Progress in Mathematics
Volume 192

Patrick Dehornoy

Braids and Self-Distributivity

Birkhäuser Verlag
Basel · Boston · Berlin

Author:

Patrick Dehornoy
Laboratoire SDAD
Mathématiques
Université Campus 2
BP 5186
14032 Caen
France
e-mail: dehornoy@math.unicaen.fr

2000 Mathematics Subject Classification 20F36, 20N02, 20M05, 06F15, 08A50, 57M25 (primary); 20F05, 20F10, 20F32, 20F60, 08B20, 68Q40, 68Q42 (secondary)

A CIP catalogue record for this book is available from the Library of Congress, Washington D.C., USA

Deutsche Bibliothek Cataloging-in-Publication Data

Dehornoy, Patrick:
Braids and self distributivity / Patrick Dehornoy. – Basel ; Boston ; Berlin : Birkhäuser, 2000
(Progress in mathematics ; Vol. 192)
ISBN 3-7643-6343-6

ISBN 3-7643-6343-6 Birkhäuser Verlag, Basel – Boston – Berlin

Printed on acid-free paper produced of chlorine-free pulp. TCF ∞
Printed in Germany
ISBN 3-7643-6343-6

9 8 7 6 5 4 3 2 1

Ferran Sunyer i Balaguer 1912–1967

* * *

This book has been awarded the Ferran Sunyer i Balaguer 1999 prize.

Each year, in honor of the memory of Ferran Sunyer i Balaguer, the Institut d'Estudis Catalans awards an international research prize for a mathematical monograph of expository nature. The prize-winning monographs are published in this series. Details about the prize can be found at http://crm.es/info/ffsb.htm

Previous winners include

- *Alexander Lubotzky*
  Discrete Groups, Expanding Graphs and Invariant Measures (vol. 125)
- *Klaus Schmidt*
  Dynamical Systems of Algebraic Origin (vol. 128)
- *M. Ram Murty & V. Kumar Murty*
  Non-vanishing of L-functions and Applications (vol. 157)
- *A. Böttcher & Yu. I. Karlovich*
  Carleson Curves, Muckenhoupt Weights, and Toeplitz Operators (vol. 154)
- *Juan J. Morales Ruiz*
  Differential Galois Theory and Non-Integrability of Hamiltonian Systems (vol. 179)

Ferran Sunyer i Balaguer 1912–1967

Born in Figueras (Gerona) with an almost fully incapacitating physical disability, Ferran Sunyer i Balaguer was confined for all his life to a wheelchair he could not move himself, and was thus constantly dependent on the care of others. His father died when Ferran was two years old, leaving his mother, Angela Balaguer, alone with the heavy burden of nursing her son. They subsequently moved in with Ferran's maternal grandmother and his cousins Maria, Angeles, and Ferran. Later, this exemplary family, which provided the environment of overflowing kindness in which our famous mathematician grew up, moved to Barcelona.

As the physician thought it advisable to keep the sickly boy away from all sorts of possible strain, such as education and teachers, Ferran was left with the option to learn either by himself or through his mother's lessons which, thanks to her love and understanding, were considered harmless to his health. Without a doubt, this education was strongly influenced by his living together with cousins who were to him much more than cousins for all his life. After a period of intense reading, arousing a first interest in astronomy and physics, his passion for mathematics emerged and dominated his further life.

In 1938, he communicated his first results to Prof. J. Hadamard of the Academy of Sciences in Paris, who published one of his papers in the Academy's "Comptes Rendus" and encouraged him to proceed in his selected course of investigation. From this moment, Ferran Sunyer i Balaguer maintained a constant interchange with the French analytical school, in particular with Mandelbrojt and his students. In the following years, his results were published regularly. The limited space here does not, unfortunately, allow for a critical analysis of his scientific achievements. In the mathematical community his work, for which he attained international recognition, is well known.

Ferran's physical handicap did not allow him to write down any of his papers by himself. He dictated them to his mother until her death in 1955, and when, after a period of grief and desperation, he resumed research with new vigor, his cousins took care of the writing. His working power, paired with exceptional talents, produced a number of results which were eventually recognized for their high scientific value and for which he was awarded various prizes. These honours not withstanding, it was difficult for him to reach the social and professional position corresponding to his scientific achievements. At times, his economic situation was not the most comfortable either. It wasn't until the 9th of December 1967, 18 days prior his death, that his confirmation as a scientific member was made public by the División de Ciencias, Médicas y de Naturaleza of the Council. Furthermore, he was elected only as "de entrada", in contrast to class membership.

Due to his physical constraints, the academic degrees for his official studies were granted rather belatedly. By the time he was given the Bachelor degree, he had already been honoured by several universities! In 1960 he finished his Master's degree and was awarded the doctorate after the requisite period of two years as a student. Although he had been a part-time employee of the Mathematical Seminar since 1948, he was not allowed to become a full member of the scientific staff until 1962. This despite his actually heading the department rather than just being a staff member.

His own papers regularly appeared in the journals of the Barcelona Seminar, *Collectanea Mathematica*, to which he was also an eminent reviewer and advisor. On several occasions, he was consulted by the Proceedings of the American Society of Mathematics as an advisor. He always participated in and supported guest lectures in Barcelona, many of them having been prepared or promoted by him. On the occasion of a conference in 1966, H. Mascart of Toulouse publicly pronounced his feeling of being honoured by the presence of F. Sunyer i Balaguer, "the first, by far, of Spanish mathematicians".

At all times, Sunyer i Balaguer felt a strong attachment to the scientific activities of his country and modestly accepted the limitations resulting from his attitude, resisting several calls from abroad, in particular from France and some institutions in the USA. In 1963 he was contracted by the US Navy, and in the following years he earned much respect for the results of his investigations. "His value to the prestige of the Spanish scientific community was outstanding and his work in mathematics of a steady excellence that makes his loss difficult to accept" (letter of condolence from T.B. Owen, Rear Admiral of the US Navy).

Twice, Sunyer i Balaguer was approached by young foreign students who wanted to write their thesis under his supervision, but he had to decline because he was unable to raise the necessary scholarship money. Many times he reviewed doctoral theses for Indian universities, on one occasion as the president of a distinguished international board. The circumstances under which Sunyer attained his scientific achievements also testify to his remarkable human qualities. Indeed, his manner was friendly and his way of conversation reflected his gift for friendship as well as enjoyment of life and work which went far beyond a mere acceptance of the situation into which he had been born. His opinions were as firm as they were cautious, and at the same time he had a deep respect for the opinion and work of others. Though modest by nature, he achieved due credit for his work, but his petitions were free of any trace of exaggeration or undue self-importance. The most surprising of his qualities was, above all, his absolute lack of preoccupation with his physical condition, which can largely be ascribed to the sensible education given by his mother and can be seen as an indicator of the integration of the disabled into our society.

On December 27, 1967, still fully active, Ferran Sunyer i Balaguer unexpectedly passed away. The memory of his remarkable personality is a constant source of stimulation for our own efforts.

Translated from Juan Augé: Ferran Sunyer Balaguer. *Gazeta Matematica*, 1.a Serie – Tomo XX – Nums. 3 y 4, 1968, where a complete bibliography can be found.

A Julien, Charlotte et Pierre

(Anglicismes) *Un dai vers middai, je tèque le beusse et je sie un jeugne manne avec une grète nèque et un hatte avec une quainnde de lèsse tressée.*

(Poor lay zanglay) *Ung joor vare meedee ger preelotobüs poor la port Changparay ... ay lar jer vee ung ohm ahvayk ung long coo ay ung chahrpo hangtooray dünn saughrt der feesel trayssay.*

R. QUENEAU, Exercices de style, (1947)

$(a+b)+c = (a+c)+(b+c)$, $(a\times b)\times c = (a\times c)\times(b\times c)$ *(...) These are other cases of the distributive principle (...) These formulae, which have hitherto escaped notice, are not without interest.*

C.S. PEIRCE, On the algebra of logic,
Amer. J. of Math., 3, 15–57, (1880)

Contents

Part B: Free LD-systems

Part C: Other LD-Systems

Preface

The aim of this book is to present recently discovered connections between Artin's braid groups B_n and left self-distributive systems (also called LD-systems), which are sets equipped with a binary operation satisfying the left self-distributivity identity

$$x(yz) = (xy)(xz). \tag{LD}$$

Such connections appeared in set theory in the 1980s and led to the discovery in 1991 of a left invariant linear order on the braid groups.

Braids and self-distributivity have been studied for a long time. Braid groups were introduced in the 1930s by E. Artin, and they have played an increasing role in mathematics in view of their connection with many fields, such as knot theory, algebraic combinatorics, quantum groups and the Yang–Baxter equation, *etc.* LD-systems have also been considered for several decades: early examples are mentioned in the beginning of the 20th century, and the first general results can be traced back to Belousov in the 1960s. The existence of a connection between braids and left self-distributivity has been observed and used in low dimensional topology for more than twenty years, in particular in work by Joyce, Brieskorn, Kauffman and their students. Brieskorn mentions that the connection is already implicit in (Hurwitz 1891).

The results we shall concentrate on here rely on a new approach developed in the late 1980s and originating from set theory. Most of the examples of self-distributive operations known at the time were connected with conjugacy in a group, and what set theory provided was a new example of a completely different type—together with the hint that something deep was hidden there, and the frustrating situation that the very existence of the example was an unprovable statement. Developing alternative constructions without relying on an unprovable statement has led to a new geometrical approach to self-distributive algebra, which has made the connection with braids more striking, and has led to a number of results about left self-distributivity, in particular many new examples and a complete description of free LD-systems; it has also led to new results about braids, the most promising so far being the existence of a natural linear order on braids.

Our story began around 1983 when the main challenge of set theory was to establish a connection between large cardinal axioms and Projective Determinacy, a structural statement that describes the Lusin hierarchy on the real line. Most large cardinal axioms involve elementary embeddings, which are a kind of endomorphism relevant in the context of set theory, and it was part of the folklore that certain elementary embeddings can be equipped with a left self-distributive operation, resulting in an intricate calculus of iterates. A proof of Projective Determinacy using large cardinals, namely Woodin cardinals, was given by Martin and Steel in 1985, and this proof, as well as a subsequent alternative proof by Woodin, requires much more than computing iterates of elementary embeddings. However, two results of the time (1986) are the author's observation that the existence of the left self-distributive operation on elementary embeddings is sufficient to prove Analytic Determinacy, a nontrivial fragment of Projective Determinacy involving the first two levels of the Lusin hierarchy, and Laver's theorem (which appeared only later) that left division in the LD-systems of elementary embeddings has no cycle. The former result shows that the self-distributive operation is nontrivial in a strong sense; the latter shows that it is quite different from the conjugacy in a group.

Motivated by the previous observations, R. Laver and the author undertook a systematic study both of the LD-systems of elementary embeddings and of free LD-systems. Contrary to related free systems, such as Joyce's free quandles or Fenn and Rourke's free racks, the free LD-systems had not received any attention so far, probably because examples were missing. The most significant result obtained during this period was the existence of a left invariant linear order on free LD-systems of rank 1 and, as a corollary, a solution to the word problem of Identity (LD), under the hypothesis that there existed an LD-system in which left division has no cycle; this was proved by R. Laver and the author independently in the Spring of 1989.

The above results created a strange situation: the existence of a left invariant order on free LD-systems, or the decidability of the word problem of (LD), are effective, finitistic properties; yet, their proof required the existence of an LD-system with a certain property, and the only known examples of such LD-systems were the LD-systems of elementary embeddings. Now, an inevitable consequence of Gödel's incompleteness theorem is that the existence of an elementary embedding of the required type is an unprovable statement, actually a higher infinity axiom. Hence the existence of an LD-system of elementary embeddings is unprovable. Therefore, the logical status of the previous results was unclear, and new developments were needed, either to construct alternative examples of LD-systems resorting to no set theoretical axiom, or to prove that such axioms were needed, as is known to be in the case of Projective Determinacy.

Here braids came into the picture, indirectly at first. The main problem was to study free LD-systems. As no concrete realization was available, the only possible approach was a syntactic approach. The main idea here has been to introduce a certain group G_{LD} which captures some geometrical properties of the identity (LD). The group G_{LD} happens to be an extension of the union B_∞ of all Artin's braid groups B_n, and it has been investigated using algebraic tools analogous to the ones Garside used for braid groups. At the end of 1991, this study led to a proof—without any set theoretical hypothesis—that left division in free LD-systems has no cycle and, as a natural by-product, to the construction of a left self-distributive operation and of a left invariant linear order on braids.

Further results both about braid ordering and about LD-systems were then proved by several authors. First, once the explicit definition of the above-mentioned self-distributive operation on braids became available, a shorter proof for existence of the braid ordering could be given, as was noted by Larue in 1992. In 1993, Laver proved that the restriction of the order to n-strand positive braids is a well-ordering. In 1995, we derived from the braid ordering a new efficient method for recognizing braid isotopy. In 1997, Fenn, Greene, Rolfsen, Rourke, and Wiest reconstructed the braid order by purely topological means, and extended the method to the mapping class group of any surface with a nonempty boundary. More recently, Short and Wiest, building on a suggestion by Thurston and work by Nielsen, gave another definition involving hyperbolic geometry and an action of braids on the real line.

As for self-distributive algebra, *i.e.*, the study of LD-systems, the main developments involve the free LD-systems, the LD-systems of elementary embeddings, and some finite LD-systems which we call the Laver tables. The results about free LD-systems consist mainly in the construction of several normal forms by Laver and the author (around 1992), and in a deepening of our understanding of the group G_{LD} (recent work). About the LD-systems of elementary embeddings, a number of results were proved by Laver, Dougherty, and Jech between 1988 and 1995 in connection with the computation of the so-called critical ordinals and attempts to construct finitistic counterparts that do not rely on large cardinal axioms. The Laver tables were introduced by Laver in the 1980s as finite quotients of the LD-systems of elementary embeddings. They are fascinating objects with a formidable combinatorial complexity. Many results about them were proved in the 1990s by Drápal. A remarkable point is that, contrary to the existence of an LD-system with an acyclic left division, some results about the Laver tables, proved using elementary embeddings, have not yet received alternative combinatorial proofs, and, therefore, they still depend on an unprovable logical axiom.

Looking at the current picture, we see that set theory is not involved in any of the braid constructions, and one may feel that its involvement in the results about the Laver tables simply implies that our understanding of these objects is far from complete. However, we would argue that both the braid order and the Laver tables are, and are to remain, applications of set theory: if the latter had not clearly shown the way, it is more than likely that most of the results we shall present in this monograph would not have been discovered yet. Similarly, the existence of the braid order can now be established quickly using Larue's short proof that left division associated with the left self-distributive operation on braids has no cycle: yet, it is not clear that this operation would have been discovered without a study of the group G_{LD}.

About this book

The current text proposes a first synthesis of this area of research. Our exposition is self-contained, and there are no prerequisites. This leads us to establish a number of basic results, about braids, self-distributive algebra, and, to some extent, set theory. However, the text is not a comprehensive course about these subjects, as we have selected only those results that are needed for our specific, mainly geometry–oriented, approach.

The text is divided into three parts, devoted to the braid order, to free LD-systems, and to general LD-systems respectively. This order does not follow the chronology of development, but it gives the braid applications first in order to motivate the study of free LD-systems that comes next. Set theory is postponed to the end, when we cannot avoid it any longer. The three parts are rather independent, and it is possible to begin with any of them, at least for a first glance.

Let us give a preview of each part of the book. The aim of Part A is to construct the linear order of braids and establish its main properties. The point is the existence of an action of braids on powers of LD-systems: if S is a given LD-system, and b is an n-strand braid, then, for each element $\vec{a}$ in S^n, we can define a new element $\vec{a} \bullet b$ by seeing $\vec{a}$ as a set of colours put on the top of b and $\vec{a} \bullet b$ as the resulting set of colours at the bottom of b when the colours flow down b according to the rule

.

When one uses classical examples of LD-systems, such as the conjugacy in a group or the barycentric mean, one deduces well-known results, such as the representation of braids inside automorphisms of a free group, and the Burau representation. New results appear when we use non-classical examples of LD-systems. In particular, a linear ordering of braids appears naturally when we use a linearly ordered LD-system $(S, <)$, *i.e.*, when the colours are ordered: the obvious idea is to extend the order to S^n lexicographically, and to define the braid b_1 to be smaller than the braid b_2, if $\vec{a} \bullet b_1 < \vec{a} \bullet b_2$ holds for every element $\vec{a}$ of S^n. The required compatibility condition turns out to be that the linear order on S satisfies $a < ab$ for all a, b, and the main technical result that makes the construction possible is that the free LD-system of rank 1 admits such a linear ordering. Technically, we avoid using abstract free LD-systems here and we resort instead to a self-distributive operation defined on braids.

Part A comprises four chapters. In Chapter I, we give an introduction to braids, and we present quickly (and somewhat informally) the connections between braids and self-distributivity, namely the above-mentioned action of braids on LD-systems, and the existence of a left self-distributive operation (called exponentiation) on braids. We also introduce what we call extended braids.

Chapter II is a preparatory chapter in which we develop a specific combinatorial method for studying those groups and monoids with a presentation of a particular syntactic form. We show that the braid groups are eligible, and deduce a number of classical properties of these groups. The method is used in Chapter VIII again.

Chapter III is devoted to the linear order of braids. Using the word reversing technique of Chapter II, we extend the action of braids on powers of LD-systems defined in Chapter I to more general LD-systems. Admitting a general result about monogenic LD-systems that will be proved in Chapter V, we deduce the existence of a left invariant order on braids. We also mention alternative definitions of this order.

In Chapter IV, we study the restriction of the braid order to positive braids, *i.e.*, to those braids that can be expressed without using the inverses σ_i^{-1} of the standard braid group generators. The main result is Laver's theorem that the restriction of the order to n-strand positive braids is a well-ordering. Here we present Burckel's construction, which computes the order type effectively and provides a new normal form for positive braids.

Part B is a general study of free LD-systems. Should the associativity identity replace left self-distributivity, the free systems would be the free semigroups, hence rather simple objects. Free LD-systems are actually much more complicated, yet they share many properties with free semigroups, and, in particular, the property that the free system of rank 1 is equipped with a unique linear order such that $a < ab$ always holds: in the case of associativity, the free semigroup of rank 1 is the set of positive integers, and the corresponding linear order is the usual order of the integers. The core of our study consists in investigating LD-equivalence of terms. The latter are abstract expressions (or words) involving variables, one binary operator, and parentheses, and we say that two terms t, t' are LD-equivalent if we can transform t into t' by applying Identity (LD) once or several times. If we think of associativity again, applying the identity to a term amounts to changing the place of parentheses, so that every equivalence class is finite, and we can select a unique representative easily, for instance by pushing all parentheses to the right. The case of LD-equivalence is much more complicated: in general, the equivalence class of a term is infinite, and finding distinguished representatives is a serious task. The main point in the approach we develop here is to put the emphasis on geometrical features*. We introduce a certain monoid $\mathcal{G}_{LD}$ consisting of partial mappings on terms, and a related group G_{LD}, so that the LD-equivalence class of a term t becomes its orbit under some partial action of G_{LD}**. All results about braids mentioned in Part A then originate from results on G_{LD}: the existence of the braid action on LD-systems comes from Artin's braid group B_∞ being a quotient of G_{LD}; braid exponentiation originates from expressing in G_{LD} a simple general property of left self-distributivity called absorption; the linear ordering of braids comes from some linear preordering on G_{LD}. The situation can be summarized in the slogan

> *The geometry of left self-distributivity is a refinement of the geometry of braids,*

which is reminiscent of the well-known observation that the geometry of braids is a refinement of the geometry of permutations: going from the symmetric group $\mathfrak{S}_\infty$ to the braid group B_∞ amounts to completing a permutation with some information about how the transpositions have been performed; similarly, going from B_∞ to G_{LD} amounts to completing a braid with some additional information about the names of the strands that have been braided, as the diagram below may suggest:

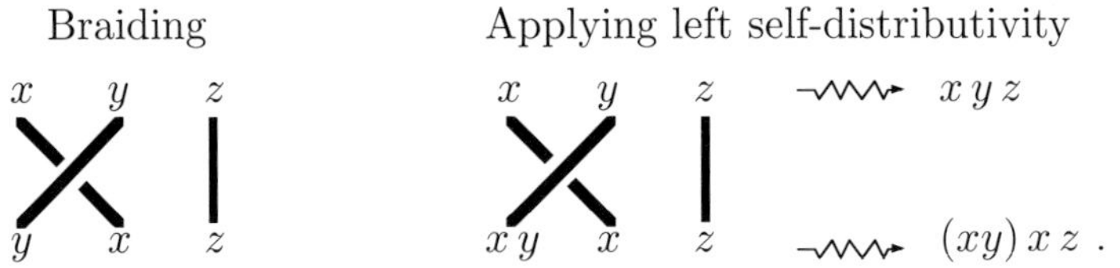

Part B is divided into five chapters. In Chapter V, we introduce free LD-systems and develop a convenient framework of finite trees. We prove the Comparison Property, which is the missing fragment in the construction of the braid order in Chapter III, and the existence of left invariant linear orders on free LD-systems, which implies the decidability of the word problem.

In Chapter VI, we prove unique normal form results for LD-equivalence, *i.e.*, we construct families of distinguished terms so that every term is LD-equivalent to exactly one distinguished term. This requires developing a precise analysis of the geometry of terms. Various applications are given, in particular about braids and their exponentiation.

---

* R. Laver has developed an alternative approach, briefly mentioned at the end of Chapter VI; neither approach is subsumed in the other, as both prove statements seemingly inaccessible to the other.

** A similar approach can be developed for every identity: in the case of associativity, the corresponding group is Richard Thompson's group F.

In Chapter VII, we introduce the monoid $\mathcal{G}_{LD}$ by considering partial operators acting on terms and specifying how and where left self-distributivity is applied. We exhibit a family of relations, called LD-relations, which hold in $\mathcal{G}_{LD}$, reminiscent of those relations involved in the MacLane–Stasheff pentagon in the case of associativity. We check that most of the results about free LD-systems established in Chapter V can be deduced from LD-relations.

In Chapter VIII, we introduce the group G_{LD} for which LD-relations yield a presentation. We describe the connection between $\mathcal{G}_{LD}$ and G_{LD}, and we construct counterparts to the braid exponentiation and to the braid order in G_{LD}. In particular, we give a purely syntactical proof of the fundamental result that left division in a free LD-system admits no cycle.

In Chapter IX, we deepen our study of the group G_{LD} and of the associated positive monoid M_{LD}. By refining the results of Chapter VI, we prove partial results about the Polish Algorithm and the Embedding Conjecture: The Polish Algorithm is a natural syntactic method for deciding LD-equivalence of terms, and its termination is one of the most puzzling open questions of the subject. The Embedding Conjecture claims that the monoid M_{LD} embeds in the group G_{LD}, a counterpart to Garside's result that the braid monoid B_n^+ embeds in the group B_n. This conjecture is equivalent to a number of properties of left self-distributivity.

Part C contains further developments about LD-systems. We did not try to be exhaustive, and we concentrated on some special families of LD-systems, in particular the Laver tables and the LD-systems of elementary embeddings. The Laver tables form an infinite family A_0, A_1, ... of finite monogenic LD-systems. The LD-system A_n has 2^n elements, and it is the unique LD-system with domain $\{1, 2, \ldots, 2^n\}$ satisfying $a*1 = a+1 \bmod 2^n$ for every a; projection *modulo* 2^n defines a surjective homomorphism of A_{n+1} onto A_n for every n, so the family $(A_n)_n$ is an inverse system. On the other hand, ranks are special sets with the weird property that every mapping of R into itself can be seen as an element of R, and elementary embeddings are defined as nontrivial injective mappings that preserve every notion that is definable from the membership relation $\in$. Let us say that a rank R is self-similar if there exists an elementary embedding of R into itself which is not the identity. The point is that, if i and j are elementary embeddings of some self-similar rank R into itself, then, as i applies to every element of R and j can be seen as an element of R, we can apply i to j, thus obtaining a new elementary embedding denoted $i[j]$. By construction, the bracket operation on elementary embeddings is left self-distributive, and the main result is that the Laver tables A_n are natural quotients of the LD-systems of elementary embeddings thus obtained. This connection results in a dictionary between elementary embeddings in set theory and values in the Laver tables. In particular, the Laver–Steel theorem, a deep well-foundedness result about elementary embeddings, translates into the result that the number of values occurring in the first row of A_n tends to infinity with n. The puzzling point is that no direct proof of the latter combinatorial statement has been discovered so far. So, as the existence of a self-similar rank is an unprovable axiom, the logical status of this statement remains open.

Part C consists of four chapters. In Chapter X, we introduce the Laver tables, we present Drápal's classification of finite monogenic LD-systems, and we mention other families of LD-systems, such as idempotent and two-sided self-distributive LD-systems.

In Chapter XI, we study LD-monoids, which are LD-systems equipped with an additional compatible associative operation. We solve the natural problem of completing a given LD-system into an LD-monoid, and we give a complete description of free LD-monoids. We also study the LD-monoid of extended braids introduced in Chapter I.

In Chapter XII, we describe the LD-systems of elementary embeddings. Here, we give a short self-contained introduction to the relevant facts needed from set theory, and show how self-distributivity naturally appears. Using a well-foundedness argument—the core of set theory—we establish the Laver–Steel theorem, which is crucial in further applications.

In Chapter XIII, we return to the Laver tables and we give a speculative conclusion to the book. We construct the above mentioned dictionary between elementary embeddings and Laver tables. Building on the unprovable hypothesis postulating the existence of a self-similar rank, we deduce results about the values in A_n, and we show why a direct combinatorial proof of the latter results has to be very complicated.

The methods used here are mostly algebraic and combinatorial in nature. The emphasis is put on words, rather than on the elements of a monoid, a group, or an LD-system they represent. Only such an approach allows us to study fine features which are not visible in the monoid, the group or the LD-system because the relations they involve are not compatible with the congruence defining the considered monoid, group, or LD-system. For instance, the notion of braid word reversing introduced in Chapter II is a refinement of the standard notion of braid word equivalence, and both induce equality of braids. However, using the former relation rather than the latter is crucial for defining the action of braids on non-classical LD-systems and deriving applications such as the braid order. This perhaps explains why such an action had not been considered before.

Exercises appear at the end of most sections, usually with the aim of mentioning in a short way further results (a lot of them correspond to unpublished material).

Historical remarks and proper credits are given in the notes at the end of each chapter. One precision is in order. The current text is deliberately centered on the author's approach to the subject—with the noticeable exception of Chapters IV, X, XII, and XIII, which rely on work by Burckel, Laver, Drápal, and Dougherty. However, we by no means claim that all unattributed results are ours: a number of them is part of the folklore, and there is no doubt that Richard Laver has known a lot of them for many years.

It is a pleasure to express my thanks to P. Ageron, G. Basset, S. Burckel, A. Drápal, I. Dynnikov, R. Fenn, A. Kanamori, R. Laver, B. Leclerc, J. Lescot, D. Rolfsen, A. Sossinsky, and B. Wiest for valuable discussions, comments and corrections. I owe special thanks to T. Kepka, who prepared the historical notes at the end of Chapter X, to M. Picantin, who found uncountably many mistakes in the manuscript, and to C. Kassel, who suggested a great number of improvements.

December 1999
Patrick Dehornoy

Part A

Ordering the Braids

I

Braids vs. Self-Distributive Systems

This chapter gives a quick introduction to our main subject, namely the connection between braids and self-distributive operations. Our presentation is based on the concrete idea of colouring the strands of a braid, and left self-distributivity arises as a natural compatibility condition. At this early stage, some of the constructions may look artificial or strange: it will be one of the aims of the subsequent chapters, in particular in Part B of this book, to explain them and hopefully make all of them natural.

Section 1 contains an elementary introduction to Artin's braid groups. We establish the standard presentation of the braid group, a basis for further algebraic developments. In Section 2, we describe an action of braids on those algebraic systems that involve a self-distributive operation. We review the classical examples of such systems, and mention the braid properties obtained using the action. We also describe a non-classical example involving injections. In Section 3, we show how a self-distributive operation, called braid exponentiation, can be constructed inside Artin's braid group B_∞ by using braid colourings. We present some related self-distributive systems, and we prove that left division associated with braid exponentiation has no cycle, a crucial result for subsequent order properties. Finally, in Section 4, we consider LD-monoids, which are structures involving both a left self-distributive operation and a compatible associative product. LD-monoids naturally appear when colouring braid diagrams. We construct such a structure on a partial topological completion of the braid group B_∞ corresponding to the intuitive idea of strands vanishing at infinity.

1. Braid Groups

There are many ways to introduce Artin's braid groups. Here, we choose the most elementary approach, namely starting from the intuitive idea of a concrete braid.

Geometric braids

Definition 1.1. (geometric braid) Assume that f is a permutation of 1, ..., n. An *n-strand geometric braid β associated with f* is defined to be the union of n disjoint polygonal arcs $\gamma_1, \ldots, \gamma_n$ in $\mathbf{R}^2 \times [0,1]$ such that γ_i connects $(f(i), 0, 0)$ to $(i, 0, 1)$ for each i and, for every z_0 in $[0,1]$, the intersection of γ_i with the plane $z = z_0$ consists of exactly one point. If β is a geometric braid associated with f, we say that f is *the* permutation of β, and we denote it by perm(β).

In Figure 1.1, we display an example of a 4-strand geometric braid. In all pictures, the z-axis will correspond to the vertical direction of the figure, with a downward orientation. The last condition in the definition implies that the z-coordinate increases monotonically along each arc.

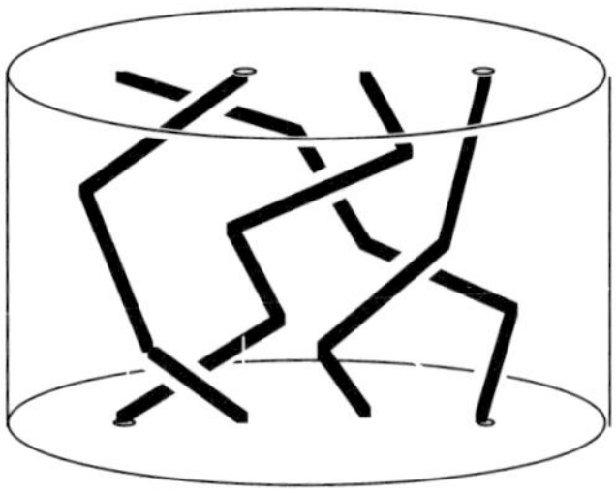

Figure 1.1. A 4-strand geometric braid

Definition 1.2. (product) (Figure 1.2) Assume that β_1 and β_2 are n-strand geometric braids. The *product* (or composition) $\beta_1\beta_2$ is defined to be the geometric braid obtained by stacking β_1 over β_2 and shrinking the result so that it lies in $\mathbf{R}^2 \times [0,1]$—so, formally, $\beta_1\beta_2$ is $c(\beta_1 \cup t(\beta_2))$, where t is translation by $(0,0,1)$ and c is the affinity that multiplies the z-coordinate by $1/2$.

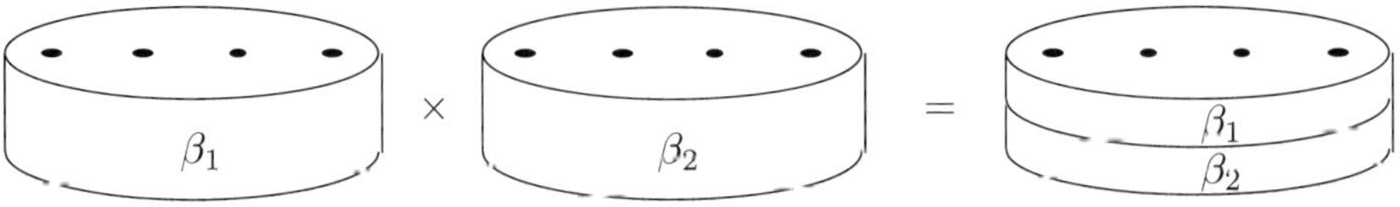

Figure 1.2. Product of geometric braids

Lemma 1.3. *The relation* $\beta = \beta_1 \beta_2$ *implies* $\mathrm{perm}(\beta) = \mathrm{perm}(\beta_1) \circ \mathrm{perm}(\beta_2)$.

The result is obvious. As it stands, the product of braids is not associative. But it becomes so provided we allow the strands to be stretched or shrunk. More generally, we shall consider as equivalent two geometric braids such that one is obtained from the other by moving strands but not allowing them to cross one another.

Definition 1.4. **(isotopy)** (Figure 1.3) Assume that β, β' are geometric braids. We say that β' is 1-*isotopic* to β if β' is obtained from β by replacing two consecutive segments $[AB]$, $[BC]$ with the segment $[AC]$, or *vice versa*, provided that the convex hull of ABC does not intersect β. We say that β' is *isotopic* to β if there exists a finite sequence $\beta_0 = \beta, \beta_1, \ldots, \beta_m = \beta'$ such that β_i is 1-isotopic to β_{i-1} for every i.

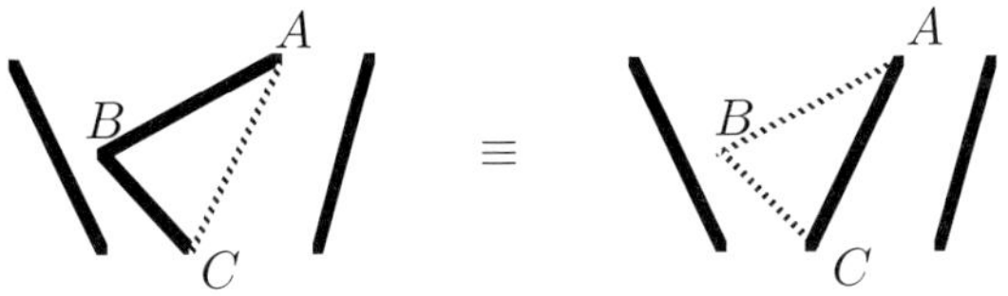

Figure 1.3. Isotopy

By construction, isotopy is an equivalence relation, and it preserves the number of strands and the permutation.

Definition 1.5. **(braid)** An n-*strand braid* is an isotopy class of geometric n-strand braids. The set of all n-strand braids is denoted by B_n.

Lemma 1.6. *Assume that* h *is a piecewise affine bijection of* $[0,1]$ *into itself. Let* $\widetilde{h}$ *be the function of* $\mathbf{R}^2 \times [0,1]$ *that maps* (x, y, z) *to* $(x, y, h(z))$. *Then every geometric braid is isotopic to its image under* $\widetilde{h}$.

Proof. By definition, isotopy is a transitive relation, so we may assume that the graph of h consists of only two segments, and, therefore, it lies either above or below the diagonal. Assume the first. Then $\widetilde{h}$ lowers each point of $\mathbf{R}^2 \times (0,1)$. Let β be a geometric braid. We define inductively an isotopy of β onto $\widetilde{h}(\beta)$: assuming that $[AB]$ and $[BC]$ are consecutive segments in β, we first replace $[BC]$ with $[B\widetilde{h}(B)] \cup [\widetilde{h}(B)C]$, and, then, we replace $[AB] \cup [B\widetilde{h}(B)]$ with $[A\widetilde{h}(B)]$. The possible problem is that some arc of β may intersect the triangles $[AB\widetilde{h}(B)]$ or $[B\widetilde{h}(B)C]$. A solution consists in ordering the segments of β as follows: we say that γ is *below* γ' if the projections of γ and γ' on the plane $z = 0$ intersect at some point $(x, y, 0)$ and, letting (x, y, z) and (x, y, z') be the liftings of $(x, y, 0)$ in γ and γ' respectively, $z > z'$ holds—according to our graphical conventions, $z = 0$ lies above $z = 1$. This relation need not be an ordering in general (cycles may exist), but it becomes an ordering provided we possibly partition the segments into smaller pieces. Then no problem will arise provided we enumerate the segments so as to begin with the lowest ones. ■

Proposition 1.7. **(braid group)** *The product of geometric braids induces a well-defined product on B_n, which gives B_n the structure of a group. The mapping* perm *induces a surjective homomorphism of B_n onto the symmetric group $\mathfrak{S}_n$.*

Proof. By construction, if β_1' is isotopic to β_1, $\beta_1'\beta_2$ is isotopic to $\beta_1\beta_2$ for every β_2. So isotopy is compatible with multiplication on the right, and, similarly, it is compatible on the left. Hence, the product on B_n is well-defined. Applying Lemma 1.6 to the function h defined by

$$h(z) = \begin{cases} 2z & \text{for } 0 \le z \le 1/4, \\ z + 1/4 & \text{for } 1/4 \le z \le 1/2, \\ z/2 + 1/2 & \text{for } 1/2 \le z \le 1 \end{cases}$$

shows that $(\beta_1\beta_2)\beta_3$ is isotopic to $\beta_1(\beta_2\beta_3)$ for all β_1, β_2, β_3. So, the product on B_n is associative. Next, define h by

$$h(z) = \begin{cases} 2z & \text{for } 0 \le z \le 1/2, \\ 1 & \text{for } 1/2 \le z \le 1. \end{cases}$$

The existence of h shows that every geometric n-strand braid β is isotopic to $\beta\varepsilon_n$, where ε_n denotes the trivial geometric braid consisting of n vertical segments (the current mapping h is not bijective, but the argument remains the same). Similarly, β is isotopic to $\varepsilon_n\beta$, and the class of ε_n is a unit for the product of B_n.

Next, let s denote the symmetry with respect to the plane $z = 1/2$. The braids $\beta s(\beta)$ and $s(\beta)\beta$ are invariant under s, and we claim that every n-strand braid that is invariant under s is isotopic to ε_n. This will prove that the class of $s(\beta)$ is the inverse of the class of β in B_n. So assume that β is a symmetric geometric braid consisting of n arcs $\gamma_1, \ldots, \gamma_n$. Let $\widehat{\gamma_i}$ denote the convex hull of γ_i. Up to using an isotopy to slightly move some segments, we may assume that the union of all intersections $\gamma_i \cap \widehat{\gamma_j}$ consists of a finite number p of vertical segments. Let us call this number p the intersection number of β. For $p = 0$, we can use an isotopy to transform each arc γ_i into a vertical segment, hence β is isotopic to ε_n. Otherwise, let γ be an intersection segment with minimal length. Assume that the ends of γ lie on γ_i and in the interior of γ_j. By construction, the part of $\widehat{\gamma_i}$ that is bounded by γ and contains no point with z-coordinate 0 or 1 cannot intersect β. Hence we can transform β into an isotopic symmetric braid with smaller intersection number by replacing slightly more than the central portion of γ_i determined by γ with a vertical segment (Figure 1.4).

Finally, the permutations of isotopic braids are equal, and Lemma 1.3 tells us that perm is a homomorphism. This homomorphism is surjective, as transpositions generate the symmetric group, and constructing a geometric braid whose permutation is a given transposition is trivial. ∎

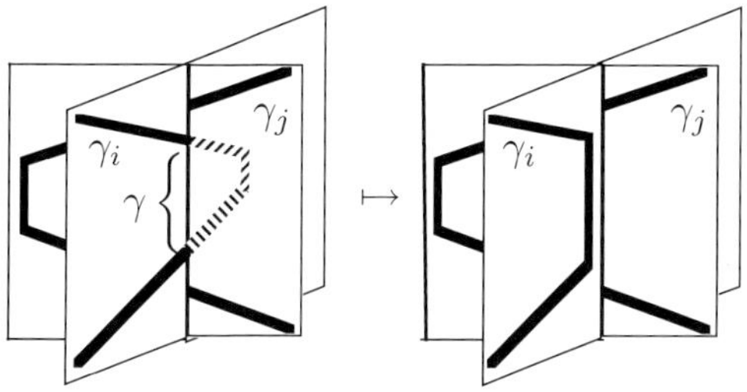

Figure 1.4. Diminishing the intersection number

For each n, we shall denote by 1 the unit of B_n, which we have seen is the class of the trivial geometric n-strand braid ε_n.

Presentation of B_n

The aim of this subsection is to prove:

Proposition 1.8. (**presentation**) *The group B_n admits the presentation*

$$\langle \sigma_1, \ldots, \sigma_{n-1} \; ; \; \sigma_i\sigma_j = \sigma_j\sigma_i \text{ for } |i-j| \geq 2, \; \sigma_i\sigma_j\sigma_i = \sigma_j\sigma_i\sigma_j \text{ for } |i-j| = 1\rangle.$$

To this end, we consider the orthogonal projections of geometric braids on the plane $y = 0$. We shall refer to this projection as the *y-projection.* The idea is to consider geometric braids with a standardized y-projection that can be fully described by a word over a finite alphabet, and, then, to characterize those words that code for isotopic braids.

Definition 1.9. (**regular**) A geometric braid is said to be *regular* if its y-projection has no triple point, it has only finitely many double points, the z-coordinates of the double points are pairwise distinct, and the double points are not the projection of the end of a segment.

Lemma 1.10. *Every geometric braid is isotopic to a regular one.*

Proof. Regular braids form a dense set with respect to the topology on geometric braids induced by the ambient topology of $\mathbf{R}^3$, while the set of all geometric braids that are isotopic to a given geometric braid is open. Hence these sets intersect. ∎

Assume that d is the y-projection of a regular braid β. Each double point $(x, 0, z)$ of d is the y-projection of two points (x, y_1, z) and (x, y_2, z) of β that belong respectively to two segments $[A_1B_1]$ and $[A_2B_2]$. We say that $[A_1B_1]$ lies *in front of* $[A_2B_2]$ if $y_1 > y_2$ holds.

Definition 1.11. (braid diagram, realization) A *braid diagram* is defined to be the y-projection of a regular braid, together with the specification, for each double point, of which segment lies in front. If d is a braid diagram, the *normal realization* of d is the geometric braid $\widetilde{d}$ obtained from d by replacing, for each double point M of d, a sufficiently short section $[AC]$ of the segment containing M and lying in front of the other with the union of two connected segments $[AB]$, $[BC]$ such that B lies in the plane $y = 1$ and M is its y-projection.

We represent a braid diagram with small interruptions on back segments, as in Figure 1.5. As it stands, the normal realization of a braid diagram is not unique, but it would be easy to make it so, the only requirement being that the y-projection of $\widetilde{d}$ be d and $\widetilde{d}$ be simple enough.

Figure 1.5. A braid diagram

Lemma 1.12. *Assume that β and β' are regular braids that project on the same braid diagram. Then β and β' are isotopic.*

Proof. Letting d be the y-projection of β and β', we show that β is isotopic to $\widetilde{d}$ using induction on the number of crossings in d. ∎

The positions of the crossings in a braid diagram can be coded by a *word*. Assume that d is a braid diagram, and $[AB]$ is a segment in d that is short enough to contain at most one double point. Assume $z_A < z_B$. We say that $[AB]$ lies at *position* i if $i-1$ points in the intersection of d with the horizontal line $z = z_A$ lie on the left of A, *i.e.*, lie on the half-line $x < x_A$.

Definition 1.13. (code, braid word) Assume that d is an n-strand braid diagram. Let $z_1 < \cdots < z_\ell$ be the z-coordinates of the double points in d. The *code* of d is defined to be the word $\sigma_{i_1}^{e_1} \cdots \sigma_{i_\ell}^{e_\ell}$, where i_k and $i_k + 1$ are the positions of the segments that cross on the line $z = z_k$, and $e_k = +1$ (*resp.* -1)

holds if the segment initially at position $i+1$ lies in front (*resp.* in back). If d contains no crossing, the code of d is defined to be the empty word ε. The code of a regular braid is defined to be the code of its y-projection. The set of all codes of n-strand braid diagrams—also called n-strand braid words—is denoted by BW_n.

For instance, the code of the braid diagram displayed in Figure 1.5—as well as the code of the geometric braid displayed in Figure 1.1—is the word $\sigma_1\sigma_2\sigma_3\sigma_1^{-1}$.

Definition 1.14. (normal diagram) Assume that w is a braid word. The *normal* n-strand diagram $D_n(w)$ is defined by the inductive rules:

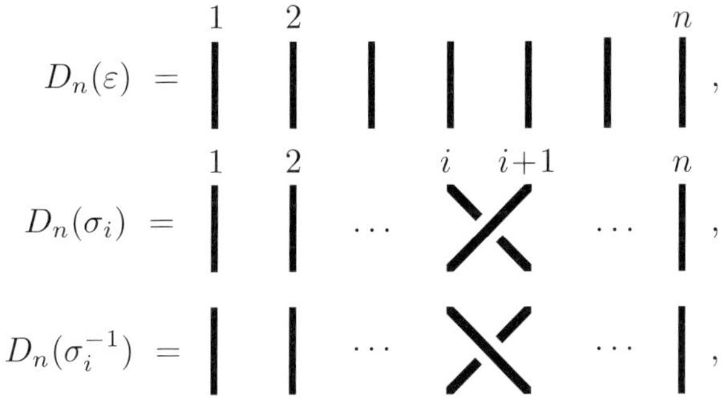

and, for $\ell \geq 2$, $D_n(\sigma_{i_1}^{e_1}\cdots\sigma_{i_\ell}^{e_\ell})$ is obtained by stacking $D_n(\sigma_{i_1}^{e_1})$, ..., $D_n(\sigma_{i_\ell}^{e_\ell})$.

For instance, the diagram $D_4(\sigma_1\sigma_2\sigma_3\sigma_1^{-1})$ is displayed in Figure 1.6.

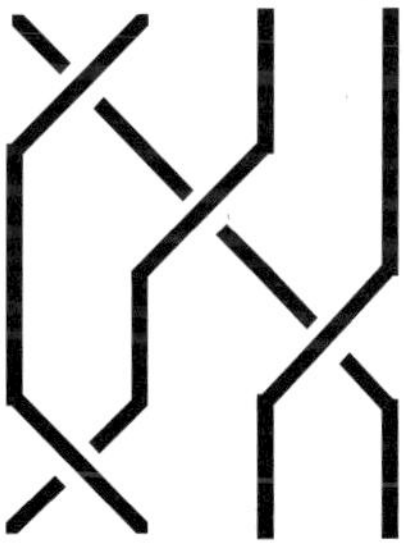

Figure 1.6. Normal braid diagram

For each n, concatenation of words turns BW_n into a monoid and, by definition, the mapping $w \mapsto \widetilde{D_n(w)}$ is a homomorphism of BW_n into geometric braids.

Lemma 1.15. *Regular geometric braids with the same code are isotopic.*

Proof. Let d be an n-strand braid diagram, and w be the code of d. An induction on the number of segments in d shows that $\widetilde{d}$ and $\widetilde{D_n(w)}$ are isotopic. ∎

Thus, if two geometric n-strand braids are coded by the same word, they are isotopic. In order to completely describe the isotopy classes of geometric braids, it remains to recognize in which cases different braid words code for isotopic braids. The following lemma gives the answer, and it completes the proof of Proposition 1.8 (presentation).

Lemma 1.16. *Let $\equiv$ be the least congruence on the monoid BW_n such that*

$$\sigma_i\ \sigma_j \equiv \sigma_j\ \sigma_i \qquad \textit{for } |i-j| \geq 2, \tag{1.1}$$

$$\sigma_i\ \sigma_j\ \sigma_i \equiv \sigma_j\ \sigma_i\ \sigma_j \qquad \textit{for } |i-j| = 1, \tag{1.2}$$

hold, and we have $\sigma_i \sigma_i^{-1} \equiv \sigma_i^{-1} \sigma_i \equiv \varepsilon$ for every i. Then two words in BW_n code for isotopic geometric braids if and only if they are $\equiv$-equivalent.

Proof. On the one hand, the geometric braids $\widetilde{D_n}(\sigma_i\sigma_j)$ and $\widetilde{D_n}(\sigma_j\sigma_i)$ are isotopic for $|i-j| \geq 2$. Similarly, the braids $\widetilde{D_n}(\sigma_i\sigma_{i+1}\sigma_i)$ and $\widetilde{D_n}(\sigma_{i+1}\sigma_i\sigma_{i+1})$ are isotopic, and so are $\widetilde{D_n}(\sigma_i\sigma_i^{-1})$, $\widetilde{D_n}(\sigma_i^{-1}\sigma_i)$ and $\widetilde{D_n}(\varepsilon)$. As braid isotopy is compatible with product and $\equiv$ is the congruence generated by the previous pairs of words, $w \equiv w'$ implies that the geometric braids $\widetilde{D_n}(w)$ and $\widetilde{D_n}(w')$ are isotopic.

On the other hand, it suffices to show that, if β' is obtained from β by replacing two consecutive segments $[AB]$, $[BC]$ with $[AC]$, then the codes of β and β' are $\equiv$-equivalent. At the expense of possibly decomposing the isotopy into several steps, we may assume that the projection of the triangle $[ABC]$ either intersects no other projection of a segment of β, or it intersects one, or it intersects two which, in addition, intersect inside the triangle.

In the first case, the codes of β and β' are equal. In the second case, there are two possibilities. Assume first that the projection of some segment γ of β intersects the projections of $[AC]$ and $[BC]$. These intersections contribute to the codes of β and β'. It follows from the definition of an isotopy that either γ lies in front both of $[AC]$ and $[BC]$, or it lies behind both of them, so the contributions are the same letter $\sigma_i^{\pm 1}$. However, the codes of β and β' need not be equal, as the z-coordinates of the intersections associated with γ, say z_0 and z_0', need not be equal. If no other intersection has z-coordinate between z_0 and z_0', the codes coincide. Otherwise, some intersections that contribute to the code before γ in β will contribute after γ in β', or *vice versa*. Now, this amounts to replacing some factor of the form $\sigma_i^e\sigma_j^{e'}$ with $|i-j| \geq 2$ by the corresponding factor $\sigma_j^{e'}\sigma_i^e$. Owing to Relation (1.1), and, possibly, to $\sigma_k\sigma_k^{-1} \equiv \sigma_k^{-1}\sigma_k \equiv \varepsilon$, the codes of β and β' are $\equiv$-equivalent.

Assume now that the projection of some segment γ of β intersects the projections of $[AB]$ and $[BC]$, but not the projection of $[AB]$. By possibly using intermediate transformations of the previous type, we may assume that no other intersection occurs between the intersections of γ with $[AB]$ and $[BC]$.

Then the latter contribute some factor $\sigma_i^e\sigma_i^{-e}$ in the code of β, which is absent in the code of β'. By definition, the codes of β and β' are $\equiv$-equivalent.

Finally, assume that the projection of $[ABC]$ contains the intersection of the projections of two segments γ, γ'. There are several cases according to the various possible positions of the segments, but, as replacing $[AB]$ and $[BC]$ with $[AC]$ is a legal isotopy, all of $[AB]$, $[BC]$ and $[AC]$ lie in front of γ, or all lie behind γ, and the same holds for γ'. Figure 1.7 displays the possible cases and the corresponding codes, up to a symmetry that replaces every code with its inverse. Using Relation (1.2) and $\sigma_k\sigma_k^{-1} \equiv \sigma_k^{-1}\sigma_k \equiv \varepsilon$, we see that the codes of β and β' are $\equiv$-equivalent. ∎

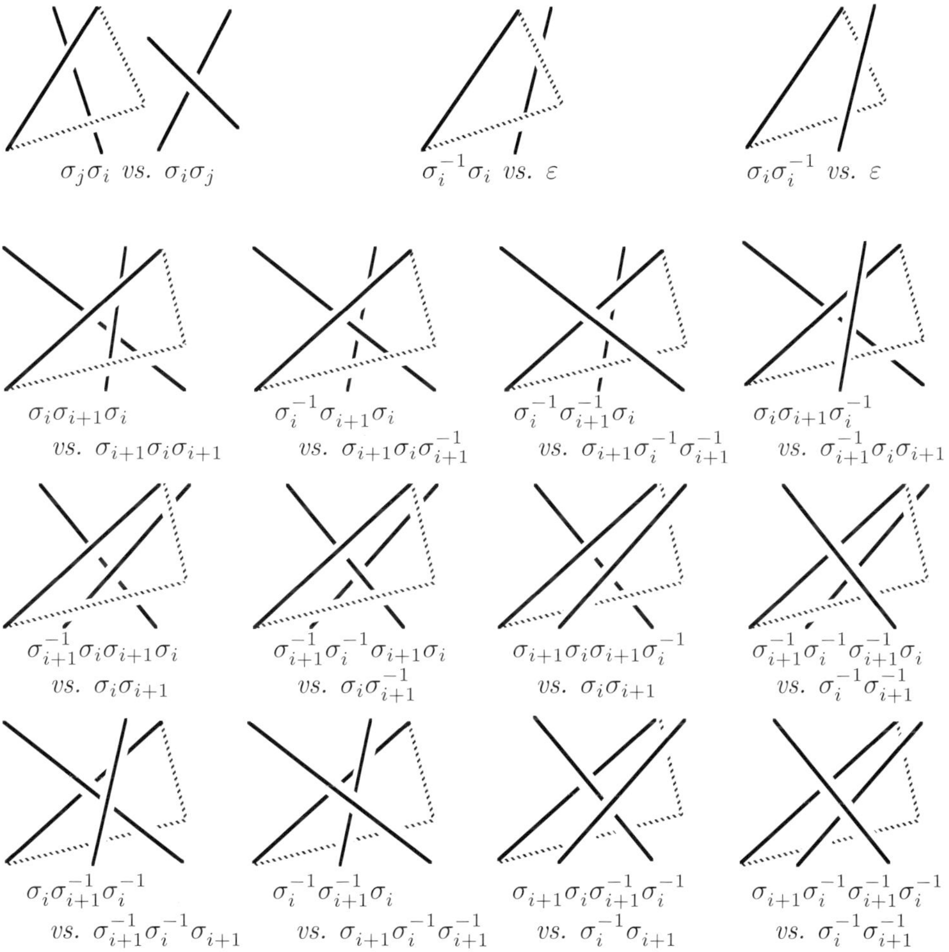

Figure 1.7. Braid relations

By the previous result, the isotopy class of a braid is determined by the braid word that codes it, and, therefore, studying braid isotopy reduces to studying the congruence $\equiv$ on braid words.

Positive braids

In the sequel, we shall often consider those braid diagrams where all crossings have the same orientation.

Definition 1.17. (positive) We say that a braid word is *positive* if it contains no letter σ_i^{-1}. The set of all n-strand positive braid words is denoted by BW_n^+. We say that a braid diagram is positive if its code is a positive braid word, and that a geometric braid is positive if its y-projection is a positive diagram. Two positive geometric braids are said to be *positively isotopic* if they are isotopic via an isotopy that involves positive braids only.

Positive braid words are closed under product, so, for each n, BW_n^+ is a submonoid of BW_n. As for isotopy classes, some care is needed: if β_1, β_2 are positive geometric braids, it is not obvious that β_1 and β_2 being isotopic force them to be positively isotopic. We shall prove this to be true in Chapter II, but, for the moment, we distinguish between isotopy and positive isotopy. So, we do not know yet that the monoid B_n^+ embeds in the group B_n. Revisiting the proof of Lemma 1.16 gives:

Lemma 1.18. *Let $\equiv^+$ be the least congruence on the monoid BW_n^+ such that*

$$\sigma_i\ \sigma_j \equiv^+ \sigma_j\ \sigma_i \qquad \text{for } |i-j| \geq 2, \tag{1.3}$$

$$\sigma_i\ \sigma_j\ \sigma_i \equiv^+ \sigma_j\ \sigma_i\ \sigma_j \qquad \text{for } |i-j| = 1, \tag{1.4}$$

holds. Then two words in BW_n^+ code for positively isotopic geometric braids if and only if they are $\equiv^+$-equivalent.

We deduce:

Proposition 1.19. (positive presentation) *The monoid B_n^+ admits—as a monoid—the presentation*

$$\langle \sigma_1, \ldots, \sigma_{n-1}\ ;\ \sigma_i\sigma_j = \sigma_j\sigma_i \text{ for } |i-j| \geq 2,\ \sigma_i\sigma_j\sigma_i = \sigma_j\sigma_i\sigma_j \text{ for } |i-j| = 1\rangle.$$

2. Braid Colourings

The first connection between braids and self-distributivity is the existence of an action of braids on sequences of elements chosen in a set equipped with a left self-distributive operation. The description of this action is natural when we use the intuition of colouring the strands of a braid.

Braid colourings

Assume that S is a fixed nonempty set, and d is an n-strand braid diagram. We use the elements of S to colour the strands of d. To this end, we attribute to each top end in d a colour from S, which we propagate along the strands of the diagram. The idea is to recover information about the braid b represented by d by comparing the final sequence of colours with the initial sequence.

If the colours are propagated without change along the strands, the final sequence is a permutation of the initial sequence, and the only datum about b we recover is the permutation $\mathrm{perm}(b)$.

Things become more interesting when we allow colours to change at crossings. Let us begin with the case of positive braid diagrams. According to our principle of propagating the colours downwards from the top of the braid diagram, it is natural to fix the rules of colouring so that the colours after a crossing (the 'new' colours) are functions of the colours before the crossings (the 'old' colours). Keeping the colours corresponds to the simplest function, namely identity. The next step is to assume that only one colour may change, say the colour of the front strand, and the new colour depends only on the two colours of the strands that have crossed. This amounts to using a function of $S \times S$ into S, *i.e.*, a binary operation $^\wedge$ on S, so that the rule for propagating colours is

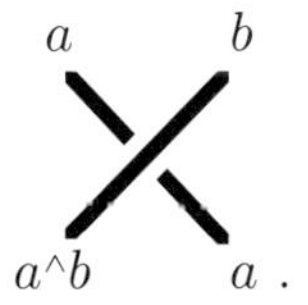

This leads to the following formal definition. Here, and everywhere in the sequel, we adopt the convention that, when $\vec{x}$ represents a (finite or infinite) sequence, then the sucessive elements of $\vec{x}$ are denoted x_1, x_2, ...

Definition 2.1. (**braid action**) Assume that $(S, {}^\wedge)$ is a binary system, *i.e.*, a set equipped with a binary operation. The (right) *action* of BW_n^+ on S^n is defined inductively by

$$\vec{a} \bullet \varepsilon = \vec{a}, \qquad \vec{a} \bullet \sigma_i w = (a_1, \dots, a_i {}^\wedge a_{i+1}, a_i, a_{i+2}, \dots, a_n) \bullet w. \tag{2.1}$$

As we are interested in braids rather than in braid words, we want the action of (2.1) to induce a well-defined action of braids, *i.e.*, the colourings have to be invariant under braid relations.

Lemma 2.2. *Assume that* $(S, ^\wedge)$ *is a binary system. The action of* BW_n^+ *on* S^n *defined by* (2.1) *is always invariant under Relations* (1.1); *it is invariant under Relations* (1.2) *if and only if* $(S, ^\wedge)$ *satisfies the left self-distributivity identity*

$$x ^\wedge (y ^\wedge z) = (x ^\wedge y) ^\wedge (x ^\wedge z). \qquad (LD)$$

Proof. Compare the diagrams:

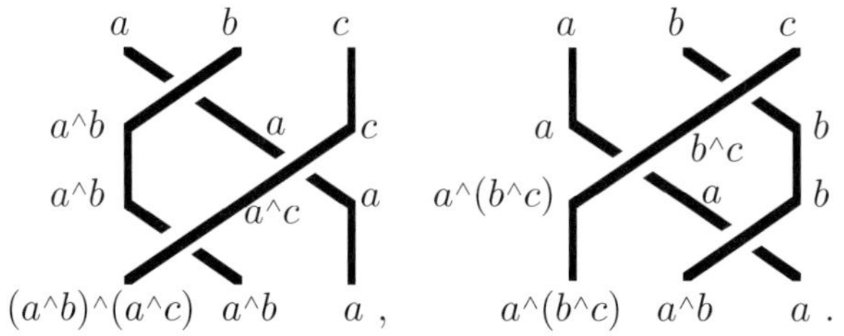

The lower colours on the left strand coincide for every initial choice of a, b, c if and only if the operation $^\wedge$ satisfies Identity (LD). ■

Let us now colour arbitrary braid diagrams, *i.e.*, let us try to define an action of arbitrary words on sequences of colours. We have to choose a rule for colouring negative crossings. In order to make the definition flexible, let us first assume that the set of colours S is equipped with two more binary operations, say $\circ$ and $\vee$. We complete the definition with the rule

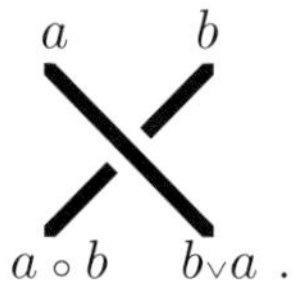

This amounts to extending the action of braid words on S^n by

$$\vec{a} \bullet \sigma_i^{-1} w = (a_1, \ldots, a_i \circ a_{i+1}, a_{i+1} \vee a_i, a_{i+2}, \ldots, a_n) \bullet w. \qquad (2.2)$$

Lemma 2.3. *Assume that* $(S, ^\wedge, \circ, \vee)$ *is a triple binary system. The right action of* BW_n *on* S^n *defined by* (2.1) *and* (2.2) *is invariant under the relations* $\sigma_i\sigma_i^{-1} = \sigma_i^{-1}\sigma_i = \varepsilon$ *if and only if* S *satisfies the identities*

$$x \circ y = y, \qquad x ^\wedge (x \vee y) = x \vee (x ^\wedge y) = y.$$

Proof. This follows from:

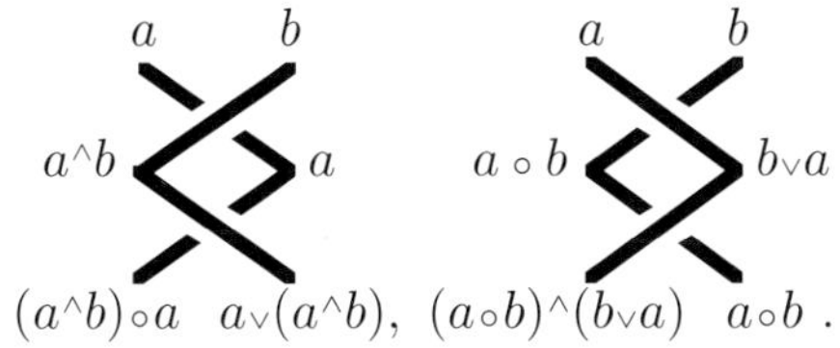

We see on the right strands that $a \circ b = b$ is required. Then looking at the left strands gives the relations between the operations ^ and ∨. ∎

We are thus led to the following definitions:

Definition 2.4. (**LD-system, LD-quasigroup**) (i) An *LD-system* is defined to be a set equipped with a left self-distributive operation, *i.e.*, a binary operation that satisfies Identity (LD).

(ii) An *LD-quasigroup* is defined to be an LD-system equipped with an additional operation ∨ that satisfies the identity

$$x \,\hat{}\, (x \vee y) = x \vee (x \,\hat{}\, y) = y. \tag{2.3}$$

The following result summarizes the above observations. Although trivial, it will be central for our purpose.

Proposition 2.5. (**action**) *(i) Assume that* $(S, \hat{})$ *is an LD-system. Then, for every* n*, Formula* (2.1) *defines an action of the braid monoid* B_n^+ *on* S^n.

(ii) Assume that $(S, \hat{}, \vee)$ *is an LD-quasigroup. Then, for every* n*, Formulas* (2.1) *and* (2.2) *define an action of the braid group* B_n *on* S^n.

Remark 2.6. In the sequel, we shall use several notations for left self-distributive operations. When only one operation is considered, there is no problem in using a "dot" notation, with the operator usually dropped. When several operations are considered simultaneously, it is often convenient to use a left exponentiation for the left self-distributive operation, thus writing ${}^{x}y$ for the "product" of x and y. In this framework, left self-distributivity is expressed by ${}^{x}({}^{y}z) = {}^{{}^{x}y}({}^{x}z)$. For better readability, we use $x\hat{}y$ as an alternative to ${}^{x}y$.

Convention 2.7. (**parentheses**) In all algebraic expressions, missing parentheses are to be added on the right: thus, everywhere in this text, $x\hat{}y\hat{}z$ stands for $x\hat{}(y\hat{}z)$, $x\hat{}y{\vee}z$ stands for $x\hat{}(y{\vee}z)$, *etc.*

Before going further, let us observe that the two operations of an LD-quasigroup determine each other. Let us recall that a binary system $(S, \hat{})$ is said to be *left cancellative* if $a\hat{}b = a\hat{}c$ implies $b = c$, and to be *left divisible* if, for all a, c in S, there exists at least one element b in S satisfying $a\hat{}b = c$. In other words, if we denote by ad_a the left translation $x \mapsto a^\wedge x$, then $(S, {}^\wedge)$ is left cancellative if and only if all mappings ad_a are injective, and it is left divisible if and only if all mappings ad_a are surjective.

Proposition 2.8. (second operation) *(i) Assume that $(Q, {}^\wedge, {}_\vee)$ is an LD-quasigroup. Then both $(Q, {}^\wedge)$ and $(Q, {}_\vee)$ are left cancellative left divisible LD-systems; the operations ${}^\wedge$ and ${}_\vee$ are mutually left distributive, and each one determines the other by:*

$$a \vee c = \text{ the unique } b \text{ satisfying } a \wedge b = c, \tag{2.4}$$

$$a \wedge c = \text{ the unique } b \text{ satisfying } a \vee b = c. \tag{2.5}$$

(ii) Conversely, assume that $(Q, {}^\wedge)$ is a left cancellative left divisible LD-system. Let ${}_\vee$ be the operation defined by (2.4). Then $(Q, {}^\wedge, {}_\vee)$ is an LD-quasigroup.

Proof. (i) By hypothesis, $(Q, {}^\wedge)$ is an LD-system, *i.e.*, the operation ${}^\wedge$ satisfies Identity (LD). If $\mathrm{ad}_a^\wedge$ and $\mathrm{ad}_a^\vee$ denote the left translations$x \mapsto a^\wedge x$ and $x \mapsto a_\vee x$ respectively, Formula (2.3) gives $\mathrm{ad}_a^\wedge \circ \mathrm{ad}_a^\vee = \mathrm{ad}_a^\vee \circ \mathrm{ad}_a^\wedge = \mathrm{id}_Q$ for every a, and, therefore, it implies that every left translation of $(Q, {}^\wedge)$ and $(Q, {}_\vee)$ is a bijection. Hence $(Q, {}^\wedge)$ and $(Q, {}_\vee)$ are left cancellative and left divisible, and (2.4) and (2.5) hold.

We prove now that $(Q, {}_\vee)$ is an LD-system. Let a, b, c be arbitrary elements of Q. We find (using Convention 2.7)

$$a \wedge (a \vee b) \wedge (a \vee b \vee c) = (a \wedge a \vee b) \wedge (a \wedge a \vee b \vee c) = b \wedge (b \vee c) = c,$$
$$a \wedge (a \vee b) \wedge ((a \vee b) \vee (a \vee c)) = a \wedge a \vee c = c,$$

which implies $a_\vee(b_\vee c) = (a_\vee b)_\vee(a_\vee c)$ as $\mathrm{ad}_{a\vee b}^\wedge$ and $\mathrm{ad}_a^\wedge$ are injective.

We prove finally that ${}^\wedge$ is left distributive with respect to ${}_\vee$. Indeed, we have

$$a \vee (a \wedge b) \vee (a \wedge (b \vee c)) = (a \vee a \wedge b) \vee (a \vee a \wedge b \vee c) = b \wedge (b \vee c) = c,$$
$$a \vee (a \wedge b) \vee ((a \wedge b) \wedge (a \wedge c)) = a \vee a \wedge c = c,$$

which implies $a^\wedge(b_\vee c) = (a^\wedge b)_\vee(a^\wedge c)$ as $\mathrm{ad}_{a\wedge b}^\vee$ and $\mathrm{ad}_a^\vee$ are injective. A symmetric argument gives $a_\vee(b^\wedge c) = (a_\vee b)^\wedge(a_\vee c)$.

Point (ii) is clear, as (2.3) follows from the definition of operation ${}_\vee$. ∎

Owing to the previous results, we shall say that a left cancellative left divisible LD-system is an implicit LD-quasigroup. There exists exactly one way to complete an implicit LD-quasigroup into an LD-quasigroup, so no ambiguity exists.

LD-quasigroups

Here we review examples of LD-quasigroups. In each case, we mention those results about braids that can be deduced from the associated action.

Example 2.9. (**trivial**) *Let S be an arbitrary set. Then the trivial operations*

$$a \wedge b = b, \qquad a \vee b = b$$

turn S into an LD-quasigroup. Using such operations to colour the strands of the braids amounts to keeping the colours unchanged. Then, for every braid b in B_n, and every sequence $\vec{a}$ in S^n, we have

$$\vec{a} \bullet b = \mathrm{perm}(b)^{-1}(\vec{a}). \tag{2.6}$$

The inversion comes from perm(b) *specifying the initial positions of the strands in terms of their final positions, i.e., going from bottom to top, as is needed for* perm *to be a homomorphism. Thus, using this particular LD-quasigroup leads to the surjective homomorphism*

$$\mathrm{perm} : B_n \twoheadrightarrow \mathfrak{S}_n.$$

Example 2.10. (**shift**) *The operations*

$$a \wedge b = b + 1, \qquad a \vee b = b - 1$$

turn $\mathbf{Z}$ into an LD-quasigroup. For every braid b in B_n, and every sequence $\vec{a}$ in $\mathbf{Z}^n$, we find

$$\sum(\vec{a} \bullet b) = \sum \vec{a} + \mathrm{es}(b), \tag{2.7}$$

where $\sum \vec{a}$ denotes $a_1 + \cdots + a_n$, and es(b) *denotes the exponent sum of b, defined to be the difference between the number of positive and negative letters in any expression of b. Thus, using this LD-quasigroup leads (in particular) to the homomorphism*

$$\mathrm{es} : B_n \twoheadrightarrow (\mathbf{Z}, +).$$

This example as well as the previous one belong to the general type $a{\wedge}b = f(b)$, $a{\vee}b = f^{-1}(b)$ with f a bijection of the considered domain.

Example 2.11. (**mean**) *Assume that E is a $\mathbf{Z}[t, t^{-1}]$-module. The binary operations*

$$a \wedge b = (1 - t)a + tb, \qquad a \vee b = (1 - t^{-1})a + t^{-1}b \tag{2.8}$$

turn E into an LD-quasigroup. For every n-strand braid b, and every sequence $\vec{a}$ in E^n, the bottom colours $\vec{a} \bullet b$ are linear combinations of the top colours $\vec{a}$. There exists an $n \times n$-matrix $r_B(b)$ satisfying

$$\vec{a} \bullet b = \vec{a} \times r_B(b), \tag{2.9}$$

and using the current LD-quasigroup leads to a linear representation

$$r_B : B_n \to \mathrm{GL}_n(\mathbf{Z}[t, t^{-1}]). \tag{2.10}$$

This linear representation is known as the (unreduced) Burau representation. Example 2.9 (trivial) corresponds to the specialization $t = 1$.

Example 2.12. **(conjugacy)** *Let G be a group. Then the binary operations defined by*

$$x \wedge y = xyx^{-1}, \qquad x \vee y = x^{-1}yx \tag{2.11}$$

turn G into an LD-quasigroup. In particular, let FG_n be the free group based on $\{x_1, \ldots, x_n\}$. For b an n-strand braid, let b^{rev} denote the image of b under the antiautomorphism of B_∞ that is the identity on the generators σ_i: thus an expression of b^{rev} is obtained from an expression of b by reversing the orders of the letters. Then define elements $y_1, \ldots, y_n$ of G by

$$(y_1, \ldots, y_n) = (x_1, \ldots, x_n) \bullet b^{rev},$$

and let $\widetilde{b}$ be the endomorphism of FG_n that maps x_i to y_i for every i. Then the mapping $\varphi : b \mapsto \widetilde{b}$ is an endomorphism of B_n into $\mathrm{End}(\mathrm{FG}_n)$, and, as $\widetilde{b^{-1}} = \widetilde{b}^{-1}$ holds by construction, the image of φ is actually included in $\mathrm{Aut}(\mathrm{FG}_n)$. As in Example 2.9 (trivial), using b^{rev} here is necessary because we consider an action on the right, while, by definition, composition of automorphisms corresponds to an action on the left. So using the current LD-quasigroup leads to a representation

$$\varphi : B_n \to \mathrm{Aut}(\mathrm{FG}_n).$$

It is known that this representation is faithful, i.e., φ is an embedding—this will be proved in Chapter III.

Using free differential calculus, one can see the mean operations (2.8) as a projection of the operations (2.11), corresponding to the fact that the Burau matrix of a braid can be deduced from the associated automorphism of a free group.

The previous list more or less exhausts classical LD-quasigroups. A few more are described in exercises below, but they are unessential variations. We see that using braid colourings and resorting to various (classical) LD-quasigroups leads to several (classical) representation results for braid groups. It is therefore natural to ask:

Question 2.13. *Can we find new examples of LD-quasigroups, and deduce further properties of braids?*

The answer to the first part is positive, but it seems that the answer to the second part is essentially negative. To explain this, let us introduce one more example.

Example 2.14. (half-conjugacy) *Let G be a group, and X be a subset of G. The binary operations defined by*

$$(a, x) \wedge (b, y) = (axa^{-1}b, y), \quad (a, x) \vee (b, y) = (ax^{-1}a^{-1}b, y) \tag{2.12}$$

turn $G \times X$ into an LD-quasigroup. This example is related to conjugacy as follows: let f be the mapping of $G \times X$ to G defined by $f((a, x)) = axa^{-1}$; then f is a surjective homomorphism of $(G \times X, \wedge, \vee)$ onto G equipped with the conjugacy operations of Example 2.12.

The point is that the previous example is, in some sense, the most general one: indeed, we shall check in Chapter V that, if FG_X is a free group based on X, then the LD-quasigroup consisting of $\mathrm{FG}_X \times X$ with the operations (2.12) is a free LD-quasigroup based on X. Hence every LD-quasigroup generated by X is a quotient of $\mathrm{FG}_X \times X$, and, therefore, the most powerful results about braids we can expect to obtain using LD-quasigroups are those coming from $\mathrm{FG}_X \times X$. The mapping f of Example 2.14 is not an embedding, and, in theory, we could obtain more using $\mathrm{FG}_X \times X$ with half-conjugacy than using FG_X with conjugacy, but the distance between the two structures is short—half-conjugacy is a sort of semidirect product of conjugacy with the operations of Example 2.10 (shift)—and no positive result in this direction is known.

On the other hand, the facts below suggest that LD-quasigroups are very special examples of LD-systems, and, in particular, they suffer strong restrictions in terms of orderability. Indeed, the LD-quasigroups of Examples 2.9 (trivial), 2.11 (mean), and 2.12 (conjugacy) all are idempotent, *i.e.*, they satisfy the identity

$$x \wedge x = x.$$

This is not true in the case of Examples 2.10 (shift) and 2.14 (half-conjugacy), but a weak form of idempotency still holds.

Lemma 2.15. *Every LD-quasigroup satisfies the identity*

$$(x \wedge x) \wedge y = x \wedge y. \tag{2.13}$$

Proof. Once we know that $\mathrm{FG}_X \times X$ equipped with half-conjugacy is a free LD-quasigroup, it suffices to check that (2.13) holds in this particular LD-quasigroup, which is straightforward. However, we can also write directly:

$$(x \wedge x) \wedge y = (x \wedge x) \wedge (x \wedge (x \vee y)) = x \wedge (x \wedge (x \vee y)) = x \wedge y. \qquad \blacksquare$$

If an LD-system embeds in an LD-quasigroup, it has to be left cancellative and, by Lemma 2.15, to satisfy Identity (2.13). Let us mention that the previous conditions turn out to be sufficient:

Proposition 2.16. (embedding) *An LD-system embeds in an LD-quasigroup if and only if it is left cancellative and it satisfies Identity* (2.13).

Proof. (Sketch) Assume that $(S, \wedge)$ is an LD-system satisfying the conditions of the proposition. Let a be an element of S such that the left translation $\mathrm{ad}_a : x \mapsto a \wedge x$ is not surjective. Letting S' be the union of S and of a disjoint copy of $S \setminus a \wedge S$, we can define on S' the structure of an LD-system in such a way that $S = a \wedge S'$ holds: this is possible as $\mathrm{ad}_a = \mathrm{ad}_{a \wedge a}$ holds. By iterating the process, one finds an extension S_a of S where the left translation ad_a is bijective. Finally, a new induction allows one to treat similarly all elements until the process comes to a saturation, *i.e.*, an LD-quasigroup is obtained. ■

Cycles for left division

Identity (2.13) implies that LD-quasigroups admit some sort of torsion.

In the sequel, if $(S, \wedge)$ is an arbitrary binary system, and a, b are elements of S, we say that a is a *left divisor* of b if $b = a \wedge x$ holds for some x. When we think about the associative context, *i.e.*, about semigroups, we see that, in the free semigroup $(\mathbf{N}_+, +)$, the left divisibility relation induces the usual order of the integers, and, in particular, it admits no cycle. On the other hand, in a finite semigroup, or in a torsion semigroup, the divisibility relation necessarily admits cycles. So, we can keep in mind the fact that the existence of cycles for left division is connected with a certain form of torsion.

If S is an idempotent LD-system, every element a of S is a left divisor of itself, and, therefore, (a) is a cycle of length 1 for left division. According to the above remark, this says that S has much torsion. The same result holds in every LD-quasigroup:

Proposition 2.17. (cycles) *In every LD-quasigroup, left division admits cycles of length* 1.

Proof. Assume that $(Q, ^\wedge)$ is an LD-quasigroup. For every a in Q, we have $(a^\wedge a)^\wedge a = a^\wedge a$, hence $(a^\wedge a)$ is a cycle of length 1 for left division. ■

Keeping our principle of using various LD-systems to colour the braids, and considering order features, we are led to the following double question:

Question 2.18. *(i) Do there exist an LD-system (necessarily not an LD-quasigroup) where left division admits no cycle—thus some sort of torsion-free LD-system?*

(ii) If so, can we use such an LD-system to colour braids?

We shall see in the sequel that the answer is positive in both cases. For the moment, let us conclude the section with one more example, which gives a (very) partial answer to Question 2.18(i).

Example 2.19. (**injection bracket**) *Let X be a nonempty set. On the set $\mathfrak{I}_X$ of all injections of X into itself, we define a new operation by*

$$f[g](n) = \begin{cases} fgf^{-1}(n) & \text{for } n \in \mathrm{Im}(f), \\ n & \text{otherwise}, \end{cases} \tag{2.14}$$

where $\mathrm{Im}(f)$ denotes the image of f. Below are displayed a few injections obtained in the case when X is $\mathbf{N}_+$ and the bracket is applied starting with the shift mapping $\mathrm{sh} : n \mapsto n+1$—in the picture, the source is the bottom line, and the target is the top one.

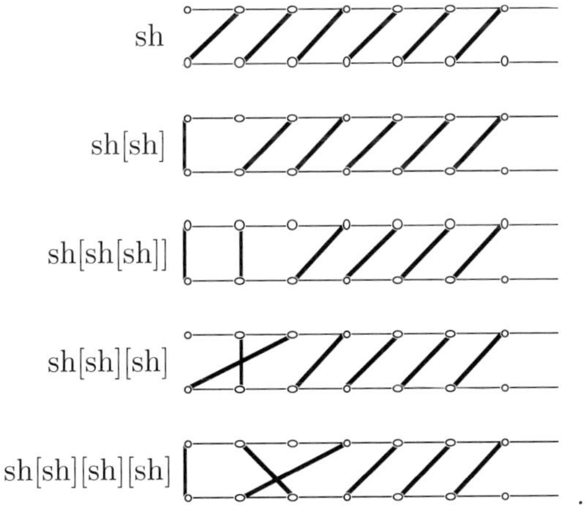

Let us denote by $\mathrm{coIm}(f)$ the coimage of f, i.e., the complement of $\mathrm{Im} f$ in X, and by $\mathrm{Gr}(f)$ the graph of f, i.e., the set of all pairs $(n, f(n))$ for $n \in X$. The equality

$$\mathrm{Gr}(f[g]) = f(\mathrm{Gr}(g)) \cup \{(n,n); n \in \mathrm{coIm}(f)\} \tag{2.15}$$

shows that $f[g]$ is in some sense the image of g under f, completed in the most trivial way when no value is forced. By definition, $fgf^{-1}(n)$ belongs to the image of f, hence $f[g]$ is an injection when f and g are, and the bracket is defined everywhere on $\mathfrak{I}_X$. Observe that, as far as bijections are concerned, the bracket operation coincides with conjugacy in the symmetric group. Thus the nontrivial part involves non-bijective injections only.

Proposition 2.20. (**injection bracket**) *(i) For every set X, the set $\mathfrak{I}_X$ equipped with the bracket of (2.14) is a left cancellative LD-system.*

(ii) Let $\mathfrak{I}_\infty$ denote the set of all injections of $\mathbf{N}_+$ into itself that eventually coincide with a shift, i.e., the set of all f's such that, for some c, $f(p) = p + c$ holds for p large enough. Then $\mathfrak{I}_\infty$ and $\mathfrak{I}_\infty \setminus \mathfrak{S}_\infty$ equipped with $[\,]$ are LD-systems, and left division in $\mathfrak{I}_\infty \setminus \mathfrak{S}_\infty$ admits no cycle of length 1—but it admits cycles of length 2.

Proof. (i) We have to check $f[g[h]] = f[g][f[h]]$ for all f, g, h. First, we have

$$\mathrm{Im}(f[g]) = \mathrm{coIm}(f) \cup f(\mathrm{Im}(g)), \tag{2.16}$$
$$\mathrm{coIm}(f[g]) = f(\mathrm{coIm}(g)). \tag{2.17}$$

Assume $p \in X$. For $p \in \mathrm{coIm}(f)$, we have $f[g[h]](p) = f[g](p) = f[h](p) = p$, hence $f[g][f[h]](p) = f[g]f[h](p) = p$. For $p = f(q)$, with $q \in \mathrm{coIm}(g)$, we get $f[g[h]](p) = f(g[h](q)) = f(q) = p$. Now, by the equality above, we have $p \in \mathrm{coIm}(f[g])$, hence $f[g][f[h]](p) = p$ as well. Finally, for $p = fg(r) = f[g]f(r)$, we have $f[g[h]](p) = f(g[h](g(r))) = fgh(r)$, and $f[g][f[h]](p) = f[g](f[h](f(r))) = f[g](fh(r)) = fgh(r)$. Hence $(\mathfrak{I}_X, [\,])$ is an LD-system.

Next, the bracket operation admits left cancellation. Indeed, assume $f[g] = f[g']$. For each p, we find $f(g(p)) = f[g](f(p)) = f[g'](f(p)) = f(g'(p))$, hence $g(p) = g'(p)$.

(ii) Assume that $f(p) = p + c_f$ and $g(p) = p + c_g$ hold for $p \geq p_0$. For $p \geq p_0 + c_f$, we have $f[g](p) = f[g](f(p - c_f)) = f(g(p - c_f)) = f(p - c_f + c_g) = p + c_g$, so $f[g]$ belongs to $\mathfrak{I}_\infty$ whenever f and g do.

Then (2.17) implies $\mathrm{coIm}(f[g]) \subseteq \mathrm{Im}(f)$, hence $\mathrm{coIm}(f[g]) \cap \mathrm{coIm}(f) = \emptyset$. So, an equality of the form $f[g] = f$ is possible only when $\mathrm{coIm}(f) = \mathrm{coIm}(g) = \emptyset$ holds, *i.e.*, for $f, g \in \mathfrak{S}_\infty$.

Writing sh for the shift mapping, we define inductively the right and left powers of sh respectively by $\mathrm{sh}^{[1]} = \mathrm{sh}_{[1]} = \mathrm{sh}$ and $\mathrm{sh}^{[p]} = \mathrm{sh}[\mathrm{sh}^{[p-1]}]$, $\mathrm{sh}^{[p]} = \mathrm{sh}_{[p-1]}[\mathrm{sh}]$ for $p \geq 2$. Then consider $f = \mathrm{sh}^{[3]}[\mathrm{sh}^{[2]}]$. The reader can check by a direct computation that $f = \mathrm{sh}_{[3]}[\mathrm{sh}_{[2]}]$ holds. Applying (LD), we find

$$\begin{aligned} f &= \mathrm{sh}[\mathrm{sh}^{[2]}][\mathrm{sh}^{[2]}] = \mathrm{sh}_{[2]}[\mathrm{sh}_{[2]}][\mathrm{sh}^{[2]}] \\ &= \mathrm{sh}_{[3]}[\mathrm{sh}_{[3]}][\mathrm{sh}^{[2]}] = \mathrm{sh}_{[3]}[\mathrm{sh}_{[2]}][\mathrm{sh}_{[3]}[\mathrm{sh}]][\mathrm{sh}^{[2]}] = f[\mathrm{sh}_{[4]}][\mathrm{sh}^{[2]}], \end{aligned}$$

which shows that $(f, f[\mathrm{sh}_{[4]}])$ is a cycle for left division in $(\mathfrak{I}_\infty \setminus \mathfrak{S}_\infty, [\,])$. ■

The previous example is our first non-classical LD-system. It has been inspired by the LD-system associated with an elementary embedding described in Chapter XII, and we shall connect it soon with other examples.

Exercise 2.21. ***(left and right equivalences)*** *(i) Assume that $(S, ^\wedge)$ is an LD-system. Write $a \equiv_L b$ for $(\forall x)(a^\wedge x = b^\wedge x)$. Prove that $\equiv_L$ is an equivalence relation on S that is compatible with $^\wedge$ on the right. Symmetrically, write $a \equiv_R b$ for $(\forall x)(x^\wedge a = x^\wedge b)$. Show that $\equiv_R$ is a congruence on $(S, ^\wedge)$, i.e., an equivalence relation on S that is compatible with $^\wedge$ on both sides.*

(ii) Assume that $(S, ^\wedge)$ is a left divisible LD-system. Show that $\equiv_L$ is a congruence on $(S, ^\wedge)$.

(iii) Assume that $(Q, ^\wedge, ^\vee)$ is an LD-quasigroup. Write $a \equiv_L b$ for $(\forall x)(a^\wedge x = b^\wedge x)$. Prove that $\equiv_L$ is a congruence on $(Q, ^\wedge, ^\vee)$, and that the quotient LD-quasigroup $Q/\equiv_L$ embeds in the LD-quasigroup made of $\mathrm{Aut}(Q)$ equipped with the conjugacy operations (hence, in particular, it is idempotent). Deduce that every right cancellative LD-quasigroup is idempotent and it embeds in a group equipped with conjugacy. Extend the result to every LD-quasigroup where $\equiv_L$ is trivial.

Exercise 2.22. ***(variations about conjugacy)*** *(i) Assume that G is a group. For $i \in \mathbf{Z}$, define $*_i$ by $x *_i y = x^i y x^{-i}$. Show that $(G, *_i)$ is an LD-quasigroup, and $*_i$ is left distributive with respect to $*_j$ for all i, j.*

(ii) For f in $\mathrm{End}(G)$, define $*_f$ on G by $x *_f y = x f(y x^{-1})$. Prove that $(G, *_f)$ is an LD-system, and an LD-quasigroup if f is an automomorphism. [A particular case is f being an inner automorphism of G: this amounts to fixing an element a of G and defining $x *_a y = x a y x^{-1} a^{-1}$.]

(iii) Define on the set $G^{\mathbf{Z}}$ of all $\mathbf{Z}$-indexed sequences from G the operation $^\wedge$ so that $\vec{x}^\wedge \vec{y} = \vec{z}$ holds for $z_i = x_i\, y_{i+1}\, x_{i+1}^{-1}$. Show that $(G^{\mathbf{Z}}, ^\wedge)$ is an LD-quasigroup.

(iv) For f in $\mathrm{End}(G)$, define $^\wedge$ on G by $x^\wedge y = x f(x^{-1} y)$. Show that $(G, ^\wedge)$ is an LD-system.

Exercise 2.23. ***(bilinear operations)*** *Define $^\wedge$ on $\mathbf{R}^3$ by*

$$(x_1, x_2, x_3)^\wedge(y_1, y_2, y_3) = (y_1, y_2, y_3 + x_1 y_2 - x_2 y_1).$$

Show that $(\mathbf{R}^3, ^\wedge)$ is an LD-quasigroup. [Hint: Observe that $^\wedge$ is the translation of conjugacy on unipotent 3×3-matrices.] Translate similarly the operations of Exercise 2.22.

Exercise 2.24. **(half-conjugacy)** *(i) Show that the relation $\equiv_L$ (Exercice 2.21) associated with half-conjugacy on $G \times A$ is the congruence associated with the projection f of Example 2.14: $(a, x) \equiv_L (a', x')$ is equivalent to $f((a, x)) = f((a', x'))$, i.e., to $axa^{-1} = a'x'a'^{-1}$. So, in particular, $(1, x) \equiv_L (x, x)$ holds.*

(ii) Prove that every element (x, a) generates under $^\wedge$ an infinite LD-system that is isomorphic to $(\mathbf{N}, {}^\wedge)$ where $p{}^\wedge q = q + 1$.

Exercise 2.25. **(symmetric)** *Say that the binary system $(S, {}^\wedge)$ is n-symmetric if it satisfies the identity $x{}^\wedge x\ op \cdots {}^\wedge x{}^\wedge y = y$, n times x. Prove that an n-symmetric LD-system is an LD-quasigroup, the second operation being defined by $x{\vee}y = x{}^\wedge x{}^\wedge \cdots {}^\wedge x{}^\wedge y$, $n - 1$ times x. [This is true in particular for $n = 1$, the second operation being then trivial.]*

Exercise 2.26. **(core)** *(i) Let G be a group. Define $^\wedge$ on G by $x{}^\wedge y = xy^{-1}x$. Prove that $(G, {}^\wedge)$ is a 2-symmetric LD-quasigroup [cf. Exercice 2.25].*

(ii) Extend the previous result to the operation defined by $x *_f y = xf(y^{-1}x)$, where f is an idempotent endomorphism of G. Show that the LD-system $(G, *_f)$ is an LD-quasigroup whenever f is an automorphism, and that, for n even (resp. odd), it is n-symmetric whenever f^n is the identity (resp. G is abelian and f^{2n} is the identity).

Exercise 2.27. **(reflections)** *Let V be a vector space with a symmetric bilinear form whose value on x, y is denoted $\langle x, y\rangle$. Let H the set of all non-isotrop vectors of V. Define $^\wedge$ on H by $x{}^\wedge y = y - 2\dfrac{\langle y, x\rangle}{\langle x, x\rangle}x$. Show that $(H, {}^\wedge)$ is a 2-symmetric LD-quasigroup. (cf. Exercise 2.25). [Thus $x{}^\wedge y$ is the image of y under the orthogonal reflection associated with x. In the case of a Euclidian space, the finite sub-LD-systems of the systems $(H, {}^\wedge)$ are the root systems, i.e., the finite families of points closed under orthogonal reflections. Another sub-LD-system is given by the n-sphera S^n: for radius 1, the binary operation induced on S^n is $x{}^\wedge y = y - 2\langle y, x\rangle x$.]*

Exercise 2.28. **(injection bracket)** *(i) Prove that, in the LD-system $(\mathfrak{I}_\infty, [\ ])$, the mapping ad : $f \mapsto (g \mapsto f[g])$ is injective. [Hint: Prove $f(\mathrm{coIm}(g)) = f'(\mathrm{coIm}(g))$ for every g, and choose for g an injection such that $\mathrm{coIm}(g)$ has exactly one element.] Prove similarly that the mapping $g \mapsto (f \mapsto f[g])$ is injective.*

(ii) Prove that $(\mathfrak{I}_\infty, [\ ])$ is neither right cancellative, nor left or right divisible. Show that $(\mathfrak{I}_\infty, [\ ])$ cannot be embedded in an LD-quasigroup.

(iii) Write $\mathrm{Fix}(g)$ *for the set of all fixed points of g. Prove that $(\exists h)(g = f[h])$ holds in $\mathfrak{I}_\infty$ if and only if $\mathrm{coIm}(f) \subseteq \mathrm{Fix}(g)$ and $g(\mathrm{Im}(f)) \subseteq \mathrm{Im}(f)$ holds Deduce that g is left divisible by f if and only if $f[g] = f[f][g]$ holds. [Hint: Assume $f[g] = f[f][g]$. For $n \in \mathrm{coIm}(f)$, we have $f[f](n) = f[g](n) = n$, so $f[g](n) = f[f][g](n)$ implies $f[f](g(n)) = n$, hence $g(n) = n$, and $\mathrm{coIm}(f) \subseteq \mathrm{Fix}(g)$. Assume $n \in \mathrm{coIm}(f)$. By the previous result, we have $g(n) = n$, hence assuming $n \in g(\mathrm{Im}(f))$ implies $n \in \mathrm{Im}(f)$, a contradiction. Hence $g(\mathrm{Im}(f)) \subseteq \mathrm{Im}(f)$ holds.]*

Exercise 2.29. **(injection bracket)** *Consider the injection bracket on $\mathbf{N}_+ \times \mathbf{N}_+$. Let a and b respectively denote the first and the second shift $(p, q) \mapsto (p+1, q)$ and $(p, q) \mapsto q+1$. Establish $a[b] \neq b[a]$ and $a[b[a[b]]] = a[a[b[b]]]$. Prove that every element in the sub-LD-system generated by a and b is a shift outside some finite rectangle.*

Exercise 2.30. **(generalized bracket)** *Let S denote the subset of $\mathfrak{I}_\infty^2$ consisting of those pairs (f_1, f_2) satisfying $f_2(\mathrm{coIm}(f_1)) = \mathrm{coIm}(f_1)$. Show that the formula $(f_1, f_2)[(g_1, g_2)] = (h_1, h_2)$, with*

$$\begin{cases} h_1(n) = f_1 g_1 f_1^{-1}(n), \quad h_2(n) = f_1 g_2 f_1^{-1}(n) & \text{for } n \in \mathrm{Im}(f_1), \\ h_1(n) = h_2(n) = f_2(n) & \text{otherwise,} \end{cases}$$

defines a left self-distributive operation on S.

Exercise 2.31. **(Wada's identities)** *(i) Assume that ^ and $\circ$ are binary operations on S. We consider colourings obeying the rule*

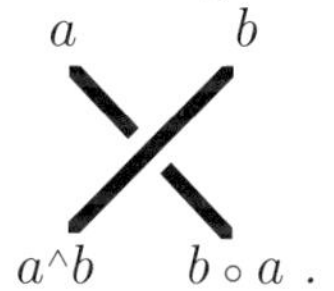

Show that the associated right action of BW_n^+ on S^n is invariant under positive braid relations if and only if $(S, {}^\wedge, \circ)$ satisfies the identities

$$x^\wedge(y^\wedge z) = (x^\wedge y)^\wedge((y \circ x)^\wedge z),$$
$$x \circ (y \circ z) = (x \circ y) \circ ((y^\wedge x) \circ z),$$
$$((y \circ x)^\wedge z) \circ (x^\wedge y) = ((y^\wedge z) \circ x)^\wedge(z \circ y).$$

(ii) Let G be a group. Show that the operations defined on G by $a^\wedge b = a(ab)^{-1}$, $a \circ b = (ab)b$ satisfy the above identities. Show that completing the construction with $a_\vee b = a(ab)$, $a \bullet b = (ab)^{-1}b$ defines an action of B_n on G^n. Generalize to $a^\wedge b = af(ab)^{-1}$, $a_\vee b = f(ab)b$, where f is an endomorphism of G satisfying $f^3 = f$.

---

3. A Self-Distributive Operation on Braids

In the previous section, we have seen how self-distributive systems can be used to obtain results about braids. In this section, we go the other way, and show how braids can be used to obtain results about self-distributivity by constructing a left self-distributive operation on braids. The two approaches look unrelated at first, but we shall see in Chapter VIII that both relie on one and the same fact, namely the connection between a certain group G_{LD} associated with self-distributivity and Artin's braid group B_∞.

Braid exponentiation

In the sequel, we consider braids with unboundedly many strands. For every integer n, the mapping that adds a trivial unbraided $(n+1)$-th strand at the right of a geometric n-strand braid induces an embedding $i_{n,n+1}$ of the group B_n into the group B_{n+1}. Then $(B_n, i_{n,n+1})_{n\geq 2}$ is a directed system.

Definition 3.1. (**group** B_∞) The direct limit of the system $(B_n, i_{n,n+1})$ is denoted B_∞.

Then B_∞ is the group generated by an infinite sequence of generators $\sigma_1, \sigma_2, \ldots$ submitted to the braid relations (1.1) and (1.2); its elements can be represented by braid diagrams with strands indexed by the positive integers. Notice that, in every such diagram, only finitely many strands are braided.

Definition 3.2. (**shift**) We write sh for the *shift* endomorphism of B_∞ that maps σ_i to σ_{i+1} for every i.

Lemma 3.3. *The endomorphism* sh *of* B_∞ *is injective.*

Proof. An algebraic argument is not obvious: if w, w' are braid words and $\mathrm{sh}(w) \equiv \mathrm{sh}(w')$ holds, we can deduce $w \equiv w'$ directly only if we know that all intermediate braid words in a sequence of elementary transformations from w to w' belong to the image of sh. This is true if we restrict to positive words, but this need not be the case in general as new factors $\sigma_1\sigma_1^{-1}$ or $\sigma_1^{-1}\sigma_1$ may appear. A geometric argument can be used here: for each braid diagram d, let us denote by $\mathrm{coll}(d)$ the diagram obtained from d by removing the strand that begins at position 1. It is easy to check that d being isotopic to d' implies $\mathrm{coll}(d)$ being isotopic to $\mathrm{coll}(d')$. Hence, with obvious definitions, $\mathrm{sh}(d) \equiv \mathrm{sh}(d')$ implies $\mathrm{coll}(\mathrm{sh}(d)) \equiv \mathrm{coll}(\mathrm{sh}(d'))$, *i.e.*, $d \equiv d'$. ■

Let us consider the question of constructing a left self-distributive operation ^ on B_∞, called braid exponentiation in the sequel. In Chapter VIII, we shall give deep arguments for such an operation to exist. For the moment, a heuristic approach will lead us to a possible definition of ^. Let us assume that the latter operation exists, and that we can use $(B_\infty, {}^\wedge)$ to colour braids—which *a priori* should require the latter structure to be an LD-quasigroup. We introduce the notion of a self-colouring braid as follows:

Definition 3.4. (self-colouring) (Assuming that a left self-distributive operation ^ has been defined on B_∞,) we say that a braid b is *self-colouring* if $(1, 1, 1, \ldots) \bullet b = (b, 1, 1, \ldots)$ holds, *i.e.*, if b produces itself starting from the colours 1, 1, ... (Figure 3.1)

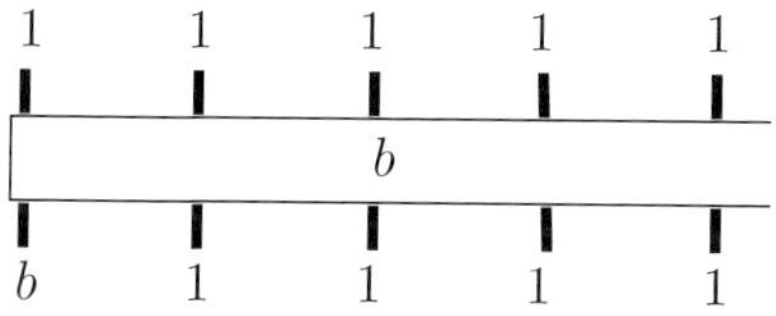

Figure 3.1. Self-colouring braid

As the existence and the definition of braid exponentiation remain open, what self-colouring braids are is not clear. However, self-colouring braids do exist, for, by definition, at least the unit braid 1 is self-colouring. Now, let us try to find a formula for operation ^ so that self-colourings braids are closed under ^. The following result comes directly from Figure 3.2:

Lemma 3.5. *Assume that a and b are self-colouring braids. Then we have* $(1, 1, \ldots) \bullet a\ \mathrm{sh}(b)\ \sigma_1\ \mathrm{sh}(a^{-1}) = (a^\wedge b, 1, 1, \ldots)$.

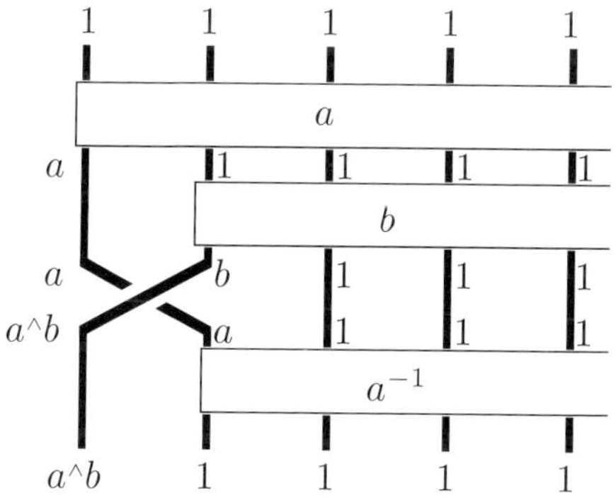

Figure 3.2. Braid exponentiation

So the family of self-colouring braids will be closed under exponentiation provided the latter is defined by the formula of Lemma 3.5. This leads us to:

Definition 3.6. (**braid exponentiation**) For a, b in B_∞, we define

$$a \wedge b = a \ \mathrm{sh}(b) \ \sigma_1 \ \mathrm{sh}(a^{-1}). \tag{3.1}$$

Example 3.7. ***(braid exponentiation)*** *The reader can check the following equalities*

$$1 \wedge 1 = \sigma_1 =$$  , $$1 \wedge (1 \wedge 1) = \sigma_2\sigma_1 =$$  ,

$$(1 \wedge 1) \wedge 1 = \sigma_1^2 \sigma_2^{-1} =$$  .

From the above arguments, we can expect exponentiation to be self-distributive on self-colouring braids. It being self-distributive on all of B_∞ is not obvious, but it is a matter of a simple computation.

Proposition 3.8. (**LD-system**) *The set B_∞ equipped with exponentiation is a left cancellative LD-system.*

Proof. We can either make a direct algebraic computation (*cf.* Exercise 3.19), or draw a picture (Figure 3.3) to check that a^$(b$^$c) = (a$^$b)$^$(a$^$c)$ holds for all a, b, c. Left cancellativity follows from sh being injective. ∎

Although the definition of braid exponentiation resembles a skew conjugacy, its properties are quite different from those of conjugacy, in particular as far as left division is concerned.

Proposition 3.9. (**not LD-quasigroup**) *No equality a^$b = 1$ holds in B_∞. Hence, in particular, $(B_\infty, \wedge)$ is not an LD-quasigroup.*

Proof. Let a, b be two braids with associated permutations f, g. Let $p = f(1) + 1$. Defining s_1 to be the transposition that exchanges 1 and 2, we find

$$\begin{aligned} \mathrm{perm}(a \wedge b)(p) &= f \circ \mathrm{sh}(g) \circ s_1 \circ \mathrm{sh}(f)^{-1}(p) \\ &= f \circ \mathrm{sh}(g) \circ s_1(2) = f \circ \mathrm{sh}(g)(1) = f(1) = p - 1, \end{aligned}$$

hence perm$(a$^$b)$ is not the identity, and a^b cannot be the unit braid. ∎

We refer to exercises for a complete study of division in $(B_\infty, \wedge)$.

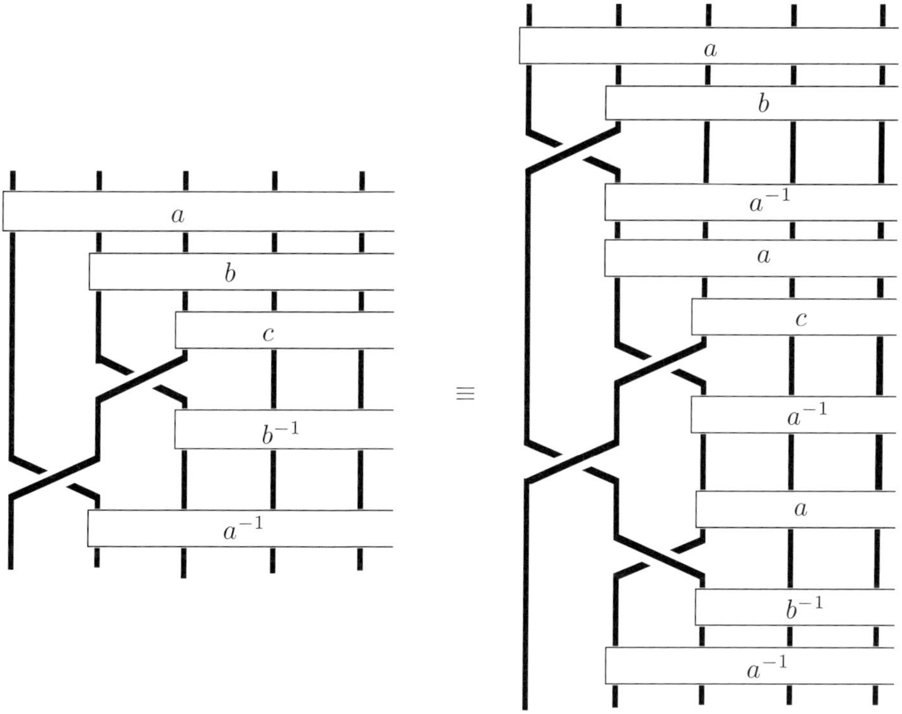

Figure 3.3. Self-distributivity of braid exponentiation

Derived operations

New left self-distributive operations can be deduced from braid exponentiation by using representations of the braid group B_∞. It need not be true that every group homomorphism of B_∞ induces a left self-distributive structure on the image, but a weak compatibility condition is sufficient.

Lemma 3.10. *Assume that r is a surjective group homomorphism defined on B_∞. A sufficient condition for the exponentiation of B_∞ to induce a left distributive operation on the image of r is the inclusion* $\mathrm{Ker}(r) \subseteq \mathrm{Ker}(r \circ \mathrm{sh})$.

The verification is left to the reader. Let us consider some examples.

Example 3.11. (permutations) *For every positive integer n, the mapping* perm *is a surjective homomorphism of B_n onto $\mathfrak{S}_n$. The symmetric groups form a directed system when equipped with the embeddings induced by the identity mapping of $\{1, \ldots, n\}$ into $\{1, \ldots, n+1\}$, and the direct limit is the group $\mathfrak{S}_\infty$ of those permutations of $\mathbf{N}_+$ that move* *finitely many integers only. Then* perm *extends into a surjective homomorphism of B_∞ onto $\mathfrak{S}_\infty$, and the compatibility relation of Lemma 3.10 is satisfied. So we obtain a left self-distributive operation on $\mathfrak{S}_\infty$ by defining*

$$f \wedge g = f \circ \mathrm{sh}(g) \circ s_1 \circ \mathrm{sh}(g^{-1}), \tag{3.2}$$

where s_1 is the transposition (1 2) and sh *is the shift endomorphism of $\mathfrak{S}_\infty$ defined by $\mathrm{sh}(f)(1) = 1$ and $\mathrm{sh}(f)(n) = f(n-1)+1$ for $n > 1$.*

Example 3.12. (**Burau matrices**) *The Burau representation of B_n into $\mathrm{GL}_n(\mathbf{Z}[t,t^{-1}])$ has been introduced in Example 2.11 (mean). It easily extends into a representation of B_∞. Let $i_{n,n+1}$ and sh_n denote the embeddings of $\mathrm{GL}_n(\mathbf{Z}[t,t^{-1}])$ into $\mathrm{GL}_{n+1}(\mathbf{Z}[t,t^{-1}])$ defined respectively by*

$$i_{n,n+1} : M \longmapsto \begin{pmatrix} & & & 0 \\ & M & & \vdots \\ & & & 0 \\ 0 & \cdots & 0 & 1 \end{pmatrix}, \qquad \mathrm{sh}_n : M \longmapsto \begin{pmatrix} 1 & 0 & \cdots & 0 \\ 0 & & & \\ \vdots & & M & \\ 0 & & & \end{pmatrix}.$$

The direct limit $\mathrm{GL}_\infty(\mathbf{Z}[t,t^{-1}])$ of the system $(\mathrm{GL}_n(\mathbf{Z}[t,t^{-1}]), i_{n,n+1})$ is the union of the groups $\mathrm{GL}_n(\mathbf{Z}[t,t^{-1}])$ turned into a group: the product of two matrices with different sizes is obtained by first completing the smaller matrix trivially on the right and the bottom. Then the mappings sh_n induce a well-defined endomorphism of $\mathrm{GL}_\infty(\mathbf{Z}[t,t^{-1}])$. The Burau representation r_B of B_∞ is defined by the rules

$$r_B(\sigma_1) = \begin{pmatrix} 1-t & t \\ 1 & 0 \end{pmatrix}, \qquad r_B(\sigma_{1+i}) = \mathrm{sh}^i(\rho(\sigma_1)).$$

The condition of Lemma 3.10 is fulfilled, so carrying the exponentiation of B_∞ gives a distributive operation on the image of r_B. The latter set is a proper subset of $\mathrm{GL}_\infty(\mathbf{Z}[t,t^{-1}])$, but we can consider the formula that defines the operation, namely

$$A \wedge B = A\ \mathrm{sh}(B)\ r_B(\sigma_1)\ \mathrm{sh}(A^{-1}), \tag{3.3}$$

for A, B ranging on all of $\mathrm{GL}_\infty(\mathbf{Z}[t,t^{-1}])$, and it is easy to check left self-distributivity directly.

So we have found a left distributive operation on matrices. Starting from the unit, which, under our conventions, is the 1,1-matrix (1), we can compute simple expressions like

$$(1) \wedge (1) = \begin{pmatrix} 1-t & t \\ 1 & 0 \end{pmatrix}, \quad (1) \wedge ((1) \wedge (1)) = \begin{pmatrix} 1-t & t & 0 \\ 1-t & 0 & t \\ 1 & 0 & 0 \end{pmatrix},$$

$$((1)\wedge(1))\wedge(1) = \begin{pmatrix} 1-t+t^2 & 0 & t-t^2 \\ 1-t & 0 & t \\ 0 & t^{-1} & 1-t^{-1} \end{pmatrix}.$$

Specializing t (with $t \neq 0$) gives a left distributive operation on matrices with complex, real or integer entries. In particular, $t = 1$ gives the exponentiation of permutations described above.

Example 3.13. (automorphisms of free groups) *We have seen in Section 2 that letting the braid group B_n act on a free group FG_n of rank n gives a homomorphism φ of B_n into the group $\mathrm{Aut}(\mathrm{FG}_n)$. Again, the construction is compatible with the canonical embedding of B_n into B_{n+1}, so, if FG_∞ denotes the free group based on the infinite sequence $\{x_1, x_2, \ldots\}$, we obtain a homomorphism, still denoted φ, of B_∞ into $\mathrm{Aut}(\mathrm{FG}_\infty)$. By construction, the automorphism $\widetilde{\sigma}_i$ is determined by*

$$\widetilde{\sigma}_i(x_k) = \begin{cases} x_k & \text{for } k < i \text{ or } k > i+1, \\ x_i x_{i+1} x_i^{-1} & \text{for } k = i, \\ x_i & \text{for } k = i+1. \end{cases} \tag{3.4}$$

We write sh for the shift endomorphism of FG_∞ that maps x_i to x_{i+1} for every i, and also for the shift endomorphism of the group $\mathrm{Aut}(\mathrm{FG}_\infty)$ defined as follows: for f in $\mathrm{Aut}(\mathrm{FG}_\infty)$, $\mathrm{sh}(f)$ is the automorphism defined by $\mathrm{sh}(f)(x_1) = x_1$ and $\mathrm{sh}(f)(x_{i+1}) = \mathrm{sh}(f(x_i))$ for $i \geq 1$. For instance, we have $\mathrm{sh}(\widetilde{\sigma}_i) = \widetilde{\sigma}_{i+1}$ for every i. Copying the definition of braid exponentiation leads to introducing for f, g in $\mathrm{Aut}(\mathrm{FG}_\infty)$

$$f \wedge g = f \circ \mathrm{sh}(g) \circ \widetilde{\sigma}_1 \circ \mathrm{sh}(f^{-1}). \tag{3.5}$$

It is easy to verify that this operation is left self-distributive on $\mathrm{Aut}(\mathrm{FG}_\infty)$, and that φ is a homomorphism of $(B_\infty, \wedge)$ into $(\mathrm{Aut}(\mathrm{FG}_\infty), \wedge)$.

Cycles for left division

As mentioned above, one of the restrictions suffered by all LD-quasigroups is a form of torsion, namely that left division admits lots of cycles, even cycles of length 1. However, we have seen that the bracket operation on injections yields an LD-system where left division admits no cycle of length 1. In this section, we prove a much stronger statement in this direction, namely that left division for the exponentations of $\mathrm{Aut}(\mathrm{FG}_\infty)$ and of B_∞ admit no cycle. Such LD-systems are therefore quite different from idempotent LD-systems, and, more generally, from all previously known examples. This fact will be crucial for the construction of a braid order in Chapter III.

Proposition 3.14. **(automorphisms acyclic)** *Left division in* $\mathrm{Aut}(\mathrm{FG}_\infty)$ *equipped with the operation of* (3.5) *admits no cycle.*

The result follows from two lemmas about possible free reductions occurring when computing the image of x_1. In the sequel, we write X for $\{x_1, x_2, \ldots\}$, and X^{-1} for $\{x_1^{-1}, x_2^{-1}, \ldots\}$. For w a word on $X \cup X^{-1}$, we say that w is *freely reduced* if it contains no pattern $x_i x_i^{-1}$ or $x_i^{-1} x_i$, and, in every case, we denote by $\mathrm{red}(w)$ the unique reduced word that can be obtained from w by iteratively deleting the patterns $x_i x_i^{-1}$ or $x_i^{-1} x_i$. Then the free group FG_∞ identifies with the set of all freely reduced words equipped with the operation $u \cdot v = \mathrm{red}(uv)$.

Definition 3.15. **(sets $S(x)$)** For x in $X \cup X^{-1}$, we define $S(x)$ to be the subset of FG_∞ consisting of all reduced words that finish with the letter x.

Let us investigate the image of the set $S(x_1^{-1})$ under the automorphism $\widetilde{\sigma}_i^{\pm 1}$.

Lemma 3.16. *For every f in* $\mathrm{Aut}(\mathrm{FG}_\infty)$, $\mathrm{sh}(f)$ *maps $S(x_1^{-1})$ into itself.*

Proof. Let us consider an arbitrary element of $S(x_1^{-1})$, say wx_1^{-1}, with $w \notin S(x_i)$. By construction, we have $\mathrm{sh}(f)(wx_1^{-1}) = \mathrm{red}(\mathrm{sh}(f)(w)\, x_1^{-1})$. Assume $\mathrm{sh}(f)(wx_1^{-1}) \notin S(x_1^{-1})$. Then the final letter x_1^{-1} in $\mathrm{sh}(f)(w)x_1^{-1}$ is cancelled by some letter x_1 occurring in $\mathrm{sh}(f)(w)$. Such a letter x_1 in $\mathrm{sh}(f)(w)$ must come from a letter x_1 in w. So there exists a decomposition $w = w_1 x_1 w_2$ satisfying $\mathrm{sh}(f)(w_2) = 1$. As $\mathrm{sh}(f)$ is injective, the latter condition implies $w_2 = 1$, hence $w \in S(x_1)$, contradicting the hypothesis. ■

Lemma 3.17. *The automorphism $\widetilde{\sigma}_1$ maps $S(x_1^{-1})$ into itself.*

Proof. Let us consider again an arbitrary element of $S(x_1^{-1})$, say wx_1^{-1}, with $w \notin S(x_i)$. We have $\widetilde{\sigma}_1(wx_1^{-1}) = \mathrm{red}(\widetilde{\sigma}_1(w)\ x_1 x_2^{-1} x_1^{-1})$. Assume $\widetilde{\sigma}_1(wx_1^{-1}) \notin S(x_1^{-1})$. Then the final x_1^{-1} in $\widetilde{\sigma}_1(w)\ x_1 x_2^{-1} x_1^{-1}$ is cancelled by some letter x_1 in $\widetilde{\sigma}_1(w)$, which comes either from some x_2 or from some $x_1^{\pm 1}$ in w.

In the first case, we display the letter x_2 involved and write $w = w_1 x_2 w_2$, with w_2 a reduced word not beginning with x_2^{-1}. We find

$$\widetilde{\sigma}_1(wx_1^{-1}) = \mathrm{red}(\widetilde{\sigma}_1(w_1)\ x_1\ \widetilde{\sigma}_1(w_2)\ x_1 x_2^{-1} x_1^{-1}),$$

and the hypothesis $\mathrm{red}(\widetilde{\sigma}_1(w_2) x_1 x_2^{-1}) = \varepsilon$ implies $\widetilde{\sigma}_1(w_2) = x_2 x_1^{-1} = \widetilde{\sigma}_1(x_2^{-1} x_1)$. As $\widetilde{\sigma}_1$ is injective, we deduce $w_2 = x_2^{-1} x_1$, contradicting the hypothesis that w_2 does not begin with x_2^{-1}.

In the second case, we write similarly $w = w_1 x_1^e w_2$ with $e = \pm 1$. We find now

$$\widetilde{\sigma}_1(wx_1^{-1}) = \mathrm{red}(\widetilde{\sigma}_1(w_1)\ x_1 x_2^e x_1^{-1}\ \widetilde{\sigma}_1(w_2)\ x_1 x_2^{-1} x_1^{-1}),$$

and the hypothesis $\mathrm{red}(x_2^e x_1^{-1} \widetilde{\sigma}_1(w_2) x_1 x_2^e) = \varepsilon$ implies $\widetilde{\sigma}_1(w_2) = x_1 x_2^{1-e} x_1^{-1} = \widetilde{\sigma}_1(x_1^{1-e})$, hence $w_2 = x_1^{1-e}$, and either $w_2 = \varepsilon$ or $w_2 = x_1^2$. In both cases, w belongs to $S(x_1)$, a contradiction. ■

Lemma 3.18. *Assume that f is an element of* $\mathrm{Aut}(\mathrm{FG}_\infty)$ *that can be expressed as a product of elements in the image of* sh *and of* $\widetilde{\sigma}_1$, *the latter occurring at least once. Then we have* $f(x_1) \in S(x_1^{-1})$, *and, therefore,* $f \neq \mathrm{id}$.

Proof. The hypothesis is that f admits a decomposition of the form

$$f = \mathrm{sh}(f_0) \circ \widetilde{\sigma}_1 \circ \mathrm{sh}(f_1) \circ \cdots \circ \widetilde{\sigma}_1 \circ \mathrm{sh}(f_p).$$

Then we have $\mathrm{sh}(f_p)(x_1) = x_1$, and $\widetilde{\sigma}_1(x_1) = x_1 x_2 x_1^{-1}$, an element of $S(x_1^{-1})$. By Lemmas 3.16 and 3.17, every subsequent factor $\mathrm{sh}(f_k)$ and $\widetilde{\sigma}_1$ maps $S(x_1^{-1})$ into $S(x_1^{-1})$. ∎

We can now prove Proposition 3.14 (automorphisms acyclic). We have to show that no equality of the form

$$f = (\cdots(f \mathbin{^\wedge} g_1) \mathbin{^\wedge} \cdots) \mathbin{^\wedge} g_p \tag{3.6}$$

is possible in $\mathrm{Aut}(\mathrm{FG}_\infty)$. By using the explicit definition of the operation $^\wedge$ on $\mathrm{Aut}(\mathrm{FG}_\infty)$, we expand (3.6) into an equality of the form

$$f = f \circ \mathrm{sh}(f_0) \circ \widetilde{\sigma}_1 \circ \mathrm{sh}(f_1) \circ \cdots \circ \mathrm{sh}(f_{p-1}) \circ \widetilde{\sigma}_1 \circ \mathrm{sh}(f_p),$$

which implies $\mathrm{id} = \mathrm{sh}(f_0) \circ \widetilde{\sigma}_1 \circ \mathrm{sh}(f_1) \circ \cdots \circ \mathrm{sh}(f_{p-1}) \circ \widetilde{\sigma}_1 \circ \mathrm{sh}(f_p)$. By Lemma 3.18, such an identity is impossible. ∎

Proposition 3.19. (braids acyclic) *(i) Left division in B_∞ equipped with braid exponentiation admits no cycle.*

(ii) Assume that the braid b admits an expression where σ_1 occurs but σ_1^{-1} does not. Then $b \neq 1$ holds.

Proof. (i) We know that there exists a homomorphism φ of $(B_\infty, {}^\wedge)$ into $(\mathrm{Aut}(\mathrm{FG}_\infty), {}^\wedge)$. The image under φ of a possible cycle in the division of $(B_\infty, {}^\wedge)$ would be a cycle in the division of $(\mathrm{Aut}(\mathrm{FG}_\infty), {}^\wedge)$, contradicting Proposition 3.14 (automorphisms acyclic).

Point (ii) follows from Lemma 3.18 directly. If b is a braid admitting an expression where σ_1 occurs but σ_1^{-1} does not, the associated automorphism $\widetilde{b}$ is eligible for Lemma 3.18, so we have $\widetilde{b} \neq \mathrm{id}$, and, therefore, $b \neq 1$. ∎

Observe that the previous arguments do not require φ to be injective (which will be proved in Chapter III). On the other hand, they do not transfer to quotients of B_∞ such as $\mathfrak{S}_\infty$, since quotienting may add cycles in the division—and it does in the case of $\mathfrak{S}_\infty$, as shown in Exercise 3.27 below.

Exercise 3.20. (braids are unavoidable) *Assume that G is a group, a is a fixed element of G, and s is an endomorphism of G. Prove that the* *formula $x{\wedge}y = x\ s(y)\ a\ s(x^{-1})$ defines a left self-distributive operation if and only if a commutes with every element in the image of s^2, and a satisfies the relation $a\ s(a)\ a = s(a)\ a\ s(a)$. [Hence the subgroup of G generated by the elements a, $s(a)$, $s^2(a)$, ... is a quotient of B_∞.]*

Exercise 3.21. (**alternative exponentiation**) *Prove that the formula $a{\vee}b = a\,\mathrm{sh}(b)\,\sigma_1^{-1}\,\mathrm{sh}(a^{-1})$ defines a left self-distributive operation on B_∞. Check that the operations $\wedge$ and $\vee$ on B_∞ are mutually left distributive.*

Exercise 3.22. (**no identity element**) *Prove that there is no braid e satisfying $b{\wedge}e = e$ for every b in B_∞. [Hint: Consider permutations.]*

Exercise 3.23. (**right powers**) *(i) For b a braid, define inductively the right power $b^{[m]}$ by $b^{[1]} = b$, $b^{[m+1]} = b{\wedge}b^{[m]}$. Write $\sigma_{m,1}$ for $\sigma_m \cdots \sigma_2\sigma_1$. Show that $b^{[2]} \neq b$ holds for every braid b. Prove $1^{[m]} = \sigma_{m-1,1}$ for $m \geq 2$.*

(ii) Show that $b \in B_n$ implies $b{\wedge}\sigma_{n-1,1} = \sigma_{n,1}$. [It will follow from the results of Chapter III that this characterizes B_n inside B_∞.]

Exercise 3.24. (**left division**) *(i) Let b be a braid. Prove that b belongs to $\mathrm{sh}(B_\infty)$ if and only if the braids $\mathrm{sh}(b)$ and σ_1 commute. [Hint: For $p \leq n$, write $\sigma_{p,n} = \sigma_p \cdots \sigma_n$. For $b \in B_n$, we have $\sigma_1^{-1}\,\mathrm{sh}(b)\,\sigma_1 = \sigma_{2,n}\,b\,\sigma_{2,n}^{-1}$, which implies $b = \mathrm{sh}(\sigma_{1,n-1}^{-1}\,b\,\sigma_{1,n-1})$ when $\mathrm{sh}(b)$ and σ_1 commute.]*

(ii) Prove that, in the LD-system $(B_\infty, {\wedge})$, $(\exists b)(a{\wedge}b = c)$ is equivalent to $a{\wedge}c = (a{\wedge}a){\wedge}c$. [Hint: Expand $a{\wedge}b = c$ and deduce that c is left divisible by a in $(B_\infty, {\wedge})$ if and only if $a^{-1}\,c\,\mathrm{sh}(a)\,\sigma_1^{-1}$ belongs to $\mathrm{sh}(B_\infty)$, hence, by (i), if and only if $\mathrm{sh}(a^{-1}\,c\,\mathrm{sh}(a)\,\sigma_1^{-1})$ commutes with σ_1. Expand $a{\wedge}c = (a{\wedge}a){\wedge}c$ and compare.]

Exercise 3.25. (**right division**) *(i) For $p \geq q$, let $\sigma_{p,q} = \sigma_p\sigma_{p-1} \cdots \sigma_q$. For $b \in B_\infty$, prove that $b \in B_n$ is equivalent to $\mathrm{sh}(b) = \sigma_{n,1}^{-1}\,b\,\sigma_{n,1}$.*

(ii) Assume $b \in B_n$, and $p < n$. Prove that $b \in B_p$ holds if and only if b commutes with $\sigma_{n,p+1}$. [Hint: Use (i) for $b \in B_n$ and $b \in B_p$.]

(iii) For b, b' in B_∞, say that b and b' are B_n-conjugate if $b' = aba^{-1}$ holds for some braid a in B_n. Prove that, for b, c in B_∞, there exists a in B_n satisfying $a{\wedge}b = c$ if and only if $c\sigma_{n,1}^{-1}$ and $\mathrm{sh}(b\sigma_{n-1,1}^{-1})$ are B_n-conjugate. Show that, if the previous condition is satisfied, the braids a in B_n satisfying $a{\wedge}b = c$ are those braids of the form a_0d, where a_0 is a particular braid satisfying $a_0{\wedge}b = c$ and d commutes with $\mathrm{sh}(b)\sigma_{n,2}^{-1}$.

(iv) Deduce that, for every p, the equality $a{\wedge}\sigma_{p-1,1} = a'{\wedge}\sigma_{p-1,1}$ in B_∞ is equivalent to $a^{-1}a' \in B_p$. In particular, the mapping $b \mapsto b{\wedge}1$ is

injective on B_∞.

Exercise 3.26. (**Burau exponentiation**) *For* M *a matrix, say* $M = (m_{i,j})$, *define* $\mathrm{Tr}^+(M)$ *to be* $\sum_i m_{i,i+1}$. *Prove* $\mathrm{Tr}^+(A^\wedge B) = \mathrm{Tr}^+(B) + t$.

Exercise 3.27. (**exponentiation of permutations**) *(i) Show that left division in* $(\mathfrak{S}_\infty, {}^\wedge)$ *admits no cycle of length* 1. *[Hint:* $f = f^\wedge g$ *would imply* $s_1 = \mathrm{sh}(g^{-1} \circ f)$.*]*

(ii) Check the equalities $\mathrm{id}^\wedge\mathrm{id} = s_1$, $(\mathrm{id}^\wedge\mathrm{id})^\wedge\mathrm{id} = s_2$, $\mathrm{id}^\wedge(\mathrm{id}^\wedge\mathrm{id}) = s_2 s_1$, $((\mathrm{id}^\wedge\mathrm{id})^\wedge\mathrm{id})^\wedge\mathrm{id} = s_2 s_1 s_3$, *where* s_i *is the transposition* $(i, i+1)$.

(iii) Let $f = s_1 s_3$ *and* $g = s_2 s_1$. *Show that* (f, g) *is a cycle of length* 2 *for left division in* $(\mathfrak{S}_\infty, {}^\wedge)$ *[Hint: prove* $f = g^\wedge s_1$ *and* $g = f^\wedge(s_2 s_1 s_3)$*]. Show that* f *and* g *belong to the sub-LD-system of* $(\mathfrak{S}_\infty, {}^\wedge)$ *generated by* id. *[Hint: We have* $f = s_2{}^\wedge s_1$.*]*

(iv) Prove that the mapping $f \mapsto f \circ \mathrm{sh}$ *defines an embedding of* $(\mathfrak{S}_\infty, {}^\wedge)$ *into* $(\mathfrak{I}_\infty, [\,])$. *Describe the image.*

---

4. Extended Braids

Several extensions of the braid groups appear naturally when the approach of the previous sections is developed. Here we introduce one such extension starting from the idea of colouring braid diagrams in such a way that both the strands and the regions between the strands receive a colour.

Region colourings

Let $(S, {}^\wedge)$ be an LD-system. For every (positive) braid diagram d, we know how to colour the strands of d using colours from S. In order to put colours on the regions between the strands, let us assume that $\cdot$ is a new binary operation on S, and consider the rule

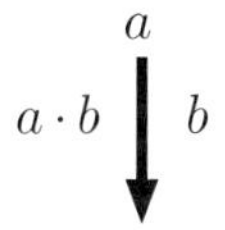

i.e., crossing a strand with colour a multiplies the colour of the region by a on the left. We look for those constraints the operations $^\wedge$ and $\cdot$ have to satisfy for the previous rule, together with the rules of Section 2, to yield a well-defined region colouring.

Lemma 4.1. *Assume that* $(S, \wedge)$ *is an LD-system,* $(S, \cdot, 1)$ *is a monoid, and the following mixed identities are satisfied:*

$$x \cdot y = (x \wedge y) \cdot x, \tag{LDM_1}$$

$$(x \wedge y) \cdot z = x \wedge (y \wedge z), \tag{LDM_2}$$

$$x \wedge (y \cdot z) = (x \wedge y) \cdot (x \wedge z), \tag{LDM_3}$$

$$1 \wedge x = x, \quad x \wedge 1 = 1. \tag{LDM_4}$$

Then, for every positive braid diagram d, and every choice of initial colours in S for the strands and the rightmost region of d, there exists a unique well-defined colouring of all strands and regions extending the initial colouring and invariant under isotopy.

Proof. Use induction on the number of crossings. The basic induction step corresponds to the scheme

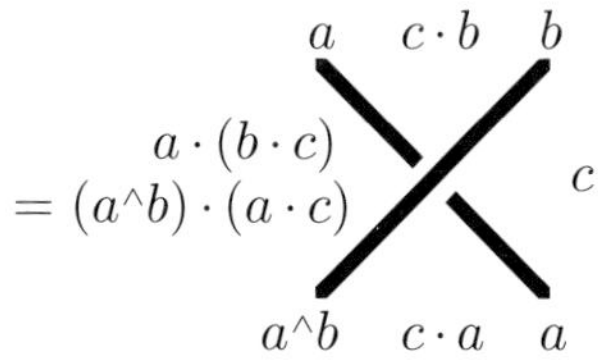

which follows from Identity (LDM_1) and associativity of operation $\cdot$. ∎

Definition 4.2. (LD-monoid) An *LD-monoid* is defined to be a monoid $(M, \cdot, 1)$ equipped with an additional left self-distributive operation $\wedge$ that satisfies the mixed identities (LDM_1), ..., (LDM_4) of Lemma 4.1.

Example 4.3. (group conjugacy) *Assume that G is a group. Then G equipped with its product and the conjugacy operation* $x{}^\wedge y = xyx^{-1}$ *is an LD-monoid.*

We shall investigate LD-monoids in Chapter XI systematically. Here, we only consider the case of braids. We have defined in Section 3 a left self-distributive operation on B_∞, namely braid exponentiation, so it is natural to ask if there exists an associative operation $\cdot$ on B_∞ so that $(B_\infty, \cdot, \wedge)$ is an LD-monoid. Actually, it is easy to check that the original product of B_∞ cannot be used here, nor can any possible product $\cdot$ that gives B_∞ the structure of a group: indeed, if the element x admits an inverse for $\cdot$, Identity (LDM_1) reads $x{}^\wedge y = x \cdot y \cdot x^{-1}$, and conjugacy is the only possible choice for the self-distributive operation. Now, as it is not idempotent, braid exponentiation cannot be the conjugacy operation of any possible exotic group structure on B_∞.

A completion of the braid group B_∞

We are thus led to looking farther, namely to some extension of B_∞. To this end, we consider the natural topology associated with the shift endomorphism of B_∞, and introduce a partial topological completion of B_∞.

Definition 4.4. (distance) For a, b braids, we define the *distance* $d(a,b)$ by $d(a,b) = 0$ for $a = b$ and $d(a,b) = 2^{-p}$ for $a^{-1}b \in \mathrm{sh}^p(B_\infty) \setminus \mathrm{sh}^{p-1}(B_\infty)$, *i.e.*, when $a^{-1}b$ can be expressed with no letter $\sigma_i^{\pm 1}$ with $i \leq p$ and p is minimal with this property.

Lemma 4.5. *The mapping d is an ultrametric distance on B_∞. The group B_∞ equipped with the associated topology is a topological group, and the shift mapping, as well as exponentiation, are continuous.*

Proof. By definition of the distance d, the open ball with radius 2^{-p} centered at b is the left coset $b\,\mathrm{sh}^p(B_\infty)$. The continuity of the product and the inverse are easily checked, as, for $a, b \in B_n$ and $p \geq n$, we have

$$a\,\mathrm{sh}^p(B_\infty) \cdot b\,\mathrm{sh}^p(B_\infty) = ab\,\mathrm{sh}^p(B_\infty), \tag{4.1}$$

$$(b\,\mathrm{sh}^p(B_\infty))^{-1} = b^{-1}\,\mathrm{sh}^p(B_\infty), \tag{4.2}$$

because a and b commute with every braid in the image of sh^p. The shift mapping is $1/2$-Lipschitz by definition, hence continuous, and so is exponentiation, which is defined by composing product, shift, and inverse. ■

As B_∞ is countable, it is not complete with respect to the above topology. We shall define an extension EB_∞ of B_∞ by adding some limits points.

Definition 4.6. (tau) For p, $q \geq 0$, we put $\tau_{p,q} = \sigma_p\sigma_{p+1} \cdots \sigma_1 \mathrm{sh}(\tau_{p,q-1})$ for $q > 0$, and $\tau_{p,0} = 1$.

Thus, $\tau_{p,q}$ is the positive braid where the q strands initially at positions $p+1$ to $p+q$ cross over the first p strands:

$\tau_{p,q}$:

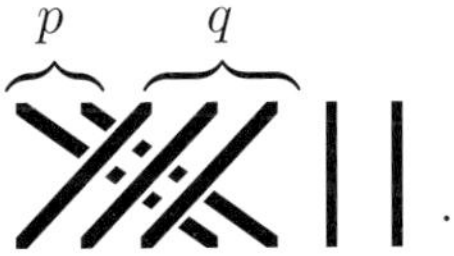

It follows from the above picture that, for every braid b in B_q, and every p, the equality

$$\tau_{p,q}\, b = \mathrm{sh}^q(b)\, \tau_{p,q} \tag{4.3}$$

holds in B_∞. We consider in the sequel sequences of the type $(a\tau_{p,n})_{n\geq 0}$. The basic observation is:

Lemma 4.7. *(i) For every braid a, and every p in $\mathbf{N}$, the sequence $(a\tau_{p,n})_{n\geq 0}$ is a Cauchy sequence in B_∞, and it has no limit in B_∞;*

(ii) The sequences $(a\tau_{p,n})_{n\geq 0}$ and $(b\tau_{q,n})_{n\geq 0}$ have the same limit in the completion of B_∞ if and only if $p = q$ and $a^{-1}b \in B_p$ hold.

This directly follows from the next criterion:

Lemma 4.8. *For a, $b \in B_\infty$, and p, $q \geq 0$, the following are equivalent:*
(i) For n large enough, we have $d(a\tau_{p,n}, b\tau_{q,n}) < 2^{-n}$;
(ii) We have $p = q$ and $a^{-1}b \in B_p$.

Proof. Assume $p = q$ and $a^{-1}b \in B_p$. Let n be large enough to guarantee that a and b belong to B_{p+n}. Then we have

$$\tau_{p,n}^{-1}\, a^{-1} b\, \tau_{q,n} = \mathrm{sh}^n(a^{-1}b),$$

and (i) holds by definition of the distance d.

Conversely, assume first $p \neq q$, say $p < q$. Choose n large enough to have $a^{-1}b \in B_{p+n-1}$ and $p + n > q$. Then the strand that begins at position n in $\tau_{p,n}^{-1}\, a^{-1} b\, \tau_{q,n}$ finishes at position $n + (q - p)$, and, therefore, $\tau_{p,n}^{-1}\, a^{-1} b\, \tau_{q,n} \in \mathrm{sh}^n(B_\infty)$ is impossible.

Finally, assume $p = q$ and $d(a\tau_{p,n}, b\tau_{q,n}) < 2^{-n}$ for n large enough. Choose n such that $a^{-1}b$ belongs to B_{p+n}. The hypothesis implies that some c in B_p satisfies

$$\tau_{p,n}^{-1}\, a^{-1} b\, \tau_{p,n} = \mathrm{sh}^n(c),$$

which, in turn, implies $a^{-1}b = c$, hence $a^{-1}b \in B_p$ (Figure 4.1). ■

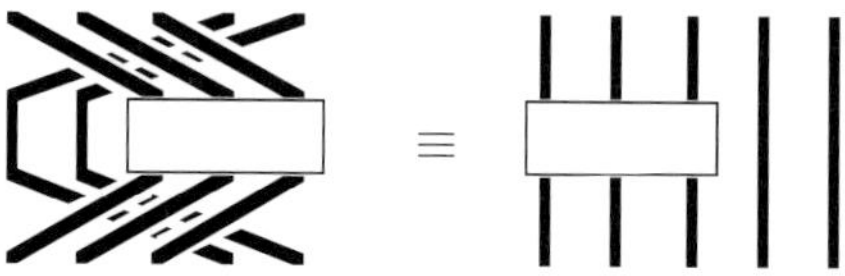

Figure 4.1. Conjugating by $\tau_{p,q}$

Thus we are led to:

Definition 4.9. (extended braid) For a, b in B_∞ and p, q in $\mathbf{N}$, we say that $(a, p) \sim (b, q)$ holds if $p = q$ and $a^{-1}b \in B_p$ do. An *extended braid* is defined to be a $\sim$-equivalence class. We write $[b, q]$ for the class of (b, q), and call q the *degree* of $[b, q]$. The set of all extended braids is denoted EB_∞.

By definition, we have

$$EB_\infty = \coprod_{p=0}^{\infty} B_\infty / B_p. \tag{4.4}$$

We define a distance d on EB_∞ by

$$d([a,p],[b,q]) = \lim_n d(a\tau_{p,n}, b\tau_{q,n}),$$

which makes sense by Lemma 4.8. By construction, the mapping $b \mapsto [b,0]$ is injective, it is an isometry, and we can identify b and $[b,0]$, and consider B_∞ as a subset of EB_∞. Then, by Lemma 4.8 again, the limit of the Cauchy sequence $(b\,\tau_{q,n})_{n\geq 0}$ in EB_∞ is $[b,q]$.

We have observed in Chapter III that multiplication is continuous on B_∞, hence there exists at most one way for extending it into a continuous operation on EB_∞. The following computation shows that the involved limits exist in EB_∞, and it gives an explicit formula.

Lemma 4.10. *For $a, b \in B_\infty$ and $p, q \geq 0$, the limit of $(a\tau_{p,m})(b\tau_{q,n})_{m,n}$ exists in EB_∞, and it is $[a\,\mathrm{sh}^p(b), p+q]$.*

Proof. Fix $r > 0$. Choose m, n satisfying $b \in B_n$, $m \geq r+q$, and $n \geq r$. Then we have $\tau_{p,m} = \tau_{p,q}\,\mathrm{sh}^q(\tau_{p,r})\,\mathrm{sh}^{r+q}(\tau_{p,m-r-q})$, and $t_{q,n} = \tau_{q,r}\,\mathrm{sh}^r(\tau_{q,n-r})$, hence

$$\begin{aligned}
a\tau_{p,m}\,b\tau_{q,n} &= a\mathrm{sh}^p(b)\tau_{p,m}\tau_{q,n}\\
&= a\,\mathrm{sh}^p(b)\tau_{p,q}\,\mathrm{sh}^q(\tau_{p,r})\,\mathrm{sh}^{r+q}(\tau_{p,m-r-q})\,\tau_{q,r}\,\mathrm{sh}^r(\tau_{q,n-r})\\
&= a\,\mathrm{sh}^p(b)\,\tau_{p,q}\,\mathrm{sh}^q(\tau_{p,r})\,\tau_{q,r}\,\mathrm{sh}^{r+q}(\tau_{p,m-r-q})\,\mathrm{sh}^r(\tau_{q,n-r})\\
&= a\,\mathrm{sh}^p(b)\,\tau_{p,q}\,\tau_{p+q,r}\,\mathrm{sh}^{r+q}(\tau_{p,m-r-q})\,\mathrm{sh}^r(\tau_{q,n-r})\\
&= a\,\mathrm{sh}^p(b)\,\tau_{p+q,r}\,\mathrm{sh}^r(\tau_{p,q})\,\mathrm{sh}^{r+q}(\tau_{p,m-r-q})\,\mathrm{sh}^r(\tau_{q,n-r}),
\end{aligned}$$

which implies $d(a\tau_{p,m}b\tau_{q,n}, a\,\mathrm{sh}^p(b)\tau_{p+q,r}) \leq 2^{-r}$, and gives the result. ∎

So, at this point, we can summarize the results as:

Proposition 4.11. **(completion)** *(i) The set EB_∞ is the partial topological completion of B_∞ obtained by adding limits to the Cauchy sequences $(b\,\tau_{q,n})_{n\geq 0}$. The isometry $b \mapsto [b,0]$ identifies B_∞ with a dense subset of EB_∞.*

(ii) The extension to EB_∞ of the multiplication on B_∞ is the operation defined by

$$[a,p]\cdot[b,q] = [a\,\mathrm{sh}^p(b), p+q]. \tag{4.5}$$

It turns EB_∞ into a monoid, and B_∞ is the set of units of EB_∞.

Definition 4.12. **(extended braid τ)** We put $\tau = [1,q]$.

By construction, the extended braid τ is the limit of the braids $\tau_{1,n}$ when n goes to ∞. It is natural to associate with τ the diagram

τ: 

where one strand vanishes at infinity. Then the diagram associated with τ^2 is

τ^2: 

but, as $\tau^2 = \sigma_1 \tau^2$ holds, we can also use instead the more simple diagram

and, more generally, associate with τ^q the similar diagram where q strands vanish at infinity.

For every extended braid $[b, q]$, we have

$$[b, q] = [b, 0] \cdot [1, q] = b \cdot \tau^q. \tag{4.6}$$

We can therefore use $b\tau^q$ as an alternative notation for $[b, q]$, and use extended braid diagrams accordingly. Such a representation is well fitted with the requirement that $b\tau^q = b'\tau^{q'}$ is equivalent to the conjunction of $q = q'$ and $b^{-1}b' \in B_q$. The extended braid diagrams associated with multiplication in EB_∞ are displayed in Figure 4.2.

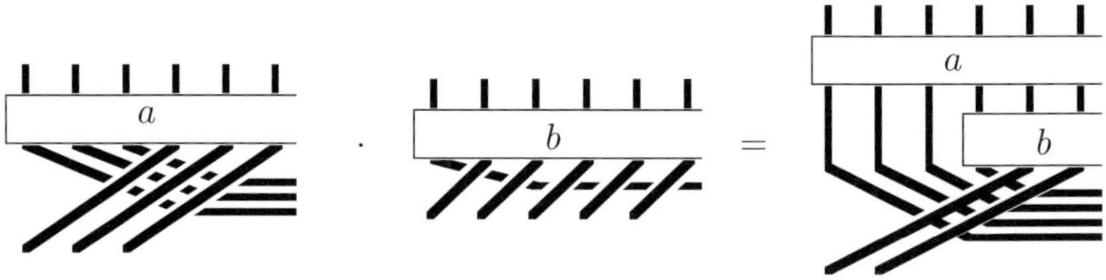

Figure 4.2. Multiplication of $a\tau^p$ and $b\tau^q$ (here $p = 3$, $q = 1$)

Proposition 4.13. (presentation) *The monoid EB_∞ is obtained from the group B_∞ by adding one generator τ subject to the relations*

$$\tau \, \sigma_i = \sigma_{i+1} \, \tau \quad \text{for } i \geq 1, \qquad \text{and} \qquad \sigma_1 \, \tau^2 = \tau^2. \tag{4.7}$$

Proof. Relations (4.7) hold in EB_∞, hence EB_∞ is a quotient of the monoid M generated by these relations. Now, every element of M admits an expression of the form $w\tau^q$ where w involves no τ, and, if w, w' are braid words satisfying $w' \equiv wu$ for some word in BW_q, then $w\tau^q$ and $w'\tau^q$ represent the same element of M. It follows that the canonical morphism of $\{\sigma_1^{\pm 1}, \sigma_2^{\pm 1}, \ldots, \tau\}^*$ onto M factors through EB_∞. ■

Self-distributive operations on EB_∞

As EB_∞ is a monoid and not a group, there may exist left self-distributive operations that are compatible with multiplication, and nevertheless do not coincide with conjugacy. As B_∞ is dense in EB_∞, it is natural to consider the topological extension of conjugacy on B_∞, as was done for multiplication. Conjugacy is continuous on B_∞, but it is not obvious that the involved limits exist in EB_∞. Actually, they do, and the counterpart of Lemma 4.10 is:

Lemma 4.14. *For $a, b \in B_\infty$ and $p, q \geq 0$, the limit of $(a\tau_{p,m})(b\tau_{q,n})(a\tau_{p,m})^{-1}$ for $m, n \to \infty$ exists in EB_∞, and it is $a\,\mathrm{sh}^p(b)\,\tau_{q,p}^{-1}\,\mathrm{sh}^q(a^{-1})\,\tau^q$.*

We are thus led to consider the operation defined on EB_∞ by

$$(a\tau^p) \vee (b\tau^q) = a\,\mathrm{sh}^p(b)\,\tau_{q,p}^{-1}\,\mathrm{sh}^q(a^{-1})\,\tau^q. \tag{4.8}$$

We shall also consider the symmetric operation (possibly) defined by

$$(a\tau^p) \wedge (b\tau^q) = a\,\mathrm{sh}^p(b)\,\tau_{p,q}\,\mathrm{sh}^q(a^{-1})\,\tau^q. \tag{4.9}$$

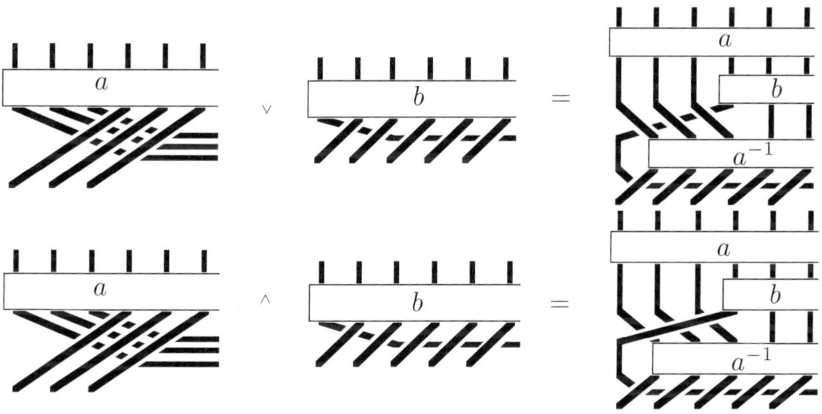

Figure 4.3. Exponentiation of $a\tau^p$ and $b\tau^q$ (here $p = 3$, $q = 1$)

Proposition 4.15. (**LD-monoid**) *The set EB_∞ equipped with multiplication and each of the operations $\wedge$ or $\vee$ is an LD-monoid.*

Proof. In the case of $\vee$, Lemma 4.14 guarantees that the definition is sound. In the case of $\wedge$, no continuity argument exists, and we have to check the compatibility of the definition with the equivalence relation $\sim$ directly. Assume $a^{-1}a' \in B_p$ and $b^{-1}b' \in B_q$. Let

$$c = (a\mathrm{sh}^p(b)\tau_{p,q}\mathrm{sh}^q(a^{-1}))^{-1}(a'\mathrm{sh}^p(b')\tau_{p,q}\mathrm{sh}^q(a'^{-1})).$$

We have to prove $c \in B_q$. Now $a^{-1}a' \in B_p$ implies that $a^{-1}a'$ commutes with $\mathrm{sh}^p(b^{-1})$, so we obtain

$$c = \mathrm{sh}^q(a)\tau_{p,q}^{-1}a^{-1}a'\mathrm{sh}^p(b^{-1}b')\tau_{p,q}\mathrm{sh}^q(a'^{-1}).$$

Then, by Formula (4.3), we have $\tau_{p,q}^{-1}a^{-1}a' = \mathrm{sh}^q(a^{-1}a')\tau_{p,q}^{-1}$, hence

$$c = \mathrm{sh}^q(a')\tau_{p,q}^{-1}\mathrm{sh}^p(b^{-1}b')\tau_{p,q}\mathrm{sh}^q(a'^{-1}).$$

By (4.3) again, $\tau_{p,q}^{-1}\mathrm{sh}^p(b^{-1}b')\tau_{p,q} = b^{-1}b'$ holds. By hypothesis, $b^{-1}b'$ belongs to B_q, hence it commutes with $\mathrm{sh}^q(a')$. Finally, we obtain $c = b^{-1}b' \in B_q$. Verifying the identities (LDM_1), (LDM_2), and (LDM_3) is easy. ■

By construction, the set EB_∞ includes *two* disjoint copies of B_∞, namely the images of the embedding $b \mapsto [b, 0]$, which has been considered already, but also the image of the "nontrivial" embedding

$$e_1 : b \mapsto [b, 1] = b\tau.$$

We see that the operations $^\wedge$ and $_\vee$ on EB_∞ both extend conjugacy on B_∞, but, on the second copy of B_∞, *i.e.*, on $\mathrm{Im}(e_1)$, we obtain different operations. The explicit formulas are

$$(a\tau) _\vee (b\tau) = a\,\mathrm{sh}(b)\,\sigma_1^{-1}\,\mathrm{sh}(a^{-1})\,\tau, \qquad (a\tau) ^\wedge (b\tau) = a\,\mathrm{sh}(b)\,\sigma_1\,\mathrm{sh}(a^{-1})\,\tau,$$

and we recognize braid exponentiation and its variant. We thus obtain a second way for introducing braid exponentiation, namely as a deformation of group conjugacy.

We finish this section with the projection of the previous operations on injections. We have associated with every braid b in B_∞ a permutation $\mathrm{perm}(b)$ of $\mathbf{N}_+$ so that, for every i, $\mathrm{perm}(b)(i)$ is the initial position of the strand that finishes at position i in b. We can apply the same construction to extended braids, according to the geometrical interpretation of Figure 4.2. This amounts to putting:

Definition 4.16. (**mapping** inj) For $b\tau^q$ in EB_∞, we define $\mathrm{inj}(b\tau^q)$ to be $\mathrm{perm}(b) \circ \mathrm{sh}^q$, where sh is the shift injection $n \mapsto n+1$ of $\mathbf{N}_+$.

The previous definition makes sense as $a^{-1}b \in B_p$ implies $\mathrm{perm}(a) \circ \mathrm{sh}^p = \mathrm{perm}(b) \circ \mathrm{sh}^p$. By construction, $\mathrm{inj}(b\tau^q)$ lies in $\mathfrak{I}_\infty$. For b a braid, $\mathrm{inj}(b)$ coincides with $\mathrm{perm}(b)$. We have $\mathrm{inj}(\tau) = \mathrm{sh}$, and, more generally, $\mathrm{inj}(\tau^q)$ is the shift mapping $n \mapsto n + q$ for every q

It is easy to establish a connection between the exponentiation of (extended) braids and the bracket operation on injections.

Proposition 4.17. (**connection**) *The mapping* inj *is a homomorphism of* $(EB_\infty, \cdot, ^\wedge)$ *into* $(\mathfrak{I}_\infty, \circ, [\,])$.

Proof. The result can be read on Figures 4.2 and 4.3. Let us consider for instance the case of ^ and [ ]. Assume $f = \mathrm{inj}(a\tau^p)$, $g = \mathrm{inj}(b\tau^q)$ and let h be the injection associated with $(a\tau^p)^\wedge(\beta\tau^q)$, *i.e.*, with $c\tau^q$, where c is $a\mathrm{sh}^p(b)\tau_{p,q}\mathrm{sh}^q(a^{-1})$.

Assume first $n \in \mathrm{Im}(f)$, say $n = f(m)$. By definition, the strand that begins at position n ends at position $m+p$ in a. Let us consider the strand γ that begins at position $f(g(m))$ in the diagram of $c\tau^q$. It ends at position n in the diagram of $c\tau^q$. Then γ moves to position $g(m)+p$ through a, then to position $m+p+q$ through $\mathrm{sh}^p(b)$. It keeps its position through $\tau_{p,q}$, and moves to position $n+q$ through $\mathrm{sh}^q(a^{-1})$. Finally, it exits at position n. So, $h(n) = f(g(f^{-1}(n)))$ holds for n in $\mathrm{Im}(f)$.

Assume now $n \notin \mathrm{Im}(f)$. This means that there exists $m \leq p$ such that the strand that begins at position n ends at position m in a. Let us consider the strand γ that begins at position n in the diagram of $c\tau^q$. Then γ exits from a at position m. It keeps the same position through $\mathrm{sh}^p(b)$, then moves to position $m+q$ through $\tau_{p,q}$. Then, γ moves to position $n+q$ through $\mathrm{sh}^q(a^{-1})$, and, finally, it exits at position n after τ^q. Hence $h(n) = n$ holds for n not in $\mathrm{Im}(f)$, and $h = f[g]$ follows.

The argument is similar (and more simple) in the case of the product. ■

We have seen that braid exponentiation is the operation induced by the operation ^ of EB_∞ on the second copy of B_∞ in EB_∞, namely on the copy associated with the embedding $e_1 : b \mapsto b\tau_{1,\infty}$. As braid exponentiation projects to permutation exponentiation, we deduce from Proposition 4.17:

Corollary 4.18. (Figure 4.5) *The mapping* $f \mapsto f \circ \mathrm{sh}$ *is an embedding of* $(\mathfrak{S}_\infty, {}^\wedge)$ *into* $(\mathfrak{I}_\infty, [\,])$, *and, therefore, the mapping* $b \mapsto \mathrm{perm}(b) \circ \mathrm{sh}$ *is a homomorphism of* $(B_\infty, {}^\wedge)$ *into* $(\mathfrak{I}_\infty, [\,])$.

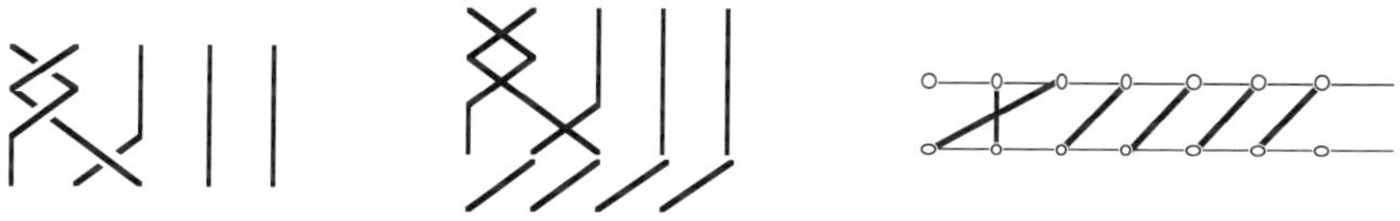

$(1^\wedge 1)^\wedge 1 \longmapsto \mathrm{perm}((1^\wedge 1)^\wedge 1) \circ \mathrm{sh} \longmapsto \mathrm{sh}[\mathrm{sh}][\mathrm{sh}]$

Figure 4.5. From braid to injection

Exercise 4.19. (**B_∞ locally abelian**) *Prove that braid group B_∞ is locally abelian in the sense that, for every braid b, there exists a neighbourhood U of b such that b commutes with every element of U.*

Exercise 4.20. (**Cauchy sequences**) *Prove that, for $p \geq 1$, the sequence $\tau_{p,1}^{-1}, \tau_{p,2}^{-1}, \tau_{p,3}^{-1}, \ldots$ is not a Cauchy sequence in B_∞.*

Exercise 4.21. **(endomorphisms of a free group)** *Extend the homomorphism φ of B_∞ into $\mathrm{Aut}(\mathrm{FG}_\infty)$ into a homomorphism of EB_∞ into the monoid of all endomorphisms of the free group FG_∞. [Hint: Map τ^q to the iterated shift endomorphism sh^q.]*

5. Notes

Braids were studied in the 1930s by E. Artin, who obtained in particular the standard presentation in terms of the generators σ_i (Artin 47). A number of equivalent definitions for Artin's braid groups can be found for instance in (Birman 75), (Burde-Zieschang 85), or (Prasolov & Sossinsky 97). Here we have chosen the most elementary one: it quickly leads to the braid relations, which, from our algebraic point of view, are central. Restricting isotopy to piecewise affine mappings (equivalence under Δ-moves) avoids in particular all pathologies such as wild knots.

The action of the braid group B_n on the n-th power of an LD-quasigroup has been used sporadically by several authors (Lyndon & Schupp 77); (Brieskorn 88) seems to be the first reference where the existence of this action is used as a unifying principle. In this paper, LD-quasigroups are called *automorphic sets*, as they are those binary systems where left translations are automorphisms. Such systems are called *racks* in (Fenn & Rourke 92). The current approach by means of colourings was used in (Dehornoy 94*a*), and it is similar to the approach of (Joyce 82), which we now briefly describe and compare.

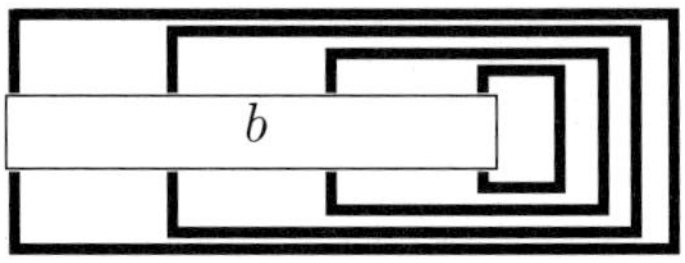

Figure 5.1. Closure of the braid b

Braids are connected with knots and links. Every oriented link can be realized as the closure of a braid in the sense of Figure 5.1. The idea of colouring the strands or the regions between the strands of a link can be traced back at least to Alexander. Let us consider link diagrams (which can be seen as closed braid diagrams). It is known that two link diagrams are the projection of isotopic three-dimensional figures if and only if they are equivalent under a sequence

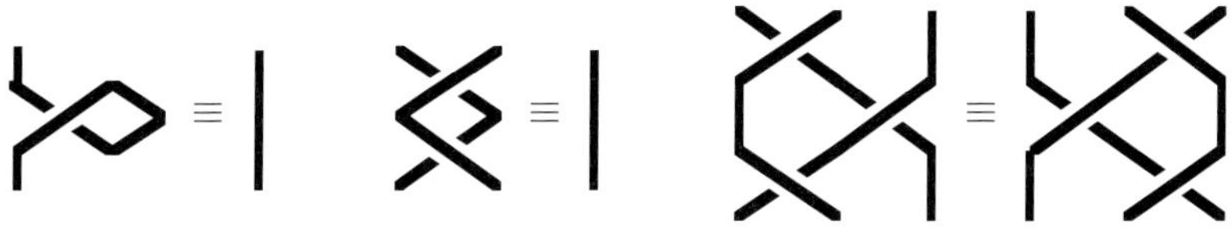

Figure 5.2. Reidemeister moves

of elementary transformations, called Reidemeister moves, of the three types illustrated in Figure 5.2 (all orientations of the strands are allowed).

Let us colour the strands of a diagram keeping the principle that, when the strand coloured with b goes over the strand coloured with a, then its colour becomes $a{}^\wedge b$ or $a{}_\vee b$ according to the orientation of the crossing. Then invariance under Reidemeister move I translates into the identities $x{}^\wedge x = x{}_\vee x = x$, invariance under Reidemeister move II translates into the identities $x{}^\wedge(x{}_\vee y) = x{}_\vee(x{}^\wedge y) = y$, while invariance under Reidemeister move III translates (owing to the previous identities) into left self-distributivity for ${}^\wedge$ and, equivalently, for ${}_\vee$. An algebraic system consisting of a set equipped with two binary operations satisfying the previous identities has been called a *quandle* in (Joyce 82). It also appears as a *crystal* in (Kauffman 91), and J.H. Conway too has considered such structures—according to the terminology used in this text, such systems could also be called idempotent LD-quasigroups. Whatever the name is, the scheme used in the context of knot theory consists in associating with the closure $\widehat{b}$ of a braid b the particular algebraic system S_b specified by the presentation

$$S_b = \langle\ \vec{x}\ ;\ \vec{x} = \vec{x} \bullet b\ \rangle.$$

For instance, if we use the action associated with conjugacy in a group, the structure S_b is the fundamental group of the complement of $\widehat{b}$; similarly using the barycentric mean gives the Alexander module (Crowell & Fox 63), while using free quandle operations leads to the quandle of $\widehat{b}$ as defined in (Joyce 82). The approach we develop here is different: we use one and the same fixed structure S for every braid b, and directly obtain information about b by comparing the sequence $\vec{x}$ and $\vec{x} \bullet b$ instead of introducing a specific quotient S_b of S.

The more general colour systems of Exercice 2.31 are considered in (Wada 92), where all solutions of small size in a free group are investigated. Various group invariants of links are deduced. One could consider further colouring rules involving more than two strands, but the complexity of the identities involved increases quickly. However, it would be interesting to investigate the free systems associated with Wada's identities.

In the context of self-distributivity, LD-quasigroups seem to play the role of groups in the context of associativity. Conjugacy as a basic example belongs to the folklore. Its many variants appear for instance in the papers of Kepka, to whom Proposition 2.16 (embedding) is also due.

Symmetric LD-quasigroups are considered in (Pierce 78, 79). See (Endres 91 to 96) for developments about the symmetric structure on the sphera S^n and its application to the study of the homotopy groups $[S^p \times S^q ; S^n]$.

The injection bracket of Example 2.19 appears in (Dehornoy 86) as a naive counterpart to the application operation on elementary embeddings in set theory (*cf.* Chapter XII). Exercice 2.30 has been proposed by R. McKenzie.

Braid exponentiation was introduced in (Dehornoy 94a) as an application for the geometrical study of self-distributivity developed in Chapter VIII below. This approach leads more naturally to Formula (3.1) than the current approach in terms of braid colourings.

Possible self-distributive systems with an acyclic left division and their ordering properties were first considered in (Dehornoy 89b) and (Laver 92) independently. The first proof that left division associated with braid exponentiation has no cycle appears in (Dehornoy 92b). The current shorter proof by means of automorphisms of a free group appears in (Larue 94).

Extended braids are introduced in (Dehornoy 98), using an alternative construction that starts with the intuition of braids involving two infinite series of strands ordered as $\mathbf{N} + \mathbf{N}$.

Most of the problems addressed in this chapter will be considered again in the sequel. However, we can mention here a few questions about exponentiation of permutations—or injection bracket, which is equivalent by Exercise 3.27(iv)—and of matrices that will not be further investigated in this book.

Question 5.1. *Give an intrinsic combinatorial characterization of those permutations that can be obtained from identity using exclusively exponentiation.*

Very little is known about this question. We shall see in Chapter VI that those braids that can be obtained from the unit braid using exponentiation are the self-colouring braids of Section 3, which is a combinatorial characterization. No similar combinatorial characterization is known for permutations.

Question 5.2. *Give a presentation of the LD-system $(\mathfrak{S}_\infty, {}^\wedge)$, and of the subsystem generated by the identity mapping.*

By describing a 2-cycle in $\mathfrak{S}_\infty$, we have obtained a relation that holds in $(\mathfrak{S}_\infty, {}^\wedge)$, but does not hold in $(B_\infty, {}^\wedge)$. It is well-known that $\mathfrak{S}_\infty$ is obtained from B_∞ by adding the Coxeter relations $\sigma_i^2 = 1$. If we give up product and keep exponentiation only, it is not known how to pass from B_∞ to $\mathfrak{S}_\infty$.

Question 5.3. *Describe those matrices that can be obtained from the identity matrix using matrix exponentiation exclusively.*

Again, very little is known, except the fact that the matrices obtained as above are Burau matrices, and therefore, they are unitary with respect to some Hermitian metric described in (Squier 84), and further used in (Dehornoy 96a).

II Word Reversing

The aim of this preparatory chapter is triple. Firstly, we wish to give a "modern" proof of the basic algebraic properties of braid groups. The second aim is to introduce a specific combinatorial technique, called word reversing, which is needed in Chapter III to extend the braid action of Chapter I on LD-systems. The third aim is to develop the techinque of word reversing in a general framework, so as to be able to apply it to a certain group G_{LD} in Chapter VIII.

The chapter is organized as follows. In Section 1, we consider monoid presentations of a particular form, of which the standard presentation of B_n is a typical example, and we define word reversing as a possible method for solving the word problem. In every case, word reversing gives a sufficient condition for two words to represent the same element of the monoid. The condition however is not necessary in general: there can exist pairs of words for which the process does not converge, or it does but it gives a wrong answer. In Sections 2 and 3, we give criteria for avoiding such problems. In Section 2, we introduce coherence, which guarantees that word reversing gives a correct answer. We show that, when an additional Noetherianity condition is satisfied, coherence is a consequence of a local assumption which can be checked effectively. When these conditions are satisfied, the considered monoid admits a nice theory of left divisibility, and word reversing is connected with the computation of right lcm's. In Section 3, we consider the termination of word reversing, and, again, we give a sufficient condition, namely the existence of a special element that we call a Garside element. Under such conditions, the monoid embeds in a group of fractions, the latter is torsion free and its word problem is solvable by a double word reversing. In Sections 4 and 5, we show that all technical hypotheses considered in Sections 2 and 3 are satisfied in the case of braid groups, and deduce a number of algebraic properties of the monoids B_n^+ and the groups B_n.

1. Complemented Presentations

We have seen in Chapter I that Artin's braid group B_n admits the presentation

$$\langle \sigma_1, \ldots, \sigma_{n-1} \; ; \; \sigma_i\sigma_j = \sigma_j\sigma_i \;\text{ for } |i-j| \geq 2,\; \sigma_i\sigma_j\sigma_i = \sigma_j\sigma_i\sigma_j \;\text{ for } |i-j| = 1 \;\rangle. \tag{1.1}$$

In this presentation, there exists exactly one relation of the form $x \cdots = y \cdots$ for each pair (x, y) of generators, and, conversely, every relation in the presentation has this form. Such relations tell us how to complete every pair of generators on the right so as to obtain a common multiple. Here we shall investigate presentations of this specific form, which we call complemented presentations.

Right complements

In the sequel, we deal with words, *i.e.*, finite sequences of letters chosen in a nonempty set, usually called an alphabet in this context. If A is an alphabet, we denote by A^* the free monoid generated by A, *i.e.*, the set of all words over A equipped with concatenation. We use ε for the empty word. The length of the word w, *i.e.*, the number of letters in w, is denoted by $\lg(w)$. For a monoid M to be generated by A means that M is a quotient of the monoid A^* under some congruence, *i.e.*, some equivalence relation that is compatible with the product on both sides. Similarly, for a group G to be generated by A means that G is a quotient of the free monoid $(A \cup A^{-1})^*$ under some congruence that contains all pairs (xx^{-1}, ε) and $(x^{-1}x, \varepsilon)$ with $x \in A$, where A^{-1} denotes a disjoint copy of A and, for x in A, x^{-1} denotes the copy of x in A^{-1}. In such a context, the words on A are called *positive* words. We shall often use u, v for positive words, and w for arbitrary words on $A \cup A^{-1}$. If the group G is the quotient $(A \cup A^{-1})^*/\equiv$, and if the element a of G is the equivalence class of the word w, we say that w *represents* a, or is *an expression of* a, and we write $a = \overline{w}$. Similar notations are used for monoids and positive words. Finally, for w a word on $A \cup A^{-1}$, we denote by w^{-1} the word obtained from w by reversing the order of the letters and exchanging everywhere x and x^{-1}. If w is an expression of a in some group G, then w^{-1} is an expression of a^{-1}.

Definition 1.1. (**complement**) Let A be an alphabet. We say that f is a *complement* on A if f is a partial mapping of $A \times A$ into A^* such that $f(x,x) = \varepsilon$ holds for every x in A, and $f(y,x)$ exists whenever $f(x,y)$ does. Then, we write $\equiv^+_{f,R}$ for the congruence on A^* generated by the pairs $(xf(x,y), yf(y,x))$ with (x,y) in the domain of f, and $\equiv_{f,R}$ for the congruence on $(A \cup A^{-1})^*$ generated by $\equiv^+_{f,R}$ together with all pairs (xx^{-1}, ε) and $(x^{-1}x, \varepsilon)$ with $x \in A$. We define the monoid and the group *associated with f on the right* to be $A^*/\equiv^+_{f,R}$ and $(A \cup A^{-1})^*/\equiv_{f,R}$ respectively.

Thus, for the monoid M (*resp.* the group G) to be associated with the complement f on the right means M (resp G) admits the presentation

$$\langle A \,;\, \{xf(x,y) = yf(y,x) \,;\, (x,y) \in \mathrm{Dom}(f)\}\rangle. \tag{1.2}$$

Convention 1.2. Most of the subsequent constructions in this chapter involve a complement f and the choice of using it on the right (a symmetric left choice is possible). It is natural to let these initial parameters appear in all derived notions, as in $\equiv_{f,R}$ or $\equiv^+_{f,R}$ above. However, we shall usually drop such subscripts to obtain more readable expressions. The convention is that, in all statements involving a complement f, all derived notions refer to that complement and, unless otherwise specified, to using it on the right.

Example 1.3. (**Artin groups**) *The braid group B_n is associated with a complement on the right, namely the complement f_B defined by*

$$f_B(\sigma_i, \sigma_j) = \begin{cases} \varepsilon & \text{for } i = j, \\ \sigma_j\sigma_i & \text{for } |i-j| = 1, \\ \sigma_j & \text{for } |i-j| \geq 2. \end{cases}$$

A similar result holds for all Artin groups. Let A be a nonempty set. A Coxeter matrix over A is a symmetric matrix $(m_{x,y})_{x,y \in A}$ with $m_{x,x} = 1$, and $m_{x,y} \in \{2, 3, \ldots, \infty\}$ for $x \neq y$. The Artin group associated with $(m_{x,y})_{x,y}$ is defined to admit the presentation

$$\langle A \,;\, \mathrm{prod}(x,y,m_{x,y}) = \mathrm{prod}(y,x,m_{y,x}) \text{ for } m_{x,y} < \infty\rangle, \tag{1.3}$$

where $\mathrm{prod}(x,y,m)$ stands for the alternating word $xyxy\ldots$ of length m. In particular, B_n is the Coxeter group associated with the matrix $(m_{i,j})_{i,j}$ defined by $m_{i,j} = 3$ for $|i-j| = 1$ and $m_{i,j} = 2$ for $|i-j| \geq 2$.

If f is a complement on A, the congruence $\equiv$ includes the congruence $\equiv^+$ on A^*: for all words w, w' on A, $w \equiv^+ w'$ implies $w \equiv w'$. This inclusion may be proper, *i.e.*, the monoid associated with f on the right need not embed in the group associated with f.

Example 1.4. (**not embedding**) *Take* $A = \{x, y\}$ *and define* f *by* $f(x,x) = f(y,y) = \varepsilon$ *and* $f(y,x) = f(x,y) = x$. *Then* $x \equiv y$ *holds, as we have* $x \equiv xxx^{-1} \equiv yxx^{-1} \equiv y$, *but* $x \equiv^+ y$ *fails, because the unique generating pair* (xx, yx) *involves no single letter. Thus the inclusion of* $\equiv^+$ *in* $\equiv$ *is proper in this case.*

Not every monoid is associated with a complement. Indeed, we have the following necessary condition:

Proposition 1.5. (**no invertible**) *Assume that* M *is a monoid associated with a complement. Then* 1 *is the only invertible element of* M.

Proof. By construction, $u \equiv^+ \varepsilon$ holds for no nonempty word u. ■

The previous condition turns out to be sufficient, at least for left cancellative monoids (*cf.* Exercise 1.20). So the class of those monoids that admit a complemented presentation is very wide, and it is not surprising that additional conditions are needed to obtain nontrivial results.

Word reversing

The main interest of complements is to make it possible to define a certain operation on words that we call *word reversing*. In good cases, this operation will allow us to recognize equivalent words and to deduce a number of properties. The principle of word reversing is as follows. If f is a complement on A, then, by definition, we have

$$x\ f(x,y) \equiv^+ y\ f(y,x)$$

for all x, y in A, which implies

$$x^{-1}\ y \equiv f(x,y)\ f(y,x)^{-1}. \tag{1.4}$$

Thus, if we replace in a word a pattern of the form $x^{-1}y$ with the corresponding pattern $f(x,y)f(y,x)^{-1}$, we obtain an equivalent word. What has been done here to replace a factor where a negative letter (a letter in A^{-1}) precedes a positive letter (a letter in A) with a factor where positive letters precede negative letters: so, we have reversed the mutual positions of negative and positive letters. *Reversing* a word w will consist in iterating this transformation so as to eventually obtain a sorted word where all positive letters lie before all negative letters—in other words, a right fraction.

Definition 1.6. (**reversing**) Assume that f is a complement on A, and w, w' are words on $A \cup A^{-1}$. We say that w is *(right) reversible* to w' in one step *(using f)*, denoted $w \curvearrowright^{(1)} w'$, if there exist x, y in A, and words w_1, w_2 on $A \cup A^{-1}$ satisfying

$$w = w_1\, x^{-1}\, y\, w_2 \quad \text{and} \quad w' = w_1\, f(x,y)\, f(y,x)^{-1}\, w_2.$$

A (finite or infinite) sequence of words w_0, w_1, ... is called a *(right) reversing sequence* if $w_i \curvearrowright^{(1)} w_{i+1}$ holds for every i. We say that $w \curvearrowright^{(n)} w'$ holds if there exists a length $n+1$ reversing sequence from w to w'—in particular, $w \curvearrowright^{(0)} w$ always holds. Finally, we say that w is *(right) reversible to w' (using f)*, denoted $w \curvearrowright w'$, if $w \curvearrowright^{(n)} w'$ holds for some n.

(Writing $\curvearrowright_{f,R}$ would remind the reader that the relation depends on the choice of f and of the right-hand side; however, according to Convention 1.2, we simply use $\curvearrowright$ here.)

Example 1.7. (**reversing**) *Let us consider the braid word $w_0 = \sigma_1^{-1}\sigma_2\sigma_2^{-1}\sigma_3$. Using the complement f_B of Example 1.3, we can reverse the factor $\sigma_1^{-1}\sigma_2$ to $\sigma_2\sigma_1\sigma_2^{-1}\sigma_1^{-1}$, and we have*

$$w_0 \curvearrowright^{(1)} w_1 = \sigma_2\sigma_1\sigma_2^{-1}\sigma_1^{-1}\sigma_2^{-1}\sigma_3.$$

Now, we can reverse the factor $\sigma_2^{-1}\sigma_3$ of w_1 into $\sigma_3\sigma_2\sigma_3^{-1}\sigma_2^{-1}$, finding

$$w_0 \curvearrowright^{(2)} w_2 = \sigma_2\sigma_1\sigma_2^{-1}\sigma_1\sigma_3\sigma_2\sigma_3^{-1}\sigma_2^{-1},$$

etc. Finally, we obtain

$$w_0 \curvearrowright^{(7)} \sigma_2\sigma_1\sigma_3\sigma_2\sigma_1\sigma_3^{-1}\sigma_2^{-1}\sigma_1^{-1}\sigma_3^{-1}\sigma_2^{-1}.$$

The last word is not reversible as it contains no factor $\sigma_i^{-1}\sigma_j$.

By Relation (1.4), we find immediately:

Lemma 1.8. *Assume that f is a complement on A, and w is a word on $A \cup A^{-1}$. Then $w \curvearrowright w'$ implies $w \equiv w'$.*

We associate now with every reversing sequence a labelled graph embedded in the plane $\mathbf{Q}^2$ which displays the word transformations. Assume that f is a complement on A, and $(w_0, w_1, \ldots)$ is a reversing sequence. The associated graph will contain three types of edges: *horizontal* A-labelled edges going from a vertex (p,q) to a vertex (p',q) with $p' > p$, *vertical* A-labelled edges going from a vertex (p,q') to a vertex (p,q) with $q < q'$, and unoriented ε-labelled edges. The construction uses induction on n. First, we associate with w_0 a graph $\Gamma(w_0)$ consisting in a single path from the point $(0,0)$ to the point (p,q), where p is the number of positive letters in w_0, and q is the number of negative letters. We use induction on w_0. The graph $\Gamma(\varepsilon)$ consists of $(0,0)$ solely. For $w_0 = wx$ with $x \in A$, the graph $\Gamma(w_0)$ is obtained from $\Gamma(w)$ by adding one horizontal x-labelled edge from $(p-1,q)$ to (p,q); for $w_0 = wx^{-1}$ with $x \in A$, the graph $\Gamma(w_0)$ is obtained from $\Gamma(w)$ by adding one vertical x-labelled edge from (p,q) to $(p,q-1)$. For instance, for $w_0 = \sigma_1^{-1}\sigma_2\sigma_2^{-1}\sigma_3$, the graph $\Gamma(w_0)$ is as follows:

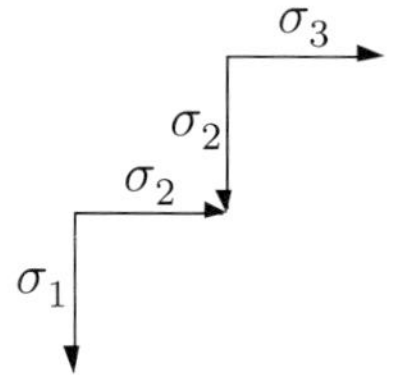

If we define the label of a path to be the word consisting of the labels of the successive edges, according to the convention that an edge labelled x crossed contrary to its orientation contributes x^{-1}, then, by construction, w_0 is the label of the unique maximal path in the graph $\Gamma(w_0)$,

Assume that the graph $\Gamma(w_0, \ldots, w_{n-1})$ has been constructed, and w_n is obtained from w_{n-1} by replacing some factor $x^{-1}y$ by the corresponding factor $f(x,y)f(y,x)^{-1}$. By induction hypothesis, we assume that w_{n-1} is traced in $\Gamma(w_0, \ldots, w_{n-1})$ from the point $(0,0)$, this meaning that there is a path labelled w_{n-1} from $(0,0)$. The factor $x^{-1}y$ involved in the reversing process labels some fragment of $\Gamma(w_0, \ldots, w_{n-1})$. Then $\Gamma(w_0, \ldots, w_n)$ is obtained by adding to $\Gamma(w_0, \ldots, w_{n-1})$ horizontal edges labelled $f(x,y)$ and vertical edges labelled $f(y,x)$ in the neighbourhood of the above fragment, according to the generic picture

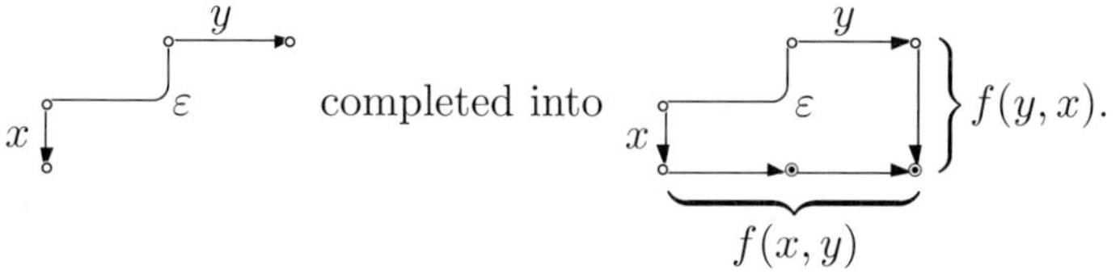

The new horizontal and vertical edges are determined so as to have equal lengths. If $f(x,y)$ or $f(y,x)$ is empty, we use instead an ε-labelled edge.

As an example, the graph associated with the reversing sequence of Example 1.7 is displayed in Figure 1.1. Notice that reversing graphs are planar, as the last edges currently added always lie on the "South-East" of all previous edges.

Remark 1.9. Reversing graphs are connected with Cayley graphs and Dehn diagrams, but, in general, a reversing graph is *not* a fragment of the Cayley graph of the corresponding group. The Cayley graph of the group G with respect to the generating set A is the graph whose vertices are the elements of G and there is an x-labelled arrow from vertex a to vertex b if and only if $b = ax$ holds in G. A reversing graph *projects* on some fragment of the

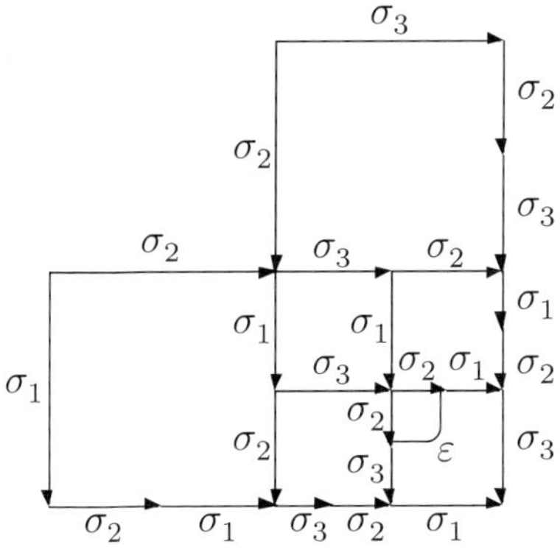

Figure 1.1. A reversing graph

Cayley graph, but several vertices in the reversing graph can represent the same element of the group, hence the same vertex of the Cayley graph. The example of the graph

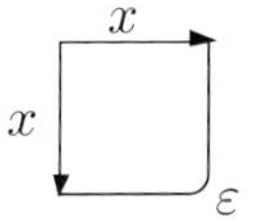

might suggest to merely collapse the ε-labelled edges, but the symmetric example of

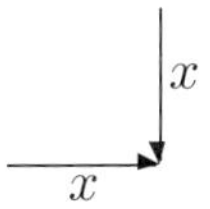

shows that this is not sufficient. Actually, identifying a reversing graph with a fragment of the Cayley graph would amount to working in the group or the monoid rather than with words, and the current approach would then fail.

Confluence of word reversing

By definition, those words that are terminal with respect to (right) reversing are of the form uv^{-1} where u and v are positive words. When we start with an arbitrary word w, there are in general several ways to reverse it, as there can exist several subfactors of the form $x^{-1}y$. Here we show that reversing is confluent, *i.e.*, the possible final result is unique when it exists.

Proposition 1.10. (**confluence**) *Assume* $w \curvearrowright^{(n')} w'$ *and* $w \curvearrowright^{(n'')} w''$. *Then there exist* n *and* w''' *satisfying* $\sup(n', n'') \le n \le n' + n''$, $w' \curvearrowright^{(n-n')} w'''$ *and* $w'' \curvearrowright^{(n-n'')} w'''$.

Proof. For $n' = 0$, we have $w = w'$, and we can take $n = n''$, $w''' = w''$. For $n'' = 0$, the argument is similar. Assume now $n' = n'' = 1$. Then there exist subfactors $x^{-1}y$ and $x'^{-1}y'$ of w that have been reversed respectively to obtain w' and w''. These factors are either equal, in which case we can take $n = 1$, $w''' = w' = w''$, or disjoint, in which case we can take $n = 2$ and define w''' to be the word obtained from w by reversing the two factors above.

We use now induction on $n' + n''$. The result has already been proved for $n' \leq 1$ and $n'' \leq 1$. Assume $n'' \geq 2$. We consider an intermediate word w_1 satisfying $w \curvearrowright^{(n''_1)} w_1 \curvearrowright^{(n''_2)} w''$ with $n''_1 + n''_2 = n''$ and $n''_1, n''_2 \geq 1$. By induction hypothesis, there exists n_1, n_2 wih $n_1 \leq n' + n''_1$ and $n_2 \leq n_1 - n''_1 + n''_2$, and w'_1, w''' satisfying $w' \curvearrowright^{(n_1 - n'_1)} w'_1$, $w_1 \curvearrowright^{(n_1 - n''_1)} w'_1$, $w'_2 \curvearrowright^{(n_2 - n_1 + n''_1)} w'''$, and $w'' \curvearrowright^{(n_2 - n''_2)} w'''$ (Figure 1.2). For $n = n''_1 + n_2$, we have $w' \curvearrowright^{(n - n')} w'''$. ∎

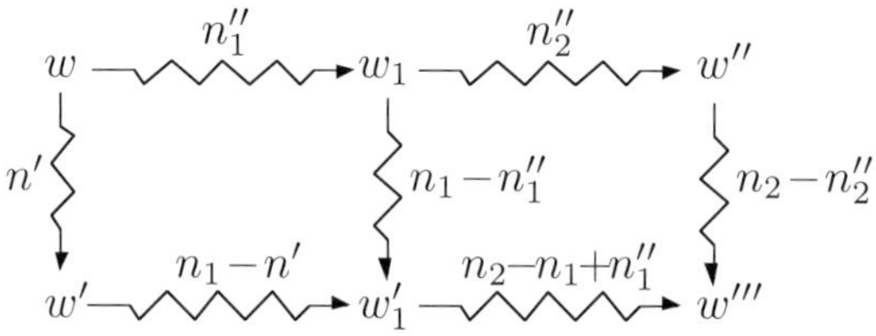

Figure 1.2. Confluence of reversing

The previous result is clear when we consider reversing graphs: it amounts to saying that, starting with a word w, there exists a unique maximal reversing graph that starts from w, whatever could be the order of the reversing steps. However, this maximal reversing graph need not be finite—in other words the reversing process need not converge, as the following counter-examples show.

Example 1.11. (not convergent, I) *Let f be the complement defined on $\{x, y\}$ by $f(x, x) = f(y, y) = f(y, x) = \varepsilon$, $f(x, y) = xy$. The associated monoid admits the presentation $\langle x, y \; ; \; x^2y = y \rangle$. Let $w = y^{-1}xy$. Then we have $w \curvearrowright^{(1)} w' = y^{-1}x^{-1}y \curvearrowright^{(1)} w$. So the reversing sequences from w are alternating sequences $(w, w', w, w', \ldots)$, and, therefore, w cannot be reversible to any word of the form uv^{-1} with u, v positive.*

Example 1.12. (not convergent, II) *Let us define a complement f on $\{\sigma_1, \sigma_2, \sigma_3\}$ by*

$$f(\sigma_1, \sigma_2) = \sigma_2\sigma_1\sigma_2, \; f(\sigma_2, \sigma_1) = \sigma_1\sigma_2\sigma_1, \; f(\sigma_2, \sigma_3) = \sigma_3\sigma_2\sigma_3,$$
$$f(\sigma_3, \sigma_2) = \sigma_2\sigma_3\sigma_2, \; f(\sigma_1, \sigma_3) = \sigma_3, \; f(\sigma_3, \sigma_1) = \sigma_1.$$

Then f generates the Artin group associated with the Coxeter matrix $m_{1,2} = m_{2,3} = 4$, $m_{1,3} = 2$—usually called Coxeter type $\widetilde{B}_2$. We see on

the graph

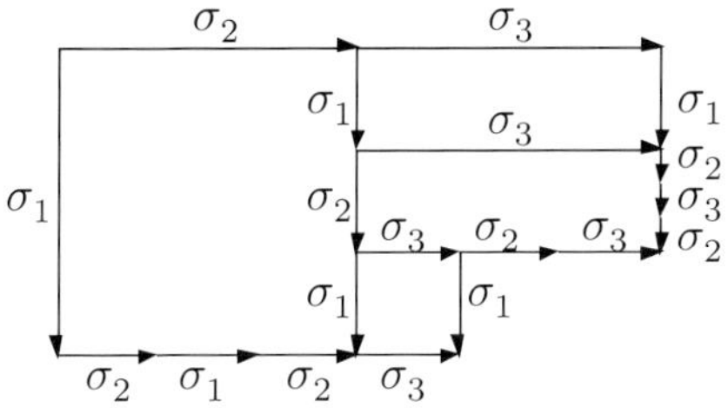

that the word $w = \sigma_1^{-1}\sigma_2\sigma_3$ *is reversible to the word*

$$(\sigma_2\sigma_1\sigma_2\sigma_3)(\sigma_1^{-1}\sigma_2\sigma_3)(\sigma_1\sigma_2\sigma_3\sigma_2)^{-1},$$

of which it is a proper factor. It follows that $\sigma_1^{-1}\sigma_2\sigma_3$ *is reversible to*

$$(\sigma_2\sigma_1\sigma_2\sigma_3)^k(\sigma_1^{-1}\sigma_2\sigma_3)(\sigma_2^{-1}\sigma_3^{-1}\sigma_2^{-1}\sigma_1^{-1})^k$$

for every nonnegative k. *Hence, reversing* w *cannot converge to any word of the form* uv^{-1} *with* u, v *positive.*

The operation \

The complement mapping we have considered so far is defined on letters only. By using word reversing, we can extend it to arbitrary positive words.

Definition 1.13. (operation \) Assume that f is a complement on A, and u, v are words on A. We define $u\backslash v$ ("u under v", or "v over u on the right") to be the unique word v_1 on A such that $u^{-1}v \curvearrowright v_1u_1^{-1}$ holds for some word u_1 on A, if such words exist.

(Once again, $\backslash_{f,R}$ would be more precise than \, but it leads to ugly formulas.)

By Proposition 1.10 (confluence), \ is a well defined partial binary operation on A^*. By construction, if $u^{-1}v$ is reversible to $v_1u_1^{-1}$, then $v^{-1}u$ is reversible to $u_1v_1^{-1}$, and, therefore, we have $u_1 = v\backslash u$. In other words, $u\backslash v$ and $v\backslash u$ exist if and only if the reversing of $u^{-1}v$ converges in a finite number of steps, and $(u\backslash v)(v\backslash u)^{-1}$ is the corresponding final word. Notice that the operation \ extends the mapping f, as, for x, $y \in A$, we have $x\backslash y = f(x, y)$ by definition.

Lemma 1.14. *Assume that* f *is a complement on* A. *Then we have*

$$(uv)\backslash w = v\backslash(u\backslash w), \tag{1.5}$$
$$w\backslash(uv) = (w\backslash u)((u\backslash w)\backslash v) \tag{1.6}$$

for all words u, v, w *on* A; *these equalities mean that either both sides are defined and they are equal, or neither of them is defined.*

Proof. If the words $u\backslash w$ and $v\backslash(u\backslash w)$ exist, then so does the word $(uv)\backslash v$, and Equalities (1.5) and (1.6) are clear on Figure 1.3. Assume conversely that $(uv)\backslash v$ exists. We have to show that $u\backslash w$ and $v\backslash(u\backslash w)$ exist, whence the result is clear. Now, let us observe that, for each pair of words (u, w), either $w\backslash u$ exists, or there exists an infinite reversing sequence starting from $u^{-1}w$. In the latter case, adding v^{-1} at the left of every word in this sequence gives an infinite reversing sequence from $v^{-1}u^{-1}w$, and $w\backslash(uv)$ cannot exist either. ■

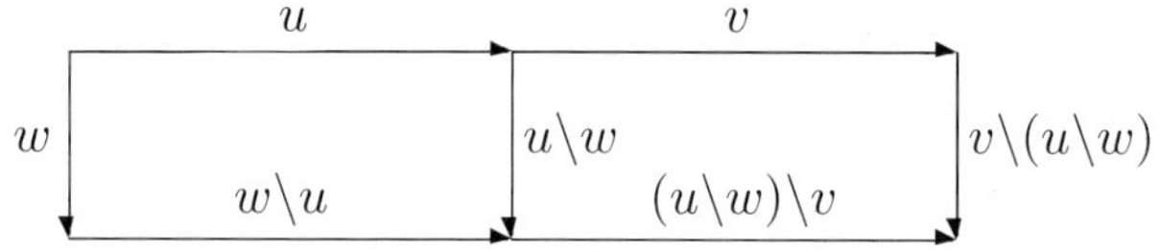

Figure 1.3. Double reversing

By Lemma 1.8, reversing produces $\equiv$-equivalent words. When we consider positive words, we have a more precise result involving the positive equivalence $\equiv^+$.

Proposition 1.15. (**positive equivalence**) *Assume that f is a complement on A, and u, v are words on A such that $u\backslash v$ exists. Then we have*

$$u\ (u\backslash v) \equiv^+ v\ (v\backslash u). \tag{1.7}$$

Proof. We use induction on the number of steps m needed to reverse the word $u^{-1}v$. For $m = 0$, at least one of u, v is empty. In this case, we have $u\backslash v = v$, $v\backslash u = u$, and the result is true. Assume $m \geq 1$. Then u and v are not empty. Let us write $u = xu_0$, $v = yv_0$, with $x, y \in A$. By repeated uses of Lemma 1.14, we obtain positive words u_1, v_1, ..., u_3, v_3 as represented in Figure 1.4. By construction, we have $m = 1 + m_1 + m_2 + m_3$, where m_1, m_2, m_3 are the numbers of steps needed to reverse $u_0^{-1}f(x,y)$, $f(y,x)^{-1}v_0$ and $u_1^{-1}v_2$ respectively. Then $m_i < m$ holds for $i = 1, 2, 3$, the induction hypothesis applies to the corresponding words, and we find

$$\begin{aligned} u\ (u\backslash v) = xu_0v_1v_3 &\equiv^+ xf(x,y)u_1v_3 \equiv^+ yf(y,x)u_1v_3 \\ &\equiv^+ yf(y,x)v_2u_3 \equiv^+ yv_0u_2u_3 = v(v\backslash u). \end{aligned}$$ ■

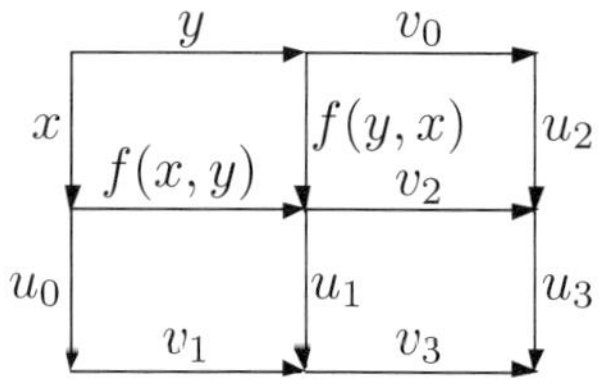

Figure 1.4. Positive equivalence

Applying the previous result, we deduce a sufficient condition for two positive words u, v to be positively equivalent:

Corollary 1.16. *Assume that f is a complement on A, and u, v are words on A. Then $u^{-1}v \quad \varepsilon$ implies $u \equiv^+ v$.*

Proof. Saying that $u^{-1}v \quad \varepsilon$ holds means that the words $u\backslash v$ and $v\backslash u$ both exist and are empty. Then apply Proposition 1.15 (positive equivalence). ■

Exercise 1.17. (quotient) *Assume that f is a complement on A, and u, v, w are words on A satisfying $w = uv$. What are $w\backslash u$ and $u\backslash w$?*

Exercise 1.18. (free and free abelian) *Prove that the free monoid A^* and the free abelian monoid $\mathbf{N}^A$ are associated with complements. What is word reversing in these cases?*

Exercise 1.19. (complemented presentation) *Assume that M is a left cancellative monoid in which 1 is the only invertible element. Let $X = M \setminus \{1\}$. Define a partial mapping f of X^2 to X by*

$$f(x,y) = \begin{cases} \varepsilon & \text{if } y \text{ is a left divisor of } x, \\ z & \text{for } y = xz \text{ with } z \neq 1. \end{cases}$$

Prove that f is a complement on X, and that M is associated with f. [Hint: Observe that M is a quotient of the monoid associated with f, and prove that, for a, a_1, ..., $a_p \in X$, $a = a_1 \cdots a_p$ implies $a \equiv^+ a_1 \cdots a_p$.] What is word reversing associated with f?

Exercise 1.20. (braid complement) *Let u_n be the braid word $\sigma_1\sigma_3 \cdots \sigma_{2n-1}$ and v_n be $\sigma_2\sigma_4 \cdots \sigma_{2n}$. Prove that reversing $u_n^{-1}v_n$ requires $\mathcal{O}(n^3)$ steps, and that the final word has length $4n^2$.*

---

2. Coherent Complements

In Section 1, we associated with every complement a combinatorial operation called reversing. This operation is connected with the associated equivalence relation $\equiv^+$, hence with the word problem of the corresponding monoid: by Corollary 1.16, if u and v are positive words and $u^{-1}v$ is reversible to the empty word, then u and v represent the same element of M. Here, we investigate whether the previous sufficient condition is necessary. This need not be the case in general, but we give effective conditions for this to be true. The main result takes the form:

Proposition 2.1. (coherent) *Assume that M is a monoid associated on the right with a complement on A that is normed and coherent on A.*

(i) Two words u, v on A represent the same element of M if and only if $u^{-1}v \quad \varepsilon$ holds.

(ii) The monoid M admits left cancellation, any two elements of M admit a unique left gcd, and they admit a unique right lcm provided they admit at least one common right multiple.

Coherence and existence of a norm are properties that will be defined below. Their interest here lies in that they can be checked easily on many examples, in particular in the case of braids, as we shall see in Section 4.

Completeness of word reversing

Definition 2.2. (complete) Assume that f is a complement on A. For u, v words on A, we write $u \equiv^{++} v$ for $u^{-1}v \quad \varepsilon$, *i.e.*, for $u\backslash v = v\backslash u = \varepsilon$. We say that (right) word reversing is *complete* for f if $u \equiv^+ v$ implies $u \equiv^{++} v$ for all words u, v on A.

Corollary 1.16 states that $u \equiv^{++} v$ always implies $u \equiv^+ v$. Saying that word reversing is complete for f means that the converse implication is true, *i.e.*, that $\equiv^+$-equivalence is always detected by word reversing. It is easy to see that this need not be the case always.

Example 2.3. (not complete) *Let f be the complement of Example 1.11 (not convergent I). We have seen that the word $y\backslash xy$ does not exist. Let us add a third letter z that absorbs x and y, i.e., let us guarantee that $z \equiv^+ xz \equiv^+ yz$ holds by defining $f(z,x) = f(z,y) = f(z,z) = \varepsilon$, $f(x,z) = f(y,z) = z$. By construction, we have $yz \equiv^+ z \equiv^+ xyz$, but $yz \equiv^{++} xyz$ does not hold, as $yz\backslash xyz$ cannot exist: for reversing $z^{-1}y^{-1}xyz$, we ought to reverse $y^{-1}xy$ first, which is impossible.*

Thus additional hypotheses are needed.

Definition 2.4. (**coherent**) Assume that f is a complement on A. For u, v, $w \in A^*$, we say that f is *coherent (on the right)* at (u, v, w) if either

$$((u\backslash v)\backslash(u\backslash w))\backslash((v\backslash u)\backslash(v\backslash w)) = \varepsilon \tag{2.1}$$

holds, or neither of the words $(u\backslash v)\backslash(u\backslash w)$, $(v\backslash u)\backslash(v\backslash w)$ exists. For $X \subseteq A^*$, we say that f is coherent *on* X if it is coherent at every triple (u, v, w) with u, v, w in X. We say that f is coherent if it is coherent on all of A^*.

Saying that f is coherent both at (u, v, w) and (v, u, w) means that either

$$(u\backslash v)\backslash(u\backslash w) \equiv^{++} (v\backslash u)\backslash(v\backslash w) \tag{2.2}$$

holds, or neither of the words above exists. Saying that f is coherent on $\{u, v, w\}$ means, when all involved words exist, that the cube represented in Figure 2.1 closes in the sense that, starting with three edges labeled u, v, w, using word reversing to close the six faces successively, and finally reversing the three remaining small triangular faces ends with empty words.

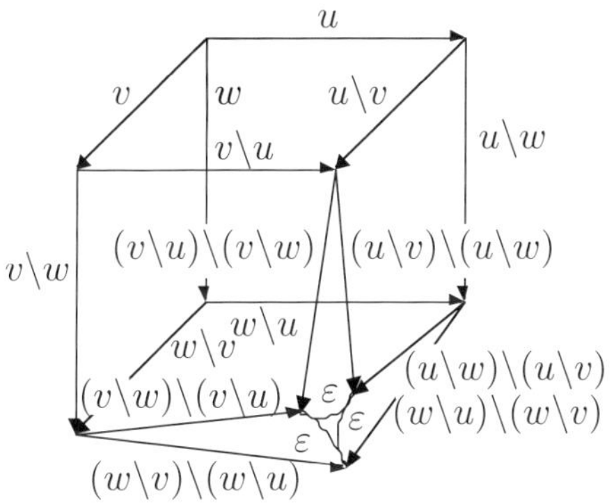

Figure 2.1. Coherence property

Proposition 2.5. (**complete**) *Assume that f is a complement on A. Then the following are equivalent:*

(i) Word reversing is complete for f;

(ii) The relation $\equiv^+$ is compatible with $\backslash$: if $u' \equiv^+ u$ and $v' \equiv^+ v$ hold, then either both $u\backslash v$ and $u'\backslash v'$ exist and $u'\backslash v' \equiv^+ u\backslash v$ holds, or neither exists;

(iii) The complement f is coherent on A^*.

Proof. Let us prove that (i) implies (ii). By symmetry and transitivity, it suffices to prove that, if $u\backslash v$ exists and $v' \equiv^+ v$ holds, then $u\backslash v'$ exists as well, and $u\backslash v' \equiv^+ u\backslash v$ and $v'\backslash u \equiv^+ v\backslash u$ hold. Assume the above hypotheses. By (1.7), we have $u\ (u\backslash v) \equiv^+ v\ (v\backslash u)$. Then $v' \equiv^+ v$ implies $u(u\backslash v) \equiv^+ v'(v\backslash u)$, hence, by completeness, $u(u\backslash v) \equiv^{++} v'(v\backslash u)$, and $(v'(v\backslash u))\backslash(u(u\backslash v)) = \varepsilon$. This implies in particular $(v'(v\backslash u))\backslash u = \varepsilon$. Using Lemma 1.14, we obtain

$$(v\backslash u)\backslash(v'\backslash u) = \varepsilon. \tag{2.3}$$

We deduce that $v'\backslash u$, and, therefore, $u\backslash v'$, exist. As v and v' now play symmetric roles, the same computation gives $(v'\backslash u)\backslash(v\backslash u) = \varepsilon$, which, together with (2.3), establishes $v'\backslash u \equiv^{++} v\backslash u$, hence $v'\backslash u \equiv^{+} v\backslash u$. Similar computations give $(u\backslash v)\backslash(u\backslash v') = \varepsilon$ and $(u\backslash v')\backslash(u\backslash v) = \varepsilon$, hence $u\backslash v \equiv^{++} u'\backslash v$, and, finally, $u\backslash v \equiv^{+} u'\backslash v$. Hence (i) implies (ii).

We prove now that (ii) implies (i). Assume $u \equiv^{+} v$. By construction, we have $u^{-1}u \curvearrowright \varepsilon$, *i.e.*, $u\backslash u$ exists and is empty. If (ii) holds, $u \equiv^{+} v$ implies that $u\backslash v$ and $v\backslash u$ exist, and that $u\backslash v \equiv^{+} v\backslash u \equiv^{+} \varepsilon$ holds, which is possible only for $u\backslash v = v\backslash u = \varepsilon$, *i.e.*, for $u \equiv^{++} v$. So, (i) and (ii) are equivalent.

Assume now (i) and (ii), and let u, v, w be words on A. If $u\backslash v$ does not exist, $v\backslash u$ does not exist either, and the words involved in (2.2) do not exist. Assume now that $u\backslash v$ and $v\backslash u$ exist. Lemma 1.14 gives $(u\backslash v)\backslash(u\backslash w) = (u(u\backslash v))\backslash w$ and $(v\backslash u)\backslash(v\backslash w) = (v(v\backslash u))\backslash w$. Assume that $(u(u\backslash v))\backslash w$ exists. By Proposition 1.15 (positive equivalence), we have $u\,(u\backslash v) \equiv^{+} v\,(v\backslash u)$. By (ii), $(v(v\backslash u))\backslash w$ exists as well, and $(v(v\backslash u))\backslash w \equiv^{+} (u(u\backslash v))\backslash w$ holds. By (i), we deduce $(v(v\backslash u))\backslash w \equiv^{++} (u(u\backslash v))\backslash w$, hence in particular $((u(u\backslash v))\backslash w)\backslash((v(v\backslash u))\backslash w) = \varepsilon$, which is (2.1).

Finally, assume (iii). We wish to prove that $\equiv^{+}$ is included in $\equiv^{++}$. By definition, $\equiv^{+}$ is the equivalence relation generated by the pairs $(uxf(x,y)v, uyf(y,x)v)$ with $x, y \in A$ and $u, v \in A^*$. So it suffices to prove that $uxf(x,y)v \equiv^{++} uyf(y,x)v$ holds, and that $\equiv^{++}$ is an equivalence relation. The first point follows from the definition. For the second point, $\equiv^{++}$ is reflexive and symmetric by definition, so it remains to prove transitivity. Let us assume $u \equiv^{++} v \equiv^{++} w$. By hypothesis, $v\backslash u$, $u\backslash v$, $v\backslash w$, and $w\backslash v$ exist and all are empty. Hence $(v\backslash u)\backslash(v\backslash w)$ exists, and it is empty. As f is coherent at (v, u, w), this implies $(u\backslash v)\backslash(u\backslash w) \equiv^{+} \varepsilon$, hence $(u\backslash v)\backslash(u\backslash w) = \varepsilon$. The existence of the latter word implies in particular that $u\backslash w$ exists. Moreover, $u\backslash v = \varepsilon$ implies $(u\backslash v)\backslash(u\backslash w) = u\backslash w$. Hence we have $u\backslash w = \varepsilon$. A symmetric argument gives $w\backslash u = \varepsilon$, hence $u \equiv^{++} w$, and $\equiv^{++}$ is transitive. This completes the proof. ∎

Local coherence

We are left with the question of recognizing coherent complements. Coherence is a matter of word reversing and, if reversing always terminates, we can decide coherence at a given triple (u, v, w) in a finite number of steps.

Example 2.6. **(coherence)** *Let us ask if the braid complement f_B is coherent at $(\sigma_1, \sigma_2, \sigma_3)$. Reversing $\sigma_3^{-1}\sigma_1\sigma_2\sigma_1$ and $\sigma_3^{-1}\sigma_2\sigma_1\sigma_2$ gives*

$$u = (\sigma_1\backslash\sigma_2)\backslash(\sigma_1\backslash\sigma_3) = \sigma_3\sigma_2\sigma_1,$$
$$u' = (\sigma_2\backslash\sigma_1)\backslash(\sigma_2\backslash\sigma_3) = \sigma_3\sigma_2\sigma_1,$$

and it remains to compute $u\backslash u'$. Here, we have $u = u'$, hence $u \equiv^{++} u'$, and coherence at $(\sigma_1, \sigma_2, \sigma_3)$—and $(\sigma_2, \sigma_1, \sigma_3)$—is true.

Now, checking coherence for every triple (u, v, w) of words is in any case an infinite process. We shall prove now that, when some additional Noetherianity condition is satisfied, it is sufficient to check coherence for triples of letters instead of triples of arbitrary words.

Definition 2.7. (**norm**) Assume that f is a complement on A. We say that a mapping ν of A^* to $\mathbf{N}$ is a *(right) norm* for f if, for all x, y in A, and all u, v in A^*, we have $\nu(xv) > \nu(v)$, $\nu(vx) \geq \nu(v)$, and $\nu(uxf(x,y)v) = \nu(uyf(y,x)v)$.

Example 2.8. (**norm**) *Assume that f is a complement on A and, for all x, y in A, the words $f(y,x)$ and $f(x,y)$ have the same length. Then the length is a norm for f. This applies in particular to the braid complement, and, more generally, to all complements of Example 1.3 (Artin).*

The monoid $\langle x, y \,;\, xyx = y^2\rangle$ is associated with the complement f defined by $f(x,y) = yx$, $f(y,x) = y$. The length is not a norm for f, as the lengths of $xf(x,y)$ and $yf(y,x)$ are not equal. However, the mapping ν defined by $\nu(u) = \#_x(u) + 2\#_y(u)$ is a norm for f, where $\#_c(u)$ denotes the number of letters c in the word u.

On the other hand, the complement f associated with the presentation $\langle x, y \,;\, y = xy\rangle$ admits no norm. Indeed, if f is the complement defined by $f(x,y) = y$, $f(y,x) = \varepsilon$, then we have $y \equiv^+ x^p y$ for every p, and ν being a norm for f would imply $\nu(y) > p$ for every p.

The interest of considering norms lies in the following result:

Proposition 2.9. (**locally coherent**) *Assume that f is a normed complement on A. Then the following are equivalent:*
(i) The complement f is coherent on A^*;
(ii) The complement f is coherent on A.

Proof. By definition, (i) implies (ii). The proof that (ii) implies (i) requires a double induction (or, equivalently, an induction of length ω^2) on convenient parameters. So, we assume (ii), and we fix a norm ν. For u, v words on A, we define $c(u,v) = \nu(u(u\backslash v))$ if $u\backslash v$ exists—possibly extended with $c(u,v) = \infty$ otherwise, but this is useless here—and we define $d(u,v)$ to be the minimal number of elementary relations $xf(x,y) = yf(y,x)$ needed to transform u into v if $u \equiv^+ v$ holds—possibly extended with $d(u,v) = \infty$ if $u \equiv^+ v$ fails, but, again, this is not needed here. We prove inductively on $c(u,v)$ that, if $u\backslash v$ exists, and if $u' \equiv^+ u$ and $v' \equiv^+ v$ hold, then $u'\backslash v'$ and $v'\backslash u'$ exist and we have $u'\backslash v' \equiv^+ u\backslash v$ and $v'\backslash u' \equiv^+ v\backslash u$. This will prove that $\equiv^+$ is compatible with $\backslash$, and, therefore, by Proposition 2.5 (complete), that f is coherent on A^*.

For $c(u,v) = 0$, the only possibility is $u = v = u' = v' = \varepsilon$, and the result is true.

Assume $c(u,v) \geq 1$. We use induction on $d(u,u') + d(v,v')$. By definition, $d(u,u') + d(v,v') = 0$ corresponds to $u' = u$, and $v' = v$, and the result is true. Assume $d(u,u')+d(v,v') = 1$. Without loss of generality, we may assume $u' = u$ and $d(v',v) = 1$, which means that there exist words v_0, v_1 and letters x, y satisfying $v = v_0x(x\backslash y)v_1$ and $v' = v_0y(y\backslash x)v_1$. Now the words $u\backslash v_0$ and $v_0\backslash u$ exist as $u\backslash v$ is supposed to exist.

Assume first $v_0\backslash u = \varepsilon$. Then we have $u'\backslash v' = (u\backslash v_0)x(x\backslash y)v_1$, and $u\backslash v = (u\backslash v_0)y(y\backslash x)v_1$, two $\equiv^+$-equivalent words, and $v'\backslash u' = v\backslash u = \varepsilon$.

Assume now $v_0\backslash u \neq \varepsilon$, say $v_0\backslash u = zu_1$ with $z \in A$ (Figure 2.2). Let $u_2 = (x(x\backslash y))\backslash z$ and $v_2 = z\backslash(x(x\backslash y))$. These words exist because $u\backslash v$ exists by hypothesis. Similarly, let $u_2' = (y(y\backslash x))\backslash z$ and $v_2' = z\backslash(y(y\backslash x))$. The hypothesis that f is coherent at each of the six permutations of (x,y,z) implies that the latter words exist and we have $u_2' \equiv^+ u_2$ and $v_2' \equiv^+ v_2$. Indeed, using coherence relations, Lemma 1.14, and Proposition 1.15 (positive equivalence), we find

$$\begin{aligned} u_2' &= (y\backslash x)\backslash(y\backslash z) \equiv^+ (x\backslash y)\backslash(x\backslash z) = u_2, \\ v_2' &= (z\backslash y)\left((y\backslash z)\backslash(y\backslash x)\right) \equiv^+ (z\backslash y)\left((z\backslash y)\backslash(z\backslash x)\right) \\ &\equiv^+ (z\backslash x)\left((z\backslash x)\backslash(z\backslash y)\right) \equiv^+ (z\backslash x)\left((x\backslash z)\backslash(x\backslash y)\right) = v_2. \end{aligned}$$

Let us introduce the words u_3, v_3, u_4, v_4, and u_5, v_5 that appear when reversing $u^{-1}v$ as shown in Figure 2.2. For instance, we have $u_3 = v_2\backslash u_1$, and $v_3 = u_1\backslash v_2$. By definition of a norm, we have

$$\begin{aligned} c(u_1,v_2) = \nu(u_1v_3) &\leq \nu(u_1v_3v_5) \\ &< \nu(zu_1v_3v_5) \leq \nu(v_0zu_1v_3v_5) = \nu(u(v\backslash u)) = c(u,v). \end{aligned}$$

Hence the induction hypothesis applies to the pairs (u_1,v_2) and (u_1,v_2'), so we deduce that the word $v_2'\backslash u_1$, which we shall call u_3', exists and is $\equiv^+$-equivalent to $v_2\backslash u_1$, *i.e.*, to u_3. A similar argument shows that the word $v_3' = u_1\backslash v_2'$ exists and is $\equiv^+$-equivalent to v_3.

The same argument shows that, with obvious notations, the words u_4' and v_4' exist and are respectively $\equiv^+$-equivalent to u_4 and v_4. Finally, the argument

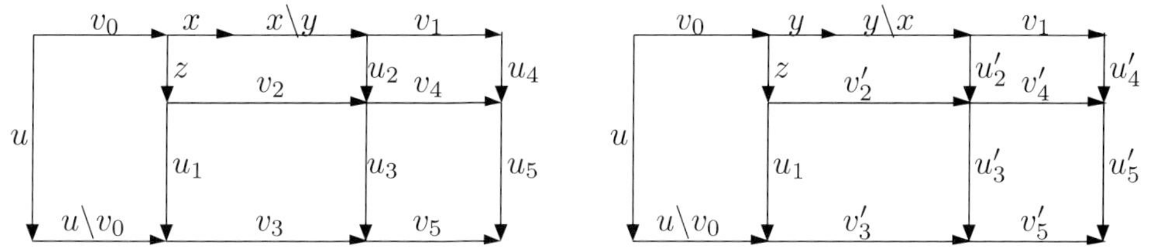

Figure 2.2. Compatibility from local coherence

is the same for u'_5 and v'_5. So, we conclude that $u'\backslash v'$ exists and we have

$$u'\backslash v' = (u\backslash v_0)v'_3v'_5 \equiv^+ (u\backslash v_0)v_3v_5 = u\backslash v,$$

and, similarly, $v'\backslash u' = u'_4u'_5 \equiv^+ u_4u_5 = v\backslash u$.

It remains to consider the case $d(u', u) + d(v', v) \geq 2$. We consider an intermediate pair of words say (u'', v'') between (u, v) and (u', v'), so that $d(u'', u) + d(v'', v) < d(u', u) + d(v', v)$ and $d(u', u'') + d(v', v'') < d(u', u) + d(v', v)$ hold. The induction hypothesis for (u, v) and (u'', v'') implies that $u''\backslash v''$ and $v''\backslash u''$ exist, and that they are $\equiv^+$-equivalent to $u\backslash v$ and $v\backslash u$ respectively. Moreover, we have $\nu(u''(u''\backslash v'')) = \nu(u(u\backslash v))$, *i.e.*, $c(u'', v'') = c(u, v)$. So, we can in turn apply the induction hypothesis to the pairs (u'', v'') and (u', v'), and conclude that $u'\backslash v'$ and $v'\backslash u'$ exist and that they are $\equiv^+$-equivalent respectively to $u\backslash v$ and $v\backslash u$. ■

Least common multiples

With the previous result, we have completed a proof of Proposition 2.1(i) (coherent). We turn now to Proposition 2.1(ii), *i.e.*, we study which properties of a monoid M follow from it being associated with a normed or coherent complement.

We have observed in Proposition 1.5 (no invertible) that 1 is the only invertible element in every monoid associated with a complement. If we add the hypothesis that the complement is normed, we can say more about the divisibility relation of the monoid. Assuming that M is a monoid, and that a, c are elements of M, we say that c is a *right multiple* of a, or that a is a *left divisor* of c if there exists b in M satisfying $ab = c$.

Definition 2.10. (**atom, atomic**) Assume that M is a monoid. For a in M, we say that a is an *atom* if $a = bc$ implies $b = 1$ or $c = 1$. We say that M is *atomic* if it is generated by its atoms and, for every a in M, the upper bound $\|a\|$ of the lengths of the decompositions of a as a product of atoms is finite.

Proposition 2.11. (**order**) *Assume that M is an atomic monoid.*

(i) The inequality

$$\|ab\| \geq \|a\| + \|b\| \tag{2.4}$$

holds for every a, b in M, and $a \neq 1$ implies $\|a\| \geq 1$.

(ii) The left and the right divisibility relations of M are partial well-founded orderings: every descending sequence below a has length at most $\|a\|$.

Proof. Formula (2.4) follows from the definition of $\|a\|$: if u and v are expressions of a and b as a products of atoms, then uv is an expression of ab, and the length of uv is the sum of the lengths of u and v. Now, (2.4) implies $\|a\| > \|b\|$ whenever b is a nontrivial (left or right) divisor of a. ■

Lemma 2.12. *Assume that M is a monoid. Then the following are equivalent:*

(i) The monoid M is atomic;

(ii) There exists a mapping $\nu : M \to \mathbf{N}$ such that $a \neq 1$ implies $\nu(ab) > \nu(b)$.

Proof. If M is atomic, then the mapping $a \mapsto \|a\|$ satisfies the conditions of (ii). Conversely, assume (ii). Let a be an element of M, and $a = a_1 \cdots a_p$ be a decomposition of a into a product of nontrivial elements of M. By hypothesis, we have

$$\nu(a) > \nu(a_2 \cdots a_p) > \nu(a_3 \cdots \alpha_p) > \cdots > \nu(a_p) > 0,$$

hence $p \leq \nu(p)$. Let $\nu'(a)$ be the upper bound of the lengths of the decompositions of a as product of nontrivial elements of M. The previous argument implies $\nu'(a) \leq \nu(a)$, so, in particular, $\nu'(a)$ is finite. Now, by construction, the inequality $\nu'(ab) \geq \nu'(a) + \nu'(b)$ holds for all a, b in M. We prove using induction on $\nu'(a)$ that a is a product of atoms. For $\nu'(a) = 0$, we must have $a = 1$, and the result is true. Otherwise, either a is an atom, or it can be decomposed as $a = a_1a_2$ with $\nu'(a_i) < \nu(a)$, and, by induction hypothesis, a_1 and a_2 are products of atoms. A product of atoms is a particular case of a product of nontrivial elements, hence, for each a, we have $\|a\| \leq \nu'(a)$, and, actually, $\|a\| = \nu'(a)$. ■

The previous criterion gives a way for recognizing that the monoid associated with a complement is possibly atomic.

Proposition 2.13. (atomic) *Assume that M is a monoid associated with a complement f. Then the following are equivalent:*

(i) The complement f admits a norm;

(ii) The monoid M is atomic.

Proof. Assume that ν is a norm for f. As the pairs $\{uxf(x,y)v), uyf(y,x)v\}$ generate the congruence $\equiv^+$ by definition, ν is compatible with $\equiv^+$, and, therefore, it induces a well-defined mapping on M, say $\overline{\nu}$. By definition, we have $\nu(xv) > \nu(v)$ for every letter x and every word v, hence, inductively, $\nu(uv) > \nu(v)$ for every nonempty word u. Thus $\overline{\nu}$ satisfies Condition (ii) of Lemma 2.12, and M is atomic. Conversely, if M is atomic, the mapping ν defined on A^* by $\nu(u) = \|\overline{u}\|$ is a norm for f (we recall that $\overline{u}$ denotes the class of the word u in M). ■

New properties appear when we introduce the hypothesis that the complement is coherent.

Proposition 2.14. **(left cancellation)** *Assume that M is a monoid associated with a coherent complement. Then M is left cancellative.*

Proof. Assume $vu \equiv^+ vu'$. Then we have $vu \equiv^{++} vu'$, *i.e.*, $u'^{-1}v^{-1}vu \quad \varepsilon$. By construction, $u'^{-1}v^{-1}vu \quad u'^{-1}u$ holds, so we deduce $u'^{-1}u \quad \varepsilon$, *i.e.*, $u \equiv^{++} u'$, which implies $u \equiv^+ u'$. ■

We consider now common multiples or divisors. Assume that M is a monoid, and a, b belong to M. We say that c is a *least common right multiple*, or *right lcm*, of a and b, if c is a right multiple of a and b, and a left divisor of every common right multiple of a and b. If c and c' are right lcm's of a and b, then, by definition, c is a left divisor of c' and c' is a left divisor of c. This implies $c = c'$ if we assume that M is left cancellative and contains no invertible element but 1. Indeed, $c = c'd$ and $c' = cd'$ imply $c = cd'd$, hence $d'd = 1$, hence $d = d' = 1$. Similarly, a *greatest common left divisor*, or *left gcd*, of a and b is an element c that is a left divisor of a and b and a right multiple of every common left divisor of a and b. Again, if M is left cancellative and has no invertible element but 1, such a left gcd of a and b is unique when it exists.

Definition 2.15. **(operations $\vee_R$, $\backslash$, $\wedge_L$)** When it exists and is unique, the right lcm of a and b is denoted by $a \vee_R b$, the element c satisfying $a \vee_R b = ac$ is denoted by $a\backslash b$, and the left gcd of a and b is denoted by $a \wedge_L b$.

The following proposition shows that reversing the word $u^{-1}v$ amounts to computing a right lcm for the elements represented by u and v.

Proposition 2.16. **(lcm)** *Assume that M is a monoid associated with a coherent complement. Let u, v be words on A, and let a, b respectively be their classes in M. Then the following are equivalent:*

(i) The word $u\backslash v$ exists;

(ii) The elements a and b admit a common right multiple.

If these conditions are satisfied, the elements a and b admit in M a (unique) right lcm, and $u\backslash v$ represents the element $a\backslash b$.

Proof. (Figure 2.3) Assume that $u\backslash v$ exists. Then $u(u\backslash v) \equiv^+ v(v\backslash u)$ holds, which induces an equality of the form $ab' = ba'$ in M. Conversely, assume that a and b have a common right multiple in M. Then there exist two words u', v' on A satisfying $uv' \equiv^+ vu'$. As word reversing is complete for f, the previous equivalence implies $uv' \equiv^{++} vu'$. Hence $(uv')\backslash(vu')$ exists, and so does $u\backslash v$. Moreover, we have $v'^{-1}u^{-1}vu' \curvearrowright v'^{-1}(u\backslash v)(v\backslash u)^{-1}u'$, so the hypothesis implies that the latter word is reversible to ε. By Lemma 1.14, this implies that there exist two words w_1, w_2 satisfying $v' \equiv^+ (u\backslash v)w_2$, $u' \equiv^+ (v\backslash u)w_1$, and $w_1 \equiv^+ w_2$. This means that $\overline{vu'}$ is a right multiple of $\overline{u\,(u\backslash v)}$. In other words, the latter element is a right lcm of a and b, and $u\backslash v$ represents $a\backslash b$. ∎

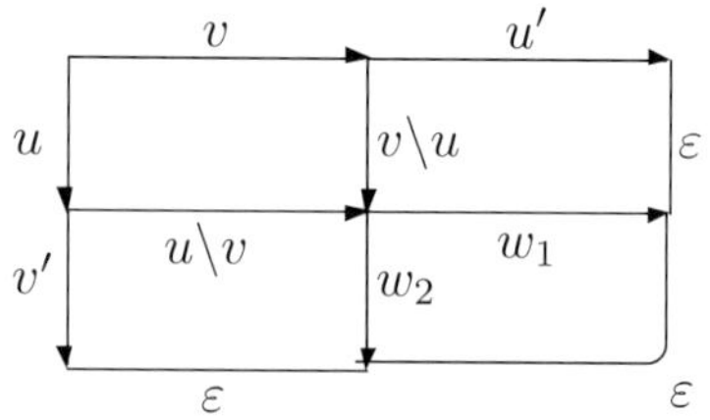

Figure 2.3. Right lcm

Proposition 2.17. (**gcd**) *Assume that M is a monoid associated with a coherent and normed complement. Then every nonempty subset of M admits a left gcd.*

Proof. By Proposition 2.13 (atomic), M is atomic. Let S be a nonempty subset of M, and D be the set of all common left divisors of S. Let a be an arbitrary element of S. Let b be an element of D such that $\|b\|$ is maximal. By (2.4), $\|b\| \le \|a\|$ holds for every b in D, so such an element exists. By construction, b is a left divisor of every element in S. On the other hand, let c be an arbitrary element of D. By construction, every element of S is a common right multiple of b and c, hence $b \vee_R c$ exists, and it belongs to D as well. By construction and by (2.4), we have

$$\|b \vee_R c\| = \|b\,(b\backslash c)\| \ge \|b\| + \|b\backslash c\|.$$

As $\|b\|$ is maximal, we find $\|b\backslash c\| = 0$, hence $b\backslash c = 1$: c is a left divisor of b. ∎

We thus have completed the proof of Proposition 2.1 (coherent).

Exercise 2.18. *(**two generators**) Prove that every complement on a 2-letter alphabet $\{x, y\}$ is coherent on $\{x, y\}$ [easy], and even on $\{x, y\}^*$ [more tricky].*

Exercise 2.19. **(two generators)** *Prove that the complement f on $\{x, y\}$ defined by $f(y, x) = y$, $f(x, y) = yxyx$ admits a norm. [This is tricky.]*

Exercise 2.20. **(norm)** *Prove that the hypothesis $\nu(vx) \geq \nu(v)$ can be removed in the definition of a norm: if f is a complement on A and there exists a mapping ν of A^* to $\mathbf{N}$ satisfying all requirements of a norm for f but the one above, then there exists another mapping ν' of A^* to $\mathbf{N}$ that is a norm for f. [Hint: Prove that the monoid $A^*/\equiv^+$ is atomic.]*

Exercise 2.21. **(homomorphism)** *(i) Assume that, for $i = 1, 2$, the monoid M_i is associated with the complement f_i on A_i, and h is a morphism of A_1^* into A_2^* such that, for every pair (x, y) in the domain of f_1, the word $h(x)\backslash h(y)$ exists and is equal to $h(x\backslash y)$. Prove that h induces a homomorphism of M_1 into M_2. Prove that the word $h(u)\backslash h(v)$ exists for every pair of words (u, v) such that $u\backslash v$ exists, and the former word is the image of the latter under h.*

(ii) Assume in addition that f_2 is normed and coherent, $h(x) \neq \varepsilon$ holds for every x in A_1, and, for x, y in A_2, $f_1(x, y)$ exists if and only if $h(x)\backslash h(y)$ does. Prove that f_1 is normed as well and that $\|\overline{u}\| \leq \|\overline{h(u)}\|$ holds for every u in A_1^*. Prove that $u\backslash v$ exists if and only if $h(u)\backslash h(v)$ does. Deduce that h induces an embedding of M_1 into M_2.

(iii) Deduce that mapping x_1 to $\sigma_1\sigma_5$, x_2 to $\sigma_2\sigma_4$, and x_3 to σ_3 defines an embedding of the Artin monoid $\langle x_1, x_2, x_3 \; ; \; x_1x_2x_1 = x_2x_1x_2, x_1x_3 = x_3x_1, x_2x_3x_2x_3 = x_3x_2x_3x_2\rangle$ into B_6^+.

Exercise 2.22. **(coherence)** *Assume that f is a complement on A, P is a subset of A^* that includes A and is closed under $\backslash$ in the sense that for all u, v in P, $u\backslash v$ exists and it belongs to P. Assume moreover that f is coherent on P.*

(i) Prove that $\backslash$ is compatible with $\equiv^{++}$ on P.

(ii) Prove that $\equiv^{++}$ is transitive on P.

(iii) Prove that word reversing is complete for f, and conclude that f is coherent on A^*. [This is tricky.]

3. Garside Elements

Going further requires that word reversing always converges. A sufficient condition is the existence of a special element that we call a Garside element. The results we shall prove here can be summarized as follows:

Proposition 3.1. (Garside element) *Assume that M and G respectively are the monoid and the group associated (on the right) with some complement f on a finite set A. Assume that f is normed and coherent on A, M is right cancellative, and there exists a Garside element Δ in M. Then:*

(i) Reversing associated with f always converges;

(ii) Any two elements in M admit a right lcm, M embeds in G, G is a group of fractions of M, and every element of M admits a unique normal form defined in terms of divisors of Δ;

(iii) The group G is torsion free, and its word problem can be solved (in quadratic time) using a double reversing.

(Under the above hypotheses, it can still be proved that the group G admits a biautomatic structure and its center is connected with the powers of Δ.)

Termination of word reversing

Definition 3.2. (convergent) Assume that f is a complement on A. We say that f is *convergent (on the right)* if $u\backslash v$ exists for all words u, v on A.

By Proposition 2.16 (lcm), convergence for a coherent complement f is a necessary and sufficient condition for every pair of elements in the associated monoid to admit a right lcm.

Definition 3.3. (lcm monoid) Assume that M is a monoid. We say that M is a *right lcm monoid* if 1 is the only invertible element in M, M is left cancellative, and any two elements of M admit a right lcm.

Thus, right lcm monoids are exactly those monoids where $\vee_R$ and $\backslash$ are well-defined binary operations. Using standard properties of lcm's and gcd's (easily extended to the current non-commutative context), we can verify that, if M is an atomic right lcm monoid, then $(M, \vee_R, \wedge_L)$ is a lattice. In this framework, the results of Section 3 imply:

Proposition 3.4. (lcm monoid) *Assume that M is a monoid associated with a normed, coherent and convergent complement. Then M is an atomic right lcm monoid, and word reversing solves its word problem.*

(We have seen in Section 2 that coherence is a sufficient condition for $u \equiv^+ v$ to be equivalent to $u^{-1}v \curvearrowright \varepsilon$; this yields an effective solution for the word problem only if we are sure that reversing $u^{-1}v$ always converges: otherwise, for non-$\equiv^+$-equivalent words u, v, we may be unable to conclude.)

Let us look at conditions implying convergence. A necessary condition is that $f(x, y)$ be defined for all letters x, y, but, clearly, this condition is not sufficient in general. Here is a sufficient condition:

Proposition 3.5. (closed) *Assume that f is a complement on A. Then the complement f is convergent if and only if there exists a subset P of A^* including A and such that $u\backslash v$ exists and belongs to P for all u, v in P.*

Moreover, if the previous condition is true and there exists k such that reversing a word $u^{-1}v$ with $u, v \in P$ requires at most k steps and ends with a word of length at most ℓ, then reversing an arbitrary word w on $A \cup A^{-1}$ requires at most $\frac{1}{4}k \cdot \lg(w)^2$ steps and ends with a word of length at most $\ell \cdot \lg(w)$.

Proof. If f is convergent, then (ii) holds with $P = A^*$. Conversely, assume (ii). Let w be a word on $A \cup A^{-1}$. We can write $w = u_1^{e_1} \cdots u_r^{e_r}$ with $u_i \in P$, $e_i = \pm 1$ for every i, and $r \le \lg(w)$. Let p be the number of positive e_i's. A simple inductive argument illustrated in Figure 3.1 shows that w is reversible to some word $v_1 \cdots v_p v_{p+1}^{-1} \cdots v_r^{-1}$ with $v_1, \ldots, v_r$ in P, and that the reversing decomposes into $p(r-p)$ reversings of words of the form $w_1^{-1}w_2$ with w_1, w_2 in P. This gives the bounds of the lemma, as $p(r-p) \le r^2/4$ holds. ∎

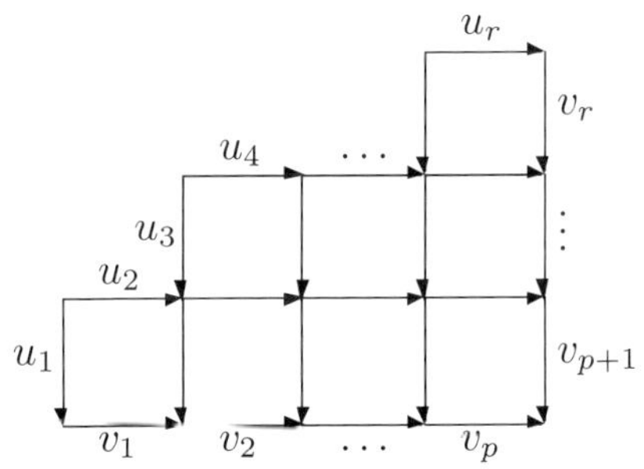

Figure 3.1. Quadratic reversing

A set of words P such that $u\backslash v$ belongs to P whenever u and v do will be said to be closed under $\backslash$. A favourable case is when a finite closed set exists. This happens if and only if the minimal possible closed set, *i.e.*, the closure of the alphabet under $\backslash$, is finite. Such a condition can be checked directly when a concrete presentation is given.

Example 3.6. (**closed**) *Let us consider the braid complement f_B on $\{\sigma_1, \sigma_2\}$. The reader can check that the closure of $\{\sigma_1, \sigma_2\}$ under operation $\backslash$ consists of the five words ε, σ_1, σ_2, $\sigma_1\sigma_2$, $\sigma_2\sigma_1$.*

The condition can also be deduced from the existence of a special element in the associated monoid, as we shall see now.

Definition 3.7. (**Garside element**) Assume that M is a monoid, and Δ is an element of M. We say that Δ is a *Garside element* in M if the set of all left divisors of Δ coincide with the set of all right divisors of Δ, and this set generates M.

Proposition 3.8. (**convergent**) *Assume that A is finite and f is a complement on A that is normed and coherent on A. Let M be the monoid associated with f. Assume that there exists a Garside element in M. Then the closure of A under $\backslash$ is finite, and f is convergent.*

Proof. Let Δ be a Garside element in M, and let S be the set of all words on A that represent a (left) divisor of Δ. We assume at first that A is exactly the set of all atoms in M. As the monoid M is atomic, the lengths of the words in S are bounded above by $\|\Delta\|$, and, as A is finite, S itself is finite. By hypothesis, S includes A. We claim that S is closed under $\backslash$. Indeed, assume $u, v \in S$. By definition, there exist words u', v' such that both uv' and vu' represent Δ. As f is coherent, this implies that $u\backslash v$ exists. By construction, $u(u\backslash v)$ represents a left divisor of the element represented by uv', *i.e.*, Δ, so $u(u\backslash v)$ belongs to S. Hence it represents also a right divisor of Δ, and so does $u\backslash v$. Therefore, $u\backslash v$ represents a right divisor of Δ, so it belongs to S. Applying Proposition 3.5 (closed), we conclude that f is convergent, and that the closure of A under $\backslash$ is included in S, so it is finite.

If we do not assume that every element of A is an atom, we can argue as follows. Let A_0 be the set of all atoms. The previous argument shows that the closure of A_0 under $\backslash$ is finite, actually included in S. There exists n such that $\|a\| \leq n$ holds for every element a of A, so every such a be expressed as the product of at most n elements of A_0, hence of S. An induction shows that, if S is closed under $\backslash$, so is every power of S. Hence the closure of A under $\backslash$ is included in S^n, which is finite. ∎

Normal form

We have seen that word reversing solves the word problem whenever a Garside element exists. Actually, we can obtain more, namely a unique normal form.

Definition 3.9. (**saturated**) Assume that M is an atomic right lcm monoid, and S is a subset of M. We say that S is *saturated (on the right)* if S contains all atoms of M, and S is closed under $\vee_R$ and $\backslash$.

Definition 3.10. (**orthogonal**) Assume that M is a monoid and S is a subset of M. For $a \in S$ and $b \in M$, we say that b is *S-orthogonal to a (on the right)* if $ac \in S$ holds for no proper left divisor c of b.

If S generates the monoid M, every element of M can be decomposed as a product of elements of S. The idea for obtaining a unique distinguished decomposition is to push the elements to the left, *i.e.*, to require the first element of the decomposition to be the maximal possible one. A saturated family is what is needed for the process to lead to a unique decomposition.

Proposition 3.11. (**decomposition**) *Assume that M is an atomic right lcm monoid, and S is a saturated family in M. Then every element of M admits a unique decomposition of the form $a_1 \cdots a_p$ where $a_1, \ldots, a_p$ belong to S and a_{k+1} is S-orthogonal to a_k for $k < p$.*

Proof. We prove the existence of a decomposition as above for every a in M using induction on $\|a\|$. For $\|a\| = 0$, *i.e.*, for $a = 1$, the result is obvious, as $1 \in S$ holds by definition. Assume $a \neq 1$. Let $\mathrm{Div}_L(a)$ denote the set of all left divisors of a. Let a_1 be an element of $\mathrm{Div}_L(a) \cap S$ such that $\|a_1\|$ is maximal. As $b \in \mathrm{Div}_L(a) \cap S$ implies $\|b\| \leq \|a\|$, such an element a_1 must exist. As a is not 1, at least one atom divides a on the left, and, as S contains all atoms, we deduce $a_1 \neq 1$. Write $a = a_1 b$. Then we have $\|b\| < \|a\|$. By induction hypothesis, b admits a decomposition $b = a_2 \cdots a_p$ that satisfies the conditions of the proposition. We have $a = a_1 a_2 \cdots a_p$, and it remains to prove that a_2 is S-orthogonal to a_1. Assume that c is a nontrivial left divisor of a_2. Then c is a left divisor of b, and $a_1 c$ is a left divisor of a. This implies $a_1 c \notin S$, for, otherwise, we would have $a_1 c \in \mathrm{Div}_L(a) \cap S$ and $\|a_1 c\| > \|a\|$, contradicting the definition of a_1.

For uniqueness, it suffices that we prove that, if $(a_1, \ldots, a_p)$ is a sequence of elements of S such that a_{k+1} is S-orthogonal to a_k for $k < p$, then a_1 is determined by the product $a_1 \cdots a_p$. Indeed, M is left cancellative, and an induction then shows that $a_2, \ldots, a_p$ are determined as well. Let $a = a_1 \cdots a_p$. By construction, we have $a_1 \in \mathrm{Div}_L(a) \cap S$. Assume $c_1 \neq 1$ and $a_1 c_1 \in S$. Define $c_k = a_k \backslash c_{k+1}$ inductively. The hypotheses $a_1 \in S$ and $a_1 c_1 \in S$ imply $c_1 \in S$, as S is closed under $\backslash$. Then, we have $a_2 c_2 = a_2 \vee_R c_1$, an element of S as S is closed under $\vee_R$. We deduce $c_2 \in S$. Similarly, we prove $a_k c_k \in S$ and $c_k \in S$ for every k using induction on k (Figure 3.2). Now, the conjunction of $c_1 \neq 1$ and a_2 being S-orthogonal to a_1 implies $c_1 \notin \mathrm{Div}_L(a_2)$, and, therefore, $c_2 \neq 1$. Repeating the argument gives inductively $c_k \neq 1$ for every k. In particular, we have $c_p \neq 1$. Now, by construction, we have $c_p = (a_2 \cdots a_p) \backslash c_1$, and $c_p \neq 1$ means $c_1 \notin \mathrm{Div}_L(a_2 \cdots a_p)$, hence $a_1 c_1 \notin \mathrm{Div}_L(a)$. So, no proper right multiple of a_1 belongs to $\mathrm{Div}_L(a) \cap S$. Now, if b belongs to $\mathrm{Div}_L(a) \cap S$, so does $a_1 \vee_R b$. The latter element is a right multiple of a, hence the above result implies $a_1 \vee_R b = a_1$, *i.e.*, b is a left divisor of a_1. It follows that a_1 is the right lcm of $\mathrm{Div}_L(a) \cap S$, which makes it clear that a_1 depends on a only. ∎

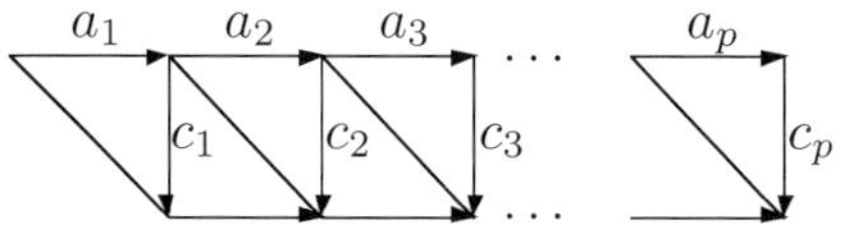

Figure 3.2. Normal decomposition

The decomposition provided by Proposition 3.11 (normal decomposition) will be called the S-*normal* decomposition of a.

Assume that M is a monoid specified by a presentation involving the alphabet A. Defining a unique normal form for this presentation means constructing a family of words N on A such that every element of M is represented by exactly one element of N. If S is a saturated family in M, using the S-normal decomposition reduces the problem of finding a normal form for the elements of M to the problem of finding a normal form for the elements of S. There is no effective solution in general (M itself is a saturated family for M), but a sufficient condition for such a solution to exist is S being finite.

Such a favourable situation occurs when a Garside element exists:

Proposition 3.12. **(saturated)** *Assume that M is a monoid associated with a complement f on a finite set A. Assume that f is normed and coherent, and there exists a Garside element Δ in M. Then*

(i) The divisors of Δ form a finite saturated family S;

(ii) For every nontrivial element a of M, the first factor of the S-normal decomposition of a is $a \wedge_L \Delta$;

(iii) For a in S and b in M, b being S-orthogonal to a is equivalent to b and $a \backslash \Delta$ being left coprime, i.e., admitting no nontrivial common left divisor.

Proof. (i) By definition, S generates M, hence it contains all atoms. Assume $a, b \in S$. Then $a \vee_R b$ is a left divisor of Δ, so it belongs to S. Moreover, there exists c satisfying $\Delta = a(a\backslash b)c$. Hence $(a\backslash b)c$ is a right divisor of Δ, hence a left divisor of Δ, and, therefore, $a\backslash b$ belongs to S.

(ii) Assume that a is an arbitrary element of M, and let $(a_1, \ldots, a_p)$ be its S-normal decomposition. By definition, a_1 is a left divisor of Δ and of a, hence of $a \wedge_L \Delta$. Now, $\|a_1\|$ has the maximal possible value among those elements of M that are left divisors of a and belong to S, *i.e.*, among the left divisors of $a \wedge_L \Delta$. Hence a_1 must be equal to $a \wedge_L \Delta$.

(iii) The element b being S-orthogonal to a means that, if x is a left divisor of b, then ax is not in S, *i.e.*, ax is not a left divisor of Δ. Now, ax being a left divisor of Δ is equivalent to x being a left divisor of $a\backslash\Delta$. ∎

Fractions

We consider now the group associated with a complement f on A, *i.e.*, the group that admits the presentation

$$\langle A \,;\, \{xf(x,y) = yf(y,x)\}\rangle.$$

Definition 3.13. (numerator, denominator) Assume that f is a complement on A. For w a word on $A \cup A^{-1}$, the *(right) numerator of* w, and the *(right) denominator of* w *(relative to* f*)* are defined to be the (unique) positive words $N(w)$ and $D(w)$ such that w is reversible to $N(w)D(w)^{-1}$—if such words exist.

Thus, $N(w)$ and $D(w)$ are the unique words satisfying $w \curvearrowright N(w)D(w)^{-1}$, and they correspond to the generic picture:

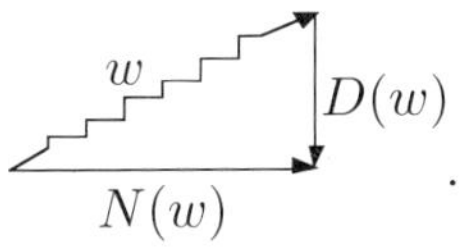

It follows from Lemma 1.8 that the equivalence

$$w \equiv N(w)\ D(w)^{-1} \tag{3.1}$$

holds for every word w whenever its right numerator and denominator exist.

Let us begin with some basic properties.

Lemma 3.14. *Assume that f is a complement on A.*

(i) For u, v words on A, we have

$$u\backslash v = N(u^{-1}v), \qquad v\backslash u = D(u^{-1}v); \tag{3.2}$$

in particular, $N(v)$ and $D(v)$ exist, and we have $N(v) = v$, $D(v) = \varepsilon$.

(ii) For w a word on $A \cup A^{-1}$, if $N(w)$ and $D(w)$ exist, so do $N(w^{-1})$ and $D(w^{-1})$, and we have $N(w^{-1}) = D(w)$, $D(w^{-1}) = N(w)$.

(iii) For w_1, w_2 words on $A \cup A^{-1}$, $N(w_1w_2)$ and $D(w_1w_2)$ exist if and only if $D(w_1)$, $N(w_2)$, and $D(w_1)\backslash N(w_2)$ exist, and, if so, we have

$$N(w_1w_2) = N(w_1)\ (D(w_1)\backslash N(w_2)),$$
$$D(w_1w_2) = D(w_2)\ (N(w_2)\backslash D(w_1)).$$

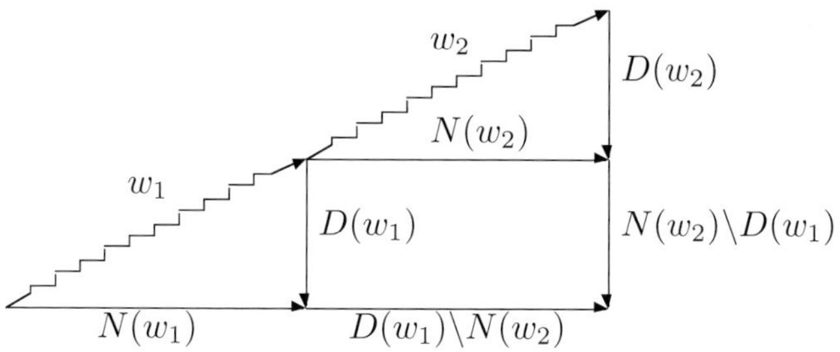

Figure 3.3. Reversing a product

Proof. Point (i) follows from the definition. For (ii), we note that $w \curvearrowright w'$ implies $w^{-1} \curvearrowright w'^{-1}$. For (iii), the computation is illustrated on Figure 3.3. ■

Proposition 3.15. **(fraction)** *Assume that G is a group associated with a convergent complement on A.*

(i) For every word w on $A \cup A^{-1}$, the words $N(w)$ and $D(w)$ exist.

(ii) Every element of G can be expressed as a fraction ab^{-1}, where a and b belong to the submonoid of G generated by A.

Proof. We prove (i) using induction on the length of w. The result is obvious if w is empty. Otherwise, assume first $w = w_0x$, with $x \in A$. By induction hypothesis, $N(w_0)$ and $D(w_0)$ exist. By Lemma 3.14, we have

$$N(w) = N(w_0)\,(D(w_0)\backslash x), \quad D(w) = x\backslash D(w_0),$$

and the latter words exist by hypothesis. The argument is similar for $w = w_0x^{-1}$ with $x \in A$. Point (ii) follows from (i) immediately. ■

The previous proposition does not tell us whether the submonoid of G generated by A is isomorphic to the monoid associated with f, *i.e.*, whether the latter monoid embeds in G. In the case of a coherent complement, we can investigate the question more closely and establish a precise connection between the relations $\equiv^+$ and $\equiv$.

Lemma 3.16. *Assume that f is a coherent and convergent complement on A.*

(i) For all words w, w' on $A \cup A^{-1}$, $w \equiv w'$ holds if and only if there exist two words v, v' on A satisfying

$$N(w)\,v \equiv^+ N(w')\,v' \qquad \text{and} \qquad D(w)\,v \equiv^+ D(w')\,v'. \tag{3.3}$$

(ii) In particular, for all words u, u' on A, $u \equiv u'$ holds if and only if $uv \equiv^+ u'v$ holds for some positive word v.

Proof. (i) Let us say that $w \sim w'$ holds when there exist v and v' satisfying (3.3). Clearly $w \sim w'$ implies $w \equiv w'$. In order to prove the converse implication, we observe first that $\sim$ is an equivalence relation on $(A \cup A^{-1})^*$. Reflexivity and symmetry are clear. So assume $w \sim w' \sim w''$. By hypothesis, there exist positive words v, v'_1 and v'_2, v'' satisfying

$$N(w)v \equiv^+ N(w')v'_1 \quad \text{and} \quad D(w)v \equiv^+ D(w')\, v'_1,$$
$$N(w')v'_2 \equiv^+ N(w'')v'' \quad \text{and} \quad D(w')v'_2 \equiv^+ D(w'')\, v''.$$

As f is convergent, we have $v'_1 u_2 \equiv^+ v'_2 u_1$ for some positive words u_1, u_2, hence $N(w)vu_2 \equiv^+ N(w'')v''u_1$ and $D(w)vu_2 \equiv^+ D(w'')\, v''u_1$, and $w \sim w''$ holds.

Next, $xf(x,y) \sim yf(y,x)$ holds for all x, y in A, and, more generally, $u \equiv^+ u'$ implies $u \sim u'$ for u, u' positive words. Similarly, $x^{-1}x \sim \varepsilon$ and $xx^{-1} \sim \varepsilon$ hold for x in A: in the latter case, we take $v = \varepsilon$, $v' = x$.

So, it remains to prove that $\sim$ is compatible with the product, *i.e.*, that $w \sim w'$ implies $x^{\pm 1}w \sim x^{\pm 1}w'$ and $wx^{\pm 1} \sim w'x^{\pm 1}$ for every x in A. Let us assume $N(w)v = N(w')v'$ and $D(w)v = D(w')v'$, and consider the question of adding $x^{\pm 1}$ on the right. The case of x^{-1} is trivial as, by Lemma 3.14, we have $N(wx^{-1}) = N(w)$ and $D(wx^{-1}) = xD(w)$, and we deduce $wx^{-1} \sim w'x^{-1}$.

In the case of x, the computation is more complicated. By hypothesis, we have $N(wx) = N(w)\,(D(w)\backslash x)$ and $D(wx) = D(w)\,x$. Let $u = (D(w)\backslash x)\backslash v$ and $u' = (D(w')\backslash x)\backslash v'$. Applying Relation (3.1), we find

$$\begin{aligned} N(wx)u &= N(w)\,(D(w)\backslash x)\,((D(w)\backslash x)\backslash v) \\ &\equiv^+ N(w)\,v\,(v\backslash(D(w)\backslash x)) = N(w)\,v\,((D(w)v)\backslash x), \\ D(wx)u &= (x\backslash D(w))\,((D(w)\backslash x)\backslash v) = x\backslash(D(w)v). \end{aligned}$$

Similar equivalences hold for the dashed variables. Now, by hypothesis, we have $N(w)v \equiv^+ N(w')v'$ and $D(w)v \equiv^+ D(w')v'$. Hence, f being coherent implies $N(wx)u \equiv^+ N(w'x)$ and $D(wx)u \equiv^+ D(w'x)$, hence $wx \sim w'x$.

The argument is symmetric for left multiplication: here multiplying by x is trivial, while multiplying by x^{-1} requires a computation as above. ■

Proposition 3.17. (**embedding**) *Assume that M and G respectively are the monoid and the group associated with a coherent and convergent complement f. Then the following are equivalent:*

(i) The monoid M is right cancellative;

(ii) The monoid M embeds in the group G, and the latter is a group of right fractions of M.

Proof. Assume that M is right cancellative. Then, by Lemma 3.16, $u \equiv u'$ implies, and, therefore, is equivalent to $u \equiv^+ u'$ for all positive words u, u'. This means that the identity mapping induces an embedding of M into G. The converse implication is straightforward. ■

Proposition 3.18. (**word problem**) *Assume that M and G respectively are the monoid and the group associated with a coherent and convergent comple-*

ment f on A Assume moreover that M is right cancellative. Then, for every word w on $A \cup A^{-1}$, the following are equivalent:

(i) The word w represents 1 in G, i.e., $w \equiv \varepsilon$ holds;

(ii) The positive words $N(w)$ and $D(w)$ are $\equiv^+$-equivalent;

(iii) The word $D(w)^{-1}N(w)$ is reversible to the empty word;

(iv) We have $N(w^2) = N(w)$ and $D(w^2) = D(w)$.

Proof. As $w \equiv N(w)D(w)^{-1}$ holds by definition, $w \equiv \varepsilon$ is equivalent to $N(w) \equiv D(w)$. Under the current hypotheses, owing to Proposition 3.17 (embedding), this is equivalent to $N(w) \equiv^+ D(w)$. As word reversing is complete for f, this in turn is equivalent to $N(w) \equiv^{++} D(w)$. So (i), (ii) and (iii) are equivalent. Now (iv) implies $w^2 \equiv w$, hence $w \equiv \varepsilon$. So it remains to prove that (ii) implies (iv). By construction, we have

$$N(w^2) = N(w)\,(N(w)\backslash D(w)), \quad D(w^2) = D(w)\,(D(w)\backslash N(w)).$$

By completeness, the hypothesis $N(w) \equiv^+ D(w)$ implies $N(w)\backslash D(w)) = D(w)\backslash N(w)) = \varepsilon$, hence $N(w^2) = N(w)$ and $D(w^2) = D(w)$. ■

So the word problem of every group associated with a convenient complement is solvable using a double word reversing: in order to decide whether a given word w represents 1 or not, we first reverse w into $N(w)D(w)^{-1}$, then we switch the factors and reverse the resulting word $D(w)^{-1}N(w)$. The initial word w represents 1 if and only if the final word is empty.

Example 3.19. (**word problem**) *Let us consider again the word $w_0 = \sigma_1^{-1}\sigma_2\sigma_2^{-1}\sigma_3$ in the braid group B_4. We shall see in Section 4 that the braid complement is eligible for the previous results. Applying the previous decision algorithm to w_0 consists in first reversing w_0 into $(\sigma_2\sigma_1\sigma_3\sigma_2\sigma_1)(\sigma_2\sigma_3\sigma_1\sigma_2\sigma_3)^{-1}$, then switching the factors, thus obtaining $(\sigma_2\sigma_3\sigma_1\sigma_2\sigma_3)^{-1}(\sigma_2\sigma_1\sigma_3\sigma_2\sigma_1)$, and reversing the latter word, which yields $\sigma_1\sigma_3^{-1}$; the latter word is not empty, so we conclude that the initial word w_0 does not represent 1 in B_4.*

Proposition 3.20. (**no torsion**) *Assume that M and G respectively are the monoid and the group associated with a coherent and convergent complement f. Assume moreover that M is right cancellative. Then the group G is torsion free.*

Proof. Assume $n \geq 2$ and the word w^n represents 1 in G. By the previous proposition, we have $N(w^{2n}) = N(w^n)$ and $D(w^{2n}) = D(w^n)$. Let us define two sequences of positive words u_i, v_i by $u_1 = D(w)$, $v_1 = N(w)$, and $u_{i+1} = v_i\backslash u_i$, $v_{i+1} = u_i\backslash v_i$. For every k, we obtain inductively

$$N(w^k) = v_1 v_2 \cdots v_k, \qquad D(w^k) = u_1 u_2 \cdots u_k.$$

Our hypothesis implies that the words $u_{n+1} \cdots u_{2n}$ and $v_{n+1} \cdots v_{2n}$ are empty. In particular, u_{n+1} and v_{n+1} are empty.

Now, assume $u_{k+1} = v_{k+1} = \varepsilon$, with $k \geq 2$. By definition, this implies $u_k \equiv^+ v_k$. By construction, we have

$$u_{k-1} v_k \equiv^+ v_{k-1} u_k,$$

so the hypothesis implies $u_{k-1} u_k \equiv^+ v_{k-1} u_k$, and, therefore, $u_{k-1} \equiv^+ v_{k-1}$ as M is right cancellative. By completeness of word reversing, this implies in turn $u_k = v_k = \varepsilon$. Using this descent argument, we conclude that $u_{n+1} = v_{n+1} = \varepsilon$ implies $u_2 = v_2 = \varepsilon$, hence $N(w^2) = N(w)$ and $D(w^2) = D(w)$, and, therefore, $w \equiv \varepsilon$. So we have shown that $w^n \equiv \varepsilon$ implies $w \equiv \varepsilon$. ■

Exercise 3.21. **(Baumslag–Solitar)** *Prove that the complement f defined on $\{x, y\}$ by $f(x, y) = x$, $f(y, x) = y^2$ is convergent, but there exists no finite set of words including $\{x, y\}$ and closed under $\backslash$.*

Exercise 3.22. **(Garside)** *Prove that y^3 is a Garside element in the monoid $\langle x, y \;;\; xyx = y^2 \rangle$. [Observe that y^3 is the smallest possible Garside element, but it is not the lcm of the generators.]*

Exercise 3.23. **(complexity)** *Assume that f is a complement on A that is normed and coherent on A. Assume u, v are words on A and a is a common right multiple of $\overline{u}$ and $\overline{v}$ in the monoid associated with f. Prove that the number of elementary steps needed to reverse the word $u^{-1}v$ is bounded above by $2^{\|a\|}$. [Hint: The length of every positive path in the reversing graph of $u^{-1}v$ is bounded above by $\|a\|$, hence the number of vertices is bounded above by $2^{\|a\|}$. Now each vertex is involved at most once as the origin of a reversing step.]*

Exercise 3.24. **(center)** *Assume that A is finite, f is a complement on A that is normed and coherent on A, and Δ is a Garside element in the monoid M associated with f. Prove that there exists an automorphism ϕ of M satisfying $a\,\Delta = \Delta\,\phi(a)$ for every a. [Hint: Define $\phi(a) = (a\backslash\Delta)\backslash\Delta$ for a a (left) divisor of Δ.] Deduce that some power of Δ belongs to the center of M.*

Exercise 3.25. **(not complete)** *(i) Consider the complement f associated with the presentation $\langle x, y, z \,;\, x = y^2 z, x = zx, yx = z \rangle$. Prove that f is coherent on $\{x, y, z\}$, but that word reversing is not complete for f. [Hint: Show that $yyx \equiv^+ zxx$ holds, but $yyx \equiv^{++} zxx$ does not.]*

(ii) Check that the set $\{\varepsilon, x, y, z, yz\}$ is closed under $\backslash$, and conclude that f is convergent. [The lack of completeness in Example 2.3 comes from the fact that certain complements do not exist. The current exercise shows that, even if word reversing converges, local coherence is not sufficient to guarantee coherence in general. On the other hand, Exercise 2.22 shows that coherence on the closure of the letters under $\backslash$ is sufficient.]

Exercise 3.26. **(lcm selector)** *Assume that M is an atomic right lcm monoid. Assume that A generates M. Define an lcm selector on A to be a mapping f of $A \times A$ to A^* such that, for every x, y in A, the word $xf(x, y)$ represents the right lcm of x and y. Prove that M is isomorphic to the monoid associated with f. Prove that the complement f is normed, coherent on A^* and convergent. [Thus the sufficient conditions of Proposition 3.4 (lcm monoid) are also necessary.]*

---

4. The Case of Braids

Here we show that Artin's braid groups B_n are eligible for the previous approach. We prove that the involved complement is normed and coherent, and that B_n^+ contains a Garside element. We deduce a number of algebraic properties of B_n^+ and B_n.

The braid complement

We have already noted that, for $2 \le n \le \infty$, the standard presentation of B_n is associated with a complement, namely the mapping f_B defined by

$$f_B(\sigma_i, \sigma_j) = \begin{cases} \varepsilon & \text{for } i = j, \\ \sigma_j \sigma_i & \text{for } |i - j| = 1, \\ \sigma_j & \text{for } |i - j| \ge 2. \end{cases}$$

There is no danger in not mentioning the index n in f_B: the complement for B_n is the restriction of the complement for B_{n+1}, and the reversing of a word w

in BW_n coincides with its reversing as a word in BW_{n+1}. This amounts to working in the direct limit B_∞, with the infinite alphabet $\{\sigma_1, \sigma_2, \ldots\}$.

In the sequel, $\curvearrowright$ and $\backslash$ refer to the braid complement f_B. However, we shall introduce below a symmetric notion of left reversing, and, therefore, we shall speak here of right reversing and use N_R and D_R for the associated numerator and denominator. We first check that the braid complement satisfies all conditions considered in Section 2.

Lemma 4.1. *The braid complement f_B admits a norm.*

Proof. For every pair of generators σ_i, σ_j, the length of the words $f_B(\sigma_i, \sigma_j)$ and $f_B(\sigma_j, \sigma_i)$ are equal, hence the length is a norm. ∎

The atoms of the monoid B_n^+ are the generators σ_i, and, for each a in B_n^+, $\|a\|$ is the common length of all positive words that represent a.

Lemma 4.2. *The braid complement f_B is coherent on $\{\sigma_1, \sigma_2, \ldots\}$.*

Proof. We claim that, for every triple (i, j, k), the words $(\sigma_i\backslash\sigma_j)\backslash(\sigma_i\backslash\sigma_k)$ and $(\sigma_j\backslash\sigma_i)\backslash(\sigma_j\backslash\sigma_k)$ exist and are $\equiv^{++}$-equivalent. In Example 2.6, we have already considered the case $i = 1$, $j = 2$, $k = 3$. A systematic study is easy. Firstly, the result is trivial if at least two integers in $\{i, j, k\}$ are equal, or if the distance between any two of them is at least 2. Up to a translation, a dilatation, or a symmetry, the remaining cases are (we do not repeat the case $i = 1$, $j = 2$, $k = 3$):

- $i = 1$, $j = 3$, $k = 4$: we find

$$\begin{cases} (\sigma_1\backslash\sigma_3)\backslash(\sigma_1\backslash\sigma_4) = \sigma_3\backslash\sigma_4 = \sigma_4\sigma_3, \\ (\sigma_3\backslash\sigma_1)\backslash(\sigma_3\backslash\sigma_4) = \sigma_1\backslash(\sigma_4\sigma_3) = \sigma_4\sigma_3; \end{cases}$$

- $i = 1$, $j = 3$, $k = 2$: we find

$$\begin{cases} (\sigma_1\backslash\sigma_3)\backslash(\sigma_1\backslash\sigma_2) = \sigma_3\backslash(\sigma_2\sigma_1) = \sigma_2\sigma_3\sigma_1\sigma_2, \\ (\sigma_3\backslash\sigma_1)\backslash(\sigma_3\backslash\sigma_2)) = \sigma_1\backslash(\sigma_2\sigma_3) = \sigma_2\sigma_1\sigma_3\sigma_2; \end{cases}$$

- $i = 1$, $j = 2$, $k = 4$: we find

$$\begin{cases} (\sigma_1\backslash\sigma_2)\backslash(\sigma_1\backslash\sigma_4) = (\sigma_2\sigma_1)\backslash\sigma_4 = \sigma_4, \\ (\sigma_2\backslash\sigma_1)\backslash(\sigma_2\backslash\sigma_4) = (\sigma_1\sigma_2)\backslash\sigma_4 = \sigma_4. \end{cases}$$

In the first and the third cases, the final words are equal, hence $\equiv_R^{++}$-equivalent. A direct verification gives $\sigma_2\sigma_3\sigma_1\sigma_2 \equiv^{++} \sigma_2\sigma_1\sigma_3\sigma_2$. ∎

By the results of Section 2, we deduce that the braid complement is everywhere coherent, and, therefore, that word reversing is complete and the operation $\backslash$ is compatible with $\equiv^+$. We thus may state:

Proposition 4.3. (**completeness**) *Assume that u, v are positive braid words. Then the following are equivalent:*

(i) The words u and v are $\equiv^+$-equivalent;

(ii) The word $u^{-1}v$ is R-reversible to the empty word.

Proposition 4.4. (**compatibility**) *Assume that u, u', v, v' are positive braid words satisfying $u' \equiv^+ u$ and $v' \equiv^+ v$. If the word $u\backslash v$ exists, then $u'\backslash v'$ exists as well, and we have $u'\backslash v' \equiv^+ u\backslash v$.*

Proposition 4.5. (**left cancellation**) *For every n, the braid monoid B_n^+ is left cancellative. Any two elements of B_n^+ admit a left gcd, and they admit a right lcm whenever they have at least one common right multiple.*

Left reversing

Besides right word reversing, where negative letters are pushed to the right, we can consider a symmetric operation of *left* word reversing, where negative letters are pushed to the left. Assume that g is a complement on an alphabet A. We say that a monoid M is *associated with g on the left* if it admits the presentation

$$\langle A\ ;\ \{g(x,y)y = g(y,x)x\}\rangle.$$

For w, w' words on $A \cup A^{-1}$, we say that w is *left reversible* to w' if we can transform w into w' by iteratively replacing factors of the form $x\,y^{-1}$ with the corresponding factors $g(y,x)^{-1}\,g(x,y)$. For u, v words on A, we introduce the words u/v and v/u as the unique positive words u_1, v_1 such that uv^{-1} is left reversible to $v_1^{-1}u_1$. The derived notions of a left numerator N_L and left denominator D_L are defined in the obvious way. In particular, the left counterparts of Formulas (1.7) and (3.1) are

$$(u/v)\,v \equiv^+ (v/u)\,u, \tag{4.1}$$

$$w \equiv D_L(w)^{-1}\,N_L(w), \tag{4.2}$$

which hold respectively for u, v words on A and w word on $A \cup A^{-1}$, when the words involved do exist.

Using the explicit form of the braid relations, we see that the braid groups are also associated on the left with a complement, namely the mapping g_B defined by $g_B(\sigma_i, \sigma_i) = \varepsilon$, $g_B(\sigma_i, \sigma_j) = \sigma_j\sigma_i$ for $|i-j| = 1$, and $g_B(\sigma_i, \sigma_j) = \sigma_i$ for $|i-j| \geq 2$. The left and right operations are closely connected.

Lemma 4.6. *For w a braid word, let w^{rev} be the word obtained from w by reversing the ordering of the letters. Then, for all w in BW_∞ and u, v in BW_∞^+, we have*

$$N_L(w^{rev}) = N_R(w)^{rev}, \quad D_L(w^{rev}) = D_R(w)^{rev}, \quad u^{rev}/v^{rev} = (v\backslash u)^{rev}. \tag{4.3}$$

Proof. A straightforward induction: if w is right reversible to w' in m steps using f_B, then w^{rev} is left reversible to w'^{rev} using g_B in m steps as well. ∎

It follows that all properties established for the complement f_B used on the right hold for g_B used on the left as well. In particular, left word reversing associated with f_B is (in the obvious sense) complete, so the counterpart of Proposition 4.5 (left cancellation) applies. We deduce:

Proposition 4.7. (right cancellation) *Assume $2 \leq n \leq \infty$. Then the braid monoid B_n^+ is right cancellative. Any two elements of B_n^+ admit a right gcd, and they admit a left lcm whenever they admit at least one common left multiple.*

For $a, b \in B_n^+$, we shall denote by $a \vee_L b$ the possible left lcm of a and b, by a/b the element satisfying $a \vee_L b = (a/b)b$, and by $a \wedge_R b$ the right gcd of a and b.

Permutation braids

We prove now that the braid complements are convergent. To this end, we show that each monoid B_n^+ contains a Garside element by introducing a special family of positive braids canonically associated with permutations.

Let us begin with some notations. First, the transposition that exchanges i and $i+1$ will be denoted s_i; thus $s_i = \mathrm{perm}(\sigma_i)$ holds.

Definition 4.8. (braid $\sigma_{i,j}$) For $i, j \geq 1$, we put $\sigma_{i,j} = \sigma_i \sigma_{i+1} \cdots \sigma_{j-1} \sigma_j$ for $i \leq j$, and $\sigma_{i,j} = \sigma_i \sigma_{i-1} \cdots \sigma_{j+1} \sigma_j$ for $i \geq j$.

Figure 4.1. The braid $\sigma_{5,3}$

For instance, $\sigma_{i,i}$ is σ_i. As in the case of σ_i, we shall use $\sigma_{i,j}$ to denote both the braid word defined above and the braid it represents.

Lemma 4.9. *Assume $i \geq j$ and u is a braid word involving only $\sigma_j, \ldots, \sigma_{i-1}$. Then we have*

$$u\, \sigma_{i,j} \equiv \sigma_{i,j}\, \mathrm{sh}(u). \tag{4.4}$$

Proof. Using induction on the length of u and the properties of sh, it suffices to establish $\sigma_i\, \sigma_{n-1,1} \equiv^+ \sigma_{n-1,1}\, \sigma_{i+1}$ for $i \leq n-2$. For such an i, we find

$$\begin{aligned} \sigma_i\, \sigma_{n-1,1} = \sigma_i\, \sigma_{n-1,i+2}\, \sigma_{i+1}\, \sigma_i\, \sigma_{i-1,1} &\equiv^+ \sigma_{n-1,i+2}\, \sigma_i \sigma_{i+1} \sigma_i\, \sigma_{i-1,1} \\ &\equiv^+ \sigma_{n-1,i+2}\, \sigma_{i+1} \sigma_i \sigma_{i+1}\, \sigma_{i-1,1} \\ &\equiv^+ \sigma_{n-1,i}\, \sigma_{i-1,1}\, \sigma_{i+1} = \sigma_{n-1,1}\, \sigma_{i+1}. \end{aligned}$$ ∎

We recall from Chapter I that a permutation of the integers $\{1, \ldots, n\}$ denoted perm(b) is associated with each braid b in B_n: perm$(b)(i) = j$ holds if and only if the strand that finishes at position i in b starts from position j. Then perm is a surjective homomorphism of the group B_n onto the symmetric group $\mathfrak{S}_n$. We introduce now a distinguished section of this surjection.

Definition 4.10. (permutation braid, word $\boldsymbol{\Delta}_n$) For f a permutation of $\{1, \ldots, n\}$, we define the positive braid word $\widetilde{f}$ by $\widetilde{\mathrm{id}} = \varepsilon$ and $\widetilde{f} = \sigma_{f(k)-1,k}\, \widetilde{g}$, where k is the smallest integer moved by f, and g is defined by

$$g(i) = \begin{cases} i & \text{for } i \le k, \\ f(i)+1 & \text{for } k \le f(i) < f(k), \\ f(i) & \text{for } f(i) > f(k). \end{cases}$$

We denote by br(f) the braid represented by $\widetilde{f}$, and call it the *permutation braid* associated with f. For $n \ge 2$, we put $\boldsymbol{\Delta}_n = \widetilde{\omega_n}$, where ω_n is the flip symmetry $i \mapsto n+1-i$ of $\{1, \ldots, n\}$, and let Δ_n be the class of $\boldsymbol{\Delta}_n$ in B_n^+.

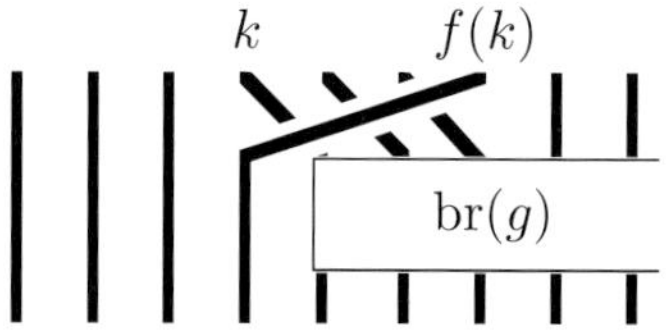

Figure 4.2. Inductive definition of br(f)

By construction, the mapping br is a section of perm, *i.e.*, perm(br(f)) $= f$ holds for every f. The braid diagram $\boldsymbol{\Delta}_n$ is displayed in Figure 4.3: it corresponds to a half-turn of a ribbon on which n strands are drawn. It follows from the definition that we have $\boldsymbol{\Delta}_2 = \sigma_1$ and, for $n \ge 3$,

$$\boldsymbol{\Delta}_n = \sigma_{n-1,1}\, \mathrm{sh}(\boldsymbol{\Delta}_{n-1}). \tag{4.5}$$

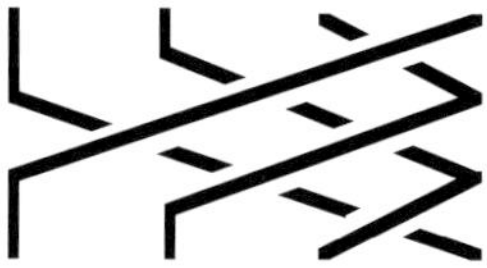

Figure 4.3. The permutation braid diagram $\boldsymbol{\Delta}_4$

Definition 4.11. (simple) For f a permutation of $\{1, \ldots, n\}$, we denote by $\ell(f)$ the *inversion number* of f, *i.e.*, the number of ordered pairs (i, j), $1 \le i < j \le n$ satisfying $f(i) > f(j)$. We say that a braid b is *simple* if b is positive and $\|b\| = \ell(\mathrm{perm}(b))$ holds.

We recall that, for b a positive braid, $\|b\|$ denotes the common length of all positive braid words that represent b. Every braid σ_i is simple, for we have $\|\sigma_i\| = \ell(\mathrm{perm}(\sigma_i)) = \ell(s_i) = 1$. The unit braid is simple as well, as $\|1\| = \ell(\mathrm{id}) = 0$ holds. On the other hand, the braid σ_1^2 is not simple, as we have $\|\sigma_1^2\| = 2$ and $\ell(\mathrm{perm}(\sigma_1^2)) = \ell(\mathrm{id}) = 0$. An induction shows that a positive braid b is simple if and only if any two strands cross at most once in any braid diagram that represents b. As $\ell(f) \le n(n+1)/2$ holds for every permutation f of n integers, the length of a simple braid in B_n^+ is bounded by $n(n+1)/2$, and, therefore, there exist only finitely many simple braids in B_n^+. The main property of simple braids for our current purpose is that they are closed under left and right divisors.

Lemma 4.12. *Assume that b_1, b_2 are positive braids and the product b_1b_2 is simple. Then both b_1 and b_2 are simple.*

Proof. For every permutation f and every integer i, we have

$$\ell(s_i f) = \begin{cases} \ell(f) + 1 & \text{if } f^{-1}(i) < f^{-1}(i+1) \text{ holds,} \\ \ell(f) - 1 & \text{if } f^{-1}(i) > f^{-1}(i+1) \text{ holds.} \end{cases} \tag{4.6}$$

This comes from the direct comparison, for each i, of the number of j's below i satisfying $f(j) > f(i)$ and the number of j's below i satisfying $f'(j) > f'(i)$, where f' is $s_i f$.

Formula (4.6) implies inductively that $\ell(\mathrm{perm}(b)) \le \|b\|$ holds for every positive braid b. It also implies the inequality

$$\ell(\mathrm{perm}(b_1b_2)) \le \|b_1\| + \ell(\mathrm{perm}(b_2)).$$

So $\ell(\mathrm{perm}(b_2)) < \|b_2\|$ implies $\ell(\mathrm{perm}(b_1b_2)) < \|b_1b_2\|$, and, therefore, the hypothesis that b_1b_2 is simple implies that b_2 is simple.

The argument is symmetric for left divisors, using the formula

$$\ell(f s_i) = \begin{cases} \ell(f) + 1 & \text{if } f(i) < f(i+1) \text{ holds,} \\ \ell(f) - 1 & \text{if } f(i) > f(i+1) \text{ holds,} \end{cases}$$

which follows from applying (4.6) to f^{-1}. ∎

Lemma 4.13. *Assume that f is a permutation of $\{1, \ldots, n\}$ and $\mathrm{br}(f)$ is a simple braid. For $1 \le i \le n-1$, two cases are possible:*

(i) Either $f^{-1}(i) < f^{-1}(i+1)$ holds, the braid $\sigma_i\,\mathrm{br}(f)$ is simple, and we have $\sigma_i\,\mathrm{br}(f) = \mathrm{br}(s_i f)$;

(ii) Or $f^{-1}(i) > f^{-1}(i+1)$ holds, and $\sigma_i\,\mathrm{br}(f)$ is not simple.

Proof. By hypothesis, $\|\mathrm{br}(f)\| = \ell(f)$ holds. In case (i), we have $\|\sigma_i\,\mathrm{br}(f)\| = \ell(f) + 1 = \ell(s_i f)$, so the braid $\sigma_i\mathrm{br}(f)$ is simple. In case (ii), we have $\|\sigma_i\,\mathrm{br}(f)\| = \ell(f) + 1$ and $\ell(s_i f) = \ell(f) - 1$, and $\sigma_i\,\mathrm{br}(f)$ is not simple.

So it remains to prove that $f^{-1}(i) < f^{-1}(i+1)$ implies $\mathrm{br}(s_i f) = \sigma_i \,\mathrm{br}(f)$. We use induction on $\ell(f)$ (or, equivalently, on the smallest integer moved by f). The result is true for $f = \mathrm{id}$, as $\mathrm{br}(f) = 1$ holds then. So, we assume $f \neq \mathrm{id}$. Let k be the smallest integer moved by f. By definition, we have $\mathrm{br}(f) = \sigma_{f(k)-1,kk}\mathrm{br}(g)$, where g is as in Definition 4.10. Five cases are to be considered.

For $i < k$, the definition gives $\mathrm{br}(s_i f) = \sigma_i \mathrm{br}(f)$ directly.

For $k \leq i < f(k) - 1$, we have $\mathrm{br}(s_i f) = \sigma_{f(k)-1,k}\mathrm{br}(s_{i+1} g)$. The hypothesis $f^{-1}(i) < f^{-1}(i+1)$ implies $g^{-1}(i+1) < g^{-1}(i+2)$. By induction hypothesis, we have $\mathrm{br}(s_{i+1} g) = \sigma_{i+1}\mathrm{br}(g)$. Using Lemma 4.9, we find

$$\mathrm{br}(s_i f) = \sigma_{f(k)-1,k}\, \sigma_{i+1}\, \mathrm{br}(g) = \sigma_i\, \sigma_{f(k)-1,k}\, \mathrm{br}(g) = \sigma_i\, \mathrm{br}(f).$$

The case $i = f(k) - 1$ is impossible, for, by construction, we would have $f^{-1}(f(k)-1) > f^{-1}(f(k)) = k$, contradicting $f^{-1}(i) < f^{-1}(i+1)$.

For $i = f(k)$, we have $s_i f(k) = f(k) + 1$, and, again, the definition gives $\mathrm{br}(s_i f) = \sigma_i \,\mathrm{br}(f)$ directly.

Finally, for $f(k) < i$, we have $\mathrm{br}(s_i f) = \sigma_{f(k)-1,k}\, \mathrm{br}(s_i g)$. The hypothesis $f^{-1}(i) < f^{-1}(i+1)$ implies $g^{-1}(i) < g^{-1}(i+1)$, and the induction hypothesis gives $\mathrm{br}(s_i g) = \sigma_i \,\mathrm{br}(g)$. Now, σ_i commutes with $\sigma_{f(k)-1}, \ldots, \sigma_k$, and we find

$$\mathrm{br}(s_i f) = \sigma_{f(k)-1,k}\, \sigma_i\, \mathrm{br}(g) = \sigma_i\, \sigma_{f(k)-1,k}\, \mathrm{br}(g) = \sigma_i \mathrm{br}(f). \qquad \blacksquare$$

Proposition 4.14. **(simple)** *Assume $n \geq 2$. For every positive braid b in B_n^+, the following are equivalent:*

(i) The braid b is a permutation braid;

(ii) The braid b is simple;

(iii) The braid b is a right divisor of Δ_n;

(iv) The braid b is a left divisor of Δ_n.

Proof. Assume that f is a permutation of $\{1, \ldots, n\}$. We prove that $\mathrm{br}(f)$ is simple using induction on the inversion number $\ell(f)$. For $\ell(f) = 0$, f is the identity, and the result is obvious. Otherwise, there exists at least one integer i satisfying $f^{-1}(i) > f^{-1}(i+1)$. Let $g = s_i f$. We have $f = \sigma_i g$, and $\ell(g) < \ell(f)$. By induction hypothesis, $\mathrm{br}(g)$ is simple. By Lemma 4.13, $\mathrm{br}(f)$ is $\sigma_i \,\mathrm{br}(g)$, and it is simple. So (i) implies (ii).

Conversely, we prove using induction on $\ell(\mathrm{perm}(b))$ that, if b is simple, then $b = \mathrm{br}(\mathrm{perm}(b))$ holds. For $\ell(\mathrm{perm}(b)) = 0$, the hypothesis that b is simple implies $\|b\| = 0$, hence $b = 1$, and b is a permutation braid. Otherwise, write $b = \sigma_i\, c$. Then we have $\mathrm{perm}(b) = s_i \,\mathrm{perm}(c)$. We have $\|b\| = \|c\| + 1$, and $\ell(\mathrm{perm}(b)) \leq \ell(\mathrm{perm}(c)) + 1$ by Lemma 4.13, hence c must be simple, with $\ell(\mathrm{perm}(c)) = \ell(\mathrm{perm}(b)) - 1$. So, by induction hypothesis, we have $c = \mathrm{br}(\mathrm{perm}(c))$. By Lemma 4.13, we deduce $b = \mathrm{br}(s_i \,\mathrm{perm}(c))$, *i.e.*, $b = \mathrm{br}(\mathrm{perm}(b))$. So (ii) implies (i).

Assume $f \neq \omega_n$. Then there exists i satisfying $f^{-1}(i) < f^{-1}(i+1)$, and we have $\ell(s_i f) > \ell(f)$ and $\sigma_i \mathrm{br}(f) = \mathrm{br}(s_i f)$. Applying the same argument to $s_i f$ and iterating, we find a permutation g satisfying $gf = \omega_n$, hence $\mathrm{br}(gf) = \mathrm{br}(g)\mathrm{br}(f)$. By construction, we have $\mathrm{br}(\omega_n) = \Delta_n$, and, therefore, Δ_n is a left multiple of $\mathrm{br}(f)$. Hence (i) and (ii) imply (iii).

The previous argument shows that, for every simple braid b in B_n^+, there exists a simple braid c that satisfies $cb = \Delta_n$. Let us denote the latter braid by $\varphi(b)$. Thus φ is a mapping of the set S consisting of all simple braids in B_n^+ into itself. As the monoid B_n^+ admits left cancellation, the mapping φ is injective. Hence, as S is finite, φ is also surjective. Hence, for every c in S, there exists b in S satisfying $\Delta_n = cb$. In particular, Δ_n is a right multiple of c. So (i) and (ii) implie (iv).

Finally, assume $\Delta_n = b_1 b_2$. Then b_1 and b_2 are simple by Lemma 4.12. ■

We immediately deduce:

Proposition 4.15. (**Garside element**) *Assume $n \geq 2$. Then Δ_n is a Garside element in B_n^+, and simple braids form a saturated family in B_n^+.*

Proof. The generators $\sigma_1, \ldots, \sigma_{n-1}$ are permutation braids by construction, so they are divisors of Δ_n. ■

By Proposition 3.8 (convergent), the braid complement f_B is convergent (on the right), and, symmetrically, g_B is convergent on the left. Thus all results of Sections 2 and 3 apply to the braid monoids and groups. In particular, the set B_n^+ equipped with the right lcm and the left gcd operations is, for every n, a lattice, and so is B_n^+ equipped with the left lcm and the right gcd operations. Applying Propositions 3.11 (decomposition) and 3.12 (saturated), we find:

Proposition 4.16. (**normal decomposition**) *Every positive n-strand braid admits a unique expression of the form $b_1 \cdots b_p$, where $b_1, \ldots, b_p$ are simple and $b_{k-1}\backslash\Delta_n$ and b_k are left coprime for $k = 2, \ldots, p$.*

Every simple braid is a permutation braid, and, therefore, it admits a distinguished representative, namely the word $\widetilde{f}$. Thus we deduce that every positive braid can be expressed in a unique way as a product of the form $\widetilde{f_1} \cdots \widetilde{f_p}$, where $f_1, \ldots, f_p$ are permutations and $\mathrm{br}(f_{k-1})\backslash\Delta_n$ and $\mathrm{br}(f_k)$ are left coprime for $k = 2, \ldots, p$.

Applying Propositions 3.17 (embedding) and 3.20 (no torsion), we find:

Proposition 4.17. (**embedding**) *For $2 \leq n \leq \infty$, the braid monoid B_n^+ embeds in the braid group B_n.*

Proposition 4.18. **(no torsion)** *For $2 \le n \le \infty$, the braid group B_n is torsion-free.*

Owing to Proposition 4.17, we need no longer distinguish between positive equivalence and general equivalence of braid words, and we use $\equiv$ everywhere in the sequel.

Exercise 4.19. **(Garside normal form)** *Prove that every braid in B_n admits a unique decomposition of the form $\Delta_n^k b$ where k is an integer and b is a positive braid of which Δ_n is not a left divisor.*

Exercise 4.20. **(center)** *(i) Consider Exercise 3.24 in the case of the monoid B_n^+. What is the corresponding automorphism? Deduce that Δ_n^2 belongs to the center of B_n.*

(ii) Assume that b is a braid whose Garside normal form (Exercise 4.19) is $\Delta_n^p a$ with $a \ne 1$. Show that b does not belong to the center of B_n. [Hint: Assume that σ_i is a left divisor of a and j is $i \pm 1$. Prove that $\Delta_n^p a$ belonging to the center of B_n implies that σ_j is a left divisor of a as well, and conclude that a is a multiple of Δ_n.] Deduce that, for $n \ge 3$, the center of B_n is the cyclic subgroup generated by Δ_n^2.

Exercise 4.21. **(conjugacy problem)** *(i) Call the integer k involved in the Garside normal form of b (Exercise 4.19) the* power *of b. Prove that, for every braid b in B_n, the powers of those braids b' that are conjugate to b have a finite upper bound $p(b)$. Define the summit set $S(b)$ of b to be the set of all braids b' such that $\Delta_n^{p(b)} b'$ is a conjugate of b. Prove that $S(b)$ is finite, and that both $p(b)$ and $S(b)$ can be determined effectively.*

(iii) Prove that two braids b_1, b_2 in B_n are conjugate if and only if their summit sets are equal.

Exercise 4.22. **(singular braids)** *Prove that the monoid obtained by adding to B_∞^+ a new infinite series of generators τ_i submitted to the relations $\tau_i\tau_j = \tau_j\tau_i$ and $\tau_i\sigma_j = \sigma_j\tau_i$ for $|i-j| \ge 2$, $\tau_i\sigma_j\sigma_i = \sigma_j\sigma_i\tau_j$ for $|i-j| = 1$, and $\tau_i\sigma_i = \sigma_i\tau_i$ for every i is associated on the right with a coherent complement.*

---

5. Double Reversing

Simultaneously using the braid complements on the right and on the left leads to additional results. The idea is that right reversing eliminates all patterns of the form $x^{-1}x$ hidden in a word, while left reversing eliminates those of the form xx^{-1}. So it is natural that combining both gives words that are minimal in some sense. In this section, we describe the braid normal form arising in this way, and we evaluate for future use the algorithmical complexity of the involved algorithms. In particular, we show how using simple braids as letters gives an algorithm which has a quadratic complexity for a fixed number of strands, and a polynomial complexity in any case.

Normal forms

By Proposition 3.18 (word problem), the word problem for the standard presentation of the braid groups is decidable, *i.e.*, there exists an effective algorithm that recognizes whether a given braid word represents 1 in the braid group. The algorithm described in Section 3 involves a double right reversing. We can instead use a more symmetric algorithm that involves both a right and a left reversing. We have observed that neither the right nor the left numerator mappings induce well-defined mappings on braids: for instance, $\sigma_1\sigma_1^{-1}$ and ε represent the same braid, but we have $N_R(\sigma_1\sigma_1^{-1}) = \sigma_1$ and $N_R(\varepsilon) = \varepsilon$. This obstruction vanishes when we combine right and left reversing.

Lemma 5.1. *Assume that w and w' are equivalent braid words. Then we have*

$$N_L(N_R(w)D_R(w)^{-1}) \equiv N_L(N_R(w')D_R(w')^{-1}), \tag{5.1}$$
$$D_L(N_R(w)D_R(w)^{-1}) \equiv D_L(N_R(w')D_R(w')^{-1}). \tag{5.2}$$

Proof. By Proposition 4.3 (completeness), $w \equiv w'$ implies the existence of two positive words v, v' satisfying

$$N_R(w)\,v \equiv N_R(w')\,v' \quad \text{and} \quad D_R(w)\,v \equiv D_R(w')\,v'.$$

Then, by construction and using the fact that braid equivalence is compatible with operation /, we find

$$N_L(N_R(w)\,D_R(w)^{-1}) = N_L(N_R(w)\,v\,v^{-1}\,D_R(w)^{-1}) = (D_R(w)\,v)/(N_R(w)\,v)$$
$$\equiv (D_R(w')v')/(N_R(w')v') = N_L(N_R(w')v'v'^{-1}D_R(w')^{-1}) = N_L(N_R(w')D_R(w')^{-1}).$$

The argument is similar for denominators. ∎

We first deduce a new solution for the word problem of B_n.

Proposition 5.2. (**word problem**) *A braid word w represents 1 if and only if the words $N_L(N_R(w)D_R(w)^{-1})$ and $D_L(N_R(w)D_R(w)^{-1})$ are empty. For each fixed value of n, there exists a constant c_n such that the number of elementary steps needed to compute the previous words is bounded above by $c_n \cdot \lg(w)^2$.*

Proof. As w is equivalent to $N_R(w)D_R(w)^{-1}$, and, therefore, to

$$D_L(N_R(w)D_R(w)^{-1})^{-1}N_L(N_R(w)D_R(w)^{-1}),$$

the condition is sufficient. Conversely, assume $w \equiv \varepsilon$. By Lemma 5.1, we have $N_L(N_R(w)D_R(w)^{-1}) \equiv N_L(N_R(\varepsilon)D_R(\varepsilon)^{-1}) \equiv \varepsilon$. Now, for a positive word u, $u \equiv \varepsilon$ implies $u = \varepsilon$. ■

Definition 5.3. (**double numerator**) For b a braid, we define $N_{RL}(b)$ and $D_{RL}(b)$ to be the braids represented by $N_L(N_R(w)D_R(w)^{-1})$ and $D_L(N_R(w)D_R(w)^{-1})$ respectively, where w is an arbitrary expression of b.

By construction, the equality

$$b = D_{RL}(b)^{-1}\, N_{RL}(b) \tag{5.3}$$

holds for every braid. Equivalently, we can consider symmetric notions involving a left reversing first and a right reversing subsequently, and, then, we have, with obvious notations,

$$b = N_{LR}(b)\, D_{LR}(b)^{-1}. \tag{5.4}$$

We shall now characterize the fractionary decompositions of (5.3) and (5.4) as minimal in some sense. We begin with auxiliary results about lcm's and gcd's.

Lemma 5.4. *Assume that a, b are positive braids. Then we have*

$$a \vee_R b = (a \wedge_L b) \cdot ((a\backslash b) \vee_L (b\backslash a)), \tag{5.5}$$

$$a = (a \wedge_R b) \cdot D_{RL}(a^{-1}b) \quad \textit{and} \quad b = (a \wedge_R b) \cdot N_{RL}(a^{-1}b). \tag{5.6}$$

Proof. Put $a' = b\backslash a$, $b' = a\backslash b$, $a'' = a'/b'$, and $b'' = b'/a'$. Then, by definition, we have

$$a \vee_R b = ab' = ba' \quad \text{and} \quad a' \vee_L b' = a''b' = b''a'.$$

By the first formula, $a \vee_R b$ is a common left multiple of a' and b', hence there exists c satisfying $a \vee_R b = c(a' \vee_L b')$. We deduce $ab' = ca''b'$, hence $a = ca''$, and, similarly, $b = cb''$. So c is a common left divisor of a and b, hence c is a left divisor of $a \wedge_L b$.

Conversely, assume $a = c_1a_1$ and $b = c_1b_1$. We have $c_1a_1b' = c_1b_1a' = a \vee_R b$, hence $a_1b' = b_1a'$. So, there exists c'' satisfying $a_1 = c''a''$ and $b_1 = c''b''$. We deduce $c_1c''a''b' = c_1c''b''a' = a \vee_R b$, hence $c_1c'' = c$. Thus c_1 is a left divisor of c, and we conclude that c is the left gcd of a and b, which proves (5.5).

With the same notations, we have

$$N_{RL}(a^{-1}b) = (b\backslash a)/(a\backslash b) = a'/b' = a'',$$

and, similarly, $D_{RL}(a^{-1}b) = b''$. Hence (5.6) is a restatement of the equalities $a = ca''$ and $b = cb''$. ■

By Proposition 2.16 (lcm), computing an lcm in B_n^+ amounts to reversing a word: if u and v represent the braids a and b respectively, then $u\,(u\backslash v)$ represents the right lcm of a and b, while $(u/v)\,v$ represents their left lcm. Using Formula (5.5) and its symmetric counterpart, we obtain the following algorithmic method for the gcd:

Proposition 5.5. (**gcd**) *Assume that the positive braids a and b are represented by u and v respectively. Then $a \wedge_R b$ and $a \wedge_L b$ are represented by $((v/u)\backslash(u/v))\backslash u$ and $u/((v\backslash u)/(u\backslash v))$ respectively.*

Proposition 5.6. (**fraction**) *For every braid c, $c = D_{RL}(c)^{-1}N_{RL}(c)$ is the unique decomposition of c as a left irreducible fraction, i.e., a fraction $a^{-1}b$ with $a \wedge_L b = 1$. If $c = a^{-1}b$ holds with a, b positive, there exists a positive braid d satisfying $a = dD_{RL}(c)$ and $b = dN_{RL}(c)$.*

Proof. The equality $c = D_{RL}(c)^{-1}N_{RL}(c)$ holds by definition. Assume that $a^{-1}b$ is a decomposition of c as a fraction. Formula (5.6) gives $a = (a \wedge_L b) \cdot D_{RL}(c)$ and $b = (a \wedge_L b) \cdot N_{RL}(c)$. Taking $d = a \wedge_L b$ gives the final result of the proposition. We deduce $N_{RL}(c) = (N_{RL}(c) \wedge_R D_{RL}(c)) \cdot N_{RL}(c)$, hence $N_{RL}(c) \wedge_L D_{RL}(c) = 1$. ■

With Proposition 5.6 (fraction), we have obtained a distinguished decomposition of every braid into a fraction. Using in addition the normal forms defined in Section 4 for positive braids, we obtain a unique normal form for every braid.

Proposition 5.7. (**normal form**) *Every braid in B_n admits a unique decomposition of the form $a_p^{-1} \cdots a_1^{-1} b_1 \cdots b_q$ where $a_p, \ldots, b_q$ are simple braids, a_1 and b_1 are left coprime, and so are a_k and $a_{k-1}\backslash\Delta_n$ for $2 \leq k \leq p$, and b_k and $b_{k-1}\backslash\Delta_n$ for $2 \leq k \leq q$.*

By exchanging the roles of right and left, we can obtain a symmetric normal decomposition of the form $b_q \cdots b_1 a_1^{-1} \cdots a_q^{-1}$ where the factors are coprime on the right.

Complementation of simple braid words

We have obtained so far two equivalent solutions for the word problem of the group B_∞: the braid word w represents 1 in B_∞ if and only if the word $D_R(w)^{-1}N_R(w)$ is right reversible to the empty word, and if and only if the word $N_R(w)D_R(w)^{-1}$ is left reversible to the empty word, *i.e.*, its normal form as defined above is empty. In both cases, we have to operate a double reversing from w. Using Proposition 3.5 (closed), we see that, for an n-strand braid word of length ℓ, this double reversing process requires at most $k_n\ell^2/2$ steps, where k_n is the maximal number of steps needed to reverse a braid word of the form $u^{-1}v$ with u and v simple n-strand braid words. For each n, the simple n-strand braid words are finite in number, and, therefore, for each fixed n, the previous algorithms have a quadratic complexity.

When we consider the word problem of B_∞, *i.e.*, when no uniform bound on the number of strands is assumed, a more precise evaluation of the parameter k_n is needed. At the moment, no good bound on k_n is known (excepted the trivial exponential bound of Exercise 3.23). We shall see now how to modify the method so as to obtain a polynomial bound.

Let BW_n^s denote the set of all n-strand permutation braid words. Thus BW_n^s consists of the $n!$ words $\widetilde{f}$ with f a permutation of $\{1, \ldots, n\}$. For u, v in BW_n^s, the words $u\backslash v$ and $v\backslash u$ need not be permutation braid words themselves, but, by Proposition 4.15 (Garside element), they are equivalent to permutation braid words.

Definition 5.8. (simple complement) For u, v permutation braid words, we denote by $f_B^s(u, v)$ the unique permutation braid word that is equivalent to $u\backslash v$.

Assume that u and v are arbitrary positive braid words. We have now an alternative way of reversing the word $u^{-1}v$, namely by considering it as a word over the alphabet BW_n^s and using the complement f_B^s. We denote by $u\backslash^s v$ and $v\backslash^s u$ the corresponding final words. The words $u\backslash v$ and $u\backslash^s v$ need not be equal in general, but an induction using Proposition 4.4 (compatibility) shows that they are equivalent. We deduce that a braid word w represents 1 in B_∞ if and only if the double f_B^s-reversing of w ends with an empty word.

We shall now evaluate the complexity of the latter algorithm. The point is to evaluate the complexity of the computation of $f_B^s(u, v)$ when u and v are permutation braid words—*i.e.*, for computing the supremum in the lattice of all simple braids.

Definition 5.9. (word $\sigma_i^{(p)}$) For $i \geq 1$, we put $\sigma_i^{(0)} = \varepsilon$ and $\sigma_i^{(p)} = \sigma_{i+p-1,i}$ for $p \geq 1$.

By construction, every permutation braid word u admits a unique decomposition of the form $\prod_1^\infty \sigma_i^{(p_i)}$ This expression will be referred to as *the* decomposition of u in the sequel.

Lemma 5.10. *Assume $i \geq j \geq 1$ and $p, q \geq 0$. Then we have*

$$\sigma_i^{(p)}\,\sigma_q^{(j)} \begin{cases} \equiv \sigma_j^{(q)}\,\sigma_i^{(p)} & \text{for } j+q<i,\\ = \sigma_j^{(p+q)} & \text{for } j+q=i,\\ \text{is not simple} & \text{for } i<j+q\leq i+p,\\ \equiv \sigma_j^{(q)}\,\sigma_{j+1}^{(p)} & \text{for } i+p<j+q. \end{cases}$$

The proof is an easy verification, which is left to the reader (for the last case, use induction on p, and then on q). The above formula emphasizes the role of the parameter "$j+q$", and it makes the following definition natural.

Definition 5.11. (admissible) Assume that u is a permutation braid word, say $u = \prod_i \sigma_i^{(p_i)}$. We define $\widehat{u}(i)$ to be $i + p_i$, and, for $i \geq 1$ and $p \geq 0$, we say that p is *i-admissible for u* if $\widehat{u}(x) < \widehat{u}(i+p)$ holds for $i \leq x < j+p$.

By definition, an integer of the form $\widehat{u}(m)$ has the form $\widehat{u}(i+p)$ for some p that is i-admissible for u if and only if the value $\widehat{u}(m)$ dominates all values of $\widehat{u}$ from $\widehat{u}(i)$.

Lemma 5.12. *Assume that u is a permutation braid word, and $j \geq 1$, $q \geq 0$ hold. Then, either q is j-admissible for u, and $u\sigma_j^{(q)}$ is equivalent to the word v determined by*

$$\widehat{v}(i) = \begin{cases} \widehat{u}(i) & \text{for } i<j \text{ and } i>j+q,\\ \widehat{u}(j+q) & \text{for } i=j,\\ \widehat{u}(i-1)+1 & \text{for } j<i\leq j+q, \end{cases}$$

or q is not j-admissible for u, and $u\sigma_j^{(q)}$ is not simple.

Proof. Apply Lemma 5.10. The condition of q being j-admissible for u is exactly what is needed to avoid case (iii). ■

Lemma 5.13. *Assume that u, v are distinct permutation braid words. Let k be the smallest integer such that σ_k occurs in uv, and $k+r$ be the smallest integer satisfying $k+r = \widehat{u}(p)$ for some p that is k-admissible for u and $k+r = \widehat{v}(q)$ for some q that is k-admissible for v. Then the first factor in the decomposition of $f_B^s(u,v)$ is $\sigma_k^{(p)}$.*

Proof. We use induction on k. As the mappings $\widehat{u}$ and $\widehat{v}$ eventually coincide with identity, every sufficiently large integer can be written as $\widehat{u}(p)$ for some p that is k-admissible for u, and, similarly, as $\widehat{v}(q)$ for some q that is k-admissible for v. Hence the integer r of the lemma exists. Assuming $k + r = \widehat{u}(p) = \widehat{v}(q)$, we deduce from Lemma 5.12 that there exist positive words u', v' containing no letter σ_k and satisfying $u\sigma_k^{(p)} \equiv \sigma_k^{(r)} u'$ and $v\sigma_k^{(q)} \equiv \sigma_k^{(r)} v'$. We then have

$$u\sigma_k^{(p)} f_B^s(u', v') \equiv \sigma_k^{(r)} u' f_B^s(u', v') \equiv \sigma_k^{(r)} v' f_B^s(v', u') \equiv v\sigma_k^{(q)} f_B^s(v', u'),$$

which shows that $\sigma_k^{(q)} f_B^s(v', u') \equiv f_B^s(v, u)w$ holds for some positive w. It remains to prove $\sigma_k^{(p)} f_B^s(u', v') \equiv u\backslash v$, *i.e.*, that $v\sigma_k^{(p)} f_B^s(u', v')$ represents the right lcm of $\overline{u}$ and $\overline{v}$. By construction, $f_B^s(u', v')$ involves no letter σ_i with $i \leq k$, so it suffices to prove that no word of the form $\sigma_k^{(r')} w'$ with $r' < r$ and w' involving no σ_i with $i \leq k$ may represent a right common multiple of $\overline{u}$ and $\overline{v}$. But, in this case, Lemma 5.12 implies that $k + r' = \widehat{u}(p')$ holds for some p' that is k-admissible for u, and, similarly, that $k + r' = \widehat{v}(q')$ holds for some q' that is k-admissible for v. Our choice of r implies $r' \leq r$. ■

Proposition 5.14. (simple complement) *There exists an algorithm that, starting with the decompositions of two permutation braid words u, v, returns the decomposition of the word $f_B^s(u, v)$ in $\mathcal{O}(m^2)$ steps, where m is the total number of factors in the decompositions of u and v.*

Proof. Write $u^{-1}v$ as $\cdots\sigma_{i_2}^{(p_2)-1}\sigma_{i_1}^{(p_1)-1}\sigma_{j_1}^{(q_1)}\sigma_{j_2}^{(q_2)}\cdots$, where m factors are involved. Writing the function $\widehat{u}$ and $\widehat{v}$ explicitly, we determine the parameters k and r of Lemma 5.13, and then reverse the words $\cdots\sigma_{i_2}^{(p_2)-1}\sigma_{i_1}^{(p_1)-1}\sigma_k^{(r)}$ and $\sigma_k^{(r)-1}\sigma_{j_1}^{(q_1)}\sigma_{j_2}^{(q_2)}\cdots$ using the partial complement formulas of Lemma 5.12: by construction, k and r have been chosen so that the complements always exist. In this way, we obtain in m steps the first factors of $f_B^s(u, v)$ and $f_B^s(v, u)$, and we are left with a word $u'^{-1}v'$ whose decomposition involves $m - 1$ factors. Iterating the process gives a quadratic algorithm. ■

Example 5.15. (computation of f_B^s) *Assume $u = \sigma_1\sigma_3$ and $v = \sigma_2\sigma_4$. The successive values of $\widehat{u}$ are $2, 2, 4, 4, 5, 6, 7, \ldots$, and those of $\widehat{v}$ are $1, 3, 3, 5, 5, 6, 7, \ldots$. The integers $\widehat{u}(p)$ such that p is 1-admissible for u, i.e., those such that $\widehat{u}(p)$ dominates all previous values, are 2, 4, 5, $\ldots$; similarly, the integers $\widehat{v}(q)$ such that q is 1-admissible for v are 1, 3, 5, 6, $\ldots$ Then the parameter $k + r$ is the least integer appearing in both lists, namely 5, i.e., the first value of r is 4. Reversing the words $u^{-1}\sigma_1^{(4)}$ and $\sigma_1^{(4)-1}v$ using the complement of Lemma 5.12 can be displayed in a single* *diagram as follows:*

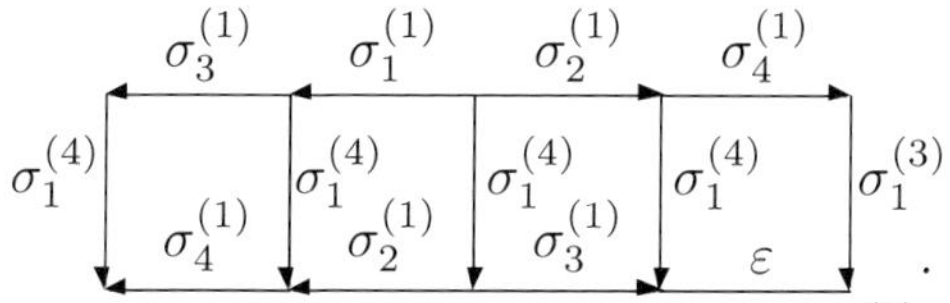

Repeating the process with the (shorter) words $u' = \sigma_3^{(1)}$ *and* $v' = \sigma_2^{(1)}\sigma_4^{(1)}$ *gives* $5 = 2 + 3$ *as the least possible next value of* $k + r$*. Finally, the complete reversing diagram is*

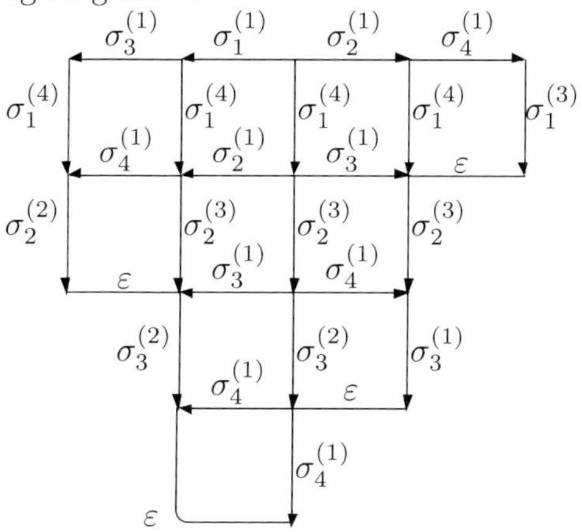

and we deduce the values $f_B^s(\sigma_1\sigma_3, \sigma_2\sigma_4) = \sigma_1^{(4)}\sigma_2^{(2)}\sigma_3^{(2)}$ *and* $f_B^s(\sigma_2\sigma_4, \sigma_1\sigma_3) = \sigma_1^{(3)}\sigma_2^{(3)}\sigma_3^{(1)}\sigma_4^{(1)}$.

We thus have completed a proof of the following result:

Proposition 5.16. (**complexity**) *There exists an algorithm that decides whether a given* n*-strand braid word* w *represents* 1 *in* $\mathcal{O}(n^2\,\mathrm{lg}(w)^2)$ *elementary steps. Hence, the word problem for the standard presentation of* B_∞ *lies in the complexity class* $\mathrm{DTIME}(n^4)$.

Exercise 5.17. (**omitting a generator**) *Say that the braid* b *omits* σ_i *if* b *has at least one expression where neither* σ_i *nor* σ_i^{-1} *occurs. Show that* b *omits* σ_i *if and only if* $\sigma_i^{\pm 1}$ *does not occur in the normal form of* b.

Exercise 5.18. (**automatic structure**) *(i) For* $b \in B_n^+$*, denote by* $F(b)$ *the set of all simple braids that are right divisors of* b*. Prove that* $b \wedge_R \Delta_n$ *is the left lcm of* $F(b)$.

(ii) Prove $F(b\,\sigma_i) = \{a\ ;\ a \text{ is simple and } a/\sigma_i \in F(b)\}$*, and deduce that, for each* i*, the braid* $(b\sigma_i) \wedge_R \Delta_n$ *depends only on* $b \wedge_R \Delta_n$ *and* σ_i.

---

6. Notes

The current approach is close to the one of (Garside 69), where the fundamental braids Δ_n and most of the results about braids established here appear. However, the main difference is the emphasis put on words, which allows one to control the situation more precisely and, in particular, to consider functions like N_R or D_R with no counterpart in the associated monoid. Such an approach will be needed both in the case of braids (to extend the action on LD-systems in Chapter III) and in the case of the group G_{LD} of Chapter VIII (where it is open whether the monoid embeds in the group).

It seems that word reversing was first considered in (Dehornoy 92*c*), for the study of the group G_{LD}. A variant was considered, in the case of Artin's groups, in (Tatsuoka 94). The basic properties of word reversing, in particular the fact that coherence on the letters implies completeness when a norm exists, are established in (Dehornoy 97). The result that the group of fractions associated with a convergent coherent complement is torsion free appears in (Dehornoy 98*b*).

The normal decomposition of positive braids in terms of simple braids—or its counterpart for Artin groups associated with finite Coxeter groups—was discovered independently by a number of authors: implicit in (Deligne 72), it appears explicitly in (Adjan 84), in circulated notes by Thurston around 1988, which appeared as Chapter 9 of (Epstein & *al.* 92), and in (ElRifai & Morton 94). A natural geometric interpretation is given in (Dehornoy 96*c*). The existence of a quadratic algorithm for the word problem of B_n relying on the computation of lattice operations is established in the notes by Thurston. He observes that computing the right gcd of two permutation braids is essentially a sorting process, so that, using a divide-and-conquer argument, one can replace the complexity n^2 with $n \log n$, thus resulting in a final braid word algorithm that requires $\mathcal{O}(n \log n \lg(w)^2)$ elementary steps to decide whether an n-strand braid word w represents 1. The equivalent treatment we have given in this chapter is from (Dehornoy 97). Let us mention here that I. Dynnikov has recently proposed a new braid word algorithm based on Teichmüller spaces and foliations which has complexity $\mathcal{O}(\lg(w)^2)$ independently on the number of strands (personal communication).

When the atomic braid complement is used, *i.e.*, when we use the generators σ_i rather than the permutation braid words, then no nontrivial complexity bound is known.

Question 6.1. *What is the upper bound for the number of reversing steps needed to reverse a braid word?*

Exercise 3.23 shows that the upper bound is at most exponential, while Exercise 1.20 shows that it is at least cubic.

In this chapter, we have restricted our study of braids to those results useful in the remainder of this text. Further classical algebraic results are Garside's solution for the conjugacy problem, which appeared in (Garside 69), and is mentioned here in Exercises 4.19 and 4.21, and the existence of a biautomatic structure, which is implicit in Exercise 5.18: the reader who knows the definitions should recognize a description of an automaton computing the right normal decomposition of a (positive) braid word using the sets of simple elements as states. Let us mention that the solution of the conjugacy problem for B_n given in (ElRifai & Morton 94) and the description of the center following (Brieskorn & Saito 72) can be extended to all groups of fractions of a monoid with a Garside element (Picantin *99*) (Picantin *00*).

By Exercise 3.26, which is from (Dehornoy & Paris 99), if some presentation of a monoid M is associated with a complement that is coherent and convergent, then every presentation of M is also associated with a coherent convergent complement. On the other hand, the group G may be the group of fractions of several non-isomorphic monoids.

Question 6.2. *Assume that G is the group of fractions of a monoid M associated with a coherent and convergent complement; what can we say about the other monoids M' of which G is a group of fractions?*

The example of B_3 shows that the involved monoids may be very different one from the other: for each k, B_3 admits the presentation $\langle x, y \; ; \; xyx = yx^k y\rangle$, and, for $k \geq 0$, it is the group of fractions of the monoid M_k with the same presentation (for instance B_3^+ is M_1). Every monoid M_k is associated with a coherent convergent complement, but, for $k \geq 2$, M_k is not atomic and no finite subset is closed under $\backslash$. In the same vein, M. Picantin has given an exotic presentation of B_3 as the group of fractions of a monoid where right lcm's need not exist. Let us also mention here the alternative presentation of B_n in terms of band generators described in (Birman, Ko & Lee 98): this presentation is associated with a complement, and most of the paper is devoted to implicitly verifying that the involved complement is coherent; see also (Bessis, Digne & Michel *00*).

Most of the results presented in this chapter can be extended to more general frameworks, and much more can be said about word reversing and its extensions. Let us only mention here three points. First, we have observed that admitting a complemented presentation is a very weak property when no additional hypothesis is assumed. However, the hypotheses we have used here are not minimal. In particular, a Garside element exists under weaker assumptions than those considered here, and, in particular, coherence can be established without using any atomicity hypothesis, a significant improvement as the existence of a norm is in general difficult to prove (Dehornoy *00*). Another rather surprising result is that, under the only hypothesis that the considered complement is coherent (but not necessarily convergent), there exists an effective upper bound for the number of reversing steps needed to reverse the word $u^{-1}u'$ when $u \equiv^+ u'$ holds, namely a double exponential of the parameter called $d(u, u')$ in the proof of Proposition 2.9 (Dehornoy *98c*). Finally, a number of results remain valid when we drop the hypothesis that the complement is functional, *i.e.*, when we consider, for each pair of letters (x, y), possibly several ways of completing them into a common right multiple of x and y. In particular, the word problem can be solved at the expense of considering non-deterministic reversing processes, and the transducers involved in an automatic structure still exist.

III
The Braid Order

In this chapter, we construct a linear ordering of the braids. This linear ordering appears canonical in several respects; in particular, the braids larger than 1 are characterized as those braids admitting an expression where the generator with smaller index appears positively only, *i.e.*, where some letter σ_i occurs, but the letter σ_i^{-1} does not, nor does any letter $\sigma_k^{\pm 1}$ with $k < i$. The order is decidable, *i.e.*, there exists an effective algorithm that compares any two given braid words, it is compatible with multiplication on one side, and the set B_∞ of all braids is order isomorphic to the rationals.

The organization of the chapter is as follows. In Section 1, we show how to define a partial action of braids on the powers of an arbitrary left cancellative LD-system using the word reversing technique of Chapter II. In Section 2, we construct the braid order. To this end, we introduce special braids as those braids that can be obtained from the unit braid using the exponentiation of Section I.3, and, by using a general property of monogenic LD-systems that will be established in Chapter V, we first construct a linear order on special braids. Then, using the partial action of braids on the LD-system consisting of special braids equipped with braid exponentiation, we extend the linear order on special braids into a linear order $<_L$ on arbitrary braids. In Section 3, we describe a geometrical algorithm that compares braids with respect to $<_L$. This algorithm is very efficient in practice. In Section 4, we give three alternative definitions of $<_L$, one in terms of automorphisms of a free group, one in terms of homeomorphisms of a punctured disk, and, finally, one in terms of an action of braids on the real line connected with hyperbolic geometry.

1. More about Braid Colourings

We have described in Section I.2 an action of the braid monoid B_n^+ on the n-th power of every LD-system by means of braid colourings, and we have observed that the action can be extended to the whole group B_n when the considered LD-system admits bijective left translations, *i.e.*, when it is what we have called an LD-quasigroup—also called an automorphic set or a rack in literature. Using the standard examples of LD-systems leads to nontrivial results about braids, and it is natural to try to use other LD-systems, such as those described in Section I.3. Unfortunately, these systems are not LD-quasigroups, and the question arises of extending the braid action to such systems. We prove in this section that such an extension is possible, at the expense of replacing an everywhere defined action with a partial action. The existence and uniqueness of this partial action relies on the braid word reversing technique of Chapter II.

Using numerators and denominators

We recall that, for $\vec{x}$ a (finite or infinite) sequence, we use x_1, x_2, ... for the successive entries of $\vec{x}$. Assume that $(S, {}^\wedge)$ is an LD-system. We have seen in Section I.2 that the rule

$$(a_1, a_2, \ldots) \bullet \sigma_i = (a_1, \ldots, a_{i-1}, a_i {}^\wedge a_{i+1}, a_i, a_{i+2}, \ldots) \tag{1.1}$$

defines an action of B_n^+ on S^n. This action can be described in terms of braid colourings by saying that, for w a positive braid word, $\vec{a} \bullet w$ is the sequence of output colours when the input colours $\vec{a}$ are applied at the top ends of the strands of (the normal braid diagram associated with) w and colours are propagated downwards according to the rule

(R_+) :

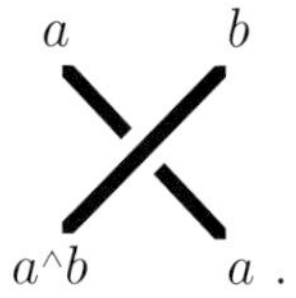

Then, for negative crossings, we must prescribe the symmetric rule

(R_-) :

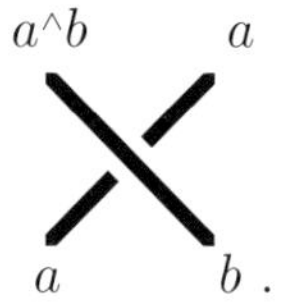

Definition 1.1. (colouring) Assume that $(S, {}^\wedge)$ is an LD-system and w is an n-strand braid word, $n \leq \infty$. We say that a pair of sequences $(\vec{a}, \vec{c})$ in S^n is an *S-colouring* for w if colours from S can be attributed to each arc in the

normal diagram $D_n(w)$ so that $\vec{a}$ are the input (top) colours, $\vec{c}$ are the output (bottom) colours, and Rules (R_+) and (R_-) are obeyed at each crossing.

If w is a positive braid word, we know that, for every sequence $\vec{a}$ in S^n, the pair $(\vec{a}, \vec{a} \bullet w)$ is an S-colouring of w, and, if S is an LD-quasigroup, the same holds for every braid word. The point is that, even if the LD-system we consider is not an LD-quasigroup, S-colourings exist for every braid word.

Lemma 1.2. *Assume that $(S, ^\wedge)$ is an LD-system and u, v are positive n-strand braid words. Then, for every sequence $\vec{x}$ in S^n, the pair $(\vec{x} \bullet u, \vec{x} \bullet v)$ is an S-colouring for $u^{-1}v$.*

Proof. (Figure 1.1) Putting the colours $\vec{x}$ in the middle of the diagram $u^{-1}v$ and propagating them upwards through u^{-1} and downwards through v gives $\vec{x} \bullet u$ on the top, and $\vec{x} \bullet v$ on the bottom. By construction, (R_+) is obeyed in the lower part of the diagram, while (R_-) is obeyed in the upper part. ∎

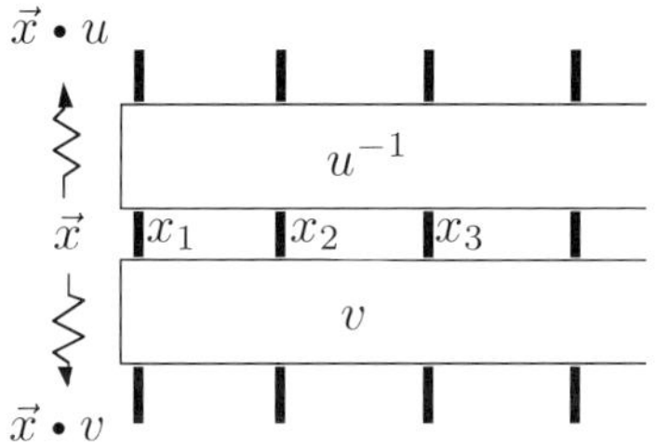

Figure 1.1. Colouring a fraction

We know that every braid can be represented by a braid word of the form $u^{-1}v$ with u, v positive braid words, so every braid word w is equivalent to a word of the form $u^{-1}v$ as above. This, however, is *not* sufficient for concluding that an S-colouring for w exists: indeed, an S-colouring for a braid word w need not be an S-colouring for every braid word w' that is equivalent to w. Now, the results of Chapter II give more than an equivalence: not only is every braid word equivalent to a fraction of the form $u^{-1}v$ with u, v positive, but also it is *left reversible* to such a word. This turns out to be exactly what we need.

Lemma 1.3. *Assume that $(S, ^\wedge)$ is an LD-system. Assume that the braid word w is left reversible to w', and that $(\vec{a}, \vec{c})$ is an S-colouring for w'. Then $(\vec{a}, \vec{c})$ is an S-colouring for w as well. In particular, if $(\vec{a}, \vec{c})$ is an S-colouring for $D_L(w)^{-1}N_L(w)$, it is an S-colouring for w as well.*

Proof. It suffices to prove the result when w is left reversible to w' in one step. If w' has been obtained from w by deleting some factor $\sigma_i\sigma_i^{-1}$, or replacing $\sigma_i\sigma_j^{-1}$ with $\sigma_j^{-1}\sigma_i$ with $|j-i| \geq 2$, the result is obvious. Assume

$|j - i| = 1$ and w' has been obtained from w by replacing some factor $\sigma_i\sigma_j^{-1}$ with $\sigma_j^{-1}\sigma_i^{-1}\sigma_j\sigma_i$. Assume for instance $i = 1$ and $j = 2$. So we start with a colouring of $\sigma_2^{-1}\sigma_1^{-1}\sigma_2\sigma_1$, and we wish to construct a colouring of $\sigma_1\sigma_2^{-1}$ with the same end colours. Let a_1, a_2, a_3 be the initial colours. Rule (R_-) for the first crossing implies that there exists c_2 in S satisfying $a_2 = a_3{}^\wedge c_2$. Similarly, Rule (R_-) for the second crossing implies that there exists c_1 in S satisfying $a_1 = a_3{}^\wedge c_1$. Then the colours must be as displayed on the left of Figure 1.2, and, then, the colours displayed on the right answer the question. ■

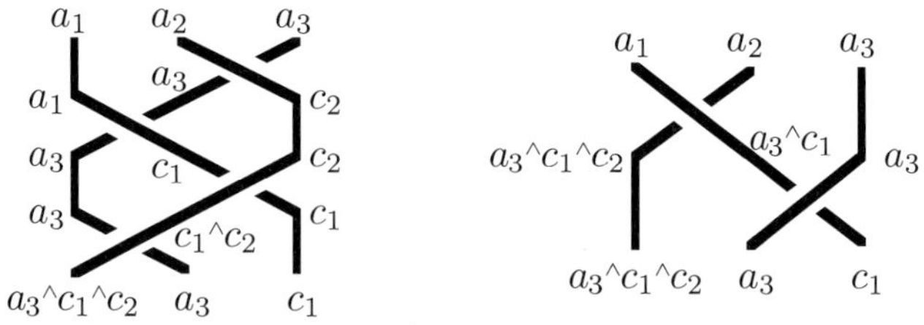

Figure 1.2. Left reversing

A similar argument gives the following result, whose proof is left to the reader.

Lemma 1.4. *Assume that $(S, {}^\wedge)$ is an LD-system. Assume that the braid word w is right reversible to w', and that $(\vec{a}, \vec{c})$ is an S-colouring for w. Then $(\vec{a}, \vec{c})$ is an S-colouring for w' as well. In particular, if $(\vec{a}, \vec{c})$ is an S-colouring for w, it is an S-colouring for $N_R(w)D_R(w)^{-1}$ as well.*

Definition 1.5. (admissible) Assume that $(S, {}^\wedge)$ is an LD-system. For $\vec{a}$ in S^∞ and w in BW_∞, we say that $\vec{a}$ is *admissible* for w if there exists at least one sequence $\vec{c}$ such that $(\vec{a}, \vec{c})$ is an S-colouring of w.

It follows from Lemmas 1.2 and 1.3 that, for every LD-system S and every braid diagram w, there exists at least one sequence in S^∞ that is admissible for w. We can extend this result to the case of several braid words simultaneously.

Lemma 1.6. *Assume that $(S, {}^\wedge)$ is an LD-system. Let $w_1, \ldots, w_p$ be a finite family of n-strand braid words. Then there exists a sequence in S^n that is admissible for each w_i.*

Proof. Lemma 1.2 implies that every sequence of the form $\vec{x} \bullet uD_L(w)$ with u a positive word is admissible for w. By the results of Section II.2, we can find positive words $u_1, \ldots, u_n$ satisfying

$$u_1 D_L(w_1) \equiv \cdots \equiv u_n D_L(w_n) :$$

we choose $u_1, \ldots, u_n$ so that $u_i D_L(w_i)$ represents the left lcm of the braids represented by $D_L(w_1), \ldots, D_L(w_n)$. Then, for every initial choice $\vec{x}$, the common value of the sequences $\vec{x} \bullet u_i D_L(w_i)$ is admissible for w_i. ■

Proposition 1.7. (action) *Assume that $(S, \wedge)$ is a left cancellative LD-system, and w, w' are equivalent braid words. Then, for each sequence $\vec{a}$ in S^∞ that is admissible both for w and w', there exists exactly one sequence $\vec{c}$ such that $(\vec{a}, \vec{c})$ is both an S-colouring of w and w'.*

Proof. Assume that $(\vec{a}, \vec{c})$ is an S-colouring of w, and $(\vec{a}, \vec{c}\,')$ is a colouring of w'. As w and w' are equivalent, by Lemma II.3.16, there exist positive words v, v' satisfying

$$N_R(w)v \equiv N_R(w')v', \qquad D_R(w)v \equiv D_R(w')v'.$$

By Lemma 1.4, $(\vec{a}, \vec{c})$ is also an S-colouring of the word $N_R(w)D_R(w)^{-1}$, and, therefore, it is an S-colouring of $N_R(w)vv^{-1}D_R(w)^{-1}$ too. By construction, we have

$$\vec{a} \bullet N_R(w)v = \vec{c} \bullet D_R(w)v.$$

The same argument gives $\vec{a} \bullet N_R(w')v' = \vec{c}\,' \bullet D_R(w')v'$, and, therefore, we have

$$\vec{c} \bullet D_R(w)v = \vec{c}\,' \bullet D_R(w')v' = \vec{c}\,' \bullet D_R(w)v.$$

This implies $\vec{c} = \vec{c}\,'$. Indeed, it suffices for an induction to show that $\vec{x} \bullet \sigma_i = \vec{x}\,' \bullet \sigma_i$ implies $\vec{x} = \vec{x}\,'$. The hypothesis implies $x_i = x'_i$ and $x_i{\wedge}x_{i+1} = x'_i{\wedge}x'_{i+1}$, hence $x_{i+1} = x'_{i+1}$ as $(S, \wedge)$ is left cancellative. ∎

The previous result applies in particular for $w = w'$. In this case, it asserts that there exists at most one colouring of w for each initial sequence of colours.

Definition 1.8. (action) Assume that $(S, \wedge)$ is a left cancellative LD-system. For w a braid word and $\vec{a}$ a sequence in S^∞, $\vec{a} \bullet w$ is defined to be the unique sequence $\vec{c}$ such that $(\vec{a}, \vec{c})$ is an S-colouring of w, if it exists. If b is a braid, $\vec{a} \bullet b$ is defined to be the sequence $\vec{a} \bullet w$, where w is an arbitrary expression of b such that $\vec{a} \bullet w$ exists, if such an expression exists.

Example 1.9. (action) *Let us consider the braid $\sigma_1^2\sigma_2^{-1}$. Assume that S is a left cancellative LD-system. The braid word $\sigma_1^2\sigma_2^{-1}$ is left reversible to $\sigma_2^{-1}\sigma_1^{-1}\sigma_2^2\sigma_1$. Applying colours a, b, c from S in the middle of the diagram of the latter word yields*

$$(a{\wedge}b, a{\wedge}c, a) \bullet \sigma_2^{-1}\sigma_1^{-1}\sigma_2^2\sigma_1 = (a{\wedge}((b{\wedge}c){\wedge}c), a, b{\wedge}c),$$

so we have $(a{\wedge}b, a{\wedge}c, a) \bullet \sigma_1^2\sigma_2^{-1} = (a{\wedge}((b{\wedge}c){\wedge}c), a, b{\wedge}c)$ as well. The sequences of the form $(a{\wedge}b, a{\wedge}c, a)$ are not the only ones for which the action of $\sigma_1^2\sigma_2^{-1}$ is defined. For instance, we have also

$$(a, b, a) \bullet \sigma_1^2\sigma_2^{-1} = ((a{\wedge}b){\wedge}a, a, b).$$

We thus have extended the action of braids to all left cancellative LD-systems, at the expense of obtaining a partial action. In the case of a left cancellative LD-system that is not an LD-quasigroup, the hypothesis that the sequence $\vec{a} \bullet b$ exists does not imply that $\vec{a} \bullet w$ exists for every braid word w that represents b. The point is that, yet the action is partial, there always exist sequences that are admissible for a given braid. More precisely, we deduce from Lemma 1.6:

Proposition 1.10. (**existence**) *Assume that $(S, {}^\wedge)$ is a left cancellative LD-system. Then, for every finite family of braids $b_1, \ldots, b_m$ in B_n, there exists at least one sequence $\vec{a}$ in S^n such that $\vec{a} \bullet b_i$ is defined for every i.*

The action of braids on braids

The LD-system consisting of B_∞ equipped with braid exponentiation is left cancellative, hence we can let B_∞ act on sequences of braids—in other words, we can use braids themselves to colour the strands of the braids. A technically significant fact is the connection between this action and multiplication on the right in B_∞, which is reminiscent of Formulas (2.7), (2.9), and (2.12) of Chapter I.

Definition 1.11. (**shifted product**) We let $B_\infty^{(\mathbf{N})}$ be the set of those sequences in $B_\infty^{\mathbf{N}}$ with finitely many entries not equal to 1. For $\vec{x} \in B_\infty^{(\mathbf{N})}$, we put

$$\prod^{sh} \vec{x} = \prod_{k=1}^{\infty} \mathrm{sh}^{k-1}(x_k) = x_1\ \mathrm{sh}(x_2)\ \mathrm{sh}^2(x_3) \cdots . \tag{1.2}$$

Lemma 1.12. *Assume $\vec{a} \in B_\infty^{(\mathbf{N})}$, $b \in B_\infty$, and $\vec{a} \bullet b$ exists. Then we have*

$$\prod^{sh}(\vec{a} \bullet b) = (\prod^{sh} \vec{a}) \cdot b. \tag{1.3}$$

Proof. We use induction on the minimal length of a braid word w representing b such that $\vec{a} \bullet w$ exists. The result is true when w is empty. Assume $w = \sigma_i w_0$. Let b_0 be the braid represented by w_0. By hypothesis, $\vec{a} \bullet \sigma_i$ and $(\vec{a} \bullet \sigma_i) \bullet b_0$ exist. We find first

$$\begin{aligned}
\prod^{sh}(\vec{a} \bullet \sigma_i) &= \prod^{sh}(a_1, \ldots, a_i{}^\wedge a_{i+1}, a_i, \ldots) \\
&= a_1\ \mathrm{sh}(a_2)\ \cdots\ \mathrm{sh}^{i-1}(a_i {}^\wedge a_{i+1})\ \mathrm{sh}^i(a_i)\ \mathrm{sh}^{i+1}(a_{i+2}) \cdots \\
&= a_1\ \mathrm{sh}(a_2)\ \cdots\ \mathrm{sh}^{i-1}(a_i)\ \mathrm{sh}^i(a_{i+1})\ \sigma_i\ \mathrm{sh}^i(a_i)^{-1}\ \mathrm{sh}^i(a_i)\ \mathrm{sh}^{i+1}(a_{i+2}) \cdots \\
&= a_1\ \mathrm{sh}(a_2)\ \cdots\ \mathrm{sh}^{i-1}(a_i)\ \mathrm{sh}^i(a_{i+1})\ \sigma_i\ \mathrm{sh}^{i+1}(a_{i+2}) \cdots \\
&= a_1\ \mathrm{sh}(a_2)\ \cdots\ \mathrm{sh}^{i-1}(a_i)\ \mathrm{sh}^i(a_{i+1})\ \mathrm{sh}^{i+1}(a_{i+2}) \cdots\ \sigma_i = (\prod^{sh} \vec{a})\ \sigma_i,
\end{aligned}$$

as σ_i commutes with every braid in the image of sh^k for $k \geq i+1$. Applying the induction hypothesis, we deduce

$$\prod^{sh}(\vec{a} \bullet b) = \prod^{sh}((\vec{a} \bullet \sigma_i) \bullet b_0) = ((\prod^{sh} \vec{a})\ \sigma_i)\ b_0 = (\prod^{sh} \vec{a})\ b. \qquad \blacksquare$$

We deduce that the action of B_∞ on $B_\infty^{(\mathbf{N})}$ is faithful:

Proposition 1.13. (faithful) *Assume that b and b' are braids in B_∞ and there exists at least one sequence $\vec{a}$ in $B_\infty^{(\mathbf{N})}$ such that $\vec{a} \bullet b$ and $\vec{a} \bullet b'$ are defined and equal. Then $b = b'$ holds.*

Proof. By Lemma 1.12, $\vec{a} \bullet b = \vec{a} \bullet b' = \vec{c}$ implies $b = (\prod^{sh} \vec{c})\,(\prod^{sh} \vec{a})^{-1} = b'$. ■

For every braid b, there exists a least sub-LD-system of $(B_\infty, {}^\wedge)$ that contains b, namely the closure of the singleton $\{b\}$ under exponentiation. A significant role is played by the closure of $\{1\}$ in the sequel.

Definition 1.14. (special braid) Assume that b is a braid. We say that b is *special* if it belongs to the closure of $\{1\}$ under exponentiation. The set of all special braids is denoted B_{sp}.

By construction, every special braid has an expression involving the trivial braid 1 and exponentiation exclusively. Thus, for instance, 1, $1^\wedge 1$, which is σ_1, $(1^\wedge 1)^\wedge 1$, which is $\sigma_1^2 \sigma_2^{-1}$, $1^\wedge(1^\wedge 1)$, which is $\sigma_2\sigma_1$, are special braids.

We have now a rigorous definition of a (partial) action of braids on sequences of braids. So there remains nothing fuzzy in the derived notion of a self-colouring braid as considered in Section I.3: we say that the braid b is self-colouring if $(1, 1, 1, \ldots) \bullet b$ exists and is equal to $(b, 1, 1, \ldots)$. As was noted in Chapter I, the unit braid is self-colouring, and self-colouring braids are closed under exponentiation. So, in particular, every special braid is self-colouring. The converse implication will be proved in Section VI.6, but it will not be used here.

By definition, special braids equipped with braid exponentiation form a left cancellative LD-system, and, therefore, we can use them to colour the strands of the braids. Restating Proposition 1.10 (existence) in this case yields:

Proposition 1.15. (existence) *For every finite family of braids $b_1, \ldots, b_m$ in B_n, there exists at least one sequence $\vec{a}$ of special braids such that $\vec{a} \bullet b_i$ is defined and consists of special braids for every i.*

Applying Formula (1.2) in the case of special braids then leads to a decomposition of every braid in terms of special braids.

Proposition 1.16. (special decomposition) *(i) Every positive braid admits a unique decomposition of the form*

$$c_1 \operatorname{sh}(c_2) \operatorname{sh}^2(c_3) \cdots \tag{1.4}$$

where $c_1, c_2, \ldots$ are special braids.

(ii) Every braid admits a decomposition of the form

$$\cdots\ \mathrm{sh}^{i-1}(a_i)^{-1}\ \cdots\ \mathrm{sh}(a_2)^{-1}\ a_1^{-1}\ c_1\ \mathrm{sh}(c_2)\ \cdots\ \mathrm{sh}^{i-1}(c_i)\ \cdots \qquad (1.5)$$

where $a_1, a_2, \ldots, c_1, c_2, \ldots$ are special braids.

Proof. (i) Assume that b is a positive n-strand braid. Then the sequence $(1, 1, 1, \ldots) \bullet b$ is defined. Let $(c_1, c_2, \ldots)$ be its value. By construction, all braids c_i are special, and, by Lemma 1.12, we have $b = \prod^{sh} \vec{c}$, which is (1.4). For uniqueness, assume that $b = \prod^{sh} \vec{c}\,'$ is an arbitrary decomposition with c'_1, c'_2, ... special. Every special braid is self-colouring. Then we find

$$\begin{aligned}(1,1,\ldots) \bullet b &= (1,1,\ldots) \bullet c'_1\ \mathrm{sh}(c'_2)\ \mathrm{sh}^2(c'_3)\cdots\\ &= (c'_1, 1, 1, \ldots) \bullet \mathrm{sh}(c'_2)\ \mathrm{sh}^2(c'_3)\cdots\\ &= (c'_1, c'_2, 1, 1, \ldots) \bullet \mathrm{sh}^2(c'_3)\cdots = \ldots = (c'_1, c'_2, c'_3, \ldots),\end{aligned}$$

so, by Proposition 1.7 (action), we deduce $c'_i = c_i$ for every i.

(ii) Every braid can be expressed as a fraction $a^{-1}b$ with a, b positive. Applying (i) to a and b gives (1.5). ■

Example 1.17. (**special decomposition**) *Let $b = \sigma_1^2\sigma_2$. We have*

$$(1,1,1) \bullet b = ((1^\wedge 1)^\wedge 1, (1^\wedge 1)^\wedge 1, 1^\wedge 1).$$

So the decomposition of b into a shifted product of special braids is

$$\sigma_1^2\sigma_2 = ((1^\wedge 1)^\wedge 1) \cdot \mathrm{sh}((1^\wedge 1)^\wedge 1) \cdot \mathrm{sh}^2(1^\wedge 1). = (\sigma_1^2\sigma_2^{-1}) \cdot (\sigma_2^2\sigma_3^{-1}) \cdot (\sigma_3).$$

If we consider positive braids with a fixed number of strands, we can say a little more, namely that, for $b \in B_n^+$, the decomposition (1.4) of b contains at most n nontrivial factors, *i.e.*, we have $b = c_1\ \mathrm{sh}(c_2) \cdots \mathrm{sh}^{n-1}(c_n)$ for some special braids $c_1, \ldots, c_n$. However, as the example above shows, or the example of σ_1^2, whose special decomposition is $(\sigma_1^2\sigma_2^{-1}) \cdot (\sigma_2)$, the involved special braids $\mathrm{sh}^{k-1}(c_k)$ need not live in B_n.

Exercise 1.18. (**left cancellative**) *Assume that $(S, ^\wedge)$ is an LD-system. For a, a' in S, let $a \equiv a'$ mean that $c_1{}^\wedge \ldots {}^\wedge c_k{}^\wedge a = c_1{}^\wedge \ldots {}^\wedge c_k{}^\wedge a'$ holds for some $c_1, \ldots, c_k$. Prove that $\equiv$ is a congruence on S, that $S/\!\equiv$ is left cancellative, and that it is the largest quotient of S with this property.*

Exercise 1.19. (**right fraction**) *Assume that S is a left cancellative LD-system, $\vec{a}$ is a sequence from S, and b is a braid such that $\vec{a} \bullet b$ exists. Let uv^{-1} be an arbitrary expression of b as a right fraction. Prove that $\vec{a} \bullet uv^{-1}$ exists. [Hint: Assume that $\vec{a} \bullet w$ exists. Argue as in the proof of Proposition 1.7 (action) using positive braid words w_1 and w_2 satisfying $N_R(w)\, w_1 \equiv u\, w_2$ and $D_R(w)\, w_1 \equiv v\, w_2$. Then $\vec{a} \bullet N_R(w) D_R(w)^{-1}$, $\vec{a} \bullet N_R(w) w_1 {w_1}^{-1} D_R(w)^{-1}$, $\vec{a} \bullet u w_2 w_2^{-1} v^{-1}$, and, finally, $\vec{a} \bullet uv^{-1}$ exist.]*

Exercise 1.20. (**special decomposition**) *For b a braid, say that $\vec{c}$ is a special decomposition for b if $\vec{c}$ is a sequence of special braids satisfying $b = \prod^{sh} \vec{c}$, i.e., $b = c_1\, \mathrm{sh}(c_2)\, \mathrm{sh}^2(c_3) \cdots$. Prove that a special decomposition is unique when it exists.*

Exercise 1.21. (**positive special**) *(i) Show that the positive braids that are special are the braids $1 {}^\wedge 1 {}^\wedge \cdots {}^\wedge 1$, i.e., those braids of the form $\sigma_{m-1} \cdots \sigma_2 \sigma_1$. [Hint: Use induction on the length of b and consider the sequence $(1, 1, \ldots) \bullet b\sigma_i$.]*

(ii) Show that the braids that admit a decomposition $\prod^{sh} \vec{c}$ with c_1, c_2, ... special positive braids are simple positive braids (in the sense of Section II.4). What is the corresponding decomposition of Δ_n?

Exercise 1.22. (**evaluation**) *Prove that, for every formal expression $t(x)$ constructed using the variable x and exponentiation, we have $t(\sigma_1^{-1}) = \mathrm{sh}(t(1))\sigma_1^{-1}$ in $(B_\infty, {}^\wedge)$, where $t(b)$ denotes the evaluation of $t(x)$ in $(B_\infty, {}^\wedge)$ when x is given the value b.*

Exercise 1.23. (**powers of special braids**) *(i) For b a braid, define inductively $b^{[1]} = b$, $b^{[m+1]} = b{}^\wedge b^{[m]}$. Prove that, for every special braid b, the equality $b{}^\wedge 1^{[m]} = 1^{[m+1]}$ holds for m large enough. Prove that there exists a positive integer h such that $b^{[m-h]} = 1^{[m]}$ holds for m large enough.*

(ii) Prove that $a{}^\wedge \sigma_{n-1,1} = \sigma_{n,1}$ holds for $a \in B_n$. Deduce that, if a is an arbitrary braid, and b is a special braid, then $a{}^\wedge b^{[m]} = b^{[m+1]}$ holds for m large enough. Show that this need not be true if b is not special.

2. The Linear Ordering of Braids

In this section, we first construct a linear order on special braids using left division and resorting to a general result about monogenic LD-systems. Then the decomposition of braids into shifted products of special braids enables us to extend the order to arbitrary braids.

The order on special braids

We have seen in Section I.3 that left division in the LD-system $(B_\infty, {}^\wedge)$ has no cycle. We can easily deduce a partial order, which turns out to be linear, on special braids.

In the case of an associative operation, the left divisibility relation is transitive. This need not be the case for a self-distributive operation, and we are led to consider the transitive closure of the divisibility relation.

Definition 2.1. (iterated divisibility) Assume that $(S, {}^\wedge)$ is a binary system. For a, b in S, we say that a is an *iterated left divisor* of b, denoted $a \sqsubset b$, if there exist $p \geq 1$ and elements $b_1, \ldots, b_p$ of S satisfying $b = (\ldots((a^\wedge b_1)^\wedge b_2)\ldots)^\wedge b_p$.

Convention 2.2. (orders) By a (partial) order, we mean a binary relation $<$ that is irreflexive—$a < a$ never holds—and transitive (hence antisymmetric, as the conjunction of $a < b$ and $b < a$ would imply $a < a$). An order $<$ on X is said to be linear if, for all a, b in X, exactly one of $a < b$, $a = b$, $a > b$ holds. If $<$ is an order, the relation "$a < b$ or $a = b$" is reflexive, antisymmetric and transitive; conversely, if $\leq$ is a reflexive, symmetric and transitive relation, the relation "$a \leq b$ and $a \neq b$" is an order. When considering relations $<$, $\sqsubset$, $\prec$, $\ldots$, we shall use $\leq$, $\sqsubseteq$, $\preceq$, $\ldots$ for the corresponding enlarged relations.

Lemma 2.3. *Assume that $(S, {}^\wedge)$ is a binary system. Then the following are equivalent:*

(i) The iterated left divisibility relation defines a partial ordering of S;

(ii) There exists a partial ordering $\prec$ on S such that $a \prec a^\wedge b$ always holds;

(iii) Left division in $(S, {}^\wedge)$ admits no cycle.

Proof. That (i) implies (ii) is obvious. If $\prec$ is a binary relation on S such that $a \prec a^\wedge b$ holds for all a, b, and $(a_1, \ldots, a_n)$ is a cycle for left division in S, then $a_1 \prec a_2 \prec \cdots \prec a_n \prec a_1$ holds, hence $\prec$ is not irreflexive. So (ii) implies (iii). Finally, (iii) implies (i), for $a \sqsubset a$ would cause (a) to be a cycle for left division. ∎

We deduce that the iterated divisibility relation $\sqsubset_{B_\infty}$ associated with braid exponentiation is a partial ordering on B_∞, and, similarly, that the iterated divisibility relation $\sqsubset_{B_{sp}}$ associated with special braid exponentiation is a partial ordering on B_{sp}. At this point, we do not know whether $\sqsubset_{B_{sp}}$ is a restriction of $\sqsubset_{B_\infty}$ as, for a, b special braids, the existence of braids b_1, b_2, ... satisfying $b = \cdots((a^\wedge b_1)^\wedge b_2)\cdots$ need not imply the existence of special braids with the same property.

Now, by definition, B_{sp} is a monogenic LD-system. The following general property will be established in Chapter V:

Proposition 2.4. (**comparison property**) *Assume that $(S, {}^\wedge)$ is a monogenic LD-system. Then any two elements of S are comparable with respect to the iterated left divisibility relation, i.e., for all a, b in S, at least one of $a \sqsubset b$, $a = b$, $a \sqsupset b$ holds.*

Applying this result to the LD-system $(B_{sp}, {}^\wedge)$ and Lemma 2.3, we conclude:

Proposition 2.5. (**order on special**) *The iterated left divisibility relation of $(B_{sp}, {}^\wedge)$ defines a linear ordering on special braids.*

Example 2.6. (**braid order**) *By definition, the sequence 1, $1_{[2]}$, $1_{[3]}$, ... is increasing in $(B_{sp}, \sqsubset)$. This corresponds to the explicit inequalities*

$$1 \sqsubset \sigma_1 \sqsubset \sigma_1^2\sigma_2^{-1} \sqsubset \sigma_1^2\sigma_2^{-1}\sigma_1\sigma_3\sigma_2^{-2} \sqsubset \cdots.$$

An immediate application of the linearity of the partial ordering $\sqsubset_{B_{sp}}$ is its coinciding with the restriction to special braids of the relation $\sqsubset_{B_\infty}$—so we may drop all subscripts without ambiguity.

Proposition 2.7. (**restriction**) *Assume that a, b are special braids. Then the following are equivalent:*

(i) The relation $a \sqsubset b$ holds in $(B_{sp}, {}^\wedge)$;

(ii) The relation $a \sqsubset b$ holds in $(B_\infty, {}^\wedge)$.

Proof. By definition, (i) implies (ii). Conversely, if a, b are special braids and $a \sqsubset_{B_{sp}} b$ does not hold, then either $a = b$ or $b \sqsubset_{B_{sp}} a$ holds: this implies $a = b$ or $b \sqsubset_{B_\infty} a$, and, in both cases, contradicts $a \sqsubset_{B_\infty} b$. (More generally, a linear ordering that is included in a partial ordering must coincide with it.) ∎

A crucial property of the linear order on B_{sp} is that it can be characterized explicitly in terms of occurrences of the generator σ_1.

Definition 2.8. (σ_i**-positive**) Let b be a braid. We say that b is *σ_i-positive* (*resp. σ_i-negative*) if it admits at least one expression where the letter σ_i occurs, but σ_i^{-1} does not (*resp.* σ_i^{-1} occurs but σ_i does not).

Proposition 2.9. (**characterization**) *Assume that a, b are special braids; Then the following are equivalent:*

(i) The relation $a \sqsubset b$ holds in $(B_{sp}, \wedge)$;

(ii) The braid $a^{-1}b$ is σ_1-positive.

Proof. Assume $a \sqsubset b$. By construction, this means that there exist special braids $b_1, \ldots, b_p$ satisfying $b = (\cdots((a^\wedge b_1)^\wedge b_2)\cdots)^\wedge b_p$. Expanding the latter expression gives

$$b = a \operatorname{sh}(b_1)\ \sigma_1 \operatorname{sh}(c_1)\ \sigma_1\ \cdots\ \sigma_1 \operatorname{sh}(c_p),$$

with $c_k = ((\cdots((a^\wedge b_1)^\wedge b_2)\cdots)^\wedge b_k)^{-1} b_{k+1}$. Hence the braid $a^{-1}b$ is explicitly σ_1-positive.

Conversely, if $a \sqsubset b$ does not hold, then we have either $a = b$, hence $a^{-1}b = 1$, or $b \sqsubset a$, and, by the previous argument, $b^{-1}a$ is σ_1-positive, hence $a^{-1}b$ is σ_1-negative. In both cases, Proposition I.3.19 (braids acyclic) implies that $a^{-1}b$ cannot be σ_1-positive. ■

The order on arbitrary braids

Assume that b and b' are braids. Let us say that a sequence of braids is a special sequence if it consists of special braids. By Proposition 1.15 (existence), there exists at least one special sequence $\vec{a}$ such that both $\vec{a} \bullet b$ and $\vec{a} \bullet b'$ exist and are special sequences. Now special braids are equipped with the linear ordering $\sqsubset$, and there exists a natural lexicographical extension of this order into a linear ordering of special sequences: for $\vec{a}$, $\vec{a}'$ special sequences, we say that $\vec{a} \sqsubset^{Lex} \vec{a}'$ is true if $a_k \sqsubset a'_k$ holds for the first index k satisfying $a_k \neq a'_k$. The obvious idea is to define the braid b to be smaller than the braid b' if the sequence $\vec{a} \bullet b$ is smaller than $\vec{a} \bullet b'$ with respect to $\sqsubset^{Lex}$.

In order to make the definition rigorous, we have to verify that the previous comparison of b and b' does not depend on the choice of the special sequence $\vec{a}$. This verification is made easy by the existence of the characterization of Proposition 2.9.

Definition 2.10. (σ**-positive**) Let b be a braid. We say that b is *σ-positive* (*resp. σ-negative*) if there exists a nonnegative integer k such that b is the image under sh^k of a σ_1-positive braid (*resp.* of a σ_1-negative braid).

Lemma 2.11. *A σ-positive braid is neither the unit braid, nor σ-negative.*

Proof. Assume $b = \mathrm{sh}^i(b_1) = \mathrm{sh}^j(b_2^{-1})$ where b_1 is σ_1-positive and b_2 is σ_1-positive or equal to 1. Assume $i \le j$. Then $b_1\ \mathrm{sh}^{j-i}(b_2)$ is σ_1-positive and equal to 1, contradicting Proposition I.3.19 (braids acyclic). ■

Lemma 2.12. *For all braids b, b', the following are equivalent:*

(i) There exists a special sequence $\vec{a}$ of special braids such that $\vec{a} \bullet b$ and $\vec{a} \bullet b'$ exist and $\vec{a} \bullet b \sqsubset^{Lex} \vec{a} \bullet b'$ holds;

(ii) The braid $b^{-1}b'$ is σ-positive.

Proof. Assume (i). Let $\vec{c} = \vec{a} \bullet b$ and $\vec{c}\,' = \vec{a} \bullet b'$. By Lemma 1.12, we have $\prod^{sh} \vec{c} = (\prod^{sh} \vec{a})\, b$, and $\prod^{sh} \vec{c}\,' = (\prod^{sh} \vec{a})\, b'$, hence

$$b^{-1}\ b' = (\textstyle\prod^{sh} \vec{c})^{-1}\ (\prod^{sh} \vec{c}\,'). \tag{2.1}$$

By definition of $\sqsubset^{Lex}$, there exists an integer i such that $c_k = c'_k$ holds for $k < i$, and $c_i \sqsubset c'_i$ holds. Thus (2.1) takes the form

$$b^{-1}\ b' = \mathrm{sh}^i(c^{-1})\ \mathrm{sh}^{i-1}(c_i^{-1}\ c'_i)\ \mathrm{sh}^i(c')$$

for some c, c'. By Proposition 2.9 (characterization), the braid $c_i^{-1}\ c'_i$ is σ_1-positive, and so is the braid $\mathrm{sh}(c^{-1})\ c_i^{-1}\ c'_i\ \mathrm{sh}(c')$. Hence the braid $b^{-1}b'$ is σ-positive, as it admits a decomposition where σ_i occurs, and neither σ_i^{-1} nor any $\sigma_k^{\pm 1}$ with $k < i$ occurs.

Conversely, assume that the braid $b^{-1}b'$ is σ-positive. Let $\vec{a}$ be an arbitrary special sequence such that both $\vec{a} \bullet b$ and $\vec{a} \bullet b'$ exist and are special sequences. As $\sqsubset^{Lex}$ is a linear ordering, it suffices to show that both $\vec{a} \bullet b = \vec{a} \bullet b'$ and $\vec{a} \bullet b' \sqsubset^{Lex} \vec{a} \bullet b$ are impossible. Now, $\vec{a} \bullet b = \vec{a} \bullet b'$ implies $b = b'$, hence $b^{-1}b' = 1$, while $\vec{a} \bullet b' \sqsubset^{Lex} \vec{a} \bullet b$ implies $b^{-1}b'$ to be σ-negative. By Lemma 2.11, both contradict $b^{-1}b'$ being σ-positive. ■

Definition 2.13. **(order)** Assume that b, b' are braids. We say that $b <_L b'$ is true if one of the equivalent conditions of Lemma 2.12 holds.

Proposition 2.14. **(order)** *(i) Every nontrivial (i.e., not equal to 1) braid is either σ-positive or σ-negative.*

(ii) The relation $<_L$ is a linear ordering on B_∞ that extends the order $\sqsubset$ of special braids.

(iii) The following relations hold for all braids a, b, c:

- *$a <_L b$ is equivalent to $\mathrm{sh}(a) <_L \mathrm{sh}(b)$;*
- *$a <_L b$ is equivalent to $ca <_L cb$, i.e., $<_L$ is left invariant on the group B_∞;*
- *$a <_L a^\wedge b$ always holds.*

Proof. Point (i) follows from the previous lemma, as the lexicographical ordering $\sqsubset^{Lex}$ on special sequences is, by construction, a linear order. For (ii), the product of two σ-positive braids is a σ-positive braid, hence the relation $<_L$ is transitive. Moreover $b <_L b$ is impossible, as the unit braid 1 is not σ-positive. Hence $<_L$ is an order. It extends the previously defined order on B_{sp}, as a σ_1-positive braid is σ-positive by definition.

For (iii), $a <_L b$ implies $\mathrm{sh}(a) <_L \mathrm{sh}(b)$ since the image under sh of a σ-positive braid is σ-positive. This gives the equivalence, as $<_L$ is a linear ordering and sh is injective. Similarly $a <_L b$ implies $ca <_L cb$, as $a^{-1}b = (ca)(cb)^{-1}$ holds. Finally, we have $a^{-1}(a{\wedge}b) = \mathrm{sh}(b)\sigma_1\mathrm{sh}(a^{-1})$, an explicitly σ_1-positive braid. ■

As shown in the following examples, the above compatibility relations are the only ones that hold without restriction.

Example 2.15. (counter-examples) *Let $a = \sigma_1$ and $b = \sigma_1^2\sigma_2^{-1}$. Then a and b both are special braids, and $a <_L b$ holds, as $a^{-1}b$ is $\sigma_1\sigma_2^{-1}$, an explicitly σ_1-positive braid. Let $c = \sigma_1^{-1}$. We have $ac = 1$ and $bc = \sigma_2^{-1}\sigma_1^{-1}\sigma_2^2$, a σ_1-negative braid. Thus $a <_L b$ does not imply $ac >_L bc$: the order $<_L$ is not compatible with multiplication on the right.*

With the same notations, we have $ab^{-1} = \sigma_1\sigma_2\sigma_1^{-2} = \sigma_2^{-2}\sigma_1\sigma_2$, again an explicitly σ_1-positive braid. Thus $a <_L b$ and $a^{-1} <_L b^{-1}$ hold: the inverse operation does not necessarily reverse the order.

Now, let $a = \sigma_1\sigma_2^{-1}$, and $b = \sigma_1\sigma_2\sigma_1$. We find $b^{-1}ab = \sigma_2\sigma_1^{-1}$, a σ_1-negative braid: a conjugate of a braid larger than 1 need not be larger than 1. With the same notation, we have $ab <_L b$: even if $a >_L 1$ holds, $ab >_L b$ need not hold.

Next, let $a = \sigma_1\sigma_3^{-2}$ and $b = \sigma_2\sigma_1$. Then we have $b^{-1}(a{\wedge}b) = \sigma_2\sigma_3^{-1}\sigma_2\sigma_1^{-1}\sigma_2^{-1}\sigma_3\sigma_2^{-1}\sigma_4^2$, a σ_1-negative braid. Thus we have $a{\wedge}b <_L b$, although $a >_L 1$ holds: $b <_L a{\wedge}b$ need not hold in general.

Finally, let $a = \sigma_3$, $b = \sigma_2$, and $c = \sigma_2\sigma_3\sigma_2$. We find $a^{-1}b = \mathrm{sh}(\sigma_2^{-1}\sigma_1)$, a σ-positive braid, so $a <_L b$ holds. Now, we have $(c{\wedge}a)^{-1}(c{\wedge}b) = \sigma_3^{-1}\sigma_4$, a σ-negative braid. So $c{\wedge}a >_L c{\wedge}b$ holds: $a <_L b$ does not imply $c{\wedge}a <_L c{\wedge}b$.

It is easy to characterize $<_L$ among all possible linear orders of B_∞:

Proposition 2.16. (uniqueness) *Assume that $\prec$ is a partial order on B_∞ that is compatible with multiplication on the left and with shift, and $1 \prec \mathrm{sh}(b)\sigma_1\mathrm{sh}(b')$ holds for all b, b'. Then $\prec$ coincides with $<_L$.*

Proof. The hypotheses imply

$$1 \prec \mathrm{sh}(b_0)\,\sigma_1\,\mathrm{sh}(b_1) \prec \mathrm{sh}(b_0)\,\sigma_1\,\mathrm{sh}(b_1)\,\sigma_1\,\mathrm{sh}(b_2) \prec \cdots$$

for all b_0, b_1, ... Hence $1 \prec b$ holds for every σ_1-positive braid b. The compatibility of $\prec$ with sh implies $1 \prec b$ for every σ-positive braid b. Hence $b <_L b'$

implies $b \prec b'$. Finally $<_L$ being a linear ordering and $\prec$ being an ordering gives the equality. ■

Applications and further results

The existence of a left invariant linear order on B_∞ implies several algebraic consequences. Let us mention one here. We have seen in Chapter II that braid groups are torsion free. The following result strengthens the property.

Proposition 2.17. (**zero divisors**) *The group algebra* $\mathbf{C}B_\infty$ *admits no zero divisor.*

Proof. Let a, b be two nonzero elements of $\mathbf{C}B_\infty$, say $a = \lambda_1 a_1 + \cdots + \lambda_p a_p$ and $b = \mu_1 b_1 + \cdots + \mu_q b_q$, where $\lambda_1, \ldots, \mu_q$ are nonzero complex numbers and $a_1, \ldots, b_q$ are braids. We may assume $a_1 <_L \cdots <_L a_p$ and $b_1 <_L \cdots <_L b_q$. Assume that $a_{i_0}b_{j_0}$ is $<_L$-maximal among all monomials a_ib_j. As $<_L$ is left invariant, $j < q$ implies $a_ib_j <_L a_ib_{j_0} \leq_L a_{i_0}b_{j_0}$ for every i, so we must have $j_0 = q$, and $j \neq j_0$ implies $a_ib_j <_L a_{i_0}b_{j_0}$. Then $i \neq i_0$ implies $a_ib_{j_0} \neq a_{i_0}b_{j_0}$, hence $a_ib_{j_0} <_L a_{i_0}b_{j_0}$. We conclude that $(i,j) \neq (i_0,j_0)$ implies $a_ib_j <_L a_{i_0}b_{j_0}$. Hence, in the product ab, the coefficient of the monomial $a_{i_0}b_{j_0}$ is exactly $\lambda_{i_0}\mu_{j_0}$, and we have $ab \neq 0$. ■

Another application is the following extension of Proposition I.3.19 (braid acyclic).

Proposition 2.18. (**σ_i-positive**) *A σ_i-positive braid is never trivial.*

Proof. Taking as an hypothesis the fact that left division in $(B_{sp}, {}^\wedge)$ has no cycle, we can reprove that a σ_1-positive braid is never trivial as follows. Assume that b is σ_1-positive. Then b admits an expression of the form $w = \mathrm{sh}(w_0)\sigma_1\mathrm{sh}(w_1)\sigma_1 \cdots \sigma_1\mathrm{sh}(w_p)$. Consider a colouring of w by special braids, let a_k denote the colour of the leftmost strand after the k-th σ_1-crossing, and b_k denote the colour of the second strand before the $k+1$-th σ_1-crossing (left diagram in Figure 2.1). By construction, we have $a_{k+1} = a_k{}^\wedge b_k$, and, therefore, $a_k \sqsubset a_{k+1}$ for every k. Thus we have $a_0 \sqsubset a_p$, hence $a_0 \neq a_p$, and, therefore, $b \neq 1$.

The previous argument extends to σ_i-positive braids as follows. Assume that b admits an expression w of the form $w_0\,\sigma_i\,w_1\,\sigma_i \cdots \sigma_i\,w_p$ with no $\sigma_i^{\pm 1}$ in $w_0, \ldots, w_p$. Consider again a colouring of w by special braids. Call the colours just before the k-th σ_i-crossing $(c_{r,1}, c_{r,2}, \ldots)$, and the colours just after this r-th σ_i-crossing $(a_{r,1}, a_{r,2}, \ldots)$ (right diagram in Figure 2.1). For $r \leq p$, define

$$a_r = a_{r,1} {}^\wedge \cdots {}^\wedge a_{r,i} {}^\wedge 1, \qquad c_r = c_{r,1} {}^\wedge \cdots {}^\wedge c_{r,i} {}^\wedge 1.$$

By construction, we have $c_r = a_r$ for every r. Indeed, it suffices to prove the result when w_r is a single letter σ_j. For $j > i$, the result is obvious, as we have $c_{r,m} = a_{r,m}$ for $m \leq i$. For $j < i$, we have

$$\begin{aligned} c_r &= c_{r,1} \wedge \cdots \wedge c_{r,i} \wedge 1 \\ &= a_{r,1} \wedge \cdots \wedge (a_{r,j} \wedge a_{r,j+1}) \wedge a_{r,j} \wedge \cdots \wedge c_{r,i} \wedge 1 = a_{r,1} \wedge \cdots \wedge a_{r,i} \wedge 1 = a_r, \end{aligned}$$

as braid exponentiation is left self-distributive. Now, using self-distributivity again, we find

$$\begin{aligned} a_{r+1} &= a_{r+1,1} \wedge \cdots \wedge a_{r+1,i} \wedge 1 \\ &= c_{r,1} \wedge \cdots \wedge c_{r,i-1} \wedge (c_{r,i} \wedge c_{r,i+1}) \wedge 1 \\ &= (c_{r,1} \wedge \cdots \wedge c_{r,i} \wedge c_{r,i+1}) \wedge (c_{r,1} \wedge \cdots \wedge c_{r,i} \wedge 1) \\ &>_L c_{r,1} \wedge \cdots \wedge c_{r,i} \wedge c_{r,i+1} \geq_L c_{r,1} \wedge \cdots \wedge c_{r,i} \wedge 1 = c_r \end{aligned}$$

As above, we deduce $c_p >_L a_0$, and we conclude that $b = 1$ is impossible. ∎

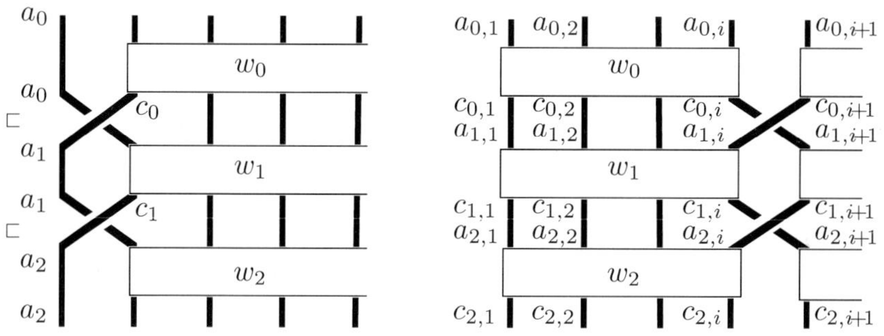

Figure 2.1. A σ_i-positive braid is never trivial

We deduce that, for every i, the relation "$a^{-1}b$ is σ_i-positive" defines a partial order on B_∞. Observe that a σ-positive braid is always σ_i-positive for some i, but the converse need not be true.

Let us now come back to the order $<_L$, and address the question of determining its order type. We begin with an auxiliary result.

Lemma 2.19. *Assume $a, b \in B_n$. Then the following are equivalent:*
(i) The relation $a <_L b$ holds;
(ii) The equality $a^{-1}b = \mathrm{sh}^k(c)$ holds for some σ_1-positive braid c and some k satisfying $k \leq n-2$.

Proof. The only new point to check is that, for $a, b \in B_n$ and $a^{-1}b = \mathrm{sh}^k(c)$ with c a σ_1-positive braid, $k \geq n-1$ is impossible. Now, for $a, b \in B_n$, $a^{-1}b$ belongs to B_n, hence it admits an expression where neither σ_{k+1} nor σ_{k+1}^{-1}

occurs. On the other hand, if c is σ_1-positive, it admits an expression where σ_1 occurs and σ_1^{-1} does not, so $\mathrm{sh}^k(c)$ admits an expression where σ_{k+1} occurs and σ_{k+1}^{-1} does not. Proposition 2.11 (σ_i-positive) then implies $a^{-1}b \neq sh^k(c)$. ■

Proposition 2.20. **(order type)** *The ordered set $(B_\infty, <_L)$ is dense without endpoints, so it is isomorphic to $(\mathbf{Q}, <)$.*

Proof. Let b be an arbitrary braid. Then, by construction, $b\sigma_1^{-1} <_L b <_L b\sigma_1$ holds, so $(B_n, <_L)$ admits no minimal or maximal element. Assume now $b \in B_n$ and $1 <_L b$. By Lemma 2.19, we have $b = \mathrm{sh}^k(c)$ with c a σ_1-positive braid c and $k \leq n-2$. Let $b' = b\sigma_n^{-1}$. Then $b\sigma_n^{-1} <_L b$ holds by definition. On the other hand, $b' = \mathrm{sh}^k(c\sigma_{n-k}^{-1})$ holds, and, by construction, $c\sigma_{n-k}^{-1}$ is σ_1-positive. Hence $1 <_L b' <_L b$ holds. So there always exists a braid strictly between 1 and b when $1 <_L b$ holds. Using left translation by a^{-1}, we deduce that there always exists a braid between a and b when $a <_L b$ holds. ■

The previous result does not answer the question of the order type of $(B_n, <_L)$ for a fixed value of n. It is obvious that $(B_2, <_L)$ is isomorphic to the integers. For $n \geq 3$, we shall come back to the question in Chapter IV.

We conclude with a description of the topology associated with $<_L$. We recall from Section I.4 that, for a, b in B_∞, the distance $d(a, b)$ is defined to be 2^{-k} where k is the least integer satisfying $a^{-1}b \in \mathrm{sh}^k(B_\infty)$.

Proposition 2.21. **(topology)** *The topology of B_∞ associated with the linear order $<_L$ is the ultrametric topology associated with d. Exponentiation is continuous on B_∞.*

Proof. By definition of d, the open disk of radius 2^{-p} centered at b is the left coset $b\,\mathrm{sh}^p(B_\infty)$, *i.e.*, the set of all braids bc such that c admits an expression where no generator $\sigma_k^{\pm 1}$ with $k \leq p$ occurs. Assume $a, b, c \in B_n$ and c belongs to the open interval (a, b), *i.e.*, $a <_L c <_L b$ holds. By Lemma 2.19, we have $a^{-1}c = sh^k(a_0)$ for some σ_1-positive braid a_0 and $k < n-1$. Then $a_0\mathrm{sh}^{n-1-k}(b')$ is σ_1-positive for every braid b', hence $a^{-1}cc'$ is σ-positive for every braid c' in $\mathrm{sh}^{n-1}(B_\infty)$. Similarly, we have $b^{-1}c = \mathrm{sh}^k(b_0)$ for some σ_1-negative braid b_0 and $k < n-1$. Then $b_0\mathrm{sh}^{n-1-k}(b')$ is σ_1-negative for every braid b', hence $b^{-1}cc'$ is σ-negative for every braid c' in $\mathrm{sh}^{n-1}(B_\infty)$. Hence, the open disk with radius 2^{n-1} centered at c is included in the interval (a, b) of $(B_\infty, <_L)$.

Conversely, let us start with an arbitrary open disk $c\,\mathrm{sh}^p(B_\infty)$. Let a, b be two points in this disk satisfying $a <_L b$. By definition, there exist a_0, b_0 satisfying $a = c\mathrm{sh}^p(a_0)$ and $b = c\,\mathrm{sh}^p(b_0)$, and $a <_L b$ implies $a_0 <_L b_0$. Let c' be an arbitrary braid in the interval (a, b). We wish to show that c' belongs to the disk $c\,\mathrm{sh}^p(B_\infty)$. The hypothesis implies $\mathrm{sh}^p(a_0) <_L c^{-1}c' <_L \mathrm{sh}^p(b_0)$. Assume that $c^{-1}c'$ does not belong to the image of sh^p. Then there exists $k < p$ and a braid c_0 which is either σ_1-positive or σ_1-negative and satisfies

$c^{-1}c' = \mathrm{sh}^k(c_0)$. If c_0 is σ_1-positive, $\mathrm{sh}^{p-k}(b_0^{-1})c_0$ is also σ_1-positive, which implies $c^{-1}c' >_L \mathrm{sh}^p(b_0)$, a contradiction. If c_0 is σ_1-negative, $\mathrm{sh}^{p-k}(a_0^{-1})c_0$ is also σ_1-negative, which implies $c^{-1}c' <_L \mathrm{sh}^p(a_0)$, again a contradiction. We deduce $c^{-1}c' \in \mathrm{sh}^p(B_\infty)$, and $c' \in c\,\mathrm{sh}^p(B_\infty)$. Hence the topologies associated with $<_L$ and with d coincide. ∎

Exercise 2.22. **(alternative proof)** *(i) Let F be a free LD-system with one generator, say x, and let π be the homomorphism of F into B_∞ that maps x to 1. We assume that F is left cancellative, hence eligible for a braid colouring. [This can be proved directly using the techniques of Chapter VIII.] Show that $(x, x, x, \ldots) \bullet \pi(a)$ exists and is equal to $(a, x, x, \ldots)$ for every a in F. Deduce that $(B_{sp}, {}^\wedge)$ is a free LD-system.*

Exercise 2.23. **(order $\ll_L$)** *(i) For a, b braids, say that $a \ll_L b$ holds if the braid $a^{-1}b$ is σ_1-positive. [Thus $a \ll_L b$ is equivalent to $a <_L b$ when a and b are special.] Prove that the relation $\ll_L$ is a partial order on B_∞, which is compatible with multiplication on the left.*

(ii) Prove that both $a \le_L b \ll_L c$ and $a \ll_L b \le_L c$ imply $a \ll_L c$.

(iii) Prove that $b \ll_L c$ implies $a{}^\wedge b \ll_L a{}^\wedge c$ for every a. [Hint: Write $(a{}^\wedge b)^{-1}\,(a{}^\wedge c) = sh(a)\,\sigma_1^{-1}\mathrm{sh}^2(d_0)\,\sigma_2\,\mathrm{sh}^2(d_1)\,\sigma_2 \cdots \sigma_2\,\mathrm{sh}^2(d_p)\,\sigma_1\,\mathrm{sh}(a^{-1})$, and observe that the latter braid is equal to

$$\mathrm{sh}(a)\,\sigma_{2,n+1}\,\mathrm{sh}(d_0)\,\sigma_1\,\mathrm{sh}(d_1)\,\sigma_1 \cdots \sigma_1\,\mathrm{sh}(d_p)\,\sigma_{2,n+1}^{-1}\,\mathrm{sh}(a^{-1}).]$$

Exercise 2.24. **(representations)** *(i) Assume that r is a linear representation of B_∞ that commutes with the shift endomorphism. Prove that r is faithful if and only if no σ_1-positive braid has a trivial r-image.*

(ii) Show that the image under the Burau representation (Example I.2.11) of a braid that admits a σ_1-positive expression with at most two σ_1's is not trivial.

Exercise 2.25. **(permutation braids)** *Assume that b_1 and b_2 are the simple braids associated with the permutations f_1 and f_2 respectively. Show that $b_1 <_L b_2$ holds if and only if the sequence $(f_1(1), \ldots, f_1(n))$ precedes the sequence $(f_2(1), \ldots, f_2(n))$ in the lexicographical order.*

Exercise 2.26. **(pure braids)** *Let PB_n denote the n-strand pure braid group, defined as the kernel of the morphism perm of B_n onto $\mathfrak{S}_n$. Let $a_{i,p} = \sigma_{p,i+1}\sigma_i^2\sigma_{p,i+1}^{-1}$. It is known that PB_n is generated by the braids $a_{i,p}$ with $1 \le i \le p \le n-1$.*

(i) Let $b = \sigma_1^2\sigma_2^2\sigma_1^{-2}$. Prove that b admits no decomposition involving σ_1^2, $\sigma_2\sigma_1^2\sigma_2^{-1}$ and $\sigma_2^{\pm 2}$ only. [Hint: Assume that b is expressed as

$w(\sigma_1^2, \sigma_2\sigma_1^2\sigma_2^{-1}, \sigma_2^2, \sigma_2^{-2})$. Write coll_i for the projection of B_3 onto B_2 corresponding to deleting the i-th strand. Using coll_3, prove that σ_1^2 does not occur in w. Using coll_2, prove that $\sigma_2\sigma_1^2\sigma_2^{-1}$ does not occur in w.]

(ii) Deduce that the restriction of the linear order $<_L$ to PB_n admits no characterization in terms of the generators $a_{i,p}$ similar to the characterization of $<_L$ in terms of the generators σ_i and σ_i-positivity.

---

3. Handle Reduction

We have established above that every nontrivial braid in B_∞ is either σ-positive or σ-negative, *i.e.*, it admits an expression where the generator with least index occurs only positively, or only negatively. The method of Section 2 is effective: starting from a braid word w, we can colour w with special braids, compare the input and the output colours with respect to the canonical order of B_{sp} by using the effective methods described in Chapter V below, and obtain a braid word w' that is equivalent to w and is either σ-positive or σ-negative.

Although effective, the previous method is intractable in practice. Moreover, it suffers a strong limitation, namely that the width of the braid is not preserved: if we start with an n-strand braid word w, the equivalent word w' so obtained need not be an n-strand braid word. Indeed, the width of w' comes from the size of the special braids involved in the colouring, and the only upper bound on this parameter coming from the results of Chapter V is some tower of exponentials in the length of w. Thus, at least two things are missing, namely a more practical algorithm for finding σ-positive or σ-negative forms, and the theoretical result that a nontrivial braid in B_n admits at least one n-strand σ-positive or σ-negative expression. In this section, we establish the following positive answers to these questions:

Proposition 3.1. (complexity) *The complexity of the linear order $<_L$ on B_∞ is at most exponential, i.e., there exists an algorithm deciding, for arbitrary braid words w_1, w_2, whether $\overline{w_1} <_L \overline{w_2}$ holds in a number of steps bounded by an exponential function of the lengths of w_1 and w_2.*

Proposition 3.2. (local version) *Every nontrivial braid in B_n can be represented either by a σ-positive n-strand braid word, or by a σ-negative n-strand braid word.*

Handles

Saying that a braid word w is neither σ_1-positive, nor σ_1-negative, nor σ_1-neutral (defined as containing neither σ_1 nor σ_1^{-1}) means that at least one subword of w has the form $\sigma_1 v \sigma_1^{-1}$ or $\sigma_1^{-1} v \sigma_1$, with neither σ_1 nor σ_1^{-1} occurring in v. Geometrically, this corresponds to the fact that the strand that starts in position 2 forms a left handle over or under the other strands, as illustrated in Figure 3.1.

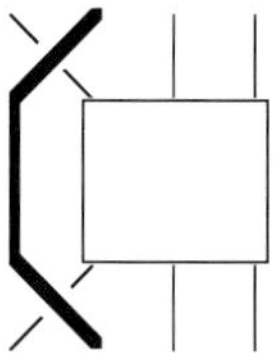

Figure 3.1. A handle

We can get rid of such a handle by replacing the initial braid diagram with a new equivalent diagram where the strand involved in the handle is pushed to the right, so as to skirt to the right of the next crossings, according to the scheme of Figure 3.2. We call this transformation *reduction* of the handle.

Figure 3.2. Handle reduction

We can then iterate handle reduction until no handle is left: if the process converges, then, by construction, the final word contains no handle, which implies that it is either σ_1-positive, or σ_1-negative, or σ_1-neutral. In the first two cases, we are done; in the third case, we can consider shifted handles $\sigma_2 v \sigma_2^{-1}$ similarly, and apply the same method again. This naive approach does not work readily: when applied to the word $w = \sigma_1\sigma_2\sigma_3\sigma_2^{-1}\sigma_3^{-1}$, it leads in one step to the word $w' = \sigma_2^{-1} w \sigma_2$: the initial handle is still there, and iterating the process leads to nothing more but longer and longer words. Now, we see that the handle in the word w' is not the original handle of the word w, but it comes from a second, shifted handle $\sigma_2\sigma_3\sigma_2^{-1}$ of w. If we reduce the latter handle into $\sigma_3^{-1}\sigma_2\sigma_3$ before reducing the main handle of w, *i.e.*, if we first go from w to $w'' = \sigma_1\sigma_3^{-1}\sigma_2\sigma_3\sigma_1^{-1}$, then applying the handle reduction yields $\sigma_3^{-1}\sigma_2^{-1}\sigma_1\sigma_2\sigma_3$, a σ-positive word equivalent to w. We shall see in the sequel that the previous problem, namely the presence of nested handles, is the only possible obstruction: provided we first reduce all nested handles, handle reduction always comes to an end in a finite number of steps.

Definition 3.3. (**handle**) A σ_i-*handle* is a braid word of the form $\sigma_i^e v \sigma_i^{-e}$ with $e = \pm 1$ and $v \in \mathrm{sh}^i(B_\infty)$.

Definition 3.4. (**reduction**) A handle $\sigma_i^e v \sigma_i^{-e}$ is said to be *permitted* if the word v includes no σ_{i+1}-handle. We say that the braid word w' is obtained from the braid word w *in one step of handle reduction* if some subword of w is a permitted σ_i-handle, say $\sigma_i^e v \sigma_i^{-e}$, and w' is obtained from w by applying in the previous handle the alphabetical homomorphism

$$\varphi_{i,e} : \sigma_k \mapsto \begin{cases} \sigma_k & \text{for } k \neq i, i+1, \\ \varepsilon & \text{for } k = i, \\ \sigma_{i+1}^{-e}\sigma_i\sigma_{i+1}^{e} & \text{for } k = i+1. \end{cases}$$

We say that w' is obtained from w *in m steps* of handle reduction if there exists a sequence of length $m+1$ from w to w' such that each term is obtained from the previous one by one step of handle reduction.

The general form of a σ_i-handle is

$$\sigma_i^e \, v_0 \, \sigma_{i+1}^{d_1} \, v_1 \, \sigma_{i+1}^{d_2} \cdots \sigma_{i+1}^{d_k} \, v_k \, \sigma_i^{-e} \tag{3.1}$$

with $d_j = \pm 1$ and $v_j \in \mathrm{sh}^{i+1}(B_\infty)$. Saying that this handle is permitted amounts to saying that all exponents d_j have a common value d. Then, reducing the handle means replacing it with

$$v_0 \, \sigma_{i+1}^{-e}\sigma_i^d\sigma_i^e \, v_1 \, \sigma_{i+1}^{-e}\sigma_i^d\sigma_i^e \cdots \sigma_{i+1}^{-e}\sigma_i^d\sigma_i^e \, v_k : \tag{3.2}$$

we remove the initial and final $\sigma_i^{\pm 1}$, and replace each σ_{i+1} with $\sigma_{i+1}^{-e}\sigma_i\sigma_{i+1}^{e}$.

Example 3.5. *(**handle reduction**) A given braid word may contain several handles, and there are in general many ways of reducing the handles. Contrary to the case of word reversing, various sequences of handle reductions may lead to different words. So handle reduction becomes an algorithm only when we fix a strategy for choosing which handle is to be reduced first. A solution is to look for the first permitted handle, i.e., for the handle whose last letter has the leftmost position. The reader can check the method on some random braid word of his choice. To give one example here, we have displayed below two reductions of the word* $\sigma_1^{-1}\sigma_2^{-1}\sigma_1\sigma_3\sigma_2^{-1}\sigma_3^{-1}\sigma_2^{-1}\sigma_1\sigma_3^{-1}\sigma_2\sigma_1^2$. *We improve readability by using the* *convention of* (Epstein & *al.* 92) *that* a, b, ... *stand for* σ_1, σ_2, ..., *and* A, B *for* σ_1^{-1}, σ_2^{-1}, *etc. So the word above becomes* ABacBCBaCbaa.

ABacBCBaCbaa	ABacBCBaCbaa
bABcBCBaCbaa	bABcBCBaCbaa
bbABcbABCbABCbaa	bbABcbABCbABCbaa
bbAcbCABCbABCbaa	bbABcbABCbAcBCaa
bbAcbCAcBCABCbaa	bbABcbABCbcbABCa
bbAcbABCABCbaa	bbABcbABCbcbbABC
bbAcbABCAcBCaa	
bbAcbABABCaa	
bbAcbAABCa	
bbAcbAbABC	

The left column shows the successive words that appear when the leftmost handle is reduced systematically, while the right column corresponds to the case where only inevitable handles are reduced, i.e., those that are nested in another handle that we wish to reduce. The handles that are reduced at each step are underlined. The final word of the first column contains no handle at all, while the final word in the second column contains no σ_1-handle, but it still contains some σ_2-handle. In both cases, the final words are σ_1-negative.

The absolute value of a braid word

As shown in the above example, handle reduction may increase the length of the braid word it is applied to. Our first task for proving convergence of handle reduction will be to show that all words obtained from a word w using handle reduction remain traced in some finite region of the Cayley graph of B_∞ depending on w. To this end, we introduce the notion of the absolute value of a braid word using the techniques of Chapter II, and we connect the operation of handle reduction with the operations of left and right reversing.

Lemma 3.6. *Assume that w is a braid word. Then we have*

$$D_L(w)\, N_R(w) \equiv N_L(w)\, D_R(w). \tag{3.3}$$

Proof. By construction, we have

$$\begin{aligned} D_L(w)\, N_R(w) &\equiv D_L(w)\, N_R(w)\, D_R(w)^{-1}\, D_R(w) \equiv D_L(w)\, w\, D_R(w) \\ &\equiv D_L(w)\, D_L(w)^{-1}\, NL(w)\, D_R(w) \equiv N_L(w)\, D_R(w). \end{aligned} \qquad \blacksquare$$

Definition 3.7. (**absolute value**) Assume that w is a braid word. We define the *absolute value* of w to be the positive braid $|w|$ represented by $D_L(w) N_R(w)$.

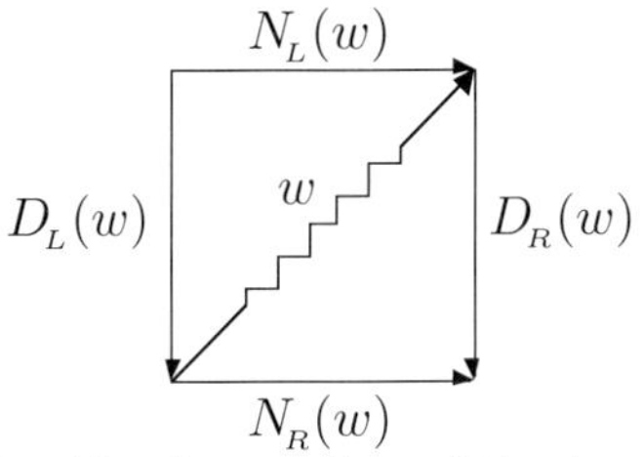

Figure 3.3. Absolute value of the braid word w

Example 3.8. **(absolute value)** *The braid word $w = \sigma_1^{-1}\sigma_2\sigma_2^{-1}\sigma_3$ is right reversible to $\sigma_2\sigma_1\sigma_3\sigma_2\sigma_1\sigma_3^{-1}\sigma_2^{-1}\sigma_1^{-1}\sigma_3^{-1}\sigma_2^{-1}$, i.e., we have $N_R(w) = \sigma_2\sigma_1\sigma_3\sigma_2\sigma_1$ and $D_R(w) = \sigma_2\sigma_3\sigma_1\sigma_2\sigma_3$. On the other hand, w is left reversible to $\sigma_1^{-1}\sigma_3$, i.e., we have $N_L(w) = \sigma_3$ and $D_L(w) = \sigma_1$. Hence the absolute value of w is the positive braid $\sigma_1\sigma_2\sigma_1\sigma_3\sigma_2\sigma_1$—which is also $\sigma_3\sigma_2\sigma_3\sigma_1\sigma_2\sigma_3$.*

In the case of B_2, *i.e.*, of $\mathbf{Z}$, the absolute value of the braid word σ_1^k is $\sigma_1^{|k|}$. Observe that, by construction, the equality $|w| = |w^{-1}|$ holds for every braid word w.

The general notion of the Cayley graph of a group is well-known—see for instance (Epstein & *al.* 92). Here we consider finite fragments of such graphs.

Definition 3.9. **(Cayley graph)** Assume that b is a positive braid. The *Cayley graph of b* is the finite labelled graph defined as follows: the vertices are the left divisors of b, and there exists an edge labelled σ_i from the vertex a_1 to the vertex a_2 if $a_2 = a_1\sigma_i$ holds.

Example 3.10. **(Cayley graph)** *The Cayley graph of the braid Δ_4 is displayed in Figure 3.4. By Proposition II.4.14 (simple), the (left) divisors of Δ_4 are in one-to-one correspondence with the 24 permutations of $\{1, \ldots, 4\}$. The Cayley graph of Δ_4 becomes the Cayley graph of the group $\mathfrak{S}_4$ if we replace each label σ_i with s_i and remove orientation. More generally, it follows from the existence of right lcm's and left gcd's in B_∞^+ that the Cayley graph of every positive braid b is a lattice with minimal element 1 and maximal element b.*

When we are given a graph Γ whose edges are labelled using letters from an alphabet A, we have the natural notion of a word *traced* in Γ: we say that w is traced in Γ from the vertex a if there exists a path γ from a in Γ labelled w, *i.e.*, w is the word obtained by concatenating the labels of the edges in γ.

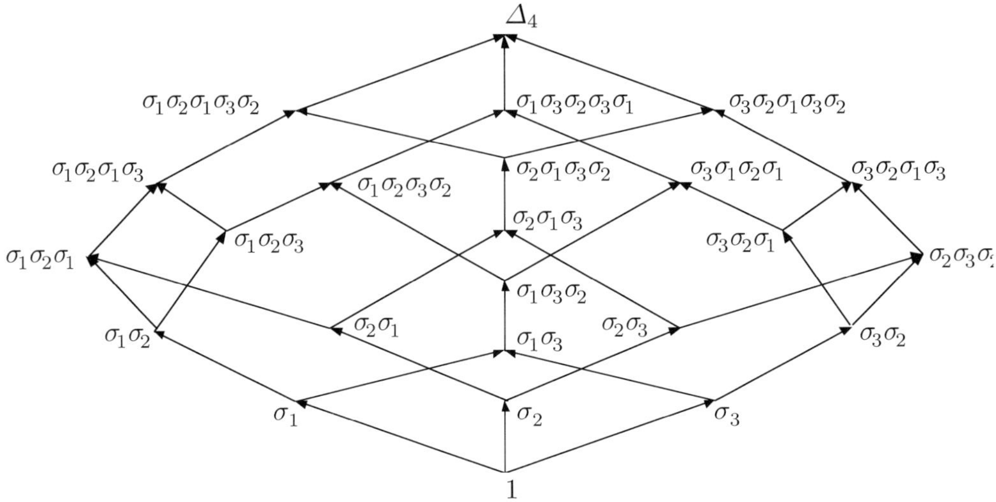

Figure 3.4. The Cayley graph of Δ_4

Definition 3.11. (**word traced**) Assume that Γ is an A-labelled graph, a is a vertex of Γ, and w is a word in $(A \cup A^{-1})^*$. We say that w is *traced in Γ from a* if either w is the empty word, or w is xv (*resp.* w is $x^{-1}v$) with x in A, there is an x-labelled edge from a to b (*resp.* from b to a) in Γ and v is traced from b in Γ.

Infinitely many words are traced in every graph Γ containing at least one edge: if x is the label of an edge from a, the word $(xx^{-1})^k$ is traced from a for every integer k. For our current purpose, the point is that, for every positive braid b, the set of all words traced in the Cayley graph of b enjoys good closure properties.

Lemma 3.12. *Assume that b is a positive braid. Then the set of all words traced in the Cayley graph of b from a given point is closed under right and left reversing.*

Proof. (Figure 3.5). Let Γ denote the Cayley graph of b. Assume that the word $v\,\sigma_i^{-1}\,\sigma_j\,v'$ is traced from a in Γ. We have to show that the word

$$v\ (\sigma_i \backslash \sigma_j)\ (\sigma_j \backslash \sigma_i)^{-1}\ v'$$

is also traced from a in Γ. The hypothesis means that there exist positive braid words u', u''_1 and u''_2 such that both $u'\sigma_i u''_1$ and $u'\sigma_j u''_2$ are traced from 1 in Γ, and $u'\sigma_i$ is equivalent to uv. By hypothesis, we have $u'\sigma_i u''_1 \equiv u'\sigma_j u''_2$, and $\sigma_i u''_1 \equiv \sigma_j u''_2$. By Proposition II.2.18 (lcm), this implies the existence of a positive word u'' satisfying

$$u''_1 \equiv (\sigma_i \backslash \sigma_j)\ u'' \quad \text{and} \quad u''_2 \equiv (\sigma_j \backslash \sigma_i)\ u''.$$

This shows that the word $(\sigma_i\backslash\sigma_j)\,(\sigma_j\backslash\sigma_i)^{-1}$ is traced in Γ from $a\overline{v}$, as was needed. The argument is symmetric for left reversing. ∎

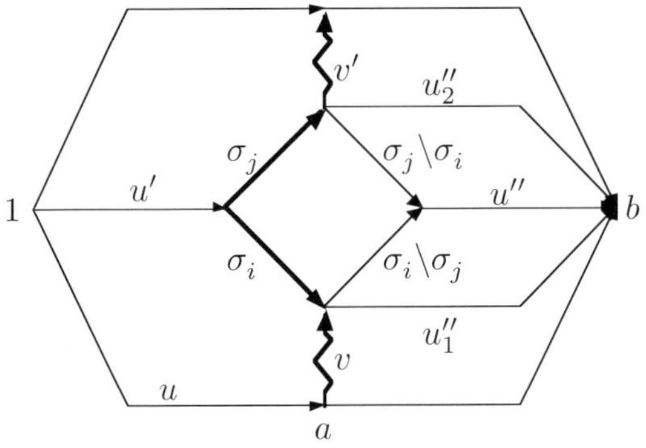

Figure 3.5. Closure of words traced under right reversing

Definition 3.13. (positive equivalence) *Positive (resp. negative) equivalence on arbitrary braid words is defined to be the congruence on* BW_∞ *generated by all pairs* (u, u') *(resp.* (u^{-1}, u'^{-1})*) where* u*,* u' *are equivalent positive braid words.*

Saying that w' is obtained from w by positive equivalence means that one can go from w to w' by applying exclusively transformations of the form $\sigma_i\sigma_j \mapsto \sigma_j\sigma_i$ with $|i-j| \geq 2$ and $\sigma_i\sigma_j\sigma_i \mapsto \sigma_j\sigma_i\sigma_j$ with $|i-j| = 1$, but *no* relation involving negative letters (not even the trivial ones).

Lemma 3.14. *Assume that b is a positive braid. Then the set of all words traced in the Cayley graph of b from a given point is closed under positive and negative equivalence.*

Proof. Let Γ denote the Cayley graph of b again. Assume, for instance, that $v\sigma_i\sigma_{i+1}\sigma_i v'$ is traced in Γ from $\overline{u}$. By definition, there exist positive words u' and u'' such that $u'\sigma_i\sigma_{i+1}\sigma_i u''$ are traced in Γ from 1 and $uv \equiv u'$ holds. Now $u'\sigma_{i+1}\sigma_i\sigma_{i+1}u''$ represents b as well, so it is traced from 1 in Γ, and, therefore, $v\sigma_{i+1}\sigma_i\sigma_{i+1}v'$ is still traced in Γ from $\overline{u}$. The case of negative equivalence is similar and corresponds to crossing the edges with reverse orientation. ∎

Proposition 3.15. (traced) *Assume that w is an arbitrary braid word. Then every word that can be obtained from w using right reversing, left reversing, positive equivalence, and negative equivalence is traced from $\overline{D_L(w)}$ in the Cayley graph of $|w|$.*

Proof. Owing to Lemmas 3.12 and 3.14, it suffices to show that w itself is traced from $\overline{D_L(w)}$ in the Cayley graph of $|w|$. We use induction on the length of w. The result is true for $w = \varepsilon$. Otherwise, assume first $w = w_0\sigma_i$. Let Γ be the Cayley graph of $|w|$, and Γ_0 be the Cayley graph of $|w_0|$. By construction, we have $D_L(w) = D_L(w_0)$ and $N_R(w) = N_R(w_0)v$, with $v = D_R(w_0)\backslash\sigma_i$ (Figure 3.6). Then, uv is traced from 1 in Γ for every word u traced from 1 in Γ_0, and, therefore, every word traced in Γ_0 from $\overline{D_L(w_0)}$ is also traced in Γ from $\overline{D_L(w_0)}$. By induction hypothesis, w_0 is traced in Γ_0 from $\overline{D_L(w_0)}$, *i.e.*, from $\overline{D_L(w)}$. So it remains to consider the final letter σ_i of w. Now we have $\overline{D_L(w)}w_0 \equiv N_L(w_0)$, hence the path starting from $\overline{D_L(w_0)}$ and labelled w_0 finishes at $N_L(w_0)$, and, by construction, σ_i is traced in Γ from $N_L(w_0)$.

Assume now $w = w_0\sigma_i^{-1}$. We have $N_R(w) = N_R(w_0)$ and $D_L(w) = vD_L(w_0)$, with $v = \sigma_i/N_R(w_0)$. With the same notation as above, the word vu is traced from 1 in Γ if u is traced from 1 in Γ_0. By induction hypothesis, w_0 is traced in Γ_0 from $\overline{D_L(w_0)}$, hence it is traced in Γ from $\overline{D_L(w_0)}$ too. Again it remains to consider the final letter σ_i^{-1} of w. As above, $\overline{D_L(w_0)}w_0$ is equivalent to $N_L(w_0)$, and the path of Γ starting from $\overline{D_L(w)}$ and labelled w_0 finishes at $\overline{N_L(w_0)}$. Now, by construction, σ_i^{-1} is traced from $\overline{N_L(w_0)}$ in Γ. ■

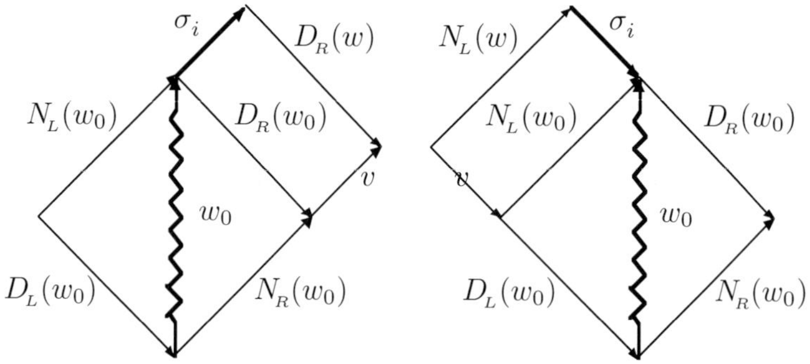

Figure 3.6. The word w is traced in the Cayley graph of $|w|$.

Our last remark deals with the size of the Cayley graph.

Definition 3.16. (width) Assume that w is a nonempty braid word. We say that w has *width* n if $n-2$ is the difference between the smallest and the largest indices i such that $\sigma_i^{\pm 1}$ occurs in w.

Every n-strand braid word has width at most n, but the inequality may be strict: for instance, the braid word $\sigma_2\sigma_3^{-1}$ has width 3. The width of w is the number of strands "really involved in w".

Proposition 3.17. (size) *Assume w is a braid word of length ℓ and width n. Then the Cayley graph of the braid $|w|$ contains at most $(n-1)^{\ell n(n-1)/2}$ edges.*

Proof. Up to a translation by a power of the shift mapping, we may assume that w belongs to BW_n. Assume that there are p positive and q negative letters in w. By Proposition II.3.5 (closed), the word $D_L(w)$ is the product of at most q braid words representing simple braids, and the word $N_R(w)$ is the product of at most p such braid words. Hence the braid $|w|$ is the product of at most ℓ simple braids. As the length of a simple n-strand braid is bounded above by $n(n-1)/2$, the length of $|w|$ is at most $\ell n(n-1)/2$. Now, at most $n-1$ edges start from a given vertex a in the Cayley graph of a word in BW_n, hence, in the Cayley graph of a word of length L, there are no more than $(n-1)^L$ edges. ∎

Let us come back to handle reduction. By construction, if the braid word w' is obtained from w using handle reduction, then w' is equivalent to w. The previous obvious fact can be refined into the following result.

Lemma 3.18. *Assume that w' is obtained from w using handle reduction. Then one can transform w into w' using right reversing, left reversing, positive equivalence, and negative equivalence.*

Proof. The point is to show that, for $v_0, \ldots, v_k$ in $\mathrm{sh}^{i+1}(B_\infty)$, we can go from

$$\sigma_i^e\, v_0\, \sigma_{i+1}^d\, v_1\, \sigma_{i+1}^d\, \cdots\, \sigma_{i+1}^d\, v_k\, \sigma_i^{-e} \tag{3.4}$$

to

$$v_0\, \sigma_{i+1}^{-e}\sigma_i^d\sigma_i^e\, v_1\, \sigma_{i+1}^{-e}\sigma_i^d\sigma_i^e\, \cdots\, \sigma_{i+1}^{-e}\sigma_i^d\sigma_i^e\, v_k\, : \tag{3.5}$$

using the transformations mentioned in the statement. Assume for instance $e=+1$ and $d=-1$. Then reduction can be done by moving to the right the initial σ_i. First transforming $\sigma_i v_0$ into $v_0\sigma_i$ can be made by a sequence of left reversings (in the case of negative letters) and positive equivalences (in the case of positive letters). Then we find the pattern $\sigma_i\sigma_{i+1}^{-1}$, which becomes $\sigma_{i+1}^{-1}\sigma_i^{-1}\sigma_{i+1}\sigma_i$ by a left reversing. So, at this point, we have transformed the initial word into

$$v_0\, \sigma_{i+1}^{-1}\sigma_i^{-1}\sigma_{i+1}\sigma_i\, v_1\, \sigma_{i+1}^{-1}\, v_2\, \cdots\, v_{k-1}\, \sigma_{i+1}^{-1}\, v_k\, \sigma_j^{-1}. \tag{3.6}$$

After k such sequences of reductions, and a last left reversing to delete the final pattern $\sigma_i\sigma_i^{-1}$, we reach the form (3.5), as we wished. The argument is similar in the case $e=-1$, $d=1$, with negative equivalences instead of positive equivalences. In the case when the exponents e and d have the same sign, we use right reversing in order to move to the left the final generator σ_i^{-e}. ∎

By applying Proposition 3.15 (traced), we deduce the following result, which is an upper complexity bound for those words obtained using handle reduction:

Proposition 3.19. (**bound**) *Assume that the braid word w' is obtained from w using handle reduction. Then w' is traced in the Cayley graph of $|w|$.*

The height of a braid word

The previous boundedness result is not sufficient for proving that handle reduction always converges. In particular, it does not discard the possibility that loops occur. To go further, we need a new parameter, the height.

Definition 3.20. (**height**) Assume that w is a braid word. The *height* of w is defined to be the maximal number of letters σ_i occurring in a word containing no σ_i^{-1} and traced in the Cayley graph of $|w|$, for any i.

Lemma 3.21. *Assume w is a braid word of length ℓ and width n. Then the height is bounded above by $(n-1)^{\ell n(n-1)/2}$.*

Proof. Assume that u is a braid word traced in the Cayley graph of $|w|$ in which the letter σ_i^{-1} does not occur. We claim that the various letters σ_i that occur in u correspond to different edges in the Cayley graph of $|w|$. Indeed, saying that the same edge is crossed twice would mean some closed path in this graph contains at least one σ_i-labelled edge and no σ_i^{-1}-labelled edge. In other words, there would exist a trivial braid word where σ_i occurs, but σ_i^{-1} does not. By Proposition 2.11 (σ_i-positive), this is impossible. Thus the number of letters σ_i in u is bounded above by the total number of edges in the Cayley graph of $|w|$, hence, according to Proposition 3.17 (size), by $(n-1)^{\ell n(n-1)/2}$. ■

Let us now consider handle reduction. Assume that $w_0 = w, w_1, \ldots$ is a sequence of handle reductions from w. The first point is that the number of σ_1-handles in w_i is not larger than the number of σ_1-handles in w, as reducing one σ_1-handle lets appear at most one new σ_1-handle. We deduce a well-defined notion of inheriting between σ_1-handles such that each σ_1-handle in the initial word w possesses at most one heir in each word w_k. We shall assume that the number of σ_1-handles in every word k is the same as in w_0. If it is not the case, *i.e.*, if some σ_1-handle vanishes without heir, say from w_k to w_{k+1}, we cut the sequence at w_k and restart from w_{k+1}. In this way, the heir of the p-th σ_1-handle of w_0 (when enumerated from the left) is the p-th σ_1-handle of w_k. Let us define the p-th *critical prefix* $\mathrm{pref}_p(w_k)$ of w_k to be the braid represented by the prefix of w_k ending at the first letter of the p-th σ_1-handle. The key point is the following observation: for every p, we have

$$\mathrm{pref}_p(w_0) \geq_L \mathrm{pref}_p(w_1) \geq_L \mathrm{pref}_p(w_2) \geq_L \cdots, \tag{3.7}$$

and, at each step where the p-th σ_1-handle is actually reduced, the inequality is strict. More precisely, we have the following key result:

Lemma 3.22. *Assume that the p-th σ_1-handle is reduced from w_k to w_{k+1}. Then there exists a braid word $u_{p,k}$ that is traced in the Cayley graph of $|w|$ from $\mathrm{pref}_p(w_k)$ to $\mathrm{pref}_p(w_{k+1})$ and contains one letter σ_1^{-1} and no letter σ_1.*

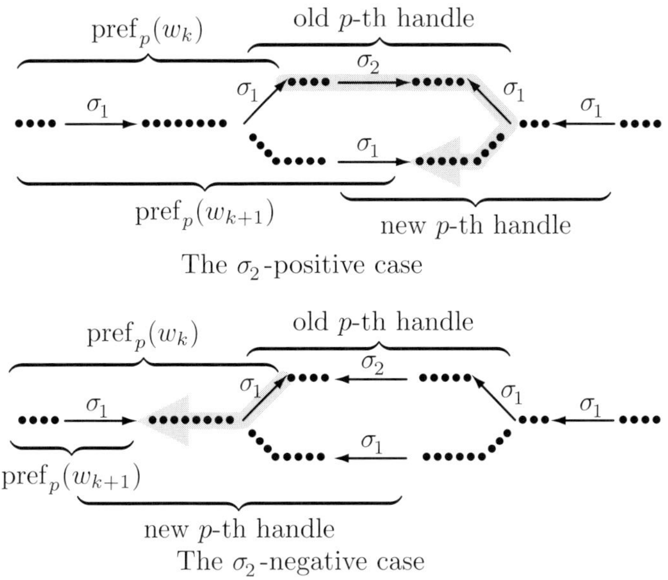

Figure 3.7. Critical prefixes

Proof. The result can be read on the diagrams of Figure 3.7, where we have represented the paths associated respectively with w_k (up) and w_{k+1} (down) in the Cayley graph of B_∞, assuming that the p-th σ_1-handle has been reduced. The word $u_{p,k}$ appears in grey, and the point is that, in every case, *i.e.*, both if σ_2 appears positively or negatively (or not at all) in the handle, this word $u_{p,k}$ contains one letter σ_1^{-1} and no letter σ_1. ∎

If the p-th σ_1-handle is not reduced from w_k to w_{k+1}, we have $\mathrm{pref}_p(w_k) \equiv \mathrm{pref}_p(w_{k+1})$, and we complete the definition with $u_{p,k} = \varepsilon$. Now, by construction, the word $u = u_{p,0}u_{p,1}u_{p,2}\cdots$ is traced in the Cayley graph of $|w|$, it is σ_1-negative, and the number N of steps in the sequence $(w_0, w_1, \ldots)$ where the p-th σ_1-handle has been reduced is equal to the number of letters σ_1^{-1} in u. It follows that the number N is bounded above by the height of w. Thus, the heirs of each σ_1-handle of the initial word w are involved in at most h reduction steps, where h is the height of w. As there are at most ℓ handles in a word of length ℓ, we can state:

Lemma 3.23. *Assume that w is a braid word of length ℓ and height h. Then the number of σ_1-handle reductions in any sequence of handle reductions from w is bounded above by ℓh.*

Assuming again that $w_0 = w, w_1, \ldots$ is a sequence of handle reductions from w, we can now iterate the result and consider the σ_2-handle reductions: the previous argument gives an upper bound for the number of σ_2-handle reductions between two successive σ_1-handle reductions, and, more generally, for the number of σ_{i+1}-handle reductions between two σ_i-handle reductions. Using a coarse upper bound on the lengths of the words w_i, one obtains the following generalization of Lemma 3.23:

Lemma 3.24. *Assume that w is a braid word of length ℓ, width n, and height h. Then the number of handle reductions in any sequence of handle reductions from w is bounded above by $\ell(2h)^{2n-1}$.*

Inserting the bound on the height given by Lemma 3.21, we finally deduce:

Proposition 3.25. (convergence) *Assume that w is a braid word of length ℓ and width n. Then every sequence of handle reductions from w converges in at most $2^{n^4\ell}$ steps.*

Propositions 3.1 (complexity) and 3.2 (local version) clearly follow.

Exercise 3.26. (absolute value) *(i) Let $v = \sigma_1\sigma_2\sigma_1^{2p}\sigma_2\sigma_1\sigma_2^{-2p}$. Prove that $|v^k| = (\sigma_1\sigma_2\sigma_1\sigma_2)^k\sigma_2^{2p}$ holds for every k. Deduce that there exists no positive constant c such that the length of $|w|$ is at least $c\cdot \lg(w)$ for every w in BW_3.*

(ii) Prove similarly that there exists no positive constant c such that the length of $|w|$ is at least $c\cdot \lg(w)^{1/2}$ in BW_4 [Hint: Consider the word $w = \sigma_1\sigma_2\sigma_3 v^q\sigma_3\sigma_2\sigma_1 \mathrm{sh}(v^{-q})$, where v is the word of (i).]

Exercise 3.27. (height) *Show that the height of the braid Δ_3^{2k} is at least $2k^2$. [Hint: Show that the word $(\sigma_1^{2k}\sigma_2\sigma_1\sigma_2^{-2k}\sigma_1\sigma_2)^k$ is traced in the Cayley graph of Δ_3^{2k}.] Show that the height of the braid Δ_n^{2k} is at least $2k^{n-1}$. [This is more tricky.]*

Exercise 3.28. (Vladivostok reduction) *Assume that n is fixed, and consider for words in BW_n the coarse reduction that consists in reducing a handle of the form $\sigma_i^e \mathrm{sh}^i(w)\sigma_i^{-e}$ into $\sigma_{n-1,i+1}^{-e}\mathrm{sh}^{i-1}(w)\sigma_{i+1,n-1}$, i.e.,*

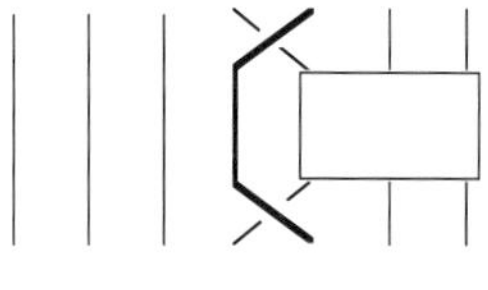 *reduces to* 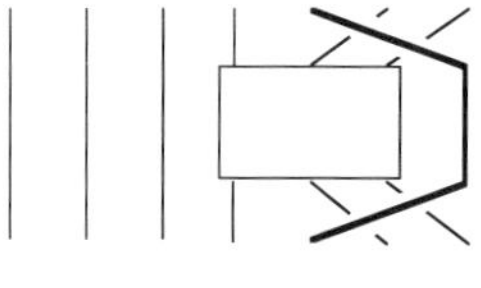

In other words, we go to the extreme right (the Far East) of the diagram systematically. Show that coarse reduction always converges for $n \leq 4$. [Hint: Show that, in this particular case, the length of the words with respect to the letters $\sigma_{p,q}$ does not increase. The general case is open, because it need not be true that all words deduced from w using coarse reduction are traced in the Cayley graph of $|w|$.]

---

4. Alternative Definitions

The linear order of braids seems to be a rather canonical object, so it is not surprising that it can be characterized or constructed using other approaches. In this section, we present three such methods. The first involves the automorphisms of a free group: it allows one to compare a braid b to 1 by looking at the images of the letters x_i under the automorphism $\widetilde{b}$. The second one uses the interpretation of braids as diffeomorphisms of a punctured disk; it compares braids by looking at the mutual positions of some associated curves, and it leads to a well-defined unique normal form result. The last one involves an action of braids on the real line connected with hyperbolic geometry.

Automorphisms of a free group

We have seen in Chapter I that there exists a homomorphism of the group B_∞ into the group $\mathrm{Aut}(\mathrm{FG}_\infty)$ of all automorphisms of the free group FG_∞ based on $\{x_1, x_2, \ldots\}$. More precisely, we have observed that the function that maps the braid σ_i to the automorphism $\widetilde{\sigma}_i$ defined by

$$\widetilde{\sigma}_i(x_k) = \begin{cases} x_k & \text{for } k \neq i, i+1, \\ x_i x_{i+1} x_i^{-1} & \text{for } k = i, \\ x_i & \text{for } k = i+1 \end{cases} \tag{4.1}$$

extends to a well-defined homomorphism φ of B_∞ into $\mathrm{Aut}(\mathrm{FG}_\infty)$ (we have not yet proved that this homomorphism is injective). By describing the images of σ-positive and σ-negative braids under φ, we shall obtain a new algorithm for deciding whether $b >_L 1$ holds.

We follow the notations and definitions of Section I.3. Thus W_∞ is the set of all words over the alphabet $\{x_1^{\pm 1}, x_2^{\pm 1}, \ldots\}$, and, for w in W_∞, we denote by $\mathrm{red}(w)$ the unique reduced word that can be obtained from w by iteratively deleting patterns $x_i x_i^{-1}$ or $x_i^{-1} x_i$. The free group FG_∞ is identified with the set of all freely reduced words. We denote by $S(x_k^e)$ the subset of FG_∞ consisting of those reduced words that finish with x_k^e.

Our first remark is that the braid order can be used to prove the following classical result.

Proposition 4.1. (**embedding**) *The homomorphism φ is an embedding of B_∞ into* $\mathrm{Aut}(\mathrm{FG}_\infty)$.

Proof. Assume that b is a nontrivial braid. Then there exists a nonnegative integer k and a σ_1-positive braid c satisfying either $b = \mathrm{sh}^k(c)$ or $b = \mathrm{sh}^k(c^{-1})$. By Lemma I.3.18, $\widetilde{c}$ is not the identity of FG_∞, hence neither $\widetilde{\mathrm{sh}^k(c)}$ nor $\widetilde{\mathrm{sh}^k(c^{-1})}$ is the identity of FG_∞. ■

In order to obtain a more precise connection between the braid order and $\mathrm{Aut}(\mathrm{FG}_\infty)$, we investigate systematically the action of $\widetilde{\sigma}_i^{\pm 1}$ on the sets $S(x_k^{\pm 1})$. First, the argument of Lemma I.3.16 gives

Lemma 4.2. *For $k \neq i, i+1$ and $e = \pm 1$, the automorphism $\widetilde{\sigma}_i$ maps $S(x_k^e)$ into itself.*

Lemma 4.3. *For each i, $\widetilde{\sigma}_i$ maps both $S(x_i)$ and $S(x_i^{-1})$ into $S(x_i^{-1})$.*

Proof. The argument is similar to the one used in the proof of Lemma I.3.17. Let us consider an arbitrary element of $S(x_i) \cup S(x_i^{-1})$, say wx_i^e with $e = \pm 1$ and $w \notin S(x_i^{-e})$. Then we have $\widetilde{\sigma}_i(wx_i^e) = \mathrm{red}(\widetilde{\sigma}_i(w)x_i x_{i+1}^e x_i^{-1})$. Assume $\widetilde{\sigma}_i(wx_i^e) \notin S(x_i^{-1})$. This means that the final x_i^{-1} in $\widetilde{\sigma}_i(wx_i^e)$ is cancelled by some letter x_i in $\widetilde{\sigma}_i(w)$. This letter comes either from some x_{i+1} or from some $x_i^{e'}$ in $\widetilde{\sigma}_i(w)$.

In the first case, we display the letter x_{i+1} involved in the cancellation by writing $w = w_1 x_{i+1} w_2$, where w_2 is a reduced word not beginning with x_{i+1}^{-1}. We find

$$\widetilde{\sigma}_i(wx_i^e) = \mathrm{red}(\widetilde{\sigma}_i(w_1)x_i\widetilde{\sigma}_i(w_2)x_i x_{i+1}^e x_i^{-1}),$$

and the hypothesis is $\mathrm{red}(\widetilde{\sigma}_i(w_2)x_i x_{i+1}^e) = \varepsilon$, which implies $\widetilde{\sigma}_i(w_2) = x_{i+1}^{-e}x_i^{-1} = \widetilde{\sigma}_i(x_{i+1}^{-1}x_i^{-e})$. We deduce $w_2 = x_{i+1}^{-1}x_i^{-e}$, contradicting $w \notin S(x_i^{-e})$.

In the second case, we write similarly $u = w_1 x_i^{e'} w_2$ with $e' = \pm 1$. So we have

$$\widetilde{\sigma}_i(wx_i^e) = \mathrm{red}(\widetilde{\sigma}_i(w_1)x_i x_{i+1}^{e'} x_i^{-1}\widetilde{\sigma}_i(w_2)x_i x_{i+1}^e x_i^{-1}),$$

and the hypothesis is $\mathrm{red}(x_{i+1}^{e'}x_i^{-1}\widetilde{\sigma}_i(w_2)x_i x_{i+1}^e) = \varepsilon$. This implies $\mathrm{red}(\widetilde{\sigma}_i(w_2)) = x_i x_{i+1}^{-e-e'}x_i^{-1} = \widetilde{\sigma}_i(x_i^{-e-e'})$, hence $w_2 = x_i^{-e-e'}$. For $e = +1$, we obtain either $w_2 = x_i^{-2}$ (for $e' = +1$) or $w_2 = 1$ (for $e' = -1$), and, in both cases, $w \in S(x_i^{-e})$, a contradiction. Similarly, for $e = -1$, we obtain either $w_2 = 1$ (for $e' = +1$) or $w_2 = x_i^2$ (for $e' = -1$), and, in both cases, $w \in S(x_i^{-e})$, again a contradiction. ■

Lemma 4.4. *For each i, $\widetilde{\sigma}_i$ maps $S(x_{i+1})$ into $S(x_i) \cup S(x_{i+1}) \cup S(x_{i+1}^{-1})$.*

Proof. Let us consider an arbitrary element of $S(x_{i+1})$, say wx_{i+1} with $w \notin S(x_{i+1}^{-1})$. Then we have $\widetilde{\sigma}_i(wx_{i+1}) = \mathrm{red}(\widetilde{\sigma}_i(w)x_i)$. Assume $\widetilde{\sigma}_i(wx_{i+1}) \notin S(x_i)$. This means that the final letter x_i is cancelled by some letter x_i^{-1} in $\widetilde{\sigma}_i(w)$. This letter comes either from some x_{i+1}^{-1} or from some $x_i^{\pm 1}$ in $\widetilde{\sigma}_i(w)$. As previously, we consider the possible cases.

The first case is $w = w_1 x_{i+1}^{-1} w_2$. We have now

$$\widetilde{\sigma}_i(wx_{i+1}) = \mathrm{red}(x_i\widetilde{\sigma}_i(w_1)x_i^{-1}\widetilde{\sigma}_i(w_2)x_i),$$

and the hypothesis is $\mathrm{red}(\widetilde{\sigma}_i(w_2)) = \varepsilon$. This implies $w_2 = \varepsilon$, and, therefore, $w \in S(x_{i+1}^{-1})$, a contradiction.

The second case is $w = w_1 x_i^e w_2$ with $e = \pm 1$ and $w_1 \notin S(x_i^{-e})$. We find

$$\widetilde{\sigma}_i(w) = \mathrm{red}(x_i\widetilde{\sigma}_i(w_1)x_i x_{i+1}^e x_i^{-1}\widetilde{\sigma}_i(w_2)x_i),$$

and the hypothesis is $\mathrm{red}(\widetilde{\sigma}_i(w_2)) = \varepsilon$. This implies $w_2 = \varepsilon$, hence $w = w_1 x_i^e$. Thus we have

$$\widetilde{\sigma}_i(wx_{i+1}) = \mathrm{red}(\widetilde{\sigma}_i(w_1)x_i x_{i+1}^e).$$

Assume that the final letter x_{i+1}^e vanishes in the reduction. It does it with some letter x_{i+1}^{-e} that necessarily comes from some letter x_i^{-e} in $\widetilde{\sigma}_i(w_1)$. So there must exist a decomposition $w_1 = w_1' x_i^{-e} w_1''$, giving

$$\widetilde{\sigma}_i(w) = \mathrm{red}(\widetilde{\sigma}_i(w_1')x_i x_{i+1}^{-e} x_i^{-1}\widetilde{\sigma}_i(w_1'')x_i)$$

with $\mathrm{red}(x_i^{-1}\widetilde{\sigma}_i(w_1'')x_i) = \varepsilon$. As above, this implies $w_1'' = \varepsilon$, and, therefore, $w_1 \in S(x_i^{-e})$, a contradiction. So $\widetilde{\sigma}_i(w) \in S(x_{i+1}^e)$ is the only possibility. ∎

Lemma 4.5. *For each i, $\widetilde{\sigma}_i$ maps $S(x_{i+1}^{-1})$ into $S(x_i^{-1})$.*

Proof. Consider an arbitrary element of $S(x_{i+1}^{-1})$, say wx_{i+1}^{-1} with $w \notin S(x_{i+1})$. We have $\widetilde{\sigma}_i(wx_{i+1}^{-1}) = \mathrm{red}(\widetilde{\sigma}_i(u)x_i^{-1})$. Assume $\widetilde{\sigma}_i(wx_{i+1}^{-1}) \notin S(x_i^{-1})$. This implies that the final letter x_i^{-1} is cancelled by some letter x_i in $\widetilde{\sigma}_i(w)$. This letter comes either from some x_{i+1} or from some $x_i^{\pm 1}$ in $\widetilde{\sigma}_i(w)$. The first case is $w = w_1 x_{i+1} w_2$. We find

$$\widetilde{\sigma}_i(w) = \mathrm{red}(x_i^{-1}\widetilde{\sigma}_i(w_1)x_i\widetilde{\sigma}_i(w_2)x_i^{-1}),$$

and the hypothesis is $\mathrm{red}(\widetilde{\sigma}_i(w_2)) = \varepsilon$. This implies $w_2 = \varepsilon$, and, therefore, $u \in S(x_{i+1})$, a contradiction.

The second case is $u = w_1 x_i^e w_2$ with $e = \pm 1$ and w_2 not beginning with x_i^{-e}. So we have

$$\widetilde{\sigma}_i(w) = \mathrm{red}(x_i^{-1}\widetilde{\sigma}_i(w_1)x_i x_{i+1}^e x_i^{-1}\widetilde{\sigma}_i(w_2)x_i^{-1}),$$

and the hypothesis is $x_{i+1}^e x_i^{-1}\widetilde{\sigma}_i(w_2) = \varepsilon$. As previously, this gives $\widetilde{\sigma}_i(w_2) = x_i x_{i+1}^{-e} = \widetilde{\sigma}_i(x_i^{-e}x_{i+1})$, and, therefore, $w_2 = x_i^{-e}x_{i+1}$, again a contradiction. ■

Lemma 4.6. *For each i, $\widetilde{\sigma}_i^{-1}$ maps $S(x_k^e)$ into itself for $k \neq i, i+1$ and $e = \pm 1$; it maps $S(x_i)$ to $S(x_{i+1})$, $S(x_i^{-1})$ to $S(x_i) \cup S(x_i^{-1}) \cup S(x_{i+1})$, and both $S(x_{i+1})$ and $S(x_{i+1}^{-1})$ to $S(x_{i+1})$.*

Proof. As the sets $S(x_i^{\pm})$ form a partition of $\mathrm{FG}_\infty \setminus \{1\}$, the results follow from the previous lemmas: it suffices to consider all possible cases and to study where an element can come from. ■

Proposition 4.7. **(trichotomy)** *Let b be an arbitrary braid.*

(i) The braid b is σ_1-positive if and only if the word $\widetilde{b}(x_1)$ finishes with x_1^{-1};

(ii) The braid b is σ_1-neutral if and only if the word $\widetilde{b}(x_1)$ is x_1;

(iii) The braid b is σ_1-negative if and only if the word $\widetilde{b}(x_1)$ finishes with $x_k^{\pm 1}$ for some k with $k \geq 2$.

Proof. As every braid is either σ_1-positive, or σ_1-neutral, or σ_1-negative, and the three conditions of the proposition each exclude one another, it is sufficient to prove that the conditions are necessary. By construction, $\widetilde{b}(x_1) = x_1$ holds if b admits a decomposition where $\sigma_1^{\pm 1}$ does not occur, *i.e.*, if b is σ_1-neutral. If b is σ_1-positive, Lemma I.3.18 gives $\widetilde{b}(x_1) \in S(x_1^{-1})$. Assume finally that b is σ_1-negative. Thus b can be written $b_1\sigma_1^{-1}\mathrm{sh}(b_2)$, where b_1 admits a decomposition where σ_1^{-1} does not occur. Then $\widetilde{\mathrm{sh}(b_2)}$ maps x_1 to itself, hence $(\sigma_1^{-1}\mathrm{sh}(b_2))^\sim$ maps x_1 to $x_2^{-1}x_1x_2$, an element of $S(x_2)$, hence of $\bigcup_{k\geq 2} S(x_k^{\pm 1})$. Then $\widetilde{\sigma}_1$ and all $\widetilde{\sigma}_k^{\pm 1}$ with $k \geq 2$ map $\bigcup_{k\geq 2} S(x_k^{\pm 1})$ into itself: indeed, $\widetilde{\sigma}_1$ is the only automorphism in the considered family that possibly maps an element of $S(x_k^{\pm 1})$ with $k \geq 2$ to an element of $S(x_1^{\pm 1})$. ■

We obtain in this way a new comparison algorithm with exponential complexity, which gives an alternative proof of Proposition 3.1 (complexity): in order to recognize whether $1 <_L \overline{w}$ holds, we compute the reduced word $\widetilde{w}(x_1)$: if it finishes with x_1^{-1}, $1 <_L \overline{w}$ holds; if it finishes with $x_k^{\pm}$ for some $k \geq 2$, $1 >_L \overline{w}$ holds. Otherwise, we know that $\widetilde{w}(x_1)$ must be x_1. In this case, we compute $\widetilde{w}(x_2)$: if the latter word finishes with x_2^{-1}, we conclude that $1 <_L \overline{w}$ holds; if it finishes with $x_k^{\pm}$ for some $k \geq 3$, we conclude that $1 >_L \overline{w}$ holds. Otherwise, $\widetilde{w}(x_2)$ must be x_2, and we continue similarly with $\widetilde{w}(x_3)$. By construction, the lengths of the words $\widetilde{w}(x_k)$ are bounded by $2^{\lg(w)}$, and they can be computed iteratively in a number of steps of the same order.

Diffeomorphisms of a disk

The linear order of braids can also be constructed using the realization of B_n as the mapping class group of a punctured disk.

There exists a natural interpretation of braids as homeomorphisms of a punctured disk of the plane. The idea is to look at the braids "from the top" instead of looking at them "from the front". In the sequel, D^2 denotes a disk in the real plane, and D_n denotes the punctured disk obtained by removing n points from the interior of D^2. To make things precise, we shall assume that our disk D^2 admits the segment $(0,0)$–$(n+1,0)$, denoted E, as a diameter, and that the n punctures are the points $(i,0)$ with $1 \leq i \leq n$. We denote by Homeo_n the set of all homeomorphisms of D_n that leave the boundary ∂D^2 pointwise fixed and permute the punctures. Two elements φ, φ' of Homeo_n are said to be isotopic if there exists a continuous mapping $I : D^2 \times [0,1] \to D^2$ such that $I(\cdot, t)$ belongs to Homeo_n for every t, and $I(\cdot, 0) = \varphi$ and $I(\cdot, 1) = \varphi'$ hold, *i.e.*, if there exists a continuous deformation of φ to φ'.

Definition 4.8. (mapping class group) The *mapping class group* of D_n is the group of all isotopy classes of homeomorphisms in Homeo_n.

Proposition 4.9. (mapping class group) *Assume $n \geq 2$. Then the mapping class group of D_n is (isomorphic to) B_n.*

Proof. (Sketch) Assume that β is a geometric n-strand braid. We associate with β a homeomorphism of D_n by defining a continuous mapping $I : D^2 \times [0,1] \to D^2$ such that $I(\cdot, 0)$ is the identity mapping of D_n, $I(\cdot, t)$ is a homeomorphism of D^2 that leaves ∂D^2 fixed for each t, and, for $1 \leq i \leq n$, the point $I((i,0), t)$ is the intersection of the i-th strand of β with the plane $z = t$ (Figure 4.1). Then the isotopy class of the restriction of $I(\cdot, 1)$ to D_n is well-defined and it depends on the isotopy class of β in B_n bijectively. ■

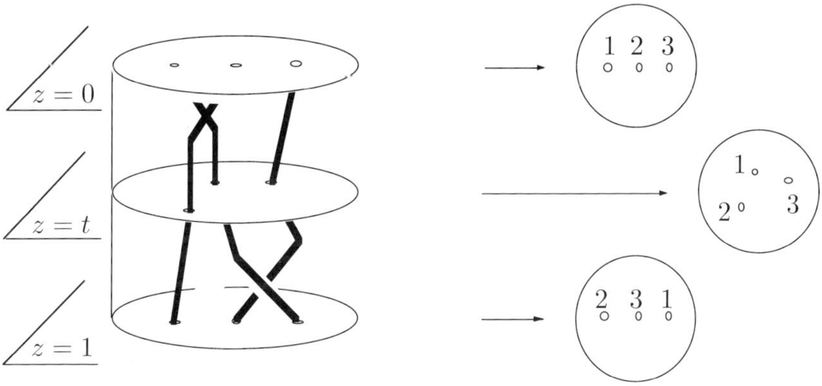

Figure 4.1. From braid to homeomorphism

Definition 4.10. (**curve diagram**) Assume that φ is an element of Homeo_n. The *curve diagram* of φ is the image of the diameter E under φ.

Example 4.11. (**curve diagram**) *By definition, the curve diagram of a homeomorphism contains the punctures. Here are some examples.*

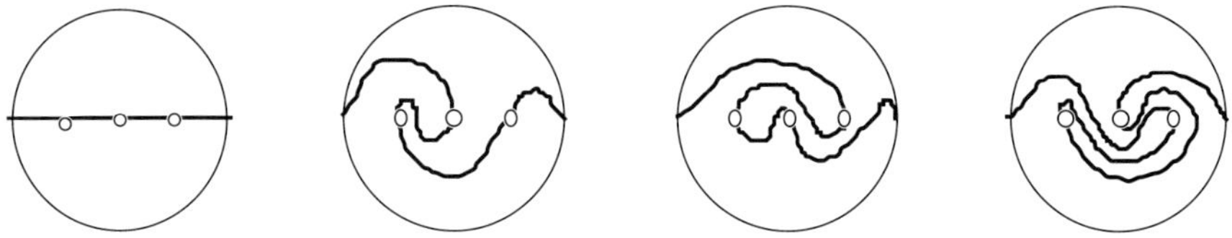

1 $\qquad$ σ_1 $\qquad$ $\sigma_1\sigma_2$ $\qquad$ $\sigma_1\sigma_2^{-1}$

Lemma 4.12. *The isotopy class of a homeomorphism φ is determined by the curve diagram of φ.*

We skip the easy verification. Isotopic homeomorphisms lead to isotopic curve diagrams, but not necessarily to the same one. The key point is that one can associate with every isotopy class of homeomorphisms, *i.e.*, with every braid, a standardized curve diagram that characterizes it. The principle of the standardization process is to minimize the cardinality of the intersection of the diameter E and its image: this can be done by pulling on the strands, and the result is well-defined and unique if we require that the diagram consist of semi-circles centered on E, contain no cusps except possibly at the punctures, and that the radii of the semi-circles are chosen so that the intersections with each segment of E are equally spaced.

Example 4.13. (**standardized diagram**) *The standardized diagrams associated with σ_1, $\sigma_1\sigma_2$, and $\sigma_1\sigma_2^{-1}$, i.e., the result of standardizing the last three curve diagrams in the previous example, are respectively:*

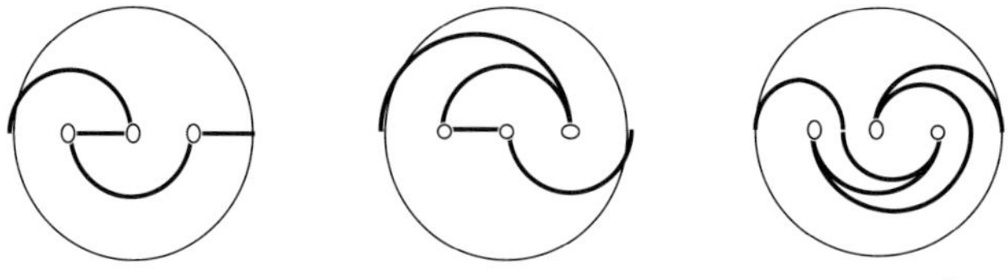

σ_1 $\qquad$ $\sigma_1\sigma_2$ $\qquad$ $\sigma_1\sigma_2^{-1}$

Proposition 4.14. (**standardization**) *The standardized curve diagrams associated with two geometric braids are equal if and only if the geometric braids are isotopic.*

We skip the proof. The main point is to prove that, if a standardized curve diagram is isotopic to E, then actually it is equal to E. We can therefore speak without ambiguity of *the* standard curve diagram associated with a braid. Once such a unique form is defined, we easily obtain a new characterization of the linear order of braids. For γ a curve diagram, we say in the sequel that γ lies *on the left* of E if γ first diverges from E in the upper half-plane.

Lemma 4.15. *Assume that b is an arbitrary braid in B_n. Then the braid b is σ-positive if and only if the standardized curve diagram associated with b lies on the left of the diameter E.*

Proof. Assume that b is σ-positive. Then there exists a σ_1-positive word w and an integer k such that b is represented by $\mathrm{sh}^k(w)$. Assume first $k = 0$. By hypothesis, w has the form

$$w = \mathrm{sh}(w_0)\,\sigma_1\,\mathrm{sh}(w_1)\,\sigma_1 \cdots .$$

Let us consider the construction of the standardized curve diagram associated with w. The first factor $\mathrm{sh}(w_0)$ leaves the first horizontal segment unchanged, while the first σ_1 then rotates it clockwise: hence, at this point, an initial segment of the first arc of the curve diagram lies in the half-plane $y > 0$.

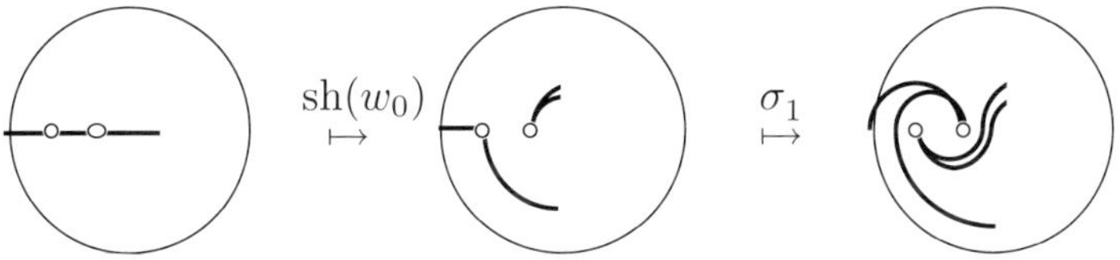

Applying a further $\sigma_i^{\pm 1}$ with $i \geq 2$ does not change the latter property, nor does applying a further σ_1 either, as this operation amounts to applying a locally clockwise rotation. The argument is the same if k is positive: the only change is that the first k horizontal segments of E remain fixed, and, therefore, the first semi-circle starts from the point $(k, 0)$ instead of from the point $(0, 0)$.

If b is σ-negative, a symmetric argument shows that an initial segment of the first non-horizontal arc in the curve diagram of b lies on the right of E. We are done: we know that exactly one of "b is σ-positive", "$b = 1$", "b is σ-negative" holds, and so does exactly one of "the curve diagram of b lies on the left of E", "the curve diagram of b is E", "the curve diagram of b lies on the right of E". Hence the above implication (i) $\Rightarrow$ (ii) is an equivalence. ■

Definition 4.16. **(order $<_R$)** For b_1, b_2 in B_∞, we say that $b_1 >_R b_2$ holds if the braid $b_1 b_2^{-1}$ is σ-positive.

The definition of $>_R$ is symmetric to that of the order $<_L$, and the connection is obvious:

Lemma 4.17. *The relation $>_R$ is a linear order on B_∞ that is compatible with multiplication on the right, and $b_1 >_R b_2$ is equivalent to $b_1^{-1} <_L b_2^{-1}$.*

Then there exists a reasonable notion of *pulling* a curve diagram γ_1 *tight with respect to* another curve diagram γ_2 so that the following holds:

Proposition 4.18. (**order**) *Assume that b_1, b_2 are braids in B_n. Then the inequality $b_1 >_R b_2$ holds if and only if any curve diagram of b_1, after being pulled tight with respect to the standardized curve diagram γ_2 of b_2, lies on the left of γ_2.*

Proof. (Sketch) Lemma 4.17 gives the result for $b_2 = 1$. Then the operation of pulling one diagram tight with respect to another one is defined so as to make the property of one curve diagram lying on the right of another one invariant under right multiplication. ■

Standardized curve diagrams can be encoded by words, and there exists an algorithm that inductively computes the code of the curve diagram of a braid word. Starting from the codes of two braids, it is easy to compare them using the previous result. One obtains in this way a new comparison algorithm for braids.

Similarly, starting with the standardized curve diagram of a braid b, and assuming for instance that this diagram lies on the left of the diameter E, it is possible to obtain a σ-positive expression for b by inductively "unraveling" the curve diagram until E is obtained. This gives a new proof of the fact that every nontrivial braid is either σ-positive or σ-negative. The method is effective, but it seems that its algorithmical complexity is higher than that of handle reduction.

Laminations

Instead of considering the image of the diameter E under the diffeomorphism associated with a braid b, we can look at the image of a family of circles L_0 starting from the origin and encircling 1, 2, ..., n punctures respectively. Such an image consists of n non-intersecting curves in the disk, and we shall call it a *lamination* (a formal intrinsic definition can be given, but we shall not need it here). Laminations can be encoded by counting their intersections with fixed lines orthogonal to the diameter E. For $i = 1, \ldots, n$, we denote by x_i and y_i the numbers of intersections of a lamination L with the half-lines $x = i, y > 0$, and $x = i, y < 0$ respectively, and by z_i be the number of intersections of L with the line $x = i + 1/2$. We encode L by the sequence $(a_1, b_1, a_2, b_2, \ldots, a_n, b_n)$ with $a_i = (x_i - y_i)/2$ and $b_i = (z_{i-1} - z_i)/2$. (In order to obtain a non-ambiguous coding, we must require x_i, y_i, and z_i to be minimal in the isotopy class of L.)

Example 4.19. (**laminations**) *The basic lamination L_0 corresponds to the values $x_i = y_i = n - i$, $z_i = 2(n+1)$, and it is therefore encoded by the sequence $(0, 1, 0, 1, \ldots)$. Below are represented the successive images of L_0 under σ_1^{-1}, σ_2, and σ_1. The top digits give the intersection numbers in the form $z_0, \frac{x_1}{y_1}, z_1, \ldots, \frac{x_n}{y_n}, z_n$; the bottom sequence is the code.*

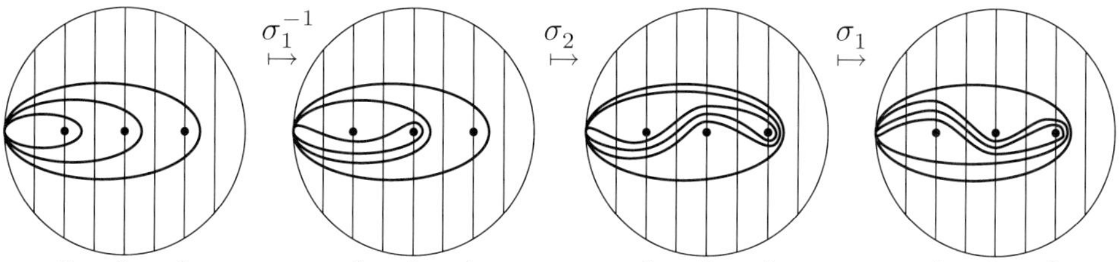

$6\ \frac{3}{3}\ 4\ \frac{2}{2}\ 2\ \frac{1}{1}\ 0$ — $(0, 1, 0, 1, 0, 1)$

$6\ \frac{2}{4}\ 6\ \frac{3}{3}\ 2\ \frac{1}{1}\ 0$ — $(-1, 0, 0, 2, 0, 1)$

$6\ \frac{2}{4}\ 6\ \frac{5}{1}\ 6\ \frac{3}{3}\ 0$ — $(-1, 0, 2, 0, 0, 3)$

$6\ \frac{4}{2}\ 6\ \frac{1}{5}\ 6\ \frac{3}{3}\ 0$ — $(1, 0, -2, 0, 0, 3)$

Proposition 4.20. (**Dynnikov's formulas**) *Assume that a lamination L is coded by $(a_i, b_i, \ldots, a_n, b_n)$. Then the image of L under σ_i (resp. σ_i^{-1}) is coded by $(a'_1, b'_1, \ldots, a'_n, b'_n)$, with $a'_k = a_k$, $b'_k = b_k$ for $k \neq i, i+1$, and, using x^+ for $\sup(x, 0)$, and x^- for $\inf(x, 0)$, and putting $c = a_i - b_i^- - a_{i+1} + b_{i+1}^+$,*

$$\begin{cases} a'_i = a_i + b_i^+ + (b_{i+1}^+ - c)^+ \\ b'_i = b_{i+1} - c^+ \end{cases} \qquad \begin{cases} a'_{i+1} = a_{i+1} + b_{i+1}^- + (b_i^- + c)^- \\ b'_{i+1} = b_i + c^+ \end{cases}$$

(resp., putting $d = a_i + b_i^- - a_{i+1} - b_{i+1}^+$,

$$\begin{cases} a'_i = a_i - b_i^+ - (b_{i+1}^+ + d)^+ \\ b'_i = b_{i+1} + d^- \end{cases} \qquad \begin{cases} a'_{i+1} = a_{i+1} - b_{i+1}^- - (b_i^- - d)^- \\ b'_{i+1} = b_i - d^- \end{cases} \quad).$$

The proof relies on studying how the intersection numbers of a normal surface with a triangulation are changed when a flip transformation is applied.

Dynnikov's formulas define a right action of B_n on $\mathbf{Z}^{2n}$. Note than, once the formulas are known, checking the compatibility with the braid relations is a tedious, but easy, verification, which amounts to checking that we have two binary operations on $\mathbf{Z}^2$ satisfying the identities of Exercise I.2.31. For $\vec{x} \in \mathbf{Z}^{2n}$, we denote by $\vec{x} * b$ the image of $\vec{x}$ under the action of b.

Lemma 4.21. *Assume that b is a σ_1-positive braid. Then the first entry in the sequence $(0, 1, 0, 1, \ldots) * b$ is positive.*

Proof. For $i \geq 2$, the action of $\sigma_i^{\pm 1}$ does not change the first two entries. Now $(a_1, b_1, a_2, \ldots) * \sigma_1$ begins with $a_1 + b_1^+ + (b_2 - c)^+$, which is positive both for

$a_1 = 0$, $b_1 = 1$, and for $a_1 > 0$. So the condition "$a_1 > 0$" is preserved whenever σ_1^{-1} is not applied. ■

Proposition 4.22. (characterization of order) *The action of B_n on $\mathbf{Z}^{2n}$ is faithful. For b in B_n, we have $b = 1$ (resp. $b >_L 1$, resp. $b <_L 1$) if and only if $(0, 1, \ldots, 0, 1) * b$ is equal to $(0, 1, \ldots, 0, 1)$ (resp. its first non-zero entry of odd index is positive, resp. is negative).*

Proof. By Lemma 4.21, if b is the k-th shift of a σ_1-positive braid, then the first k entries of odd index in $(0, 1, 0, 1, \ldots) * b$ are zero, and the next one is positive. A symmetric computation shows that, if b is the k-th shift of a σ_1-negative braid, then the first k entries of odd index in $(0, 1, 0, 1, \ldots) * b$ are zero, and the next one is negative. As every braid not equal to 1 is either σ-positive or σ-negative, the odd entries in $(0, 1, 0, 1, \ldots) * b$ can be zero only for $b = 1$. ■

For instance, Example 4.19 gives $(0, 1, 0, 1, \ldots) * \sigma_1^{-1}\sigma_2 = (-1, 0, 2, 0, 0, 3)$ and $(0, 1, 0, 1, \ldots) * \sigma_1^{-1}\sigma_2\sigma_1 = (1, 0, -2, 0, 0, 3)$: $\sigma_1^{-1}\sigma_2$ is σ-negative, while $\sigma_1^{-1}\sigma_2\sigma_1$ is σ-positive.

Proposition 4.23. (complexity) *The action of Proposition 4.20 solves the word problem of the standard presentation of B_∞ and decides the order $<_L$ in quadratic time: for every braid word w, we decide whether $\overline{w} = 1$, $\overline{w} >_L 1$, or $\overline{w} <_L 1$ holds in $C \cdot \lg(w)^2$ steps, where C is some constant not depending on the number of strands.*

Proof. Computing the action of $\sigma_i^{\pm 1}$ on a sequence of integers of total length ℓ requires $\mathcal{O}(\ell)$ steps, and it ends with a new sequence of length $\ell + \mathcal{O}(1)$. So, for each braid word w, computing the sequence $(0, 1, 0, 1, \ldots) * w$ requires $\mathcal{O}(\lg(w)^2)$ steps. ■

Hyperbolic geometry

We conclude with still another definition of the braid order. This characterization involves an action of the braids on the real line, which derives an order on braids—actually, an infinite family of orders—from the order of the reals.

Proposition 4.24. (action) *There exists an action of B_n on the real line consisting of order preserving homeomorphisms, and the action is faithful on an uncountable Borel set U (actually a dense G_δ), i.e., for every x in U, $b_1 \neq b_2$ implies $b_1(x) \neq b_2(x)$.*

Proof. (Sketch) The principle for defining the action of B_n, again identified with the mapping class group of the punctured disk D_n, consists in considering a universal covering $\widetilde{D_n}$ of D_n in the hyperbolic plane H_2. The action associated with a homeomorphism φ of D_n is obtained by introducing a distinguished lifting $\widetilde{\varphi}$ of φ, namely the one that fixes an arbitrary chosen basepoint on the boundary of $\widetilde{D_n}$, and looking at the action of $\widetilde{\varphi}$ on the boundary of $\widetilde{D_n}$: if we attach to $\widetilde{D_n}$ its limit points on the circle at infinity, we obtain a space which is homeomorphic to a closed disk, and we look at the action induced on the boundary of this disk. By construction, the action on the basepoint is trivial, so we obtain an action on the boundary circle of $\widetilde{D_n}$ with one point removed, hence on the real line. The problem now is to prove that, for many points, the action is free. One can show that this is the case at least for those points in $\partial\widetilde{D_n}$ that are associated with a geodesic γ in D_n such that every path component of $D_n \setminus \gamma$ contains at most one puncture. ■

We then obtain, for each x in U, a linear ordering of the braids by defining

$$b_1 <_x b_2 \quad \Longleftrightarrow \quad b_1(x) < b_2(x), \tag{4.2}$$

and, by construction, this order is compatible with multiplication on the left as the action of braids is order preserving: $b_1(x) < b_2(x)$ implies $b(b_1(x)) < b(b_2(x))$ for every b.

Thus, we have defined a linear order $<_x$ on B_n for every point x in U. The particular linear order $<_L$ considered in this chapter corresponds (for instance) to the orbit associated with a geodesic of D_n that encircles each puncture once.

5. Notes

The linear order of braids and the result that every nontrivial braid admits either a σ-positive or a σ-negative expression appear in (Dehornoy 92*c*) and (Dehornoy 94*a*). The current exposition puts the emphasis on special braids and braid exponentiation, and it is simpler than the original one, where the braid order is constructed as a projection of a certain preorder on the group G_{LD} of Chapter VIII.

The result that every n-strand braid admits a σ-positive or a σ-negative expression with no more than n strands was first proved by R. Laver (unpublished) using results about free LD-systems. A more effective proof was then proposed in (Larue 94*a*), using a topological interpretation reminiscent of the approach of Section 4. The algorithmical complexity of Larue's construction seems to be essentially exponential.

Handle reduction is described in (Dehornoy 97*b*). Several variants are introduced there. In particular, one can use more general σ_i-handles of the form $\sigma_i^e w \sigma_i^{-e}$ where we only assume that σ_i and σ_{i-1} do not occur in w, but we allow the generators σ_k with $k \le i-2$. The principle of the convergence proof remains the same, yet some details are more complicated.

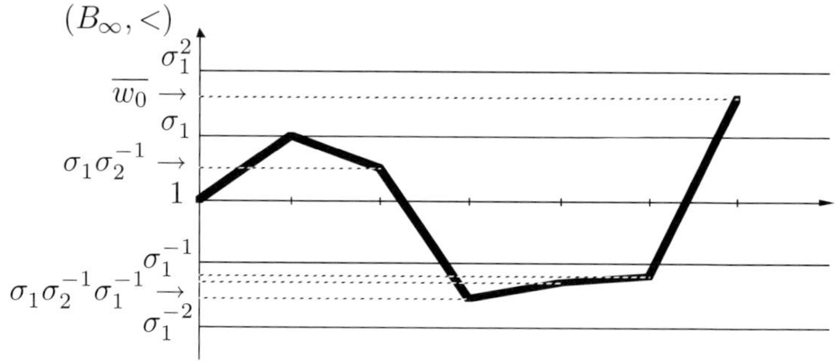

Figure 5.1. The characteristic graph of $w_0 = \sigma_1\sigma_2^{-1}\sigma_1^{-1}\sigma_2\sigma_3\sigma_1$

Handle reduction is deeply connected with the linear ordering of braids, and a valuable intuition can be given using characteristic graphs of the following type: the braids equipped with $<_L$ are order isomorphic to the rationals, and, therefore, we can associate with every braid word w of length ℓ some characteristic graph displaying the position of the successive prefixes of w in the braid order, on the shape of Figure 5.1. With this representation, reducing handles in a braid word w amounts to smoothing the characteristic graph by erasing peaks or filling gaps.

Handle reduction is very efficient in practice, especially when one uses a divide-and-conquer strategy consisting in iteratively cutting the considered word into two shorter words w_1, w_2 of nearly equal length, reducing w_1 and w_2 separately to w_1', w_2', and then reducing $w_1'w_2'$. In this way, reducing random braids involving several thousands of crossings on any number of strands typically requires less than one second on a computer. Experiments show that the average number of braid relations needed for deciding whether a random braid word is trivial using handle reduction is by several orders smaller than when using left–right reversing as in Chapter II (less than 10^2 relations *vs.* more than 10^4 for 10-strand braid words of length 100). Thus the gap between the practical efficiency and the complexity bounds of Section 3 is large.

Conjecture 5.1. *Assume that w is a braid word of length ℓ and width n. Then the number of possible handle reduction steps from w is bounded above by some quadratic function of ℓ (and of n?).*

A better estimate of the height of a braid word could improve the results of Section 3. Another approach could be to study handle reduction starting with words of a particular form, for instance of the form uv^{-1} with u, v positive. B. Wiest has obtained preliminary results, but the problem remains open.

Other open questions involve the length of the words obtained by handle reduction: to date, the only known bound is exponential.

Conjecture 5.2. *Assume that w is an n-strand braid word, and w' is a freely reduced word obtained from w by handle reduction. Then we have* $\lg(w') \leq n(n-1)/2 \cdot \lg(w)$.

By Lemma 3.18, this statement follows from the more general conjecture:

Conjecture 5.3. *Assume that w is an n-strand braid word, and w' is a freely reduced braid word obtained from w using right and left reversing, and positive and negative equivalence. Then we have* $\lg(w') \leq n(n-1)/2 \cdot \lg(w)$.

(We say that a word is freely reduced if it contains no pair $\sigma_i\sigma_i^{-1}$ or $\sigma_i^{-1}\sigma_i$.) The point that makes the previous conjecture plausible is that the hypotheses discard any possibility of creating a pair $\sigma_i\sigma_i^{-1}$ or $\sigma_i^{-1}\sigma_i$ *ex nihilo*. Presumably, Conjecture 5.3 is not specific of braids, and a similar property is likely to hold for every group of fractions of an atomic lcm monoid admitting a Garside element, at the expense of replacing the length with the norm.

Nothing is proved about the average complexity of handle reduction in B_n, in particular about the way it depends on n.

The problem of determining the shortest expression of a given braid is difficult (Paterson & Razborov 91). Handle reduction does not solve the question. However, a short expression for a braid b can be obtained by first reducing an arbitrary word w representing b to w' using handle reduction, and then reducing w' to w'' using a flipped handle reduction consisting in pushing the right handles to the left, *i.e.*, applying usual handle reduction to the image of w' under a flip, and flipping the result.

A possibly related question is to find, for a given σ_1-positive braid b, an expression of b with the minimal number of σ_1's (and no σ_1^{-1}). Again, handle reduction need not give the optimal solution. Trying to minimize the number of letters σ_1 leads to the following conjecture (of which one direction is trivial):

Conjecture 5.4. *The braid $\sigma_1\mathrm{sh}(b)\sigma_1$ admits a σ_1-positive expression with exactly one σ_1 if and only if the braid b does.*

Since every braid admits a σ-positive or a σ-negative expression, we are led to investigate the images of σ-positive braid words under linear representations. In the case of the Burau representation (Exercise 2.24), which is known not to be faithful (Moody 91), starting with the counterexamples of (Long & Paton 93), one can find braid words with five σ_1's and a trivial Burau image. On the other hand, 7-strand braid words with three σ_1's, and 4-strand braid words with four σ_1's have a nontrivial Burau image (Dehornoy 96*a*). Nothing seems to be known for the other linear representations.

The characterization of the braid order in terms of automorphisms of a free group is an easy application of the ideas used in (Larue 94).

The reconstruction of the braid order by means of diffeomorphisms of a punctured disk and curve diagrams appears in (Fenn, Greene, Rolfsen, Rourke & Wiest 99). (Rourke & Wiest *98*) extends the technique to all mapping class groups of surfaces with a nonempty boundary.

The approach involving laminations is due to I. Dynnikov (personal communication). It is connected with work on the Teichmüller space. The formulas of Proposition 4.20 are remarkable. Observe that using them gives an extremely simple proof for the acyclicity of the braid ordering—and, therefore, of left division in free LD-systems: Lemma 4.21 is similar to Proposition I.3.14 (automorphisms acyclic), but much more straightforward. Let us mention that $\mathbf{Z}, +, 0, \sup$ in Proposition 4.20 could be replaced with $\mathbf{Q}$ (or $\mathbf{C}$), $*, 1, +$ respectively (then x^+ becomes $x+1$ and x^- becomes $1/(x+1)$).

The characterization of the braid order in terms of hyperbolic geometry relies on a suggestion by W. Thurston; the existence of an action of the braid group on the circle S^1 can be traced back to (Nielsen 27). The current exposition follows (Short & Wiest *99*); see also (Kassel 99).

Define the *pure braid group* PB_n to be the kernel of the projection of B_n onto the symmetric group. The question of whether PB_n is orderable turns out to be quite different from the question for B_n. Indeed, PB_n is a semi-direct product of free groups—see (Birman 75). Using the Magnus representation of the free group FG_n into $\mathbf{Z}[[X_1, \ldots, X_n]]$ that maps x_i to $1 - X_i$ and x_i^{-1} to $1 + X_i + X_i^2 + \cdots$, one deduces a linear order on FG_n from the lexicographical order on formal power series. It turns out that the automorphisms involved in the semi-direct decomposition of PB_n are compatible with the previous order, and, from there, one can define a linear order on PB_n which is compatible both with left and right multiplication (Kim & Rolfsen *98*). This linear order cannot be extended into a linear order of B_n compatible with multiplication on one side, and, more generally, the same negative result holds for any linear order on PB_n compatible with multiplication on both sides for $n \geq 5$ (Rhemtulla & Rolfsen *99*).

Well-known generalizations of the braid groups B_n are Artin groups of Example II.1.3—and, more generally, all groups associated with a coherent complement. For each such a group, and each generator x, we can ask whether an element admitting an expression where x occurs and x^{-1} does not can be equal to 1, and whether every element admits at least one expression where not both x and x^{-1} occur. Artin groups of type B embed in Artin group of type A, so the answers to the above questions are the same as for type A. The first open case is type D_4: in this case, the symmetry of the three external generators could result in the failure of any order property.

IV
The Order on Positive Braids

We investigate now the restriction of the linear braid order of Chapter III to positive braids, *i.e.*, to those braids that admit an expression where no letter σ_i^{-1} occurs. The main result is Laver's theorem that the restriction of $<_L$ to B_n^+ is a well ordering. Here we follow Burckel's approach, which associates with every positive braid a normal form consisting in a finite tree; the order of positive braids is then a lexicographical ordering for the associated trees, and one deduces that the order type of $(B_n^+, <_L)$ is the ordinal $\omega^{\omega^{n-2}}$.

As the general construction is intricate, we first consider the special case of 3-strand braids in Section 1, and give a complete proof in this case. Here, things are very simple, as the involved trees can be identified with sequences of integers. In Section 2, we describe the general case, which is similar yet combinatorially more complicated. In Section 3, we give some applications of the properties of the order on positive braids, in particular, we prove that the linear order $<_L$ extends the subword order of positive words, and we present a conjecture of Laver about the action of braids on braids. In an appendix, we give the first elements in the well ordering on B_4^+, together with the associated trees and their ordinal rank.

1. The Case of Three Strands

Here we study the restriction of the linear order $<_L$ to B_3^+. We are going to prove that $(B_3^+, <_L)$ is isomorphic to sequences of integers ordered by the "ShortLex" ordering of sequences which compares the lengths first and, in case of equality, resorts to the lexicographical ordering.

Coding braid words by sequences of integers

We recall that BW_n^+ denotes the set of all positive n-strand braid words. The general idea is to decompose a positive braid word involving n different generators into a product of braid words each of which involves at most $n-1$ generators. Then, assuming that we have defined an ordering on the latter words, we shall define an ordering on words with n generators by using a lexicographical extension of the order on words with $n-1$ generators. In the particular case of 3-strand braid words, only two generators are used, namely σ_1 and σ_2, and the decomposition consists in parsing the word into an alternating sequence of blocks of σ_1's and σ_2's. Such a decomposition is unique if we require the last word to be a block of σ_2's and every block, except possibly the last one, to be nonempty.

Definition 1.1. (**code**) Assume $u \in BW_3^+$. The *code* of u is defined to be the unique sequence of positive integers $(e_1, \ldots, e_p)$ satisfying $u = \sigma_1^{e_1}\sigma_2^{e_2} \cdots \sigma_1^{e_{p-1}}\sigma_2^{e_p - 1}$ for p even, and $u = \sigma_2^{e_1}\sigma_1^{e_2} \cdots \sigma_1^{e_{p-1}}\sigma_2^{e_p - 1}$ for p odd. The integer p is called the *breadth* of u.

For instance, the code of $\sigma_2^2\sigma_1\sigma_2^3$ is $(2, 1, 4)$, and its breadth is 3, while the code of $\sigma_1^2\sigma_2\sigma_1^3$ is $(2, 1, 3, 1)$, and its breadth is 4. It is clear that the code of a word determines that word completely, and that, conversely, each finite sequence of positive integers is the code of a unique positive 3-strand braid word.

We introduce now our main tool, namely a linear ordering of braid words. For X an arbitrary set, we denote by X^* the set of all finite sequences on X, and, for $\vec{x}$ in X^*, we denote by $\lg(\vec{x})$ the length of $\vec{x}$.

Definition 1.2. (**ShortLex extension**) (i) Assume that $(X, <)$ is an ordered set. We denote by $<^{ShortLex}$ the order of X^* such that $\vec{x} <^{ShortLex} \vec{y}$ holds if and only if either $\lg(\vec{x}) < \lg(\vec{y})$ holds, or we have $\lg(\vec{x}) = \lg(\vec{y})$ and $\vec{x}$ precedes $\vec{y}$ with respect to the lexicographical extension of $<$, *i.e.*, there exists an integer i such that $x_i < y_i$ holds, and $x_k = y_k$ holds for $k < i$.

(ii) For u, v in BW_3^+, we say that $u \prec v$ holds if the code of u precedes the code of v with respect to the order $<^{ShortLex}$ deduced from the standard order of the integers.

The ShortLex-extension of a linear order is a linear order, hence, by construction, $\prec$ is a linear order on BW_3^+.

Example 1.3. (**order**) *Let p and q be the breadths of u and v respectively. Then $u \prec v$ holds if we have either $p < q$, or $p = q$ and, for some i, the $i-1$ first entries in the code of u and v coincide, but the i-th entry in the code of u is smaller than the i-th entry in the code of v.*

For instance, we have $\sigma_2^2\sigma_1\sigma_2^3 \prec \sigma_1^2\sigma_2\sigma_1^3$, which corresponds to $(2,1,4) <^{ShortLex} (2,1,3,1)$: here the breadth of the first word is smaller than the breadth of the second one. On the other hand, we have $\sigma_2^2\sigma_1\sigma_2^2 \prec \sigma_2^2\sigma_1\sigma_2^3$, which corresponds to $(2,1,3) <^{ShortLex} (2,1,4)$: here, the breadths are the same, the first two components of the parsings coincide, but not the third one.

Although easy, the following well ordering property will be crucial in the sequel. We refer to Chapter XII for a quick introduction to ordinals.

Proposition 1.4. (**well ordering**) *(i) The linear order $\prec$ on BW_3^+ is compatible with multiplication on the left, and it includes the subword order: $u \prec \sigma_i u$ and $u \prec u\sigma_i$ hold for every u and every i. Every word u admits an immediate successor with respect to $\prec$, namely $u\sigma_2$.*

(ii) The linear order $\prec$ on BW_3^+ is a well ordering of type ω^ω. For every u in BW_3^+, the rank of u in $(BW_3^+, \prec)$, i.e., the order type of the set of all predecessors of u, is

$$\omega^{p-1} \cdot e_1 + \omega^{p-2} \cdot (e_2 - 1) + \cdots + \omega \cdot (e_{p-1} - 1) + (e_p - 1), \tag{1.1}$$

where $(e_1, \ldots, e_p)$ is the code of u.

Proof. (i) Assume $u \prec v$. We wish to show that $\sigma_i u \prec \sigma_i v$ holds for $i = 1, 2$. Let $(e_1, \ldots, e_p)$ be the code of u, and $(e'_1, \ldots, e'_q)$ be the code of v. By hypothesis, we have

$$(e_1, \ldots, e_p) <^{ShortLex} (e'_1, \ldots, e'_q). \tag{1.2}$$

If both p and q are even, we have seen above that the codes of $\sigma_1 u$ and $\sigma_1 v$ are respectively $(e_1 + 1, e_2, \ldots, e_p)$ and $(e'_1 + 1, e'_2, \ldots, e'_q)$, and it is clear that (1.2) implies

$$(e_1 + 1, e_2, \ldots, e_p) <^{ShortLex} (e'_1 + 1, e'_2, \ldots, e'_q),$$

i.e., $\sigma_1 u \prec \sigma_1 v$. Similarly, the codes of $\sigma_2 u$ and $\sigma_2 v$ are respectively $(1, e_1, \ldots, e_p)$ and $(1, e'_1, \ldots, e'_q)$, and (1.2) implies also

$$(1, e_1, \ldots, e_p) <^{ShortLex} (1, e'_1, \ldots, e'_q),$$

which implies $\sigma_2 u \prec \sigma_2 v$. The argument is symmetric if both p and q are odd. Assume now that p and q do not have the same parity. Necessarily $p < q$ holds. The inequalities we have to establish are now

$$(e_1 + 1, e_2, \dots, e_p) <^{ShortLex} (1, e'_1, e'_2, \dots, e'_q),$$
$$(1, e_1, e_2, \dots, e_p) <^{ShortLex} (e'_1 + 1, e'_2, \dots, e'_q).$$

The first follows from $p < q$. The second holds because $q \neq 0$ implies $e'_1 \geq 1$.

The relation $u \prec \sigma_i u$ holds for all u and i if and only if the inequalities

$$(e_1, e_2, \dots, e_p) <^{ShortLex} (1, e_1, e_2, \dots, e_p),$$
$$(e_1, e_2, \dots, e_p) <^{ShortLex} (e_1 + 1, e_2, \dots, e_p)$$

are true, which follows from the definition of $<^{ShortLex}$.

Finally, every sequence $(e_1, \dots, e_p)$ admits $(e_1, \dots, e_p + 1)$ as an immediate successor with respect to $<^{ShortLex}$. If $(e_1, \dots, e_p)$ is the code of u, then $(e_1, \dots, e_p + 1)$ is the code of $u\sigma_2$. Hence $u\sigma_2$ is the immediate successor of u with respect to $\prec$.

(ii) For each p, the lexicographical order on the length p sequences of positive integers is a well ordering of type ω^p. Hence, by definition of $<^{ShortLex}$, the set of all codes is the disjoint sum of a well ordering of type ω, followed by a well ordering of type ω^2, followed by a well ordering of type ω^3, *etc.* Therefore, the codes ordered by $<^{ShortLex}$ and, equivalently, the words in BW_3^+ ordered by $\prec$, make a well ordering of type $\omega + \omega^2 + \omega^3 + \cdots$, *i.e.*, of type ω^ω.

Assume that u admits the code $(e_1, \dots, e_p)$. The order type of the sequences of length at most $p-1$ is ω^{p-1}, so the rank of u in BW_3^+ is $\omega^{p-1} + \alpha$, where α is the rank of the sequence $(e_1, \dots, e_p)$ among all possible sequences of length exactly p ordered lexicographically. As we consider only sequences with positive entries, α is equal to

$$\omega^{p-1} \cdot (e_1 - 1) + \omega^{p-2} \cdot (e_2 - 1) + \cdots + \omega \cdot (e_{p-1} - 1) + (e_p - 1),$$

which gives (1.1). ∎

Example 1.5. **(rank)** *The rank of the word $\sigma_2^2 \sigma_1 \sigma_2^3$ is the ordinal $\omega^2 \cdot 2 + 3$, as its code is the sequence $(2, 1, 4)$, while the rank of $\sigma_1^2 \sigma_2 \sigma_1^3$ is $\omega^3 \cdot 2 + \omega \cdot 2$, as its code is $(2, 1, 3, 1)$. By definition of rank, $u \prec v$ holds if and only if the rank of u is less than the rank of v.*

Observe that the well ordering $\prec$ is *not* compatible with multiplication on the right. For instance, we have $\sigma_2 \prec \sigma_1$ and $\sigma_1^2 \prec \sigma_2 \sigma_1$, since $(2) <^{ShortLex} (1, 1)$ and $(2, 1) <^{ShortLex} (1, 1, 1)$ hold.

Reducible and irreducible braid words

As the relation $\prec$ is a linear ordering on words, it cannot be compatible with braid word equivalence: for instance, we have $\sigma_2\sigma_1\sigma_2 \prec \sigma_1\sigma_2\sigma_1$ as the respective codes are $(1,1,2)$ and $(1,1,1,1)$. In the sequel, the idea is to consider for each positive braid b the $\prec$-minimal word that represents b. The main result is that there exists an algorithmic process, called reduction here, that leads from any braid word representing b to its $\prec$-minimal expression, and that, for such minimal expressions, the linear order $<_L$ corresponds to the word order $\prec$. More precisely, we shall introduce a convenient notion of irreducible braid word and prove (we recall that, for u a braid word, $\overline{u}$ denotes the braid represented by u):

Proposition 1.6. (**Burckel's normal form**) *(i) Every word in BW_3^+ is equivalent to exactly one irreducible word, namely the $\prec$-minimal element of its equivalent class;*

(ii) If u and v are irreducible words in BW_3^+, then $\overline{u} <_L \overline{v}$ holds in B_3^+ if and only if $u \prec v$ holds in BW_3^+.

Let us first introduce the reduction process explicitly.

Definition 1.7. (**reduction**) Let u be a positive 3-strand braid word, and $(e_1, \ldots, e_p)$ be its code. We say that u is *reducible* if $e_k = 1$ holds for at least one index k satisfying $2 \le k \le p-2$. In this case, assuming that k is maximal with the previous property, we define $\mathrm{red}(u)$ to be the word with code

$$(e_1, \ldots, e_{k-2}, e_{k-1}-1, e_{k+1}, 1, e_{k+2}+1, e_{k+3}), \quad \text{for } e_{k-1} \ge 2, \tag{1.3}$$
$$(e_1, \ldots, e_{k-2}+e_{k+1}, 1, e_{k+2}+1, e_{k+3}, \ldots, e_p), \quad \text{for } e_{k-1} = 1. \tag{1.4}$$

Observe that (1.3) and (1.4) are not really different: for $e_{k-1} = 1$, (1.3) gives the sequence $(e_1, \ldots, e_{k-2}, 0, e_{k+1}, 1, e_{k+2}+1, e_{k+3}, \ldots, e_p)$, which is not a code, but becomes (1.4) when we gather the two blocks that border the empty block.

Example 1.8. (**reduction**) *Applying reduction consists in replacing the rightmost pattern of the form $(p,1,q,r)$ in the code with $(p-1,q,1,r+1)$, and contracting for $p=0$. Let $u = \sigma_1\sigma_2\sigma_1\sigma_2^2\sigma_1$. Its code is $(1,1,1,2,1,1)$, and u is reducible, as the second and the third entries in the code are 1. For computing $\mathrm{red}(u)$, we consider the rightmost reducible pattern in the code of u, here $(1,1,2,1)$, which we replace with $(0,2,1,2)$, thus obtaining $(1,0,2,1,2,1)$, which contracts into $(3,1,2,1)$. Thus $\mathrm{red}(u)$ is the word $\sigma_1^3\sigma_2\sigma_1^2$.*

The latter word is reducible as well, since the second entry in its code is 1. Reduction leads to the code $(2,2,1,2)$, i.e., the word $\mathrm{red}^2(u)$ is

$\sigma_1^2\sigma_2^2\sigma_1\sigma_2$. *The latter word is irreducible, i.e., no reduction may be applied to it.*

By definition, the word u is reducible if and only if it contains the pattern $\sigma_1\sigma_2\sigma_1$, or a pattern of the form $\sigma_2\sigma_1\sigma_2^e\sigma_1$ with $e \geq 1$. The technical interest of reduction appears in the following result.

Lemma 1.9. *Assume that u is a reducible positive 3-strand braid word. Then we have* $\mathrm{red}(u) \equiv u$ *and* $\mathrm{red}(u) \prec u$.

Proof. Constructing $\mathrm{red}(u)$ from u consists in replacing in u a subword of the form $\sigma_i^p\sigma_j\sigma_i^q\sigma_j^r$, with $p, q > 0$ and $\{i, j\} = \{1, 2\}$, by the word $\sigma_i^{p-1}\sigma_j^q\sigma_i\sigma_j^{r+1}$. In order to prove $u \equiv \mathrm{red}(u)$, it suffices that we prove that the latter patterns are equivalent. Now we have

$$\begin{aligned}\sigma_i^p\sigma_j\sigma_i^q\sigma_j^r &= \sigma_i^{p-1}\sigma_i\sigma_j\sigma_i\sigma_i^{q-1}\sigma_j^r \equiv \sigma_i^{p-1}\sigma_j\sigma_i\sigma_j\sigma_i^{q-1}\sigma_j^r\\ &\equiv \sigma_i^{p-1}\sigma_j^2\sigma_i\sigma_j\sigma_i^{q-2}\sigma_j^r \equiv \ldots \equiv \sigma_i^{p-1}\sigma_j^q\sigma_i\sigma_j\sigma_j^r = \sigma_i^{p-1}\sigma_j^q\sigma_i\sigma_j^{r+1}.\end{aligned}$$

It remains to prove $\mathrm{red}(u) \prec u$. We keep the notation of Definition 1.7. In the special case $k = 2$, $e_{k-1} = 1$, the breadth of $\mathrm{red}(u)$ is strictly less than the breadth of u, so the result is true. Otherwise, the breadths are equal, the $k-2$ first entries of the codes coincide, and e_{k-1} in u is replaced with $e_{k-1} - 1$ in $\mathrm{red}(u)$. So $u \prec \mathrm{red}(u)$ holds. ∎

Lemma 1.10. *Assume that u is a reducible positive 3-strand braid word. Then there exists an integer m such that the word* $\mathrm{red}^m(u)$ *is irreducible.*

Proof. By Lemma 1.9, the sequence of words u, $\mathrm{red}(u)$, $\mathrm{red}^2(u)$, ... is $\prec$-decreasing. By Proposition 1.4 (well ordering), $(BW_3^+, \prec)$ is a well ordered set, so it contains no infinite descending chain. Hence the previous sequence must be finite, *i.e.*, there exists a finite integer m such that $\mathrm{red}^m(u)$ is irreducible. ∎

Definition 1.11. (iterated reduction) For u in BW_3^+, the word $\mathrm{red}^*(u)$ is defined to be the unique irreducible word of the form $\mathrm{red}^m(u)$.

For instance, we have seen in Example 1.8 that, if u is the word $\sigma_1\sigma_2\sigma_1\sigma_2^2\sigma_1$, then $\mathrm{red}^2(u)$ is irreducible. So $\mathrm{red}^*(u)$ is defined to be the latter word.

The connection between $\prec$ and $<_L$

The key point of the construction is the result that, if u and v are irreducible words, then $u \prec v$ holds if and only if the inequality $\overline{u} <_L \overline{v}$ holds in the monoid B_3^+. In other words, the linear order of positive braids is induced by the order $\prec$ on irreducible words, *i.e.*, by the ShortLex ordering on finite sequences of positive integers. The main technical point for proving the above result is the following statement:

Lemma 1.12. *Assume that u, v are 3-strand positive braid words, $u \prec v$ holds, and v is irreducible. Then $\overline{u} <_L \overline{v}$ holds.*

We prove the result using ordinal induction on the rank $\mathrm{rk}(v)$ of v in the well ordered set $(BW_3^+, \prec)$. For ρ an ordinal, we denote by (S_ρ) the statement:

If v is irreducible and $\mathrm{rk}(v) \leq \rho$ holds, then $u \prec v$ implies $\overline{u} <_L \overline{v}$.

Our aim is to prove (S_ρ) for every ordinal ρ with $\rho < \omega^\omega$. First, (S_0) is vacuously true, as, if v has rank 0, then v is the empty word, and $u \prec v$ never holds. So the point is to prove that, if (S_ξ) holds for $\xi < \rho$, which we shall denote by $(S_{<\rho})$, then (S_ρ) holds as well. Let us assume that v is irreducible with rank ρ, and that $u \prec v$ holds. We wish to prove $\overline{u} <_L \overline{v}$. It is not easy to establish the result directly, for we have few hypotheses about u. The idea will be to consider only some special words u below v for which the inequality $\overline{u} <_L \overline{v}$ can be proved easily, and to show that this is sufficient to deduce the inequality in the general case. In the current case of 3-strand braid words, the task will be easy.

Definition 1.13. (**companion**) Assume $u, v \in BW_3^+$, and $p > 0$. Let q be the breadth of v. We say that u is a *p-companion of v* if u has breadth $q - 1$ and its code begins with p.

By definition, if u is a companion of v, then $u \prec v$ holds. The converse need not be true, but we can show that the companions of v are almost cofinal in v, in the sense that every predecessor of v is a predecessor of some companion of v, or it has a special form.

Lemma 1.14. *Assume that u, v are 3-strand positive braid words and $u \prec v$ holds. Then at least one of the following holds:*
(i) We have $u = \sigma_i u_0$, $v = \sigma_i v_0$ and $u_0 \prec v_0$ for some i, u_0, v_0;
(ii) For p large enough, we have $u \prec u' \prec v$ for every p-companion u' of v.

Proof. Assume that u has breadth p, and v has breadth q. If $p = q$ holds, necessarily v begins with the same letter as u, namely σ_1 for p odd, and σ_2 for

p even. So we can write $u = \sigma_i u_0$, $v = \sigma_i v_0$. As $v_0 \prec u_0$ would imply $v \prec u$, $u_0 \prec v_0$ holds.

Assume now $p \le q-1$. For $p \le q-2$, then $u \prec u'$ holds for every companion u' of v. For $p = q-1$, $u \prec u'$ holds whenever u' is a p'-companion of v with $p' > e$, where e is the first entry in the code of u. ■

We arrive at the key point of the argument.

Lemma 1.15. *Assume that v is an irreducible 3-strand positive braid word with breadth at least 2. Then, for every p, there exists a p-companion u of v that is irreducible and satisfies $\overline{u} <_L \overline{v}$.*

Proof. Let $(e_1, \dots, e_q)$ be the code of v. We define u to be the word with code $(p, 2, 2, \dots, 2, 1, 0)$ and breadth $q-1$. By definition, u is a p-companion of v, and it is irreducible. We prove $\overline{u} <_L \overline{v}$ by considering the possible values of q.

Assume first $q = 2$, *i.e.*, $v = \sigma_1^{e_1}\sigma_2^{e_2-1}$. Then u is σ_2^p, and, $\overline{u} <_L \overline{v}$ holds since $u^{-1}v$ is a σ_1-positive braid word.

Assume now $q = 3$. We have $v = \sigma_2^{e_1}\sigma_1^{e_2}\sigma_2^{e_3-1}$, and $u = \sigma_1^p$. Then $\overline{u} <_L \overline{v}$ holds as we have

$$u^{-1}v = \sigma_1^{-p}\sigma_2^{e_1}\sigma_1^{e_2}\sigma_2^{e_3-1} \equiv \sigma_2\sigma_1\sigma_2^{-(p-1)}\sigma_1^{e_1-1}\sigma_2^{-1}\sigma_1^{e_2-1}\sigma_2^{e_3-1},$$

and the latter word is σ_1-positive.

Assume next $q = 4$. Then we have $v = \sigma_1^{e_1}\sigma_2^{e_2}\sigma_1^{e_3}\sigma_2^{e_4-1}$ and $u = \sigma_2^p\sigma_1$. Then $\overline{u} <_L \overline{v}$ holds as we have

$$u^{-1}v = \sigma_1^{-1}\sigma_2^{-p}\sigma_1^{e_1}\sigma_2^{e_2}\sigma_1^{e_3}\sigma_2^{e_4-1} \equiv \sigma_2\sigma_1^{-p}\sigma_2^{-1}\sigma_1^{e_1-1}\sigma_2^{e_2}\sigma_1^{e_3}\sigma_2^{e_4-1},$$

and the latter braid word is equivalent to the σ_1-positive word

$$\sigma_2^2\sigma_1\sigma_2^{-(p-2)}\sigma_1^{e_1-1}\sigma_2^{-1}\sigma_1^{e_2-2}\sigma_2^{-1}\sigma_1^{e_3-1}\sigma_2^{e_4-1}.$$

Assume $q \ge 5$ with q odd. Write $v = \sigma_2 v_0$, and $u = \sigma_1^p\sigma_2^2 u_0$. Then we have

$$\sigma_1\sigma_2^{e+2}u_0 \prec v_0 \prec v.$$

Indeed, either v begins with at least two σ_2's, and v_0 has breadth q while $\sigma_1\sigma_2^{p+2}u_0$ has breadth $q-1$, or v begins with only one σ_2, and both $\sigma_1\sigma_2^{p+2}u_0$ and v_0 have breadth $q-1$, but the first entry in the code of $\sigma_1\sigma_2^{p+2}u_0$ is 1, while the first entry in the code of v_0, which is e_2, is at least 2 by hypothesis. Now, v_0 is irreducible, hence the induction hypothesis gives in B_3^+ the inequality $\sigma_1\sigma_2^{p+2}\overline{u_0} <_L \overline{v_0}$, and, therefore, multiplying on the left by σ_2,

$$\sigma_2\sigma_1\sigma_2^{p+2}\overline{u_0} <_L \overline{v}. \tag{1.5}$$

Now $\sigma_2\sigma_1\sigma_2^{p+2}u_0 \equiv \sigma_1^p\sigma_2\sigma_1\sigma_2^2u_0$ holds. Consider the word $\sigma_1\sigma_2^2u_0$: it is irreducible by construction, and $\sigma_1\sigma_2^2u_0 \prec v$ holds. Now $\sigma_2u_0 \prec \sigma_1\sigma_2^2u_0$ holds, so the induction hypothesis gives $\sigma_2\overline{u_0} <_L \sigma_1\sigma_2^2\overline{u_0}$, and, therefore, using (1.5),

$$\overline{u} = \sigma_1^p\sigma_2^2\overline{u_0} <_L \sigma_1^p\sigma_2\sigma_1\sigma_2^2\overline{u_0} <_L \overline{v}.$$

The argument is similar when q is even, exchanging σ_1 and σ_2. ∎

We are now ready to complete the proof of Lemma 1.12.

Proof. (Lemma 1.12) Assume that $(S_{<\rho})$ is true, and v is an irreducible word with rank ρ. Assume $u \prec v$. By Lemma 1.14, two cases are possible. Assume first that $u = \sigma_i u_0$ and $v = \sigma_i v_0$ hold for some i, u_0, v_0. Then, by definition, v_0 is irreducible, and $v_0 \prec v$ holds. Hence, by induction hypothesis, we have $\overline{u_0} <_L \overline{v_0}$, and therefore

$$\overline{u} = \sigma_i\overline{u_0} <_L \sigma_i\overline{v_0} = \overline{v}$$

as the order $<_L$ is compatible with multiplication on the left. Assume now that $u \prec u'$ holds for every p-companion u' of v. By Lemma 1.15, there exists such a companion u' that is irreducible and satisfies $\overline{u'} <_L \overline{v}$. The induction hypothesis implies $\overline{u} <_L \overline{u'}$, as $u' \prec v$ holds. So we find $\overline{u} <_L \overline{u'} <_L \overline{v}$. Hence $\overline{u} <_L \overline{v}$ holds in every case, and the property (S_ρ) is true. ∎

It is now easy to deduce:

Lemma 1.16. *Assume that v is 3-strand positive braid word. Then the following are equivalent:*

(i) The word v is irreducible;

(ii) The word v is the $\prec$-least element in its $\equiv$-equivalence class.

Proof. Assume that v is irreducible. By Lemma 1.12, $\overline{u} <_L \overline{v}$, and, therefore, $u \not\equiv v$, holds for every word u that satisfies $u \prec v$. Hence v is the $\prec$-least element in its equivalence class. Conversely, assume that v is the $\prec$-least element in its equivalence class. If v is not irreducible, then, by Lemma 1.9, the word $\mathrm{red}(v)$ satisfies $\mathrm{red}(v) \prec v$ and $\mathrm{red}(v) \equiv v$, a contradiction. ∎

We have now completed the proof of Proposition 1.6 (Burckel's normal form): Point (i) is the previous lemma, and, then, Point (ii) follows from Lemma 1.12 directly.

With this result at hand, we can deduce the following fundamental property of the linear order $<_L$ on B_3^+:

Proposition 1.17. **(well ordering)** *The restriction of the braid order $<_L$ to B_3^+ is a well ordering of type ω^ω.*

Proof. Let BW_3^{irred} denote the set of all irreducible words in BW_3^+. As $\prec$ is a well ordering on BW_3^+, its restriction to BW_3^{irred} is also a well ordering, and the order type of $(BW_3^{\text{irred}}, \prec)$ is at most the type of $(BW_3^+, \prec)$, which is ω^ω. As some words in BW_3^+ are reducible, the previous inequality could be strict. This however does not happen. Indeed, the mapping $d : (e_1, \ldots, e_p) \mapsto (2e_1, \ldots, 2e_p)$ defines a $\prec$-increasing mapping of BW_3^+ into a subset of BW_3^{irred}. Hence the order type of the latter subset is ω^ω, and so is that of BW_3^{irred}. ■

Exercise 1.18. (Garside) *Compute the rank of Δ_3^k in $(B_3^+, <_L)$.*

---

2. The General Case

We extend now the previous constructions and results to the case of braids with an arbitrary number of strands. The principle remains the same. We associate with each braid word a code, which is now an iterated sequence of sequences of integers. Such codes can be described as trees in a natural way. They are still equipped with a ShortLex-ordering, now a well ordering of type ω^{ω^ω}, and we use this order to define a well ordering $\prec$ on braid words. Then, we introduce the notion of a reducible word, and define a reduction that maps every braid word that is not irreducible to a $\prec$-smaller equivalent braid word. The point again is to show, using an induction on the ordinal rank, that an irreducible word is the $\prec$-least word in its equivalence class and that the restriction of the order $\prec$ to irreducible words induces the linear order $<_L$ of the braids. As above, we shall deduce that the restriction of $<_L$ to positive n-strand braids is a well ordering for each n.

Coding positive braid words by trees

We iterate the coding of 3-strand braid words by sequences of integers to any number of strands, thus using sequences of sequences of integers for 4-strand braid words, sequences of sequences of sequences of integers for 5-strand braid words, *etc.* Such iterated sequences naturally appear as finite trees.

Definition 2.1. (uniform tree) An *n-uniform tree* is defined to be a positive integer for $n = 2$, and to be a finite sequence of $(n-1)$-uniform trees for $n \geq 3$.

We associate a graph with every uniform tree as follows: for $t = (t_1, \ldots, t_p)$, the graph of t consists of a single root connected with the p graphs of t_1, ...,

t_p enumerated from left to right; if t is the integer e, the graph of t is a root with e leaves below it. We use in the sequel the standard vocabulary for trees, thus speaking of root, leaves, nodes, successors of a node (here successor always means immediate successor), *etc.* Observe that every branch in an n-uniform tree, defined as a path from the root to a leaf, has length $n-1$, *i.e.*, it contains n nodes.

Example 2.2. (**tree**) *A typical* 4*-uniform tree is* $((1,1),(2),(1,2,2))$. *The associated graph is*

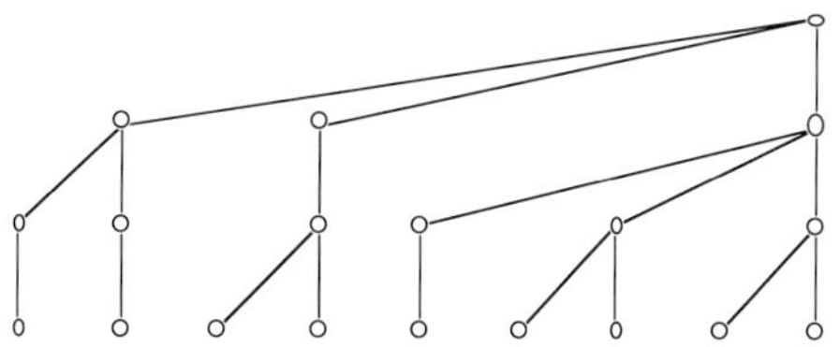

In order to construct a coding of braid words by uniform trees, we first define the braid word associated with a uniform tree, *i.e.*, we begin with the decoding process.

Definition 2.3. (**decoding**) Assume that t is an n-uniform tree. We attribute *labels* to the inner nodes of t as follows. First, we attribute the label $(1, n-1)$ to the root. Then, if the label (i,j) with $i < j$ has been attributed to a node, we alternatively attribute the labels $(i+1, j)$ and $(j-1, i)$ to the successors of this node starting from the right. Symmetrically, if the label (i,j) with $i > j$ has been attributed to a node, we alternatively attribute the labels $(i-1, j)$ and $(j+1, i)$ to the successors of this node starting from the right. Finally, we attribute to each leaf the label σ_i where (i,i) is the label of its predecessor, except for the rightmost leaf, which receives the label ε. The word *coded by* t is the word obtained by concatenating the labels of the leaves of t enumerated from left to right.

Example 2.4. (**decoding**) *Labelling the tree considered above gives*

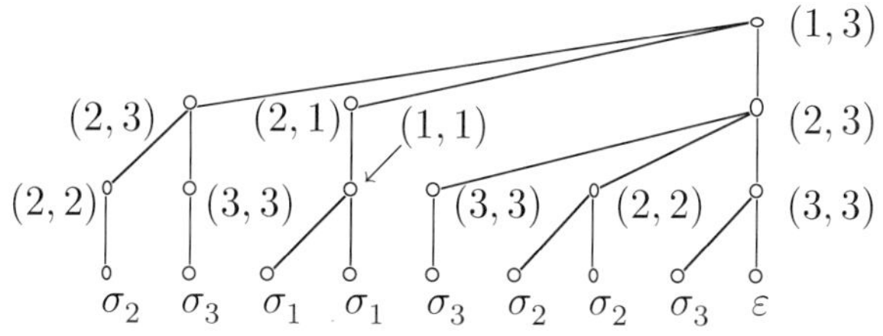

Thus the word coded by this tree is $\sigma_2\sigma_3\sigma_1^2\sigma_3\sigma_2^2\sigma_3$.

Lemma 2.5. *Every positive n-strand braid word is coded by a unique n-uniform tree, i.e., coding establishes a bijective correspondence between n-uniform trees and positive n-strand braid words.*

Proof. We describe a coding process that associates with each braid word u in BW_n^+ an n-uniform tree t, and prove that u is coded by t and t is the unique tree that codes u. The argument is inductive, and consists in defining the code of $\sigma_i u$ from the code of u.

First, we define the code of the empty braid word to be the tree $(\ldots(1)\ldots)$, $n-2$ pairs of parentheses. The graph associated with this tree consists of a single branch, the label of its unique leaf is ε by definition, the word coded is ε, and ε is coded by no other tree. Assume now that the code t of u has been constructed, and let i be an integer between 1 and $n-1$. Let t' be the tree obtained by adding to t a new leaf on the left, and a branch that connects this leaf to the lowest left node N in t whose label is an interval that contains i. Such a node exists, as the label of the root of t is $(1, n-1)$ by definition. We claim that the word coded by t' is $\sigma_i u$. Indeed, the hypothesis that we take the lowest possible left node N means that the label of the left successor of N in t does not contain i. This implies that the label of the left successor of N in t' has the form (j, i) for some j, and, therefore, the label of the leftmost leaf in t' is σ_i. Now we claim that t' is the only tree coding for $\sigma_i u$. Indeed, if t'' codes for $\sigma_i u$, then the tree obtained from t'' by removing the leftmost branch codes u, so, by induction hypothesis, it is equal to t. So t'' is obtained from t by adding one new branch. Let N' be the node of t where this additional branch is grafted. By construction, the label of N' must contain i, so N' cannot be below N. On the other hand, assume that N' lies strictly above N. Then i is one end of the label of N' and it belongs to the label of its left successor in t'', so, by construction, it cannot belong to the label of the second successor of N' in t'', which is the left successor of N' in t. Hence i cannot belong to the label of N in t, a contradiction. ∎

Definition 2.6. (code) For u in BW_n^+, the *n-code* of u is defined to be the unique n-uniform tree t such that u is coded by t.

Example 2.7. (coding) *Let* u *be the word* $\sigma_2\sigma_3\sigma_1^2\sigma_3\sigma_2^2\sigma_3$. *We have seen above that the* 4*-code of* u *is the* 4*-uniform tree* $((1,1),(2),(1,2,2))$. *As* u *is an element of* BW_n^+ *for every* n *above* 4*, it admits also an* n*-code for all such values of* n. *An induction shows that this* n*-code is obtained by adding* $n-4$ *pairs of parentheses around the* 4*-code of* u*, i.e., by adding a single branch of length* $n-4$ *above the root of the* 4*-code of* u.

Let us look now for the code of $\sigma_i u$. This code is constructed by adding a new branch to the tree t that codes u. For $i = 1$, the lowest left node of t whose label contains 1 is the root of t. So the coding of $\sigma_1 u$ is

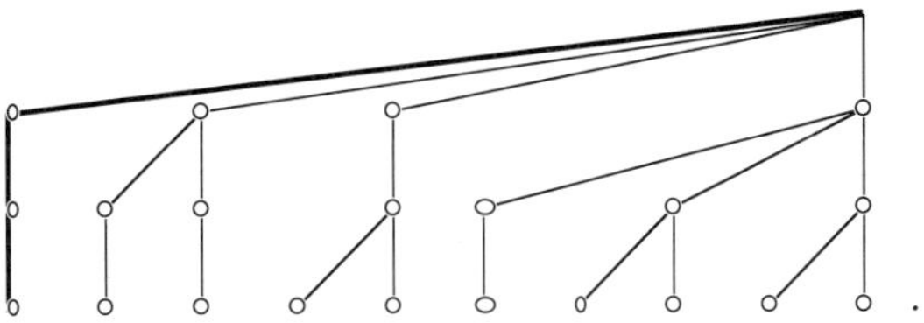

For $i = 2$, the lowest left node of t whose label contains 2 lies on level 2 from the root (just above the leaves). So the coding of $\sigma_2 u$ is

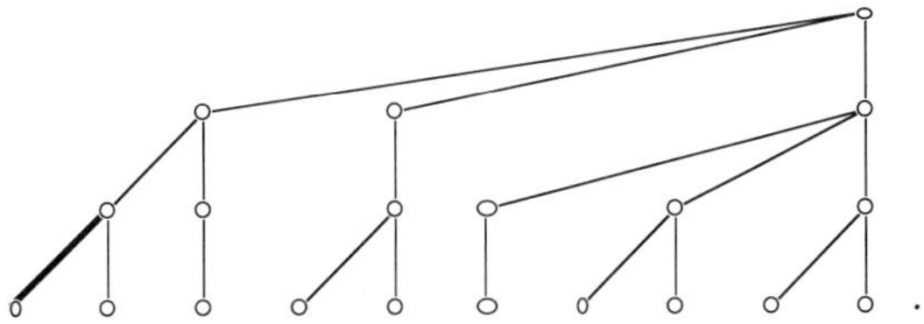

Finally, for $i = 3$, the lowest left node of t whose label contains 3 lies on level 1 from the root. So the coding of $\sigma_3 u$ is

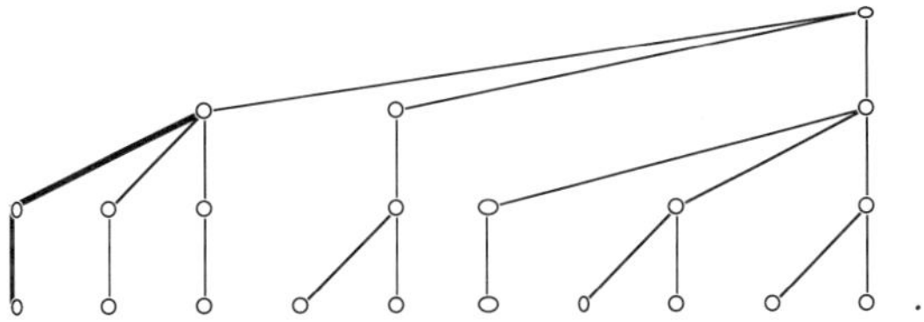

Observe that, by construction, adding the $n - 1$ possible generators to a word in BW_n^+ corresponds to adding a left branch in the $n - 1$ possible ways to a uniform tree of height n.

In the case of 3-strand braid words, the current coding by trees coincides with our previous coding by sequences of positive integers. Coding a braid word by a tree amounts to gathering the letters into blocks of close letters, inductively: for instance, the bracketed expression of the word $\sigma_2\sigma_3\sigma_1^2\sigma_3\sigma_2^2\sigma_3$ is

$$(((\sigma_2)(\sigma_3))((\sigma_1\sigma_1))((\sigma_3)(\sigma_2\sigma_2)(\sigma_3))).$$

In particular, there exists a main decomposition that corresponds to isolating the successors of the root, for instance

$$(\sigma_2\sigma_3)(\sigma_1\sigma_1)(\sigma_3\sigma_2\sigma_2\sigma_3)$$

in the previous case.

We introduce now an order on the trees, and, from there, on braid words.

Definition 2.8. (order) (i) Assume that t and t' are n-uniform trees. For $n = 2$, we say that $t <^{ShortLex} t'$ holds if $t < t'$ holds (in this case, t and t' are positive integers). Otherwise, assume $t = (t_1, \dots, t_p)$ and $t' = (t'_1, \dots t'_{p'})$ where $t_1, \dots, t'_{p'}$ are $(n-1)$-uniform trees. We say that $t <^{ShortLex} t'$ holds if either $p < p'$ holds, or there exists i such that $t_k = t'_k$ holds for $k < i$ and $t_i <^{ShortLex} t'_i$ holds.

(ii) Assume that u and v are positive n-strand braid words. We say that $u \prec v$ holds if the code of u is $<^{ShortLex}$-smaller than the code of v.

Observe that, in the case of 3-strand braid words, the relations defined above coincide with those defined in Section 1.

Example 2.9. *(order) Let us consider the words $u = \sigma_2\sigma_3\sigma_1^2\sigma_3\sigma_2^2\sigma_3$ and $v = \sigma_2\sigma_3\sigma_2\sigma_1$. The codes of u and v are*

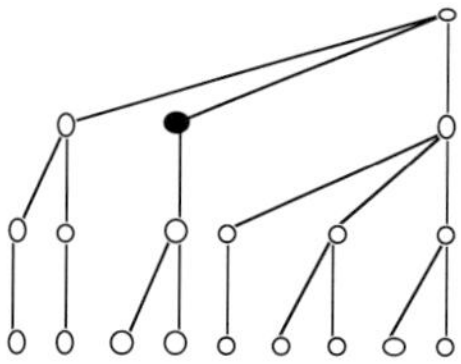 *and* 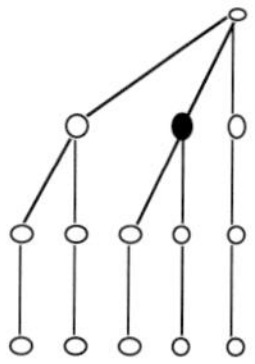 *respectively.*

Then $u \prec v$ holds (although the tree of v appears as thinner than the tree of u): in both cases the roots have three successors, and the left successors of the root are equal, but the second successor of the root in u has only one successor, while the one in v has two successors (the corresponding nodes appear in black).

Proposition 2.10. (well ordering) *(i) The linear order $\prec$ on BW_n^+ is a well ordering of type $\omega^{\omega^{n-2}}$. It is compatible with multiplication on the left, and it includes the subword order: $u \prec \sigma_i u$ and $u \prec u\sigma_i$ hold for every u and every i.*

(ii) For u in BW_n^+ coded by t, the rank $\mathrm{rk}(u)$ of u in $(BW_n^+, \prec)$ is $t-1$ for $n = 2$, and, otherwise, it is

$$\omega^{\omega^{n-3}\cdot(p-1)} \cdot (1 + \mathrm{rk}(u_1)) + \omega^{\omega^{n-3}\cdot(p-2)} \cdot \mathrm{rk}(u_2) + \cdots + \omega^{\omega^{n-3}} \cdot \mathrm{rk}(u_{p-1}) + \mathrm{rk}(u_p), \tag{2.1}$$

where t is $(t_1, \dots, t_p)$, and $u_1, \dots, u_p$ are the words of BW_{n-1}^+ coded by $t_1, \dots, t_p$ respectively.

Proof. The argument is the same as in the case of three strands. An induction on n shows that the order type of $(BW_n^+, \prec)$ is $\omega^{\omega^{n-2}}$, as the order on n-uniform trees is the ShortLex extension of the order on $(n-1)$-uniform trees. Similarly, Formula (2.1) asserts that the rank of u is the rank of $(u_1, \ldots, u_p)$ in the ShortLex extension of the order $\prec$ on BW_{n-1}^+. Indeed, the order type of BW_{n-1}^+ is $\omega^{\omega^{n-3}}$, hence there are $\omega^{\omega^{n-3}\cdot(p-1)}$ words in BW_n^+ with breadth at most $p-1$, and the rank of u is the latter ordinal augmented by the rank of u among all words of breadth p, which is the rank of $(u_1, \ldots, u_p)$ in the lexicographical extension of $\prec$ on BW_{n-1}^+. ∎

Example 2.11. **(rank)** *The rank of the word* $u = \sigma_2\sigma_3\sigma_1^2\sigma_3\sigma_2^2\sigma_3$ *(considered in the previous examples) is*

$$\omega^{\omega\cdot 2}\cdot(1+\mathrm{rk}(u_1)) + \omega^{\omega}\cdot\mathrm{rk}(u_2) + \mathrm{rk}(u_3),$$

where u_1, u_2, u_3 *are the 3-strand braid words with respective codes* $(1,1)$, (2), *and* $(1,2,2)$. *Applying the same argument to* u_1, u_2, *and* u_3, *which amounts to using Formula* (1.5), *we obtain* $\mathrm{rk}(u_1) = \omega$, $\mathrm{rk}(u_2) = 1$, *and* $\mathrm{rk}(u_3) = \omega^2 + \omega + 1$, *so, finally, the rank of* u *is*

$$\omega^{\omega\cdot 2}\cdot(1+\omega)) + \omega^{\omega}\cdot 1 + \omega^2 + \omega + 1 = \omega^{\omega\cdot 2+1} + \omega^{\omega} + \omega^2 + \omega + 1.$$

Observe that, because of the empty contribution of the rightmost branch in a tree and of the shifting of the letters, the word u *is not the product of the words* u_1, $\ldots$, u_3.

Reducible and irreducible braid words

As in the case of three strands, we introduce now a convenient notion of reduction. In order to state the definition, we need to introduce a few geometric notions connected with trees. First, we attribute to every node in a uniform tree an *address* that describes the path from the root to that node. Addresses are finite sequences of nonnegative integers.

Definition 2.12. **(address)** Assume that t is a uniform tree. The empty address is attributed to the root of t, and, assuming that the address α has been given to a node, we attribute the addresses $\alpha 0$, $\alpha 1$, ... to its successors enumerated from *right to left*.

So, for instance, 0^{n-1} is always the address of the rightmost leaf in an n-uniform tree. The reader can check that the addresses of the leaves in the tree of Example 2.2 respectively are 210, 200, 101, 100, 020, 011, 010, 001, 000, when enumerated from left to right.

Definition 2.13. **(neighbour)** Assume that t is a uniform tree, and N is a node of t that does not lie on the rightmost branch. Let $\alpha(p+1)0^m$ be the address of N in t. We define the *right neighbour* of N in t to be the node with address αp in t.

Thus the right neighbour of N is the closest node N_1 on the right of N: we reach N_1 from N by going up in the tree toward the root, and we go one step on the right as soon as possible, according to the figure below:

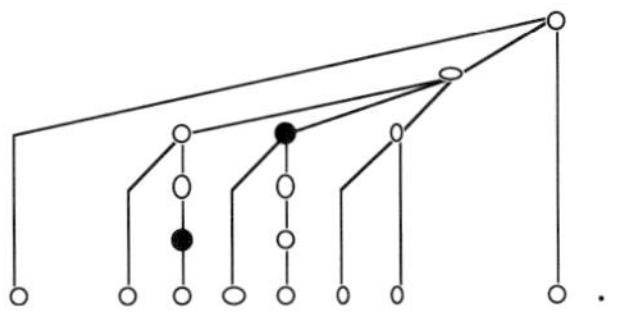

If N is a node in a uniform tree t, the word obtained by concatenating the labels of the leaves below N enumerated from left to right will be called the word *below* N. So, for instance, if t codes for u, then, by definition, u is the word below the root of t.

Definition 2.14. **(set $W_{i,j}$, $W^\bullet_{i,j}$)** Let i, j be positive integers. We denote by $W_{i,j}$ the set of all positive braid words that involve only letters from σ_i to σ_j; we denote by $W^\bullet_{i,j}$ the subset of $W_{i,j}$ consisting of all words in $W_{i,j}$ that finish with σ_j.

(With this notation, the set BW_n^+ of all positive n-strand braid words coincides with $W_{n+1,1}$ and $W_{1,n+1}$.) An induction shows that the braid word that lies below a node labelled (i,j) in a uniform tree belongs to $W_{i,j}$, and even to $W^\bullet_{i,j}$, except possibly in the case of a node on the rightmost branch: because the last leaf is labelled ε, we cannot be sure in this case that the considered word finishes with σ_j.

Definition 2.15. **(decomposition)** Assume that u is a positive braid word, and t is the code of u. The *main decomposition* of u is the sequence $(\sigma_i, u_1, u_2, \ldots)$, where σ_i is the first letter in u, *i.e.*, it is the word below the leftmost leaf in t, and, for every i, u_i is the word below N_i in t, where N_0 is the leftmost leaf in t and, for every i, N_{i+1} is the right neighbour of N_i in t.

Example 2.16. **(decomposition)** *Let us consider again the word $\sigma_2\sigma_3\sigma_1^2\sigma_3\sigma_2^2\sigma_3$. Starting from the left leaf in the associated tree, we find*

the successive right neighbours

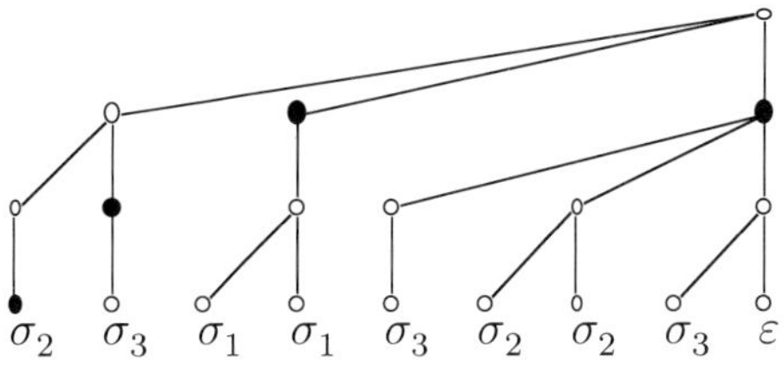

Hence the main decomposition of u is $(\sigma_2, \sigma_3, \sigma_1^2, \sigma_3\sigma_2^2\sigma_3)$.

By construction, the main decomposition of a word u is a decomposition of u in the sense that, if it is $(\sigma_i, u_1, \ldots, u_\ell)$, then $u = \sigma_i u_1 \cdots u_\ell$ holds. Observe that the length of the main decomposition of a word u is at least 3, unless u is empty, or u has the form $\sigma_i u_1$ with u_1 in $W_{i+1,n-1}$. Indeed, saying that the main decomposition has length 2 means that the right neighbour of the left leaf belongs to the right branch in the tree that codes u. This happens when there exists i such that the tree takes the form

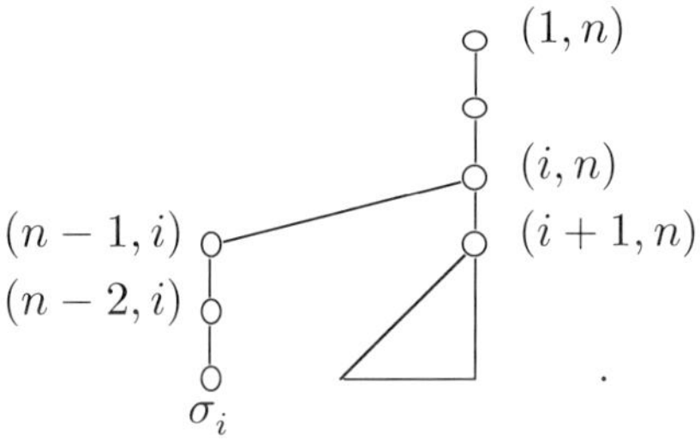

Lemma 2.17. *Assume that u is a positive n-strand braid word and the main decomposition of u begins with (σ_i, u_1, u_2). Then exactly one of the following cases occurs:*

(i) The word u begins with σ_i^2;

(ii) We have $u_1 \in W^\bullet_{i+1,j+1}$ and $u_2 \in W^\bullet_{j,i'}$ for some $j \geq i$ and $i' \leq i$:

(iii) We have $u_1 \in W^\bullet_{i-1,j-1}$ and $u_2 \in W_{j,i'}$ for some $j \leq i$ and $i' \geq i$; moreover, we have $u_2 \in W^\bullet_{j,i'}$ if $\sigma_i u_1 u_2$ is a proper prefix of u, i.e., if the main decomposition of u has length 4 at least.

Proof. Let t be the code of u, N_0 be the leftmost leaf in t, N_1 and N_2 be the right neighbour and second right neighbour of N_0 in t. If N_1 is a leaf, the label of this leaf must be σ_i, and we are in case (i). Otherwise, the label of the left "brother" of N_1, of whom N_0 is the only iterated successor, has the form (j, i) for some j. In this case, the label of N_1 is either $(i-1, j-1)$ or $(i+1, j+1)$, and, by hypothesis, i does not belong to the involved interval $(i-1, j-1)$ or $(i+1, j+1)$. So the only possibilities are $(i+1, j+1)$ for $j \geq i$, and $(i-1, j-1)$ for $j \leq i$. In the first case, the word u_1 involves letters between σ_{j+1} and σ_{i+1}, and it must finish with σ_{j+1}; in the second case, it involves letters between σ_{i-1} and σ_{j-1}, and it must finish with σ_{j-1}. A similar argument shows that the label of N_2 must contain i and j but not $j+1$ in the case $j \geq i$, so it has the form (j, i') or (i', j) for some $i' \leq i$, and, in this case, u_2 belongs to $\{\sigma_{i'}, \ldots, \sigma_j\}$, and it must finish with $\sigma_{i'}$. The argument is symmetric for $i \geq j$, except that, if there is no letter after u_2 in u, we cannot be sure that u_2 ends with $\sigma_{i'}$ because of the empty label of the last leaf. ∎

We are now ready to define reducible words and the associated reduction mapping. We recall from Chapter II that, for i, j positive integers, $\sigma_{i,j}$ denotes the braid word $\sigma_i \ldots \sigma_j$. Then, for every braid word u involving the generators $\sigma_i^{\pm 1}$, ..., $\sigma_{j-1}^{\pm 1}$ only, we have

$$\sigma_{i,j}\, u \equiv \mathrm{sh}^e(u)\, \sigma_{i,j} \tag{2.2}$$

with $e = 1$ for $i < j$, and $e = -1$ for $i > j$ (Lemma II.4.9) .

Definition 2.18. (reduction) Assume that u is a positive n-strand braid word. We say that u is *reducible* if there exists a decomposition $u_0 \sigma_i u_1 u_2 u_3$ of u such that (σ_i, u_1, u_2) is the beginning of the main decomposition of $\sigma_i u_1 u_2 u_3$ and one of the following occurs:

(i) The word u_1 contains neither σ_i nor σ_{i-1} nor σ_{i+1}; then we put

$$\mathrm{red}(u) = u_0 u_1 \sigma_i u_2 u_3;$$

(ii) The word u_1 is $\sigma_{i+1,j+1}$ for some $j \geq i$, and u_2 can be written as $u_2' u_2''$, with $u_2' = \varepsilon$ or $u_2' \in W_{i,j}^{\bullet}$ and u_2'' not containing σ_j; then we put

$$\mathrm{red}(u) = u_0 \mathrm{sh}(u_2') \sigma_{i,j} u_2'' \sigma_{j+1} u_3;$$

(iii) The word u_1 is $\sigma_{i-1,j-1}$ for some $j \leq i$, and u_2 can be written as $u_2' u_2''$, with $u_2' = \varepsilon$ or $u_2' \in W_{i,j}^{\bullet}$, u_2'' not containing σ_j, and $u_3 \neq \varepsilon$; then we put

$$\mathrm{red}(u) = u_0 \mathrm{sh}^{-1}(u_2') \sigma_{i,j} u_2'' \sigma_{j-1} u_3.$$

Before considering an example, let us immediately observe the compatibility of the current definition with that of Section 1.

Lemma 2.19. *For 3-strand braid words, the reduction of Definition 2.18 coincides with that considered in Section 1.*

Proof. Assume that u is a positive 3-strand braid word. Then u is reducible if it admits a suffix $\sigma_i u_1 u_2 u_3$ as in the above definition. Now case (i) is impossible as only σ_1 and σ_2 occur in u. Case (ii) occurs for $i = 1$ and $u_1 = \sigma_2$. Then u is reducible in the sense of Section 1, and the current word red(u) coincides with the one defined there. Indeed, for $u_2 = \sigma_1^e$, red(u) is obtained in both case by replacing $\sigma_1\sigma_2\sigma_1^e$ with $\sigma_2^e\sigma_1\sigma_2$. The argument is similar in case (iii), owing in addition to the fact that the condition $u_3 \neq \varepsilon$ corresponds to the additional requirement that some σ_1 lies on the right of the pattern $\sigma_2\sigma_1$ involved in the reduction. ■

Example 2.20. (reduction) *Let us consider $u = \sigma_2\sigma_3\sigma_1^2\sigma_3\sigma_2^2\sigma_3$, as in Example 2.16. In order to recognize whether u is reducible, we look at the main decompositions of the suffixes of u.*

First, the suffix $\sigma_3\sigma_2^2\sigma_3$ of u and its suffixes are irreducible, as they are the image under the shift mapping of 3-strand braid words that are irreducible in the sense of Section 1.

Next, the main decomposition of the suffix $\sigma_1\sigma_3\sigma_2^2\sigma_3$ of u is $(\sigma_1, \sigma_3\sigma_2^2\sigma_3)$. It has length 2 only, hence this suffix is irreducible.

Then, the main decomposition of the suffix $\sigma_1^2\sigma_3\sigma_2^2\sigma_3$ of u is $(\sigma_1, \sigma_1, \sigma_3\sigma_2^2\sigma_3)$, it has length 3, but no reduction can be applied.

Similarly, the main decomposition of the suffix $\sigma_3\sigma_1^2\sigma_3\sigma_2^2\sigma_3$ of u is $(\sigma_3, \sigma_1^2, \sigma_3\sigma_2^2\sigma_3)$, and no reduction can be applied.

So, it remains to consider u itself. The main decomposition begins with $(\sigma_2, \sigma_3, \sigma_1^2)$. With the notation of Definition 2.18, we have $i = 2$, $u_1 = \sigma_3$, so we are in case (iii) with $j = 2$. Now u_2 is $\varepsilon\sigma_1^2$, σ_1^2 does not contain σ_2, and u_3 is not empty. Hence the word u is reducible, and we have

$$\mathrm{red}(u) = \sigma_2\sigma_1^2\sigma_3^2\sigma_2^2\sigma_3,$$

with associated tree

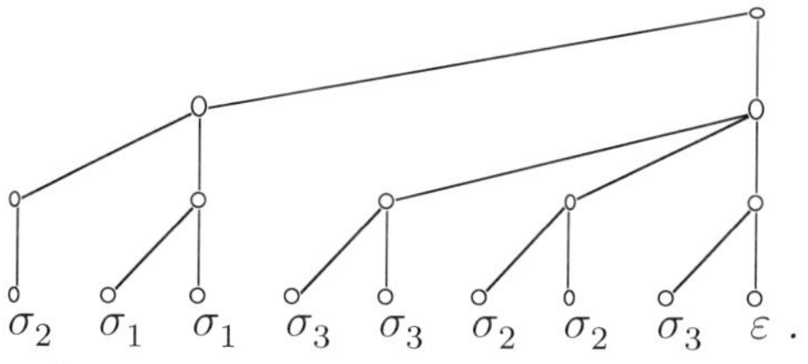

Reduction can then be iterated. As above, we look at the main decomposition of the suffixes of the word red(u)—which do not coincide with the suffixes of u, at least for some of them. In the current case, the reader can check that no suffix is reducible. In particular, the main decomposition of red(u) is $(\sigma_2, \sigma_1^2, \sigma_3\sigma_2^2\sigma_3)$, and no reduction can be applied. We conclude that the word red(u) is irreducible.

In the current framework, we still have the crucial result of Section 1 that applying reduction to a word u yields a new word that is equivalent to u and is $\prec$-smaller than u:

Lemma 2.21. *Assume that u is a reducible positive braid word. Then we have* $\mathrm{red}(u) \equiv u$ *and* $\mathrm{red}(u) \prec u$.

Proof. Equivalence follows from the definition. In case (i), we push the generator σ_i through a word u_1 which, by hypothesis, contains only generators that commute with σ_i. In case (ii), the equivalence $\sigma_{i,j-1}u_2' \equiv \mathrm{sh}^{-1}(u_2')\sigma_{i,j-1}$ is an occurrence of (2.2) as, by hypothesis, u_2' involves only letters among σ_j, ..., σ_i. Then σ_{j-1} commutes with u_2'' as we assume that u_2'' contains no σ_j (by construction, it contains no σ_{j-1} either). The argument is the same in case (iii).

As for the order, the latter is compatible with multiplication on the left, so it suffices that we show $\mathrm{red}(u) \prec u$ in the case where u itself (and not a proper suffix) is involved in the reduction. In case (i), the code t' of $\mathrm{red}(u)$ is obtained from the code t of u by transferring the leftmost branch to a position more on the right. So there exists some iterated left subtree s' of t' that is strictly less than the corresponding subtree in t. By definition of the ShortLex ordering, this implies $t' <^{ShortLex} t$, and, therefore, $\mathrm{red}(u) \prec u$ (Figure 2.1). In case (ii), we observe that $\mathrm{sh}(u_2') \prec \sigma_{i,j}$ holds by construction. This means that the tree associated with $\mathrm{red}(u)$ is obtained from the tree associated with u by diminishing some iterated left subtree. In order to conclude, it remains to see that the breadths do not increase: this is true because, by construction, the final character σ_{j+1} that remains after u_2'' can always be integrated in the tail u_3 without augmenting any breadth. The argument is symmetric in case (iii), and the hypothesis $u_3 \neq \varepsilon$ guarantees again that the letter σ_{j-1} can be integrated without augmenting the breadth (Figure 2.2). ■

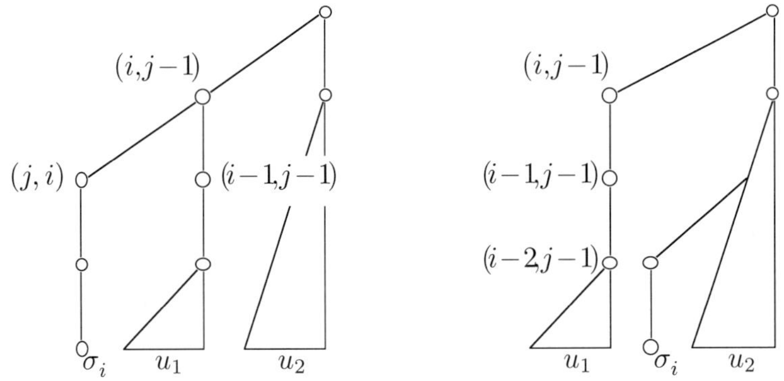

Figure 2.1. Reduction, case (i) with $j \leq i$

Reduction consists in pushing to the right a generator or a sequence of consecutive generators through a block it commutes with, possibly at the expense of shifting the indices. The problem is to recognize where we stop, and this can be decided only when the geometry of the tree associated with the word is

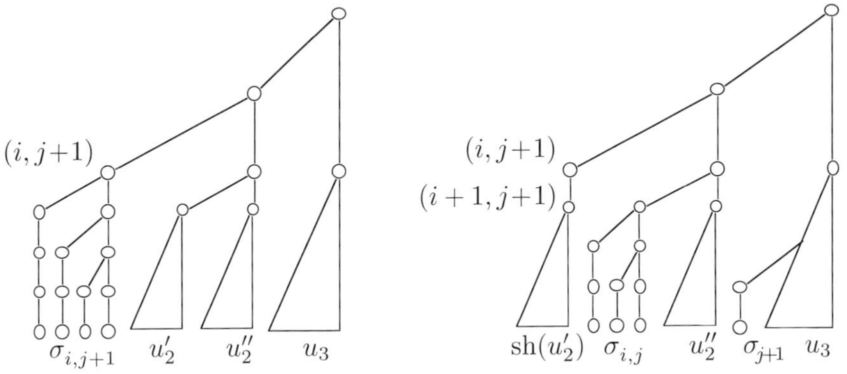

Figure 2.2. Reduction, case (ii)

known. In particular, in the case of a pattern $\sigma_1\sigma_3$, there is no uniform rule to decide whether the generator σ_1 has to be pushed through σ_3 or not: the word $\sigma_1\sigma_3$ is irreducible, but the word $\sigma_1\sigma_3\sigma_1$ is reducible to $\sigma_1^2\sigma_3$. This is why the definition of a reducible word does not state something like "assume that u has a subword of the form...", but involves the more complicated framework of the main decomposition.

As in the case of 3-strand braid words, the fact that the order $\prec$ is a well ordering implies that, for every braid word u, there exists a finite integer k such that the word $\mathrm{red}^k(u)$ is irreducible. The latter word will still be denoted by $\mathrm{red}^*(u)$.

The connection between $\prec$ and $<_L$

As in Section 1, the key of the construction is the result that, if u and v are irreducible words, then $u \prec v$ is equivalent to $\overline{u} <_L \overline{v}$. In order to prove the result, we follow the same scheme as above. So, we denote by (S_ρ) the statement

If v is irreducible and $\mathrm{rk}(v) \le \rho$ holds, then $u \prec v$ implies $\overline{u} <_L \overline{v}$,

and our aim is to prove (S_ρ) inductively for every ordinal ρ.

As before, we have to prove that $u \prec v$ implies $\overline{u} <_L \overline{v}$ when v is irreducible. To this end, we again introduce some special words, called the companions of v, for which the result is easier, and we show that almost every predecessor of v is a predecessor of some companion of v.

Definition 2.22. (companion) Assume that v is a nonempty positive n-strand braid word. Let $\alpha(q+1)0^m$ be the address of the leftmost leaf in the code of t. We say that u is a *p-companion* of v if the leftmost address in the code of u begins with αqp.

By definition, $u \prec v$ holds for every companion u of v: the main parts of the shapes of the codes of u and v coincide, but we removed the leftmost branch from u to v.

Lemma 2.23. *Assume that u, v are positive n-strand braid words and $u \prec v$ holds. Then at least one of the following holds:*

(i) We have $u = \sigma_i u_0$, $v = \sigma_i v_0$ and $u_0 \prec v_0$ for some i, u_0, v_0;

(ii) We have $v = \sigma_i^2 v_0$ and $u \prec \sigma_i v_0$ for some i and v_0;

(iii) The word v begins with σ_i^2 for no i, and, for p large enough, $u \prec u'$ holds for every p-companion u' of v.

Proof. Let α be the leftmost address in (the tree associated with) u, and β be that of v. The hypothesis $u \prec v$ implies that α cannot be bigger than β in the sense of the lexicographical order of addresses. We consider three cases. For $\alpha = \beta$, the first letter in u is the first letter in v, and we are in case (i). Assume now that α is smaller than β and β has the form $\beta'(q+1)$. Then the labels of $\beta'(q+1)$ and $\beta' q$ are the same, and v begins with twice the same letter, say $v = \sigma_i^2 v_0$. Then we have either $\sigma_i v \preceq u\sigma_i^2 v_0$, which implies that u begins with σ_i, and we are in case (i) again, or we have $u \prec \sigma_i v_0$, and we are in case (ii). Assume finally that α is smaller than β and β has the form $\beta'(q+1)0^m$ with $m \geq 1$. Then either α lies on the right of βq, and $u \prec u'$ holds for every companion of w, or α has a prefix of the form βqp, and $u \prec u'$ holds for every p'-companion u' of v with $p' > p$. Hence we are in case (iii). ∎

We arrive at the key result. The reader will observe that the property $(S_{<\rho})$ appears in the hypotheses of the lemma. This was not needed in the case of 3 strands, but this makes the proof of the auxiliary result easier here, and introduces no problem in the final argument.

Lemma 2.24. *Assume that v is an irreducible word in BW_n^+ with the following properties:*

(i) the main decomposition of v has length at least 3,

(ii) the word v begins with σ_i^2 for no i,

(iii) the property $(S_{<\rho})$ holds, where ρ is the rank of v in $(BW_n^+, \prec)$.

Then, for every p, there exists a $2p$-companion u of v that is irreducible and satisfies $\overline{u} <_L \overline{v}$.

Proof. Write $v = \sigma_i v_1 v_2 v_3$, where (σ_i, v_1, v_2) is the beginning of the main decomposition of v. Assume first $v_1 \in W^\bullet_{i+1,j+1}$ and $v_2 \in W^\bullet_{j,i'}$ with $i' \leq i$. Then we define $u = u_1^p u_2 v_3$, with

$$u_1 = \sigma_j^2 \sigma_{j+1}^2 \cdots \sigma_{i+1}^2 \sigma_{i+2}^2 \cdots \sigma_j^2 \sigma_{j+1}^2,$$
$$u_2 = (\sigma_i^2 \sigma_{i+1}^2 \cdots \sigma_j^2)\,(\sigma_{j-1}^2 \sigma_{j-2}^2 \cdots \sigma_{i'}^2).$$

We claim that u has the required properties. First the explicit definition of u shows that it is a $2p$-companion of v. Then u is irreducible, for v_3 is irreducible as v is, and, again, the explicit definition of u_1 and u_2 makes any reduction impossible in the left part of u. So the point is to prove $\overline{u} <_L \overline{v}$. Now, as v is irreducible, the word v_1 necessarily contains the letter σ_{i+1}, and, by Lemma 2.17 and by definition of an irreducible word, one of the following two cases occur:

(i) The word $\sigma_{i+1,j+1}$ is a proper factor of v_1;

(ii) We have $v_1 = \sigma_{i+1,j+1}$ and $v_2 \in W_{i,j}W_{i',j-1}$.

In both cases, we have $\sigma_{i+1,j+1}wu_2v_3 \prec v_1v_2v_3$ for every w in $W_{i,j}$. Indeed, in case (i), $\sigma_{i+1,j+1}w'v_3 \prec v_1v_2v_3$ holds for every w' in $W_{i',j}$, and wu_2 is such a w'; In case (ii), $wu_2v_3 \prec v_2v_3$ holds, which implies $v_1wu_2v_3 \prec v_1v_2v_3$..

Now $v_1v_2v_3 \prec v$ holds, hence, using the hypothesis $(S_{<\rho})$, we deduce

$$\sigma_{i+1,j+1}\overline{w}\,\overline{u_2}\,\overline{v_3} <_L \overline{v_1}\,\overline{v_2}\,\overline{v_3},$$

and, therefore, $\sigma_{i,j+1}\,\overline{w}\,\overline{u_2}\,\overline{v_3} <_L \overline{v}$ for every w as above. By (2.2), we have $\sigma_{i,j+1}w \equiv \mathrm{sh}^{-1}(w)\sigma_{i,j+1}$, so we have obtained the relation

$$\mathrm{sh}(\overline{w})\sigma_{i,j+1}\overline{u_2}\,\overline{v_3} <_L \overline{v}. \tag{2.3}$$

Applying (2.3) with $w = u_1^p$ gives the desired result provided we can eliminate the factor $\sigma_{i,j+1}$. To this end, let u_2' be the factor $\sigma_i\sigma_{i+1}^2 \cdots \sigma_j^2$ of u_2, and u_2'' be $\sigma_{j-1}^2\sigma_{j-2}^2 \cdots \sigma_{i'}^2$. So we have $u_2 = \sigma_i u_2' u_2''$, and

$$u_2'u_2''v_3 \prec \sigma_{i+1,j+1}u_2v_3. \tag{2.4}$$

Then $\sigma_i u_2'$ does not belong to $W_{i+1,j}$, hence the word $\sigma_{i,j+1}u_2v_3$ is irreducible, and, as $\sigma_{i,j+1}u_2v_3 \prec v$ holds, the hypothesis $(S_{<\rho})$ together with (2.4) implies

$$\overline{u_2'}\,\overline{u_2''}\,\overline{v_3} <_L \sigma_{i+1,j+1}\,\overline{u_2}\,\overline{v_3}.$$

Multiplying on the left by $u_1^p\sigma_i$ yields $\overline{u} <_L \overline{v}$.

The argument for $v_1 \in W_{j-1,i-1}$ with $j \le i$ is symmetric. ∎

We are now ready to prove the counterpart of Lemma 1.12.

Lemma 2.25. *Assume that u, v are positive braid words, $u \prec v$ holds, and v is irreducible. Then $\overline{u} <_L \overline{v}$ holds.*

Proof. Our aim is to prove (S_ρ) for every ordinal ρ less than ω^{ω^ω}. First, as in Section 1, (S_0) is vacuously true, and the point is to prove that $(S_{<\rho})$ implies (S_ρ). So, we assume that v is irreducible with rank ρ, and that $u \prec v$ holds.

If the main decomposition of v has length 3 at least, then, by Lemma 2.23, three cases are possible. Assume first $u = \sigma_i u_0$, $v = \sigma_i v_0$ and $u_0 \prec v_0$. Then, by definition, v_0 is irreducible, and $v_0 \prec v$ holds. Hence, by induction hypothesis, we have $\overline{u_0} <_L \overline{v_0}$, and therefore

$$\overline{u} = \sigma_i \, \overline{u_0} <_L \sigma_i \, \overline{v_0} = \overline{v}$$

as $<_L$ is compatible with multiplication on the left.

Assume then $v = \sigma_i^2 v_0$ and $u \prec \sigma_i v_0$. Then, by hypothesis, $\sigma_i v_0$ is irreducible, and $\sigma_i v_0 \prec v$ holds. Hence, by induction hypothesis, we have $\overline{u} <_L \sigma_i \, \overline{v_0}$. For a similar reason, we have $\overline{v_0} <_L \sigma_i \, \overline{v_0}$, which implies $\sigma_i \, \overline{v_0} <_L \sigma_i^2 \overline{v_0} = \overline{v}$. By transitivity of $<_L$, we deduce $\overline{u} <_L \overline{v}$.

Assume now that $u \prec u'$ holds for every p-companion u' of v with p large enough. By Lemma 2.24, and, because $(S_{<\rho})$ is assumed to be true by induction hypothesis, there exists such a companion u' that is irreducible and satisfies $\overline{u'} <_L \overline{v}$. As $u' \prec v$ holds, the induction hypothesis implies $\overline{u} <_L \overline{u'}$. So we find $\overline{u} <_L \overline{u'} <_L \overline{v}$, and $\overline{u} <_L \overline{v}$ holds.

Finally, assume that the main decomposition of v has length at most 2, *i.e.*, v has the form $\sigma_i v_1$ with v_1 in $W_{i+1,n-1}$. Then $u \prec v$ implies either $u = \sigma_i u_1$ with $u_1 \prec v_1$, and we prove $\overline{u} <_L \overline{v}$ as above, or $u \in W_{i+1,n-1}$, and, in this case, $\overline{u} <_L \overline{v}$ holds since $\overline{u}^{-1}\overline{v}$ admits an expression where σ_i occurs, but neither σ_i^{-1} nor any $\sigma_k^{\pm 1}$ with $k < i$ does. We conclude that (S_ρ) is true, which completes the proof. ■

As in Section 1, we deduce the following results:

Lemma 2.26. *Assume that v is an n-strand positive braid word. Then the following are equivalent:*

(i) The word v is irreducible;

(ii) The word v is the $\prec$-least element in its $\equiv$-equivalence class.

Proposition 2.27. (**Burckel's normal form**) *(i) Every positive braid word is equivalent to exactly one irreducible word, namely the $\prec$-minimal element of its equivalent class;*

(ii) If u and v are irreducible words, then $\overline{u} <_L \overline{v}$ is equivalent to $u \prec v$.

Proposition 2.28. (**well ordering**) *For every integer n, the restriction of the braid order $<_L$ to B_n^+ is a well ordering of type $\omega^{\omega^{n-2}}$.*

Proof. As $(BW_n^+, \prec)$ is a well ordering of order type $\omega^{\omega^{n-2}}$, Proposition 2.27(ii) implies that $(B_n^+, <_L)$ is a well ordering of type at most $\omega^{\omega^{n-2}}$. In order to prove that the order type is $\omega^{\omega^{n-2}}$ exactly, we have to check as in the case of B_3^+ that there are enough irreducible words. To this end, we denote by d the mapping of BW_n^+ into itself that maps every word u to the word u' such that the tree associated with u' is obtained from the tree associated with u by doubling the degree of every inner node. For instance, the tree associated with $\sigma_3\sigma_1$ is $((1),(1),(1))$, hence the tree associated with $d(u)$ is $((2,2),(2,2),(2,2),(2,2),(2,2),(2,2))$, and we have $d(u) = (\sigma_2^2\sigma_1^2\sigma_2^2\sigma_3^2)^3$. Then the image of d consists of irreducible words only, and d is $\prec$-increasing. Thus, with obvious notation, the order type of $(d(B_n^+), <_L)$ is $\omega^{\omega^{n-2}}$, and, therefore, the order type of $(B_n^+, <_L)$ is at least, hence is exactly, this ordinal. ■

Exercise 2.29. **(successor)** *Show that, for every braid word u in BW_n^+, the immediate successor of u with respect to $\prec$ is $u\sigma_{n-1}$.*

Exercise 2.30. **(decreasing sequences)** *Prove that, if $b_1 >_L b_2 >_L \cdots$ is a strictly decreasing sequence in B_n, then necessarily the lengths of the denominators $D_{RL}(b_1)$, $D_{RL}(b_2)$, ... tend to infinity.*

Exercise 2.31. **(Garside)** *Compute the rank of Δ_4 in $(B_4^+, <_L)$, and, more generally, the rank of Δ_n in $(B_n^+, <_L)$.*

---

3. Applications

Here we gather several further results involving the braid order. Firstly, using irreducible words, we prove that $<_L$ extends the subword partial ordering of braids. This implies that $b >_L 1$ holds for every conjugate of a braid belonging to B_∞^+. Then, we pose Laver's conjecture, a deep open well foundedness question. Finally, we show how to reverse the ordering of the generators σ_i so as to obtain a well ordering on the whole of B_∞^+, which leads to associating with every braid a unique well defined pair of ordinals.

The subword property

Let b be an arbitrary braid. By construction, $b <_L b\sigma_i$ holds for every i, but we have so far no symmetric result for comparing b and $\sigma_i b$. Here we establish such a result.

Lemma 3.1. *Assume that u is an irreducible n-strand positive braid word. Then the inequality $\overline{u} <_L \sigma_i\overline{u}$ holds for $i = 1, \ldots, n-1$.*

Proof. We use induction on the rank of u. By definition, we have $u \prec \sigma_i u$, hence, if $\sigma_i u$ is irreducible, $\overline{u} <_L \sigma_i \overline{u}$ follows from Proposition 2.27 (Burckel's normal form). Assume now that $\sigma_i u$ is reducible, say $u = u_1 u_2 u_3$, where (σ_i, u_1, u_2) is the beginning of the main decomposition of $\sigma_i u$. Assume first that u_1 does not contain $\sigma_{i\pm 1}$. Then we know that the word $u_2 u_3$ is irreducible, and $u_2 u_3 \prec u$ holds. Hence, by induction hypothesis, we have $\overline{u_2}\,\overline{u_3} <_L \sigma_i\,\overline{u_2}\,\overline{u_3}$, which implies in B_n^+

$$\overline{u} = \overline{u_1}\,\overline{u_2}\,\overline{u_3} <_L \overline{u_1}\,\sigma_i\,\overline{u_2}\,\overline{u_3} = \sigma_i\,\overline{u_1}\,\overline{u_2}\,\overline{u_3} = \sigma_i\,\overline{u}.$$

Assume now $u_1 = \sigma_{i+1,j+1}$ and $u_2 = u_2' u_2''$ with $u_2' \in W_{i,j}^{\bullet} \cup \{\varepsilon\}$ and u_2'' not containing σ_j. As u is assumed to be irreducible, the letter σ_i occurs in u_2'. So let us write $u_2'' = v\sigma_i w$, where v does not contain σ_i. As $w\,u_2''\,u_3 \prec u_2\,u_3$ holds, we have $\sigma_{i+2,j+1}\,w\,u_2''\,u_3 \prec \sigma_{i+2,j+1}\,u_2\,u_3 \prec u$. Applying the induction hypothesis, we find in B_n^+

$$\sigma_{i+2,j+1}\,\overline{w}\,\overline{u_2''}\,\overline{u_3} <_L \sigma_{i+1,j+1}\,\overline{w}\,\overline{u_2''}\,\overline{u_3},$$

and, therefore,

$$\overline{v}\,\sigma_{i+1}\,\sigma_i\,\sigma_{i+2,j+1}\,\overline{w}\,\overline{u_2''}\,\overline{u_3} <_L \overline{v}\,\sigma_{i+1}\,\sigma_{i,j+1}\,\overline{w}\,\overline{u_2''}\,\overline{u_3}. \tag{3.1}$$

Applying braid relations, we see that the left factor of (3.1) is $\overline{u_1}\,\overline{u_2}\,\overline{u_3}$, *i.e.*, $\overline{u}$, while the second one is $\sigma_i\,\overline{u_1}\,\overline{u_2}\,\overline{u_3}$, *i.e.*, $\sigma_i\overline{u}$. So $\overline{u} <_L \sigma_i\overline{u}$ holds. ■

Remark 3.2. In the case of B_3^+, an inductive argument is not necessary. Indeed, if $(e_1, \ldots, e_p)$ is the code of an irreducible word u, then the code of $\mathrm{red}^*(\sigma_i u)$ is $(e_1 + 1, e_2, \ldots, e_p)$ if $p - i$ is odd, $(1, e_1, \ldots, e_p)$ if $p - i$ is even and $e_1 \geq 2$ or $p \leq 2$ holds, and $(e_2 - 1, e_3, \ldots, e_{p-2}, e_{p-1} + 1, 1, e_p + 1)$ if $p - i$ is even, and $e_1 = 1$ and $p \geq 3$ hold. In all three cases, we deduce $u \prec \mathrm{red}^*(\sigma_i u)$, hence, $\overline{u} <_L \sigma_i\overline{u}$, directly.

The previous lemma leads to the following general result:

Proposition 3.3. (**subword property**) *Assume that b_1 and b_2 are braids and some expression of b_2 is obtained by inserting positive generators in some expression of b_1. Then $b_1 <_L b_2$ holds.*

Proof. For an inductive argument, it suffices to prove the result when there exist two braids a_1, a_2 and a generator σ_i satisfying $b_1 = a_1 a_2$ and $b_2 = a_1 \sigma_i a_2$. Owing to the compatibility of $<_L$ with multiplication on the left, it suffices to prove that $b <_L \sigma_i b$ holds for every braid b.

Assume $b \in B_n$. The braid Δ_n^2 belongs to the center of B_n, and, for $i < n$, $\Delta_n^2 \sigma_i^{-1}$ belongs to B_n^+. Hence, we can choose an integer k such that $\Delta_n^{2k} b$ belongs to B_n^+. Let u be the (unique) irreducible word that represents $\Delta_n^{2k} b$. By construction, the word $\mathrm{red}^*(\sigma_i u)$ is irreducible, and it represents $\sigma_i b$. By Lemma 3.1, we deduce $\Delta_n^{2k}\,b <_L \sigma_i\,\Delta_n^{2k}\,b = \Delta_n^{2k}\,\sigma_i\,b$, hence $b <_L \sigma_i b$. ■

Proposition 3.4. (conjugate) *Assume that b is a positive braid. Then $aba^{-1} >_L 1$ holds for every braid a.*

Proof. Proposition 3.3 (subword) gives $a^{-1} <_L ba^{-1}$, hence $1 <_L aba^{-1}$ by compatibility with multiplication on the left. ∎

This result should be compared to our previous remark that a conjugate aba^{-1} of a braid b that satisfies $b >_L 1$ need not satisfy $aba^{-1} >_L 1$ in general, as shown for instance by the example $b = \sigma_1\sigma_2^{-1}$, $a = \sigma_1\sigma_2$.

Laver's conjecture

We have seen in Section III.2 that, for every braid b in B_n and every left cancellative LD-system S, there exists at least one sequence $\vec{a}$ in S^n such that the action of b on $\vec{a}$ is defined, *i.e.*, $\vec{a} \bullet b$ exists. Here, we reverse the point of view, and, starting with a sequence $\vec{a}$, we consider the set of all braids b such that $\vec{a} \bullet b$ exists.

Definition 3.5. (set $D(S, \vec{a})$) Assume that S is a left cancellative LD-system, and $\vec{a}$ belongs to S^n. We put

$$D(S, \vec{a}) = \{b \in B_n \,;\, \vec{a} \bullet b \text{ exists}\}.$$

Thus b belongs to $D(S, \vec{a})$ if and only if $\vec{a} \bullet w$ exists for at least one braid word w representing b.

***Example 3.6.** (set $D(S, \vec{a})$) If S is an LD-quasigroup, then, for every sequence $\vec{a}$ in S^n, $D(S, \vec{a})$ is all of B_n. On the other hand, if S is the LD-system $(B_\infty, {}^\wedge)$, we can for instance consider the set $D(B_\infty, (1, \ldots, 1))$ consisting of those braids b for which $(1, \ldots, 1) \bullet b$ is defined. It is easy to see that $D(B_\infty, (1, \ldots, 1))$ is a proper subset of B_n: for instance, σ_i^{-1} belongs to $D(B_\infty, (1, \ldots, 1))$ for no i. We have seen in Chapter III that every braid in $D(B_\infty, (1, \ldots, 1))$ admits a special decomposition, and it will follow from the results of Chapter VI that this condition is also sufficient.*

Assume that S is a left cancellative LD-system. As the action of positive braids on S^n is always defined, every set $D(S, \vec{a})$ with $\vec{a} \in S^n$ includes B_n^+. We shall see in Chapter V that, in some cases, $D(S, \vec{a})$ may coincide with B_n^+, and, in such cases, the results of Section 2 show that the restriction of the braid order $<_L$ to $D(S, \vec{a})$ is a well ordering. This leads to the following

Conjecture 3.7. (Laver's conjecture, braid form) *For every sequence $\vec{a}$ in B_∞^n, the subset $D(B_\infty, \vec{a})$ of B_n is well ordered by $<_L$.*

A well ordering on B_∞^+

The result that $<_L$ is a well ordering on B_n^+ does not extend to the whole of B_∞^+, as, by construction, we have

$$\sigma_1 >_L \sigma_2 >_L \cdots,$$

a strictly descending sequence. However, this sequence is the only obstruction.

Definition 3.8. (flip) For $n \geq 2$, we denote by ϕ_n the flip automorphism of B_n that maps σ_i to σ_{n-i} for every i.

Lemma 3.9. *Assume b_1, $b_2 \in B_n$, and $p \geq n$. Then $\phi_n(b_1) <_L \phi_n(b_2)$ is equivalent to $\phi_p(b_1) <_L \phi_p(b_2)$.*

Proof. By definition, we have $\phi_p(b) = \mathrm{sh}^{p-n}(\phi_n(b))$ for every braid b in B_n. Now the braid order $<_L$ is invariant under sh. ■

Definition 3.10. (order $<_L^\phi$) Assume that b_1, b_2 are braids. We say that $b_1 <_L^\phi b_2$ holds if $\phi_n(b_1) <_L \phi_n(b_2)$ whenever b_1 and b_2 belong to B_n.

Proposition 3.11. (well ordering) *(i) The relation $<_L^\phi$ is a left invariant linear order on the group B_∞.*

(ii) The restriction of $<_L^\phi$ to B_∞^+ is a well ordering of type ω^{ω^ω}. For every n, B_n^+ is the initial segment of $(B_\infty^+, <_L^\phi)$ associated with σ_n, i.e., for $b \in B_\infty^+$, $b \in B_n^+$ is equivalent to $b <_L^\phi \sigma_n$. For every positive braid b in B_n^+, the rank of b in $(B_\infty^+, <_L^\phi)$ is equal to the rank of $\phi_n(b)$ in $(B_n^+, <_L)$. In particular, the rank of σ_n in $(B_\infty^+, <_L^\phi)$ is 1 for $n = 1$, and is $\omega^{\omega^{n-2}}$ for $n \geq 2$.

Proof. Point (i) is clear. As for (ii), owing to the previous results, it suffices to prove that $b \in B_n^+$ is equivalent to $b <_L^\phi \sigma_n$. Assume $b \in B_n^+$. Then we have

$$\phi_{n+1}(b) = \mathrm{sh}(\phi_n(b)), \qquad \phi_n(\sigma_n) = \sigma_1,$$

and $\mathrm{sh}(\phi_n(b)) <_L \sigma_1$ implies $b <_L^\phi \sigma_n$. Conversely, assume $b <_L^\phi \sigma_n$, with $b \in B_p^+$, $p \geq n$. Then we have $\phi_p(b) <_L \sigma_{p-n}$. By Proposition 3.3 (subword), $b_1 \sigma_{p-n} b_2 \geq_L \sigma_{p-n}$ holds for all positive braids b_1, b_2, hence $\phi_p(b) <_L \sigma_{p-n}$ implies that $\phi_p(b)$ admits a decomposition involving σ_{p-n+1}, ..., σ_{p-1} only, *i.e.*, we have $\phi_p(b) \in \mathrm{sh}^{p-n}(B_n)$, and $b \in B_n$. ■

The interest of considering the order $<^{\phi}_{L}$ rather than its counterpart $<_{L}$ is that the rank of a positive braid b with respect to $<^{\phi}_{L}$ does not depend on a reference set B_n in which b lies.

We have seen in Chapter II how to associate with every braid b a unique distinguished pair of positive braids $N_{RL}(b)$, $D_{RL}(b)$ satisfying $b = D_{RL}(b)^{-1}N_{RL}(b)$—and a symmetric decomposition $b = N_{LR}(b)D_{LR}(b)^{-1}$. By the previous results, every positive braid is specified by a unique ordinal α with $0 \le \alpha < \omega^{\omega^{\omega}}$, namely its rank in the well ordered set $(B^{+}_{\infty}, <^{\phi}_{L})$. It follows that every braid is specified by a pair of such ordinals.

Definition 3.12. (ordinal decomposition) Assume that b is a braid in B_n. The *ordinal decomposition* of b is defined to be the ordered pair $o(b)$ consisting of the ranks of $D_{RL}(b)$ and $N_{RL}(b)$ in $(B^{+}_{\infty}, <^{\phi}_{L})$.

Example 3.13. (ordinal decomposition) *Let b be the braid $\sigma_2\sigma_1^{-1}$. The normal decomposition of b as a negative–positive fraction is $(\sigma_2\sigma_1)^{-1}(\sigma_1\sigma_2)$. The rank of $\sigma_1\sigma_2$ in $(B^{+}_{\infty}, <^{\phi}_{L})$, which is the rank of $\sigma_2\sigma_1$ in $(B^{+}_{3}, <_{L})$, is $\omega + 1$, while the rank of $\sigma_2\sigma_1$ is ω^2. So, we have $o(\sigma_2\sigma_1^{-1}) = (\omega + 1, \omega^2)$.*

Proposition 3.14. (order) *Assume that b is a braid and we have $o(b) = (\alpha, \beta)$. Then $b >_{L} 1$ holds in B_{∞} if and only if $\beta > \alpha$ holds in the ordinals.*

Proof. As $<_{L}$ is left invariant, $1 <_{L} b$ is equivalent to $D_{RL}(b) <_{L} N_{RL}(b)$, hence, by definition of the ordinal rank, to $\alpha < \beta$. ■

It is not true that every pair of positive braids is the normal decomposition of a braid: by construction, the pair (b_1, b_2) is the normal decomposition of the braid $b_1^{-1}b_2$ if and only if the positive braids b_1, b_2 are left coprime in B^{+}_{∞}.

Example 3.15. (case of B_2) *The elements of B^{+}_{2} have the form σ_1^{p} with $p \ge 0$. Now σ_1^{p} and σ_1^{q} are left coprime if and only if at least one of $p = 0$, $q = 0$ holds, and, therefore, the ordinal decomposition in B_2 is the pair $(p, 0)$ for σ_1^{p}, $p \ge 0$, and $(0, q)$ for σ_1^{-q}, $q \ge 0$. The group B_2 is isomorphic to $(\mathbf{Z}, +)$, and we recognize the standard representation of integers by pairs of natural numbers.*

Exercise 3.16. (quasipositive braids) *Define a quasipositive braid to be a braid with an expression of the form* $(b_1\sigma_{i_1}b_1^{-1})(b_2\sigma_{i_2}b_2^{-1})\ldots(b_k\sigma_{i_k}b_k^{-1})$. *Prove that every quasipositive braid* b *satisfies* $b >_L 1$.

Exercise 3.17. (compatibility) *(i) Show that, for every braid* b, *the braid* $b^{-1}\mathrm{sh}(b)\sigma_1$ *is* σ_1*-positive.*

(ii) Assume $a = a_1a_2\mathrm{sh}(a_1^{-1})$ *with* a_2 *positive. Prove that* $b <_L a\hat{}b$ *holds for every* b. *Observe that this applies when* a *is positive or special.*

Exercise 3.18. (Laver's conjecture for B_2**)** *(i) For* b *a braid, define* $\mathrm{ht}_R(b)$ *to be the number of positive integers* p *satisfying* $\mathrm{perm}(b)(p+1) = p$. *Prove that* $\mathrm{ht}_R(a\hat{}b) = \mathrm{ht}_R(b) + 1$ *holds for all braids* a, b.

(ii) Deduce Laver's conjecture for B_2.

---

4. Notes

The result that the restriction of the linear order $<_L$ to B_n^+ is a well ordering for every n has been established in (Laver 96) using a normal form result for the free LD-monoid FLDM_∞ (see Chapter XI). Laver proves Proposition 3.3 (subword) first, and he deduces that $<_L$ is a well ordering using a theorem of Higman asserting that every linear order on words over a finite alphabet with the subword property is a well ordering. This method does not give the order type, nor does it allow one to determine the rank of a given positive braid in the well ordering. The current approach, which is more effective, is due to S. Burckel, and the material of Sections 1 and 2, as well as the proof of Lemma 3.1, appears in (Burckel *94*) and (Burckel 97). The subword property has been reproved in (Wiest *98*) using the reconstruction of the linear order of braids in terms of the curve diagrams of Section III.4. Laver's Conjecture (in the free LD form mentioned in Chapter V) was proposed around 1995.

We think that Burckel's normal form of positive braids in terms of trees is interesting in itself, and we expect further applications in the future. The question of practically determining the ordinal rank of an arbitrary positive braid remains partly open in the general case; partial results appear in (Burckel *96*).

Laver's Conjecture seems to be a very difficult question. In the case of braids, the set $D(B_\infty, (1, 1, \ldots, 1))$, n times 1, contains all special braids that live in B_n. It easily follows from the results of Chapter V that the restriction of the linear order $<_L$ to special braids is not a well ordering. This does not contradict Laver's Conjecture, as the set of those special braids that live in a fixed braid group B_n may be finite. This however is not obvious: the inductive construction of special braids implies that, if $t(x)$ is a term of height n, then the braid $t(1)$ belongs to B_{n+1}, but this does not dismiss the possibility that $t(1)$ belongs to B_p for some $p \leq n$. We propose the following conjecture.

Conjecture 4.1. *The number of special braids in B_n is bounded above by 2^n.*

The existence of a canonical representation of braids by pairs of ordinals could lead to further applications. In particular, each time we have an equivalence relation $\sim$ on braids, we can define for every braid b a distinguished pair of ordinals

$$o_\sim(b) = \inf\{o(b') \ ; \ b' \sim b\},$$

where the pairs of ordinals are ordered lexicographically. From what we mentioned above, computing the value of $o(b)$ in the general case is not obvious, and, so, computing $o_\sim(b)$ for various equivalence relations $\sim$ is likely to be still more difficult. Nevertheless, we can raise the following question:

Question 4.2. *Does some algorithm compute the value of $o_\sim(b)$ when $\sim$ is conjugacy? And when $\sim$ is the Markov equivalence such that $b \sim b'$ holds if and only if the closures $\widehat{b}$ and $\widehat{b'}$ are isotopic links?*

In theory, considering the latter equivalence leads us to associate with every link a well defined pair of ordinals that exactly characterizes the isotopy class of that link—presumably an observation of little interest as long as no clue about how to compute this pair of ordinals is given.

Instead of using the symmetric normal form of braids, we could also consider Garside forms $\Delta_n^{-k}b$ with $k \geq 0$ and b positive, which amounts to using pairs of ordinals in which the first factor is the rank of some braid Δ_n^k.

Let us conclude with another hypothetical development. We have seen that the mapping $\varphi : p \mapsto \sigma_1^p$ defines an isomorphism of the ordered group $(\mathbf{Z}, +, <)$ onto $(B_2, \cdot, <_L)$, and, in this sense, braids appear as a natural extension of integers.

Question 4.3. *Does some binary operation $\times$ on B_n correspond to multiplication of integers and, in particular, satisfy $\sigma_1^p \times \sigma_1^q = \sigma_1^{pq}$?*

In other words, does there exist a convenient (necessarily non-commutative) arithmetic of braids? Perhaps the braid order could help us to guess what a possible definition of $\times$ should be.

Appendix. Rank Tables

We give below a list of the first braids in the well orderings $(B_3^+, <_L)$, $(B_4^+, <_L)$, and $(B_\infty^+, <_L^\phi)$, together with their rank, and, in the first two cases, their code. Every braid appears there exactly once, and its expression is the $\prec$-minimal word in its equivalence class, *i.e.*, its Burckel's normal form. Thus, for instance, $\sigma_2\sigma_1\sigma_2$ appears in Table 1, but $\sigma_1\sigma_2\sigma_1$ does not appear, as it is not an irreducible expression. Note that the rank mentioned here is the rank with respect to $<_L$, which is not the rank with respect to $\prec$, as reducible words as skipped: the braid $\sigma_1\sigma_2^2\sigma_1$ is the ω^3-th element of $(B_3^+, <_L)$, but, by Proposition 1.4 (well ordering), the word $\sigma_1\sigma_2^2\sigma_1$ is the $\omega^3+\omega^2$-th element of $(BW_3^+, \prec)$: the missing words are the words $\sigma_1\sigma_2\sigma_1^{p+1}\sigma_2^q$, which are reducible.

braid	1	σ_2	σ_2^2	...	σ_1	$\sigma_1\sigma_2$	$\sigma_1\sigma_2^2$	...
code	(1)	(2)	(3)		(1, 1)	(1, 2)	(1, 3)	
rank	0	1	2	...	ω	$\omega+1$	$\omega+2$	...

braid	...	σ_1^2	$\sigma_1^2\sigma_2$	...	$\sigma_2\sigma_1$	$\sigma_2\sigma_1\sigma_2$	...
code		(2, 1)	(2, 2)		(1, 1, 1)	(1, 1, 2)	
rank	...	$\omega{\cdot}2$	$\omega{\cdot}2+1$	...	ω^2	ω^2+1	...

braid	...	$\sigma_2\sigma_1^2$	$\sigma_2\sigma_1^2\sigma_2$	...	$\sigma_2\sigma_1^3$	$\sigma_2\sigma_1^3\sigma_2$
code		(1, 2, 1)	(1, 2, 2)		(1, 3, 1)	(1, 3, 2)
rank	...	$\omega^2+\omega$	$\omega^2+\omega+1$	...	$\omega^2+\omega{\cdot}2$	$\omega^2+\omega{\cdot}2+1$

braid	...	$\sigma_2^2\sigma_1$	...	$\sigma_2^2\sigma_1^2$	...	$\sigma_1\sigma_2^2\sigma_1$	$\sigma_1\sigma_2^2\sigma_1\sigma_2$
code		(2, 1, 1)		(2, 2, 1)		(1, 2, 1, 1)	(1, 2, 1, 2)
rank	...	$\omega^2{\cdot}2$	...	$\omega^2{\cdot}2+\omega$	...	ω^3	ω^3+1

braid	...	$\sigma_1\sigma_2^2\sigma_1^2$	...	$\sigma_1\sigma_2^3\sigma_1$	...	$\sigma_1^2\sigma_2^2\sigma_1$
code		(1, 2, 2, 1)		(1, 3, 1, 1)		(2, 2, 1, 1)
rank	...	$\omega^3+\omega$	...	$\omega^3+\omega^2$	...	$\omega^3{\cdot}2$

braid	...	$\sigma_2\sigma_1^2\sigma_2^2\sigma_1$	...	$\sigma_1\sigma_2^2\sigma_1^2\sigma_2^2\sigma_1$	...
code		(1, 2, 2, 1, 1)		(1, 2, 2, 2, 1, 1)	
rank	...	ω^4	...	ω^5	...

Table 1. The well ordering $(B_3^+, <_L)$

braid	1	σ_3	σ_3^2	...	σ_2	$\sigma_2\sigma_3$	...	σ_2^2	$\sigma_2^2\sigma_3$
code									
rank	0	1	2	...	ω	$\omega+1$	...	$\omega\cdot 2$	$\omega\cdot 2+1$

braid	...	σ_2^3	...	$\sigma_3\sigma_2$	...	$\sigma_3\sigma_2^2$	...	$\sigma_3\sigma_2^3$
code						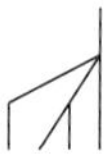		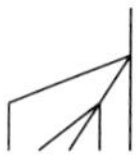
rank	...	$\omega\cdot 3$	...	ω^2	...	$\omega^2+\omega$	...	$\omega^2+\omega\cdot 2$

braid	...	$\sigma_3^2\sigma_2$	...	$\sigma_2\sigma_3^2\sigma_2$	$\sigma_2\sigma_3^2\sigma_2\sigma_3$	...	$\sigma_2\sigma_3^2\sigma_2^2$
code				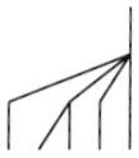	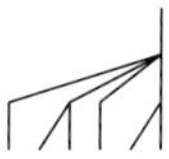		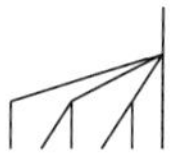
rank	...	$\omega^2\cdot 2$	...	ω^3	ω^3+1	...	$\omega^3+\omega$

braid	...	$\sigma_2\sigma_3^3\sigma_2$	...	σ_1	$\sigma_1\sigma_3$	...	$\sigma_1\sigma_2$	...
code		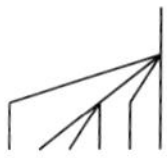						
rank	...	$\omega^3+\omega^2$	...	ω^ω	$\omega^\omega+1$	...	$\omega^\omega+\omega$	...

braid	...	$\sigma_1\sigma_3\sigma_2$	...	$\sigma_1\sigma_2\sigma_3^2\sigma_2$	...	σ_1^2	$\sigma_1^2\sigma_3$
code		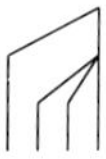		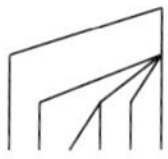			
rank	...	$\omega^\omega+\omega^2$	...	$\omega^\omega+\omega^3$	...	$\omega^\omega\cdot 2$	$\omega^\omega\cdot 2+1$

braid	...	$\sigma_1^2\sigma_2$	...	$\sigma_2\sigma_1$	...	$\sigma_2\sigma_1^2$	...	$\sigma_2^2\sigma_1$
code								
rank	...	$\omega^\omega\cdot 2+\omega$	...	$\omega^{\omega+1}$	...	$\omega^{\omega+1}+\omega^\omega$	...	$\omega^{\omega+1}\cdot 2$

braid	...	$\sigma_1\sigma_2\sigma_1^2$	...	$\sigma_3\sigma_2\sigma_1$	...
code		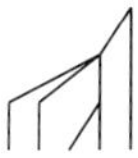			
rank	...	$\omega^{\omega+2}$	...	$\omega^{\omega\cdot 2}$	...

Table 2. The well ordering $(B_4^+, <_L)$

braid	1	σ_1	σ_1^2	...	σ_2	$\sigma_2\sigma_1$	$\sigma_2\sigma_1^2$	...
rank	0	1	2	...	ω	$\omega+1$	$\omega+2$	...

braid	σ_2^2	$\sigma_2^2\sigma_1$	...	$\sigma_1\sigma_2$	$\sigma_1\sigma_2\sigma_1$	...	$\sigma_1\sigma_2^2$
rank	$\omega{\cdot}2$	$\omega{\cdot}2+1$	...	ω^2	ω^2+1	...	$\omega^2+\omega$

braid	...	$\sigma_1\sigma_2^2\sigma_1$	...	$\sigma_1\sigma_2^3$	$\sigma_1\sigma_2^3\sigma_1$	...
rank	...	$\omega^2+\omega+1$	...	$\omega^2+\omega{\cdot}2$	$\omega^2+\omega{\cdot}2+1$	...

braid	$\sigma_1^2\sigma_2$	...	$\sigma_1^2\sigma_2^2$	...	$\sigma_2\sigma_1^2\sigma_2$	$\sigma_2\sigma_1^2\sigma_2\sigma_1$	...
rank	$\omega^2{\cdot}2$	...	$\omega^2{\cdot}2+\omega$	...	ω^3	ω^3+1	...

braid	$\sigma_2\sigma_1^2\sigma_2^2$	...	$\sigma_2\sigma_1^3\sigma_2$	...	$\sigma_2^2\sigma_1^2\sigma_2$	...	$\sigma_1\sigma_2^2\sigma_1^2\sigma_2$
rank	$\omega^3+\omega$	...	$\omega^3+\omega^2$	...	$\omega^3{\cdot}2$	...	ω^4

braid	...	$\sigma_2\sigma_1^2\sigma_2^2\sigma_1^2\sigma_2$	...	σ_3	$\sigma_3\sigma_1$	...	$\sigma_3\sigma_2$
rank	...	ω^5	...	ω^ω	$\omega^\omega+1$	...	$\omega^\omega+\omega$

braid	...	$\sigma_3\sigma_1\sigma_2$	...	$\sigma_3\sigma_2\sigma_1^2\sigma_2$	...	σ_3^2	$\sigma_3^2\sigma_1$
rank	...	$\omega^\omega+\omega^2$	...	$\omega^\omega+\omega^3$	...	$\omega^\omega{\cdot}2$	$\omega^\omega{\cdot}2+1$

braid	...	$\sigma_3^2\sigma_2$	...	$\sigma_2\sigma_3$	...	$\sigma_2\sigma_3^2$	...
rank	...	$\omega^\omega{\cdot}2+\omega$	...	$\omega^{\omega+1}$	...	$\omega^{\omega+1}+\omega^\omega$	...

braid	$\sigma_2^2\sigma_3$	...	$\sigma_3\sigma_2\sigma_3^2$	...	$\sigma_1\sigma_2\sigma_3$	...	σ_4
rank	$\omega^{\omega+1}{\cdot}2$	...	$\omega^{\omega+2}$	...	$\omega^{\omega\cdot 2}$	...	ω^{ω^2}

braid	$\sigma_4\sigma_1$	...	$\sigma_4\sigma_2$	...	$\sigma_4\sigma_3$	...	σ_5
rank	$\omega^{\omega^2}+1$	...	$\omega^{\omega^2}+\omega$	...	$\omega^{\omega^2}+\omega^\omega$	...	ω^{ω^3}

braid	...	σ_6	...
rank	...	ω^{ω^4}	...

Table 3. The well ordering $(B_\infty^+, <_L^\phi)$

Part B

Free LD-systems

V
Orders on Free LD-systems

Here, we begin our study of free LD-systems, which will be continued in the following four chapters. The main result of this chapter is that every free LD-system admits a canonical linear ordering. We deduce a solution for the word problem of left self-distributivity, and a simple criterion for recognizing free LD-systems.

The chapter is organized as follows. In Section 1, we recall the general construction of free algebraic systems, and we describe free LD-quasigroups. In Section 2, we introduce the notion of LD-equivalent terms, and we establish a general property of monogenic LD-systems called absorption by right powers. In Section 3, we introduce the notion of being an LD-expansion, a refinement of being LD-equivalent, and we show that two terms are LD-equivalent if and only if they admit a common LD-expansion (confluence property). This result is used in Section 4 to establish the comparison property used in Chapter III when constructing the braid order. In Section 5, we use the iterated left divisibility relation to construct a linear order on every free LD-system. Finally, in Section 6, we deduce that the word problem of the left self-distributivity identity is solvable and we establish Laver's criterion for a given LD-system to be free. This leads to realizations of the free LD-system of rank 1 inside the braid group B_∞, and of the free LD-systems of any rank inside some extension of B_∞.

1. Free Systems

Here we recall some basic facts about free systems. As an example, we shall describe free LD-quasigroups.

Assume that $\vec{I}$ is a (finite or infinite) family of algebraic identities. An algebraic system that satisfies all identities in $\vec{I}$ will be called an *$\vec{I}$-system*, which is compatible with our definition of an LD-system. The class of all $\vec{I}$-systems is called the *equational variety* associated with $\vec{I}$. An equational variety can be characterized as a class of structures that is closed under product, substructure and quotient, but we shall not use this result here as we are only interested in varieties with a given family of defining identities.

Definition 1.1. (**free, basis**) Assume that V is an equational variety, S belongs to V, and X is a subset of S. We say that X is *free* in S if, for every S' in V and every map f of X to S', there exists a homomorphism of the subsystem of S generated by X into S'. We say that X is a *basis* of S if it is free in S and generates S. We say that S is *free* if it admits at least one basis.

Proposition 1.2. (**free**) *Assume that V is an equational variety.*

(i) Assume that S belongs to V and X is a free subset of S. If f is a mapping of X into a system S' of V, then the morphism that extends f to the subsystem of S generated by X is unique.

(ii) Assume that F belongs to V and X is a basis of F. Then every free system based on X in V is isomorphic to F, and every system in V that is generated by X is a quotient of F.

Proof. (i) Let S_X denote the subsystem of S generated by X. Assume that $\widehat{f}$ and $\widehat{f}'$ are two extensions of f into a morphism of S_X into S', and let S_1 be the subset of S made of those elements a satisfying $\widehat{f}(a) = \widehat{f}'(a)$. By hypothesis, S_1 includes X, and it is closed under the operations of S, as $\widehat{f}$ and $\widehat{f}'$ are morphisms. So S_1 includes the subsystem of S generated by X, which is S_X.

(ii) Assume that f is a bijection of X into X', F is free based on X and F' is free based on X'. There exist an extension $\widehat{f}$ of f from F to F', and an extension $\widehat{f^{-1}}$ of f^{-1} from F' to F. The composed map $\widehat{f^{-1}} \circ \widehat{f}$ is an endomorphism of F that extends the identity mapping on X: by (i) this extension is the identity endomorphism, as the latter is another such extension. Similarly $\widehat{f} \circ \widehat{f^{-1}}$ has to the identity of F', hence $\widehat{f}$ and $\widehat{f^{-1}}$ are mutually inverse isomorphisms.

Assume now that S is an arbitrary element of V generated by X. The identity mapping of X extends into a morphism $\widehat{\mathrm{id}}$ of F into S. As in (i), the image of $\widehat{\mathrm{id}}$ includes the subsystem of S generated by X, which is S by hypothesis. Hence, $\widehat{\mathrm{id}}$ is surjective, and S is isomorphic to $F/\equiv$, where $\equiv$ is the congruence $\widehat{\mathrm{id}}(a) = \widehat{\mathrm{id}}(b)$. ∎

Corollary 1.3. *If V is an equational variety, then a given algebraic identity holds in every system of V if and only if it holds in every free system of V.*

Proof. Identities are preserved under quotient. ■

By Proposition 1.2(ii) (free), free systems based on equipotent sets are isomorphic. A free system based on a set of cardinality k is usually called a free system of *rank* k. (We do not claim that the rank is unique in general.)

Example 1.4. (**free monoids**) *For X a nonempty set, the monoid X^* consisting of all words over the alphabet X equipped with concatenation is a free monoid based on X. Indeed, every word is the concatenation of its letters, hence X generates X^*. If M is a monoid and f is any mapping of X into M, there is a unique way to extend f into a morphism f^* of X^* into M, namely to define $f^*(\varepsilon)$ to be the unit of M, and $f^*(x_1 \cdots x_\ell)$ to be the product $f(x_1) \cdots f(x_\ell)$, which makes sense as the decomposition is unique.*

Example 1.5. (**free groups**) *Let X be a nonempty set, and X^{-1} be a disjoint copy of X. For w in the free monoid $(X \cup X^{-1})^*$, let* red(w) *be the word obtained from w by iteratively deleting all possible patterns of the form xx^{-1} or $x^{-1}x$. Such a mapping is well defined, i.e., the final result is independent of the way the reductions are operated. The relation* red(w) = red(w') *defines a congruence $\equiv$ on $(X \cup X^{-1})^*$, so concatenation induces a well defined associative operation on the quotient-set* FG_X. *For w in $(X \cup X^{-1})^*$, the class of the word w^{-1} (the word obtained from w by exchanging x and x^{-1} everywhere and reversing their ordering) is an inverse for the class of w, and* FG_X *is a group generated by X. For f a mapping of X into a group G, there exists a unique extension of f into a morphism f^* of the free monoid $(X \cup X^{-1})^*$ into G satisfying $f(x^{-1}) = f(x)^{-1}$ for x in X. Then f^* is compatible with $\equiv$, and it induces a morphism of the group* FG_X *into G. So* FG_X *is a free group based on X.*

As we shall see below, it is essentially trivial to prove that free systems exist for every family of algebraic identities. As our main interest is in LD-systems, we could consider this special case only. However, the general argument is even more easily stated. The only problem is to start with a sufficiently flexible system. In the case of associative structures, free monoids are convenient: for instance, free groups (and, subsequently, free systems in varieties of groups or semigroups) can be constructed as quotients of free monoids. This approach cannot work for non-associative systems, and we need to start with some universal system playing the role of free monoids. Such systems are the free systems in the variety associated with an empty family of identities, called *absolutely free systems*. Here we consider the case of binary systems, *i.e.*, of sets equipped with a single binary operation. The extension to other cases is straightforward. We recall that, for X a set, X^* denotes the set of all words over the alphabet X.

Definition 1.6. (term, size) Assume that X is a nonempty set. Let $\bullet$ be a fixed letter not in X. For $k \geq 1$, a *term on X of size k* is defined to be a nonempty word on $X \cup \{\bullet\}$ that obeys the following inductive rules:
- a term of size 1 is an element of X;
- a term of size k with $k \geq 2$ is a word of the form $t_1 t_2 \bullet$, where t_1 and t_2 are terms of sizes i and j respectively and $i + j = k$ holds.

The set of all terms on X is denoted by T_X. For t_1, t_2 in T_X, we define $t_1 \cdot t_2$ to be the word $t_1 t_2 \bullet$.

For instance, if x, y, z are elements of X—called *variables* in this context—then the words $xy\bullet x\bullet$ and $xyz\bullet\bullet$ are terms of size 3 on X, while $x\bullet y$ is not a term, since every binary term that is not a single letter has to end with the letter $\bullet$. The following lemma characterizes terms among words over $X \cup \{\bullet\}$—in terms of formal grammars, T_X is the algebraic language generated by the productions $S \to SS\bullet$ and $S \to x$ for $x \in X$.

Lemma 1.7. *Assume that w is a word over $X \cup \{\bullet\}$. Then w is a term if and only if, letting $\#_X(v)$ and $\#_\bullet(v)$ denote the number of letters from X and of $\bullet$ in v, we have*
- $\#_X(w) = \#_\bullet(w) + 1$, *and*
- $\#_X(u) \geq \#_\bullet(u) + 1$ *for every nonempty prefix u of w.*

Proof. The above conditions hold when w is a single variable, and they are preserved under the operation $(w_1, w_2) \mapsto w_1 w_2 \bullet$. Conversely, if w satisfies the conditions, then either w is a variable, or it admits a unique decomposition as $w_1 w_2 \bullet$ where w_1 and w_2 still satisfy the conditions. Indeed, w_1 is the longest proper prefix u of w that satisfies $\#_X(u) = \#_\bullet(u) + 1$. We then apply induction on the length. ∎

Using the previous criterion, we deduce the fundamental property of terms:

Lemma 1.8. *Assume that t is a term of size n, with $n \geq 2$. Then t admits a unique decomposition $t = t_1 \cdot t_2$, with t_1, t_2 terms of size $< n$.*

Proof. Existence is in the definition. Uniqueness comes from the uniqueness of the possible decomposition of a word over $X \cup \{\bullet\}$ as $w_1 w_2 \bullet$ where w_1 and w_2 satisfy the conditions of Lemma 1.7. ∎

Proposition 1.9. (absolutely free) *Assume that X is a nonempty set. Then $(T_X, \cdot)$ is a free binary system based on X.*

Proof. The product of two terms is still a term, so the subsystem of T_X generated by X must be all of T_X. Let S be a binary system, and f be a mapping of X into S. We show inductively on the size of t that there exists a unique way to define $\widehat{f}(t)$ so that $\widehat{f}$ is a homomorphism of T_X into S extending f. If t is a variable, $\widehat{f}(t) = f(t)$ holds by hypothesis. Otherwise, there exists a decomposition of t as $t_1 \cdot t_2$, and the sizes of t_1 and t_2 are smaller than the size of t. We define $\widehat{f}(t) = \widehat{f}(t_1) \cdot \widehat{f}(t_2)$. Because the decomposition of t as the product of t_1 and t_2 is unique, this definition is non-ambiguous, and, by construction, $\widehat{f}$ is a homomorphism. ∎

Definition 1.10. (evaluation) In the above framework, $\widehat{f}(t)$ is called the *f-evaluation* of t in S. We shall occasionally write $t(x_1, \ldots, x_n)$ for a term involving the variables $x_1, \ldots, x_n$, and then use $t(a_1, \ldots, a_n)^S$ for the evaluation of t in S when the variable x_i is given the value a_i, *i.e.*, for the image of t under the morphism $\widehat{f}$ defined by $\widehat{f}(x_i) = a_i$.

Construction of terms easily extends to arbitrary systems: for instance, for systems involving two binary operations, it suffices to consider terms constructed using two new letters (or operators). The case of unary (or ternary) operations is similar. We shall occasionally denote by $T_{(X,I)}$ the set of all terms constructed using variables in X and operators in I (or indexed by I). In this framework, the current set T_X should be denoted $T_{(X,\cdot)}$, or, even, $T_{(X,\cdot_2)}$, in order to indicate that $\cdot$ is a binary operation.

Let us come back to the existence of free systems in equational varieties. Such systems can be constructed as quotients of absolutely free systems under a convenient congruence. Let us observe that, syntactically, an algebraic identity is *a pair of terms*: for instance, the left distributivity identity (LD) is (or can be defined as) the pair of terms $(x \cdot (y \cdot z), (x \cdot y) \cdot (x \cdot z))$.

Definition 1.11. (substitution, instance) (i) Assume that X, Y are sets and f maps X into T_Y. The *substitution* associated with f is the extension of f into a morphism of T_X to T_Y; the image of a term t on X under this substitution is denoted t^f, and it is called a *Y-substitute* of t.

(ii) Assume that (t, t') is a pair of terms on the set X, and S is a binary system. Every pair of the form $(\widehat{f}(t), \widehat{f}(t'))$ with f a mapping of X to S is called an *S-instance* of (t, t').

Saying that the identity (t, t') (or "$t = t'$") is satisfied in a system S means that every S-instance of this pair is a diagonal pair of the form (a, a).

Lemma 1.12. *Assume that R is a set of pairs on an algebraic system S. Then there exists a least congruence on S that includes R, namely the relation $\widehat{R}$ such that, for a, a' in S, $a\ \widehat{R}\ a'$ holds if and only if there exists a finite sequence $a_0 = a$, a_1, ..., $a_n = a'$ such that, for every i, we have $\{a_{i-1}, a_i\} = \{cbd, cb'd\}$ for some pair $\{b, b'\}$ in R and some c, d in S.*

Proof. If there exists a sequence a_0, a_1, ... as above, (a_{i-1}, a_i) belongs to every congruence on S that includes R. Conversely, $\widehat{R}$ is a congruence that includes R. ∎

In the above situation, the congruence $\widehat{R}$ is said to be *generated by* R.

Proposition 1.13. (free exist) *Assume that $\vec{I}$ is a family of algebraic identities, and X is any nonempty set. Then there exists a free $\vec{I}$-system based on X, namely the system $T_X/\equiv_{X,\vec{I}}$, where $\equiv_{X,\vec{I}}$ is the congruence on T_X generated by all T_X-instances of identities in $\vec{I}$.*

Proof. Let $\equiv_{X,\vec{I}}$ be the congruence generated on T_X by all T_X-instances of identities in $\vec{I}$. By definition, the operations of T_X induce well defined operations on the quotient set F_X. First, we claim that F_X is an $\vec{I}$-system. This amounts to verifying that, for each identity (t, t') in the sequence $\vec{I}$, every instance of (t, t') in F_X is a pair of the form (a, a). Now, an instance of (t, t') in F_X is the projection of an instance of (t, t') in T_X, *i.e.*, it it is (t^f, t'^f) for some substitution f. By definition, the congruence $\equiv_{X,\vec{I}}$ contains all such instances, and, therefore, the image of any such pair in F_X has the form (a, a).

It is clear that X generates the system F_X, since X generates the system T_X.

Assume that S is an $\vec{I}$-system, and f is a mapping of X into S. Let $\widehat{f}$ denote the unique extension of f into a morphism of T_X into S. In order to show that $\widehat{f}$ induces a well defined mapping of F_X into S, we have to verify that $t \equiv_{X,\vec{I}} t'$ implies $\widehat{f}(t) = \widehat{f}(t')$. The relation $\widehat{f}(t) = \widehat{f}(t')$ is a congruence on T_X, and $\equiv_{X,\vec{I}}$ is, by definition, the smallest congruence that includes all T_X-instances of the identities in $\vec{I}$. Hence, in order to show that the first congruence includes the latter, it suffices to show that $\widehat{f}(t) = \widehat{f}(t')$ holds when (t, t') is an instance of an identity in $\vec{I}$. But, in this case, $(\widehat{f}(t), \widehat{f}(t'))$ is an S-instance of the considered identity, and, because S is supposed to be an $\vec{I}$-system, this pair has the form (a, a). So the mapping $\widehat{f}$ induces a well defined morphism of F_X into S, which extends f by construction. ∎

Remark 1.14. Nothing in our hypotheses forbids that some identity of the form $x = y$ appears in $\vec{I}$, or, more generally, that $x \equiv_{X,\vec{I}} y$ holds for two distinct variables x, y. In this case the projection of X onto $T_X/\equiv_{X,\vec{I}}$ is not injective: all elements of X are collapsed to a single element, and we could wonder how an arbitrary mapping of X into S could be extended when f is not constant on X. This actually cannot happen, for, in this degenerated case, all $\vec{I}$-systems have one element, as they satisfy the identity $x = y$.

Free LD-quasigroups

The previous general argument implies that free LD-systems exist. However, the description given by Proposition 1.13 (free exist) is not effective. Indeed, the absolutely free systems T_X are concrete objects, but defining the congruence $\equiv_{X,\vec{I}}$ as the congruence generated by the T_X-instances of identities in $\vec{I}$ does not suffice in general to describe this congruence *effectively*, in the sense of specifying an algorithmic procedure that decides whether two terms are or not equivalent. This question is called the *word problem* for $\vec{I}$ (on X). Answering it in the case of the identity (LD) is one of the aims of this chapter.

At the moment, we study the case of free LD-quasigroups. Describing such systems turns out to be easy. As in the case of free groups, we shall define a concrete structure and show that it is free using a family of distinguished representatives. We shall use the following criterion.

Lemma 1.15. *Assume that NT_X is a subset of T_X such that every term in T_X is $\vec{I}$-equivalent to at least one term in NT_X, and that S is an $\vec{I}$-system generated by X such that distinct terms in NT_X receive distinct evaluations in S. Then S is free based on X.*

Proof. By Proposition 1.13 (free exist), S is a quotient of the free $\vec{I}$-system $T_X/{\equiv_{X,\vec{I}}}$. In order to show that S is free, it suffices to show that the associated morphism is injective. Assume that t_1, t_2 are non-$\vec{I}$-equivalent terms. Our aim is to show $t_1^S \neq t_2^S$. By hypothesis, there exist terms t'_1, t'_2 in NT_X equivalent to t_1 and t_2 respectively. Because S is an $\vec{I}$-system, we have $t_1^S = {t'_1}^S$ and $t_2^S = {t'_2}^S$. As t_1 and t_2 are not $\vec{I}$-equivalent, t'_1 and t'_2 are also not $\vec{I}$-equivalent. So, by hypothesis, we have ${t'_1}^S \neq {t'_2}^S$, hence $t_1^S \neq t_2^S$. ∎

In the situation of Lemma 1.15, the distinguished terms are usually called normal terms. We show now that such normal terms exist in the case of free LD-quasigroups.

We denote by (LDQ) the family of those three identities that define LD-quasigroups, namely (LD) and the two mixed identities $x{\wedge}(x{\vee}y) = y$, $x{\vee}(x{\wedge}y) = y$. We use $T_{(X,\wedge,\vee)}$ for the corresponding system of terms, *i.e.*, those terms on X constructed using two binary operations denoted $\wedge$ and $\vee$. We recall that $a{\wedge}b{\wedge}c$ stands for $a{\wedge}(b{\wedge}c)$. We shall use $*^{\pm 1}$ to denote $\wedge$ in the case of $+1$, and $\vee$ in the case of -1.

Lemma 1.16. *Let NT_X be the set of those terms in $T_{(X,\wedge,\vee)}$ of the form $x_1 *^{e_1} x_2 *^{e_2} \cdots *^{e_{p-1}} x_p *^{e_p} x$, with $x_1, \ldots, x_p, x \in X$, $e_1, \ldots, e_p = \pm 1$ and $x_i = x_{i+1}$ implying $e_i = e_{i+1}$. Then every term in $T_{(X,\wedge,\vee)}$ is LDQ-equivalent to a term in NT_X.*

Proof. We use induction on the size of the term t. If t is a variable, the result is obvious. Otherwise, assume $t = t_1 *^e t_2$. By induction hypothesis, t_1 is *LDQ*-equivalent to some term t_1' in NT_X, say $y_1 *^{d_1} \cdots y_q *^{d_q} y$, and t_2 is *LDQ*-equivalent to some term t_2' in NT_X, say $z_1 *^{c_1} \cdots z_r *^{c_r} z$, where $y_1, \ldots, z$ are variables. Hence t is *LDQ*-equivalent to $t_1' *^e t_2'$. We use induction on q. For $q = 0$, *i.e.*, $t_1' = y$, if we have $y \neq z_1$, or $y = z_1$ and $e = c_1$, then $t_1' *^e t_2'$ belongs to NT_X; if we have $y = z_1$ and $e = -c_1$, then $t_1' *^e t_2'$ is LDQ-equivalent to $z_2 *^{c_2} \cdots z_r *^{c_r} z$, an element of NT_X. Assume now $q > 0$. We have $t_1' = y_1 *^{d_1} t_1''$, with $t_1'' = y_2 *^{d_2} \cdots y_q *^{d_q} y$, and t is LDQ-equivalent to $(y_1 *^{d_1} t_1'') *^e t_2'$. Now we have

$$(y_1 *^{d_1} t_1'') *^e t_2' = (y_1 *^{d_1} t_1'') *^e y_1 *^{d_1} y_1 *^{-d_1} t_2' =_{\scriptscriptstyle LDQ} y_1 *^{d_1} t_1'' *^e y_1 *^{-d_1} t_2'.$$

By induction hypothesis, the term $t_1'' *^e y_1 *^{-d_1} t_2'$, which corresponds to the value $q-1$ of the induction parameter, is *LDQ*-equivalent to a term t'' in NT_X, say $t'' = x_1 *^{e_1} \cdots x_p *^{e_p} x$. Hence t is LDQ-equivalent to $y_1 *^{d_1} t''$. For $y_1 \neq x_1$, and for $y_1 = x_1$ with $d_1 = e_1$, t'' belongs to NT_X. Otherwise, *i.e.*, for $y_1 = x_1$ with $d_1 = -e_1$, t'' is LDQ-equivalent to $x_2 *^{e_2} \cdots x_p *^{e_p} x$, an element of NT_X. ∎

Proposition 1.17. (free LD-quasigroups) *Assume that X is a nonempty set. Let FG_X be a free group based on X, and $^\wedge$ be the half-conjugacy operation defined on $\mathrm{FG}_X \times X$ by $(a, x)^\wedge(b, y) = (axa^{-1}b, y)$. Then $(\mathrm{FG}_X \times X, ^\wedge)$ is a free LD-quasigroup based on $\{1\} \times X$.*

Proof. First $(\mathrm{FG}_X \times X, ^\wedge)$ is an LD-quasigroup, as was seen in Example I.2.14. The fact that the elements $(1, x)$ generate all of $\mathrm{FG}_X \times X$ when both $^\wedge$ and $_\vee$ are used follows from the formulas

$$(1, x) \wedge (a, y) = (xa, y) \qquad \text{and} \qquad (1, x) \vee (a, y) = (x^{-1}a, y). \tag{1.1}$$

To show that $\mathrm{FG}_X \times X$ is free, we apply the criterion of Lemma 1.15 with the normal terms of Lemma 1.16. By the above formula the evaluation of the term $x_1{}^{\wedge e_1} \cdots x_p{}^{\wedge e_p} x$ in $\mathrm{FG}_X \times X$ is $(x_1^{e_1} \cdots x_p^{e_p}, x)$, and therefore different terms have different evaluations. ∎

In the particular case of rank 1, the free group FG_X is the group $\mathbf{Z}$, and the free LD-quasigroup is $\mathbf{Z}$ equipped with the two operations $x^\wedge y = y+1$, $x_\vee y = y-1$.

Observe that the previous result implies that the normal form of Lemma 1.16 is unique, *i.e.*, different normal terms cannot be equivalent: this was not proved in the existence lemma and follows only from our having found a system where different terms receive different values.

The word problem for a family of identities $\vec{I}$ is the question of algorithmically recognizing whether two given terms are or not $\vec{I}$-equivalent. In the case of (LDQ), the answer is immediate:

Proposition 1.18. (**word problem**) *The word problem for (LDQ) is decidable.*

Proof. For t, t' in $T_{(X,^\wedge,_\vee)}$, we decide whether $t =_{LDQ} t'$ holds by evaluating t and t' in $\mathrm{FG}_X \times X$ equipped with the half-conjugacy operations (1.1)—or, equivalently, by computing the normal terms LDQ-equivalent to t and t' respectively, and checking whether the latter coincide. ■

Exercise 1.19. (**length**) *Show that the length of a size k term (as a word) is $2k - 1$.*

Exercise 1.20. (**polynomials**) *Assume that R is a ring. Verify that commutative associative R-algebras form an equational variety with respect to the set of operations comprising addition, multiplication, and multiplication by λ for each scalar λ (the latter being unary operations). Let X be a nonempty set. Prove that the algebra $R[X]$ of all polynomials with coefficients in R and variables in X is free based on X in this variety.*

Exercise 1.21. (**free n-symmetric LD-quasigroup**) *Adapt the construction of free LD-quasigroups to the case of the free n-symmetric LD-quasigroups of Exercise I.2.25. [Hint: Replace the free group FG_X with the group generated by X subject to the relation $x^n = 1$.]*

Exercise 1.22. (**free quandles**) *Define a quandle (or LDI-quasigroup) to be an LD-quasigroup $(Q, ^\wedge, _\vee)$ satisfying in addition the identities $x^\wedge x = x_\vee x = x$. Prove that the free quandle based on X is the free group FG_X equipped with the two conjugacy operations $a^\wedge b = aba^{-1}$, $a_\vee b = a^{-1}ba$.*

Exercise 1.23. (**completely free**) *(i) Assume that V is an equational variety, and S belongs to V. Say that a subset X of S is completely free in S if, for every system S' of V and every map f of X to S', some homomorphism of S to S' extends f. Prove that every free family in a vector space is completely free.*

(ii) Show that the previous implication fails for groups and monoids. [Hint: Consider for instance the subset $\{2\}$ in the group $(\mathbf{Z}, +)$, and the subset $\{xy, xyy\}$ in the monoid $\{x, y\}^*$.]

---

2. The Absorption Property

We turn to the case of free LD-systems. Convenient for LD-quasigroups, a naive normal form approach is not sufficient here. We shall see in Chapter VI that normal forms can be constructed in the case of LD-systems, but at the expense of using more sophisticated results. However, we shall see first that other approaches can be developed for describing free systems.

The congruence $=_{LD}$ on terms

Since being an LD-system means satisfying the algebraic identity (LD), the general framework of Section 1 applies, and, in particular, Proposition 1.13 (free exist) tells us that free LD-systems exist. More precisely, if X is any nonempty set, and if $=_{LD,X}$ denotes the congruence on T_X generated by all T_X-instances of Identity (LD), then the binary system $T_X/{=_{LD,X}}$ is a free LD-system based on X. We recall that T_X denotes the set of all well formed terms constructed from variables in X using a binary operation. As only one binary operation is involved here, as well as (almost) everywhere in Part B of this text, there is no problem in using the symbol $\cdot$ for the operation, with the standard convention that the symbol is often skipped. However, we shall not skip it in terms, so as to avoid the possible confusion between the term $t_1{\cdot}t_2$, which, by definition, is the word $t_1t_2\bullet$, and the word t_1t_2 obtained by concatenating t_1 and t_2—which is not a term.

Notation 2.1. (free LD-systems) (i) For X a nonempty set, the free LD-system based on X will be denoted by FLD_X. For t in T_X, the equivalence class of t with respect to $=_{LD,X}$ is called its *LD-class*, and it is denoted by $\bar{t}$. Terms in T_X that are equivalent with respect to the relation $=_{LD,X}$ will be said to be *LD-equivalent*.

(ii) We fix an infinite sequence $x_1, x_2, \ldots$, and, for $1 \le n \le \infty$, we write T_n and FLD_n for the set $T_{\{x_i\ ;\ 1 \le i \le n\}}$ and for the free LD-system based on $\{x_i\ ;\ 1 \le i \le n\}$ respectively. When dealing with FLD_1 and T_1, we write x for x_1.

By construction, every element of FLD_X is an LD-class. For a in FLD_X, every term t in T_X such that a is the LD-class of t will be called an *expression* of a. By Proposition 1.2 (free), every free LD-system of finite or countable rank is isomorphic to a system FLD_n with $1 \le n \le \infty$.

The first result we shall establish here is that the congruence $=_{LD,X}$ does not depend on the set X. Indeed, we have the following result:

Proposition 2.2. (independence) *Assume that the set X is included in the set X'. Then two terms in T_X are equivalent with respect to $=_{LD,X'}$ if and only if they are equivalent with respect to $=_{LD,X}$.*

Once this is proved, we shall write $=_{LD}$ instead of $=_{LD,X}$. For proving Proposition 2.2—and for many future uses—we introduce the relations corresponding to applying (LD) once.

Definition 2.3. (basic expansion) We say that the term t' is a *basic LD-expansion* of the term t if t' is obtained from t by replacing some subterm of the form $t_1{\cdot}(t_2{\cdot}t_3)$ with the corresponding term $(t_1{\cdot}t_2){\cdot}(t_1{\cdot}t_3)$.

***Example 2.4.** (**basic expansion**) Complicated terms are not easy to read. We improve readability by associating with every term a labelled binary tree using the convention that the tree associated with the variable x reduces to a single vertex labelled x, and that the tree associated with $t_1{\cdot}t_2$ consists of a root labelled $\cdot$ with two successors, namely a left one which is the tree associated with t_1, and a right one which is the tree associated with t_2. In this way, the tree*

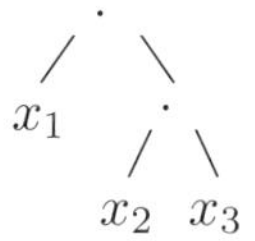

represents the term $x_1{\cdot}(x_2{\cdot}x_3)$. Since only one operator is involved here, we shall usually not mention it, thus writing 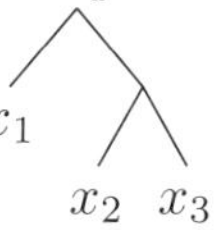

for the latter term. Similarly, we can forget about the variable when terms with one variable are involved, thus writing  *for $x{\cdot}(x{\cdot}x)$.*

With this representation, performing a basic LD-expansion means replacing some subterm, i.e., some subtree, of the form

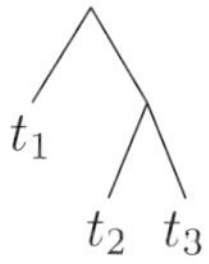

by the corresponding

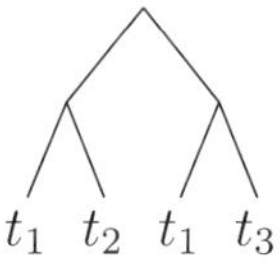

.

For instance, here is a sequence of successive basic LD-expansions:

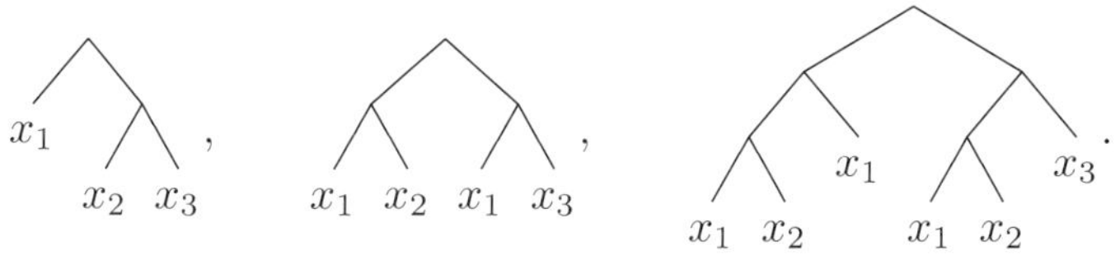

By construction, if the term t' is a basic LD-expansion of the term t, then t' is LD-equivalent to t. The converse need not be true in general, but we deduce from Lemma 1.12 the following criterion:

Lemma 2.5. *(i) The congruence $=_{LD,X}$ on T_X is the equivalence relation generated by those pairs (t, t') such that t' is a basic LD-expansion of t.*

(ii) Assume that t, t' are terms in T_X. Then $t =_{LD,X} t'$ holds if and only if there exists a finite sequence $t_0, \ldots, t_k$ with $t_0 = t$, $t_k = t'$ such that, for $i < k$, either t_{i+1} is a basic LD-expansion of t_i or t_i is a basic LD-expansion of t_{i+1}.

It is then easy to prove Proposition 2.2 (independence).

Proof. Assume that t, t' belong to T_X and $t =_{LD,X'} t'$ holds. By Lemma 2.5, there exists a finite sequence of terms $t_0 = t$, t_1, ..., $t_k = t'$ in $T_{X'}$ such that, for every intermediate index i, the term t_{i+1} is an LD-expansion of t_i or t_i is an LD-expansion of t_{i+1}. Now, inductively, the variables occurring in the terms t_1, t_2, ... are the same as those occurring in t. Hence all terms t_1, t_2, ... belong to T_X, and, from the definition of a basic LD-expansion, $t =_{LD,X} t'$ holds. The converse implication is obvious. ∎

It should be clear that the previous result is not specific to Identity (LD), and that a similar result holds for every algebraic identity such that the same variables occur on both sides.

Another application of Lemma 2.5 is the following result.

Proposition 2.6. (**basis**) *A free LD-system has a unique basis.*

Proof. Let X be an arbitrary set. By Lemma 2.5, the elements of X are alone in their LD-equivalence class, as, by construction, a variable admits no basic LD-expansion. Hence every set that generates FLD_X must include X. Now a subset of FLD_X that properly includes X cannot be free. ∎

By the previous result, the *rank* of a free LD-system is unique. Thus FLD_n is, up to isomorphism, the unique free LD-system of rank n, and, for $n \neq p$, FLD_n is not isomorphic to FLD_p.

Powers in left self-distributive algebra

Before turning to the general study of LD-equivalence, we establish here some computational rules for powers in LD-systems. This study leads us to a phenomenon that we call absorption by right powers. It plays a crucial role in the sequel, in particular in the proof of the comparison property.

In the framework of associativity, there is one notion of power only: for every element g of an associative system, the powers g^p of g may be defined using multiplication on the left or on the right with the same result. Moreover, every element in the semigroup generated by g is a power of g, which means that every term $t(x)$ is equivalent up to associativity to a term of the form x^p. When left self-distributivity replaces associativity, several types of powers can be considered, and it is not true in general that the sub-LD-system generated by an element g consists of powers of g only.

Definition 2.7. **(powers)** Assume that $(S, \cdot)$ is a binary system, and a is an element of S. For $p \geq 1$, we respectively define the p-th *left power* $a_{[p]}$ and the p-th *right power* $a^{[p]}$ of a by

$$a_{[1]} = a^{[1]} = a, \quad \text{and} \quad a_{[p]} = a_{[p-1]} \cdot a, \quad a^{[p]} = a \cdot a^{[p-1]} \quad \text{for } p \geq 2.$$

In particular, we use the previous notation for terms, *i.e.*, in the case of the absolutely free systems T_X. For instance, if x is a variable, $x^{[3]}$ denotes the term $x{\cdot}(x{\cdot}x)$, while $x_{[3]}$ denotes the term $(x{\cdot}x){\cdot}x$.

Powers obey some rules in the left self-distributive framework. In each case we give two equivalent forms of the result, one involving LD-equivalence of terms, and one involving elements of LD-systems.

Lemma 2.8. *(i) For all terms t_1, t_2 in T_X, we have*

$$t_1 \cdot t_{2[p]} =_{LD} (t_1 \cdot t_2)_{[p]}, \qquad t_1 \cdot t_2^{[p]} =_{LD} (t_1 \cdot t_2)^{[p]} \qquad \text{for } p \geq 1. \tag{2.1}$$

(ii) For every LD-system S, and for all a, b in S, we have

$$a \cdot b_{[p]} = (a \cdot b)_{[p]}, \qquad a \cdot b^{[p]} = (a \cdot b)^{[p]} \qquad \text{for } p \geq 1. \tag{2.2}$$

Proof. Powers are defined from the operation, and, therefore, they are preserved under homomorphism. As left translations are homomorphisms in LD-systems, we deduce (2.2). Applying the formulas in FLD_X gives (2.1). ∎

Lemma 2.9. *(i) For every term t in T_X, we have*

$$t^{[p]} \cdot t^{[q]} =_{LD} t^{[q+1]}, \qquad \text{for } q \geq p \geq 1, \tag{2.3}$$

$$(t^{[p]})^{[q]} =_{LD} t^{[p+q-1]}, \quad \text{for } p,\ q \geq 1. \tag{2.4}$$

(ii) For every LD-system S, and for every a in S, we have

$$a^{[p]} \cdot a^{[q]} = a^{[q+1]}, \qquad \text{for } q \geq p \geq 1, \tag{2.5}$$

$$(a^{[p]})^{[q]} = a^{[p+q-1]}, \quad \text{for } p,\ q \geq 1. \tag{2.6}$$

Proof. For (2.3), we use induction on $p \geq 1$. For $p = 1$, it is the definition of the right power. Otherwise, we find

$$t^{[p]} \cdot t^{[q]} = (t \cdot t^{[p-1]}) \cdot (t \cdot t^{[q-1]}) =_{LD} t \cdot (t^{[p-1]} \cdot t^{[q-1]}) =_{LD} t \cdot t^{[q]} = t^{[q+1]}.$$

Relation (2.4) is established using induction on $q \geq 1$. For $q = 1$, the equality is trivial. Otherwise, using (2.3), we find

$$(t^{[p]})^{[q]} = t^{[p]} \cdot (t^{[p]})^{[q-1]} =_{LD} t^{[p]} \cdot t^{[p+q-2]} =_{LD} t^{[p+q-1]}.$$

The previous relations give (2.5) and (2.6) in FLD_X, and, from there, in every LD-system. ■

We mentioned that the sub-LD-system S_g of an LD-system S generated by an element g need not consist of powers of g only (*cf.* Exercise 2.15). However, it turns out that the right powers of g with sufficiently high exponent absorb every element of S_g in the sense below. In order to state the result precisely, we need to introduce two parameters measuring the complexity of a term.

Definition 2.10. (height, right height) For t a term, the *height* $\mathrm{ht}(t)$ and the *right height* $\mathrm{ht}_R(t)$ are defined inductively by $\mathrm{ht}(t) = \mathrm{ht}_R(t) = 1$ for t a variable, and $\mathrm{ht}(t) = \sup(\mathrm{ht}(t_1), \mathrm{ht}(t_2)) + 1$, $\mathrm{ht}_R(t) = \mathrm{ht}_R(t_2) + 1$ for $t = t_1 \cdot t_2$.

For instance, the height of the term $(x \cdot x) \cdot x$ is 3, while its right height is 2. Observe that the height of a term is the height of the associated binary tree, *i.e.*, the maximal length of a branch, while its right height is the length of its rightmost branch. By a trivial induction, the inequality $\mathrm{ht}_R(t) \leq \mathrm{ht}(t)$ holds for every term t.

Proposition 2.11. (absorption property) *(i) For every term t in T_1, we have*

$$t \cdot x^{[p]} =_{LD} x^{[p+1]} \qquad \text{for } p \geq \mathrm{ht}(t). \tag{2.7}$$

(ii) Assume that S is a monogenic LD-system, and g is a generator of S. Then, for every a in S, we have

$$a \cdot g^{[p]} = g^{[p+1]} \qquad \text{for } p \text{ large enough.} \tag{2.8}$$

Proof. (i) We use induction on the size of t. For $t = x$, (2.7) is true by definition, the LD-equivalence being an equality. Assume $t = t_1 \cdot t_2$, and $p \geq \mathrm{ht}(t)$, hence $p - 1 \geq \mathrm{ht}(t_1)$ and $p - 1 \geq \mathrm{ht}(t_2)$. We have $t \cdot x^{[p]} = (t_1 \cdot t_2) \cdot x^{[p]}$, and, applying the induction hypothesis, we find

$$t \cdot x^{[p]} =_{LD} (t_1 \cdot t_2) \cdot (t_1 \cdot x^{[p-1]}) =_{LD} t_1 \cdot (t_2 \cdot x^{[p-1]}) =_{LD} t_1 \cdot x^{[p]} =_{LD} x^{[p+1]}.$$

Point (ii) for FLD_1 is a restatement of (i), and, by projection, the result follows for every monogenic LD-system. ■

Proposition 2.12. (attraction property) *(i) For every term t in T_1, we have*

$$t^{[p - \mathrm{ht}_R(t) + 1]} =_{LD} x^{[p]} \qquad \text{for } p \geq \mathrm{ht}(t). \tag{2.9}$$

(ii) Assume that S is a monogenic LD-system, and g is a generator of S. Then, for every a in S, there exists a nonnegative integer h satisfying

$$a^{[p-h]} = g^{[p]} \qquad \text{for } p \text{ large enough.} \tag{2.10}$$

Proof. (i) We use induction on t again. For $t = x$, (2.9) is an equality for $p \geq 1$. Otherwise, assume $t = t_1{\cdot}t_2$, and $p \geq \mathrm{ht}(t)$. Using (2.1), we find

$$t^{[p-\mathrm{ht}_R(t)+1]} = (t_1 \cdot t_2)^{[p-\mathrm{ht}_R(t)+1]} =_{LD} t_1 \cdot t_2^{[p-\mathrm{ht}_R(t)+1]} = t_1 \cdot t_2^{[p-1-\mathrm{ht}_R(t_2)+1]},$$

hence, by applying the induction hypothesis for t_2 using $p - 1 \geq \mathrm{ht}(t_2)$, and for t_1 using $p - 1 \geq \mathrm{ht}(t_2)$ twice,

$$t^{[p-\mathrm{ht}_R(t)+1]} =_{LD} t_1 \cdot x^{[p-1]} =_{LD} t_1 \cdot t_1^{[p-1-\mathrm{ht}_R(t_1)+1]} = t_1^{[p-\mathrm{ht}_R(t_1)+1]} =_{LD} x^{[p]}$$

As usual, Point (ii) follows by projection. ∎

Exercise 2.13. (powers) *Prove the equivalences*

$x_{[p]}{\cdot}x^{[q]} =_{LD} x^{[q+1]}$ *and* $(x_{[p+1]})^{[p]} =_{LD} x^{[p+1]}$

for $q \geq p \geq 1$. [Hint: Apply Proposition 2.11 (absorption).]

Exercise 2.14. (mixed left powers) *Assume that S is an LD-system. For a, b in S, define the p-th mixed left power of a and b by $a{\cdot}_{[0]}b = a$ and $a{\cdot}_{[p]}b = (a{\cdot}_{[p-1]}b){\cdot}b$ for $p \geq 1$. Prove that the identity $x{\cdot}(y{\cdot}_{[p]}z) = (x{\cdot}y){\cdot}_{[p]}(x{\cdot}z)$ holds in S for every p.*

Exercise 2.15. (tables T_m) *(i) Define, for $m \geq 1$, T_m to be the set $\{1, \ldots, m+1\}$ equipped with the operation*

$$a * b = \begin{cases} m+1 & \text{for } a \leq m, \text{ or } b = m+1, \\ b+1 & \text{for } a = m+1 \text{ and } b < m, \\ 1 & \text{for } a = m+1 \text{ and } b = m. \end{cases}$$

Show that T_m is an LD-system and that every element a with $1 \leq a \leq m$ is a generator of T_m.

(ii) Assume $1 \leq a \leq m$, and let $b = (m+1) * a$, i.e., $b = a + 1$ for $a \neq m$, and $b = 1$ for $a = m$. Prove the equalities $a^{[p]} = a_{[2q]} = m + 1$ and $a_{[2q+1]} = b$ for $p \geq 2$ and $q \geq 1$. [For $m \geq 3$, this gives an example of a monogenic LD-system containing elements that are neither left nor right powers of the generator(s).]

Exercise 2.16. (**free infinite**) *Assume that X is a nonempty set. Show that all terms of the form $(\cdots((x_1{\cdot}x_2){\cdot}x_3)\cdots){\cdot}x_p$ with $x_1, \ldots, x_p$ in X are alone in their LD-class. Deduce that every free LD-system is infinite.*

Exercise 2.17. (**substitution**) *Prove that every substitution $f : X \to T_X$ induces an endomorphism of the free LD-system* FLD_X.

Exercise 2.18. (**LD-invariants**) *For f a mapping of T_X into a set S, say that f is an LD-invariant if $t =_{LD} t'$ implies $f(t) = f(t')$, and that f is a strong LD-invariant if it is an LD-invariant and, in addition, the value of $f(t_1{\cdot}t_2)$ depends only on the values of $f(t_1)$ and $f(t_2)$.*

(i) Prove that f is an LD-invariant if and only if $f(t') = f(t)$ holds whenever t' is a basic LD-expansion of t.

(ii)Prove that f is a strong LD-invariant if and only if $f(t_1{\cdot}(t_2{\cdot}t_3)) = f((t_1{\cdot}t_2){\cdot}(t_1{\cdot}t_3))$ holds for all terms t_1, t_2, t_3, and the conjunction of $f(t_1) = f(t'_1)$ and $f(t_2) = f(t'_2)$ implies $f(t_1{\cdot}t_2) = f(t'_1{\cdot}t'_2)$.

(iii) Assume that S is an LD-system, and φ is a homomorphism of FLD_X *into S. Show that the mapping f defined on T_X by $f(t) = \varphi(\overline{t})$ is a strong LD-invariant. Conversely, assume that f is a strong LD-invariant, and let S be the image of f. Show that $f(t_1){\cdot}f(t_2) = f(t_1{\cdot}t_2)$ defines a well defined left self-distributive operation on S, and f induces a homomorphism of* FLD_X *onto S. [In the situation above, we shall say that the LD-system S is associated with the LD-invariant f.]*

Exercise 2.19. (**right and left variable**) *(i) Let $\mathrm{var}_R(t)$ denote the rightmost variable of the term t. Check that var_R is a strong LD-invariant on T_X (Exercise 2.18). What is the associated LD-system?*

(ii) Same questions with the leftmost variable var_L.

Exercise 2.20. (**right height**) *Show that the right height is a strong LD-invariant (Exercise 2.18). What is the associated LD-system? Is the height an LD-invariant?*

Exercise 2.21. (**medial**) *Let λ be a fixed complex number. For t in T_X, define $f(t)$ in the* **C**-*span of X inductively by $f(t) = t$ for t a variable, and $ft) = (1-\lambda)f(t_1) + \lambda f(t_2)$ for $t = t_1{\cdot}t_2$. Check that f is an LD-invariant (Exercise 2.18), and give a direct geometric definition of the value of $f(t)$ in terms of the tree associated with t.*

Exercise 2.22. **(list of variables)** *(i) Prove that the set of all variables occurring in a term is a strong LD-invariant (Exercise 2.18).*

(ii) For t a term, define $\mathrm{Var}(t)$ to be x if t is the variable x, and to be the concatenation of $\mathrm{Var}(t_1)$ and of the subsequence of $\mathrm{Var}(t_2)$ made of those variables that do not occur in $\mathrm{Var}(t_1)$ for $t = t_1{\cdot}t_2$. For instance, we have $\mathrm{Var}((x{\cdot}y){\cdot}(y{\cdot}z)) = xyz$. Prove that Var is a strong LD-invariant. Show that the table

	x	y	xy	yx
x	x	xy	xy	xy
y	yx	y	yx	yx
xy	xy	xy	xy	xy
yx	yx	yx	yx	yx

specifies the associated operation for $X = \{x, y\}$.

Exercise 2.23. **(Burau LD-invariant)** *For t in T_1, define $f(t)$ to be the evaluation of t in $\mathrm{GL}_\infty(\mathbf{Z}[t, t^{-1}])$ that maps x to the identity matrix (1) and uses the exponentiation of matrices defined in Section I.3. For instance, we have*

$$f(x) = (1), \quad f(x_{[2]}) = \begin{pmatrix} 1-t & t \\ 1 & 0 \end{pmatrix}, \text{ etc.}$$

Let $f^i_j(t)$ be the (i,j)-entry of the matrix $f(t)$. Show that f^i_j is an LD-invariant for all i, j, but not a strong LD-invariant (Exercise 2.18).

Exercise 2.24. **(polynomial invariants)** *Assume that $P(x,y)$ is a degree 1 polynomial. Show that there exists an LD-invariant f satisfying $f(t_1{\cdot}t_2) = P(f(t_1), f(t_2))$ if and only if $P(x,y)$ has the form $(1-a)x + ay$ (Exercise 2.18). Extend the result to degree 2 polynomials.*

3. The Confluence Property

The main notion in the sequel is the notion of an LD-expansion. It refines LD-equivalence: the term t' is said to be an LD-expansion of t if t' is LD-equivalent to t and, in addition, one can transform t to t' using Identity (LD) in the expanding direction only, *i.e.*, composing finitely many basic LD-expansions. The aim of this section is to prove the following result:

Proposition 3.1. (**confluence property**) *Two terms are LD-equivalent if and only if they admit a common LD-expansion.*

This property, which we shall see in Chapter VII expresses that every element in a certain group can be written as a fraction, is also connected with the existence of right lcm's. We shall establish it by introducing, for every term t, a distinguished family of LD-expansions $\partial^k t$ of t, $k \geq 0$, and proving that, if t' is LD-equivalent to t, then $\partial^k t$ is an LD-expansion of t' for k large enough, *i.e.*, we prove that the family of all $\partial^k t$ is cofinal in the LD-class of t with respect to the relation of being an LD-expansion.

LD-Expansions

Definition 3.2. (**LD-expansion**) Assume $k \geq 0$. We say that the term t' is a k-*LD-expansion* of the term t if there exists a finite sequence of terms $t_0 = t$, t_1, $\ldots$, $t_k = t'$ such that, for each i, t_i is a basic LD-expansion of t_{i-1} or $t_i = t_{i-1}$ holds. We say that t' is an *LD-expansion* of t if it is a k-LD-expansion of t for some k.

Thus, the 1-LD-expansions of t consist of t itself and of all basic LD-expansions of t. The following property is clear from the definition:

Lemma 3.3. *Assume that t'_1 is a k_1-LD-expansion of t_1, and t'_2 is a k_2-LD-expansion of t_2. Then $t'_1 \cdot t'_2$ is a $k_1 + k_2$-LD-expansion of $t_1 \cdot t_2$.*

Our aim is to find some structure in the family of all LD-expansions of a given term t, which is infinite in general. The question is difficult, and, as we shall see in Chapter IX, it remains partially open. At the moment, we introduce some distinguished LD-expansions of t, by using a certain iterated distribution process.

Definition 3.4. (**uniform distribution**) Let t_0, t be terms. The result of *uniformly distributing* t_0 in t, denoted $t_0 * t$, is defined to be t^f, where f is the substitution that maps every variable x to $t_0 \cdot x$.

By definition of a substitution, the following inductive relation holds:

$$t_0 * t = \begin{cases} t_0 \cdot t & \text{if } t \text{ is a variable,} \\ (t_0 * t_1) \cdot (t_0 * t_2) & \text{for } t = t_1 \cdot t_2. \end{cases}$$

Example 3.5. *(**uniform distribution**) Assume* $t_0 =$  *and* $t =$ 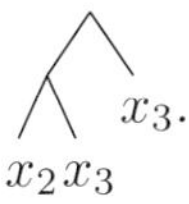 *Then* $t_0 * t$ *is* 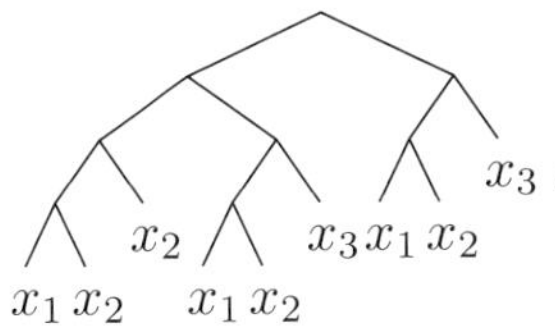 *:* $t_0 * t$ *is obtained from* $t_0{\cdot}t$ *by distributing* t_0 *until the variables of* t *are reached, but not beyond this point.*

Lemma 3.6. *(i) For all* t_0, t, *the term* $t_0 * t$ *is an LD-expansion of* $t_0{\cdot}t$.

(ii) Assume that t_0' *is an LD-expansion of* t_0, *and* t' *is an LD-expansion of* t. *Then* $t_0' * t'$ *is an LD-expansion of* $t_0 * t$.

(iii) For all t_0, t_1, t, *the term* $(t_0 * t_1) * (t_0 * t)$ *is an LD-expansion of* $t_0 * (t_1 * t)$.

Proof. (i) Use induction on t. If t is a variable, then $t_0 * t = t_0{\cdot}t$ holds. Otherwise, assume $t = t_1{\cdot}t_2$. Then we have $t_0 * t = (t_0 * t_1){\cdot}(t_0 * t_2)$. By induction hypothesis, $t_0 * t_e$ is an LD-expansion of $t_0{\cdot}t_e$ for $e = 1, 2$, and, therefore, $t_0 * t$ is an LD-expansion of $(t_0{\cdot}t_1){\cdot}(t_0{\cdot}t_2)$, which is by definition an LD-expansion of $t_0{\cdot}(t_1{\cdot}t_2)$.

(ii) It suffices to prove the result when t' is either t or a basic LD-expansion of t. We do it using induction on t. If t is a variable, then we have $t' = t$, $t_0' * t' = t_0' * t$, and, by Lemma 3.3, the latter term is an LD-expansion of $t_0 * t$. Otherwise, assume $t = t_1{\cdot}t_2$. By induction hypothesis, $t_0' * t_e$ is an LD-expansion of $t_0 * t_e$ for $e = 1, 2$. For $t' = t$, we have $t_0' * t' = (t_0' * t_1){\cdot}(t_0' * t_2)$, an LD-expansion of $(t_0 * t_1){\cdot}(t_0 * t_2)$, *i.e.*, of $t_0 * t$. Assume now that t' is a basic LD-expansion of t, *i.e.*, is obtained from t by applying distributivity to some subterm s of t. Three cases may occur. If s is a subterm of t_1, we have $t' = t_1'{\cdot}t_2$, where t_1' is the LD-expansion of t_1 obtained by applying distributivity to s. By induction hypothesis, $t_0' * t_1'$ is an LD-expansion of $t_0 * t_1$, and, as above, $t_0' * t'$ is an LD-expansion of $t_0 * t$. The argument is similar if s is a subterm of t_2. The remaining case is $s = t$. Then we have $t = t_1{\cdot}(t_2{\cdot}t_3)$, and $t' = (t_1{\cdot}t_2){\cdot}(t_1{\cdot}t_3)$ for some t_1, t_2, t_3. By induction hypothesis, $t_0' * (t_1{\cdot}t_e)$ is an LD-expansion of $t_0 * (t_1{\cdot}t_e)$ for $e = 2, 3$. Hence $t_0' * t'$, which is $(t_0' * (t_1{\cdot}t_2)){\cdot}(t_0' * (t_1{\cdot}t_3))$, is an LD-expansion of $(t_0 * (t_1{\cdot}t_2)){\cdot}(t_0 * (t_1{\cdot}t_3))$. The latter term is

$$((t_0 * t_1) \cdot (t_0 * t_2)) \cdot ((t_0 * t_1) \cdot (t_0 * t_3)),$$

an LD-expansion of $(t_0 * t_1){\cdot}((t_0 * t_2){\cdot}(t_0 * t_3))$, *i.e.*, of $t_0 * t$.

(iii) We use induction on t. For t a variable, we find

$$(t_0 * t_1) * (t_0 * t) = (t_0 * t_2) * (t_0 \cdot t),$$
$$t_0 * (t_1 * t) = t_0 * (t_1 \cdot t) = (t_0 * t_2) \cdot (t_0 \cdot t),$$

and the result follows from (i). For $t = t_2{\cdot}t_3$, we have

$$\begin{aligned}(t_0 * t_1) * (t_0 * t) &= (t_0 * t_1) * (t_0 * (t_2 \cdot t_3)) = (t_0 * t_1) * ((t_0 * t_2) \cdot (t_0 * t_3))) \\ &= ((t_0 * t_1) * (t_0 * t_2)) \cdot ((t_0 * t_1) * (t_0 * t_3)), \\ t_0 * (t_1 * t) &= t_0 * (t_1 * (t_2 \cdot t_3) \\ &= t_0 * ((t_1 * t_2) \cdot (t_1 * t_3)) = (t_0 * (t_1 * t_2)) \cdot (t_0 * (t_1 * t_3)).\end{aligned}$$

By induction hypothesis, $(t_0 * t_1) * (t_0 * t_e)$ is an LD-expansion of $t_0 * (t_1 * t_e)$ for $e = 2, 3$, and the induction goes through. ∎

A new induction leads to the *dilatation* operator ∂, crucial in the sequel.

Definition 3.7. (**operator** ∂) Assume that t is a term. Then ∂t is the term defined inductively by

$$\partial t = \begin{cases} t & \text{if } t \text{ is a variable,} \\ \partial t_1 * \partial t_2 & \text{for } t = t_1 * t_2. \end{cases}$$

Example 3.8. ***(operator*** ∂***)*** *For* $t = t_1{\cdot}t_2$, *the term* ∂t *is the term obtained by uniformly distributing* ∂t_1 *in* ∂t_2. *For instance, let* $t = x_1{\cdot}(x_2{\cdot}x_3)$. *Since* x_2 *and* x_3 *are variables, we have* $\partial x_2 = x_2$ *and* $\partial x_3 = x_3$, *hence* $\partial(x_2{\cdot}x_3) = \partial x_2 * \partial x_3 = x_2 * x_3 = x_2{\cdot}x_3$. *Then we find*

$$\partial t = \partial x_1 * \partial(x_2 \cdot x_3) = x_1 * (x_2 \cdot x_3) = (x_1 \cdot x_2) \cdot (x_1 \cdot x_3).$$

The reader can check similarly

$\partial^2 t =$ 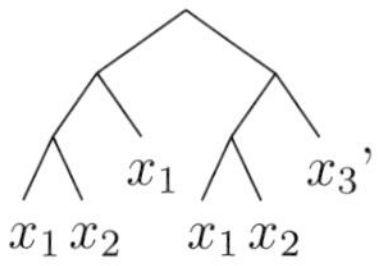

, $\partial^3 t =$ 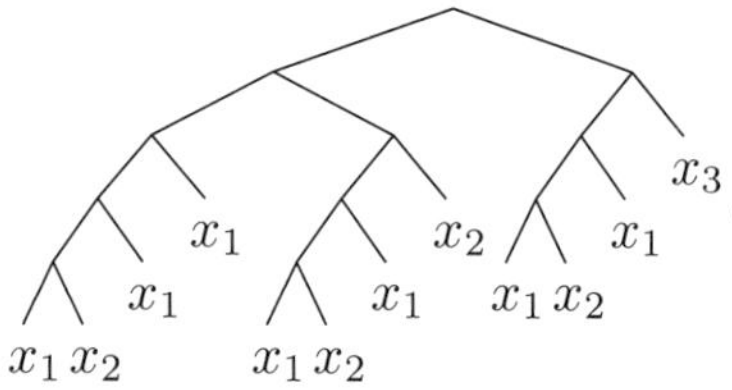

.

In general, the size of $\partial^k t$ *quickly increases with* k*—except for those terms of the form* $x_{[p]}$, *which are fixed under* ∂.

Lemma 3.9. *For every term* t, *we have* $\lg(\partial t) \leq 2^{\lg(t)}$.

Proof. First, we have the general relation

$$\lg(t_0 * t) = (\lg(t_0) + 3)(\lg(t) + 1)/2 - 1. \tag{3.1}$$

Indeed, if t_0 has size p and t has size q, then $t_0 * t$ has size $(p+1)q$, since it is obtained by replacing each variable of t with a term of size $p+1$. The result follows as the length of a term of size r is $2r - 1$.

We now establish the lemma using induction on t. If t is a variable, we have $\partial t = t$, and the result is true. Otherwise, assume $t = t_1 \cdot t_2$, hence $\partial t = \partial t_1 * \partial t_2$. Using (3.1) and the induction hypothesis, we find

$$\lg(\partial t) = (\lg(\partial t_1) + 3)(\lg(\partial t_2) + 1)/2 - 1 \leq (2^{\lg(t_1)} + 3)(2^{\lg(t_2)} + 1)/2 - 1,$$

hence $\lg(\partial t) \leq 2^{\lg(t_1)+\lg(t_2)+1} = 2^{\lg(t)}$. ∎

Lemma 3.10. *For every k, $\partial^k t$ is an LD-expansion of t.*

Proof. It suffices to prove the result for $k = 1$. We use induction on t. If t is a variable, we have $\partial t = t$, and the result is true. Otherwise, assume $t = t_1 \cdot t_2$. By induction hypothesis, ∂t_e is an LD-expansion of t_e for $e = 1, 2$, so $\partial t_1 \cdot \partial t_2$ is an LD-expansion of t, and so is ∂t by Lemma 3.6. ∎

The point now is that ∂t is an LD-expansion of all 1-expansions of t.

Lemma 3.11. *Assume that t' is a basic LD-expansion of t. Then ∂t is an LD-expansion of t'.*

Proof. We use induction on t. The result is vacuously true if t is a variable. Assume $t = t_1 \cdot t_2$. By definition, t' is obtained from t by expanding some subterm s of t into a new subterm s'. Three cases may occur according to the position of the subterm s in t. If s is a subterm of t_1, we have $t' = t'_1 \cdot t_2$, where t'_1 is the basic LD-expansion of t_1 obtained by replacing the considered subterm s with s'. By induction hypothesis, ∂t_1 is an LD-expansion of t'_1, and, by Lemma 3.10, ∂t_2 is an LD-expansion of t_2. Hence $\partial t_1 \cdot \partial t_2$ is an LD-expansion of t', and, by Lemma 3.6, so is ∂t, which is $\partial t_1 * \partial t_2$. The argument is similar if t' is obtained from t by expanding some subterm of t_2. There remains the case $s = t$. In this case, t_2 can written as $t_3 \cdot t_4$. By Lemma 3.10, ∂t is an LD-expansion of $t_1 * t_2$, which is $(t_1 * t_3) \cdot (t_1 * t_4)$, and, therefore, it is an LD-expansion of $(t_1 \cdot t_3) \cdot (t_1 \cdot t_4)$, *i.e.*, of t'. ∎

The second point is that ∂ is increasing with respect to the LD-expansion relation.

Lemma 3.12. *Assume that t' is an LD-expansion of t. Then $\partial t'$ is an LD-expansion of ∂t.*

Proof. It suffices to prove the result when t' is a basic LD-expansion of t. We use induction on t again. The result is vacuously true if t is a variable. Otherwise, assume $t = t_1 \cdot t_2$, and assume that t' is obtained from t by applying distributivity to some subterm s of t. If s is a subterm of t_1, let t'_1 be the term obtained from t_1 by applying distributivity to the considered subterm s. By induction hypothesis, $\partial t'_1$ is an LD-expansion of ∂t_1, and therefore, by

Lemma 3.6, $\partial t'$, which is $\partial t'_1 * \partial t_2$, is an LD-expansion of ∂t, which is $\partial t_1 * \partial t_2$. The argument is similar if s is a subterm of t_1. Finally, for $s = t$, we have $t_2 = t_3{\cdot}t_4$ and $t' = (t_1{\cdot}t_3){\cdot}(t_1{\cdot}t_4)$ for some t_3, t_4. By construction, we have

$$\partial t = \partial t_1 * (\partial t_3 * \partial t_4), \quad \text{and} \quad \partial t' = (\partial t_1 * \partial t_3) * (\partial t_1 * \partial t_4),$$

and we conclude using Lemma 3.6(iii). ■

We are now ready to prove the confluence property, actually a more precise quantitative version of this property. Owing to Lemma 2.5, we can introduce the following complexity measure for LD-equivalence.

Definition 3.13. (LD-distance) For t, t' terms, the *LD-distance* of t and t' is defined to be the minimal number $d_{LD}(t, t')$ of instances of (LD) needed to transform t into t', if t and t' are LD-equivalent, and to be ∞ otherwise.

So, for instance, we have $d_{LD}(t, t') = 0$ for $t = t'$, and $d_{LD}(t, t') = 1$ if t' is a basic LD-expansion of t, or t is a basic LD-expansion of t'.

Proposition 3.14. (quantitative confluence) *Assume $d_{LD}(t, t') = k < \infty$. Then $\partial^k t$ is an LD-expansion of t'.*

Proof. We use induction on k. The result is clear for $k = 0$. Otherwise, there exists a term t'' satisfying $d_{LD}(t, t'') = k - 1$ and $d_{LD}(t'', t') = 1$. By induction hypothesis, $\partial^{k-1} t$ is an LD-expansion of t''. Two cases may occur. Either t'' is a basic LD-expansion of t': in this case, $\partial^{k-1} t$ is an LD-expansion of t', and so is $\partial^k t$; Or t' is a basic LD-expansion of t'': in this case, by Lemma 3.11, $\partial t''$ is an LD-expansion of t'; by Lemma 3.12, $\partial^k t$ is an LD-expansion of $\partial t''$ since $\partial^{k-1} t$ is an LD-expansion of t'', and $\partial^k t$ is an LD-expansion of t'. ■

If t' is a k-LD-expansion of the term t, then $d_{LD}(t, t') \leq k$ holds, so we deduce:

Corollary 3.15. *Assume that the term t' is a k-LD-expansion of the term t. Then $\partial^k t$ is an LD-expansion of t'.*

Remark 3.16. Being an LD-expansion is a decidable relation: when we are given two terms t, t', we can algorithmically decide whether t' is an LD-expansion of t or not by systematically enumerating all LD-expansions of t with length $\lg(t')$ at most. However, the confluence result does not give any effective way for recognizing similarly whether two given terms t, t' are LD-equivalent or not, as it provides no upper bound on the size of their possible common LD-expansion.

Exercise 3.17. (LD-contraction) *Prove that the confluence property fails for the LD-contraction relation, defined to be the inverse of LD-expansion. [Hint: Verify that $x_1{\cdot}x_2{\cdot}x_3{\cdot}x_4$ is LD-equivalent to*

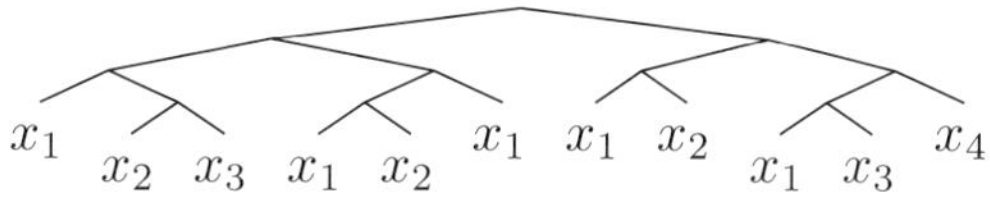

and check that the latter term is an LD-expansion of no smaller term.]

Exercise 3.18. (operator ∂) *Prove $\partial x^{[p+1]} = (\partial x^{[p]})^{[2]}$ for every p. Deduce that $\partial x^{[p]}$ is a complete binary tree of height p. Compare the lengths of $x^{[p]}$ and $\partial x^{[p]}$.*

Exercise 3.19. (subinjective terms) *Define an injective term to be a term where no variable occurs twice, and a subinjective term to be a subterm of an LD-expansion of an injective term. Prove that, if t is injective and s is a subterm of an LD-expansion of t, then the sequence $\mathrm{Var}(s)$ defined in Exercise 2.22 is a subsequence of the sequence $\mathrm{Var}(t)$. Deduce that the term* 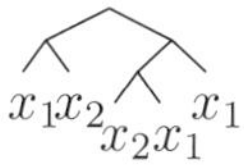

is not subinjective.

Exercise 3.20. (substitution) *Assume that h is a substitution, and t is a term. Prove that the term $\partial(t^h)$ is an LD-expansion of the term $(\partial t)^h$.*

Exercise 3.21. (injectivity) *Prove that the mapping ∂ is injective on terms, and give an effective way for computing $\partial^{-1}t$ when t belongs to the image of ∂. [Hint: For $t = t_1 * t_2$ with $\mathrm{ht}_R(t) = n$, t_1 is the left subterm of the $(n-1)$-th iterated right subterm of t.]*

Exercise 3.22. (not minimal) *Show that the terms* 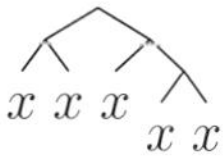

and 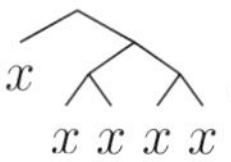

admit 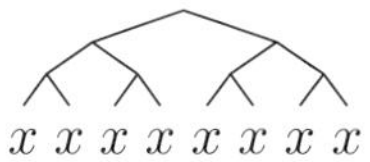

as a least common LD-expansion. Deduce that ∂t need not be the least common LD-expansion of all basic LD-expansions of t.

4. The Comparison Property

The interest of considering LD-expansions rather than general LD-equivalence is that every subterm of a term t possesses at least one LD-equivalent copy in every LD-expansion of t. This fact, together with the absorption property of Section 2, leads us to the comparison property of Chapter III.

Left subterms

When a term is not a variable, it has well defined left and right subterms: by definition, t_1 is the left subterm of a term of the form $t_1 \cdot t_2$, and t_2 is its right subterm. Similarly, if the left subterm t_1 of a term t is not a variable, the second left subterm of t is the left subterm of t_1, so we can continue and consider iterated left subterms of t.

Definition 4.1. (left subterm) For t a term that is not a variable, we denote by $\mathrm{left}(t)$ the left subterm of t. For $p \geq 0$, we denote by $\mathrm{left}^p(t)$ the iterated left subterm of t, inductively defined for p small enough by $\mathrm{left}^0(t) = t$ and $\mathrm{left}^p(t) = \mathrm{left}(\mathrm{left}^{p-1}(t))$ for $p \geq 1$.

Example 4.2. *(iterated left subterm) Let* $t = (x_1 \cdot x_2) \cdot x_3$. *Then the terms* $\mathrm{left}(t, 0)$, $\mathrm{left}(t, 1)$ *and* $\mathrm{left}(t, 2)$ *are defined, and they are equal to* t, $x_1 \cdot x_2$ *and* x_1 *respectively. When terms are viewed as trees, the iterated left subterms correspond to the leftmost subtrees, as suggested below:*

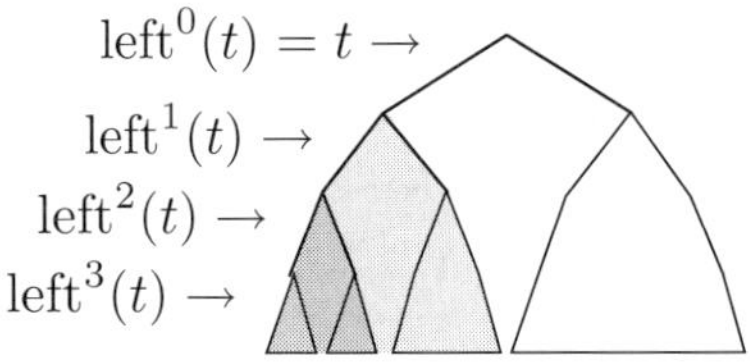

Using the criterion of Lemma 1.7, we see that the iterated left subterms of a term t are those prefixes of the word t that are themselves well formed terms.

Definition 4.3. (relations $\sqsubset$ and $\sqsubset_{LD}$) For t_1, t_2 terms, we say that $t_1 \sqsubset t_2$ holds if t_1 is a proper iterated left subterm of t, *i.e.*, if $t_1 = \mathrm{left}^p(t_2)$ holds for some positive p. We say that $t_1 \sqsubset_{LD} t_2$ holds if there exist two terms t'_1, t'_2 satisfying $t'_1 =_{LD} t_1$, $t'_2 =_{LD} t_2$, and $t'_1 \sqsubset t'_2$. We write $t_1 \sqsubseteq_{LD} t_2$ for "$t_1 \sqsubset_{LD} t_2$ or $t_1 =_{LD} t_2$", *i.e.*, if there exist t'_1, t'_2 satisfying $t'_1 =_{LD} t_1$, $t'_2 =_{LD} t_2$, and $t'_1 \sqsubseteq t'_2$.

Example 4.4. **(relation $\sqsubset_{LD}$)** *We have $x_1{\cdot}x_2 \sqsubset_{LD} x_1{\cdot}(x_2{\cdot}x_3)$, since the second term is LD-equivalent to $(x_1{\cdot}x_2){\cdot}(x_1{\cdot}x_3)$, of which $x_1{\cdot}x_2$ is the left subterm. Similarly $(x_1{\cdot}x_2){\cdot}x_1 \sqsubset_{LD} x_1{\cdot}(x_2{\cdot}x_3)$ holds, since the latter term is LD-equivalent to $((x_1{\cdot}x_2){\cdot}x_1){\cdot}((x_1{\cdot}x_2){\cdot}x_3))$.*

By definition, LD-equivalence is compatible with the product. Hence $t_1 \sqsubset_{LD} t_2$ holds if and only if there exists a positive integer p and terms $s_1, \ldots, s_p$ satisfying

$$t_2 =_{LD} (\cdots((t_1 \cdot s_1) \cdot s_2)\cdots) \cdot s_p.$$

Thus the relation $\sqsubset_{LD}$ on T_X induces the iterated left divisibility relation $\sqsubset$ of the LD-system FLD_X, and the relation $\sqsubset_{LD}$ (as well as $\sqsubseteq_{LD}$) is transitive.

Our aim in this section is to prove the following result, which has been used in Chapter III:

Proposition 4.5. **(comparison property)** *(i) Let t_1, t_2 be arbitrary terms in T_1. Then at least one of $t_1 \sqsubset_{LD} t_2$, $t_1 =_{LD} t_2$, $t_2 \sqsubset_{LD} t_1$ holds.*

(ii) Assume that S is a monogenic LD-system. Then any two elements of S are comparable with respect to the iterated left divisibility relation, i.e., for all a, b in S, at least one of $a \sqsubset b$, $a = b$, $a \sqsupset b$ holds.

By the above remark, (ii) for FLD_1 is a restatement of (i), and the property for an arbitrary monogenic LD-system then follows by projection. So the point is to prove (i). The proof uses the confluence property (Proposition 3.1), the absorption property (Proposition 2.11), and easy relations connecting the left subterms in a term and in its LD-expansions.

Left subterms in LD-expansions

It follows from the particular form of the left self-distributivity identity that LD-expansions are nearly compatible with left subterms: if s is the left subterm of t and t' is an LD-expansion of t, then the left subterm of t' need not be an LD-expansion of s, but some iterated left subterm of t' is an LD-expansion of s.

Proposition 4.6. **(left subterm)** *Assume that t' is a k-LD-expansion of t, and $\mathrm{left}^p(t)$ is defined. Then there exists p' and k' with $p' \geq p$ and $k' \leq k$ such that $\mathrm{left}^{p'}(t')$ is a k'-LD-expansion of $\mathrm{left}^p(t)$.*

Proof. (Figure 4.1) We may assume that t' is a basic LD-expansion of t. Then we use induction on t. If t is a variable, the result is true. Otherwise, we use induction on p. For $p = 0$, the result is clear. For $p > 0$, three cases may occur.

If t is $t_1{\cdot}t_2$ and t' is obtained from t by expanding some subterm of t_1, then the left subterm t'_1 of t' is a basic LD-expansion of t_1, and, for every p such that $\mathrm{left}^p(t)$ exists, $\mathrm{left}^{p-1}(t_1)$ exists and, by induction hypothesis, there exists $p' \geq p$ such that $\mathrm{left}^{p'-1}(t'_1)$ is a 1-LD-expansion of $\mathrm{left}^{p-1}(t_1)$, *i.e.*, $\mathrm{left}^{p'}(t')$ is a 1-LD-expansion of $\mathrm{left}^p(t)$. If t is $t_1{\cdot}t_2$ and t' is obtained from t by expanding some subterm of t_2, then we have $\mathrm{left}^p(t') = \mathrm{left}^p(t)$ for $p > 0$. Finally, assume $t = t_1{\cdot}(t_2{\cdot}t_3)$ and $t' = (t_1{\cdot}t_2){\cdot}(t_1{\cdot}t_3)$. Then, for every p such that $\mathrm{left}^p(t)$, *i.e.*, $\mathrm{left}^{p-1}(t_1)$, is defined, $\mathrm{left}^{p+1}(t')$ is defined and it equals $\mathrm{left}^p(t)$. ∎

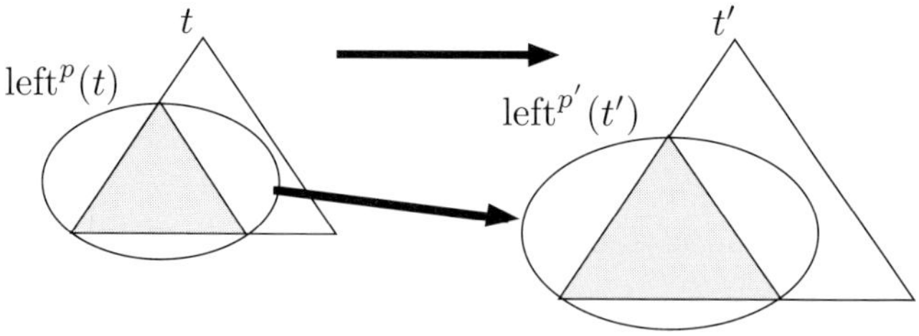

Figure 4.1. Expansion of iterated left subterm

For future use, let us state two more technical results. The first one is obvious.

Lemma 4.7. *Assume that s is a subterm of t, and s' is an LD-expansion of s. Let t' be the term obtained from t by replacing the subterm s with s'. Then t' is an LD-expansion of t.*

Lemma 4.8. *For t_1, t_2 terms, $t_1 \sqsubset_{LD} t_2$ (resp. $t_1 \sqsubseteq_{LD} t_2$) holds if and only if there exist an LD-expansion t'_1 of t_1 and an LD-expansion t'_2 of t_2 satisfying $t'_1 \sqsubset t'_2$ (resp. $t'_1 \sqsubseteq t'_2$).*

Proof. Being an LD-expansion implies being LD-equivalent, hence the condition is sufficient. Conversely, assume $t_1 \sqsubset_{LD} t_2$. By definition, there exist t''_1 and t''_2 satisfying $t''_1 =_{LD} t_1$, $t''_2 =_{LD} t_2$, and $t''_1 \sqsubset t''_2$, *i.e.*, $t''_1 = \mathrm{left}^p(t''_2)$ for some positive p. By Proposition 3.1 (confluence), the terms t_1 and t''_1 have a common LD-expansion t'''_1. Let t'''_2 be the term obtained from t''_2 by replacing the iterated left subterm t''_1 with its LD-expansion t'''_1. By Lemma 4.7, t'''_2 is an LD-expansion of t''_2, hence is LD-equivalent to t_2. By applying Proposition 3.1 again, we obtain a common LD-expansion t'_2 for t_2 and t'''_2. By Proposition 4.6 (left subterm), some iterated left subterm t'_1 of t'_2 is an LD-expansion of t'''_1, hence of t_1. The argument is the same for $\sqsubseteq_{LD}$. ∎

The case of one variable

We are ready to prove the comparison property (Proposition 4.5).

Proof. (Figure 4.2) Assume that t_1 and t_2 are arbitrary terms in T_1. By Proposition 2.11 (absorption), there exists an integer p satisfying

$$t_1 \cdot x^{[p]} =_{LD} t_2 \cdot x^{[p]} =_{LD} x^{[p+1]}.$$

By Proposition 3.1 (confluence), the terms $t_1{\cdot}x^{[p]}$ and $t_2{\cdot}x^{[p]}$ admit a common LD-expansion t. Now, t_1 is the left subterm of $t_1{\cdot}x^{[p]}$, hence, by Proposition 4.6 (left subterm), there exists some positive integer p_1 such that $\mathrm{left}^{p_1}(t)$ is an LD-expansion of t_1. Similarly, there exists some positive integer p_2 such that $\mathrm{left}^{p_2}(t)$ is an LD-expansion of t_2. The integers p_1 and p_2 are comparable: for $p_1 > p_2$, $\mathrm{left}^{p_1}(t)$ is an iterated left subterm of $\mathrm{left}^{p_2}(t)$, so we have $\mathrm{left}^{p_1}(t) \sqsubset \mathrm{left}^{p_2}(t)$, and $t_1 \sqsubset_{LD} t_2$; for $p_1 < p_2$, we obtain $t_1 \sqsupset_{LD} t_2$ similarly; finally, for $p_1 = p_2$, we obtain $t_1 =_{LD} t_2$. ■

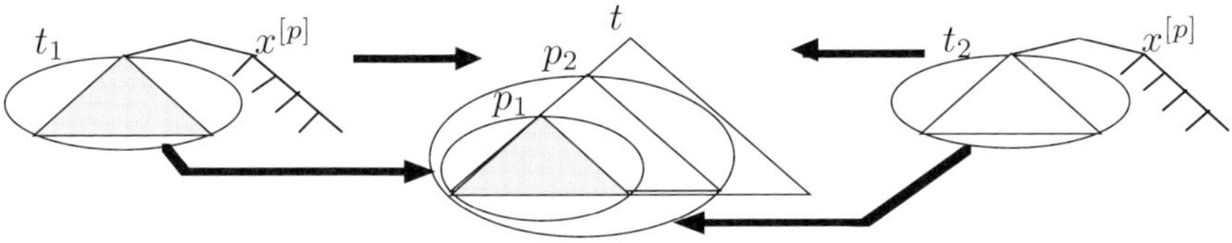

Figure 4.2. Comparison property

Example 4.9. (comparison) *Let $t_1 = (x{\cdot}x){\cdot}x$ and $t_2 = x{\cdot}(x{\cdot}x)$. Then both $t_1{\cdot}x^{[3]}$ and $t_2{\cdot}x^{[3]}$ are LD-equivalent to $x^{[4]}$, and*

$t =$ 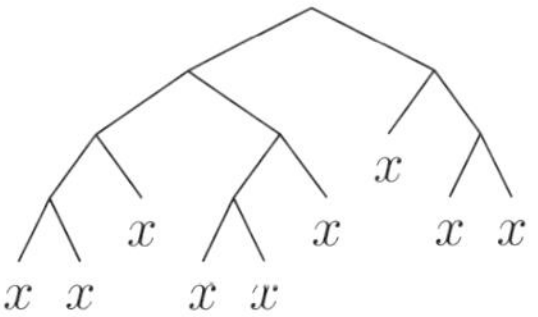

is an LD-expansion of $t_1{\cdot}x^{[3]}$ and of $t_2{\cdot}x^{[3]}$. We have $\mathrm{left}^2(t) = t_1$, while $\mathrm{left}(t)$ is an LD-expansion of t_2: so $t_1 \sqsubset_{LD} t_2$ holds.

The argument used to prove the comparison property implies that, for every term t in T_1, some LD-expansion of t is an iterated left subterm of some LD-expansion of $x^{[p]}$ for p large enough. We can choose the latter LD-expansion to be of the particular form $\partial^k x^{[p]}$, and, by a bookkeeping argument, we can obtain effective upper bounds on p and k, which will used in Chapter VI.

Definition 4.10. (complexity) The *complexity* $\mathrm{cp}(t)$ of the term t is defined to be 0 for t a variable, and to be $2\mathrm{cp}(t_1) + \mathrm{cp}(t_2) + 1$ for $t = t_1{\cdot}t_2$.

Proposition 4.11. (quantitative absorption) *For every term t in T_1, and for $p \geq \mathrm{ht}(t)$, the LD-distance between $x^{[p+1]}$ and $t{\cdot}x^{[p]}$ is at most $\mathrm{cp}(t)$.*

Proof. We copy the proof of Proposition 2.11, taking the complexity into account. For $t = x$, the result is clear. For $t = t_1{\cdot}t_2$ and $p \geq \mathrm{ht}(t)$, we find

$$\begin{aligned} d_{LD}(x^{[p+1]}, t \cdot x^{[p]}) &= d_{LD}(x^{[p+1]}, (t_1 \cdot t_2) \cdot x^{[p]}) \\ &\leq d_{LD}(x^{[p+1]}, (t_1 \cdot t_2) \cdot (t_1 \cdot x^{[p-1]})) + \mathrm{cp}(t_1) \\ &\leq d_{LD}(x^{[p+1]}, t_1 \cdot (t_2 \cdot x^{[p-1]})) + 1 + \mathrm{cp}(t_1) \\ &\leq d_{LD}(x^{[p+1]}, t_1 \cdot x^{[p]}) + \mathrm{cp}(t_2) + 1 + \mathrm{cp}(t_1) \\ &\leq d_{LD}(x^{[p+1]}, x^{[p+1]}) + \mathrm{cp}(t_1) + \mathrm{cp}(t_2) + 1 + \mathrm{cp}(t_1) \\ &= d_{LD}(x^{[p+1]}, x^{[p+1]}) + \mathrm{cp}(t). \end{aligned}$$ ■

Proposition 4.12. (distinguished expansion) *Assume that t is a term in T_1. Then, for all p, k satisfying $p \geq \mathrm{ht}(t)$ and $k \geq \mathrm{cp}(t)$, some iterated left subterm of $\partial^k x^{[p+1]}$ is an LD-expansion of t.*

Proof. By Proposition 4.11 (quantitative absorption), the LD-distance between $x^{[p+1]}$ and $t{\cdot}x^{[p]}$ is at most k. By Proposition 3.14 (quantitative confluence), it follows that $\partial^k x^{[p+1]}$ is an LD-expansion of $t{\cdot}x^{[p]}$. Hence, by Proposition 4.6 (left subterm), some iterated left subterm of $\partial^k x^{[p+1]}$ is an LD-expansion of the left subterm of $t{\cdot}x^{[p]}$, which is t. ■

The case of several variables

In the case of terms involving several variables, the comparison property as stated above is no longer true: if x and y are distinct variables, they are not LD-equivalent, and neither $x \sqsubset_{LD} y$ nor $y \sqsubset_{LD} x$ holds, since variables have no proper LD-expansion, and no proper iterated left subterm either. However, a comparison property still holds, provided a linear ordering of the variables is given and we use a lexicographical extension of the relation $\sqsubset$.

Definition 4.13. (relation $<_{LD}$) Assume that $<$ is a linear order on X. For t_1, t_2 in T_X, we say that $t_1 < t_2$ holds if t_1 precedes t_2 in the lexicographic extension of $<$ to words over $X \cup \{\bullet\}$, *i.e.*, if

- either the word t_1 is a proper prefix of the word t_2, *i.e.*, $t_1 \sqsubset t_2$ holds,
- or there exist words w, w_1, w_2 and elements x, y of X with $x < y$ satisfying $t_1 = wxw_1$ and $t_2 = wyw_2$.

We say that $t_1 <_{LD} t_2$ holds if $t_1' < t_2'$ holds for some terms t_1', t_2' satisfying $t_1' =_{LD} t_1$ and $t_2' =_{LD} t_2$.

Note that, for x, y in X, $x < y$ holds in T_X if and only if it does in X, so no ambiguity can arise. By definition, the relation $<$ extends the relation $\sqsubset$. The second case for $t_1 < t_2$ to hold occurs when the words t_1 and t_2 have a variable clash when one reads them from left to right. By construction, both $\sqsubset$ and $<$ are partial orders on terms. As no order condition connects the variables and the operator, the order $<$ is not linear: the terms $x{\cdot}x$ and $x{\cdot}(x{\cdot}x)$, *i.e.*, the words $xx\bullet$ and $xxx\bullet\bullet$, are not $<$-comparable, since their clash, which occurs at the third letter, is between a variable and an operator. In the case of terms with one variable, the relations $\sqsubset$ and $<$ coincide since a variable clash is impossible.

We shall now extend to $<$ the results established above for $\sqsubset$. In particular, we shall prove:

Proposition 4.14. (**extended comparison**) *Assume that $<$ is a linear order on X. For t_1, t_2 in T_X, at least one of $t_1 <_{LD} t_2$, $t_1 =_{LD} t_2$, $t_2 <_{LD} t_1$ holds.*

The problem is to handle clashes. The main point is that we can always suppose them to occur at certain positions at the expense of replacing the initial terms by convenient LD-expansions.

Definition 4.15. (**relations $\ll$ and $\ll_{LD}$**) Assume that $<$ is a linear order on X. For t_1, t_2 in T_X, we say that $t_1 \ll t_2$ holds if there exist a word w, two variables x, y in X with $x < y$, and an integer p satisfying $wx\bullet^p \sqsubseteq t_1$ and $wy\bullet^p \sqsubseteq t_2$. We say that $t_1 \ll_{LD} t_2$ holds if $t'_1 \ll_{LD} t'_2$ holds for some terms t'_1, t'_2 satisfying $t'_1 =_{LD} t_1$ and $t'_2 =_{LD} t_2$.

By definition, $t_1 \ll t_2$ implies $t_1 < t_2$, but the converse implication is not true in general: $t_1 \ll t_2$ requires that the variable clash between t_1 and t_2 occurs at the last position in some iterated left subterm of t_1 and t_2. For instance, $x_1x_1x_3\bullet\bullet < x_1x_2x_3\bullet\bullet$ holds, but $x_1x_1x_3\bullet\bullet \ll x_1x_2x_3\bullet\bullet$ does not, since x_1x_1 and x_1x_2 are not well formed terms. However, it is always possible to complete prefixes of a term in order to obtain well formed terms.

Lemma 4.16. *Assume that t is a term, and w is a nonempty word that is a prefix of t. Then there exists a unique integer p such that the word $w\bullet^p$ is a term. Moreover, the latter term is an iterated left subterm of some LD-expansion of t.*

Proof. The uniqueness of p follows from Lemma 1.7. For the existence, we use induction on t. If t is a variable, the only possibility is $w = t$, and the result is true with $p = 0$. Otherwise, assume $t = t_1{\cdot}t_2$. Three cases are possible.

Assume first that the word w is a prefix of t_1. By induction hypothesis, there exists an integer p such that $w\bullet^p$ is a term, and there exists an LD-expansion t'_1 of t_1 such that $w\bullet^p$ is an iterated left subterm of t'_1. Then $w\bullet^p$ is an iterated left subterm of $t'_1t_2\bullet$ as well, and this term is an LD-expansion of t.

Assume now $w = t_1 w_2$ with w_2 a prefix of t_2. By induction hypothesis, $w_2 \bullet^q$ is a well formed term for some q, and there exists an LD-expansion t_2' of t_2 satisfying $w_2 \bullet^q = \mathrm{left}^r(t_2')$ for some r. Let $t' = t_1 * t_2'$: by construction, t' is an LD-expansion of t, and we have $\mathrm{left}^r(t') = t_1 * (w_2 \bullet^q)$. Hence $\mathrm{left}^r(t')$ is an LD-expansion of $t_1 \cdot (w_2 \bullet^q)$, *i.e.*, of $w \bullet^{q+1}$, and $p = q + 1$ works.

Finally, the result is trivial when $w = t$ holds. ∎

We extend now Lemma 4.8 and show that, when dealing with $<_{LD}$, we can, as in the case of $\sqsubset_{LD}$, use LD-expansions instead of arbitrary LD-equivalent terms; moreover, we may assume that the clashes occur at the end of iterated left subterms.

Lemma 4.17. *Assume that $<$ is a linear order on X. Then $t_1 <_{LD} t_2$ holds if and only if there exist LD-expansions t_1' of t_1 and t_2' of t_2 satisfying $t_1' \sqsubset t_2'$ or $t_1' \ll t_2'$.*

Proof. The condition is sufficient, since both $t_1' \sqsubset t_2'$ and $t_1' \ll t_2'$ imply $t_1' < t_2'$, and, therefore, $t_1 <_{LD} t_2$. Conversely, assume $t_1 <_{LD} t_2$. By definition, there exist two terms t_1'', t_2'' satisfying $t_1'' =_{LD} t_1$, $t_2'' =_{LD} t_2$ and $t_1'' < t_2''$. If $t_1'' \sqsubset t_2''$ holds, then, by Lemma 4.8, there exist an LD-expansion t_1' of t_1 and an LD-expansion t_2' of t_2 satisfying $t_1' \sqsubset t_2'$. So, it remains to consider the case when t_1'' and t_2'' have a variable clash. Assume $t_1'' = wxu_1$, $t_2'' = wyu_2$, where x, y are variables satisfying $x < y$. By Lemma 4.16, there exists a (unique) integer p such that $wx\bullet^p$ and $wy\bullet^p$ are well formed terms, and, moreover, there exist LD-expansions t_1''', t_2''' of t_1'' and t_2'' respectively satisfying $wx\bullet^p \sqsubseteq t_1'''$ and $wy\bullet^p \sqsubseteq t_2'''$ (Figure 4.3). For $e = 1, 2$, let t_e'''' be a common LD-expansion for t_e and t_e'''. By Proposition 4.6 (left subterm), some iterated left subterm of t_1'''' is an LD-expansion of $wx\bullet^p$, and, by LD-invariance of the right height and of the right variable, this LD-expansion has the form $w_1 x \bullet^p$. Similarly, some iterated left subterm of t_2'''' of the form $w_2 y \bullet^p$ is an LD-expansion of $wy\bullet^p$. Now, since the rightmost variable of a term is never duplicated or modified when (*LD*) is applied, $w_2 y \bullet^p$ being an LD-expansion of $wy\bullet^p$ implies $w_2 x \bullet^p$ being an LD-expansion of $wx\bullet^p$. Let $w' x \bullet^p$ be a common LD-expansion of $w_1 x \bullet^p$ and $w_2 x \bullet^p$. Then $w' y \bullet^p$ is an LD-expansion of $w_2 y \bullet^p$. Let t_1' be the LD-expansion of t_1'''' obtained by replacing the iterated left subterm $w_1 x \bullet^p$ with $w' x \bullet^p$, and t_2' be the LD-expansion of t_2'''' obtained by replacing the iterated left subterm $w_2 y \bullet^p$ with $w' y \bullet^p$. Then, by construction, t_1' is an LD-expansion of t_1, t_2' is an LD-expansion of t_2, and $t_1' \ll t_2'$ holds since t_1' begins with $w' x \bullet^p$ while t_2' begins with $w' y \bullet^p$. ∎

Lemma 4.18. *Let R be one of $\sqsubset$, $\sqsupset$, $\ll$ or $\gg$. Assume that the terms t_1, t_2 satisfy $t_1 \mathrel{R} t_2$, and that t_2' is an LD-expansion of t_2. Then there exists an LD-expansion t_1' of t_1 satisfying $t_1' \mathrel{R} t_2'$.*

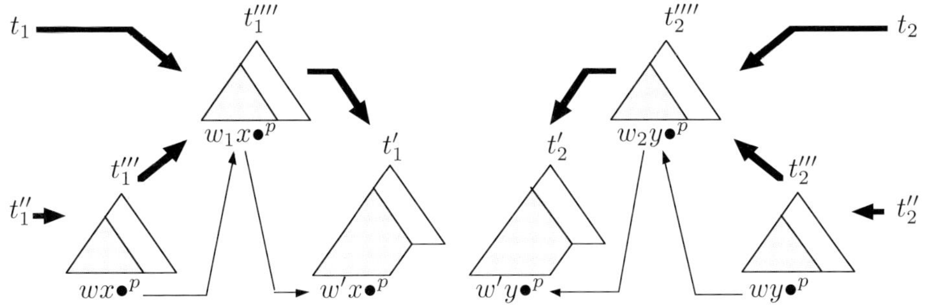

Figure 4.3: Pushing the clash

Proof. If R is $\sqsubset$, the result follows from Proposition 4.6 (left subterm). If R is $\sqsupset$, it follows from Lemma 4.7. Assume now that R is $\ll$: by definition, there exist iterated left subterms $wx\bullet^p$ of t_1 and $wy\bullet^p$ of t_2 with $x < y$. Since t_2' is an LD-expansion of t_2, some iterated left subterm of t' is an LD-expansion of $wy\bullet^p$, hence of the form $w'y\bullet^p$. Now consider the LD-expansion t_1' of t_1 obtained by replacing the iterated left subterm $wx\bullet^p$ with $w'x\bullet^p$: then $t_1' \ll t_2'$ holds. The argument is similar when R is $\gg$. ■

Lemma 4.19. *The relation $<_{LD}$ is transitive.*

Proof. Assume that t_1, t_2, t_3 are terms, and $t_1 <_{LD} t_2 <_{LD} t_3$ holds. By Lemma 4.17, some LD-expansions t_1' of t_1, t_2' and t_2'' of t_2, and t_3'' of t_3 satisfy $t_1'\ R'\ t_2'$ and $t_2''\ R''\ t_3''$ hold, where R' and R'' are $\sqsubset$ or $\ll$. We first reduce to the case when t_2' and t_2'' coincide. To this end, choose a common LD-expansion t_2''' of t_2' and t_2''. By applying Lemma 4.18 to $t_1'\ R'\ t_2'$ and t_2''', we obtain an LD-expansion t_1''' of t_1' satisfying $t_1'''\ R'\ t_2'''$. Similarly, by applying the lemma to $t_2''\ R''\ t_3''$ and t_2''', we obtain an LD-expansion t_3''' of t_3'' satisfying $t_2'''\ R''\ t_3'''$.

There remains to consider the four possible cases. If R' and R'' both are $\sqsubset$, or both are $\ll$, $t_1''' < t_3'''$ holds since $\sqsubset$ and $\ll$ are transitive. Assume $t_1''' \sqsubset t_2''' \ll t_3'''$. Then $t_1''' \sqsubseteq t_3'''$ or $t_1''' \ll t_3'''$ holds respectively, according to whether the variable clash between t_2''' and t_3''' occurs after or before the end of the iterated left subterm t_1'''. Finally, if $t_1''' \ll t_2''' \sqsubset t_3'''$ holds, then $t_1''' \ll t_3'''$ occurs. So, in each case, $t_1''' < t_3'''$ holds, and this implies $t_1 <_{LD} t_3$. ■

Definition 4.20. (**ignoring variables**) For t in T_∞, we denote by $t^\dagger$ the image of t under the substitution that maps every variable to x_1.

Lemma 4.21. *Assume $t_0 = t^\dagger$, and t_0' is an LD-expansion of t_0. Then there exists an LD-expansion t' of t satisfying $t_0' = t'^\dagger$.*

Proof. It suffices to prove the result when t_0' is a basic LD-expansion of t_0. So, assume that t_0' is obtained from t_0 by replacing some subterm $s_1{\cdot}(s_2{\cdot}s_3)$ with

$(s_1{\cdot}s_2){\cdot}(s_1{\cdot}s_3)$. Then the corresponding subterm of t has the form $t_1{\cdot}(t_2{\cdot}t_3)$, and we define t' to be the term obtained from t by replacing the latter subterm with $(t_1{\cdot}t_2){\cdot}(t_1{\cdot}t_3)$ (see Chapter VII for more precise notation). ■

We are now ready to prove Proposition 4.14 (extended comparison)

Proof. Assume that t_1, t_2 are terms in T_X. By Proposition 4.5 (comparison), the terms $t_1^\dagger$ and $t_2^\dagger$ are comparable with respect to $\sqsubset_{LD}$ or $=_{LD}$. Assume for instance $t_1^\dagger \sqsubset_{LD} t_2^\dagger$. By Lemma 4.8, there exist two terms t_1'', t_2'' in T_1 that are LD-expansions of $t_1^\dagger$ and $t_2^\dagger$ respectively and $t_1'' \sqsubset t_2''$ holds. By Lemma 4.21, there exists an LD-expansion t_1' of t_1 satisfying $(t_1')^\dagger = t_1''$, and an LD-expansion t_2' of t_2 satisfying $(t_2')^\dagger = t_2''$. Two cases may occur. If t_1' and t_2' have no variable clash, then $t_1' \sqsubset t_2'$ holds, which implies $t_1 \sqsubset_{LD} t_2$, and, therefore, $t_1 <_{LD} t_2$. Otherwise, there must exist distinct variables x, y and a word w such that the word t_1' begins with wx while t_2' begins with wy. Indeed, a clash of the type "variable *vs.* operator" is impossible, because it would imply a similar clash between t_1'' and t_2''. Hence, by definition, $t_1' \ll t_2'$ or $t_1' \gg t_2'$ holds, which implies $t_1 <_{LD} t_2$, or $t_1 >_{LD} t_2$. The argument is similar for $t_1^\dagger =_{LD} t_2^\dagger$ and $t_1^\dagger \sqsupset_{LD} t_2^\dagger$. ■

Example 4.22. (extended comparison) *Consider the terms*

$t_1 = x_1x_2x_3{\bullet}{\bullet}$, $t_2 = x_1x_2{\bullet}x_3x_3{\bullet}{\bullet}$.

They are not comparable with respect to $<$, *but the term* $x_1x_2{\bullet}x_1x_3{\bullet}{\bullet}$ *is an LD-expansion of* t_1, *and, assuming* $x_1 < x_3$, $x_1x_2{\bullet}x_1x_3{\bullet}{\bullet} < t_2$ *holds, and so does* $t_1 <_{LD} t_2$. *For obtaining variable clashes occurring at right-most positions, we must expand the terms further:* $x_1x_2{\bullet}x_1{\bullet}x_1x_2{\bullet}x_3{\bullet}{\bullet}$ *is an LD-expansion of* t_1, $x_1x_2{\bullet}x_3{\bullet}x_1x_2{\bullet}x_3{\bullet}{\bullet}$ *is an LD-expansion of* t_2, *and we find* $x_1x_2{\bullet}x_1{\bullet}x_1x_2{\bullet}x_3{\bullet}{\bullet} \ll x_1x_2{\bullet}x_3{\bullet}x_1x_2{\bullet}x_3{\bullet}{\bullet}$.

Exercise 4.23. (complexity) *Show that* $\mathrm{cp}(t) \le 3^{\mathrm{ht}(t)}/6$ *always holds.*

Exercise 4.24. (right powers) *(i) Prove that* $t \sqsubset_{LD} x^{[\mathrm{ht}(t)+1]}$ *holds for every term* t *in* T_1 *satisfying* $\mathrm{ht}(t) > \mathrm{ht}_R(t)$.

(ii) Deduce that, if t *is a term in* T_∞ *satisfying* $\mathrm{ht}(t) > \mathrm{ht}_R(t)$, *the term* $t^{[\mathrm{ht}(t)-\mathrm{ht}_R(t)+1]}$ *is* $\sqsubset_{LD}$*-comparable with no term* $x_1{\cdot}x_2 \cdots x_n$.

---

5. Canonical Orders

It is now easy to prove that, for every linear order $<$ on the set X, the associated relation $<_{LD}$ induces a linear order on the free LD-system based on X.

Construction and characterization

The only point still missing for proving that the relation $<_{LD}$ induces a linear order on FLD_X is that $<_{LD}$ has no cycle, or that left division in FLD_X has no cycle, which turns out to be equivalent. We shall give a direct proof of this result in Chapter VIII. However, we can also establish it by exhibiting any example of an LD-system where left division has no cycle. Such an example has been constructed in Chapter I.

Proposition 5.1. (**acyclicity**) *(i) Left division in free LD-systems has no cycle.*

(ii) For every set X, the relation $\sqsubset_{LD}$ on T_X has no cycle.

Proof. (i) By Proposition I.3.14 (automorphisms acyclicit), left division in the LD-system $(\mathrm{Aut}(\mathrm{FG}_\infty), {}^\wedge)$ has no cycle. Let X be an arbitrary set. By definition, there exists a homomorphism π of FLD_X into $(\mathrm{Aut}(\mathrm{FG}_\infty), {}^\wedge)$. If $(a_1, \ldots, a_p)$ is a cycle for left division in FLD_X, $(\pi(a_1), \ldots, \pi(a_p))$ is a cycle for left division in $\mathrm{Aut}(\mathrm{FG}_\infty)$, which is impossible.

(ii) As $\sqsubset_{LD}$ is transitive, it suffices to show that $t \sqsubset_{LD} t$ is impossible. Now, by definition, the latter relation in T_X is equivalent to $\bar{t} \sqsubset \bar{t}$ in FLD_X, contradicting (i)—we recall that $\bar{t}$ denotes the LD-class of t in FLD_X. ∎

The previous acyclicity result extends to the lexicographical relations $<_{LD}$. Although simple, the proof below is remarkable, as it deduces a cycle for $\sqsubset_{LD}$ from a possible cycle for $<_{LD}$, a rather surprising connection.

Proposition 5.2. (**extended acyclicity**) *Assume that $<$ is a linear order on X. Then the relation $<_{LD}$ on T_X has no cycle.*

Proof. (Figure 5.1) By Lemma 4.19, $<_{LD}$ is transitive, and it suffices to show that $t <_{LD} t$ is impossible. Assume the latter relation. By Proposition 5.1 (acyclicity), $t \sqsubset_{LD} t$ is impossible, hence, by Lemma 4.17, there exist two LD-expansions t', t'' of t satisfying $t' \ll t''$. Thus there exist a word w, two distinct variables x, y, and integers r, p', and p'' satisfying $\mathrm{left}^{p'}(t') = wx\bullet^r$ and $\mathrm{left}^{p''}(t') = wy\bullet^r$. Let t''' be a common LD-expansion of t' and t''. By Proposition 4.6 (left subterm), some subterm $\mathrm{left}^{q'}(t''')$ of t''' is an LD-expansion of $\mathrm{left}^{p'}(t')$, and, similarly, some subterm $\mathrm{left}^{q''}(t''')$ is an

LD-expansion of $\mathrm{left}^{p''}(t'')$. As $\mathrm{left}^{p'}(t')$ is $wx\bullet^r$, its LD-expansion $\mathrm{left}^{q'}(t''')$ must have the form $w'x\bullet^r$, and, similarly, $\mathrm{left}^{q''}(t''')$ has the form $w''y\bullet^r$. The integers q' and q'' cannot be equal, since the right variables of $\mathrm{left}^{q''}(t''')$ and $\mathrm{left}^{q''}(t''')$ are x and y respectively. Assume for instance $q' < q''$, and let t_0 be the term $w'y\bullet^r$. Then t_0 is an LD-expansion of $wy\bullet^r$, hence $t_0 =_{LD} \mathrm{left}^{q''}(t''')$ holds. Now, by construction, we have $\mathrm{left}^{q''-q'}(t_0) = \mathrm{left}^{q''}(t''')$. We deduce $t_0 =_{LD} \mathrm{left}^{q''-q'}(t_0)$, hence $t_0 \sqsubset_{LD} t_0$, contradicting Proposition 5.1. ∎

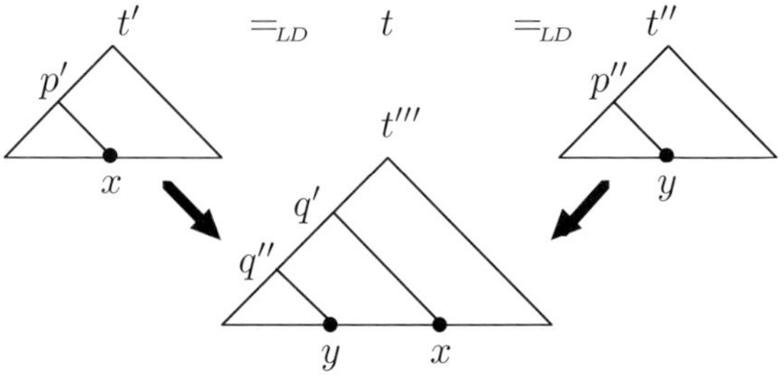

Figure 5.1. No cycle

A consequence of the previous result is that the three cases in the comparison properties exclude each other:

Proposition 5.3. (trichotomy) *(i) For all t_1, t_2 in T_1, exactly one of $t_1 \sqsubset_{LD} t_2$, $t_1 =_{LD} t_2$, $t_1 \sqsupset_{LD} t_2$ holds.*

(ii) Assume that $<$ is a linear order on X. For all t_1, t_2 in T_X, exactly one of $t_1 <_{LD} t_2$, $t_1 =_{LD} t_2$, $t_1 >_{LD} t_2$ holds.

We can now state the main results about linear orders on free LD-systems.

Proposition 5.4. (order, rank 1) *Assume that S is a ree LD-system of rank 1. Then there exists a unique partial order $<$ on S that extends the left divisibility relation of S, i.e., $a < ab$ always holds. This partial order is linear, it is compatible with multiplication on the left, and it coincides with the iterated left divisibility relation, i.e., $a < b$ holds if and only if there exists a positive integer p and elements $b_1, \ldots, b_p$ satisfying*

$$b = (\cdots((ab_1)b_2)\cdots)b_p. \tag{5.1}$$

Proof. By definition, the relation $\sqsubset_{LD}$ on T_1 is compatible with the equivalence relation $=_{LD}$, so it induces a well defined relation $<$ on FLD_1. By construction, $a < b$ holds if and only if (5.1) does. By definition, $<$ is transitive, and, as $\sqsubset_{LD}$ has no cycle, $<$ is a partial order. The comparison property states that this order is linear. Assuming $a < b$, hence (5.1), we deduce for every c using left self-distributivity

$$cb = (\cdots(((ca)(cb_1))\cdots)(cb_p),$$

hence $ca < cb$.

Assume that $\prec$ is a partial order on S that extends left divisibility. Assume that a, b are elements of S that satisfy $a < b$. Then there exist a positive integer p and elements $b_1, \ldots, b_p$ satisfying $b = (\cdots((ab_1)b_2)\cdots)b_p$. By hypothesis, this implies

$$a \prec ab_1 \prec (ab_1)b_2 \prec \cdots \prec (\cdots((ab_1)b_2)\cdots)b_p = b,$$

hence $a \prec b$. Thus $a < b$ implies $a \prec b$. As $<$ is a linear order and $\prec$ is a partial order, we deduce that $\prec$ coincides with $<$. ■

Example 5.5. (order, rank 1) *In* FLD_1, *we have the inequalities*

$$x < x_{[2]} < x_{[3]} < x_{[4]} < \cdots < x^{[3]} < x^{[4]} < \ldots$$

Indeed, in T_1, $x_{[p]} \sqsubset x_{[p+1]} = x_{[p]}{\cdot}x$ *holds for every* p. *Then, for* $p \geq 2$, *we find*

$$x_{[p]} \cdot x_{[2]} =_{LD} (x_{[p]} \cdot x) \cdot (x_{[p]} \cdot x) = x_{[p+1]} \cdot x_{[p+1]} \sqsupset_{LD} x_{[p+1]} \cdot x_{[2]},$$

and we deduce $x_{[p+1]} \sqsubset_{LD} x_{[p+1]}{\cdot}x_{[2]} \sqsubset_{LD} x_{[2]}{\cdot}x_{[2]} = x^{[3]}$. *Finally, for every* p, *we have* $x^{[p]} \sqsubset x^{[p]}{\cdot}x^{[p]} =_{LD} x^{[p+1]}$.

Proposition 5.6. (order) *Assume that* S *is free LD-system based on* X, *and* $<$ *is a linear order on* X. *Then there exists a unique partial order* $<$ *on* S *that extends the left divisibility relation of* S *and is such that that* $x < y$ *in* X *implies*

$$(\cdots((c_1 \cdots c_r x)a_1)\cdots)a_p < (\cdots((c_1 \cdots c_r y)b_1)\cdots)b_q \tag{5.2}$$

for all $a_1, \ldots, c_r$. *This partial order is linear; it is compatible with multiplication on the left, and* $a < b$ *holds if and only if either* $a \sqsubset b$ *holds, i.e., there exists a positive integer* p *and elements* $b_1, \ldots, b_p$ *satisfying*

$$b = (\cdots((ab_1)b_2)\cdots)b_p, \tag{5.3}$$

or $a \ll b$ *holds, this meaning that there exist* x, y *in* X *with* $x < y$ *and* $a_1, \ldots, c_r$ *satisfying*

$$\begin{cases} a = (\cdots((c_1 \cdots c_r x)a_1)\cdots)a_p, \\ b = (\cdots((c_1 \cdots c_r y)b_1)\cdots)b_q. \end{cases} \tag{5.4}$$

Each of the relations $\sqsubset$ *and* $\ll$ *is a partial order on* FLD_X.

Proof. By definition, the relations $\sqsubset_{LD}$, $\lll_{LD}$, and $<_{LD}$ on T_X are compatible with the equivalence relation $=_{LD}$, so they induce well defined relations $\sqsubset$, $\lll$, and $<$ on FLD_X. Then, by Lemma 4.17, $a < b$ holds in FLD_X if and only if either $a \sqsubset b$, *i.e.*, (5.3), holds, or $a \lll b$ holds, which, by definition, means that there exist elements $a_1, \ldots, c_r$ satisfying (5.4). By Lemma 4.19, the relation $<$ is transitive on FLD_X, and, by Proposition 5.2 (extended acyclicity), it is irreflexive. Hence $<$ is a partial order on FLD_X. By Proposition 4.14 (extended comparison), this order is a linear order.

Assume that $\prec$ is a partial order on S that extends left divisibility and satisfies the condition of (5.2). Assume that a, b are elements of S that satisfy $a < b$. Then either $a \sqsubset b$ or $a \lll b$ holds. In the first case, there exists a positive integer p and elements $b_1, \ldots, b_p$ satisfying $b = (\cdots((ab_1)b_2)\cdots)b_p$, which implies

$$a \prec ab_1 \prec (ab_1)b_2 \prec \cdots \prec (\cdots((ab_1)b_2)\cdots)b_p = b,$$

hence $a \sqsubset b$. In the second case, a and b admit decompositions of the form (5.4) with $x < y$, hence Condition (5.2) implies $a \prec b$ again. So $a < b$ implies $a \prec b$. As $<$ is a linear order and $\prec$ is a partial order, we deduce that $\prec$ coincides with $<$. ∎

Convention 5.7. In the case of the free LD-system FLD_n of rank n, we use $\lll$ and $<$ for the orders that extend the ordering $x_1 < x_2 < \cdots$ of the variables.

Example 5.8. (**order**) *In* FLD_2, *we have* $x_1(x_2x_2) \lll (x_1x_2)x_2$, *and, therefore,* $x_1(x_2x_2) < (x_1x_2)x_2$, *as, in* T_2, *we have*

$$x_1 \cdot (x_2 \cdot x_2) =_{LD} ((x_1 \cdot x_2) \cdot x_1) \cdot (x_1 \cdot x_2) \cdot x_2 \lll_{LD} (x_1 \cdot x_2) \cdot x_2.$$

Properties of the orders on free LD-systems

Here we list some basic properties of the orders on free LD-systems. It will be convenient to introduce the following notion.

Definition 5.9. (**canonical order**) Assume that S is a free LD-system, and $<$ is an order on S. We say that $<$ is *canonical* if, letting X be the (unique) basis of S, $<$ is the (unique) linear order on S that extends $< \restriction X$ as in Proposition 5.6 (order).

A free LD-system of rank n admits $n!$ canonical orders corresponding to the $n!$ linear orderings of its basis. In particular, a free LD-system of rank 1 admits exactly one canonical order, namely its iterated left divisibility relation.

Proposition 5.10. (compatibility) *Assume that S is a free LD-system and $<$ is a canonical order on S. Let $\prec$ be $<$, $\ll$, or $\sqsubset$. Then, for all a, b, c, d,*

(i) $a \prec b$ is equivalent to $ca \prec cb$,
(ii) $a \preceq b$ implies $a \prec bc$,
(iii) $a \ll b$ implies $ac \ll bd$,
(iv) $ac \ll b$ implies $a < b$,
(v) $a \ll b \leq c$ implies $a \ll c$.

Proof. Point (i), (ii), and (iii) follow from the definition of $\sqsubset$ and $\ll$ directly. For (iv), assume $ac \ll b$. If $a < b$ does not hold, then $b \leq a$ holds, and we deduce $b \leq ac$ from (ii), a contradiction. For (v), assume $a \ll b \leq c$. If $b \ll c$ holds, then $a \ll c$ does since $\ll$ is transitive. If $b \sqsubset c$ holds, there exist $c_1, \ldots, c_p$ satisfying $c = (\cdots((bc_1)c_2\cdots)c_p$, and repeated uses of (ii) give $a \ll (\cdots((bc_1)c_2\cdots)c_k$ for every k inductively. ∎

Proposition 5.11. (order type) *In* $(\mathrm{FLD}_1, <)$, *x is minimal and every element a admits ax as an immediate successor. The order* $(\mathrm{FLD}_1, <)$ *is isomorphic to the standard order on* $\mathbf{N} \times (1 + \mathbf{Q})$.

Proof. The element x is a minimum in $(\mathrm{FLD}_1, <)$, since the term x is an iterated left subterm of each term in T_1. For every term t in T_1, the term $t{\cdot}x$ satisfies $t \sqsubset t{\cdot}x$, so $a < ax$ holds in FLD_1 for every a. Now, let t be an arbitrary term such that $s \sqsubset_{LD} t$ holds. Without loss of generality, we may assume that s is a proper iterated left subterm of t, say $t = (\cdots(s{\cdot}t_1)\cdots){\cdot}t_p$ with $p \geq 1$. By construction, there exist a nonnegative number ℓ and terms $s_1, \ldots, s_\ell$ satisfying $t_1 = (\cdots(x{\cdot}s_1)\cdots){\cdot}s_\ell$. This implies

$$s \cdot t_1 =_{LD} (\cdots((s \cdot x) \cdot (s \cdot s_1))\cdots) \cdot (s \cdot s_\ell),$$

so $s{\cdot}x$ is an iterated left subterm of some term that is LD-equivalent to t. In other words, $s{\cdot}x \sqsubseteq_{LD} t$ holds. Translated in FLD_1, this means that $ax \leq b$ holds for all elements b satisfying $a < b$.

Let T_1' denote the subset of T_1 comprising x and all terms with right height 2 or more. Every term t in T_1 can be written in a unique way as $(\cdots((s{\cdot}x){\cdot}x\ldots){\cdot}x$, and $s \in T_1'$. In this case, the terms that are LD-equivalent to t are exactly those of the form $(\cdots((s{\cdot}x){\cdot}x\cdots){\cdot}x$ with $s' =_{LD} s$. It follows that the order $(\mathrm{FLD}_1, <)$ is the product of the discrete order $(\mathbf{N}, <)$ by the restriction of the order $<$ to those classes that do not contain any term of the form $s{\cdot}x$, *i.e.*, to the LD-classes of terms in T_1'. So, it remains to show that, if t_2 is a term with right height at least 2, and if $t_1 \sqsubset_{LD} t_2$ holds, then one can find a term t satisfying $t_1 \sqsubset_{LD} t \sqsubset_{LD} t_2$. Without loss of generality, we may assume that t_1 is a proper iterated left subterm of t_2, and even that t_1 is the left subterm of t_2 since $\mathrm{left}^k(t) \sqsubseteq_{LD} \mathrm{left}(t_2) \sqsubset_{LD} t_2$ holds for $k \geq 1$. Then t_2 has the form $t_1{\cdot}(s_1{\cdot}s_2)$, and $t = t_1{\cdot}s_1$ is a solution. ∎

Example 5.12. **(order)** *The first elements of* $(\mathrm{FLD}_1, <)$ *are*

$$x < x_{[2]} = x^{[2]} < x_{[3]} < x_{[4]} < \cdots$$

We have seen in Example 5.8 (order, rank 1) that the element $x^{[3]}$ *is beyond all the above elements, but it is not the least such element. Actually, we have seen in Example 5.8 that the sequence* $x^{[3]} = x_{[2]}x_{[2]}$, $x_{[3]}x_{[2]}$, $x_{[4]}x_{[2]}$, ... *is decreasing, and we have therefore*

$$\sup_p x_{[p]} \le \inf_q (x_{[q]}x_{[2]}) < x^{[3]}.$$

The argument showing that, for every a in FLD_1, ax is the immediate successor of a also shows that, for every a in FLD_n, ax_1 is the immediate successor of a. Describing the predecessors of a requires more care.

Definition 5.13. **(iteration)** Assume that S is a binary system. For $(a_1, \ldots, a_p)$ a finite sequence of elements of S, we put

$$\partial(a_1, \ldots, a_p) = (a_1 \cdots a_p, a_1 \cdots a_{p-1}, \ldots, a_1a_2, a_1). \tag{5.5}$$

The n-*th iteration* of $(a_1, \ldots, a_p)$ is defined to be the first component, denoted $\mathrm{iter}_n(a_1, \ldots, a_p)$, of the p-tuple $\partial^n(a_1, \ldots, a_p)$. We write $\mathrm{iter}(a_1, \ldots, a_p)$ for the sequence of all $\mathrm{iter}_n(a_1, \ldots, a_p)$ with $n \ge 0$.

Example 5.14. **(iteration)** *For a 1-tuple* (a), *we have* $\partial^n(a) = a$ *for every* n, *so the* n-*th iterate of* a *is* a. *For a pair* (a, b), *the iterates are defined by the inductive clauses*

$$\mathrm{iter}_n(a,b) = \begin{cases} a & \text{for } n = 0, \\ ab & \text{for } n = 1, \\ \mathrm{iter}_{n-1}(a,b)\mathrm{iter}_{n-2}(a,b) & \text{for } n \ge 2. \end{cases}$$

For instance, we have $\mathrm{iter}_2(a,b) = (ab)a$, $\mathrm{iter}_3(a,b) = ((ab)a)(ab)$, *a kind of Fibonacci sequence.*

Lemma 5.15. *Assume that* a_1, ..., a_p *belong to a free LD-system based on* X, *and that* t_1, ..., t_p *are terms in* T_X *representing* a_1, ..., a_p. *Let* x *be an arbitrary variable in* X. *Then, for every* n, *the term* $\mathrm{left}(\partial^n(t_1 \cdots t_p{\cdot}x))$ *represents* $\mathrm{iter}_n(a_1, \ldots, a_p)$.

Proof. The result is obvious for $n = 0$. It suffices to prove it for $n = 1$, and an induction then completes the proof. Now, it suffices to prove that $t = t_1 \cdots t_p{\cdot}x$ implies $\partial t = t'_1 \cdots t'_p{\cdot}x$ with $t'_i = \partial(t_1 \cdots t_i)$. This equality is proved using induction on $p \geq 1$. For $p = 1$, we have $\partial(t_1{\cdot}x) = \partial t_1{\cdot}\partial x = \partial t_1{\cdot}x$ by definition. Assume $p \geq 2$. Using the induction hypothesis, we find

$$\begin{aligned}\partial(t_1 \cdot \ldots t_p \cdot x) &= \partial t_1 * \partial(t_2 \cdots t_p \cdot x)\\ &= \partial t_1 * (\partial(t_2 \cdots t_p) \cdots \partial t_2 \cdot x)\\ &= (\partial t_1 * \partial(t_2 \cdots t_p)) \cdots (\partial t_1 * \partial t_2) \cdot \partial t_1 \cdot x\\ &= (\partial(t_1 \cdots t_p)) \cdots \partial(t_1 \cdot t_2) \cdot \partial t_1 \cdot x. \end{aligned}$$ ∎

Proposition 5.16. **(predecessors)** *Assume that S is a free LD-system based on X, and $<$ is a canonical order on S. Assume that a is an element of S with right height at least 3, say $a = a_1 \cdots a_p x$ with $x \in X$. Then the elements b of S satisfying $b < a$ are:*
- *those elements that satisfy $b \leq \mathrm{iter}_n(a_1, \ldots, a_p)$ for some n, and*
- *those elements of the form $a_1 \cdots a_p y$ with y in X satisfying $y < x$.*

Proof. Choose terms $t_1, \ldots, t_p$ that represent $a_1, \ldots, a_p$ respectively. Thus $t = t_1 \cdots t_p{\cdot}x$ represents a. By Lemma 5.15, the successive elements of the sequence $\mathrm{iter}(a_1, \ldots, a_p)$ are represented by the left subterms of the terms $\partial^n t$. By construction, and because $p \geq 2$ holds, the relation $\mathrm{left}(\partial^{n-1}t) \sqsubset \mathrm{left}(\partial^n t)$ is true, so we have $\mathrm{iter}_{n-1}(a_1, \ldots, a_p) \sqsubset \mathrm{iter}_n(a_1, \ldots, a_p)$.

Assume $b \sqsubset a$. By construction, there exist a term s representing b and a term t' representing a such that s is a proper iterated left subterm of t', hence s is a (possibly improper) iterated left subterm of $\mathrm{left}(t')$. By Proposition 3.1 (confluence), $\partial^n t$ is an LD-expansion of t' for some n. Hence, by Proposition 4.6 (left subterm), there exists $k \geq 1$ such that $\mathrm{left}^k(\partial^n t)$ is an LD-expansion of $\mathrm{left}(t')$. So $s \sqsubseteq_{LD} \mathrm{left}(\partial^n t)$ holds, which, by the previous lemma, implies $b \sqsubseteq \mathrm{iter}_n(a_1, \ldots, a_p)$.

Assume now $b \ll a$. By the same argument as above, there exist s representing b and n such that $s \ll \partial^n t$ holds. If the variable clash between s and $\partial^n t$ does not involve the rightmost variable in $\partial^n t$, then, by construction, we have $s \ll_{LD} \mathrm{left}(\partial^{n+1}t)$, and, therefore, $b \ll \mathrm{iter}_{n+1}(a_1, \ldots, a_p)$. Otherwise, the only possibility is s being the term obtained from $\partial^n t$ by replacing the final variable x with some smaller variable y. Then $b = a_1 \cdots a_p y$ holds. ∎

So, for instance, if S is a free LD-system based on X, and x is an element of X, the element $x^{[3]}$ is the $\sqsubset$-lowest upper bound of the "Fibonacci" sequence $\mathrm{iter}(x, x)$: x, $x_{[2]}$, $x_{[3]}$, $x_{[3]}x_{[2]}(= x_{[4]}{}^{[2]})$, $(x_{[3]}x_{[2]})x_{[3]}$, ...

Exercise 5.17. **(divisibility)** *Assume that S is a free LD-system of rank 1, and g is the generator of S. Let a, b belong to S. Prove that*

$(ab)g$ is not left divisible by a, i.e., $(ab)g = ac$ is impossible. [Hint: Use $(ab)g \sqsubset abg$.]

Exercise 5.18. (commensurability) *Assume that S is a free LD-system. For a, b in S, say that a and b are commensurable, denoted $a \asymp b$, if $a^{[p]} = b^{[q]}$ holds for some p, q.*

(i) Prove that $\asymp$ is an equivalence relation that is compatible with $\cdot$ on the left, but not compatible with $\cdot$ on the right if S has rank at least 2. [Hint: Compare x_1x_2 and $x_1^{[2]}x_2$.] What is $\asymp$ if S has rank 1?

(ii) Prove that $a \asymp b$ holds if and only if the sub-LD-systems of S generated by a and b are not disjoint, and that, in this case, their intersection includes a free LD-system.

(iii) Let E be an equivalence class for $\asymp$. Show that E is a sub-LD-system of S, and that the restriction of $\sqsubset$ to E is a linear order. [Hint: Introduce a projection π of S onto FLD_1, and show that the conjunction of $a \sqsubset b^{[q]}$ and $\pi(a) \sqsubset \pi(b)$ implies $a \sqsubset b$.]

---

6. Applications

The existence of canonical orders on free LD-systems leads to several significant applications. Here we list some of them. In particular, we prove that special braids equipped with braid exponentiation yield a free LD-system of rank 1, and that the word problem for left self-distributivity, *i.e.*, the problem of algorithmically deciding whether two given terms are LD-equivalent, is algorithmically solvable.

Cancellation

Proposition 6.1. (left cancellativity) *Free LD-systems are left cancellative.*

Proof. Assume that S is a free LD-system, and a, b are distinct elements of S. Choose a canonical order $<$ on S. Then either $a < b$ or $b < a$ holds. By compatibility of $<$ with left translations, these relations imply $ca < cb$ and $cb < ca$ respectively, so both imply $ca \neq cb$. ∎

Observe that free LD-systems are not right cancellative: if S is an arbitrary LD-system and a is an element of S, $aa^{[2]} = a^{[2]}a^{[2]}$ holds, and, if S is free, a and $a^{[2]}$ are not equal, since S is acyclic.

An application is the following characterization of the right powers.

Proposition 6.2. (**right powers**) *Assume that S is a free LD-system of rank 1. For all a, b in S, the following are equivalent:*
(i) The element b is a right power of a;
(ii) The relation $ab = b^{[2]}$ holds.

Proof. If $b = a^{[m]}$ holds, then we have $bb^{[m]} = b^{[m+1]} = (b^{[m]})^{[2]}$, so (i) implies (ii). Conversely, assume $ab = b^{[2]}$. Let us write $x^k b$ for $x \cdots xb$, k times x. We claim that $a^k b = b^k b$ holds for every $k \geq 1$. For $k = 1$, it is our hypothesis. Assume $k \geq 2$. We have

$$a^k b = a^{k-1}ab = a^{k-1}b^{[2]} = (a^{k-1}b)^{[2]} = (b^{k-1}b)^{[2]} = b^{k-1}bb = b^k b. \tag{6.1}$$

Now, by absorption, there exist positive integers p, q satisfying $a^{[p]} = b^{[q]}$. Applying (6.1), we find

$$a^{q-1}b = b^{q-1}b = b^{[q]} = a^{[p]} = a^{p-1}a. \tag{6.2}$$

Assume $q > p$. Applying left cancellativity p times, we deduce $a = a^{q-p}b$ from (6.2), which gives a cycle for left division in S. Hence we must have $q \leq p$, and, applying left cancellativity, we obtain $b = a^{p-q}a = a^{[p-q+1]}$. ■

Corollary 6.3. *Assume that S is a free LD-system of rank 1, g is the generator of S, and a is an element of S. Then, for every m, the following are equivalent:*
(i) There exists p satisfying $a^{[p]} = g^{[m]}$;
(ii) We have $g^{[m+1]} = ag^{[m]}$.

Criteria of freeness

Proposition 6.4. (**Laver's criterion**) *Assume that S is a monogenic LD-system. Then the following are equivalent:*
(i) The LD-system S is free;
(ii) Left division in S has no cycle, i.e., no equality of the form

$$a = (\cdots((ab_1)b_2)\cdots)b_p \tag{6.3}$$

with $p \geq 1$ holds in S;
(iii) There exists a partial order on S that includes left divisibility.

Proof. We already know that (i) implies (iii), and (iii) implies (ii) as, if $\prec$ is a transitive relation that includes left divisibility, $b = (\cdots((ab_1)b_2)\cdots)b_p$ implies $a \prec b$, hence $a \neq b$ if $\prec$ is acyclic. It remains to prove that (ii) implies (i). So assume that S is an LD-system generated by an element g and left division in S has no cycle. There exists a (unique) surjective homomorphism π of FLD_1 onto S that maps x to g. We claim that π is injective. Indeed, assume that a, b are distinct elements of FLD_1. Then $a < b$ or $b < a$ holds. Assume for instance $a < b$. Then, by definition, b admits in FLD_1 a decomposition of the form

$$b = (\cdots((ab_1)b_2)\cdots)b_p.$$

Applying the homomorphism π, we find in S the equality

$$\pi(b) = (\cdots((\pi(a)\pi(b_1))\pi(b_2))\cdots)\pi(b_p).$$

Now $\pi(a) = \pi(b)$ is impossible: this would give a cycle for left division in S. ■

Let us consider the injection bracket of Example I.2.19. We have seen that left division in the monogenic sub-LD-system $\mathfrak{I}_{sp}$ of $\mathfrak{I}_\infty$ generated by the shift injection sh admits cycles of length 2. Applying Laver's criterion, we deduce that $(\mathfrak{I}_{sp}, [\,])$ is not free.

Extending the previous result to the case of an arbitrary rank is easy.

Definition 6.5. (quasi-free) Assume that S is an LD-system, and X is a subset of S. We say that X is *quasi-free* in S if no equality of the form

$$(\cdots((c_1 \cdots c_r x)a_1)\cdots)a_p = (\cdots((c_1 \cdots c_r y)b_1)\cdots)b_q \tag{6.4}$$

with x, y distinct elements of X holds in S.

Proposition 6.6. (criterion) *Assume that S is an LD-system generated by X. Then the following are equivalent:*

(i) The LD-system S is free based on X;

(ii) Left division in S admits no cycle, and X is quasi-free in S.

(iii) There exists a partial order $\prec$ on S including left divisibility and such that $x \prec y$ in X implies

$$(\cdots((c_1 \cdots c_r x)a_1)\cdots)a_p \prec (\cdots((c_1 \cdots c_r y)b_1)\cdots)b_q$$

for all $a_1, \ldots, c_r$.

Proof. We already know that (i) implies (iii), and, as above, (iii) implies (ii). We show now that (ii) implies (i). So we assume that S and X satisfy the conditions of (ii). There exists a (unique) surjective homomorphism π of FLD_X onto S that is the identity on X. Let $<$ be a canonical order of S. Let a, b be distinct elements of FLD_X. Then exactly one of $a < b$, $b < a$ holds. Assume for instance $a < b$. Then, by Proposition 5.6 (order), either b admits in FLD_X a decomposition

$$b = (\cdots((ab_1)b_2)\cdots)b_p,$$

or a and b admit decompositions

$$a = (\cdots((c_1 \cdots c_r x)a_1)\cdots)a_p, \quad b = (\cdots((c_1 \cdots c_r y)b_1)\cdots)b_q$$

with x, y in X satisfying $x < y$. Applying the homomorphism π, we find in S either

$$\pi(b) = (\cdots((\pi(a)\pi(b_1))\pi(b_2))\cdots)\pi(b_p),$$

or

$$\begin{cases} \pi(a) = (\cdots((\pi(c_1)\cdots\pi(c_r)x)\pi(a_1))\cdots)\pi(a_p) \\ \pi(b) = (\cdots((\pi(c_1)\cdots\pi(c_r)y)\pi(b_1))\cdots)\pi(b_q). \end{cases}$$

The hypothesis that left division in S has no cycle implies that the first type equality is incompatible with $\pi(a) = \pi(b)$. The hypothesis that X is quasi-free in S implies that the second type equalities are incompatible with $\pi(a) = \pi(b)$ as well. So the mapping π is injective. ∎

As an application, let us consider subsystems of a free LD-system.

Lemma 6.7. *Assume that S is a free LD-system, and $<$ is a canonical order on S. Then every subset of S that is a chain for $\ll$ is free.*

Proof. Let X be a subset of S. As S is free, left division in S has no cycle, and, therefore, X is free in S if and only if it is quasi-free. Assume $x, y \in X$ and $x < y$. For all $c_1, \ldots, c_r$ in S, we have $c_1 \cdots c_r x \ll c_1 \cdots c_r y$, hence, for all $a_1, \ldots, b_q$,

$$(\cdots((c_1 \cdots c_r x)a_1)\cdots)a_p \ll (\cdots((c_1 \cdots c_r y)b_1)\cdots)b_q,$$

and the above two elements are not equal. Thus, a sufficient condition for X to be free is that any two elements of X are comparable with respect to $\ll$. ∎

Proposition 6.8. (**subsystems**) *(i) Every monogenic subsystem of a free LD-system is free.*

(ii) A free LD-system of rank 2 or more admits free sub-LD-systems of any finite or countable rank.

Proof. For (i), we observe that Lemma 6.7 applies to every family with one element. For (ii), assume that S is a free LD-system of rank at least 2, and $\{x_1, x_2, \ldots\}$ is a basis of S. Then $(x_1, x_2x_1, x_3x_2x_1, \ldots)$ is a countable subset of S that is a chain for $\ll$. ∎

In contradistinction, we have

Proposition 6.9. (**not free**) *A free LD-system of rank 1 admits no free sub-LD-system of rank 2.*

Proof. Assume that S is a free LD-system generated by x. Let a, b be distinct elements of S. By Proposition 2.11 (absorption property), we have $a^{[p]} = b^{[q]} = x^{[r]}$ for some p, q, r. The first equality contradicts the freeness criterion of Proposition 6.6 (criterion). So $\{a, b\}$ is not free in S. ∎

We have no complete characterization of the free subfamilies in a free LD-system. See exercises below for a few additional remarks.

Realizations of free LD-systems

A natural question is then to find a concrete realization of free LD-systems in the spirit of the realization of free groups in terms of reduced words. We shall see that braids equipped with exponentiation provide such realizations.

Let us begin with the case of rank 1. Laver's criterion tells us that every monogenic LD-system where left division has no cycle is free. Now, we have seen in Section I.3 that left division in B_∞ and $\mathrm{Aut}(\mathrm{FG}_\infty)$ equipped with exponentiation has no cycle. We deduce:

Proposition 6.10. (**braids free**) *(i) For every braid b, the sub-LD-system of $(B_\infty, {}^\wedge)$ generated by b is a free LD-system of rank 1. In particular, special braids equipped with braid exponentiation yield a free LD-system of rank 1.*

(ii) For every automorphism φ of FG_∞, the sub-LD-system of $(\mathrm{Aut}(\mathrm{FG}_\infty), {}^\wedge)$ generated by φ is a free LD-system.

For rank 2 and above, we introduce two extensions of braid groups, called respectively CB_∞ and LB_∞. A natural explanation for our introducing CB_∞ will be given in Chapter VIII.

Definition 6.11. (**charged braids**) For $n \le \infty$, the group CB_n is defined as the extension of B_n obtained by adding a sequence of mutually commuting generators $\rho_1, \ldots, \rho_n$ submitted to the relations

$$\sigma_i \rho_j = \rho_j \sigma_i \quad \text{for } j < i \text{ or } j > i+1, \qquad \sigma_i \rho_i \rho_{i+1} = \rho_i \rho_{i+1} \sigma_i \quad \text{for every } i.$$

A concrete interpretation for the group CB_n as a group of charged braids can be given as follows: we associate with the generators σ_i braid diagrams as usual, and we associate with the generator ρ_j an elementary charge ('$\oplus$') sitting on the j-th strand of the braid. Then the relations of CB_n express that charges freely move on the strands, but a charge is allowed to pass a crossing on one strand only when another charge passes on the other strand of the crossing, as below:

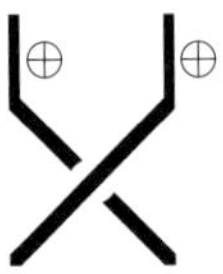

is equivalent to

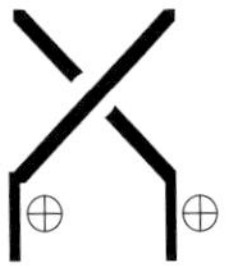

.

The new generators ρ_j that have been added are free enough to extend the construction of braid exponentiation. First, we extend the shift endomomorphism of B_∞ to CB_∞ by defining $\mathrm{sh}(\rho_j) = \rho_{j+1}$ for every j.

Proposition 6.12. (**exponentiation on** CB_∞) *The operation* $\wedge$ *defined on* CB_∞ *by* $a{\wedge}b = a\ \mathrm{sh}(b)\ \sigma_1\ \mathrm{sh}(a)^{-1}$ *is left self-distributive.*

The easy verification is left to the reader, as well as the verification of the following result:

Proposition 6.13. (**Larue's group**) *Let* LB_∞ *be the extension of* B_∞ *obtained by adding a double infinite sequence of generators* $\theta_{j,k}$, $j, k \geq 1$ *subject to the relations*

$$\sigma_i\theta_{j,k} = \theta_{j,k}\sigma_i \quad \text{for } j > i+1.$$

(i) Let sh *be the endomorphism of* LB_∞ *that maps* σ_i *to* σ_{i+1} *and* $\theta_{j,k}$ *to* $\theta_{j+1,k}$ *for all* i, j, k. *Then the operation* $\wedge$ *defined on* LB_∞ *by* $a{\wedge}b = a\,\mathrm{sh}(b)\,\sigma_1\,\mathrm{sh}(a^{-1})$ *is left self-distributive.*

(ii) Mapping σ_i *to* σ_i *and* $\theta_{j,k}$ *to* ρ_j^{k-1} *induces a surjective homomorphism of* LB_∞ *onto* CB_∞.

Observe that the relation $\sigma_i\rho_j^{k-1} = \rho_j^{k-1}\sigma_i$ holds in CB_∞ for $j > i+1$, hence the surjective homomorphism alluded to in (ii) above is well defined.

We show now that the LD-systems $(CB_\infty, \wedge)$ and $(LB_\infty, \wedge)$ are eligible for the criterion of Proposition 6.6.

Proposition 6.14. (**charged braids free**) *The elements* 1, ρ_1, ρ_1^2, $\ldots$ *generate a free LD-system in* $(CB_\infty, \wedge)$.

Proof. Proving that left division in $(CB_\infty, \wedge)$ admits no cycle is easy. Indeed, there exists a surjective homomorphism of CB_∞ onto B_∞ that keeps every σ_i unchanged, and collapses every ρ_j to 1. Under this homomorphism, a possible cycle for left divisibility in CB_∞ would project onto a cycle for left divisibility in B_∞, which is impossible.

There remains to prove that the family $(1, \rho_1, \rho_1^2, \ldots)$ is quasi-free in $(CB_\infty, \wedge)$, *i.e.*, to prove that

$$\begin{cases} a = (\cdots((c_1{\wedge}\cdots{\wedge}c_r{\wedge}\rho_1^i){\wedge}a_1){\wedge}\cdots){\wedge}a_p, \\ b = (\cdots((c_1{\wedge}\cdots{\wedge}c_r{\wedge}\rho_1^j){\wedge}b_1){\wedge}\cdots){\wedge}b_q \end{cases} \tag{6.5}$$

with $i \neq j$ implies $a \neq b$. So, assume that a and b satisfy (6.5) with $j = i+k$, $k > 0$. Expanding (6.5) gives $a = c\sigma_{r,1}a_0$ and $b = c\rho_r^k\sigma_{r,1}b_0$, where a_0 and b_0 admit expressions where neither σ_1^{-1} nor $\rho_1^{\pm 1}$ occurs.

Let $\mathrm{FG}_{\infty,\infty}$ denote a free group based on a double sequence of generators $x_{p,q}$, $p \geq 1$, $q \in \mathbf{Z}$. Using the defining relations of CB_∞, we see that the formulas

$$\widetilde{\sigma}_i(x_{p,q}) = \begin{cases} x_{p,q} & \text{for } p \neq i, i+1, \\ x_{i,q} x_{i+1,q} x_{i,q}^{-1} & \text{for } p = i, \\ x_{i,q} & \text{for } p = i+1, \end{cases}$$

$$\widetilde{\rho}_j(x_{p,q}) = \begin{cases} x_{p,q} & \text{for } p \neq j, \\ x_{p,q+1} & \text{for } p = j, \end{cases}$$

define an antihomomorphism ψ of CB_∞ into $\mathrm{Aut}(\mathrm{FG}_{\infty,\infty})$ (resorting to an antihomomorphism rather than to a homomorphism here amounts to using reversed composition rather than composition, *i.e.*, thinking of CB_∞ as acting on the right). Let $x = \psi(c\sigma_{r,1})^{-1}(x_{1,0})$. By definition, we have $\psi(c\sigma_{r,1})(x) = x_{1,0}$, and we find

$$\begin{cases} \psi(a)(x) = \psi(a_0)(x_{1,0}), \\ \psi(b)(x) = \psi(b_0)(\psi(\sigma_{r,1}^{-1} \rho_r^k \sigma_{r,1})(x_{1,0})), = \psi(b_0)(x_{1,k}) \end{cases}$$

We use now the argument of Lemma I.3.17. The hypothesis that a_0 admits an expression without σ_1^{-1} and $\rho_1^{\pm 1}$ implies that $\psi(a_0)(x_{1,0})$ is either $x_{1,0}$, or it finishes with $x_{1,0}^{-1}$. Indeed, the only new fact that could possibly occur due to the presence of charges $\rho_j^{\pm 1}$ would be some indices q in variables $x_{p,q}^{\pm 1}$ to be modified. Now, the only letters $\rho_j^{\pm 1}$ whose action may modify a variable $x_{1,k}^{\pm 1}$ are $\rho_1^{\pm 1}$, which we assume do not occur in a_0. Similarly, the hypothesis that b_0 admits an expression without σ_1^{-1} and $\rho_1^{\pm 1}$ implies that $\psi(b_0)(x_{1,k})$ is either $x_{1,k}$, or it finishes with $x_{1,k}^{-1}$. We deduce $\psi(a)(x) \neq \psi(b)(x)$, hence $\psi(a) \neq \psi(b)$, and, finally, $a \neq b$. ∎

Proposition 6.15. (free LD from Larue's group) *The elements $\theta_{1,1}$, $\theta_{1,2}$, ... generate a free LD-system in $(LB_\infty, ^\wedge)$.*

Proof. The surjective homomorphism of the group LB_∞ onto the group CB_∞ that maps $\theta_{j,k}$ to ρ_j^{k-1} is compatible with the shift mappings, and, therefore, it is also a surjective homomorphism of the LD-system $(LB_\infty, ^\wedge)$ onto the LD-system $(CB_\infty, ^\wedge)$. A possible cycle for left division in $(LB_\infty, ^\wedge)$ would induce a cycle for left division in $(CB_\infty, ^\wedge)$, which is impossible. Similarly, a possible equality for elements satisfying equalities of type (6.5) in LB_∞ and involving the elements $\theta_{1,k}$ would project onto a similar equality in CB_∞ involving the elements ρ_1^{k-1}, which we have seen above is impossible. ∎

The word problem

As was already mentioned in Section 1, the word problem for a sequence of algebraic identities $\vec{I}$ is the question of algorithmically deciding whether two given terms t, t' become equal when the identities are obeyed. So, in the case of left self-distributivity, the question is to recognize whether $t =_{LD} t'$ holds. Such a question is obviously semi-decidable: starting from t, we can enumerate all terms LD-equivalent to t systematically by repeatedly applying the identity (LD) at each possible position. If t' is LD-equivalent to t, it will appear in the previous list at some (finite) step, and we can then certify that $t' =_{LD} t$ holds. On the other hand, the method gives no answer: if $t' =_{LD} t$ fails, after any finite number of steps, we can only see that t' has not yet appeared, which does not prove that it will not appear later.

We give now a full answer to the question—and not only a half-answer as above.

Proposition 6.16. (**word problem**) *The relations $=_{LD}$, $\sqsubset_{LD}$ and $<_{LD}$ are decidable, i.e., there exist algorithms that decide whether two given terms t_1, t_2 satisfy $t_1 =_{LD} t_2$, $t_1 \sqsubset_{LD} t_2$, or $t_1 <_{LD} t_2$. More precisely, let $\exp^*$ denote the iterated exponential defined by $\exp^*(0) = 1$ and $\exp^*(n+1) = 2^{\exp^*(n)}$. Then the previous relations lie in the complexity class $\mathrm{DSPACE}(\exp^*(2^\ell)))$, i.e., the algorithms can be constructed so that, for ℓ large enough and for terms of length at most ℓ, the total decision requires at most $\exp^*(2^\ell)$ space resources.*

Proof. Assume first that t_1 and t_2 are terms in T_1 with length at most ℓ. Let m be the supremum of their heights, and k be the supremum of their complexities. We first construct the term $\partial^k x^{[m+1]}$. We have $m \le (\ell+1)/2$, and $k \le (3^m - 1)/2$ (Exercise 4.23), hence $k < 2^{a\ell}$ with $a = \log_2(\sqrt{3})$. By Lemma 3.9, the length of the term $\partial^k x^{[m]}$ is bounded by the tower of exponentials $2^{2^{\cdot^{\cdot^{\cdot^{2^\ell}}}}}$ with k times 2. On the other hand, $\ell \le \exp^*((1-a)\ell)$ holds for ℓ large enough. So, finally, the length of the term $\partial^k x^{[m+1]}$ is bounded above by $\exp^*(2^\ell)$.

Then we consider each iterated left subterm $\mathrm{left}^p(\partial^k x^{[m+1]})$ and we check whether it is an LD-expansion of t_1 or t_2. Recognizing whether t' is an LD-expansion of t can be done by enumerating all LD-expansions of t whose length is at most that of t': as expanding a term increases its length, the considered LD-expansions are finite in number, and the decision can be made using $\lg(t')$ space resources at most. By Proposition 4.5 (comparison), we shall eventually find integers p_1 and p_2 such that $\mathrm{left}^{p_1}(\partial^k x^{[m+1]})$ is an LD-expansion of t_1, and $\mathrm{left}^{p_2}(\partial^k x^{[m+1]})$ is an LD-expansion of t_2. We conclude that $t_1 =_{LD} t_2$ holds if and only if $p_1 = p_2$ holds, and that $t_1 \sqsubset_{LD} t_2$ holds if and only if $p_1 > p_2$ holds.

Let us now consider the general case of t_1 and t_2 involving an arbitrary number of variables. Letting again m be the supremum of the heights of t_1 and t_2, and k be the supremum of their complexities, we apply the previous method to the terms $t_1^\dagger$ and $t_2^\dagger$, and find integers p_1 and p_2 such that $\mathrm{left}^{p_i}(\partial^k x^{[m+1]})$ is an LD-expansion of t_i for $i = 1, 2$. Now, by Lemma 4.21, and always using the same space resources, we find two terms t'_1, t'_2 expanding t_1 and t_2 respectively and such that $t'^{\dagger}_i$ is $\mathrm{left}^{p_i}(\partial^k x^{[m+1]})$. Then $t_1 =_{LD} t_2$ is equivalent to $t'_1 = t'_2$. Similarly, $t_1 \sqsubset_{LD} t_2$ is equivalent to $t'_1 \prec t'_2$. Finally $t_1 \ll_{LD} t_2$ is equivalent to $t'_1 \ll t'_2$. ∎

Example 6.17. (word problem) *Let $t_1 = (x_1{\cdot}x_2){\cdot}(x_1{\cdot}x_3){\cdot}x_4$ and $t_2 = x_1{\cdot}(x_2{\cdot}x_3){\cdot}(x_1{\cdot}x_4)$. The height of t_1 and t_2 is 4 and their complexities are 6 and 5 respectively. It follows that some iterated left subterms of the term $\partial^6 x^{[5]}$ expand the terms $t_1^\dagger$ and $t_2^\dagger$. The term $\partial^6 x^{[5]}$ is a huge term involving thousands of letters. However, considering the term $\partial x^{[5]}$ is already sufficient here: indeed its left subterm, which is $\partial x^{[4]}$, turns out to be an LD-expansion both of $t_1^\dagger$ and of $t_2^\dagger$. Now, applying to t_1 and t_2 those sequences of basic LD-expansions that lead from $t_1^\dagger$ and $t_2^\dagger$ to $\partial x^{[4]}$ yields in both cases the term $((x_1{\cdot}x_2){\cdot}(x_1{\cdot}x_3)){\cdot}((x_1{\cdot}x_2){\cdot}(x_1{\cdot}x_4))$. Thus, we conclude that t_1 and t_2 are LD-equivalent.*

The previous algorithms are very inefficient in practice. For the word problem, more tractable methods can be obtained by using braids or charged braids.

Proposition 6.18. (complexity I) *The restriction of LD-equivalence to terms in T_1 has at most an exponential complexity.*

Proof. Let t, t' be terms in T_1. As the sub-LD-system of $(B_\infty, {}^\wedge)$ generated by the unit braid 1 is free, the terms t and t' are LD-equivalent if and only if they receive equal evaluations in $(B_\infty, {}^\wedge)$ when the variable x is mapped to the unit braid 1. By the results of Section II.5, two braid words w, w' represent the same braid if and only if the double reversing of the word $w^{-1}w'$ ends with an empty word. So, starting with two terms t, t' in T_1, we can decide $t =_{LD} t'$ as follows:

(i) Compute the images t and t' under the mapping F of T_1 into BW_∞ inductively defined by $F(x) = \varepsilon$, $F(t_1{\cdot}t_2) = F(t_1)\,\mathrm{sh}(F(t_2))\,\sigma_1\,\mathrm{sh}(F(t_1)^{-1})$;
(ii) Operate the double reversing of the braid word $F(t)^{-1}F(t')$.
Then the terms t and t' are LD-equivalent if and only if the final braid word after step (ii) is empty.

If t has length at most n (as a word), the length of the braid word $F(t)$ is bounded above by $\ell = 2^{(n+1)/2} - 1$ (which is obtained in the case of the left power $x_{[(n-1)/2]}$), and its width is bounded above by $m = (n-1)/2$. We have seen in Chapter II that the double reversing of an m-strand braid word of length ℓ requires $\mathcal{O}(m^2\ell^2)$ steps, which leads to the upper bound of the proposition. ∎

Example 6.19. **(word problem)** *Let us apply the previous method to the terms $t = (x{\cdot}x){\cdot}x{\cdot}x{\cdot}x$ and $t' = x{\cdot}(x{\cdot}x){\cdot}(x{\cdot}x)$. We find*

$$F(t) = \sigma_1\sigma_3\sigma_2\sigma_1\sigma_2^{-1}, \quad F(t') = \sigma_2\sigma_3\sigma_2\sigma_3^{-1}\sigma_1.$$

The rightreversing of $F(t)^{-1}F(t')$ ends with the word $\sigma_2\sigma_3\sigma_2\sigma_3^{-1}\sigma_2^{-1}\sigma_3^{-1}$, and the left reversing of the latter word ends with the empty word. We conclude that t and t' are LD-equivalent.

Extending the previous method to the case of rank 2 and above requires that we first solve the word problem of the group CB_∞ or LB_∞. Here we shall consider the latter group. Let us denote by LW the free monoid of all words over the alphabet $\{\sigma_i^{\pm1}\ ;\ i \geq 1\} \cup \{\theta_{j,k}^{\pm1}\ ;\ j,k \geq 1\}$. We introduce two word homomorphisms, namely π_B that preserves the letters σ_i and collapses all letters $\theta_{j,k}$ to ε, and π_F that collapses the letters σ_i and preserves $\theta_{j,k}$. These mappings induce well defined group homomorphisms of LB_∞ onto B_∞ and onto the free group $\mathrm{FG}_{\infty,\infty}$ generated by the double sequence of all $\theta_{j,k}$'s respectively. If we are given a word w in LW, a necessary condition for w to represent 1 in LB_∞ is that the projections $\pi_B(w)$ and $\pi_F(w)$ both represent 1, which, by the results of the previous sections, can be decided in polynomial time. So, the only problem is to recognize whether w represents 1 in LB_∞ when $\pi_B(w)$ and $\pi_F(w)$ represent 1 respectively in B_∞ and FG_∞.

Definition 6.20. **(reduction)** Assume that w is a word in LW. We say that w is *reducible* if it contains a subword of the form $\theta_{j,k}^{e}u\theta_{j,k}^{-e}$ with $e = \pm1$ and u a braid word satisfying $u \equiv u'$ for some u' in BW_{j-2}. In this case, we say that w *reduces* to w' if w' is obtained from w by replacing a factor $\theta_{j,k}^{e}u\theta_{j,k}^{-e}$ as above with the corresponding factor u.

The question of whether a given braid word u is equivalent to a braid word in BW_m for some given m is algorithmically decidable by a double reversing process: indeed, $u \equiv u'$ implies $N_{RL}(u) \equiv N_{RL}(u')$ and $D_{RL}(u) \equiv D_{RL}(u')$, and, therefore, if u' involves only letters $\sigma_i^{\pm1}$ with $i \leq m$, so does $D_{RL}(u)^{-1}N_{RL}(u)$.

Assume that w is a word in LW such that $\pi_F(w)$ is equivalent to ε. Then there exist sequences of free reductions reducing the word $\pi_F(w)$ to ε. We define a *pairing* for w to be a set of pairs of positions (p, q) in w such that the letter at p vanishes with the letter at q in such a sequence of free reductions (there is no uniqueness in general). When a pairing is fixed, we shall indicate that two letters are paired by underlining the subword they determine. If a word w as above and a pairing for w are given, we define a *special pattern* in w to be a subword of w of the form $\underline{\theta_{j,k}^{e}u\theta_{j,k}^{-e}}$, where the first and the last letters are paired, e is ±1 and $\overline{\pi_B(u)}$ does not belong to B_{j-1}.

Lemma 6.21. *Assume that $\pi_F(w)$ is trivial and every pairing for w contains a special pattern. Then the same is true for every word w' equivalent to w.*

Proof. Assume that w' is obtained from w by applying one of the defining relations of LB_∞. Consider an arbitrary pairing for w'. If w' has been obtained using a braid relation, then the pairing of w' comes from a pairing of w, the latter contains by hypothesis a special pattern, say $\underline{\theta^e_{j,k} u \theta^{-e}_{j,k}}$. This pattern of w induces a pattern of w' of the form $\underline{\theta^e_{j,k} u' \theta^{-e}_{j,k}}$ with $\pi_B(u') \equiv \pi_B(u)$. The latter is a special pattern for the considered pairing of w'. Similarly, if w' has been obtained from w using some commutation relation $\sigma_i^{\pm 1} \theta^{\pm 1}_{j,k} \equiv \theta^{\pm 1}_{j,k} \sigma_i^{\pm 1}$, we can lift the pairing of w' into a pairing of w; the latter contains a special pattern $\underline{\theta^e_{j,k} u \theta^{-e}_{j,k}}$, which in turn induces a pattern of w' which is either equal to the latter (if it is not involved in the commutation relation that has been applied), or has the form $\underline{\theta^e_{j,k} u \sigma_i^{\pm 1} \theta^{-e}_{j,k}}$ or $\underline{\theta^e_{j,k} \sigma_i^{\pm 1} u \theta^{-e}_{j,k}}$. This gives the result, as, by hypothesis, $i \le j-2$ holds in this case, so $\overline{u} \in B_{j-1}$ is equivalent to $\overline{u}\sigma_i^{\pm 1} \in B_{j-1}$ and $\sigma_i^{\pm 1}\overline{u} \in B_{j-1}$.

Assume that w' has been obtained from w by deleting a pair $\theta^e_{j,k}\theta^{-e}_{j,k}$. Every pairing for w' extends into a pairing of w by re-adding the pair $(\theta^{e1}_{j,k}, \theta^{-e}_{j,k})$ that has been removed. Then w contains at least one special pattern, which cannot be the additional pattern $\underline{\theta^e_{j,k}\theta^{-e}_{j,k}}$ that has been re-added at the end, and, so, it gives a special pattern in the pairing of w'.

There remains the case when w' is obtained from w by adding a pair $\theta^e_{j,k}\theta^{-e}_{j,k}$. Let us consider an arbitrary pairing of w'. If the newly created pair $\theta^e_{j,k}\theta^{-e}_{j,k}$ appears in the pairing, the pairs of w remain present in w', and the result is true. Assume now that the pairing takes the form $\underline{\theta^{-e}_{j,k} w_1 \theta^e_{j,k}}\ \underline{\theta^{-e}_{j,k} w_2 \theta^e_{j,k}}$, where the newly created pair corresponds to the central two letters. Then the word $\theta^{-e}_{j,k} w_1 w_2 \theta^e_{j,k}$ is a subword of w, and the hypothesis that there exists a pairing as indicated for w' implies that there exists a pairing for w where the first and the last letters in the above subword are paired. If w contains a special pattern that is disjoint from the above considered pattern, this pattern belongs to w'. Otherwise, $\underline{\theta^{-e}_{j,k} w_1 w_2 \theta^e_{j,k}}$ is special in w. This implies that at least one of $\underline{\theta^{-e}_{j,k} w_1 \theta^e_{j,k}}$, $\underline{\theta^{-e}_{j,k} w_2 \theta^e_{j,k}}$ is special in w', for $\overline{w_1 w_2} \notin B_{j-1}$ implies $\overline{w_1} \notin B_{j-1}$ or $\overline{w_2} \notin B_{j-1}$. The last case is a pairing of the form $\theta^e_{j,k}\theta^{-e}_{j,k} w_1 \theta^e_{j,k} w_2 \theta^{-e}_{j,k}$, *i.e.*, the new letters vanish with letters both on their right—or, symmetrically, on their left. The argument is similar. We obtain in w a pairing with the fragment $\underline{w_1 \theta^e_{j,k} w_2 \theta^{-e}_{j,k}}$. If w contains another special pattern, so does w'. Otherwise, we must have $\overline{w_2} \notin B_{j-1}$. Then either $\overline{w_1} \in B_{j-1}$ holds, so we have $\overline{w_1 w_2} \notin B_{j-1}$, and $\underline{\theta^e_{j,k}\theta^{-e}_{j,k} w_1 \theta^e_{j,k} w_2 \theta^{-e}_{j,k}}$ is the special pattern we need, or $\overline{w_1} \notin B_{j-1}$ holds, and $\underline{\theta^{-e}_{j,k} w_1 \theta^e_{j,k}}$ is the special pattern we need. ■

It is now easy to conclude the argument.

Proposition 6.22. (**word problem**) *The word problem for the standard presentation of the group LB_∞ is solvable in polynomial time.*

Proof. Let w be a word in LW. We can decide whether w represents 1 in the group LB_∞ as follows:

(i) Reduce w until an irreducible word w' is obtained. If w' contains at least one letter $\theta_{j,k}^{\pm 1}$, then w does not represent 1;
(ii) Otherwise, w represents 1 in LB_∞ if and only if w' represents 1 in B_∞.

As reduction diminishes the length, it terminates in a finite number of steps, bounded above by the length of w. Owing to the defining relations of LB_∞, reduction transforms a word into an equivalent word, since $\theta_{j,k}$ commutes with every braid of B_{j-1}. Hence, the word w' of the algorithm exists, and it represents the same element of LB_∞ as w. So, if w' contains no θ letter and represents 1 in B_∞, then w represents 1 in LB_∞. Conversely, assume that w' contains at least one θ letter. If $\pi_F(w')$ is not trivial, then w does not represent 1 in LB_∞. So let us assume that $\pi_F(w')$ is trivial. Fix a pairing in w', and consider a pattern $\underline{\theta_{j,k}^{e} u \theta_{j,k}^{-e}}$ of minimal length in the pairing. Then u contains no letter θ, *i.e.*, it is a braid word. The hypothesis that w' is irreducible implies $\overline{u} \notin B_{j-1}$. This means that the pattern above is special in w'. By Lemma 6.21, every word that is equivalent to w' contains a special pattern, and, therefore, it cannot be the empty word, which contains no special pattern. ∎

Proposition 6.23. (**complexity** II) *LD-equivalence has at most an exponential complexity.*

Proof. For comparing two terms t, t', we compute their images $F(t)$, $F(t')$ in the group LB_∞ when the variable x_i is mapped to the generator $\theta_{1,i}$. By Proposition 6.15 (free from Larue's group), the terms t and t' are LD-equivalent if and only if the word $F(t)^{-1}F(t')$ represents 1 in LB_∞. By Proposition 6.22 (word problem), the latter question can be solved algorithmically using time resources that are polynomial in the length of $F(t)^{-1}F(t')$, hence exponential with respect to the parameter $\lg(t) + \lg(t')$. ∎

Laver's Conjecture for free LD-systems

We have mentioned in Section IV.3 the braid form of Laver's Conjecture about the action of braids on the left cancellative LD-system $(B_\infty, {}^\wedge)$. Here we state a similar conjecture involving free LD-systems.

We recall that, for S a cancellative LD-system, and $\vec{a}$ a sequence of elements of S, we denote by $D(S, \vec{a})$ the set of all braids such that $\vec{a} \bullet b$ exists, where the action is the one defined in Section III.1. Here we consider the action of braids on powers of the free LD-system FLD_∞, which we have seen is left cancellative, hence eligible.

Conjecture 6.24. (**Laver's conjecture, LD form**) *For every sequence $\vec{a}$ in FLD_∞^n, the subset $D(\mathrm{FLD}_\infty, \vec{a})$ of B_n is well ordered by $<_L$.*

We establish below a result that makes the previous conjecture plausible. As the action of positive braids is always defined, every set of the form $D(S, \vec{a})$ includes the set B_n^+, which we know is well ordered by $<_L$. We show now that, in some cases, the set $D(\mathrm{FLD}_\infty, \vec{a})$ coincides with B_n^+, so that, in such cases, the conjecture is true.

Definition 6.25. (**strongly decreasing**) Assume that $(a_1, \dots, a_n)$ is a finite sequence in FLD_∞^n. We say that $(a_1, \dots, a_n)$ is *strongly decreasing* if the inequality $a_k \gg a_{k+1}a_{k+2} \cdots a_n$ holds for $k = 1, \dots, n-1$.

A strongly decreasing sequence is $<$-decreasing, as, according to Proposition 5.10 (compatibility), $c \gg ab$ implies $c > a$.

Lemma 6.26. *Assume that $\vec{a}$ is a strongly decreasing sequence in FLD_∞^n and b belongs to B_n^+. Then $\vec{a} \bullet b$ is strongly decreasing as well.*

Proof. For an induction on the length of b, it suffices to consider the case of a single crossing σ_i. Write $(b_1, \dots, b_n) = (a_1, \dots, a_n) \bullet \sigma_i$. For $k < i$, we have

$$\begin{aligned} b_k = a_k &\gg a_{k+1} \cdots a_i a_{i+1} \cdots a_n \\ &= a_{k+1} \cdots (a_i a_{i+1}) a_i \cdots a_n = b_{k+1} \cdots b_i b_{i+1} \cdots b_n. \end{aligned}$$

For $k = i$, we find, using the compatibility of $\gg$ with multiplication on the left,

$$b_k = a_k a_{k+1} \gg a_k(a_{k+2} \cdots a_n) = b_{k+1} b_{k+2} \cdots b_n.$$

For $k = i+1$, we have $b_k = a_{k-1}$. By Proposition 5.10 (compatibility), the conjunction of $a_{k-1} \gg a_k b$ and $a_k \gg c$ implies $a_{k-1} \gg c$. Applying this with $c = b = a_{k+1} \cdots a_n$, we obtain $a_{k-1} \gg a_{k+1} \cdots a_n$, and

$$b_k = a_{k-1} \gg a_{k+1} \cdots a_n = b_{k+1} \cdots b_n.$$

Finally, for $k > i+1$, we find $b_k = a_k \gg a_{k+1} \cdots a_n = b_{k+1} \cdots b_n$. ∎

Lemma 6.27. *Assume that $\vec{a}$ is a strongly decreasing sequence in FLD_∞^n and b is a braid in B_n^+ such that, letting $(c_1, \dots, c_n) = \vec{a} \bullet b$, the inequality $c_q \sqsubset c_p$ holds for some p, q with $p < q$. Then $\sigma_{p,q-1}$ is a right divisor of b in B_n^+.*

Proof. We use induction on the length of a positive word u representing b. If u is empty, the result is vacuously true, as $a_q \ll a_p$ holds for $p < q$, and, therefore, $a_q \sqsubset a_p$ does not. Otherwise, let $u = u'\sigma_i$ and $(c'_1, \dots, c'_n) = \vec{a} \bullet u'$. We assume $p < q$ and $c_q \sqsubset c_p$, and consider the various possible positions of i. Our aim is to construct a positive braid word v satisfying $u \equiv v\, \sigma_{p,q-1}$.

Assume first $i+1 < p$ or $i > q$. Then, by definition, we have $c_p = c'_p$, $c_q = c'_q$. The induction hypothesis implies that $\sigma_{p,q-1}$ is a right divisor of $\overline{u'}$, *i.e.*, there exists a positive braid word v' satisfying $u' \equiv v'\sigma_{p,q-1}$. We deduce $u \equiv v\sigma_{p,q-1}$ for $v = v'\sigma_i$, as σ_i commutes with $\sigma_{p,q-1}$.

Assume now $i+1 = p$. Then, we have $c_p = c'_{p-1}$ and $c_q = c'_q$, so the hypothesis $c_q \sqsubset c_p$ gives $c'_q \sqsubset c'_{p-1}$, the induction hypothesis implies that there exists v' satisfying $u' \equiv v'\sigma_{p-1,q-1}$. Using $\sigma_{p-1,q-1}\sigma_{p-1} \equiv \sigma_{p,q-1}$, we deduce $u \equiv v\sigma_{p,q-1}$ with $v = v'\sigma_{p,p-1}$.

Assume $i = p$. For $q = p+1$, the result is true, as u finishes with σ_p by hypothesis. So, assume $p+1 < q$. Then, we have $c_p = c'_p c'_{p+1}$ and $c_q = c'_q$. The hypothesis is $c'_q \sqsubset c'_p c'_{p+1}$. By Lemma 6.26, $c'_q < c'_p$ holds, and we deduce $c'_q \sqsubset c'_p$. Indeed, the other possibility $c'_q \ll c'_p$ would imply $c'_q \ll c'_p c'_{p+1}$, contradicting $c'_q \sqsubset c'_p c'_{p+1}$. By induction hypothesis, there exists v' satisfying $u' \equiv v'\sigma_{p,q-1}$. Using $\sigma_{p,q-1}\sigma_p \equiv \sigma_{p+1}\sigma_{p,q-1}$, we deduce $u \equiv v\sigma_{p,q-1}$ with $v = v'\sigma_{p+1}$.

Assume $p < i < q-1$. Here again, we have $c_p = c'_p$, $c_q = c'_q$. Applying the induction hypothesis, we find $u' \equiv v'\sigma_{p,q-1}$, whence $u \equiv v\sigma_{p,q-1}$ with $v = v'\sigma_{i+1}$, using $\sigma_{p,q-1}\sigma_i \equiv \sigma_{i+1}\sigma_{p,q-1}$.

Assume then $i = q-1$. We have $c_p = c'_p$, $c_q = c'_{q-1}$. By induction hypothesis, we have $u' \equiv v'\sigma_{p,q-2}$, and we deduce $u \equiv v'\sigma_{p,q-1}$.

Assume finally $i = q$. Here we have $c_p = c'_p$, and $c_q = c'_q c'_{q+1}$. The hypothesis is $c'_q c'_{q+1} \sqsubset c'_p$, which implies $c'_q \sqsubset c'_p$ by definition of $\sqsubset$. By induction hypothesis, there exists v' satisfying $u' \equiv v'\sigma_{p,q-1}$. Let $(c''_1, \ldots, c''_n) = \vec{a} \bullet v'$. Applying the definition, we obtain $c_p = c''_p c''_{p+1}$ and $c_q = c''_p c''_{q+1}$. Hence, the hypothesis implies $c''_p c''_{q+1} \sqsubset c''_p c''_{p+1}$, which in turn implies $c''_{q+1} \sqsubset c''_{p+1}$. By induction hypothesis, there exists v'' satisfying $v' \equiv v''\sigma_{p+1,q}$, we find

$$u \equiv v''\ \sigma_{p+1,q}\ \sigma_{p,q-1}\ \sigma_q \equiv v''\ \sigma_p\ \sigma_{p+1,q}\ \sigma_{p,q-1} = v''\ \sigma_{p,q}\ \sigma_{p,q-1},$$

and taking $v = v''\sigma_{p,q}$ gives the result. All cases have been considered. ∎

Proposition 6.28. (**only positive braids**) *Assume that $\vec{a}$ is a strongly decreasing sequence in* FLD^n_∞. *Then the set $D(\mathrm{FLD}_\infty, \vec{a})$ is exactly B_n^+.*

Proof. Assume that w is a braid word such that $\vec{a} \bullet w$ exists. The word $D_L(w)^{-1}N_R(w)$ is equivalent to w, and, therefore, by Proposition III.1.7 (action) and Lemma III.1.4, the sequence $\vec{a} \bullet N_{LR}(w)D_{LR}(w)^{-1}$ exists as well. Assume that $D_{LR}(w)$ is not empty, and let σ_i be its last letter. By hypothesis, the sequence $\vec{a} \bullet N_{LR}(w)\sigma_i^{-1}$ is defined. Applying Lemma 6.27, we conclude that σ_i is a right divisor of $N_{LR}(w)$ in B_n^+, and, therefore, σ_i is a common right divisor of $N_{LR}(w)$ and $D_{LR}(w)$ in B_n^+. This contradicts Proposition II.5.6 fraction). ∎

Thus, if $\vec{a}$ is a strongly decreasing sequence in FLD^n_∞, the subset $D(\mathrm{FLD}_\infty, \vec{a})$ of B_n is well ordered by $<_L$.

Exercise 6.29. **(powers)** *Assume that S is a free LD-system of rank 1, and the elements a, b of S satisfy $ab = b^{[2]}$ and $ba = a^{[2]}$. Prove $a = b$.*

Exercise 6.30. **(free subsets)** *(i) Assume that S is a free LD-system and $<$ is a canonical order on S. Show that the completely free subsets of S (cf. Exercise 1.23) are the chains of $\ll$.*

(ii) Prove that a subset $\{a, b\}$ with $b = ac$ or $b = (ac)a$ is not free in S. [Hint: We have $ab = a^{[2]}b$ and $ab^{[2]} = a^{[2]}b^{[2]}$ respectively.]

Exercise 6.31. **(braid exponentiation)** *Prove that the sets $\{1, \sigma_1\}$ and $\{\sigma_1, \sigma_2\}$ are not free in $(B_\infty, {}^\wedge)$. [Hint: $\sigma_1^{[2]} = \sigma_2^{[2]}$ holds.]*

Exercise 6.32. **(action of charged braids)** *(i) Assume that S is an LD-system, and f is an endomorphism of S. Show that the action of B_n^+ on S^n defined in Section I.2 can be extended to CB_n^+ (the monoid defined by the charged braids relations) by adding the rule*

which amounts to using $(a_1, \dots, a_j, \dots)\, \rho_j = (a_1, \dots, a_{j-1}, f(a_j), a_{j+1}, \dots)$.

(ii) Show that the action extends to the group CB_n if S is an LD-quasigroup and f is an automorphism. Can we define a partial action when S is only assumed to be left cancellative and f to be injective?

Exercise 6.33. **(charged braids)** *(i) Prove that the presentation of the group CB_∞ is associated on the right with a complement that is coherent but not convergent.*

(ii) Show that the group CB_∞ is a semidirect product of B_∞ by $\mathrm{Ker}(\phi)$, where $\phi(\sigma_i) = \sigma_i$ and $\phi(\rho_j) = 1$.

(iii) Extend the Burau representation of braids into a linear representation r_B of CB_∞ into $\mathrm{GL}_\infty \mathbf{Z}[t, t^{-1}, u, u^{-1}]$ by mapping ρ_1 to (u).

(iv) Deduce that B_∞ is not a normal subgroup in CB_∞. [Hint: Compute $r_B(\rho_1 \sigma_1 \rho_1^{-1})$, and observe that the sum of the entries in the i-th row is t^{i-1} in the image of a braid under r_B.]

(v) Adapt the results to the group LB_∞ using the surjective homomorphism of Proposition 6.28.

Exercise 6.34. **(Laver's conjecture for** B_2) *Prove the LD form of Laver's conjecture for* B_2. *[Hint: Applying a negative braid to a sequence* $\vec{a}$ *in* FLD_∞^n *diminishes the sum* $\sum_i \mathrm{ht}_R(a_i)$, *where* ht_R *is the mapping induced by the right height of the terms.]*

Exercise 6.35. **(declining sequences)** *Say that a finite sequence* $(a_1, \ldots, a_n)$ *in* FLD_∞^n *is declining if* $a_k > a_{k+1} \cdots a_n$ *holds for* $k = 1, \ldots, n-1$. *Prove that, if* $\vec{a}$ *is a declining sequence in* FLD_∞^n *and* b *is a positive braid, then* $\vec{a} \bullet b$ *is declining.*

7. Notes

The general results about free algebraic systems are part of folklore. A comprehensive source for a number of results in universal algebra is (McKenzie, McNulty & Taylor 86). The terms we use here are right Polish terms where product is postponed. The only requirement in the construction of terms is that each term that is not a variable admits a unique decomposition as the product of two terms, and we can use any binary system with this property. The present construction however turns out to be convenient for the specific case of left self-distributivity, as it makes the iterated left subterms of a term be its prefixes—which will be crucial in Chapter VI. As far as right Polish terms are used, introducing the binary operation $\cdot$ besides $\bullet$ can appear strange, but it greatly improves readability.

The description of free LD-quasigroups is mentioned in (Fenn & Rourke 92)—it was also probably part of folklore among algebraists of Eastern Europ.

It seems that free LD-systems had not been investigated for themselves until the first results of set theory about iterations of elementary embeddings (*cf.* Chapter XII) gave a specific motivation. LD-expansions of terms and the operator ∂ are introduced in (Dehornoy 89*a*), where the confluence property is established. The comparison property and its application to the word problem are established in (Dehornoy 89*b*). An independent approach was developed at the same time by R. Laver, starting from results in set theory. In particular, the use of the order for proving freeness appears in (Laver 92). The canonical order for the case of several generators is described in (Dehornoy 93). These results were obtained around 1989–90, when the existence of an LD-system with no cycle for left division was not known—or was known only under some unprovable set-theoretical hypothesis: both the braid exponentiation of Section I.3 and the syntactic proof of acyclicity of Chapter VIII below were discovered later. Thus, in early papers, the results always take the form "Assume that there exists an LD-system where left division has no cycle. Then *etc.*".

Charged braids are introduced in (Dehornoy *94c*), while the group LB_∞ is defined in (Larue *94a*), where the solution to the word problem is established. The group CB_∞ is smaller than the group LB_∞, and it would be natural to try to use CB_∞ everywhere. However, we still have no proof of

Conjecture 7.1. *The word problem of the group CB_∞ is solvable in polynomial time.*

It is likely that CB_∞ embeds is a group of automorphisms of a free group, which would result in an exponential algorithm for the word problem. Using for CB_∞ the argument developed by Larue for LB_∞ is not obvious because of the partial commutation relations between the generators ρ_j.

Laver's Conjecture (for free LD-systems) is implicit in (Laver 96); it was proposed first around 1995. It seems to be one of the deepest open questions about the braid order. It is not obvious that the two forms of Laver's Conjecture, the one involving braids, and the one involving free LD-systems, are equivalent, but both are natural extensions of the same restricted statement, namely the conjecture that, for every sequence $\vec{a}$ in FLD_1^n, or, equivalently, in $(B_{sp})^n$, the subset $D(B_{sp}, \vec{a})$ of B_n is well ordered by $<_L$. Proposition 6.28 (only positive braids) appears in (Larue *94a*).

Observe that, as far as orders are involved, the case of LD-systems is reminiscent of the case of semigroups, *i.e.*, the case when associativity replaces left self-distributivity. In both cases, left divisibility defines a partial order on free objects, which is linear in the case of rank 1, and extends to a linear order by means of a lexicographical extension in the general case. The existence of such orders characterizes free structures in both cases.

A number of natural questions about free LD-systems and LD-equivalence remain open. Let us mention a few of them here. Firstly, we have seen that the right height of a term is a trivial example of a strong LD-invariant, which is associated with the LD-system made of $\mathbf{N}$ equipped with $x{\wedge}y = y + 1$. This turns out to be the only known example of a (strong) LD-invariant defined using geometrical parameters of the terms and not constant on one variable terms. We shall construct in Chapter VI families of LD-invariants with the latter property, but defining them by purely geometrical properties of the terms remains an open question.

We have no general criterion for recognizing free subsets in a free LD-system. A promising approach could be to use realizations of free LD-systems inside CB_∞ or LB_∞. However, little is known about these groups. On the one hand, they seem to share a number of properties with B_∞, but, on the other hand, decompositions into fractions fail to exist in CB_∞ and LB_∞.

Question 7.2. *Does the LD-system $(B_\infty, {\wedge})$ include a free sub-LD-system of rank 2?*

Partial results such as in Exercise 6.31 support a negative answer.

Other problems deal with the equational theory and the first-order theory of free LD-systems, *i.e.*, the family of all identities and the family of all first-order sentences satisfied in FLD_n.

Question 7.3. (Drápal) *Is every identity satisfied in* FLD_1 *a consequence of (LD), i.e., does every identity that holds in* FLD_1 *hold in every free LD-system* FLD_∞*, and, therefore, in every LD-system?*

In the case of associativity, the answer is negative: the free semigroup of rank 1, *i.e.*, $(\mathbf{N}, +)$, is commutative, but there exist non-commutative semigroups. So commutativity is a typical example of an identity that is not a consequence of associativity, but holds in the free system of rank 1 in the corresponding variety. In the case of left self-distributivity, no such example is known.

Question 7.4. *Are the first-order theories of* FLD_1 *and* FLD_∞ *decidable, i.e., does there exist an algorithm that, starting with an arbitrary sentence involving one binary operation, recognizes whether this sentence is true in* FLD_1 *or* FLD_∞*?*

In the case of the equational theory of FLD_∞, *i.e.*, when we restrict to sentences with universal quantifiers only, the answer is positive, as deciding whether $(\forall x_1)\dots(\forall x_n)(t_1(x_1,\dots,x_n) = t_2(x_1,\dots,x_k))$ holds in FLD_∞ amounts to deciding whether the terms t_1 and t_2 are LD-equivalent, what we know is decidable by Proposition 6.23 (word problem).

Question 7.5. *Does the theory of* $(\mathrm{FLD}_1, \cdot, 1)$ *admit quantifier elimination, i.e., is there an algorithm that, starting with an arbitrary sentence* F *involving one binary operation and one name for the generator, returns a quantifier-free sentence* F' *such that* F *is true in* $(\mathrm{FLD}_1, \cdot)$ *if and only if* F' *is?*

A quantifier-free formula of the type above is a Boolean combination of equalities $t_1(x) = t_2(x)$, and, therefore, it can be decided. Thus, a positive answer to Question 7.5 would imply a positive answer to Question 7.4 (in the case of FLD_1). Considering quantifier elimination here is natural: this is the method used by Ackerman in the 1920s for proving that the first-order theory of the free monogenic semigroup $(\mathbf{N}_+, +, 1)$ is decidable (the theory of free semigroups with at least two generators is undecidable). Some results in Chapter VI may support a positive answer to Question 7.5, but we are certainly far from a complete solution.

VI
Normal Forms

In Chapter V, we solved the word problem of LD-equivalence by constructing an effective algorithm directly recognizing whether two given terms are LD-equivalent. Here, we prove normal form results: we construct several families of distinguished terms such that every term is LD-equivalent to exactly one term in the family. This gives alternative solutions to the word problem, as two terms are LD-equivalent if and only if their normal forms are equal, but also new applications.

The main technical notion is that of a cut of a term. For t_0 a fixed term, we define a cut of t_0 to be a term obtained from t_0 by ignoring what lies at the right of some variable. So, there are as many cuts in t_0 as there are occurrences of variables. Then, we define a fractional cut of t_0 to be an iterated product of cuts of t_0 that, in some sense, interpolates between the cuts of t_0. If there are n occurrences of variables in t_0, the cuts of t_0 can be numbered $1, \ldots, n$, and, then, the fractional cuts of t_0 can be specified by rational numbers between 1 and n, typically 2.1 or 3.101. The main result of the chapter states that every term t satisfying $t \sqsubseteq_{LD} t_0$ is LD-equivalent to a unique fractional cut of t_0, naturally called the t_0-normal form of t, and, therefore, the LD-class of t can be specified by a rational number as above. The result extends to convenient infinite terms t_0, and, in particular, every term in T_1 admits a well-defined x^{∞}-normal form.

The chapter comprises five sections. In Section 1, we develop a geometrical framework for working with terms viewed as binary trees, and, in particular, for specifying the subterms by using addresses. In Section 2, we introduce the cuts of a term, and establish their basic properties, in particular their behaviour in LD-expansions. In Section 3, we prove a first normal form result. We introduce the notion of a ∂-normal term, and prove that every term is LD-equivalent to a ∂-normal term. The remarkably simple argument relies on the absorption property of Section V.2. In Section 4, we introduce a new family of normal terms connected with the fractional cuts alluded above. The proof that every term is LD-equivalent to a normal term starts from the existence of the ∂-normal form. Section 5 is devoted to applications of the previous results: we prove that the inequality $b \sqsubset ab$ holds for all a, b in the free LD-system FLD_1, we study left division in FLD_1, and we deduce a complete description of special braids.

1. Terms as Trees

In this and the next chapters, we need an efficient treatment of terms and, in particular, we must be able to work easily with the subterms of a term. A natural approach consists in considering terms as binary trees, as was done in Section V.2. In this preliminary section, we introduce additional notions.

Addresses

Let X be a nonempty set. We recall that T_X denotes the set of all terms constructed using variables in X and a binary operator, here denoted $\cdot$. The names of the variables do not matter. In order to obtain uniqueness properties, we fix an infinite sequence of variables x_1, x_2, ..., and, as in Chapter V, we use T_n for the set of all terms constructed using x_1, ..., x_n, and T_∞ for the union of all T_n's. In the case of T_1, we usually write x for x_1.

Definition 1.1. (canonical) For t a term, we denote by $\mathrm{Var}(t)$ the sequence obtained by enumerating the variables of t from left with repetitions omitted. The term t is said to be *canonical* if $\mathrm{Var}(t)$ is an initial segment of $x_1 x_2 \cdot \cdots$.

For instance, the term $x_1{\cdot}(x_2{\cdot}x_3)$—also denoted $x_1{\cdot}x_2{\cdot}x_3$ according to our general convention that missing brackets are to be added on the right—is canonical. Similarly, $(x_1{\cdot}x_2){\cdot}x_1$ is canonical, but $x_2{\cdot}x_1$ is not.

Every term that is not a variable admits a left and a right subterm. In Chapter V, we considered the iterated left subterms $\mathrm{left}^p(t)$ of t. Here, we need notations for arbitrary subterms of t. The idea is to iterate the "left–right" specification, and to speak for instance of the left-left-right subterm of a term. To get shorter notations, we shall use 0 for "left" and 1 for "right".

Definition 1.2. (**address**) An *address* is a finite sequence of 0's and 1's. The empty address is denoted ϕ. The set of all addresses is denoted $\mathbf{A}$.

When writing addresses, we simply juxtapose the digits: so 00 and 1011 are addresses, corresponding respectively to "left-left" and to "right-left-right-right". If α, β are addresses, $\alpha\beta$ is the address obtained by concatenating α and β. The set $\mathbf{A}$ equipped with concatenation is a free monoid based on $\{0, 1\}$. It is also equipped with several natural orderings.

Definition 1.3. (**prefix**) For $\alpha, \beta \in \mathbf{A}$, we say that α is a *prefix* of β, denoted $\alpha \sqsubseteq \beta$, if $\beta = \alpha\gamma$ holds for some γ.

In the present context, saying that the address α is a prefix of the address β means that the node specified by β lies below the node specified by α in the considered tree. So, we shall also say that β lies *below* α.

Lemma 1.4. *The relation $\sqsubset$ is a partial order on $\mathbf{A}$; the address ϕ is a minimum for $\sqsubset$, and, for every address α, the prefixes of α are linearly ordered by $\sqsubset$.*

The trivial verification is left as an exercise. Another way to state the latter property—which means that $(\mathbf{A}, \sqsubset)$ is a tree in the language of orders—is to say that, if two addresses α, β are not $\sqsubset$-comparable, *i.e.*, if neither $\alpha \sqsubseteq \beta$ nor $\beta \sqsubseteq \alpha$ holds, then they are $\sqsubset$-incompatible, *i.e.*, they have no common successor. By construction, every address γ admits two minimal successors, namely $\gamma 0$ and $\gamma 1$, and, therefore, α and β are incomparable for $\sqsubseteq$ if and only if there exists an address γ satisfying $\gamma 0 \sqsubseteq \alpha$ and $\gamma 1 \sqsubseteq \beta$, or *vice versa*.

Definition 1.5. (**left–right, orthogonal**) For $\alpha, \beta \in \mathbf{A}$, we say that α lies *on the right of* β, or that β lies *on the left of* α, and write $\alpha >_{LR} \beta$, if there exists an address γ satisfying $\gamma 1 \sqsubseteq \alpha$ and $\gamma 0 \sqsubseteq \beta$. We say that α and β are *orthogonal*, denoted $\alpha \perp \beta$, if $\alpha >_{LR} \beta$ or $\beta >_{LR} \alpha$ holds.

The relation $>_{LR}$ is a partial ordering on $\mathbf{A}$.

Definition 1.6. (**precedence**) For $\alpha, \beta \in \mathbf{A}$, we say that $\alpha > \beta$ holds if either $\alpha >_{LR} \beta$ or $\alpha \sqsubset \beta$ holds.

Lemma 1.7. *The relation $<$ is a linear order on $\mathbf{A}$, and ϕ is the maximum of $(\mathbf{A}, <)$. Every address of the form $\alpha 1$ admits α as an immediate successor, and every address of the form $\alpha 0$ is the lower bound of the decreasing sequence $\alpha 1$, $\alpha 10$, $\alpha 100$, ...*

The proof is left as an (easy) exercise.

The subterms of a term

Addresses are convenient for specifying the subterms of a term.

Definition 1.8. (**subterm**) Assume that t is a term. For α an address, the *subterm of t at α*, or α-subterm of t, is the term $\mathrm{sub}(t, \alpha)$ possibly specified by:

$$\mathrm{sub}(t,\alpha) = \begin{cases} t & \text{for } \alpha = \phi, \\ \mathrm{sub}(t_1, \beta) & \text{for } \alpha = 0\beta \text{ and } t = t_1 \cdot t_2, \\ \mathrm{sub}(t_2, \beta) & \text{for } \alpha = 1\beta \text{ and } t = t_1 \cdot t_2. \end{cases}$$

If $\mathrm{sub}(t, \alpha)$ exists and is a variable, we also denote it by $\mathrm{var}(t, \alpha)$.

Example 1.9. (**subterms**) *When a term t is viewed as a tree, $\mathrm{sub}(t, \alpha)$ is the subtree of t lying below the node N such that α describes the path from the root to N. Let for instance $t = x_1 \cdot (x_2 \cdot x_3 \cdot x_4) \cdot x_5$. Viewed as a tree, t is*

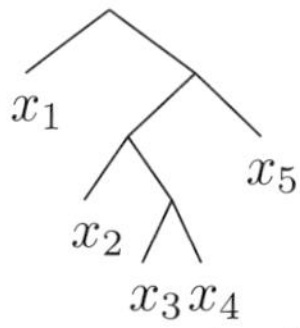

. As can be read in this expression, we have $\mathrm{sub}(t, 0) = x_1$ and $\mathrm{sub}(t, 1) = (x_2 \cdot x_3 \cdot x_4) \cdot x_5$. Similarly, we have $\mathrm{sub}(t, 10) = x_2 \cdot x_3 \cdot x_4$, $\mathrm{sub}(t, 101) = x_3 \cdot x_4$, and $\mathrm{sub}(t, 1010) = x_3$, while $\mathrm{sub}(t, 10100)$ does not exist.

Observe that $\mathrm{left}^p(t) = \mathrm{sub}(t, 0^p)$ holds for all t and p.

Lemma 1.10. *Assume that t is a term, and α, β are addresses. Then we have $\mathrm{sub}(t, \alpha\beta) = \mathrm{sub}((\mathrm{sub}(t, \alpha)), \beta)$: either both terms exist and they are equal, or neither of them is defined.*

Proof. We use induction on the length of the address α. For $\alpha = \phi$, the result is true as $\mathrm{sub}(t, \alpha) = t$ holds by definition. Otherwise, assume $\alpha = e\alpha'$ with $e \in \{0, 1\}$. Assume for instance $e = 0$. By definition, we have $\mathrm{sub}(t, \alpha\beta) = \mathrm{sub}(\mathrm{left}(t), \alpha'\beta)$. By induction hypothesis, the latter term is equal to $\mathrm{sub}(\mathrm{sub}(\mathrm{left}(t), \alpha'), \beta)$, hence, by definition, to $\mathrm{sub}(\mathrm{sub}(t, \alpha), \beta)$. The argument is similar for $e = 1$. ∎

Definition 1.11. (**skeleton, outline**) Assume that t is a term and α is an address. We say that α is an address *in* t if the subterm $\mathrm{sub}(t, \alpha)$ exists. In this case, we say that α is *external* in t if $\mathrm{sub}(t, \alpha)$ is a variable, and is *internal* otherwise. The *skeleton* of t is defined to be the set $\mathrm{Skel}(t)$ of all addresses in t; the *outline* of t is defined to be the set $\mathrm{Out}(t)$ of all external addresses in t.

Example 1.12. (**skeleton**) *Let t be, as above, $x_1{\cdot}(x_2{\cdot}x_3{\cdot}x_4){\cdot}x_5$. The skeleton of t comprises nine addresses, namely ϕ, 0, 1, 10, 100, 101, 1010, 1011, and 11. Among them, ϕ, 1, 10, 101 are internal, and 0, 100, 1010, 1011, 11 are external. The latter form the outline of t.*

Definition 1.13. (**shift**) Assume that α is an address, and B is a set of addresses. We define $\mathrm{sh}_\alpha(B)$ to be $\{\alpha\beta\ ;\ \beta \in B\}$.

Lemma 1.14. *For $t = t_1{\cdot}t_2$, we have*

$$\mathrm{Skel}(t) = \{\phi\} \cup \mathrm{sh}_0(\mathrm{Skel}(t_1)) \cup \mathrm{sh}_1(\mathrm{Skel}(t_2)), \tag{1.1}$$

$$\mathrm{Out}(t) = \mathrm{sh}_0(\mathrm{Out}(t_1)) \cup \mathrm{sh}_1(\mathrm{Out}(t_2)). \tag{1.2}$$

Proof. The first equality follows from the definition of $\mathrm{sub}(t, \alpha)$. The second one follows from Lemma 1.10: $\mathrm{sub}(t, 0\alpha) = \mathrm{sub}(t_1, \alpha)$ holds, and, in particular, the former is a variable if and only if the latter is. ■

Lemma 1.15. *Assume that t is a term. Then α belongs to the skeleton of t if and only if there exists β such that $\alpha\beta$ belongs to the outline of t.*

Proof. We use induction on α. For $\alpha = \phi$, α always belongs to the skeleton of t, and α is a prefix of every address in t, hence of every address in the outline of t. Assume $\alpha = e\beta$, with $e = 0$ or $e = 1$. Then α belongs to the skeleton of t if and only if t is not a variable, say $t = t_1{\cdot}t_2$, and β belongs to the skeleton of t_e. Applying the induction hypothesis to t_e and β gives the result. ■

Notation 1.16. (**0^* and 1^***) Assume that t is a term, and α belongs to the skeleton of t. Then there exist unique nonnegative integers p, q such that $\alpha 0^p$ and $\alpha 1^q$ belong to the outline of t. When no confusion is possible, we write $\alpha 0^*$ and $\alpha 1^*$ for the latter addresses, thus avoiding an introduction of p and q.

Observe that two terms t_1, t_2 have the same outline if and only if they have the same image under the substitution that maps every variable to x_1, *i.e.*, if and only if $t_1^\dagger = t_2^\dagger$ holds.

The existence of the canonical orders on free LD-systems implies the following property, which is the basis of all subsequent normal form results:

Proposition 1.17. (**uniqueness**) *Assume that t_1, t_2 are terms with the same outline. Then t_1 and t_2 are LD-equivalent if and only if they are equal.*

Proof. Assume that t_1 and t_2 have the same outline. Fix a linear order $<$ on the variables that occur in t_1 and t_2, and consider the extension $<_{LD}$ of $<$ defined in Section V.4. If t_1 and t_2 do not coincide, one of $t_1 \ll t_2$, $t_2 \ll t_1$ holds, and, therefore, one of $t_1 <_{LD} t_2$, $t_2 <_{LD} t_1$ does. By Proposition V.5.3 (trichotomy), both forbid $t_1 =_{LD} t_2$. ∎

Trees *vs.* words

In Section V.1, we constructed the terms in T_X so that a term on X is a word over the alphabet $X \cup \{\bullet\}$. We used a right Polish notation, so that $t_1 \cdot t_2$ is the word $t_1 t_2 \bullet$. We recall that we consider a word over the alphabet X as a mapping of $\{1, 2, \ldots, \ell\}$ into X, where ℓ is the length $\lg(w)$ of w. For such a word w, we call $\{1, \ldots, \ell\}$ the *domain* of w, and we denote by $w(p)$ the p-th letter in w. If $w(p) = x$ holds, we say that p is a *position* of x in w. For instance, the letter x_1 has two positions in $x_1 x_2 x_3 x_4 \bullet\bullet x_1 \bullet\bullet$—which is the term $x_1 \cdot ((x_2 \cdot x_3 \cdot x_4) \cdot x_1)$—namely 1 and 7, while the letter $\bullet$ has four positions, namely 5, 6, 8, and 9. By definition, a word of length ℓ contains ℓ positions, namely 1, ..., ℓ. Such numbers are called positions *in* t.

Two representations of terms are available, namely by trees, the nodes of which are specified using addresses, and by words, the letters of which are specified using positions. There exists a natural one-to-one correspondence between positions in a term and addresses in the associated tree.

Definition 1.18. (address, position) Assume that t is a term and p is a position in t. The *address of* p *in* t is inductively specified by the clauses

$$\operatorname{add}(p, t) = \begin{cases} \phi & \text{for } t \text{ a variable or } p = \lg(t), \\ 0 \operatorname{add}(p, t_1) & \text{for } t = t_1 \cdot t_2 \text{ and } p \le \lg(t_1), \\ 1 \operatorname{add}(q, t_2) & \text{for } t = t_1 \cdot t_2 \text{ and } p = \lg(t_1) + q \text{ with } q \le \lg(t_2). \end{cases}$$

Assume now that α is an address in t. The *position of* α *in* t is inductively specified by the clauses

$$\operatorname{pos}(\alpha, t) = \begin{cases} \lg(t) & \text{for } \alpha = \phi, \\ \operatorname{pos}(\beta, t_1) & \text{for } t = t_1 \cdot t_2 \text{ and } \alpha = 0\beta, \\ \lg(t_1) + \operatorname{pos}(\alpha_1, t_2) & \text{for } t = t_1 \cdot t_2 \text{ and } \alpha = 1\beta. \end{cases}$$

Proposition 1.19. (address and position) *Assume that t is a term. Then the map* $\operatorname{add}(., t)$ *is an increasing bijection of* $(\operatorname{Dom}(t), <)$ *onto* $(\operatorname{Skel}(t), <)$*, and* $\operatorname{pos}(., t)$ *is the inverse bijection. For every position p in t, the subterm* $\operatorname{sub}(t, \operatorname{add}(p, t))$ *is the unique subterm of t that ends at position p; in particular, if the letter $t(p)$ is a variable x, then we have* $\operatorname{sub}(t, \operatorname{add}(p, t)) = x$.

Proof. We use induction on t. Everything is true when t is a variable, since then 1 is the unique position in t, and ϕ is the unique address in t. Assume

$t = t_1 \cdot t_2$. The $<$-increasing enumeration of $\mathrm{Dom}(t)$ is, by definition of the right Polish notation, the $<$-increasing enumeration of $\mathrm{Dom}(t_1)$ followed by the $<$-increasing enumeration of $\mathrm{Dom}(t_2)$ shifted by the number $\mathrm{lg}(t_1)$, followed by the number $\mathrm{lg}(t)$. Similarly, the $<$-increasing enumeration of the skeleton of t is, by definition of the order $<$ on addresses, the $<$-increasing enumeration of the skeleton of t_1 with an initial 0 added to every address, followed by the $<$-increasing enumeration of the skeleton of t_2 with an initial 1 added, followed by the empty address. Since, by induction hypothesis, the mapping 'add' establishes an increasing bijection in the case of t_1 and t_2, it establishes an increasing bijection in the case of t. Similarly, the property that the mapping 'pos' is the inverse of the mapping 'add' in the case of t_1 and t_2 implies the property for t. ∎

Example 1.20. **(addresses** *vs.* **positions)** *Let t be the term $x_1 \cdot (x_2 \cdot (x_3 \cdot x_4)) \cdot x_5$, i.e., the word $x_1 x_2 x_3 x_4 \bullet\bullet x_5 \bullet\bullet$. There are 9 positions in t, and there are 9 nodes in the associated binary tree*

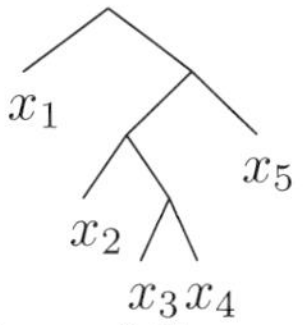

.

The correspondence between positions and addresses in t is as follows:

$$
\begin{array}{lll}
1 & \mapsto 0, & \text{associated subtree: } x_1, \\
2 & \mapsto 100, & \text{associated subtree: } x_2, \\
3 & \mapsto 1010, & \text{associated subtree: } x_3, \\
4 & \mapsto 1011, & \text{associated subtree: } x_4, \\
5 & \mapsto 101, & \text{associated subtree: } x_3 \cdot x_4, \\
6 & \mapsto 10, & \text{associated subtree: } x_2 \cdot x_3 \cdot x_4, \\
7 & \mapsto 11, & \text{associated subtree: } x_5, \\
8 & \mapsto 1, & \text{associated subtree: } (x_2 \cdot (x_3 \cdot x_4)) \cdot x_5, \\
9 & \mapsto \emptyset, & \text{associated subtree: } t.
\end{array}
$$

Observe that, for every term t, the letter at position p in t viewed as a word coincides with the letter at address $\mathrm{add}(p, t)$ provided we label every inner node of the tree t with the operator $\bullet$.

Exercise 1.21. **(antichain)** *Prove that the outline of a term is a maximal antichain of $(\mathbf{A}, \sqsubset)$, i.e., a maximal set of pairwise $\sqsubset$-incomparable addresses. Conversely, prove that every finite maximal antichain of $(\mathbf{A}, \sqsubset)$ is the outline of a unique term in T_1. [$(\mathbf{A}, \sqsubset)$ also has infinite antichains.]*

Exercise 1.22. (number of 1's) *Assume that t is a term, and α is an address in the outline of t. Let p be the position of α in t. Let $\#_1(\alpha)$ and $\#_1^{fin}(\alpha)$ denote respectively the total number of 1's in α, and the number of final 1's in α, i.e., the number of letters 1 not followed by any 0. Prove that $\#_1(\alpha)$ is equal to the difference between the number of variables and the number of letters $\bullet$ before p in t. Prove that $\#_1^{fin}(\alpha)$ is equal to the number of letters $\bullet$ between p and the next position where a variable occurs in t.*

---

2. The Cuts of a Term

We have associated with every term the family of its subterms. In this section, we associate with every term a new family of terms, that we call its *cuts*. These terms are natural generalizations of iterated left subterms.

Cuts

The cuts of a term t are defined from the prefixes of t viewed as a word. It is not true that every word prefix w of t is a term, but we have seen in Lemma V.4.16 that there exists a unique nonnegative integer p such that $w\bullet^p$ is a term.

Definition 2.1. (cut) For t a term and α in the skeleton of t, the *cut of t at α* is the term $\mathrm{cut}(t, \alpha)$ inductively specified by the clauses:

$$\mathrm{cut}(t,\alpha) = \begin{cases} t & \text{for } \alpha = \phi, \\ \mathrm{cut}(t_1, \beta) & \text{for } \alpha = 0\beta \text{ and } t = t_1{\cdot}t_2, \\ t_1{\cdot}\mathrm{cut}(t_2, \beta) & \text{for } \alpha = 1\beta \text{ and } t = t_1{\cdot}t_2. \end{cases}$$

Example 2.2. (cuts) *Let $t = x_1{\cdot}(x_2{\cdot}x_3{\cdot}x_4){\cdot}x_5$ as in Example 1.20. There are 9 addresses in the skeleton of t, and the corresponding cuts (according to the increasing enumeration of addresses) are*

$\mathrm{cut}(t, 0) = x_1$,
$\mathrm{cut}(t, 100) = x_1{\cdot}x_2$,
$\mathrm{cut}(t, 1010) = x_1{\cdot}x_2{\cdot}x_3$,
$\mathrm{cut}(t, 1011) = \mathrm{cut}(t, 101) = \mathrm{cut}(t, 10) = x_1{\cdot}x_2{\cdot}x_3{\cdot}x_4$,
$\mathrm{cut}(t, 11) = \mathrm{cut}(t, 1) = \mathrm{cut}(t, \phi) = t$.

Lemma 2.3. *Assume that t is a term, and α belongs to the skeleton of t. Then $\mathrm{cut}(t, \alpha)$ is the unique term of the form $w\bullet^p$, where w is the prefix of t that ends at position $\mathrm{pos}(\alpha, t)$.*

Proof. Induction on t. For t a variable, we have $\alpha = \phi$, $\mathrm{pos}(\alpha, t) = 1$, and the result is true. Otherwise, assume $t = t_1 \cdot t_2$. For $\alpha = 0\beta$ with β in the skeleton of t_1, we have $\mathrm{pos}(\alpha, t) = \mathrm{pos}(\beta, t_1)$, and $\mathrm{cut}(t, \alpha) = \mathrm{cut}(t_1, \beta)$, so the result for t follows from the result for t_1. For $\alpha = 1\beta$ with β in the skeleton of t_2, we have $\mathrm{pos}(\alpha, t) = \mathrm{lg}(t_1) + \mathrm{pos}(\beta, t_2)$, and the prefix of t of length $\mathrm{pos}(\beta, t)$ is $t_1 w$, where w is the prefix of t_2 of length $\mathrm{pos}(\beta, t_2)$. By definition, we have $\mathrm{cut}(t, \alpha) = t_1 \cdot \mathrm{cut}(t_2, \beta)$. Using the induction hypothesis, we deduce $\mathrm{cut}(t_2, \beta) = w\bullet^{p_1}$, hence $\mathrm{cut}(t, \alpha) = t_1 w \bullet^{p_1+1}$. ■

Lemma 2.4. *Assume that t is a term. Then, for every α in the skeleton of t, we have $\mathrm{cut}(t, \alpha) = \mathrm{cut}(t, \alpha 1^*)$.*

Proof. Lemma 2.3 implies that $\mathrm{cut}(t, \alpha) = \mathrm{cut}(t, \alpha 1)$ holds whenever $\alpha 1$ belongs to the skeleton of t. ■

It follows that a complete list of cuts of a term t is obtained by considering those cuts associated with external addresses in t.

The following lemma shows that the iterated left subterms of a term t are particular cuts of t. Its proof is obvious from Lemma V.4.16, and it is left to the reader.

Lemma 2.5. *Let t be an arbitrary term, and α an address in the skeleton of t. Then the term $\mathrm{cut}(t, \alpha)$ is a prefix of t if and only if $\alpha = 0^p$ holds for some p. In this case, $\mathrm{cut}(t, \alpha)$ is the p-th iterated subterm of t.*

Arbitrary cuts of t need not be subterms of t, but they can always be expressed as products of subterms. We describe the correspondence precisely.

Definition 2.6. (**left edge**) For α an address, the *left edge of* α is defined to be the finite sequence $(\gamma_1 0, \ldots, \gamma_p 0)$, where $\gamma_1, \ldots, \gamma_p$ is the $\sqsubset$-increasing enumeration of those prefixes γ of α satisfying $\gamma 1 \sqsubseteq \alpha$.

For instance, the left edge of 010011 is the sequence $(00, 01000, 010010)$. As an induction shows, the length of the left edge of the address α is equal to the number of 1's in α.

Lemma 2.7. *Assume that t is a term, and α is an address in the skeleton of t. Let $(\alpha_1, \ldots, \alpha_p)$ be the left edge of α. Then we have*

$$\mathrm{cut}(t, \alpha) = \mathrm{sub}(t, \alpha_1) \cdot \cdots \cdot \mathrm{sub}(t, \alpha_p) \cdot \mathrm{sub}(t, \alpha), \tag{2.1}$$

$$\mathrm{cut}(t, \alpha) =_{LD} \mathrm{cut}(t, \alpha_p) \cdot \cdots \cdot \mathrm{cut}(t, \alpha_1) \cdot \mathrm{sub}(t, \alpha). \tag{2.2}$$

Proof. We use induction on the length of α. The result is trivial for $\alpha = \phi$. Otherwise, assume $t = t_1 \cdot t_2$, and $\alpha = 0\beta$, where the left edge of β is $(\beta_1, \dots, \beta_q)$. Then the left edge of α is $(0\beta_1, \dots, 0\beta_q)$. We have $\mathrm{cut}(t, \alpha) = \mathrm{cut}(t_1, \beta)$ and $\mathrm{sub}(t, 0\gamma) = \mathrm{sub}(t_1, \gamma)$ for every γ, so, using the induction hypothesis, we obtain

$$\begin{aligned}\mathrm{cut}(t,\alpha) = \mathrm{cut}(t_1,\beta) &= \mathrm{sub}(t_1,\beta_1)\cdot \dots \cdot \mathrm{sub}(t_1,\beta_q)\cdot \mathrm{sub}(t_1,\beta)\\ &= \mathrm{sub}(t,0\beta_0)\cdot \dots \cdot \mathrm{sub}(t,0\beta_q)\cdot \mathrm{sub}(t,\alpha),\\ \mathrm{cut}(t,\alpha) = \mathrm{cut}(t_1,\beta) &=_{LD} \mathrm{cut}(t_1,\beta_q)\cdot \dots \cdot \mathrm{cut}(t_1,\beta_1)\cdot \mathrm{sub}(t_1,\beta)\\ &= \mathrm{cut}(t,0\beta_q)\cdot \dots \cdot \mathrm{cut}(t,0\beta_1)\cdot \mathrm{sub}(t,\alpha).\end{aligned}$$

Assume now $\alpha = 1\beta$, where the left edge of β is $(\beta_1, \dots, \beta_q)$ again. Then the left edge of α is $(0, 1\beta_1, \dots, 1\beta_q)$. Applying the induction hypothesis, we obtain

$$\begin{aligned}\mathrm{cut}(t,\alpha) &= t_1\cdot \mathrm{cut}(t_2,\beta) = t_1\cdot \mathrm{sub}(t_2,\beta_1)\cdot \dots \cdot \mathrm{sub}(t_2,\beta_q)\cdot \mathrm{sub}(t_2,\beta)\\ &= \mathrm{sub}(t,0)\cdot \mathrm{sub}(t,1\beta_0)\cdot \dots \cdot \mathrm{sub}(t,1\beta_q)\cdot \mathrm{sub}(t,\alpha),\\ \mathrm{cut}(t,\alpha) &= t_1\cdot \mathrm{cut}(t_2,\beta) =_{LD} t_1\cdot \mathrm{cut}(t_2,\beta_q)\cdot \dots \cdot \mathrm{cut}(t_2,\beta_1)\cdot \mathrm{sub}(t_2,\beta)\\ &=_{LD} (t_1\cdot \mathrm{cut}(t_2,\beta_q))\cdot \dots \cdot (t_1\cdot \mathrm{cut}(t_2,\beta_1))\cdot t_1\cdot \mathrm{sub}(t_2,\beta)\\ &= \mathrm{cut}(t,1\beta_q)\cdot \dots \cdot \mathrm{cut}(t,1\beta_1)\cdot \mathrm{cut}(t,0)\cdot \mathrm{sub}(t,\alpha).\end{aligned}$$ ∎

It follows that the tree associated with $\mathrm{cut}(t, \alpha)$ is obtained from the tree associated with t by deleting everything on the right of α, as shown in Figure 2.1.

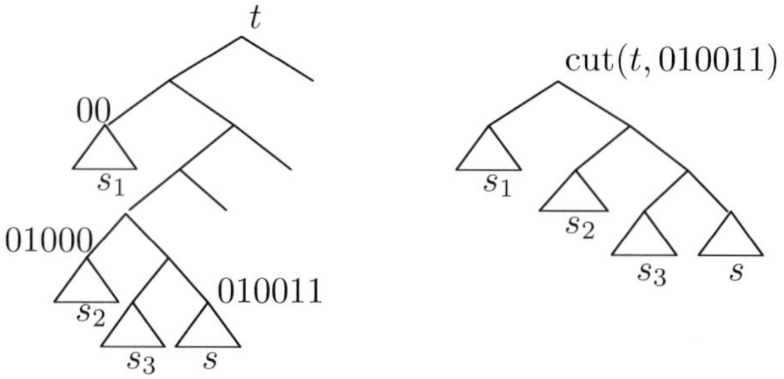

Figure 2.1. The term $\mathrm{cut}(t, 010011)$ is $\mathrm{sub}(t, 00)\cdot\mathrm{sub}(t, 01000)\cdot\mathrm{sub}(t, 010010)\cdot\mathrm{sub}(t, 010011)$

We show now that distinct cuts of a term are never LD-equivalent. More precisely, the cuts of a term make a $\sqsubset_{LD}$-increasing sequence.

Proposition 2.8. (**cut order**) *Assume that t is a term, and α, β are addresses in the outline of t. Then $\alpha >_{LR} \beta$ is equivalent to $\mathrm{cut}(t, \alpha) \sqsupset_{LD} \mathrm{cut}(t, \beta)$. So, in particular, distinct cuts of a given term are never LD-equivalent.*

Proof. The outline of t consists of pairwise incompatible addresses, hence the restriction of the left–right ordering of addresses to $\mathrm{Out}(t)$ is a linear order.

On the other hand, $\sqsupset_{LD}$ is a partial order. Hence, it suffices to prove that $\alpha >_{LR} \beta$ implies $\mathrm{cut}(t,\alpha) \sqsupset_{LD} \mathrm{cut}(t,\beta)$. As $\sqsupset_{LD}$ is transitive, we may assume moreover that α is the immediate successor of β in $\mathrm{Out}(t)$. We use induction on t. If t is a variable, the result is vacuously true. Otherwise, assume $t = t_1{\cdot}t_2$. By construction, there exist an address γ and two nonnegative integers p, q satisfying $\alpha = \gamma 10^q$ and $\beta = \gamma 01^p$. Assume first $\gamma = 0\gamma_0$. Then we have $\mathrm{cut}(t,\alpha) = \mathrm{cut}(t_1, \gamma_0 10^q)$, $\mathrm{cut}(t,\beta) = \mathrm{cut}(t_1, \gamma_0 01^p)$, and the result follows from the induction hypothesis. Assume now $\gamma = 1\gamma_1$. Then we have $\mathrm{cut}(t,\alpha) = t_1{\cdot}\mathrm{cut}(t_2, \gamma_1 10^q)$, and $\mathrm{cut}(t,\beta) = t_1{\cdot}\mathrm{cut}(t_2, \gamma_1 01^p)$. The result follows from the induction hypothesis, as $t'_2 \sqsupset_{LD} t''_2$ implies $t_1{\cdot}t'_2 \sqsupset_{LD} t_1{\cdot}t''_2$ by left self-distributivity. Finally, for $\gamma = \phi$, we have $\mathrm{cut}(t,\alpha) = t_1{\cdot}x$, and $\mathrm{cut}(t,\beta) = t_1$, where x is the leftmost variable in t_2. The result is then true, as $t_1{\cdot}x \sqsupset t_1$ holds. ∎

The successor of a cut

We establish now an explicit connection between every non-final cut of a term t and the cut associated with the next address in the outline of t. To this end, we partition the left–right order of addresses into two partial suborders, called covering and uncovering respectively. These relations are crucial in the sequel of this chapter, and in Chapter IX.

Definition 2.9. (covering, uncovering) Assume that α and β are addresses and $\alpha >_{LR} \beta$ holds. We say that α *covers* β if there exist an address γ and a positive integer q satisfying $\alpha = \gamma 1^q$ and $\gamma 0 \sqsubseteq \beta$. Otherwise, we say that α *uncovers* β.

For instance, 1011 covers 1001, while 1010 uncovers 00. Covering and uncovering are partial orders. For every address α, the set of those addresses covered by α form a final segment in the addresses on the left of α. More precisely, if we have $\alpha = \gamma 1^q$ where γ does not end with 1, and $\alpha >_{LR} \beta$, then α covers β if and only if $\beta \not<_{LR} \gamma 0$ holds. For a given address α, there is no unique leftmost address covered by α, but, if we restrict to the outline of a term, such an address exists.

Definition 2.10. (function μ_t) Assume that t is a term, and α is an address in the outline of t. Assuming $\alpha = \gamma 1^*$ with γ not ending with 1, we define $\mu_t(\alpha)$ to be the unique address of the form $\gamma 0^*$ in $\mathrm{Out}(t)$.

Example 2.11. (function μ_t) *Let $t = x_1{\cdot}(x_2{\cdot}x_3{\cdot}x_4){\cdot}x_5$. Then the outline of t comprises five addresses, among which four contain at least one 0.*

The corresponding values of the function μ_t are $\mu_t(0) = 0$, $\mu_t(100) = 100$, $\mu_t(1010) = 1010$, $\mu_t(1011) = 100$, $\mu_t(11) = 0$.

The previous remarks can be summarized as follows:

Lemma 2.12. *Assume that t is a term, and α, β belong to the outline of t. Then α covers β if and only if $\alpha >_{LR} \beta \geq_{LR} \mu_t(\alpha)$ holds.*

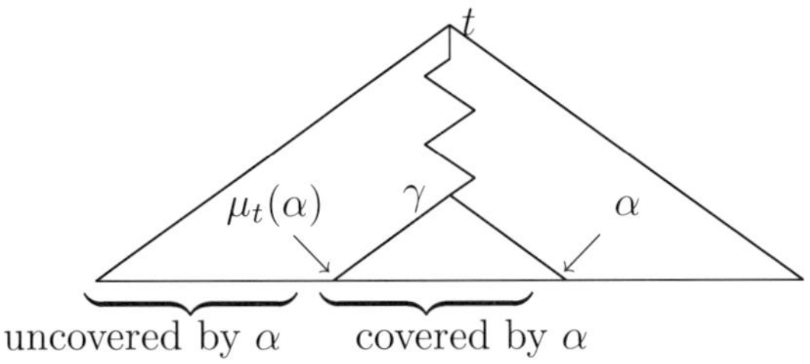

Figure 2.2. The covering relation in t

We can now describe the immediate successor of a cut explicitly.

Definition 2.13. (**almost LD-equivalent**) Assume that t, t' are terms. We say that t and t' are *almost LD-equivalent*, denoted $t =_{LD}^{almost} t'$, if t' is LD-equivalent to a term obtained from t by possibly changing the rightmost variable.

Proposition 2.14. (**next cut**) *Assume that t is a term, α is a non-final address in the outline of t, α^+ is the immediate successor of α in the outline of t, and x is the variable that occurs at α^+ in t.*

(i) Assume $\mathrm{cut}(t,\alpha) = t_1 \cdot \cdots \cdot t_p$. Then we have $\mathrm{cut}(t,\alpha^+) = t_1 \cdot \cdots \cdot t_p \cdot x$.

(ii) The term $\mathrm{cut}(t,\alpha^+)$ is LD-equivalent to the term obtained from $\mathrm{cut}(t,\alpha) \cdot \mathrm{cut}(t,\mu_t(\alpha))$ by replacing the rightmost variable with x. Thus, in particular, we have

$$\mathrm{cut}(t,\alpha^+) =_{LD}^{almost} \mathrm{cut}(t,\alpha) \cdot \mathrm{cut}(t,\mu_t(\alpha)). \tag{2.3}$$

Proof. There exist an address γ and nonnegative integers p, q satisfying $\alpha = \gamma 0 1^p$ and $\alpha^+ = \gamma 1 0^q$. Then we have $\mu_t(\alpha) = \gamma 0^r$ for some r. Assume that the left edge of γ is $(\gamma_1, \ldots, \gamma_r)$. Then the left edge of $\gamma 0$ and of $\mu_t(\alpha)$ is $(\gamma_1, \ldots, \gamma_r)$ as well, while the left edge of α^+ is $(\gamma_1, \ldots, \gamma_r, \gamma 0)$. By Lemma 2.7, we have

$$\begin{aligned}
\mathrm{cut}(t,\alpha) &= \mathrm{sub}(t,\gamma_1) \cdot \cdots \cdot \mathrm{sub}(t,\gamma_r) \cdot \mathrm{sub}(t,\gamma 0)\\
\mathrm{cut}(t,\mu_t(\alpha)) &= \mathrm{sub}(t,\gamma_1) \cdot \cdots \cdot \mathrm{sub}(t,\gamma_r) \cdot \mathrm{sub}(t,\mu_t(\alpha))\\
\mathrm{cut}(t,\alpha^+) &= \mathrm{sub}(t,\gamma_1) \cdot \cdots \cdot \mathrm{sub}(t,\gamma_r) \cdot \mathrm{sub}(t,\gamma 0) \cdot \mathrm{sub}(t,\alpha^+).
\end{aligned}$$

This gives (i) directly, and (ii) by applying left self-distributivity. ■

A consequence of the previous result is that, if α is an address in the skeleton of t, then no cut s of t may satisfy the double inequality

$$\mathrm{cut}(t,\alpha)\cdot\mathrm{cut}(t,\mu_t(\alpha)) \sqsupset_{LD} s \sqsupset_{LD} \mathrm{cut}(t,\alpha). \tag{2.4}$$

Definition 2.15. (descent) Assume that t is a term. A *descent* in t is a finite sequence of addresses $(\alpha_1,\ldots,\alpha_p)$ in the outline of t such that α_i uncovers α_{i+1} for $i<p$. We write $\mathrm{Desc}(t)$ for the set of all descents in t.

Thus a descent in t is a rapidly decreasing sequence in the outline of t.

Example 2.16. (**descent**) *Let* $t = x_1{\cdot}(x_2{\cdot}x_3{\cdot}x_4){\cdot}x_5$. *There are ten descents in* t, *namely* (0), (100), (100, 0), (1010), (1010, 0), (1010, 100), (1010, 100, 0), (1011), (1011, 0), *and* (11).

Proposition 2.17. (obstruction) *Assume that* t *is a term, and* $(\alpha_1,\ldots,\alpha_p)$ *is a descent in* t *with* $p\geq 2$. *Then we have*

$$\mathrm{cut}(t,\alpha_1^+) \sqsupset_{LD} \mathrm{cut}(t,\alpha_1)\cdot\cdots\cdot\mathrm{cut}(t,\alpha_p) \sqsupset_{LD} \mathrm{cut}(t,\alpha_1) \tag{2.5}$$

where α_1^+ *is the immediate successor of* α_1 *in* $\mathrm{Out}(t)$. *Hence, no cut of* t *may be LD-equivalent to* $\mathrm{cut}(t,\alpha_1)\cdot\cdots\cdot\mathrm{cut}(t,\alpha_p)$.

Proof. (The statement makes sense as the hypothesis that α_1 uncovers α_2 implies that α_1 is not the final address in t.) For $p\geq k>1$, we show by a descending induction on k the relation

$$\mathrm{cut}(t,\alpha_k^+) \sqsupset_{LD} \mathrm{cut}(t,\alpha_k)\cdot\cdots\cdot\mathrm{cut}(t,\alpha_p), \tag{2.6}$$

where α_k^+ is the immediate successor of α_k in the outline of t. For $k=p$, the result is obvious. Assume $k<p$. Using the induction hypothesis and $\mu_t(\alpha_k) >_{LR} \alpha_{k+1}$, which implies $\mu_t(\alpha_k) \geq_{LR} \alpha_{k+1}^+$, we find

$$\mathrm{cut}(t,\mu_t(\alpha_k)) \sqsupseteq_{LD} \mathrm{cut}(t,\alpha_{k+1}^+) \sqsupset_{LD} \mathrm{cut}(t,\alpha_{k+1})\cdot\cdots\cdot\mathrm{cut}(t,\alpha_p).$$

Multiplying by $\mathrm{cut}(t,\alpha_k)$ on the left and applying Proposition 2.14 (next cut) gives

$$\mathrm{cut}(t,\alpha_k^+) =_{LD}^{\text{almost}} \mathrm{cut}(t,\alpha_k)\cdot\mathrm{cut}(t,\mu_t(\alpha_k)) \sqsupset_{LD} \mathrm{cut}(t,\alpha_k)\cdot\cdots\cdot\mathrm{cut}(t,\alpha_p).$$

This implies (2.6) as, for $t_1 =_{LD}^{\text{almost}} t_1'$, $t_1 \sqsupset_{LD} t_0$ is equivalent to $t_1' \sqsupset_{LD} t_0$. ■

Cuts of LD-expansions

We recall from Chapter V that an LD-expansion of a term t is a term t' obtained from t by applying the left self-distributivity identity in the expanding direction, *i.e.*, from $x{\cdot}(y{\cdot}z)$ to $(x{\cdot}y){\cdot}(x{\cdot}z)$. An important phenomenon is that, if the term t' is an LD-expansion of the term t, then each cut s of t has a copy in t', more exactly some cut of t' is an LD-expansion of s—as is the case for iterated left subterms.

In order to determine the cuts of an LD-expansion, it will be convenient to take into account the address(es) where the LD-expansion process has been performed—an approach that we shall develop in Chapter VII systematically.

Definition 2.18. (**LD-expansion at** α) Assume that α is an address, and t is a term such that $\mathrm{sub}(t, \alpha)$ exists and has the form $t_1{\cdot}(t_2{\cdot}t_3)$. The LD-expansion of t *at* α is defined to be the basic LD-expansion of t obtained by replacing $\mathrm{sub}(t, \alpha)$ with $(t_1{\cdot}t_2){\cdot}(t_1{\cdot}t_3)$.

Thus, expanding t at α means applying the left self-distributivity transformation $x{\cdot}(y{\cdot}z) \mapsto (x{\cdot}y){\cdot}(x{\cdot}z)$ to the α-subterm of t. By construction, every basic LD-expansion of t is the LD-expansion of t at some unique address.

Lemma 2.19. *Assume that t' is the LD-expansion of t at α. Then the outline of t' comprises*
- *all addresses β in the outline of t that satisfy $\beta \perp \alpha$ or $\alpha 11 \sqsubseteq \beta$,*
- *all addresses $\alpha 00\gamma$ and $\alpha 10\gamma$ such that $\alpha 0\gamma$ belongs to the outline of t,*
- *all addresses $\alpha 01\gamma$ such that $\alpha 10\gamma$ belongs to the outline of t.*

Moreover,
- *for $\beta <_{LR} \alpha$, we have $\mathrm{cut}(t', \beta) = \mathrm{cut}(t, \beta)$;*
- *for $\beta = \alpha 00\gamma$, we have $\mathrm{cut}(t', \beta) = \mathrm{cut}(t, \alpha 0\gamma)$;*
- *for $\beta = \alpha 01\gamma$, we have $\mathrm{cut}(t', \beta) = \mathrm{cut}(t, \alpha 10\gamma)$;*
- *for $\beta = \alpha 10\gamma$, we have $\mathrm{cut}(t', \beta) =_{LD} \mathrm{cut}(t, \alpha 10){\cdot}\mathrm{cut}(t, \alpha 0\gamma)$;*
- *for $\beta = \alpha 11\gamma$ or $\alpha <_{LR} \beta$, $\mathrm{cut}(t', \beta)$ is a basic LD-expansion of $\mathrm{cut}(t, \beta)$.*

Proof. We use induction on α. Assume first $\alpha = \phi$. By definition, we have $t = t_1{\cdot}(t_2{\cdot}t_3)$ and $t' = (t_1{\cdot}t_2){\cdot}(t_1{\cdot}t_3)$ for some t_1, t_2, t_3. Applying Lemma 1.14, we obtain

$$\mathrm{Out}(t) = \mathrm{sh}_0(\mathrm{Out}(t_1)) \cup \mathrm{sh}_{10}(\mathrm{Out}(t_2)) \cup \mathrm{sh}_{11}(\mathrm{Out}(t_3)),$$
$$\mathrm{Out}(t') = \mathrm{sh}_{00}(\mathrm{Out}(t_1)) \cup \mathrm{sh}_{01}(\mathrm{Out}(t_2)) \cup \mathrm{sh}_{10}(\mathrm{Out}(t_1)) \cup \mathrm{sh}_{11}(\mathrm{Out}(t_3)),$$

from which the description of the outline of t' follows. For $0\gamma \in \mathrm{Out}(t)$, we find $\mathrm{cut}(t, 0\gamma) = \mathrm{cut}(t', 00\gamma) = \mathrm{cut}(t_1, \gamma)$, and

$$\begin{aligned}\mathrm{cut}(t', 10\gamma) &= \mathrm{sub}(t', 0) \cdot \mathrm{cut}(\mathrm{sub}(t', 1), 0\gamma)\\ &= (t_1 \cdot t_2) \cdot \mathrm{cut}(t, 0\gamma) = \mathrm{cut}(t, 10) \cdot \mathrm{cut}(t, 0\gamma).\end{aligned}$$

For $10\gamma \in \mathrm{Out}(t)$, we find $\mathrm{cut}(t, 10\gamma) = \mathrm{cut}(t', 01\gamma) = t_1{\cdot}\mathrm{cut}(t_2, \gamma)$. Finally, for $11\gamma \in \mathrm{Out}(t)$, we find $\mathrm{cut}(t, 11\gamma) = t_1{\cdot}t_2{\cdot}\mathrm{cut}(t_3, \gamma)$, and $\mathrm{cut}(t', 11\gamma) = (t_1{\cdot}t_2){\cdot}(t_1{\cdot}\mathrm{cut}(t_3, \gamma))$. The latter term is a basic LD-expansion of the former.

Assume now $\alpha = 0\beta$. Write $t = t_1{\cdot}t_2$. Then we have $t' = t_1'{\cdot}t_2$, where t_1' is the LD-expansion of t_1 at β. Applying the induction hypothesis to t_1' gives the result for all addresses beginning with 0. For those addresses that begin with 1, we have $\mathrm{cut}(t', 1\gamma) = t_1'{\cdot}\mathrm{cut}(t_2, \gamma)$, and the latter is a basic LD-expansion of $\mathrm{cut}(t, 1\gamma)$, *i.e.*, of $t_1{\cdot}\mathrm{cut}(t_2, \gamma)$.

Finally, assume $\alpha = 1\beta$. For $t = t_1{\cdot}t_2$, we have $t' = t_1{\cdot}t_2'$ where t_2' is the LD-expansion of t_2 at β. Everything is obvious as far as addresses that begin with 0 are concerned. For those addresses that begin with 1, we apply the induction hypothesis to t_2'. Then coming back to t_2 amounts to multiplying everywhere by t_1 on the left: in this way, an equality $\mathrm{cut}(t_2', \gamma') = \mathrm{cut}(t_2, \gamma)$ gives a similar equality $\mathrm{cut}(t', 1\gamma') = \mathrm{cut}(t, 1\gamma)$, but an equality of the form $\mathrm{cut}(t_2', \gamma') = \mathrm{cut}(t_2, \beta 10){\cdot}\mathrm{cut}(t_2, \gamma)$ only gives $\mathrm{cut}(t', 1\gamma') =_{LD} \mathrm{cut}(t, \alpha 10){\cdot}\mathrm{cut}(t, 1\gamma)$ as we have to distribute t_1. ■

The previous statement shows in particular that, if t' is a basic LD-expansion of t, then, for every cut s of t, there exists a cut s' of t' that expands s. Thus, a straightforward iteration gives the following result:

Proposition 2.20. (cuts of expansion) *Assume that t' is an LD-expansion of t. Then, for every address α in the outline of t, there exists an address α' in the outline of t' such that $\mathrm{cut}(t', a')$ is an LD-expansion of $\mathrm{cut}(t, \alpha)$.*

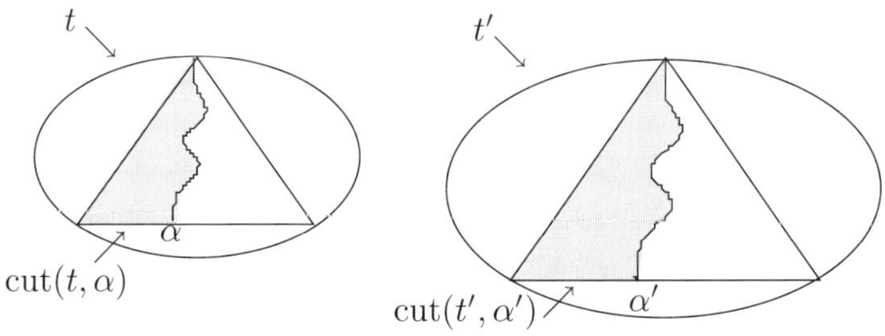

Figure 2.3. Cuts of expansion

As every left iterated subterm of a term t is a particular cut of t, a special case of the previous result is Proposition V.4.6 (left subterm). However both results are not comparable, as the current one only asserts that some cut of the LD-expansion expands the considered iterated subterm, while Proposition V.4.6 tells us in addition that this cut is an iterated left subterm—which actually could be also derived from the current proof.

Exercise 2.21. (**decomposition**) *Prove that the prefix of length p of a term t is the word $\mathrm{sub}(t, \alpha_1) \ldots \mathrm{sub}(t, \alpha_m)\mathrm{sub}(t, \alpha)$ where α is the address of p in t and $(\alpha_1, \ldots, \alpha_m)$ is the left edge of α.*

Exercise 2.22. (**cuts**) *If t' is an LD-expansion of t, then every cut of t is LD-equivalent to some cut of t'. Prove that the converse implication is false. [Hint: Consider $(x{\cdot}x){\cdot}x{\cdot}x{\cdot}x$ and $((x{\cdot}x){\cdot}(x{\cdot}x)){\cdot}x{\cdot}x{\cdot}x$.]*

Exercise 2.23. (**order $\sqsupset^{\ell in}$**) *Say that $t_1 \sqsupset^{\ell in} t_2$ holds if $t_1 \sqsupset t_2$ holds or there exist a word w and a variable x such that t_1 begins with wx and t_2 begins with $w\bullet$. Then $\sqsupset^{\ell in}$ is a partial order not compatible with $=_{LD}$, and its restriction to T_1 is linear. Prove that, for t a term and α, β in the outline of t, $\mathrm{cut}(t, \alpha) \sqsupset^{\ell in} \mathrm{cut}(t, \beta)$ is equivalent to $\alpha >_{LR} \beta$.*

3. The ∂-Normal Form

In this section, we address the question of constructing a normal form for LD-equivalence, *i.e.*, of defining a distinguished element in every LD-equivalence class. Here, we introduce the notion of a ∂-normal term, and we establish the following answer:

Proposition 3.1. (∂-**normal**) *Every term in T_∞ is LD-equivalent to a unique ∂-normal term. The function that maps every term to the unique LD-equivalent ∂-normal term lies in the complexity class* $\mathrm{DSPACE}(\exp^*(\mathcal{O}(2^n)))$.

(We recall that $\exp^*$ denotes the iterated exponential defined by $\exp^*(0) = 1$ and $\exp^*(n+1) = 2^{\exp^*(n)}$.) This solution is not optimal, and a better solution will be given in the next section. However, the current construction is useful as a preliminary result, and its proof requires little work.

The t_0-∂-normal terms

We start from the following result:

Lemma 3.2. *For all terms t_0, t, we have $t \sqsubseteq_{LD} t_0$ if and only if there exists $k \geq 0$ such that some cut of $\partial^k t_0$ is an LD-expansion of t.*

Proof. Assume $t \sqsubseteq_{LD} t_0$. By definition, there exist a term t' LD-equivalent to t and a term t'_0 LD-equivalent to t_0 such that t' is an iterated left subterm of t'_0. Let t''_0 be the term obtained from t'_0 by replacing the iterated left subterm t' with t. Then we have $t''_0 =_{LD} t'_0 =_{LD} t_0$. By Proposition V.3.14 (quantitative confluence), there exists an integer k such that the term $\partial^k t_0$ is an LD-expansion of t''_0. By construction, t is an iterated left subterm, hence a cut, of t''_0. By Proposition 2.20 (cuts of expansion), some cut of $\partial^k t_0$ is an LD-expansion of t.

Conversely, if some LD-expansion t' of t is a cut of $\partial^k t_0$, then we have $t' \sqsubseteq_{LD} \partial^k t_0$, hence $t \sqsubseteq_{LD} t_0$ as $\partial^k t_0$ is LD-equivalent to t_0. ■

Remark 3.3. We shall see below that, if t is a cut of t_0, then ∂t is an iterated left subterm of ∂t_0. Thus we could replace cuts with iterated left subterms in the previous result. However, technical reasons that should become clear below make it more natural to use cuts.

Our principle will be to construct normal forms that depend on a reference base term t_0—which can be seen as reminiscent of the base b expansions of the integers.

Definition 3.4. (**t_0-∂-normal**) Let t_0 be a term. For $k \geq 0$, we say that a term t is a *t_0-∂-normal term of degree* k if t is a cut of the term $\partial^k t_0$ and, if $k > 0$ holds, t is LD-equivalent to no cut of $\partial^{k-1} t_0$.

By Proposition 2.20 (cuts of expansion), every cut of t_0 has an LD-expansion that is a cut of ∂t_0. Thus, the t_0-∂-normal terms of degree 1 are the "new" cuts of ∂t_0, those that are not copies (= LD-expansions) of cuts of t_0. Similarly, the t_0-∂-normal terms of degree 2 are those cuts of $\partial^2 t_0$ that are not copies of t_0-∂-normal terms of degree ≤ 1, *etc.*

Example 3.5. (**t_0-∂-normal**) *Let $t_0 = x^{[3]}$. Then t has three cuts, namely x, $x^{[2]}$ and $x^{[3]}$. These are the $x^{[3]}$-∂-normal terms of degree 0.*

Next, we have $\partial x^{[3]} = (x \cdot x) \cdot (x \cdot x)$. This term has four cuts, namely x and $x \cdot x$, which are already $x^{[3]}$-∂-normal terms of degree 0, $(x \cdot x) \cdot x$, which is new, and $(x \cdot x) \cdot (x \cdot x)$ itself, which is LD-equivalent to $x^{[3]}$, an $x^{[3]}$-∂-normal term of degree 0. Thus, there exists only one $x^{[3]}$-∂-normal term of degree 1, namely $(x \cdot x) \cdot x$.

Then, we find $\partial^2 x^{[3]} = ((x \cdot x) \cdot x) \cdot (x \cdot x) \cdot x$, a term with six cuts, among which four are LD-expansions of $x^{[3]}$-∂-normal terms of degree ≤ 1, and two are new, namely $\mathrm{cut}(\partial^2 x^{[3]}, 100) = ((x \cdot x) \cdot x) \cdot x$, and $\mathrm{cut}(\partial^2 x^{[3]}, 101) = ((x \cdot x) \cdot x) \cdot (x \cdot x)$. The latter terms are therefore the two $x^{[3]}$-∂-normal terms of degree 2.

Writing complicated terms soon becomes cumbersome (we shall introduce below more convenient notation). In the figure below we display the terms $\partial^k x^{[3]}$ for $k \leq 3$: the addresses with a dot are those that correspond to new cuts, hence to $x^{[3]}$-∂-normal terms.

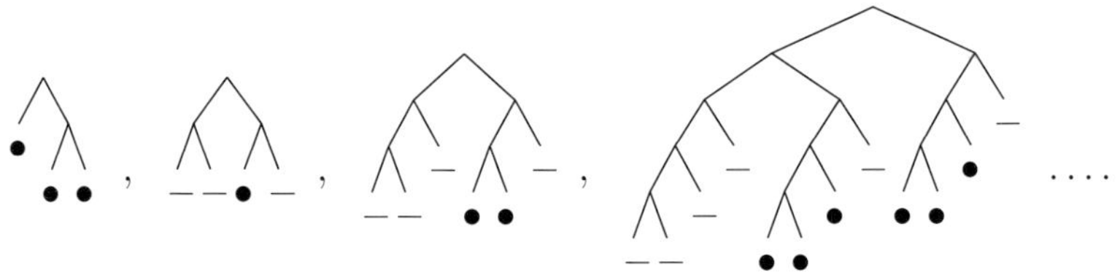

We see that there are six $x^{[3]}$-∂-normal terms of degree 3.

We obtain a unique normal form property immediately:

Proposition 3.6. **(t_0-∂-normal form)** *Let t_0 be a fixed term. Then every term t satisfying $t \sqsubseteq_{LD} t_0$ is LD-equivalent to a unique t_0-∂-normal term.*

Proof. Assume $t \sqsubseteq_{LD} t_0$. By Lemma 3.2, there exists an integer k such that some cut of $\partial^k t_0$ is an LD-expansion of t. By construction, every cut of $\partial^k t_0$ is LD-equivalent to—actually, is an LD-expansion of—a t_0-∂-normal term: either this cut is a new cut and is itself a t_0-∂-normal term, or it is an LD-expansion of some cut of $\partial^{k-1} t_0$, which is inductively an LD-expansion of a t_0-∂-normal term. Hence t is LD-equivalent to a t_0-∂-normal term. This t_0-∂-normal term is unique, as Proposition 2.8 (cut order) tells us that different cuts of a term are LD-unequivalent, and, therefore, so are distinct t_0-∂-normal terms. ■

Extended terms

By construction, the relation "being a cut" is transitive. It is easy to verify that, if t_0 is a cut of t'_0, then every t_0-∂-normal term is also a ∂-t'_0-normal term. For instance, every $x^{[m]}$-∂-normal term is also $x^{[m']}$-∂-normal for $m' \geq m$. In order to avoid unessential limitations, it will be convenient to extend the definition of t_0-∂-normal terms so as to include the possibility of t_0 being an infinite word.

Definition 3.7. **(extended term)** Assume that X is a nonempty set. An *extended term over* X is defined to be either a term, or an infinite word t_0 over the alphabet $X \cup \{\bullet\}$ such that the inequality $\#_X(w) \geq \#_\bullet(w) + 1$ holds for every finite prefix w of t_0. For t_0 an extended term, a *cut of* t_0 is defined to be a term of the form $w\bullet^p$ with w a word prefix of t_0. If t is a term, and t_0 is an extended term, we say that $t \sqsubseteq_{LD} t_0$ holds if $t \sqsubseteq s$ holds for some cut s of t_0.

Example 3.8. (**extended term**) *The infinite word $x^\infty = xxx\cdots$ is an extended term. Its cuts are the terms $x^m \bullet^{m-1}$, i.e., $x^{[m]}$. More generally, if t_0 is a term, then $t_0^\infty = t_0 t_0 t_0 \cdots$ is an extended term, whose cuts are those terms of the form $t_0 \cdot \cdots \cdot t_0 \cdot t$ with t a cut of t_0.*

Similarly, the word $t_{0[\infty]} = t_0(t_0\bullet)^\infty = t_0 t_0 \bullet t_0 \bullet t_0 \bullet \cdots$ is an extended term. Its cuts are those terms of the form $(\cdots((t_0 \cdot t_0) \cdot t_0) \cdots) \cdot t$, with t a cut of t_0. In particular, the cuts of $x_{[\infty]}$ are the terms $x_{[m]}$.

In some cases, an (infinite) binary tree may be associated with an extended term. For instance, it is natural to associate with the extended terms x^∞ and $x_{[\infty]}$ respectively the trees  *and*  *. More complicated examples like $xx(x\bullet)^\infty$ show that this is not always possible.*

Observe that, for t a term, the cuts of t considered as an extended term coincide with the cuts of t as defined in Section 2.

Lemma 3.9. *Assume that s is a cut of t. Then ∂s is a cut of ∂t.*

Proof. Assume $s = \mathrm{cut}(t, \alpha)$. We prove $\partial s \sqsubseteq \partial t$ using induction on α. For $\alpha = \phi$, we have $s = t$, hence $\partial s = \partial t$. Otherwise, assume $t = t_1 \cdot t_2$. For $\alpha = 0\beta$, we have $s = \mathrm{cut}(t_1, \beta)$. By induction hypothesis, we have $\partial s \sqsubseteq \partial t_1$, hence $\partial s \sqsubset \partial t$, as $\partial t_1 \sqsubset \partial t_1 * \partial t_2$ always holds. For $\alpha = 1\beta$, we have $s = t_1 \cdot \mathrm{cut}(t_2, \beta)$. By induction hypothesis, we have $\partial \mathrm{cut}(t_2, \beta) \sqsubseteq \partial t_2$, and we deduce $\partial s = \partial t_1 * \partial(cut(t_2, \beta)) \sqsubseteq \partial t_1 * \partial t_2 = \partial t$, as an easy induction on s_2 shows that $s_1 \sqsubseteq s_2$ implies $s_0 * s_1 \sqsubseteq s_0 * s_2$ for every s_0. ∎

Lemma 3.10. *Assume that t_0' is a term and t_0 is a cut of t_0'. Then every t_0-∂-normal term of degree k is a t_0'-∂-normal term of degree k as well.*

The result follows from Lemma 3.9 immediately. So, there is no problem in considering the notion of a t_0-∂-normal term for an extended term t_0: we say that the term t is t_0-∂-normal of degree k if t is a cut of $\partial^k t_0$ that expands no cut of $\partial^{k-1} t_0$. Then the following result is a consequence of Proposition 3.6:

Proposition 3.11. (t_0-∂-**normal form**) *Assume that t_0 is an extended term. Then every term t satisfying $t \sqsubseteq_{LD} t_0$ is LD-equivalent to a unique t_0-∂-normal term.*

In the previous situation, the unique t_0-∂-normal term that is LD-equivalent to a given term t will be called the t_0-∂-*normal form* of t.

The case of one variable

So far, we have not obtained a unique normal form for every term, since, for each (extended) term t_0, only those terms t satisfying $t \sqsubseteq_{LD} t_0$ admit a t_0-∂-normal form. This limitation however is vacuous in the case of one variable.

Lemma 3.12. *Assume that t_0 is a term in T_1. Then the extended term t_0^∞ is cofinal in $(T_1, \sqsubset_{LD})$ in the sense that $t \sqsubset_{LD} t_0^\infty$ holds for every term t in T_1. The same result holds for every extended term of the form $t_1 \cdots t_k t_0^\infty$.*

Proof. Assume $t \in T_1$, $m \geq \mathrm{ht}(t)$, and $k \geq \mathrm{cp}(t)$. By Proposition V.4.12 (distinguished expansion), some iterated left subterm, hence some cut, of $\partial^k x^{[m+1]}$ is an LD-expansion of t. Hence $t \sqsubseteq_{LD} x^{[m+1]}$ holds, and so does $t \sqsubset_{LD} x^\infty$. Now, by Proposition V.2.11 (absorption), there exists an integer h such that $t_0^{[m]} =_{LD} x^{[m+h]}$ holds for m large enough. So $t \sqsubseteq_{LD} x^{[m]}$ holding for m large enough implies $t \sqsubseteq_{LD} t_0^{[m]}$ holding as well for m large enough. Finally, we observe that, if t_0' denotes the term $t_1 \cdot \cdots \cdot t_k \cdot t_0$, then ${t_0'}^{[m]}$ is LD-equivalent to $t_1 \cdot \cdots \cdot t_k \cdot t_0^{[m]}$, hence $t \sqsubseteq_{LD} t_1 \cdot \cdots \cdot t_k \cdot t_0^{[m]}$ holds for m large enough. ■

It follows that, for every base term t_0 in T_1, every term in T_1 admits a t_0^∞-∂-normal form. Moreover, we have an effective algorithm for computing this normal form, and the quantitative lemmas of Chapter V give us an upper bound for the complexity of this algorithm.

Proposition 3.13. (∂**-normal form**) *(i) Let $t_1, \ldots, t_k, t_0$ be arbitrary terms in T_1. Then every term in T_1 is LD-equivalent to a unique $t_1 \cdots t_k t_0^\infty$-$\partial$-normal term. In particular, every term in T_1 is LD-equivalent to a unique x^∞-∂-normal term.*

(ii) The function that maps every term in T_1 to its $t_1 \cdots t_k t_0^\infty$-$\partial$-normal form lies in the complexity class $\mathrm{DSPACE}(\exp^*(\mathcal{O}(2^n)))$.

Proof. Existence is clear from Lemma 3.12. For complexity, as in the proof of Proposition V.6.16 (word problem), we observe that, if t has length ℓ, then the whole computation can be made inside the term $\partial^{\mathrm{cp}(t)} x^{[\mathrm{ht}(t)+1]}$, and that the length of the latter term is at most $\exp^*(\ell)$. This is sufficient in the case of the x^∞-∂-normal form. In the general case of the $t_1 \cdots t_k t_0^\infty$-$\partial$-normal form, the result is similar as we are interested only in large sizes, and, for m large enough, $t_1 \cdots t_k t_0^{[m]}$ is LD-equivalent to $x^{[m+c]}$ for some constant c. ■

The case of several variables

In the case of several variables, no extended term is cofinal with respect to the $\sqsubset$-ordering. However, we can find a unique normal form by first ignoring the variables, *i.e.*, by considering the projection on T_1, and reintroducing the variables directly in the ∂-normal form.

Definition 3.14. (**variant**) Assume that t_1 and t_2 are terms. We say that t_1 is a *variant* of t_2 if $t_1^\dagger = t_2^\dagger$ holds, where we recall $t^\dagger$ is the projection of t on T_1. In other words, t_1 is a variant of t_2 if and only if they have the same skeleton.

Definition 3.15. (**∂-normal**) Assume that t is a term in T_∞. We say that t is a *∂-normal term of degree k* if t is a variant of a cut of $\partial^k x^\infty$ and t is an LD-expansion of no ∂-normal term of degree $k-1$.

Example 3.16. *(∂-**normal**) The ∂-normal terms of degree 0 are all variants of cuts of x^∞: the cuts of x^∞ are the terms $x^{[m]}$, so their variants are all terms of the form $x_{i_1}\cdot \cdots \cdot x_{i_m}$.*

For degree 1, we have seen that $(x{\cdot}x){\cdot}x$ is a cut of $\partial x^{[3]}$ that is an LD-expansion of no cut of x^∞. It follows that no variant of $(x{\cdot}x){\cdot}x$ may be an LD-expansion of a ∂-normal term of degree 0. So every term of the form $(x_{i_1}{\cdot}x_{i_2}){\cdot}x_{i_3}$ is a ∂-normal term of degree 1. Now consider the cut $(x{\cdot}x){\cdot}(x{\cdot}x)$ of ∂x^∞. This term is not an x^∞-∂-normal term, since it is an LD-expansion of $x{\cdot}(x{\cdot}x)$. However, some variants of this term are ∂-normal. Indeed, such a variant has the form $(x_{i_1}{\cdot}x_{i_2}){\cdot}(x_{i_3}{\cdot}x_{i_4})$. For $i_3 = i_1$, the term is an LD-expansion of $x_{i_1}{\cdot}(x_{i_2}{\cdot}x_{i_4})$, and, therefore, it is not ∂-normal. But for $i_1 \neq i_3$, the term is a proper LD-expansion of no term, and it is therefore a ∂-normal term.

By definition, a term in T_1 is ∂-normal if and only if it is ∂-x^∞-normal.

We can now easily complete the proof of Proposition 3.1 (∂-normal):

Proof. Let t be an arbitrary term in T_∞. Then the term $t^\dagger$ belongs to T_1, so, by Proposition V.4.12 (distinguished expansion), some iterated left subterm t_1 of $\partial^{\mathrm{cp}(t)}x^{[\mathrm{ht}(t)+1]}$ (hence a cut of that term) is an LD-expansion of $t^\dagger$. By Lemma V.4.21, some variant t' of t_1 is an LD-expansion of t. Now, by construction, either t' is ∂-normal, or it is an LD-expansion of a ∂-normal term. As for complexity, the argument is the same as in the case of one variable, since, once again, the whole construction takes place inside the term $\partial^k x^{[m+1]}$. ∎

Exercise 3.17. (**∂-normal form**) *Show that the ∂-normal form of the term $((x_1{\cdot}x_2){\cdot}x_3){\cdot}((x_1{\cdot}x_2){\cdot}x_4)$ is $(x_1{\cdot}x_2){\cdot}(x_3{\cdot}x_4)$, while the one of $((x_1{\cdot}x_2){\cdot}x_1){\cdot}((x_1{\cdot}x_2){\cdot}x_4)$ is $x_1{\cdot}(x_2{\cdot}x_4)$.*

4. The Right Normal Form

The existence of the ∂-normal form has been established easily, but its properties are not outstanding: the size of the terms $\partial^k x^{[m]}$ increases quickly with k, making practical computation impossible, no intrinsic characterization of ∂-normal terms is known, comparing ∂-normal terms with respect to $<_{LD}$ is uneasy, *etc.* In this section, starting from the ∂-normal form, we construct a new normal form with better properties. As for t_0-∂-normal terms, we use a base term t_0 and construct t_0-normal terms. The result is once again that every term t satisfying $t \sqsubseteq_{LD} t_0$ is LD-equivalent to a unique representative, here a fractional cut of t_0 of the form $\mathrm{cut}_{\boldsymbol{c}}(t_0)$, where $\boldsymbol{c}$ can be seen as a rational number of a certain type that we call a code. Defining a t_0-normal term to be one of the latter form, we shall obtain:

Proposition 4.1. (**normal form**) *Assume that t_0 is an extended term.*

(i) Every term t satisfying $t \sqsubseteq_{LD} t_0$ is LD-equivalent to a unique t_0-normal term. In particular, every term in T_1 is LD-equivalent to a unique x^∞-normal term.

(ii) Every t_0-normal term is specified by a rational number called its code. The function that maps every term t with $t \sqsubseteq_{LD} t_0$ to its t_0-code lies in the complexity class DSPACE$(\exp^*(\mathcal{O}(2^n)))$.

(iii) For $t_1, t_2 \sqsubseteq_{LD} t_0$, the relation $t_1 \sqsubset_{LD} t_2$ is equivalent to $\boldsymbol{c}_1 < \boldsymbol{c}_2$, where $\boldsymbol{c}_e$ is the t_0-code of t_e and $<$ is the usual ordering of the rationals.

The geometry of ∂t

The idea for constructing t_0-normal terms is to start with t_0-∂-normal terms, and to apply the operator ∂^{-1} as many times as possible to the subterms of the considered term. The key point is the existence of a bijective correspondence between the outline of ∂t and the descents of t, which results in a complete description of the cuts of ∂t in terms of the cuts of t.

The first step is to study the geometry of the terms $t_0 * t$.

Lemma 4.2. *Assume that t_0, t are terms. The outline of $t_0 * t$ consists of:*
- *all addresses $\alpha 0 \alpha_0$ with $\alpha_0 \in \mathrm{Out}(t_0)$ and $\alpha \in \mathrm{Out}(t)$,*
- *all addresses $\alpha 1$ with $\alpha \in \mathrm{Out}(t)$.*

Assume $\alpha_0 \in \mathrm{Out}(t_0)$, $\alpha \in \mathrm{Out}(t)$, and $\mathrm{cut}(t, \beta) = s_1 \cdot \cdots \cdot s_p \cdot x$ with x a variable. Then we have

$$\mathrm{cut}(t_0 * t, \alpha 0 \alpha_0) = (t_0 * s_1) \cdot \cdots \cdot (t_0 * s_p) \cdot \mathrm{cut}(t_0, \alpha_0),$$
$$\mathrm{cut}(t_0 * t, \alpha 1) = t_0 * \mathrm{cut}(t, \beta).$$

Proof. The result about the outline is clear owing to the definition of $t_0 * t$ as the term obtained from t by substituting every variable x with the term $t_0 \cdot x$. We obtain a formal proof, as well as a proof for the formulas for the cuts, by using induction on t. For t a variable, everything is obvious. Assume $t = t_1 \cdot t_2$. By definition, we have $t_0 * t = (t_0 * t_1) \cdot (t_0 * t_2)$. Applying the induction hypothesis to t_1 and t_2 gives the result about $\mathrm{Out}(t_0 * t)$. Assume $\alpha_0 \in \mathrm{Out}(t_0)$ and $\alpha \in \mathrm{Out}(t)$. Assume first $\alpha = 0\alpha_1$ with $\alpha_1 \in \mathrm{Out}(t_1)$. Write $\mathrm{cut}(t_1, \alpha_1) = s_1 \cdot \cdots \cdot s_p \cdot x$ with x a variable. The induction hypothesis gives

$$\mathrm{cut}(t_0 * t, 0\alpha_1 0 \alpha_0) = \mathrm{cut}(t_0 * t_1, \alpha_1 0 \alpha_0) = (t_0 * s_1) \cdot \cdots \cdot (t_0 * s_p) \cdot \mathrm{cut}(t_0, \alpha_0),$$

which is the expected result as $s_1 \cdot \cdots \cdot s_p \cdot x = \mathrm{cut}(t, 0\alpha_1)$ holds. Similarly, we have

$$\mathrm{cut}(t_0 * t, 0\alpha_1 1) = \mathrm{cut}(t_0 * t_1, \alpha_1 1) = t_0 * \mathrm{cut}(t_1, \alpha_1).$$

Assume now $\alpha = 1\alpha_1$, with $\alpha_1 \in \mathrm{Out}(t_2)$. Write $\mathrm{cut}(t_2, \alpha_1) = s_1 \cdot \cdots \cdot s_p \cdot x$ with x a variable. We find

$$\begin{aligned}\mathrm{cut}(t_0 * t, 1\alpha_1 0 \alpha_0) &= (t_0 * t_1) \cdot \mathrm{cut}(t_0 * t_2, \alpha_1 0 \alpha_0)\\ &= (t_0 * t_1) \cdot (t_0 * s_1) \cdot \cdots \cdot (t_0 * s_p) \cdot \mathrm{cut}(t_0, \alpha_0),\end{aligned}$$

the desired formula as $\mathrm{cut}(t, \alpha) = t_1 \cdot s_1 \cdot \cdots \cdot s_p \cdot x$ holds. Finally, we have

$$\begin{aligned}\mathrm{cut}(t_0 * t, 1\alpha_1 1) &= (t_0 * t_1) \cdot \mathrm{cut}(t_0 * t_2, \alpha_1 1)\\ &= t_0 * (t_1 \cdot \mathrm{cut}(t_2, \alpha_1)), = t_0 * \mathrm{cut}(t, \alpha).\end{aligned}$$ ∎

(Observe that the argument in the previous proof has to be essentially trivial as the formulas are equalities and not equivalences.)

We turn to the study of the cuts of the term ∂t, which we shall see are connected with the descents of t. We order the sequences of addresses, hence in particular the descents, using the lexicographical extension of the left–right ordering. For every term t, the largest descent in t is the descent (1^*), *i.e.*, the unique descent of the form (1^q) with 1^q in the outline of t. We call the latter descent the *final* descent of t. For $\vec{\alpha} = (\alpha_1, \ldots, \alpha_p)$ a sequence of addresses, and γ an address, we write $\gamma\vec{\alpha}$ for the sequence $(\gamma\alpha_1, \ldots, \gamma\alpha_p)$, and, similarly, $\vec{\alpha}\gamma$ for the sequence $(\alpha_1\gamma, \ldots, \alpha_p\gamma)$. Finally, if $\vec{\alpha}$ and $\vec{\beta}$ are two sequences of addresses, say $\vec{\alpha} = (\alpha_1, \dots, \alpha_p)$ and $\vec{\beta} = (\beta_1, \dots, \beta_q)$, we denote by $\vec{\alpha}^\frown\vec{\beta}$ the concatenated sequence $(\alpha_1, \dots, \alpha_p, \beta_1, \dots, \beta_q)$.

Definition 4.3. (functions π_t and π_t^+) Assume that t is a term, and $\vec{\alpha}$ is a non-final descent in t, say $\vec{\alpha} = (\alpha_1, \dots, \alpha_p)$. We define

$$\pi_t(\vec{\alpha}) = 0^{\nu_t(+\infty,\alpha_1)}10^{\nu_t(\alpha_1,\alpha_2)}10^{\nu_t(\alpha_2,\alpha_3)}1\dots 0^{\nu_t(\alpha_{p-1},\alpha_p)}01^{\#_1(\alpha_p)},$$
$$\pi_t^+(\vec{\alpha}) = 0^{\nu_t(+\infty,\alpha_1)}10^{\nu_t(\alpha_1,\alpha_2)}10^{\nu_t(\alpha_2,\alpha_3)}1\dots 0^{\nu_t(\alpha_{p-1},\alpha_p)}10^{\nu_t(\alpha_p,-\infty)},$$

where $\nu_t(\alpha, \beta)$ denotes the number of addresses γ in the outline of t that satisfy $\mu_t(\alpha) >_{LR} \gamma >_{LR} \beta$ and $\pm\infty$ are two virtual addresses with the convention that $\mu_t(+\infty) >_{LR} \gamma >_{LR} -\infty$ holds for every address γ not of the form 1^q; $\#_1(\alpha)$ denotes the number of 1's in α. If (1^q) is the final descent in t, we extend π_t with $\pi_t((1^q)) = 1^q$.

Proposition 4.4. (decomposition) *(i) For every term t, the mapping π_t is an increasing bijection of* Desc(t) *ordered by the lexicographical extension of $<_{LR}$ onto* Out(∂t) *ordered by $<_{LR}$. For $\vec{\alpha}$ a non-final descent in t, the immediate successor of $\pi_t(\vec{\alpha})$ in the outline of ∂t is $\pi_t^+(\vec{\alpha})$.*

(ii) Assume that $\vec{\alpha}$ is a descent in t, say $\vec{\alpha} = (\alpha_1, \dots, \alpha_p)$. Then we have

$$\mathrm{cut}(\partial t, \pi_t(\vec{\alpha})) = \partial\mathrm{cut}(t, \alpha_1) \cdot \dots \cdot \partial\mathrm{cut}(t, \alpha_p), \tag{4.1}$$

and, if $\vec{\alpha}$ is not final,

$$\mathrm{cut}(\partial t, \pi_t^+(\vec{\alpha})) = \partial\mathrm{cut}(t, \alpha_1) \cdot \dots \cdot \partial\mathrm{cut}(t, \alpha_p) \cdot x, \tag{4.2}$$

where x is a variable.

Proof. (i) We use induction on t. For t a variable, the only descent of t is (ϕ), and we have $\partial t = t$ and $\pi_t((\phi)) = \phi$, so the results are true. We assume now $t = t_1 \cdot t_2$, hence $\partial t = \partial t_1 * \partial t_2$. By definition, the descents of t consist of the sequences

- $1\vec{\gamma}^\frown 0\vec{\beta}$, with β a descent in t_1 and γ a non-final descent in t_2;
- $1\vec{\gamma}$, with γ a descent in t_2;
- $0\vec{\beta}$, with β a descent in t_1.

Assume first that $\vec{\beta}$ is a non-final descent in t_1 and $\vec{\gamma}$ is a non-final descent in t_2. For α_2 non-final in Out(t_2) and α_1 in Out(t_1), we have

$$\nu_t(1\alpha_2, 0\alpha_1) = \nu_{t_2}(\alpha_2, -\infty) + 1 + \nu_{t_1}(+\infty, \alpha_1), \tag{4.3}$$

which remains valid for $\alpha_2 = +\infty$ or $\alpha_1 = -\infty$. Using the explicit definition and (4.3), we obtain

$$\pi_t(1\vec{\gamma}^\frown 0\vec{\beta}) = \pi_{t_2}^+(\vec{\gamma})\,0\,\pi_{t_1}(\vec{\beta}), \qquad \pi_t^+(1\vec{\gamma}^\frown 0\vec{\beta}) = \pi_{t_2}^+(\vec{\gamma})\,0\,\pi_{t_1}^+(\vec{\beta}),$$

the latter holding for $\vec{\beta}$ non-final only. With similar notation, we find

$$\pi_t(1\vec{\gamma}) = \pi_{t_2}(\vec{\gamma})\,1, \qquad \pi_t^+(1\vec{\gamma}) = \pi_{t_2}^+(\vec{\gamma})\,0^{m_0},$$
$$\pi_t(0\vec{\beta}) = 0^{m_1}\pi_{t_1}(\vec{\beta})1, \qquad \pi_t^+(0\beta) = 0^{m_1}\pi_{t_1}^+(\vec{\beta}),$$

where m_e denotes the size of t_e.

Let now (1^q) be the final descent in t_1. For every address γ, we have $\nu_t(1\gamma, 01^q) = \nu_{t_2}(\gamma, -\infty)$. Then, for $\vec{\gamma}$ a non-final descent in t_2, we obtain

$$\pi_t(1\vec{\gamma}^\frown(01^q)) = \pi_{t_2}^+(\vec{\gamma})\,0\,1^q = \pi_{t_2}^+(\vec{\gamma})\,0\,\pi_{t_1}((1^q)), \quad \pi_t^+(1\vec{\gamma}^\frown(01^q)) = \pi_{t_2}^+(\vec{\gamma})\,1,$$
$$\pi_t((01^q)) = 0^{m_1}1^q = 0^{m_1}\pi_{t_1}((1^q)), \qquad \pi_t^+((01^q)) = 0^{m_1-1}1.$$

Finally, let (1^r) be the final descent in t_2. Then (1^{r+1}) is the final descent in t, and we have

$$\pi_t((1^{r+1})) = 1\pi_{t_2}((1^r)).$$

We conclude that the image of the mapping π_t consists of all sequences $\pi_{t_2}^+(\vec{\gamma})0\pi_{t_1}(\vec{\beta})$ for $\vec{\beta}$ a descent in t_1 and $\vec{\gamma}$ a non-final descent in t_2, together with all sequences $\pi_{t_2}^+(\vec{\gamma})1$ with $\vec{\gamma}$ a descent in t_2 and all sequences $0^{m_1}\pi_{t_1}(\vec{\beta})$ for $\vec{\beta}$ a descent in t_1. The induction hypothesis implies that the set of all sequences of the form $\pi_{t_2}^+(\vec{\gamma})$ with $\vec{\gamma}$ a non-final descent in t_2 is the outline of ∂t_2 with 0^* omitted, and that the set of all sequences of the form $\pi_{t_1}(\vec{\beta})$ with $\vec{\beta}$ a descent in t_1 is the outline of ∂t_1. Hence, by Lemma 4.2, the image of π_t is the outline of the term $\partial t_1 * \partial t_2$, which is ∂t. A similar argument shows that the image of π_t^+ is the outline of ∂t with the leftmost element omitted.

By construction, $\pi_t(\vec{\alpha}) <_{LR} \pi_t^+(\vec{\alpha})$ holds for every non-final descent in t. Moreover, $\pi_t(\vec{\alpha})$ and $\pi_t^+(\vec{\alpha})$ have the form $\gamma 01^*$, $\gamma 10^*$. Hence, no address β may satisfy $\pi_t(\vec{\alpha}) <_{LR} \beta <_{LR} \pi_t^+(\vec{\alpha})$. This shows that $\pi_t^+(\vec{\alpha})$ is the immediate successor of $\pi_t(\vec{\alpha})$ in the outline of ∂t.

Finally, the following relations hold by construction:
- $\pi_t(0\vec{\beta}) <_{LR} \pi_t(1\vec{\gamma}) <_{LR} \pi_t(1\vec{\gamma}^\frown 0\vec{\beta})$ for $\vec{\beta}$ a descent in t_1 and $\vec{\gamma}$ a descent in t_2;
- $\pi_t(1\vec{\gamma}^\frown 0\vec{\beta}) <_{LR} \pi_t(1\vec{\gamma}'^\frown 0\vec{\beta}')$ for $\vec{\beta}$, $\vec{\beta}'$ descents in t_1 and $\vec{\gamma}$, $\vec{\gamma}'$ non-final descents in t_2 satisfying $\pi_{t_2}(\vec{\gamma}) <_{LR} \pi_{t_2}(\vec{\gamma}')$;
- $\pi_t(1\vec{\gamma}^\frown 0\vec{\beta}) <_{LR} \pi_t(1\vec{\gamma}^\frown 0\vec{\beta}')$ for $\vec{\beta}$, $\vec{\beta}'$ descents in t_1 satisfying $\pi_{t_1}(\vec{\beta}) <_{LR} \pi_{t_1}(\vec{\beta}')$ and $\vec{\gamma}$ a non-final descent in t_2.

It follows that the mapping π_t is increasing, since, by induction hypothesis, the mappings π_{t_1} and π_{t_2} are increasing. This completes the proof of (i).

(ii) Assume now that $\vec{\beta}$ is a descent in t_1 and $\vec{\gamma}$ is a non-final descent in t_2, say $\vec{\beta} = (\beta_1, \dots, \beta_q)$ and $\vec{\gamma} = (\gamma_1, \dots, \gamma_r)$. Then $\pi_t(1\vec{\gamma}^\frown 0\vec{\beta})$, which is $\pi_{t_2}^+(\vec{\gamma})0\pi_{t_1}(\vec{\beta})$, belongs to the outline of ∂t. Lemma 4.2 then gives

$$\begin{aligned}\mathrm{cut}(\partial t, \pi_t(1\vec{\gamma}^\frown 0\vec{\beta})) &= \mathrm{cut}(\partial t_1 * \partial t_2, \pi_{t_2}^+(\vec{\gamma})\,0\,\pi_{t_1}(\vec{\beta}))\\ &= (\partial t_1 * s_1)\cdot\dots\cdot(\partial t_1 * s_m)\cdot \mathrm{cut}(\partial t_1, \pi_{t_1}(\vec{\beta})),\end{aligned}$$

where $\mathrm{cut}(\partial t_2, \pi^+_{t_2}(\vec\gamma))$ is supposed to be $s_1\cdot \cdots \cdot s_m\cdot x$ for some variable x. By induction hypothesis, we have $m = r$, and $s_k = \partial\mathrm{cut}(t_2, \gamma_k)$ for $k \le r$. Using $\mathrm{cut}(t, 1\gamma_k) = t_1\cdot\mathrm{cut}(t_2, \gamma_k)$ and $\partial\mathrm{cut}(t, 1\gamma_k) = \partial t_1 * \partial\mathrm{cut}(t_2, \gamma_k)$, we find

$$\mathrm{cut}(\partial t, \pi_t(1\vec\gamma{}^\frown 0\vec\beta)) = \partial\mathrm{cut}(t, 1\gamma_1)\cdot\cdots\cdot\partial\mathrm{cut}(t, 1\gamma_r)\cdot\partial\mathrm{cut}(t, 0\beta_1)\cdot\cdots\cdot\partial\mathrm{cut}(t, 0\beta_q),$$

which is Formula (4.1). The case of $0\vec\beta$ alone (no $1\vec\gamma$) is similar.

Let us now consider the case of $1\vec\gamma$, where $\vec\gamma$ is a descent in t_2, say $\vec\gamma = (\gamma_1, \ldots, \gamma_r)$. Then we have $\pi_t(1\vec\gamma) = \pi_{t_2}(\vec\gamma)1$, and, by Lemma 4.2,

$$\mathrm{cut}(\partial t, \pi_t(1\vec\gamma)) = \mathrm{cut}(\partial t_1 * \partial t_2, \pi^+_{t_2}(\vec\gamma)\,1) = \partial t_1 * \mathrm{cut}(\partial t_2, \pi^+_{t_2}(\vec\gamma)).$$

By induction hypothesis, $\mathrm{cut}(\partial t_2, \pi^+_{t_2}(\vec\gamma)) = \partial\mathrm{cut}(t_2, \gamma_1)\cdot \cdots \cdot\partial\mathrm{cut}(t_2, \gamma_r)$ holds, so, finally, we find

$$\begin{aligned}\mathrm{cut}(\partial t, \pi_t(1\vec\gamma)) &= \partial t_1 * (\partial\mathrm{cut}(t_2, \gamma_1)\cdot\cdots\cdot\partial\mathrm{cut}(t_2, \gamma_r))\\ &= (\partial t_1 * \partial\mathrm{cut}(t_2, \gamma_1))\cdot\cdots\cdot(\partial t_1\cdot\partial\mathrm{cut}(t_2, \gamma_r))\\ &= (\partial(t_1\cdot\mathrm{cut}(t_2, \gamma_1)))\cdot\cdots\cdot(\partial(t_1\cdot\mathrm{cut}(t_2, \gamma_r)))\\ &= \partial\mathrm{cut}(t, 1\gamma_1))\cdot\cdots\cdot\partial\mathrm{cut}(t, 1\gamma_r)),\end{aligned}$$

which gives Formula (4.1) again.

As $\pi^+_t(\vec\alpha))$ is the successor of $\pi_t(\vec\alpha)$ in the outline of ∂t, Formula (4.2) follows by applying Proposition 2.14 (next cut). ∎

Example 4.5. **(cuts of** ∂t) *Let t be (again) the term $x_1\cdot(x_2\cdot x_3\cdot x_4)\cdot x_5$. We have seen that there are ten descents in t. Thus, there are ten elements in the outline of ∂t, which is the term*

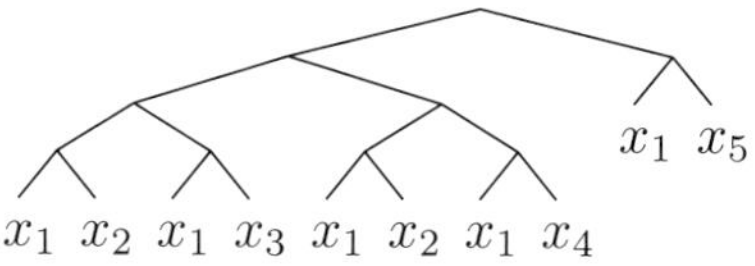

Let us consider for instance the descent $(1010, 100)$ in t. We have $\nu_t(+\infty, 1010) = 1$ since there is one non-final address in the outline of t on the right of 1010, $\nu_t(1010, 100) = 0$ since no point in the outline of t lies between 1010 and 100, and $\nu_t(100, -\infty) = 1$ as $\mu_t(100) = 100 >_{LR} 0$ holds. We deduce

$$\pi_t((1010, 100)) = 0^1 10^0 01^1 = 0101,$$
$$\pi^+_t((1010, 100)) = 0^1 10^0 10^1 = 0110,$$

and, according to Formulas (4.1) *and* (4.2),

$$\begin{aligned}\mathrm{cut}(\partial t, 0101) &= x_1x_2\bullet x_1x_3\bullet\bullet x_1x_2\bullet\bullet = \partial(x_1x_2x_3\bullet\bullet)\cdot\partial(x_1x_2\bullet)\\ &= \partial\mathrm{cut}(t,1010)\cdot\partial\mathrm{cut}(t,100),\\ \mathrm{cut}(\partial t, 0110) &= x_1x_2\bullet x_1x_3\bullet\bullet x_1x_2\bullet x_1\bullet\bullet = \partial(x_1x_2x_3\bullet\bullet)\cdot\partial(x_1x_2\bullet)\cdot x_1\\ &= \partial\mathrm{cut}(t,1010)\cdot\partial\mathrm{cut}(t,100)\cdot x_1.\end{aligned}$$

It is now easy to characterize those cuts of ∂t that expand some cut of t.

Lemma 4.6. *Assume that t is a term, and α belongs to the outline of ∂t. Then the following are equivalent:*

(i) The term $\mathrm{cut}(\partial t, \alpha)$ *is LD-equivalent to some cut of t;*
(ii) The term $\mathrm{cut}(\partial t, \alpha)$ *is equal to ∂s for some cut s of t;*
(iii) The address α is $\pi_t((\beta))$ for some address β in the outline of t;
(iv) The address α has the form 0^i1^j.

Proof. It is obvious that (ii) implies (i). By Proposition 4.4 (decomposition), $\alpha = \pi_t((\beta))$ implies $\mathrm{cut}(\partial t, \alpha) = \partial\mathrm{cut}(t, \beta)$, so (iii) implies (ii). Conversely, if $\alpha = \pi_t(\vec{\beta})$ holds for some descent $\vec{\beta}$ in t of length $q \geq 2$, then Proposition 4.4 gives $\mathrm{cut}(\partial t, \alpha) =_{LD} \mathrm{cut}(t, \beta_1)\cdot\dots\cdot\mathrm{cut}(t, \beta_q)$. By Proposition 2.17 (obstruction), no cut of t may be LD-equivalent to such a term, hence (i) cannot be true.

The equivalence of (iii) and (iv) follows from the explicit definition of the function π_t. If the outline of t contains m addresses and β is the i-th address from the left in this outline, then the definition gives $\pi_t((\beta)) = 0^{m-i}1^{\#_1(\beta)}$. On the other hand, the factor 10 necessarily appears in the address $\pi_t(\vec{\beta})$ when $\vec{\beta}$ comprises at least two components. ■

We conclude this preparatory section with the description of the covering relation for a term ∂t.

Lemma 4.7. *Assume that t is a term and $\vec{\alpha}$ is a descent in t, say $\vec{\alpha} = (\alpha_1, \dots, \alpha_p)$. Then we have*

$$\mu_{\partial t}(\pi_t(\vec{\alpha})) = \pi_t((\alpha_1, \dots, \alpha_{p-1}, 0^*)). \tag{4.4}$$

Proof. If $\alpha_p = 0^*$ holds, then, by construction, $\pi_t(\vec{\alpha})$ ends with 0, hence it covers nothing, and we have $\mu_{\partial t}(\pi_t(\vec{\alpha})) = \vec{\alpha}$. Assume now that α_p contains at least one 1, and $p \geq 2$ holds. By definition, we have

$$\pi_t(\vec{\alpha}) = 0^{\nu_t(+\infty,\alpha_1)}10^{\nu_t(\alpha_1,\alpha_2)}1\dots10^{\nu_t(\alpha_{p-2},\alpha_{p-1})}10^{\nu_t(\alpha_{p-1},\alpha_p)}01^{\#_1(\alpha_p)},$$

so, by definition of $\mu_{\partial t}$, there must exist a positive integer i such that $\mu_{\partial t}(\pi_t(\vec{\alpha}))$ has the form $0^{\nu_t(+\infty,\alpha_1)}10^{\nu_t(\alpha_1,\alpha_2)}1\dots10^{\nu_t(\alpha_{p-2},\alpha_{p-1})}10^i$. Now, we have

$$\pi_t((\alpha_1, \dots, \alpha_{p-1}, 0^*)) = 0^{\nu_t(+\infty,\alpha_1)}10^{\nu_t(\alpha_1,\alpha_2)}1\dots10^{\nu_t(\alpha_{p-2},\alpha_{p-1})}10^{\nu_t(\alpha_{p-1},0^*)+1},$$

which has the desired form, and the result follows by uniqueness. Assume finally that $\vec{\alpha}$ has length 1. Then, by the previous lemma, $\pi_t(\vec{\alpha})$ has the form 0^i1^j, $\mu_{\partial t}(\pi_t(\vec{\alpha}))$ has the form 0^*, and, again, $\pi_t((0^*))$ has this form. ∎

Fractional cuts

We shall now construct new terms from the cuts of a fixed (extended) base term t_0. Such terms are called the fractional cuts of t_0, as they lie between the (ordinary) cuts of t_0 with respect to the relation $\sqsubset_{LD}$.

Let us consider the question of constructing normal terms again. As every term t satisfying $t \sqsubseteq_{LD} t_0$ is LD-equivalent to a t_0-∂-normal term, it suffices that we construct normal representatives for t_0-∂-normal terms, *i.e.*, for the cuts of the terms $\partial^k t_0$. Proposition 4.4 (decomposition) tells us that every cut of ∂t is LD-equivalent to a unique descending product of cuts of t. So, if we number the cuts of t from left to right, say $\mathrm{cut}_1(t)$, $\mathrm{cut}_2(t)$, *etc.*, every cut s of ∂t can be specified by the decreasing sequence of integers $(i_1, \ldots, i_p)$ such that s is LD-equivalent to $\mathrm{cut}_{i_1}(t) \cdot \cdots \cdot \mathrm{cut}_{i_p}(t)$. This is the normal form for degree 1 terms, and the general case just consists in iterating the construction: every cut of $\partial^2 t$ can be specified by a sequence of sequence of integers, *etc.* This amounts to specifying the cuts of $\partial^k t$ using k-uniform trees as in Chapter IV. Actually, technical reasons make it more convenient to use here binary trees (with branches of various length, but only two successors for each node).

Definition 4.8. (**precode**) For $1 \leq n \leq \infty$, an *n-precode* is defined to be a term on $\{1, 2, \ldots, n\}$—hence, in particular, a word over the alphabet $\{\bullet, 1, 2, \ldots n\}$. For $\boldsymbol{c}_1$, $\boldsymbol{c}_2$ precodes, we say that $\boldsymbol{c}_1 \prec \boldsymbol{c}_2$ holds if $\boldsymbol{c}_1$ precedes $\boldsymbol{c}_2$ in the lexicographical extension of $\bullet < 1 < 2 < \cdots$.

The linear ordering $\prec$ does not coincide with the partial lexicographical ordering considered in Section V.4: here the letter $\bullet$ is comparable with the other letters.

By definition, natural numbers are special precodes. It is convenient to view general precodes as rational numbers interpolating the integers.

Notation 4.9. (**precode**) For $\boldsymbol{c}$ an n-precode, we write $(\boldsymbol{c})_n$ for the word obtained from $\boldsymbol{c}$ by removing the final $\bullet$'s, replacing the other $\bullet$'s with 0, and, for $\lg(\boldsymbol{c}) \geq 2$, adding one dot after the first letter.

In practice, all precodes we shall mention are 10-precodes, and we use $(\boldsymbol{c})_{10}$ to represent $\boldsymbol{c}$. By construction, this word is the decimal expansion of a rational number, and the interest of our convention lies in the following straightforward compatibility result:

Lemma 4.10. *Assume that $\boldsymbol{c}_1$ and $\boldsymbol{c}_2$ are n-precodes. For $e = 1, 2$, let q_e be the rational number with base n expansion $(\boldsymbol{c}_e)_n$. Then we have $\boldsymbol{c}_1 \prec \boldsymbol{c}_2$ if and only if $q_1 < q_2$ holds in $\mathbf{Q}$.*

Example 4.11. (**precode**) *Typical precodes are 2, 31•, and 31•21••. According to Notation 4.9, they will be denoted 2, 3.1, and 3.1021 respectively. The associated trees are 2,*  *,* 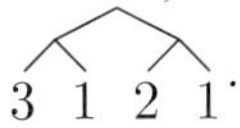 *. The inequalities $2 \prec 3.1 \prec 3.1021$ hold both when the expressions are considered as precodes and as rational numbers.*

Up to replacing i with x_i, a precode is a term in T_∞, and we could use the latter terms everywhere. However, it seems more convenient for the intuition to keep different names and notations here, as precodes are not used as terms in the sequel, but as frames for building complex cuts.

Definition 4.12. (**fractional cut**) Assume that t_0 is an extended term with n cuts. For $\boldsymbol{c}$ an n-precode, the *fractional cut* $\mathrm{cut}_{\boldsymbol{c}}(t_0)$ is defined to be the image of $\boldsymbol{c}$ under the homomorphism that maps i to the i-th cut of t from the left.

Example 4.13. (**fractional cut**) *The extended term x^∞ has infinitely many cuts, so $\mathrm{cut}_{\boldsymbol{c}}(x^\infty)$ is defined for every precode $\boldsymbol{c}$. We have for instance*

$$\mathrm{cut}_2(x^\infty) = x^{[2]},$$
$$\mathrm{cut}_{3.1}(x^\infty) = \mathrm{cut}_3(x^\infty)\cdot\mathrm{cut}_1(x^\infty) = x^{[3]}\cdot x,$$
$$\mathrm{cut}_{3.1021}(x^\infty) = \mathrm{cut}_{3.1}(x^\infty)\cdot\mathrm{cut}_{2.1}(x^\infty) = (x^{[3]}\cdot x)\cdot(x^{[2]}\cdot x).$$

In the sequel, we shall be interested in special precodes only, which are called codes. We introduce them now.

Definition 4.14. (**degree**) For $\boldsymbol{c}$ a precode, the *degree* $\deg(\boldsymbol{c})$ of $\boldsymbol{c}$ is the maximal number of 0's in an address in $\boldsymbol{c}$.

For instance, the outline of the precode 3.1021 is $\{00, 01, 10, 11\}$, so its degree is 2. By construction, the precodes of degree 0 are the integers.

For $\deg(\boldsymbol{c}) \geq 1$, $\boldsymbol{c}$ admits a decomposition of the form $\boldsymbol{c} = \boldsymbol{c}_1 \cdot \cdots \cdot \boldsymbol{c}_p \cdot i$, with i an integer. By definition, we have $\deg(\boldsymbol{c}) = \sup_i \deg(\boldsymbol{c}_i) + 1$, so there exists a largest index k satisfying $\deg(\boldsymbol{c}_k) = \deg(\boldsymbol{c}) - 1$.

Definition 4.15. (**head**) For $\boldsymbol{c}$ a precode, the *head* of $\boldsymbol{c}$ is the precode $\mathrm{hd}(\boldsymbol{c})$ defined as follows: for $\boldsymbol{c}$ an integer, we put $\mathrm{hd}(\boldsymbol{c}) = \boldsymbol{c}$; for $\boldsymbol{c} = \boldsymbol{c}_1 \cdot \cdots \cdot \boldsymbol{c}_p \cdot i$ with $p \geq 1$ and i an integer, we put $\mathrm{hd}(\boldsymbol{c}) = \boldsymbol{c}_1 \cdot \cdots \cdot \boldsymbol{c}_k \cdot 1$, where k is the largest index k satisfying $\deg(\boldsymbol{c}_k) = \deg(\boldsymbol{c}) - 1$.

The head of $\boldsymbol{c}$ is obtained from $\boldsymbol{c}$ by keeping the complicated initial part of $\boldsymbol{c}$, but replacing the associated tail with 1. Notice that $\deg(\mathrm{hd}(\boldsymbol{c})) = \deg(\boldsymbol{c})$ always holds, since the fragment we remove does not contribute to the degree.

Example 4.16. (**head**) *The head of* 3.2 *is* 3.1, *while the head of* 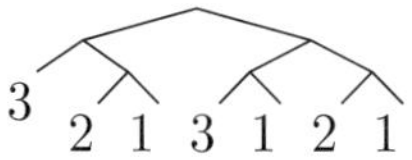

, *i.e., of* 3.210031021, *is* 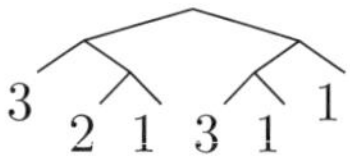

, *i.e.,* 3.21003101. *Observe that the head of* $\boldsymbol{c}$ *need not be a cut of* $\boldsymbol{c}$*: the projection of* $\mathrm{hd}(\boldsymbol{c})$ *on* T_1 *is a cut of the projection of* $\boldsymbol{c}$*, but the labels need not be the same.*

Definition 4.17. (**code**) Assume that $\boldsymbol{c}$ is a precode and α is an internal address in $\boldsymbol{c}$. We say that $\boldsymbol{c}$ is *descending at* α if we have

$$\deg(\mathrm{sub}(\boldsymbol{c}, \alpha 0)) + 1 \geq \deg(\mathrm{sub}(\boldsymbol{c}, \alpha 1)), \text{and } \mathrm{hd}(\mathrm{sub}(\boldsymbol{c}, \alpha 0)) \succ \mathrm{sub}(\boldsymbol{c}, \alpha 1). \quad (4.5)$$

We say that $\boldsymbol{c}$ is a *code* if it is descending at every internal address.

Notice that every subterm of a code is a code, as being a code is defined by local conditions.

Example 4.18. (**code**) *Every positive integer is a code, as Condition* (4.5) *is vacuously true.*

Let $\boldsymbol{c}$ *be a precode of degree* 1. *Then* $\boldsymbol{c}$ *can be expressed as* $i_1 \cdot \cdots \cdot i_{p+1}$. *The internal addresses of* $\boldsymbol{c}$ *are* ϕ, 1, ..., 1^{p-1}. *For* $\alpha = 1^k$, *the precodes* $\mathrm{sub}(\boldsymbol{c}, \alpha 0)$ *and* $\mathrm{sub}(\boldsymbol{c}, \alpha 1)$ *involved in* (4.5) *are respectively* i_{k+1} *and* $i_{k+2} \cdot \cdots \cdot i_{p+1}$. *The degree condition always holds, and the head condition holds if and only if* $i_k > i_{k+1}$ *does. Thus, the codes of degree* 1 *are those precodes* $i_1 \cdot \cdots \cdot i_{p+1}$ *with* $i_1 > \cdots > i_{p+1}$. *Thus the first codes of degree* 1 *are* 2.1, 3.1, 3.21, 4.1, 4.2, 4.21, 4.3, *etc.*

For degree 2, *things are more complicated. For instance,* 3.2031 *is not descending at* ϕ. *Indeed, the condition on degrees holds, but we have* $\mathrm{hd}(3.2) = 3.1$, *so the head condition does not hold.*

Lemma 4.19. *Assume that $\boldsymbol{c}$ is a code of positive degree.*

(i) Let i be the leftmost integer in $\boldsymbol{c}$. Then, for every integer j occurring in $\boldsymbol{c}$, we have $i \geq j$, and even $i > j$ if the left subterm of $\boldsymbol{c}$ has size 1.

(ii) Assume that i occurs at $\alpha 0^p$ in $\boldsymbol{c}$, and j occurs at $\alpha\beta$ in $\boldsymbol{c}$, with $\beta \neq 0^p$. Then, $i \geq j$ holds. If β begins with 0^{p-1}, then $i > j$ holds.

Proof. We show (i) using induction on $\boldsymbol{c}$. If $\boldsymbol{c}$ is an integer, the result is vacuously true. Otherwise, assume $\boldsymbol{c} = \boldsymbol{c}_1 \cdot \boldsymbol{c}_2$. Let i the leftmost integer in $\boldsymbol{c}_1$, and j be the leftmost integer in $\boldsymbol{c}_2$. The hypothesis that $\boldsymbol{c}$ is descending at ϕ implies $i \geq j$. Now, we have observed that every subterm of a code is a code, so $\boldsymbol{c}_1$ and $\boldsymbol{c}_2$ are codes, and, by induction hypothesis, every integer occurring in $\boldsymbol{c}_1$ is at most i, and every integer occurring in $\boldsymbol{c}_2$ is at most j, hence at most i as well. If $\boldsymbol{c}_1$ is an integer, the hypothesis that $\boldsymbol{c}$ is descending at ϕ implies in addition $i > j$, for, otherwise, $\boldsymbol{c}_1$ would be a prefix of $\boldsymbol{c}_2$, contradicting the head condition. Point (ii) follows from applying (i) to the α-th subterm of $\boldsymbol{c}$. ∎

The key point for our subsequent use of codes is that every code of degree k can be expressed as a product of codes of degree at most $k-1$, .

Proposition 4.20. (induction) *Assume that k is a positive integer and $\boldsymbol{c}$ is a precode of degree k. Then the following are equivalent:*

(i) The precode $\boldsymbol{c}$ is a code;

(ii) The precode $\boldsymbol{c}$ admits a decomposition $\boldsymbol{c} = \boldsymbol{c}_1 \cdot \cdots \cdot \boldsymbol{c}_{p+1}$ where $\boldsymbol{c}_1, \ldots, \boldsymbol{c}_{p+1}$ are codes satisfying $\deg(\boldsymbol{c}_1) = \cdots = \deg(\boldsymbol{c}_p) = k-1$, $\deg(\boldsymbol{c}_{p+1}) \leq k-1$, and $\mathrm{hd}(\boldsymbol{c}_i) \succ \boldsymbol{c}_{i+1}$ for $i \leq p$.

Moreover the decomposition of (ii) is unique when it exists.

Proof. We use induction on k. For $k = 1$, we have obtained the characterization in Example 4.18 (code). Assume $k \geq 2$. Assume that $\boldsymbol{c}$ is a code of degree k. By construction, it admits a decomposition $\boldsymbol{c} = \boldsymbol{c}_1 \cdot \cdots \cdot \boldsymbol{c}_{p+1}$ with $\deg(\boldsymbol{c}_p) = k-1$ and $\deg(\boldsymbol{c}_{p+1}) \leq k-1$. As $\boldsymbol{c}_1, \ldots, \boldsymbol{c}_{p+1}$ are subterms of $\boldsymbol{c}$, they are codes. Moreover, $\boldsymbol{c}$ is descending at 1^i for $i \leq p$. This implies $\deg(\boldsymbol{c}_i) \geq \deg(\boldsymbol{c}_{i+1} \cdot \cdots \cdot \boldsymbol{c}_{p+1}) - 1 = k-1$. Moreover, the condition $\mathrm{hd}(\boldsymbol{c}_i) \succ \boldsymbol{c}_{i+1}$ holds by definition, so (i) implies (ii). As for uniqueness, p must be the largest integer such that the subterm of $\boldsymbol{c}$ with address $1^p 0$ has degree $k-1$, so the only possible decomposition is the one described above.

Conversely, assume (ii). By hypothesis, the precode $\boldsymbol{c}$ is descending at every address which is not of the form 1^i with $i \leq p$. So it suffices to consider the latter addresses. Now, we have $\deg(\boldsymbol{c}_i) = k-1$, and $\deg(\boldsymbol{c}_{i+1} \cdot \cdots \cdot \boldsymbol{c}_{p+1}) = k$ for $i < p$. So the degree condition holds. On the other hand, the condition $\mathrm{hd}(\boldsymbol{c}_i) \succ \boldsymbol{c}_{i+1}$ holds by hypothesis. Hence $\boldsymbol{c}$ is a code. ∎

Using the previous characterization, we can enumerate systematically the first codes of every given degree (see Appendix).

The right normal form

We are ready to describe the right normal form. We start with the ∂-normal form, and prove that every t_0-∂-normal term is LD-equivalent to a unique fractional cut associated with a convenient code.

Definition 4.21. (**uncover, cover**) Assume that t is a term, and s_1, s_2 are cuts of t, say $s_1 = \mathrm{cut}(t, \alpha)$, $s_2 = \mathrm{cut}(t, \beta)$. We say that s_1 *covers* s_2 *in* t (*resp. uncovers*) if α covers β (*resp.* uncovers).

Lemma 4.22. *Assume that t_0 is a cut of t'_0, and s_1, s_2 are non-final cuts of t_0. Then s_1 uncovers (resp. covers) s_2 in t_0 if and only if it does in t'_0.*

Proof. Assume $t_0 = \mathrm{cut}(t'_0, \gamma)$. Let $(\gamma_1, \ldots, \gamma_r)$ be the left edge of γ. Assume $s_1 = \mathrm{cut}(t_0, \alpha_1) = \mathrm{cut}(t'_0, \alpha'_1)$, and $s_2 = cut(t_0, \alpha_2) = \mathrm{cut}(t'_0, \alpha'_2)$. We assume $s_1 \sqsupset_{LD} s_2$. Now assume that α'_1 covers α'_2. Then there exists γ', with $\gamma' = \gamma$ or $\gamma' = \gamma_i$ for some i, such that both α'_1 and α'_2 lie below γ', say $\alpha'_1 = \gamma'\beta_1$, $\alpha'_2 = \gamma'\beta_2$, and β_1 covers β_2. Now, in this case, we have $\alpha_1 = 1^r\beta_1$ and $\alpha_2 = 1^r\beta_2$ (case $\gamma' = \gamma$), or $\alpha_1 = 1^{i-1}\beta_1$ and $\alpha_2 = 1^{i-1}\beta_2$ (case $\gamma' = \gamma_i$). Hence α_1 covers α_2 in t_0. The converse implication is proved similarly. ∎

It follows that we can define without ambiguity the notion of a cut uncovering another cut in every extended term t_0: if t_0 is infinite, we say that s_1 uncovers s_2 in t_0 if it does so in every sufficiently large (finite) cut of t_0.

Definition 4.23. (**t_0-head, t_0-code, t_0-normal**) Assume that t_0 is an extended term of size n. For $\boldsymbol{c}$ a precode, the t_0-*head* $\mathrm{hd}_{t_0}(\boldsymbol{c})$ is defined as the head of $\boldsymbol{c}$, but starting with $\mathrm{hd}_{t_0}(\boldsymbol{c}) = j$ if $\boldsymbol{c}$ is the integer i and $\mathrm{cut}_j(t_0)$ is the least cut of t_0 covered by $\mathrm{cut}_i(t_0)$, if any, and $j = i$ otherwise. A t_0-*code* is defined to be a code that involves no integer bigger than n, and that satisfies the t_0-descending condition obtained from (4.5) by replacing hd with hd_{t_0}. Finally, we say that t is t_0-*normal* if $t = \mathrm{cut}_{\boldsymbol{c}}(t_0)$ holds for some t_0-code $\boldsymbol{c}$.

Example 4.24. (*t_0-***code**) *Let $t_0 = x^\infty$. There are infinitely many cuts in t_0, and, moreover, $\mathrm{cut}_i(t_0)$ uncovers $\mathrm{cut}_j(t_0)$ in x^∞ whenever $i > j$ holds. Hence, by Lemma 4.19, every code is an x^∞-code.*

On the other hand, let $t_0 = (x{\cdot}x){\cdot}(x{\cdot}x)$. Then t_0 has four cuts, $\mathrm{cut}_3(t_0)$ uncovers $\mathrm{cut}_1(t_0)$ and $\mathrm{cut}_2(t_0)$ in t_0, and these are the only uncovering cases. Assume that $\boldsymbol{c}$ is a code. The requirements for $\boldsymbol{c}$ to be a t_0-code are that $\boldsymbol{c}$ contains no integer i with $i > 4$, and

- 2 does not occur at $\alpha 0$: indeed, we have $\mathrm{hd}_{t_0}(2) = 1$, and the t_0-descending condition at α would require $1 \succ \mathrm{sub}(\boldsymbol{c}, \alpha 1)$, an impossible

condition; so, for instance, 2.1 is a code, but not a t_0-code—but 3.2 is a t_0-code;

- 4 does not occur in $\mathbf{c}$ *unless for* $\mathbf{c} = 4$*: indeed, we have* $\mathrm{hd}_{t_0}(4) = 1$*, so, as above, there is no possible value at the right of 4, and, on the other hand, no possible value at the left either, as 4 is maximal.*

So, there are only two t_0-codes of degree 1, namely 3.1 and 3.2.

For a code to be a t_0-code is a local condition:

Proposition 4.25. (t_0**-code**) *Assume that t_0 is an extended term, and* $\mathbf{c}$ *is a code of degree at least 1. Then the following are equivalent:*

(i) The code $\mathbf{c}$ *is a t_0-code;*

(ii) Every degree 1 subterm of $\mathbf{c}$ *is a t_0-code;*

(iii) If i occurs at $\alpha 0$ and j occurs below $\alpha 1$ in $\mathbf{c}$*, then* $\mathrm{cut}_i(t_0)$ *uncovers* $\mathrm{cut}_j(t_0)$ *in t_0.*

Proof. By definition, (i) implies (ii). Assume (ii). Assume that i occurs at $\alpha 0$ and j occurs below $\alpha 1$ in $\mathbf{c}$. As $\mathbf{c}$ is descending at α, $\mathrm{sub}(\mathbf{c}, \alpha)$ must have degree 1, say $\mathrm{sub}(\mathbf{c}, \alpha) = i \cdot j_1 \cdot \cdots \cdot j_{p+1}$. The t_0-descending condition at α requires that $\mathrm{cut}_i(t_0)$ uncovers $\mathrm{cut}_{j_1}(t_0)$ in t_0. Applying the same argument at $\alpha 1$, we see that $\mathrm{cut}_{j_1}(t_0)$ uncovers $\mathrm{cut}_{j_2}(t_0)$ in t_0, and, by transitivity, $\mathrm{cut}_i(t_0)$ uncovers $\mathrm{cut}_{j_2}(t_0)$ in t_0, and, inductively, it uncovers $\mathrm{cut}_{j_\ell}(t_0)$ for every ℓ, and (iii) holds.

Conversely, (iii) implies (ii) by definition, and (ii) implies (i) as the only difference between the descending condition defining codes and the t_0-descending condition defining t_0-codes involves the head and the t_0-head, and the latter coincide for codes of positive degree. ■

We are ready to establish the correspondence between t_0-∂-normal terms and t_0-normal terms.

Definition 4.26. (t_0**-coding**) Assume that t_0 is an extended term, and t is LD-equivalent to some cut t' of $\partial^k t_0$. For $k = 0$, we have $t' = \mathrm{cut}_i(t_0)$ for some integer i; then the *t_0-code of t* is defined to be i. For $k \geq 1$, the term t' admits a unique decomposition $t = \partial t_1 \cdot \cdots \cdot \partial t_p$ where $t_1, \ldots, t_p$ are cuts of $\partial^{k-1} s$; then the *t_0-code* of t is defined to be $\mathbf{c}_1 \cdot \cdots \cdot \mathbf{c}_p$, where $\mathbf{c}_1, \ldots, \mathbf{c}_p$ are the t_0-codes of $t_1, \ldots, t_p$ respectively.

Example 4.27. *(t_0-coding) Let $t_0 = x^\infty$. The term $t =$* 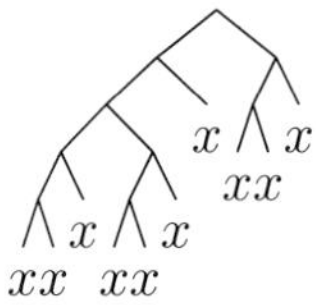

is a cut of $\partial^2 t_0$. Its decomposition in terms of cuts of ∂t_0 is $t = \partial t_1 \cdot \partial t_2$, with $t_1 =$  *and $t_2 =$* *. Now the decomposition of t_1 in terms of cuts of t_0 is $t_1 = \partial(x \cdot x \cdot x) \cdot \partial(x)$, and the one of t_2 is $t_2 = \partial(x \cdot x) \cdot \partial(x)$. We have $\mathrm{cut}_1(t_0) = x$, $\mathrm{cut}_2(t_0) = x \cdot x$, and $\mathrm{cut}_3(t_0) = x \cdot x \cdot x$. Hence the t_0-code of t_1 is 3.1, the t_0-code of t_2 is 2.1, and, finally, the t_0-code of t is 3.1021.*

The main technical result is as follows:

Lemma 4.28. *Assume that t_0 is an extended term. Then, for $k \geq 0$, the t_0-code mapping is an increasing bijection of the cuts of $\partial^k t_0$ ordered by $\sqsubset_{LD}$ onto the t_0-codes of degree at most k ordered by $\prec$. This bijection maps the t_0-∂-normal term of degree k onto the t_0-codes of degree k. If t is a cut of $\partial^k t_0$, and $\boldsymbol{c}$ is the t_0-code of t, then t is an LD-expansion of $\mathrm{cut}_{\boldsymbol{c}}(t_0)$, and the t_0-code of the least cut t' of $\partial^k t_0$ covered by t in $\partial^k t_0$ is $\mathrm{hd}_{t_0}(\boldsymbol{c})$.*

Proof. We use induction on k. For $k = 0$, everything is true as, by definition, the t_0-code mapping establishes an increasing bijection between the cuts of t_0 and the integers $1, \ldots, n$. By Proposition 2.8 (cut order), $\mathrm{cut}_i(t_0) \sqsubset_{LD} \mathrm{cut}_j(t_0)$ is equivalent to $i < j$, and, by definition, $\mathrm{hd}_{t_0}(i)$ is the least cut of t_0 covered by $\mathrm{cut}_i(t_0)$

Assume now $k > 0$. Let t be a cut of $\partial^k t_0$. By Proposition 4.4 (decomposition), there exists a (unique) descent $\vec{\alpha}$ in $\partial^{k-1} t_0$, say $\vec{\alpha} = (\alpha_1, \ldots, \alpha_{p+1})$, such that, letting $t_i = \mathrm{cut}(\partial^{k-1} t_0, \alpha_i)$, we have $t = \partial t_1 \cdot \cdots \cdot \partial t_{p+1}$. Hence t is an LD-expansion of $t_1 \cdot \cdots \cdot t_{p+1}$. By induction hypothesis, for each i, the t_0-code $\boldsymbol{c}_i$ of t_i is a t_0-code of degree at most $k-1$, and t_i is an LD-expansion of $\mathrm{cut}_{\boldsymbol{c}_i}(t_0)$. Let $\boldsymbol{c}$ be the t_0-code of t. By construction, we have $\boldsymbol{c} = \boldsymbol{c}_1 \cdot \cdots \cdot \boldsymbol{c}_{p+1}$, and, therefore, t is an LD-expansion of $\mathrm{cut}_{\boldsymbol{c}}(t_0)$. By induction hypothesis, $\boldsymbol{c}_1, \ldots, \boldsymbol{c}_{p+1}$ are t_0-codes of degree at most $k-1$, hence we have $\deg(\boldsymbol{c}) \leq k$. Owing to Proposition 4.20 (induction), in order to show that $\boldsymbol{c}$ is a t_0-code, it remains to show that the degree of $\boldsymbol{c}_i$ is exactly $k-1$ for $i \leq p$, and that, for $k = 1$, $\boldsymbol{c}$ is a t_0-code, and, for $k \geq 2$, $\mathrm{hd}_{t_0}(\boldsymbol{c}_i) \succ \boldsymbol{c}_{i+1}$ holds.

For $k = 1$, we have $\deg(\boldsymbol{c}_i) \leq 0$ by construction, hence $\deg(\boldsymbol{c}_i) = 0$, and the hypothesis that $\vec{\alpha}$ is a descent in t_0 implies that $\boldsymbol{c}_1 \cdot \cdots \cdot \boldsymbol{c}_{p+1}$ is a t_0-code. So we are done in this case.

Assume now $k \geq 2$. We claim that, for $i \leq p$, the address α_i cannot be of the form 0^*1^*. Indeed, such an address covers all lower addresses in $\mathrm{Out}(\partial^k t_0)$, and, therefore, it cannot appear in a descent, except at the last position. By Lemma 4.6, this means that the cut of $\partial^{k-1}t_0$ at α_i is a new cut, and, by induction hypothesis, this proves that the degree of $\boldsymbol{c}_i$ is exactly $k-1$. Hence, if $p > 0$ holds, we have $\deg(\boldsymbol{c}) = k$. Moreover, the hypothesis that $\vec{\alpha}$ is a descent implies

$$\mu_{\partial^{k-1}t_0}(\alpha_i) >_{LR} \alpha_{i+1}. \tag{4.6}$$

By induction hypothesis, the t_0-code of $\mathrm{cut}(\partial^k t_0, \mu_{\partial^{k-1}t_0}(\alpha_i))$ is $\mathrm{hd}_{t_0}(\boldsymbol{c}_i)$, while the t_0-code of $\mathrm{cut}(\partial^k t_0, \alpha_{i+1})$ is $\boldsymbol{c}_{i+1}$. Applying the induction hypothesis, we deduce that (4.6) implies $\mathrm{hd}_{t_0}(\boldsymbol{c}_i) \succ \boldsymbol{c}_{i+1}$. Hence $\boldsymbol{c}$ is a t_0-code of degree at most k.

By construction, t_0-coding preserves orders: $\prec$ on codes of degree k is the lexicographical extension of $\prec$ on codes of degree $k-1$ at most, while $\sqsubset_{LD}$ on cuts of $\partial^k t_0$ corresponds to the ordering on descents of $\partial^{k-1}t_0$, which is the lexicographical extension of the left-right ordering on the outline of $\partial^{k-1}t_0$, itself isomorphic to the $\sqsubset_{LD}$-ordering on the cuts of $\partial^{k-1}t_0$, hence, by induction hypothesis, to $\prec$ on codes of degree $k-1$ at most.

It follows that t_0-coding is injective. Assuming in the above argument that t is a new cut of $\partial^k t_0$, *i.e.*, one that is not an LD-expansion of a cut of $\partial^{k-1}t_0$, amounts to assuming $p \geq 1$, and we have seen that, in this case, $\boldsymbol{c}$ has degree k exactly.

Let us now consider the surjectivity of t_0-coding. Let $\boldsymbol{c}$ be a t_0-code of degree k. By definition, we have $\boldsymbol{c} = \boldsymbol{c}_1 \cdot \cdots \cdot \boldsymbol{c}_{p+1}$, where $\boldsymbol{c}_1, \ldots, \boldsymbol{c}_p$ are t_0-codes of degree $k-1$ and $\boldsymbol{c}_{p+1}$ is a t_0-code of degree $k-1$ at most. By induction hypothesis, there exists for every i a cut t_i of $\partial^{k-1}t_0$ such that $\boldsymbol{c}_i$ is the t_0-code of t_i. Let $t = \partial t_1 \cdot \cdots \cdot \partial t_{p+1}$. Then t is a cut of $\partial^k t_0$, and the t_0-code of t is $\boldsymbol{c}$. Indeed, if α_i denotes the address t_i comes from in $\partial^{k-1}t_0$, $(\alpha_1, \ldots, \alpha_p)$ being a descent in $\partial^{k-1}t_0$ follows from the induction hypothesis and the assumption that $\boldsymbol{c}$ is a code. Thus t_0-coding induces an increasing bijection of the cuts of $\partial^k t_0$ onto the t_0-codes of degree k at most, and, by the same argument, a bijection of the new cuts of $\partial^k t_0$ onto the t_0-codes of degree k exactly.

Assume finally that t is a cut of $\partial^k t_0$, and let $\boldsymbol{c}$ be its t_0-code. It remains to compute the t_0-code $\boldsymbol{c}'$ of the least cut t' of $\partial^k t_0$ covered by t in $\partial^k t_0$. Now, assuming again $t = \partial t_1 \cdot \cdots \cdot \partial t_{p+1}$, where t_i uncovers t_{i+1} in $\partial^{k-1}t_0$ for every i, we have $t' = \partial t_1 \cdot \cdots \cdot \partial t_p \cdot x$ by Lemma 4.7, where x is the left variable of t. This gives $\boldsymbol{c}' = \boldsymbol{c}_1 \cdot \cdots \cdot \boldsymbol{c}_p \cdot 1$, where $\boldsymbol{c}_i$ is the t_0-code of t_i. Thus $\boldsymbol{c}'$ is the head of $\boldsymbol{c}$, and the proof is complete. ■

Proving Proposition 4.1 (normal form) is now straightforward:

Proof. For (i), assume that t_0 is a fixed extended term, and t is a term satisfying $t \sqsubseteq_{LD} t_0$. By Proposition 3.11 (t_0-∂-normal form), there exists a unique

t_0-∂-normal term, *i.e.*, a unique new cut of some term $\partial^k t_0$, say t', that is LD-equivalent to t. By the previous lemma, there exists a unique t_0-code $\boldsymbol{c}$, namely the t_0-code of t and t', such that $\mathrm{cut}_{\boldsymbol{c}}(t_0)$ is LD-equivalent to t', hence to t. In the case of T_1, we notice that every code is an x^∞-code.

For (ii), we observe that computing $\boldsymbol{c}$ from t' requires no more space resource than computing the associated term $\partial^k t_0$, so the complexity upper bound is the same as for the ∂-normal form.

Point (iii) follows from Lemma 4.28 directly. ∎

Although all steps are effective, computing the normal form is a difficult task in practice. We shall return to the question in Chapter IX and describe a simple method that works in good cases. Let us observe here that one of the difficulties lies in the fact that a fractional cut of t_0 need not be of minimal size in its LD-equivalence class (see Exercise 4.32).

Exercise 4.29. (addresses) *Assume that* $\boldsymbol{c}$ *is a code. Prove that* 1 *may occur in* $\boldsymbol{c}$ *only at addresses that do not finish with* 0.

Exercise 4.30. (successor) *Show that, if* $\boldsymbol{c}$ *is a code, then* $\boldsymbol{c}{\cdot}1$ *is also a code, and it is the immediate successor of* $\boldsymbol{c}$.

Exercise 4.31. (minimal) *(i) As precodes are terms, we can speak of LD-equivalent precodes. Show that, if* $\boldsymbol{c}$ *and* $\boldsymbol{c}'$ *are LD-equivalent precodes, then* $\mathrm{cut}_{\boldsymbol{c}}(t_0)$ *and* $\mathrm{cut}_{\boldsymbol{c}'}(t_0)$ *are LD-equivalent for every* t_0. *Deduce that each LD-class of precodes contains at most one code.*

(ii) Show that every degree 1 *code is of minimal size in its LD-class, but that this is not true for degree* 2 *or more. [Hint: The code* 3.210031, *of size* 5, *is LD-equivalent to the precode* 3.2101, *of size* 4.*]*

Exercise 4.32. (LD-expansion) *Assume that* t_0' *is an LD-expansion of* t_0. *Prove that, for every term* t, *the* t_0'*-normal form of* t *exists if and only if the* t_0*-normal form of* t *exists, and the degree of the former is at most equal to the degree of the latter. [Hint: Observe that* $\partial^k t_0'$ *is an LD-expansion of* $\partial^k t_0$ *for every* k.*]*

---

5. Applications

Once unique normal forms are available, new results about free LD-systems can be established. Here we study left division in free LD-systems, and we deduce a geometrical characterization of special braids.

More about canonical orders

By construction, the iterated left divisibility relation $a \sqsubset ab$ holds for all a, b in the free LD-system FLD_∞. On the other hand, the relation $b \sqsubset ab$ need not hold in general: for instance, in FLD_2, we have $x_2 \gg x_1x_2$, and, therefore, $x_2 \sqsubset x_1x_2$ is false. However, we can prove a positive result when a and b are $\sqsubset$-comparable. Let us begin with an auxiliary result.

Lemma 5.1. *Assume that t_0 is an extended term, t_1 is a term, and there exists a (necessarily increasing) function f of $\mathbf{N}_+$ into itself such that*

(i) for every i, we have $t_1 \cdot \mathrm{cut}_i(t_0) =_{LD} \mathrm{cut}_{f(i)}(t_0)$, and

(ii) for all i, j, if $\mathrm{cut}_j(t_0)$ uncovers $\mathrm{cut}_i(t_0)$ in t_0, then $\mathrm{cut}_{f(j)}(t_0)$ uncovers $\mathrm{cut}_{f(i)}(t_0)$ in t_0.

Then, for each term t satisfying $t \sqsubseteq_{LD} t_0$, the t_0-code of $t_1 \cdot t$ exists, and it is obtained from the t_0-code of t by substituting i with $f(i)$ everywhere.

Proof. For $\boldsymbol{c}$ a precode, let $f(\boldsymbol{c})$ denote the precode obtained from $\boldsymbol{c}$ by substituting i with $f(i)$ everywhere. It follows from (i) and an induction that, for each term t, we have

$$t_1 \cdot \mathrm{cut}_{\boldsymbol{c}}(t_0) =_{LD} \mathrm{cut}_{f(\boldsymbol{c})}(t_0).$$

So the point is to show that $f(\boldsymbol{c})$ is a t_0-code whenever $\boldsymbol{c}$ is.

We first observe that $f(\boldsymbol{c})$ is a code whenever $\boldsymbol{c}$ is. This follows from the fact that f is increasing. Indeed, the skeleton of the term $f(\boldsymbol{c})$ coincides with the skeleton of $\boldsymbol{c}$. Assume that α is an internal address in t. The hypothesis $\deg(\mathrm{sub}(\boldsymbol{c}, \alpha 0)) + 1 \geq \deg(\mathrm{sub}(\boldsymbol{c}, \alpha 1))$ implies

$$\deg(\mathrm{sub}(f(\boldsymbol{c}), \alpha 0)) + 1 \geq \deg(\mathrm{sub}(f(\boldsymbol{c}), \alpha 1))$$

and, similarly, the hypothesis $\mathrm{hd}_{t_0}(\mathrm{sub}(\boldsymbol{c}, \alpha 0)) \succ \mathrm{sub}(\boldsymbol{c}, \alpha 1)$ implies

$$f(\mathrm{hd}_{t_0}(\mathrm{sub}(\boldsymbol{c}, \alpha 0))) \succ \mathrm{sub}(f(\boldsymbol{c}), \alpha 1). \tag{5.1}$$

The equality $f(\mathrm{hd}_{t_0}(\mathrm{sub}(\boldsymbol{c}, \alpha 0))) = \mathrm{hd}_{t_0}(\mathrm{sub}(f(\boldsymbol{c}), \alpha 0))$ need not be true in general, for the last integer of the latter precode is 1 while the last integer of the former is $f(1)$. However, for $\boldsymbol{c}'$ in the image of f, satisfying $f(\mathrm{hd}_{t_0}(\mathrm{sub}(\boldsymbol{c}, \alpha 0))) \succ \boldsymbol{c}'$ and $\mathrm{hd}_{t_0}(\mathrm{sub}(f(\boldsymbol{c}), \alpha 0)) \succ \boldsymbol{c}'$ are equivalent conditions, so (5.1) implies

$$\mathrm{hd}_{t_0}(\mathrm{sub}(f(\boldsymbol{c}), \alpha 0)) \succ \mathrm{sub}(f(\boldsymbol{c}), \alpha 1),$$

and we conclude that $f(\boldsymbol{c})$ is a code. By Proposition 4.25 (t_0-code), it suffices now to verify that every degree 1 subterm of $f(\boldsymbol{c})$ is t_0-normal: this is what Condition (ii) asserts. ■

Proposition 5.2. (**translation**) *Assume that t_0, t are finite terms, and $t \sqsubset_{LD} t_0^\infty$ holds. Then the t_0^∞-code of $t_0 \cdot t$ exists as well, and it is obtained from that of t by shifting every factor by the size of t_0.*

Proof. We apply the previous lemma, with t_0^∞ in place of t_0, and f being the mapping $i \mapsto i + n$, where n is the number of cuts in t_0, *i.e.*, its size. By construction, the cuts of t_0^∞ are $\mathrm{cut}_1(t_0)$, ..., $\mathrm{cut}_n(t_0)$, followed by $t_0 \cdot \mathrm{cut}_1(t_0)$, ..., $t_0 \cdot \mathrm{cut}_n(t_0)$, then by $t_0^{[2]} \cdot \mathrm{cut}_1(t_0)$, ..., $t_0^{[2]} \cdot \mathrm{cut}_n(t_0)$, *etc.* Condition (i) of Lemma 5.1 holds. As for Condition (ii), we observe that $\mathrm{cut}_{pn+i}(t_0^\infty)$ uncovers $\mathrm{cut}_{qn+j}(t_0^\infty)$ in t_0^∞ if either $p > q$ holds, or $p = q$ holds and $\mathrm{cut}_i(t_0)$ uncovers $\mathrm{cut}_j(t_0)$ in t_0. Hence Condition (ii) of Lemma 5.1 holds as well. Applying the latter lemma gives the result. ■

Example 5.3. (**translation**) *Let $t = ((x \cdot x \cdot x) \cdot x) \cdot (x \cdot x)$. We have $t = \mathrm{cut}_{3.102}(x^\infty)$, 3.102 is a code, and t is an x^∞-normal term. The size of x is* 1. *Applying Proposition* 5.2 *(translation) shows that the x^∞-code of $x \cdot t$ is* 4.203, *and the x^∞-code of $x \cdot x \cdot t$ is* 5.304.

Proposition 5.4. (**compatibility**) *Assume that a and b belong to* FLD_∞, *and $b \sqsubseteq a^\infty$ holds. Then $b \sqsubset ab$ holds.*

Proof. Let t_0 be a term representing a, and t be a term representing b. The hypothesis $b \sqsubseteq a^\infty$ implies that the t_0^∞-code of t exists. By Proposition 5.2 (translation), the t_0^∞-code $\boldsymbol{c}'$ of $t_0 \cdot t$ exists as well, and it is obtained from the t_0^∞-code $\boldsymbol{c}$ of t by shifting the indices of the variables by the size n of t_0. Then $\boldsymbol{c} < \boldsymbol{c}'$ holds, for, if $\boldsymbol{c}$ begins with i, $\boldsymbol{c}'$ begins with $i + n$. By Proposition 4.1(ii) (normal form), this implies $t \sqsubset_{LD} t_0 \cdot t$. ■

Corollary 5.5. *The relation $b \sqsubset ab$ holds for all a, b in* FLD_1.

Observe that the inequality $a \sqsubset ab$ is also clear using normal forms: with the previous notation, the t_0^∞-code of t_0 is n, while we have seen that the t_0^∞-code of $t_0 \cdot t$ begins with a variable at least equal to $n + 1$.

Left division

We consider now left division in free LD-systems, *i.e.*, we address the question of finding an element b that satisfies $ab = c$ when a and c are given. Because free LD-systems are left cancellative, the solution, when it exists, is unique. However, the solution need not exist in general. For instance, if c belongs to the (unique) basis of the considered free LD-system, the LD-invariance of the right height implies that c has no expression of the form ab.

We have already observed that, if S is an LD-system and $ab = c$ holds in S, then $a^{[2]}c = ac$ holds necessarily. This makes the following definition natural:

Definition 5.6. (**closed**) Assume that S is an LD-system. We say that S is *closed* if the previous necessary condition is sufficient in S, *i.e.*, if, for a, c in S, the equality $a^{[2]}c = ac$ implies the existence of b satisfying $ab = c$.

It was proved in Exercises I.2.28 and I.3.24 that the LD-systems $(B_\infty, {}^\wedge)$ and $(\mathfrak{I}_\infty, [\;])$ are closed. Also, by definition, every LD-quasigroup is closed. Here we investigate the case of free LD-systems.

Lemma 5.7. *Assume that t_0, t are terms, t_0 has size n, and $t \sqsubseteq_{LD} t_0^\infty$ holds. Then $t_0^{[2]}{\cdot}t \sqsubseteq_{LD} t_0^\infty$ holds as well, and the t_0^∞-code of $t_0^{[2]}{\cdot}t$ is obtained from that of t by applying the transformation f defined on degree 1 t_0^∞-codes by*

$$f(i_1 \cdot \dots \cdot i_p) = \begin{cases} (i_1+n)\cdot \dots \cdot(i_p+n) & \text{for } i_p > n,\\ (i_1+n)\cdot \dots \cdot(i_q+n)\cdot(2n)\cdot i_{q+1}\cdot \dots \cdot i_p & \\ & \text{for } i_p \le n \text{ with } i_q > n \ge i_{q+1}. \end{cases}$$

Proof. The formula above is more complicated than the formula of Proposition 5.2 (translation), because multiplying by $t_0^{[2]}$ does not map every cut of t_0^∞ into a cut of t_0^∞, so we cannot simply use a substitution on degree 0 codes, but we must consider degree 1 codes. However, the principle of the proof remains the same. First, for $\boldsymbol{c}$ a t_0^∞-code of degree 1, we have $\mathrm{cut}_{f(\boldsymbol{c})}(t_0^\infty) =_{LD} t_0^{[2]}{\cdot}\mathrm{cut}_{\boldsymbol{c}}(t_0^\infty)$. Indeed, for $i \ge n$, we have $\mathrm{cut}_i(t_0^\infty) =_{LD} t_0{\cdot}\mathrm{cut}_{i-n}(t_0^\infty)$, hence $t_0^{[2]}{\cdot}\mathrm{cut}_i(t_0^\infty) =_{LD} \mathrm{cut}_{i+n}(t_0^\infty)$, and, on the other hand, $t_0^{[2]} = \mathrm{cut}_{2n}(t_0^\infty)$. So it remains to prove that $f(\boldsymbol{c})$ is a t_0^∞-code whenever $\boldsymbol{c}$ is. We consider precodes with degree 1 at most. For degree 0, say $\boldsymbol{c} = i$, we have $f(\boldsymbol{c}) = i+n$ for $i > n$, and $f(\boldsymbol{c}) = (2n){\cdot}i$ for $i \le n$: the latter precode is a t_0^∞-code, as $\mathrm{cut}_{2n}(t_0^\infty)$, which is $t_0^{[2]}$, uncovers $\mathrm{cut}_i(t_0^\infty)$ in t_0^∞. The argument is similar for degree 1: the point is that, if every factor uncovers the next one in t_0^∞ in the sequence $(\mathrm{cut}_{i_1}(t_0^\infty), \dots, \mathrm{cut}_{i_p}(t_0^\infty))$ and $i_q > n \ge i_{q+1}$ holds, then the same is true for the sequence

$$(\mathrm{cut}_{i_1+n}(t_0^\infty), \dots, \mathrm{cut}_{i_q+n}(t_0^\infty), \mathrm{cut}_{2n}(t_0^\infty), \mathrm{cut}_{i_{q+1}}(t_0^\infty), \dots, \mathrm{cut}_{i_p}(t_0^\infty)).$$

The general case follows, as f preserves codes of positive degree and $\boldsymbol{c}_1 \prec \boldsymbol{c}_2$ implies $f(\boldsymbol{c}_1) \prec f(\boldsymbol{c}_2)$ for such codes. It is not true that $\mathrm{hd}_{t_0}(f(\boldsymbol{c})) = f(\mathrm{hd}_{t_0}(\boldsymbol{c}))$ must hold, but, by the previous remark and the fact that no precode $\boldsymbol{c}$ satisfies $\boldsymbol{c} \prec 1$, it is true that $\boldsymbol{c}_1 \prec \mathrm{hd}_{t_0}(\boldsymbol{c}_2)$ implies $f(\boldsymbol{c}_1) \prec \mathrm{hd}_{t_0}(f(\boldsymbol{c}_2))$. ■

Proposition 5.8. **(left division)** *Assume that S is a free LD-system, and $c \sqsubset a^{[p]}$ holds for some p. Then the following are equivalent:*
(i) There exists b satisfying $ab = c$;
(ii) The equality $a^{[2]}c = ac$ holds.
If the previous conditions hold, the element b is unique, and its t_0^∞-code can be determined by choosing an expression t_0 of a, computing the t_0^∞-code of c, and diminishing by the size of t_0 all integers in this code.

Proof. We prove that (ii) implies (i). Assume that t_0 represents a and has size n. By hypothesis, the t_0^∞-normal form of c exists. So we may assume that c is represented by $\mathrm{cut}_{\boldsymbol{c}}(t_0^\infty)$, where $\boldsymbol{c}$ is some t_0^∞-code. By Proposition 5.2 (translation), ac is represented by $\mathrm{cut}_{\boldsymbol{c}'}(t_0^\infty)$, where $\boldsymbol{c}'$ is obtained from $\boldsymbol{c}$ by shifting all integers by n. By the previous lemma, $a^{[2]}c$ is represented by $\mathrm{cut}_{\boldsymbol{c}''}(t_0^\infty)$, where $\boldsymbol{c}''$ is obtained from $\boldsymbol{c}$ by shifting by n the integers above n, keeping the integers below n and inserting some $2n$'s. Then $\boldsymbol{c}' = \boldsymbol{c}''$ is possible only if no integer i with $i \leq n$ occurs in $\boldsymbol{c}$. In this case, we obtain $c = ab$, where b is the element represented by $\mathrm{cut}_{\tilde{\boldsymbol{c}}}(t_0^\infty)$, and $\tilde{\boldsymbol{c}}$ is the code obtained from $\boldsymbol{c}$ by diminishing all integers by n. ■

In particular, we deduce:

Proposition 5.9. **(closed)** *Free LD-systems of rank 1 are closed.*

Characterization of special braids

As an application, we can now give a geometrical characterization of the special braids considered in Chapter III. We recall that a braid b is said to be special if it can be obtained from the unit braid using braid exponentiation exclusively.

Lemma 5.10. *Assume that $\vec{a}$ is a sequence of special braids and $\vec{a} \bullet b$ is defined. Then $\vec{a} \bullet b$ consists of special braids.*

Proof. It suffices to show that the new braids introduced at each elementary step of the action are special. The case of positive crossings is trivial: the new braid has the form a^b where a and b occur at the previous step, and, by definition, a^b is special whenever a and b are. Let us now consider a negative crossing. The new braid that appears is a quotient: we have to show that, if a

and c are special, and b is a braid satisfying $a^\wedge b = c$, then b is special. The hypothesis $a^\wedge b = c$ implies $a^\wedge c = a^{[2]\wedge} c$. By Proposition III.2.5 (order on special), special braids form a free LD-system of rank 1. By Proposition 5.9 (closed), the equality $a^\wedge c = a^{[2]\wedge} c$ in B_{sp} implies the existence of b' in B_{sp} satisfying $a^\wedge b' = c$. By left cancellation, we must have $b' = b$, *i.e.*, b is special. ∎

Proposition 5.11. (**special**) *A braid is special if and only if it is self-colouring.*

Proof. We have seen in Chapter III that every special braid is self-colouring. Conversely, assume that b is self-colouring, *i.e.*, the sequence $\vec{1} \bullet b$ exists, and it is equal to $(b, 1, 1, \ldots)$. By Lemma 5.10, all braids in the latter sequence are special, and so is b in particular. ∎

Proposition 5.12. (**parsing**) *There exists an algorithm that recognizes whether a given braid word represents a special braid, and, if so, gives an expression of this braid in terms of the unit braid and exponentiation.*

Proof. Let w be an arbitrary braid word. We decide whether w represents a special braid as follows. First, we reverse w into $N_R(w)D_R(w)^{-1}$. Then, we compute $(1, 1, 1, \ldots) \bullet N_R(w)D_R(w)^{-1}$, which, by Lemma III.1.4, exists whenever $(1, 1, 1, \ldots) \bullet \overline{w}$ does. Then w represents a special braid if and only if the previous computation is successful and it ends with a sequence of the form $(b, 1, 1, \ldots)$, *i.e.*, all components but the first are trivial. This can be tested effectively using one of the algorithms that solve the word problem of braids. Moreover, we have obtained above a left division algorithm in FLD_1. Hence, we can obtain an effective expression of the special braids involved in $(1, 1, 1, \ldots) \bullet N_R(w)D_R(w)^{-1}$ in terms of 1 and exponentiation. ∎

Example 5.13. (**parsing**) *Let $w = \sigma_2^{-1}\sigma_1^{-1}\sigma_2^2\sigma_1$. We first reverse w into $\sigma_1^2\sigma_2^{-1}$. Then we compute $(1, 1, 1) \bullet \sigma_1^2\sigma_2^{-1}$. We find $(1, 1, 1) \bullet \sigma_1^2 = ((1^\wedge 1)^\wedge 1, 1^\wedge 1, 1)$. Then, in order to apply σ_2^{-1}, we have to divide $1^\wedge 1$ by 1 on the left. In the present case, the result is obvious: division is possible and the quotient is 1. So we obtain that $(1, 1, 1) \bullet \sigma_1^2\sigma_2^{-1} = ((1^\wedge 1)^\wedge 1, 1, 1)$. We conclude that w represents the special braid $(1^\wedge 1)^\wedge 1$, i.e., $1_{[3]}$.*

Exercise 5.14. (**left division**) *Assume that S is a free LD-system of rank 1. For $a, b, c \in S$, prove that a is a left divisor of c if and only if ba is a left divisor of bc. [Hint: Use Proposition 5.9 (closed).] Prove similarly that a is a left divisor of c if and only if a is a left divisor of $(ab)c$.*

Exercise 5.15. (**cuts of product**) *(i) Assume $t = t_1{\cdot}t_2$. Prove $t_1{\cdot}s \neq_{LD} t_1^{[2]}{\cdot}s$ for every proper cut s of t_1. Deduce that $t_1{\cdot}s$ is LD-equivalent to a cut of t if and only if s is LD-equivalent to a cut of t_2.*

(ii) Prove that $t_1{\cdot}t$ is LD-equivalent to a cut of $t_1 * t_2$ if and only if t is LD-equivalent to a cut of t_2. Deduce that $t_1{\cdot}t$ is LD-equivalent to a cut of $\partial(t_1{\cdot}t_2)$ if and only if t is LD-equivalent to a cut of ∂t_2.

Exercise 5.16. (**index**) *(i) For a in FLD_1, define the index of a to be the last integer in the x^∞-code of an (arbitrary) expression of a. Compute the index of $x^{[n]}$ and $x_{[n]}$ for every n. Show that the index of $a^{[2]}$ need not be strictly larger than the index of a. [Hint: Prove that the index of $(x_{[5]})^{[m]}$ is 2 for $m = 2, 3$.]*

(ii) Prove that the index i of a can be determined as follows: start with an expression t of a, find an LD-expansion of $x^{[n]}$ containing a cut s that expands t: then $1^{i-1}0$ is the address of the origin in $x^{[n]}$ of the rightmost variable in s.

Exercise 5.17. (**special decomposition**) *Let b be a braid. Prove that $b = b_1\,\mathrm{sh}(b_2)\,\mathrm{sh}^2(b_3)\cdots$ holds for some special braids b_1, b_2, ..., i.e., b admits a special decomposition, if and only if the sequence $(1, 1, 1, \ldots) \bullet b$ is defined (using the action of braids on $(B_\infty, {}^\wedge)$), and, in this case, the sequence $(b_1, b_2, \ldots)$ is unique.*

6. Notes

Most of the results of this chapter appear in (Dehornoy 94*b*), in the case of one variable. This paper includes the study of the terms ∂t and of their geometry, in particular the correspondence between the outline of ∂t and the descents of t. The presentation in terms of fractional cuts is new. The characterization of special braids as self-colouring braids appears in (Dehornoy 99*a*).

The question of finding unique normal forms for the elements of the free LD-systems is natural, but what a good answer is remains vague: a good normal form is one that leads to applications. We know that special braids form a free LD-system of rank 1, and we can obtain a unique normal form by using the unique normal form of braids constructed in Chapter II. As computing the normal form of a braid has a polynomial complexity, computing the "braid normal form" of a term in T_1 is simply exponential with respect to the size of that term—thus much simpler than computing the normal forms of the current chapter. However, it is not clear that the braid normal form has much interest: for instance, proving the results of Section 5 by means of this normal form seems problematic. More generally, the connection between special braids and the current normal forms remains unclear. For instance, nothing is known about the following problem:

Question 6.1. *Define the index of a special braid as in Exercise* 5.16. *Does there exist a geometrical characterization of this parameter?*

Many combinatorial questions about the terms $\partial^k t$ remain open. In particular, the size of the term $\partial x^{[n]}$ is 2^n, as the tree associated with the latter term is a complete binary tree of height n. We have observed that, if t has size n, then the size of $\partial^k t$ is bounded above by $2^{2^{\cdot^{\cdot^{2^n}}}}$, a tower of height k. The exact value is unknown:

Question 6.2. *What is the size of the term* $\partial^k x^{[n]}$*?*

More difficult problems involve computing the normal form of a term in practice, which, as we mentioned above, remains essentially open. For an induction, it would be sufficient to know how to determine the t_0-code of the product of two t_0-normal terms, which amounts to solving the following problem:

Question 6.3. *Assume that* $\mathbf{c}_1$ *and* $\mathbf{c}_2$ *are* t_0*-codes. Which* t_0*-code is LD-equivalent to* $\mathbf{c}_1 \cdot \mathbf{c}_2$*?*

In the case of degrees 0 or 1, the answer can be given explicitly. The general case is difficult. In particular, it need not be true that the product of two codes of degree 2 is equivalent to a code with degree ≤ 3. For instance, it is proved in (Dehornoy 97*a*) that the product of the degree 2 codes 3.21003103 and 3.210031021 has degree 4.

Question 6.4. *Are all free LD-systems closed?*

Proposition 5.9 (closed) gives a positive answer for rank 1, and we conjecture a similar result in the general case.

Laver's normal forms

R. Laver has introduced in (Laver 92), (Laver 93) and (Laver 96) several normal forms which do not coincide with the ones constructed here.

Let us briefly describe the principle of Laver's construction. It takes place in the context of LD-monoids, *i.e.*, when a second, associative operation is considered besides the left self-distributive operation. We refer to Section I.4 and to Chapter XI for precise definitions. Here, it suffices to know that LD-monoids are associated with a family of identities (LDM) which contains (LD). We work inside the set $T_1^{(2)}$ of terms involving one variable x and two operators $\cdot$ and $\circ$. The aim is to construct a family of terms NF_L such that every term is LDM-equivalent to exactly one term in NF_L. Normal terms will have the form

$$t = (\cdots((t_0 \cdot t_1) \cdot t_2) \cdot \cdots) * t_p, \tag{6.1}$$

where t_0 is a term of the form $x^{[i]}$ and $*$ denotes either $\cdot$ or $\circ$. Call every term of this form *prenormal*. For t a prenormal term of the form (6.1), we define the *unfolding* $t^\#$ of t to be the finite sequence $(t_0, t_1, t_2, \ldots, t_p)$ if $*$ is $\cdot$, and to be the infinite sequence $(t_0, t_1, t_2, \ldots, t_p, \ldots)$ where $t_k = (\cdots((t_0 {\cdot} t_1){\cdot} t_2){\cdot}\cdots) * t_{k-2}$ holds for $k > p$ otherwise. Assuming that $\prec$ is some ordering on terms, we denote by $\prec^{Lex}$ the lexicographical extension of $\prec$ to sequences of terms. Then the set NF_L of all left normal terms and a linear ordering $\prec$ on this set are defined by a simultaneous induction.

Definition 6.5. (left normal) (i) The terms $x^{[i]}$ are in NF_L, and $x^{[i]} \prec x^{[j]}$ holds if and only if $i < j$ does;

(ii) The term t belongs to NF_L if it is prenormal with the form of (6.1), the terms $t_0, \ldots, t_p$ belong to NF_L, $t_1 \prec t_0$ holds, $t_k \preceq (\cdots((t_0{\cdot}t_1){\cdot}t_2){\cdot}\cdots) * t_{k-2}$ holds for $2 \le k \le p$, with $\prec$ instead of $\preceq$ for $* = \circ$ and $k = p$;

(iii) If $t \circ t'$ belongs to NF_L, so do all iterates of (t, t'), *i.e.*, all terms t_n in the Fibonacci sequence $t_0 = t$, $t_1 = t{\cdot}t'$, $t_{k+2} = t_{k+1}{\cdot}t_k$.

(iv) If t, t' belong to NF_L, then $t \prec t'$ holds if and only if $t^\# \prec^{Lex} t'^\#$ does.

It is true, but not trivial, that the previous definition makes sense. Calling this normal form left normal is natural, as its construction relies on an induction along the *left* branch of a binary tree—while Proposition 4.20 (induction) gives a similar induction along the *right* branch in the case of right normal terms. (Here our convention that default association is on the right is not convenient, and, actually, Laver chooses the opposite convention.)

Left normal terms do not coincide with the right normal terms of Section 4. For instance, the term $(x^{[3]}{\cdot}x^{[2]}{\cdot}x){\cdot}(x^{[3]}{\cdot}x^{[2]}){\cdot}x$ is an x^∞-normal term of degree 2—the associated code is 3.210032—, but its left normal form is the (LD-equivalent!) term $(x^{[3]}{\cdot}x^{[3]}){\cdot}(x^{[2]}{\cdot}x){\cdot}x$—associated with the precode 3.2102, which we have seen is not a code.

Proposition 6.6. (**Laver's normal form**) *Every term in $T_1^{(2)}$ is LDM-equivalent to exactly one term in NF_L.*

The proof of this result is difficult. It consists in proving directly that the normal forms of $t_1 \cdot t_2$ and $t_1 \circ t_2$ exist when t_1 and t_2 are normal. This is done by introducing the auxiliary notion of a t_1-normal term, and proving first that the t_1-normal forms of $t_1 \cdot t_2$ and $t_1 \circ t_2$ exist. Because of this indirect procedure, no explicit bound on the complexity of normalization is known: there exists an effective algorithm that determines the normal form of every term in $T_1^{(2)}$, but no primitive recursive bound on its complexity is known.

The left normal form is very powerful in terms of applications. For instance, the result that every nontrivial braid in B_n admits an expression by an n'-strand σ-positive or σ-negative braid word with $n' \leq n$ was first proved by R. Laver using the left normal form for LD-systems. Also, this normal form can be used to solve equations in free LD-systems (Laver, personal communication). No text is currently available.

Still other normal forms with respect to LD-equivalence can be considered. Recently, J. Moody has proposed to start from the following principle, which R. Laver announced he can prove using the left normal form in the case of one variable (personal communication):

Proposition 6.7. (**Moody's normal form**) *If $t_1 \cdot t_2 =_{LD} t_1' \cdot t_2'$ holds, then either there exists t satisfying $t \sqsubset_{LD} t_1$ and $t \sqsubset_{LD} t_1'$, or there exists t satisfying $t \sqsubset_{LD} t_2$ and $t \sqsubset_{LD} t_2'$, or we have $t_1' =_{LD} t_1$ and $t_2' =_{LD} t_2$.*

A natural task would be to establish a connection between the left and the right normal forms, which, in particular, could led to complexity bounds for the left normal form. Nothing is known so far in this direction.

Appendix. The First Codes

We give below a list of the first codes, in increasing order. For each code $\boldsymbol{c}$, the list comprises

- the decimal representation of $\boldsymbol{c}$;
- the associated tree;
- the degree of $\boldsymbol{c}$;
- the skeleton of the associated x^∞-normal term, *i.e.*, $\mathrm{cut}_{\boldsymbol{c}}(x^\infty)$;
- the evaluation of $\boldsymbol{c}$ in the free LD-quasigroup based on $\{\mathtt{a}, \mathtt{b}, \ldots\}$ when 1 is mapped to a, 2 to b, *etc.*; we write A, B, ... for $\mathtt{a}^{-1}$, $\mathtt{b}^{-1}$, ... Such data are given in connection with Conjecture X.6.2, which claims that this evaluation is injective.

code	tree	degree	x^∞-normal term	half-conjugate
1	1	0	·	1, a
2	2	0		1, b
2.1		1		b, a
2.101		2		baB, a
2.10101	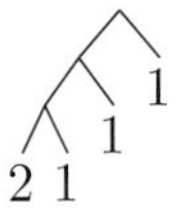	3		baBabAB, a
2.1010101	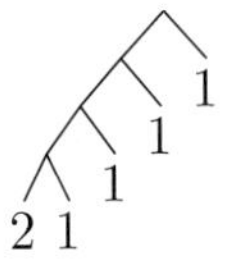	4	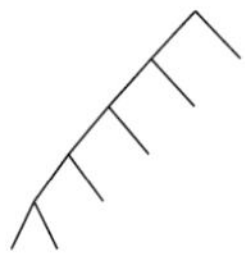	baBabABabaBAbAB, a
...				
2.10102	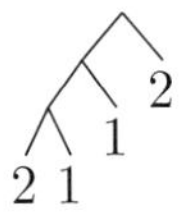	3		baBabAB, b
...				
2.101021		3	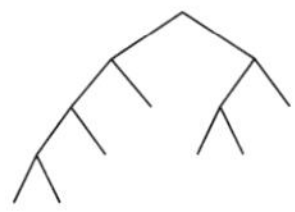	baBabA, a

...				
2.102	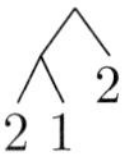	2	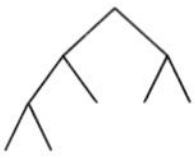	baB, b
2.10201	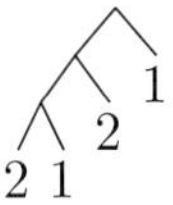	3		babAB, a
...				
2.10202	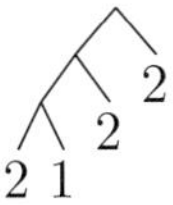	3	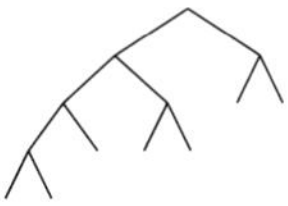	babAB, b
...				
2.102021	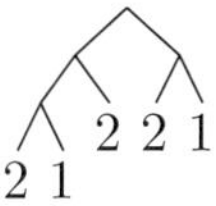	3	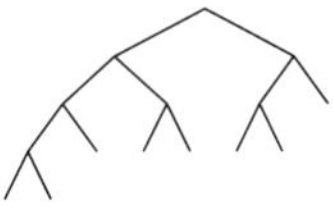	babA, a
...				
3	3	0		1, c
3.1		1		c, a
3.101	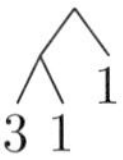	2		caC, a
3.10101	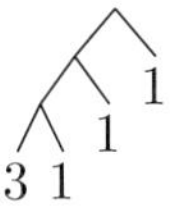	3	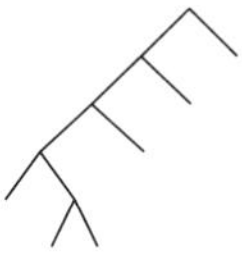	caCacAC, a
...				
3.10102	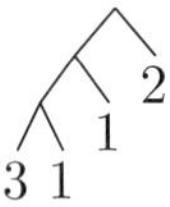	3		caCacAC, b

...				
3.101021	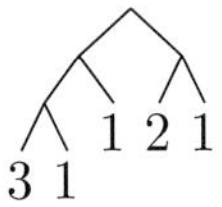	3	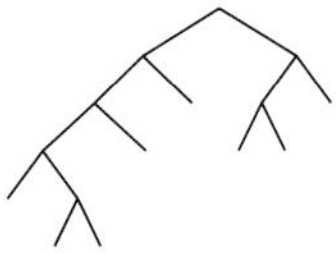	caCacACb, a
...				
3.102		2	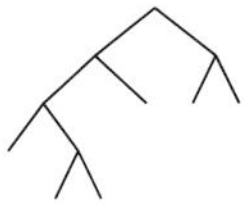	caC, b
3.10201	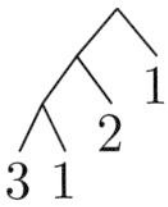	3	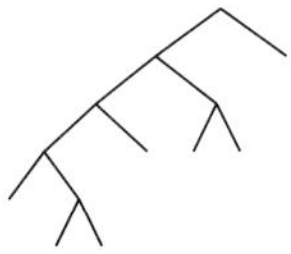	caCbcAC, a
...				
3.10202	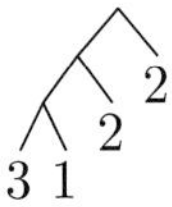	3	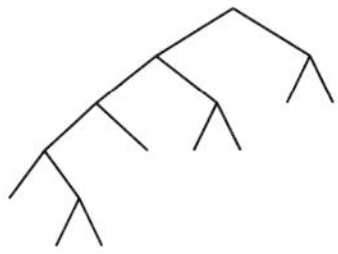	caCbcAC, b
...				
3.102021		3	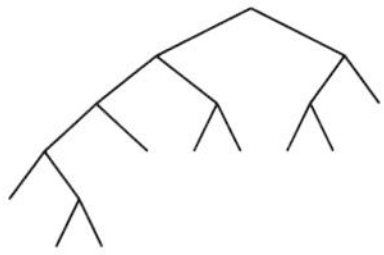	caCbcACbcaCBcAC, a
...				
3.103	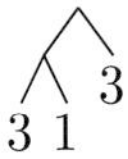	2	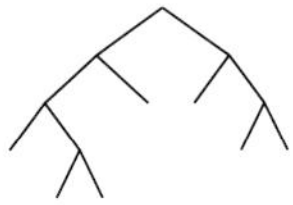	caC, c
3.10301	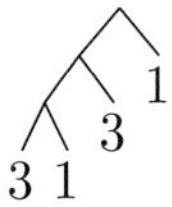	3	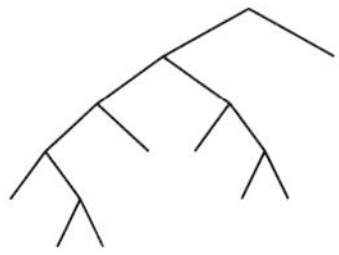	cacAC, a
...				
3.2		1		c, b

3.201	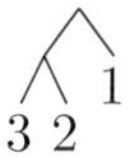	2		cbC, a
3.20101	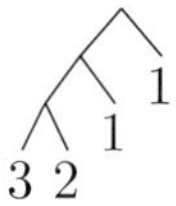	3	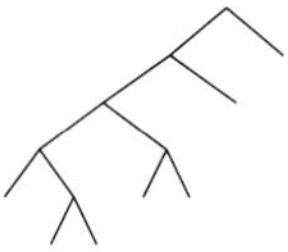	cbCacBC, a
...				
3.202	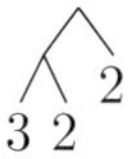	2	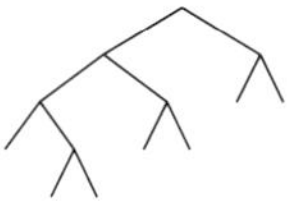	cbC, b
3.20201	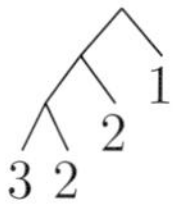	3	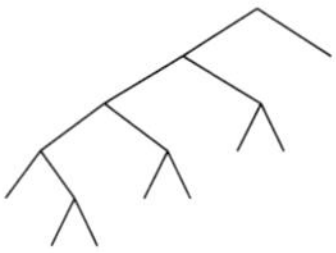	cbCbcBC, a
...				
3.2021		2	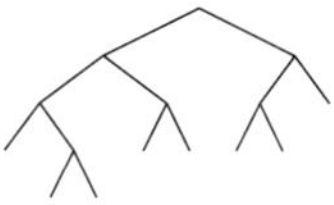	cbCb, a
...				
3.203	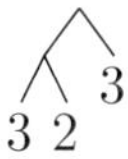	2	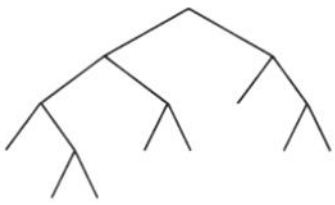	cbC, c
...				
3.21	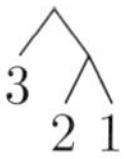	1		cb, a
3.21001	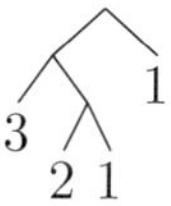	2	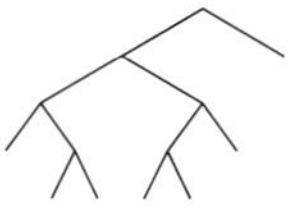	cbaBC, a
...				
4	4	0		1, d
...				

VII
The Geometry Monoid

Studying free LD-systems amounts to studying LD-equivalence of terms. Now $t =_{LD} t'$ holds if t can be transformed into t' by iteratively applying the left self-distributivity identity, which can be seen as applying some operator that specifies where and how (from left to right or from right to left) the identity is applied. Applying the identity several times amounts to composing the associated operators. So we obtain a monoid of (partial) operators acting on terms, so that the LD-equivalence class of a term is its orbit under the action. The aim of this chapter is to study the monoid $\mathcal{G}_{LD}$ involved in this action, which we call the geometry monoid of (LD) as it captures a number of geometrical relations involving left self-distributivity.

The chapter is organized as follows. In Section 1, we introduce the operators LD_w and the monoid $\mathcal{G}_{LD}$ they generate. We describe the domain and image of LD_w, and we interpret the product in $\mathcal{G}_{LD}$ as a term unification process. In Section 2, we use geometric arguments to build a list of relations, called LD-relations, that hold in the monoid $\mathcal{G}_{LD}$. We do not prove that LD-relations form a presentation of $\mathcal{G}_{LD}$, but we show in Section 3 and 4 how to use them to re-prove a number of previously known properties of left self-distributivity. In particular, we give a syntactic proof of the confluence property, *i.e.*, one that resorts to LD-relations only. To this end, we introduce for each term t a distinguished word $\boldsymbol{\Delta}_t$ which describes the passage from t to ∂t, and which is directly reminiscent of Garside's braid word $\boldsymbol{\Delta}_n$..

1. Left Self-Distributivity Operators

Transforming a term t using Identity (LD) amounts to applying to t some operator that specifies how the identity is used: there may exist several ways to apply (LD) to t, but things become functional when we fix an orientation for the identity, *i.e.*, when we choose definite left and right sides, and when we specify where the identity is applied in t. Here, we establish the basic properties of the involved operators. In particular, we show that the monoid they generate does not depend on the considered set of variables: it depends on left self-distributivity only, and it will be called the geometry monoid of (LD).

The operators LD_α

We recall that T_X denotes the set of all terms constructed using variables from X and one binary operation. As in Chapter VI, we shall view every term t as a finite binary tree. Then each subterm of t is specified by an address, *i.e.*, a finite sequence of 0's and 1's, that describes the path from the root of the tree to the considered node; we write $\mathrm{sub}(t, \alpha)$ for the α-subterm of t, *i.e.*, for the subterm whose root lies at address α in t, if it exists, *i.e.*, if α is short enough.

Definition 1.1. (**operator** $\mathrm{LD}_{\alpha,X}$) For X a nonempty set, and α an address, $\mathrm{LD}_{\alpha,X}$ is defined to be the partial function on T_X that maps every term t to its LD-expansion at α, if it exists.

Thus, the operator $\mathrm{LD}_{\alpha,X}$ applies to t if and only if the subterm $\mathrm{sub}(t, \alpha)$ exists and it can be decomposed as $t_1{\cdot}(t_2{\cdot}t_3)$, in which case, $\mathrm{LD}_{\alpha,X}$ maps t to the term t' obtained from t by replacing its α-subterm with $(t_1{\cdot}t_2){\cdot}(t_1{\cdot}t_3)$. As the possible decomposition of a term into $t_1{\cdot}(t_2{\cdot}t_3)$ is unique, the partial operator $\mathrm{LD}_{\alpha,X}$ is functional, and our definition makes sense. The construction of $\mathrm{LD}_{\alpha,X}(t)$ is local in the sense that it involves the variables occurring in t only. Hence, if X is included in X' and t belongs to T_X, the values of $\mathrm{LD}_{\alpha,X}(t)$ and $\mathrm{LD}_{\alpha,X'}(t)$ coincide. Taking this remark into account, we shall usually drop the subscript X in $\mathrm{LD}_{\alpha,X}$.

Example 1.2. (**operator** LD_α) *Let* $t = x_1{\cdot}((x_2{\cdot}(x_3{\cdot}x_4)){\cdot}x_5)$. *Then we have* $t = t_1{\cdot}(t_2{\cdot}t_3)$, *with* $t_1 = x_1$, $t_2 = x_2{\cdot}(x_3{\cdot}x_4)$ *and* $t_3 = x_5$. *Hence* t *lies in the domain of* LD_ϕ, *and its image is the term* $(t_1{\cdot}t_2){\cdot}(t_1{\cdot}t_3)$, *i.e.*, $(x_1{\cdot}(x_2{\cdot}(x_3{\cdot}x_4))){\cdot}(x_1{\cdot}x_5)$.

Similarly, we have $\mathrm{sub}(t, 10) = t_1{\cdot}(t_2{\cdot}t_3)$ *with* $t_1 = x_2$, $t_2 = x_3$, *and* $t_3 = x_4$. *Hence* t *lies in the domain of* LD_{10}, *and its image is the term* $x_1{\cdot}((t_1{\cdot}t_2){\cdot}(t_1{\cdot}t_3)){\cdot}x_5$, *i.e.*, $x_1{\cdot}(((x_2{\cdot}x_3){\cdot}(x_2{\cdot}x_4)){\cdot}x_5)$.

It is easier to read the transformations when terms are displayed in tree form. In the current case, we have

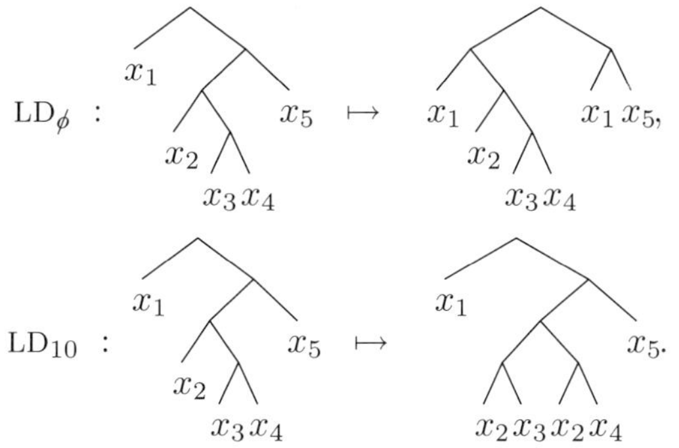

The following property follows from the definition readily:

Lemma 1.3. *For every address α, the operator $\mathrm{LD}_{\alpha,X}$ is a one-to-one partial function on T_X; its inverse is the operator $\mathrm{LD}_{\alpha,X}^{-1}$ defined as $\mathrm{LD}_{\alpha,X}$ exchanging the roles of $t_1{\cdot}(t_2{\cdot}t_3)$ and $(t_1{\cdot}t_2){\cdot}(t_1{\cdot}t_3)$.*

The geometry monoid

We consider now the (partial) operators obtained by composing operators $\mathrm{LD}_\alpha^{\pm 1}$.

Definition 1.4. (geometry monoid) For X a nonempty set, the *geometry monoid* of (LD) relative to X is defined to be the monoid $\mathcal{G}_{LD,X}$ generated by all operators $\mathrm{LD}_{\alpha,X}^{\pm 1}$ using composition; similarly, the *positive geometry monoid* $\mathcal{G}_{LD,X}^{+}$ is the monoid generated by all operators $\mathrm{LD}_{\alpha,X}$.

The monoid $\mathcal{G}_{LD,X}$ consists of one-to-one partial functions, so, by construction, it is an inverse semigroup: for every f in $\mathcal{G}_{LD,X}$, there exists a unique g in $\mathcal{G}_{LD,X}$, namely f^{-1}, satisfying $fgf = f$ and $gfg = g$. But $\mathcal{G}_{LD,X}$ is not a group, as $f^{-1}f$ is the identity mapping on the domain of f, and not on all terms.

A bijection between X and X' induces an isomorphism of the monoids $\mathcal{G}_{LD,X}$ and $\mathcal{G}_{LD,X'}$, hence the isomorphism class of $\mathcal{G}_{LD,X}$ may depend on the cardinality of X only. We shall see below that it actually does not depend on this cardinality, but the proof requires preliminary results, and we shall keep the subscript X in $\mathcal{G}_{LD,X}$ for the moment.

By definition, the elements of $\mathcal{G}_{LD,X}^{+}$ are finite products of operators LD_α, thus of the generic form

$$\mathrm{LD}_{\alpha_1} \circ \cdots \circ \mathrm{LD}_{\alpha_\ell}.$$

Such elements are indexed by finite sequences of addresses, *i.e.*, by words on the alphabet $\mathbf{A}$. The set of all such words is denoted by $\mathbf{A}^*$. The product (concatenation) of two such words u, v is denoted by uv, or $u \cdot v$: because addresses themselves are words, we often keep the dot to avoid ambiguity. For instance, $0 \cdot 10 \cdot \phi$ is a typical length 3 word on $\mathbf{A}$. We use ε for the empty word of $\mathbf{A}^*$, which should not be confused with the length 1 word ϕ consisting of the empty address.

In the sequel, we think of the transformations LD_α as acting on terms on the right. In such a framework, it is natural to replace composition, which is convenient for an action on the left, by the opposed operation, denoted $\bullet$ here.

Definition 1.5. (operator LD_u, I) For $u = \alpha_1 \cdot \cdots \cdot \alpha_\ell \in \mathbf{A}^*$, the operator LD_u is defined to be $\mathrm{LD}_{\alpha_1} \bullet \cdots \bullet \mathrm{LD}_{\alpha_\ell}$, *i.e.*, $\mathrm{LD}_{\alpha_\ell} \circ \cdots \circ \mathrm{LD}_{\alpha_1}$. The operator LD_ε is defined to be the identity mapping.

Let us consider now the monoid $\mathcal{G}_{LD,X}$. Its elements have the form

$$\mathrm{LD}_{\alpha_1}^{e_1} \bullet \cdots \bullet \mathrm{LD}_{\alpha_\ell}^{e_\ell}$$

with $\alpha_1, \ldots, \alpha_\ell \in \mathbf{A}$ and $e_1, \ldots, e_\ell = \pm 1$. To specify such products, it is natural to introduce a formal inverse α^{-1} for each address α, and to consider words on $\mathbf{A} \cup \mathbf{A}^{-1}$, where $\mathbf{A}^{-1}$ denotes the set of all α^{-1}.

Definition 1.6. (operator LD_w, II) For $w = \alpha_1^{e_1} \cdot \cdots \cdot \alpha_\ell^{e_\ell}$ in $(\mathbf{A} \cup \mathbf{A}^{-1})^*$, the operator LD_w is defined to be $\mathrm{LD}_{\alpha_1}^{e_1} \bullet \cdots \bullet \mathrm{LD}_{\alpha_\ell}^{e_\ell}$.

With such notations, the mapping $u \mapsto \mathrm{LD}_u \restriction T_X$ defines a surjective homomorphism of the free monoid $\mathbf{A}^*$ onto the monoid $(\mathcal{G}_{LD,X}^+, \bullet)$, and, similarly, $w \mapsto \mathrm{LD}_w \restriction T_X$ defines a surjective homomorphism of the free monoid $(\mathbf{A} \cup \mathbf{A}^{-1})^*$ onto $(\mathcal{G}_{LD,X}, \bullet)$. The elements of $\mathbf{A}^*$ will be called positive words, while the elements of $(\mathbf{A} \cup \mathbf{A}^{-1})^*$ are simply called words. As in Chapter II, we shall often use u and v for positive words, and w for arbitrary words.

Definition 1.7. (action) For t a term, and w a word on $\mathbf{A} \cup \mathbf{A}^{-1}$, the term $t \bullet w$ is defined to be the image of t under LD_w, if it exists.

By definition, the previous action is a monoid action on the right, as we have

$$t \bullet \varepsilon = t, \qquad t \bullet w_1 w_2 = (t \bullet w_1) \bullet w_2. \tag{1.1}$$

Example 1.8. (action) *Let* $w = 10 \cdot \phi \cdot 01^{-1}$. *By definition,* LD_w *is the product* $\mathrm{LD}_{10} \bullet \mathrm{LD}_\phi \bullet \mathrm{LD}_{01}^{-1}$, *i.e., the composition* $\mathrm{LD}_{01}^{-1} \circ \mathrm{LD}_\phi \circ \mathrm{LD}_{10}$. *The reader can check that the term* $x_1 \cdot (x_2 \cdot x_3 \cdot x_4) \cdot x_5$ *belongs to the domain of* LD_w, *and its image is the term* $(x_1 \cdot x_2 \cdot x_3 \cdot x_4) \cdot x_1 \cdot x_5$, *as shown by the*

diagram

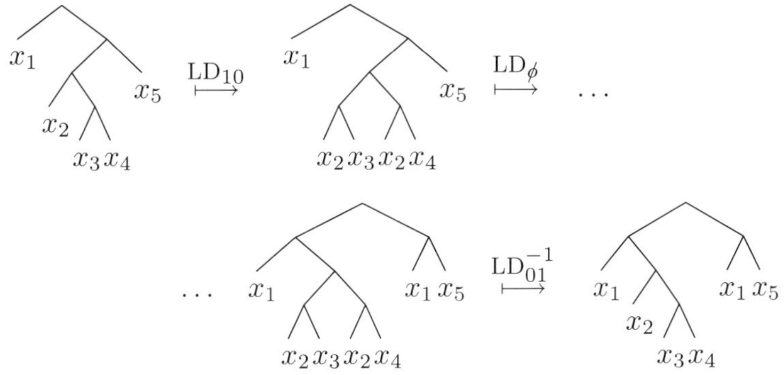

By construction, $t \bullet w =_{LD} t$ holds when $t \bullet w$ exists, and, for u a positive word, $t \bullet u$ is an LD-expansion of t when defined. These implications are equivalences.

Proposition 1.9. (**action**) *Assume that t, t' are terms in T_X.*

(i) The terms t and t' are LD-equivalent if and only if some operator in $\mathcal{G}_{LD,X}$ maps t to t', i.e., if $t' = t \bullet w$ holds for some word w on $\mathbf{A} \cup \mathbf{A}^{-1}$.

(ii) The term t' is an LD-expansion of the term t if and only if some operator in $\mathcal{G}^+_{LD,X}$ maps t to t', i.e., if $t' = t \bullet u$ holds for some word u on $\mathbf{A}$.

Proof. By definition, the basic LD-expansions of t are exactly those terms of the form $t \bullet \alpha$ with α an address. Now, by definition, an LD-expansion of t is a term obtained from t by repeatedly taking basic LD-expansions, which amounts to composing several operators LD_α, *i.e.*, to applying an element of $\mathcal{G}^+_{LD,X}$. Similarly, Lemma V.2.5 tells us that $t' =_{LD} t$ holds if and only if there exists a finite sequence of terms t_i from t to t' such that every term in the sequence is either a basic LD-expansion of the previous term, or a term of which the previous term is a basic LD-expansion. This means that t_i is the image of t_{i-1} under some operator $\mathrm{LD}_\alpha^{\pm 1}$. So t' is LD-equivalent to t if and only if some finite composition of operators $\mathrm{LD}_\alpha^{\pm 1}$, *i.e.*, some element of $\mathcal{G}_{LD,X}$, maps t to t'. ■

It follows that studying the free LD-system based on X, studying the equivalence relation $=_{LD}$ on T_X and studying the action of the monoids $\mathcal{G}_{LD,X}$ are essentially one and the same thing. The advantage of the latter approach is to make a more precise analysis possible, as it takes into account the specific geometrical properties of (LD) and make them explicit.

Domains and images

The operators $\mathrm{LD}_\alpha^{\pm 1}$ are partial functions. For instance, a variable is too small a term to lie in the domain of any operator LD_w with $w \neq \varepsilon$. In this section, we describe the domain and the image of LD_w.

We recall from Section V.1 that, if s is a term, every term of the form s^h with h a substitution with values in T_X is called an X-substitute of s, and that, if (s, s') is a pair of terms—also called an identity—, every pair of terms of the form (s^h, s'^h) with h a substitution with values in T_X is called an X-instance of (s, s'). With this terminology, we see that t' is the image of t under LD_ϕ if and only if the pair (t, t') is an X-instance of the identity $(x{\cdot}(y{\cdot}z), (x{\cdot}y){\cdot}(x{\cdot}z))$. This implies that the domain of $\mathrm{LD}_{\phi,X}$ is the set of all X-substitutes of the term $x{\cdot}(y{\cdot}z)$, while its image is the set of all X-substitutes of the term $(x{\cdot}y){\cdot}(x{\cdot}z)$. The situation is similar for every operator LD_w, as we shall prove:

Proposition 1.10. (domain) *(i) For every word w on $\mathbf{A} \cup \mathbf{A}^{-1}$, either the operator LD_w is empty, or there exists a unique pair of LD-equivalent canonical terms (t_w^L, t_w^R) such that the operator LD_w maps t to t' if and only if the pair (t, t') is an instance of (t_w^L, t_w^R).*

(ii) For every word u on $\mathbf{A}$, the operator LD_u is nonempty, and the term t_u^L is injective; in this case, a term t lies in the domain of the operator LD_u if and only if its skeleton includes the skeleton of t_u^L.

(We recall that t is canonical if the variables of t make an initial segment of $x_1 x_2 \cdots$ when enumerated from left to right with repetitions omitted.) Let us begin with basic LD-expansions.

Definition 1.11. (injective) The term t is said to be *injective* if no variable occurs twice in t. The substitution h is said to be *injective* if, for every variable z, the term $h(z)$ is injective, and, in addition, the variables occurring in $h(z)$ are disjoint from those occurring in $h(y)$ for $y \neq z$.

Lemma 1.12. *For every address α, there exists a unique pair of LD-equivalent canonical terms (t_α^L, t_α^R) such that the operator LD_α maps t to t' if and only if the pair (t, t') is an instance of (t_α^L, t_α^R). Moreover, t_α^L is injective.*

Proof. We use induction on the length of α. For $\alpha = \phi$, the choice $t_\phi^L = x_1{\cdot}(x_2{\cdot}x_3)$, $t_\phi^R = (x_1{\cdot}x_2){\cdot}(x_1{\cdot}x_3)$ is convenient, and it is the only possible choice for t_α^L and t_α^R to be canonical. Assume now $\alpha = 0\beta$. Then LD_α applies to t if and only if t is not a variable and LD_β applies to the left subterm of t. Assume $t = t_1{\cdot}t_2$. The image of t under LD_α is $t_1'{\cdot}t_2$, where t_1' is the image of t_1 under LD_β. Assuming that LD_β, viewed as a set of pairs of terms, is the set of all instances of some pair (s, s'), we deduce that LD_α is the set of all instances of the pair $(s{\cdot}z, s'{\cdot}z)$, where z is an arbitrary variable which does not occur in s or s'. With this choice, the term $s{\cdot}z$ is injective provided s is, and, if (s, s') is canonical, $(s{\cdot}z, s'{\cdot}z)$ is canonical provided we take for z the first variable x_i that does not occur in s. The argument is similar for $\alpha = 1\beta$. ■

Example 1.13. **(domain)** *The previous argument is effective, and we can write the terms* t^L_α *and* t^R_α *explicitly. We have found* $t^L_\phi = x_1{\cdot}(x_2{\cdot}x_3)$, $t^R_\phi = (x_1{\cdot}x_2){\cdot}(x_1{\cdot}x_3)$, *i.e.,*

$(t^L_\phi, t^R_\phi) =$ 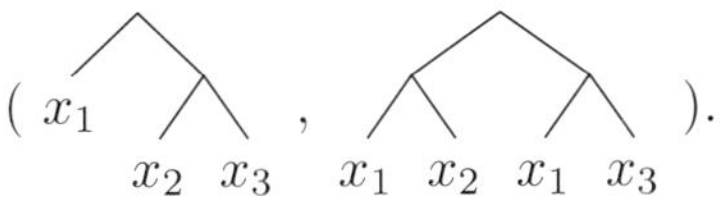

We obtain inductively

$(t^L_1, t^R_1) =$ 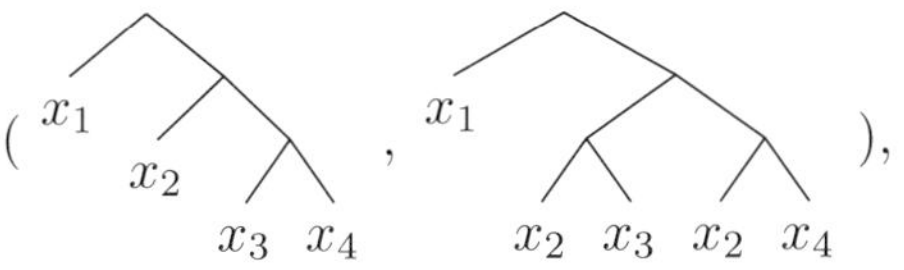

$(t^L_{10}, t^R_{10}) =$ 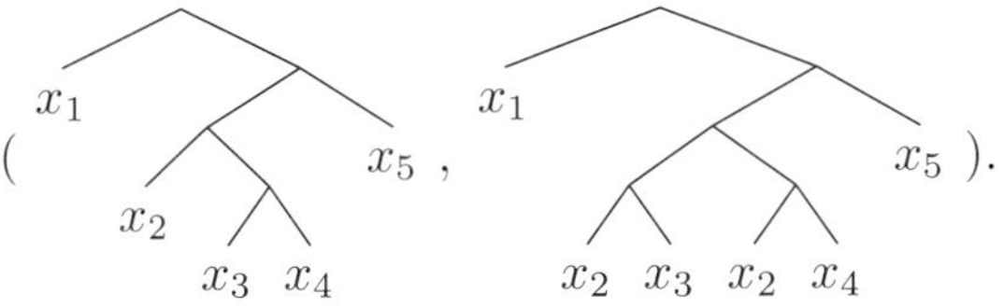

The previous result implies that a term belongs to the domain of LD_α if and only if its skeleton is large enough. More precisely, we have:

Lemma 1.14. *(i) The term* t *lies in the domain of* LD_α *if and only if the address* $\alpha 10$ *belongs to the skeleton of* t.

(ii) The term t *lies in the domain of* LD_α^{-1}, *i.e., in the image of* LD_α, *if and only if the addresses* $\alpha 00$ *and* $\alpha 10$ *belong to the skeleton of* t, *and, in addition,* $\mathrm{sub}(t, \alpha 00) = \mathrm{sub}(t, \alpha 10)$ *holds.*

We leave the easy inductive proof to the reader, and consider composition now.

Definition 1.15. **(unifier, specialization)** Assume that s, t are terms, and f, g are substitutions. We say that (f, g) is a *unifier* for s and t if $s^f = t^g$ holds. If both (f, g) and (f', g') are unifiers for s and t, we say that (f', g') is a *specialization* of (f, g) if there exists h such that $f'(x) = h(f(x))$ holds for every variable x occurring in s, and $g'(y) = h(g(y))$ holds for every variable y occurring in t.

The point is that there exists an effective method that recognizes whether two given terms can be unified and, if so, determines a distinguished unifier.

Proposition 1.16. (**unification**) *Assume that the terms s and t are unifiable. Then they have a most general unifier, i.e., one of which every unifier of s and t is a specialization.*

Proof. Let us first address the special question of finding one single substitution f satisfying $s^f = t^f$, *i.e.*, let us consider unifiers of the form (f, f). For $s \neq t$, we define the *disagreement set* $D(s,t)$ inductively by

$$D(s,t) = \begin{cases} \{\phi\} & \text{if } s \text{ or } t \text{ is a variable,} \\ \mathrm{sh}_0(D(s_1,t_1)) \cup \mathrm{sh}_1(D(s_2,t_2)) & \text{for } s = s_1{\cdot}s_2,\ t = t_1{\cdot}t_2, \end{cases}$$

and we call the leftmost address in $D(s,t)$ the discrepancy between s and t. Assume that s and t admit a discrepancy at α. By construction, at least one of the terms $\mathrm{sub}(s,\alpha)$, $\mathrm{sub}(t,\alpha)$ is a variable. Assume for instance that $\mathrm{sub}(t,\alpha)$ is the variable z. Two cases may occur. Either z occurs in $\mathrm{sub}(t,\alpha)$. Then there cannot exist f satisfying $s^f = t^f$, as this would imply $z^f = (\mathrm{sub}(t,\alpha))^f$, whereas a term never equals one of its proper subterms. Or z does not occur in $\mathrm{sub}(t,\alpha)$. Then we define f_1 by $f_1(z) = \mathrm{sub}(t,\alpha)$, and $f_1(x) = x$ for $x \neq z$. By construction, the terms s^{f_1} and t^{f_1} have no discrepancy at α. Moreover, if $s^f = t^f$ holds, we have $\mathrm{sub}(s^f,\alpha) = \mathrm{sub}(t^f,\alpha)$, *i.e.*, $f(z) = (\mathrm{sub}(t,\alpha))^f$. Let us define h by $h(z) = z$, and $h(x) = f(x)$ for $x \neq z$. Then, for $x \neq z$, we have $h(f_1(x)) = h(x) = f(x)$, and

$$h(f_1(z)) = h(\mathrm{sub}(t,\alpha)) = f(\mathrm{sub}(t,\alpha)) = f(z),$$

since, by hypothesis, z does not occur in $\mathrm{sub}(t,\alpha)$. This shows that every possible substitution f satisfying $s^f = t^f$ factors through f_1. We then iterate the process, and consider the discrepancy between s^{f_1} and t^{f_1}. The iteration converges, as each step diminishes by one the total number of variables occurring in s and t. So, if there exists f satisfying $s^f = t^f$, then the above algorithm provides such a substitution f_0, and every substitution f satisfying $s^f = t^f$ factors through f_0.

Let us now return to the general question of unifying s and t. Assume $s^f = t^g$. Fix a permutation of the variables π such that s and t^π have no variable in common. By construction, we have $s^f = (t^\pi)^{\pi^{-1}g}$, hence $s^{f'} = (t^\pi)^{f'}$ where f' coincides with f on $\mathrm{Var}(s)$ and with $g \circ \pi^{-1}$ on $\mathrm{Var}(t^\pi)$. Then the above algorithm running on s and t^π succeeds, it yields a substitution f_0, and f' factors through f_0, *i.e.*, $f' = h \circ f_0$ holds for some h. Define $g_0 = f_0 \circ \pi$. By construction we have $s^{f_0} = t^{g_0}$, $f(x) = h(f_0(x))$ for x occurring in s, and $g(y) = h(g_0(y))$ for y occurring in t. ∎

Example 1.17. (**unification**) Let $s =$ 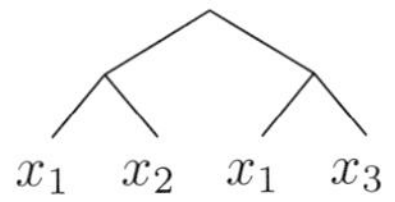

and $t =$ 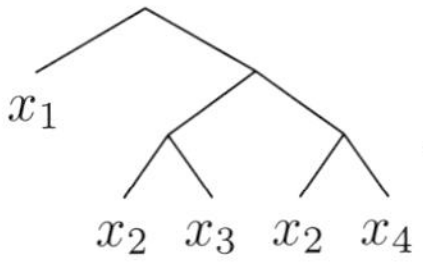

. Unifying s and t amounts to finding a substitution f satisfying $s^f = (t')^f$, with $t' =$ 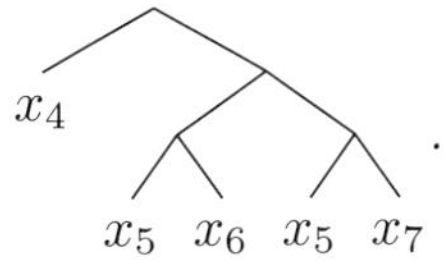

. We have $D(s,t') = \{0, 10, 11\}$. The discrepancy is 0, between $x_1{\cdot}x_2$ and x_4. So, we first apply f_1 that maps x_4 to $x_1{\cdot}x_2$, and keeps the other variables unchanged. The resulting terms are

$s^{f_1} =$ 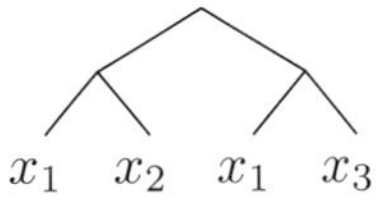

, $(t')^{f_1} =$ 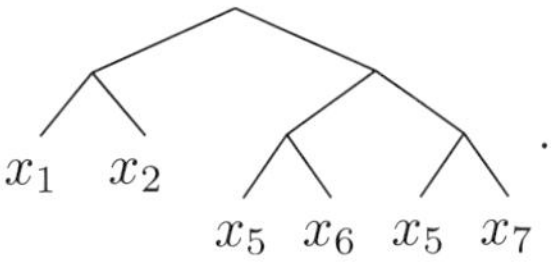

.

Now, the discrepancy is in 10, between x_1 and $x_5{\cdot}x_6$. We apply f_2, which maps x_1 to $x_5{\cdot}x_6$, and we obtain

$s^{f_1 f_2} =$ 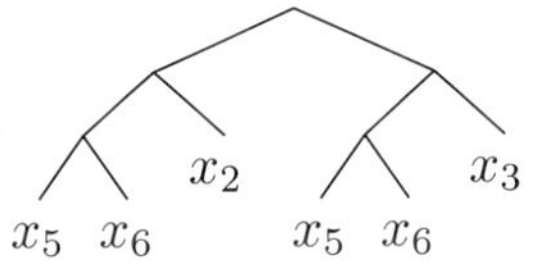

, $(t')^{f_1 f_2} =$ 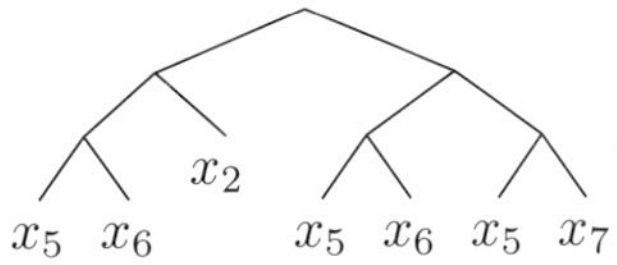

.

The discrepancy is now 11, between x_3 and $x_5{\cdot}x_7$. Applying f_3 that maps x_3 to $x_5{\cdot}x_7$, we find

$s^{f_1 f_2 f_3} =$ 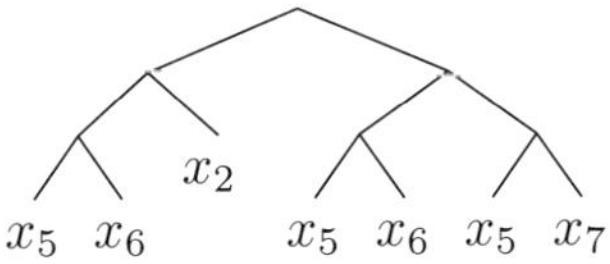

$= (t')^{f_1 f_2 f_3}$.

We conclude that f: $x_1 \mapsto x_5{\cdot}x_6$, $x_3 \mapsto x_5{\cdot}x_7$, $x_4 \mapsto (x_5{\cdot}x_6){\cdot}x_2$, $x_i \mapsto x_i$ for $i \neq 1, 3, 4$ is a most general substitution satisfying $s^f = (t')^f$. Finally, s and t are unifiable, and a most general unifier is the pair (f, g), where f is as above and g is the composition of f and the variable permutation used to transform t into t', i.e.,

f: $x_1 \mapsto x_5{\cdot}x_6$, $x_2 \mapsto x_2$, $x_3 \mapsto x_5{\cdot}x_7$,

g: $x_1 \mapsto (x_5{\cdot}x_6){\cdot}x_2$, $x_2 \mapsto x_5$, $x_3 \mapsto x_6$, $x_4 \mapsto x_7$.

(It suffices to specify the substitutions on the variables that actually appear in the considered terms.)

Similarly, let us consider the unification of

$s =$ 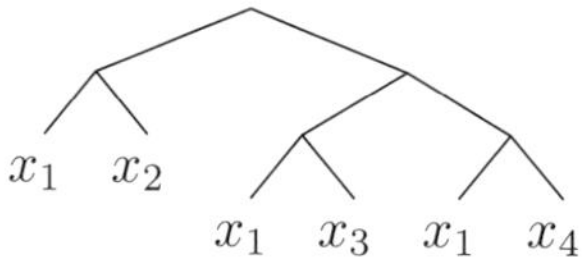

and $t =$ 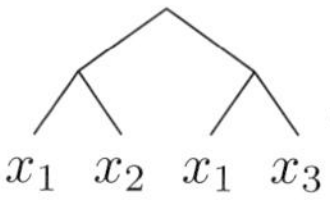

.

We translate t into $t' =$ 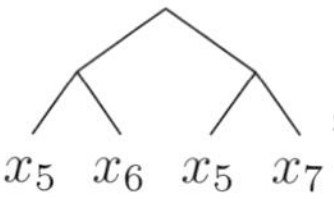

, and try to find f satisfying $s^f = (t')^f$. We first apply f_1 that maps x_5 to x_1, and f_2 that maps x_6 to x_2, thus obtaining

$s^{f_1 f_2} =$ 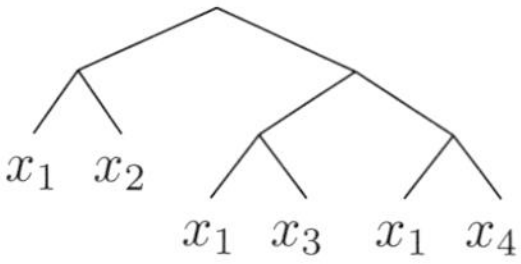

, $t^{f_1 f_2} =$ 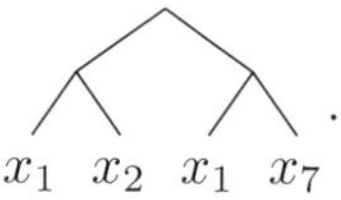

.

The discrepancy between $s^{f_1 f_2}$ and $t^{f_1 f_2}$ is in 10, between $x_1{\cdot}x_3$ and x_1. Since the variable x_1 occurs in $x_1{\cdot}x_3$, we cannot find f satisfying $s^f = (t')^f$, and, therefore, the terms s and t cannot be unified.

With the previous technique, it is easy to determine the product of mappings associated with term substitutions.

Lemma 1.18. *(i) Assume that F, G are partial mappings defined on terms, that F (viewed as a set of pairs) is the family of all instances of the pair (s, s') and G is the family of all instances of the pair (t, t'). Then*

- either the terms s' and t are unifiable; then, if (f, g) is a most general unifier for s' and t, the map $F \bullet G$ is the family of all instances of the pair $(s^f, (t')^g)$;

- or the terms s' and t are not unifiable, and the map $F \bullet G$ is empty.

(ii) Assume in addition that s and t are injective. Then s' and t are unifiable, and, letting h be a most general unifier of s' and t, the term s^h is injective.

Proof. (i) Assume that the mapping $F \bullet G$, *i.e.*, $G \circ F$, is not empty, and let t be an arbitrary term in its domain. Then t belongs in particular to the domain of F, and, therefore, we have $t = s^{f'}$ for some substitution f'. Then $F(t) = s'^{f'}$ holds, and, because the latter term belongs by hypothesis to the domain of G, it is also a substitute of t, say $t^{g'}$. Thus the terms s' and t are unifiable. Let (f, g) be a most general unifier of s' and t. Then there exists a substitution h satisfying $f'(x) = h(f(x))$ for x occurring in s', hence in s, and $g'(y) = h(g(y))$ for y occurring in t, hence in t'. So, we have $t = (s^f)^h$, and $G(F(t)) = ((t')^g)^h$, *i.e.*, the pair $(t, G(F(t))$ is an instance of the pair $(s^f, (t')^g)$.

(ii) Let us consider the successive steps of the algorithm of Proposition 1.16 (unification) constructing a substitution f satisfying $(s')^f = (t'')^f$, where t''

is a copy of t such that s' and t'' have no variable in common. Let f_i be the substitution obtained at the i-th step. We write s'_i for $s'^{f_1 \cdots f_{i-1}}$, and t''_i for $(t'')^{f_1 \cdots f_{i-1}}$. We show inductively on i that f_i exists if s'_i and t''_i are not yet equal, and that s_i is injective. Let α be the discrepancy between s'_i and t''_i. Assume first that $\mathrm{sub}(s'_i, \alpha)$ is a variable, say z. We have to check that z does not occur in $\mathrm{sub}(t''_i, \alpha)$. Now, the term t'' is injective, so the variables occurring below α in t'' do not occur on the left of α, and, therefore, they are not involved in the previous substitutions $f_1, \ldots, f_{i-1}$. So we have $\mathrm{sub}(t''_i, \alpha) = \mathrm{sub}(t'', \alpha)$. Now the variable z is either a variable of s', or a variable that comes from some substitution f_j with $j < i$: such a variable comes from the left of α, and, therefore, it cannot occur in $\mathrm{sub}(t'', \alpha)$. Hence f_i exists, and s_{i+1}, which is $s_i^{f_i}$, is obtained by replacing one variable of s_i (necessarily occurring at most once) by an injective term $\mathrm{sub}(t'', \alpha)$ involving only new variables, therefore it is injective. Assume now that $\mathrm{sub}(t''_i, \alpha)$ is a variable, say z. As before, z cannot occur on the left of α in t, and therefore it was not involved in the previous substitutions. In particular, z cannot occur in s'_i. So f_i exists in this case also, and s_{i+1}, which is simply s_i, is injective. So the induction goes on. ∎

Proposition 1.10 (domain) can now be proved easily.

Proof. We use induction on the length of the word w: if w has length 1, we apply Lemma 1.12, otherwise, we propagate the induction hypothesis using Lemma 1.18(i). For (ii) we resort to Lemma 1.18(ii). Finally, we observe that, if t is an injective term, every term whose skeleton includes the skeleton of t is a substitute of t—a property that fails if t is not injective, as variable clashes may occur in the construction of the substitution. ∎

Example 1.19. **(domain)** *The pair (t^L_w, t^R_w) can be computed effectively using the method of Lemma 1.18: assuming that $(t^L_{w_1}, t^R_{w_1})$ and $(t^L_{w_2}, t^R_{w_2})$ are known, we compute $(t^L_{w_1 w_2}, t^R_{w_1 w_2})$ by unifying $t^R_{w_1}$ and $t^L_{w_2}$; at the end the variables are possibly renamed in order to obtain canonical terms.*

For instance, let us consider the word $\phi \cdot 1^{-1}$. We know that (t^L_ϕ, t^R_ϕ) is the pair (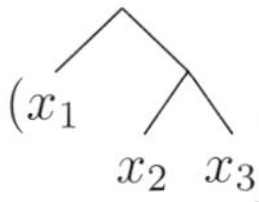

, 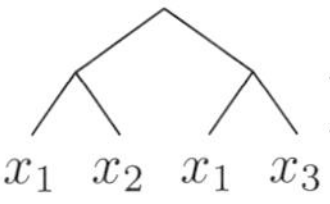

), while $(t^L_{1^{-1}}, t^R_{1^{-1}})$ is the pair (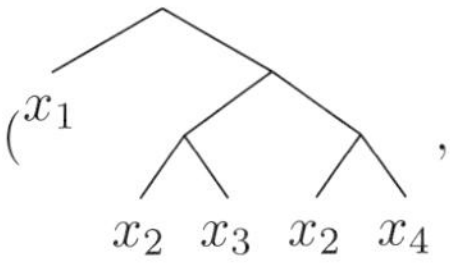

, 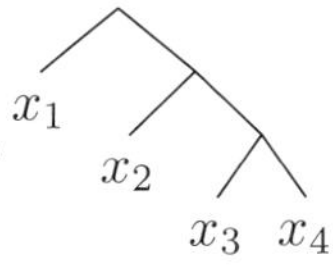

). In order to determine $(t^L_{\phi \cdot 1^{-1}}, t^R_{\phi \cdot 1^{-1}})$ (if it exists), we unify t^R_ϕ and $t^L_{1^{-1}}$. We have seen in Example 1.17 (unification) that the unification is possible, and, using the values computed

there, we conclude that the product of LD_ϕ *and* LD_1^{-1} *consists of all instances of the pair* (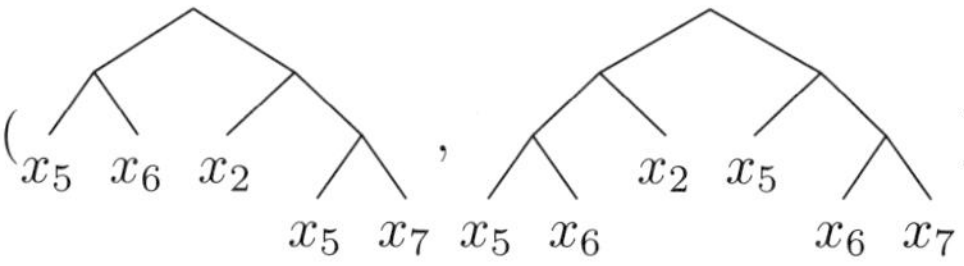

). *It remains to rename the variables in order to obtain canonical terms, and we find finally*

$(t^L_{\phi\cdot 1^{-1}}, t^R_{\phi\cdot 1^{-1}}) = ($ 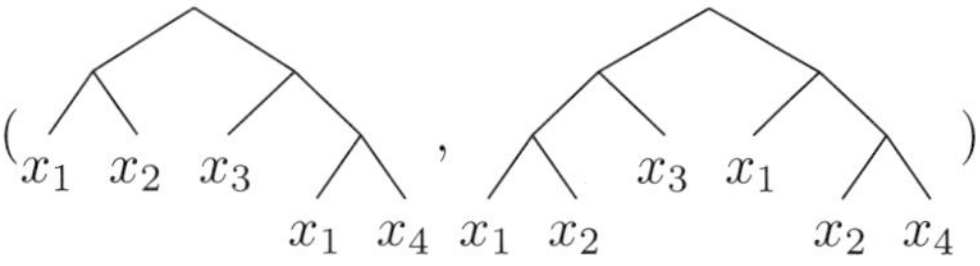

).

Let us now consider $\phi \cdot 1 \cdot \phi^{-1}$. *As the reader can check, we have*

$(t^L_{\phi\cdot 1}, t^R_{\phi\cdot 1}) = ($ 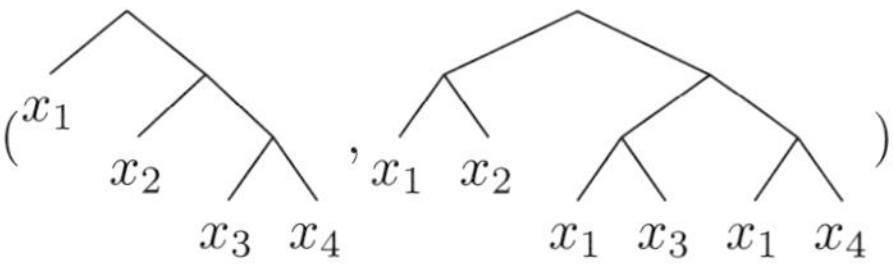

),

and $(t^L_{\phi^{-1}}, t^R_{\phi^{-1}}) = ($ 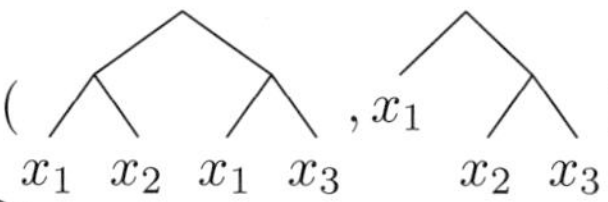

). *So we are led to unify the terms* 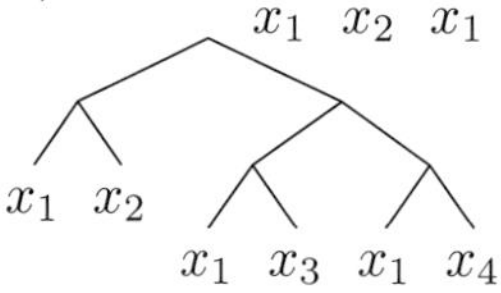

and 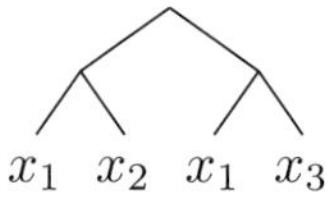

, *which we have seen is impossible. We conclude that the operator* $\mathrm{LD}_{\phi\cdot 1\cdot \phi^{-1}}$ *is empty.*

Remark 1.20. Let us write $(LD)_w$ for the pair (t^L_w, t^R_w). The terms t^L_w and t^R_w are LD-equivalent when they exist, as, by definition, LD_w maps t^L_w to t^R_w. It is usual to refer to pairs of terms that are equivalent under a given identity I as to *consequences* of I. So, each identity $(LD)_w$ is a consequence of (LD). Conversely, for every consequence (t, t') of (LD), there exists a finite sequence w such that the operator LD_w maps t to t', hence (t, t') is an instance of $(LD)_w$. Thus we can summarize our approach by saying that we have selected a family of consequences of (LD), namely the family of all $(LD)_w$ for w a word on $\mathbf{A} \cup \mathbf{A}^{-1}$, which is complete in the sense that every consequence of (LD) is an instance of an indentity $(LD)_w$. In addition, this family is equipped with an associative product—but a partial one only: for instance, the product of $(LD)_{\phi\cdot 1}$ and $(LD)_{\phi^{-1}}$ is not defined. In terms of operators, composition is always defined, but it can be the empty mapping.

The previous results about the domain of LD_w can be extended to several operators simultaneously.

Proposition 1.21. (**intersection**) *Assume that* $w_1, \ldots, w_m$ *are words on* $\mathbf{A} \cup \mathbf{A}^{-1}$. *Then either the intersection of the domains of the operators* $\mathrm{LD}_{w_1}, \ldots, \mathrm{LD}_{w_m}$ *is empty, or it is the set of all substitutes of some unique canonical term* $t^L_{w_1,\ldots,w_m}$. *If* $w_1, \ldots, w_m$ *are words on* $\mathbf{A}$, *only the second case may occur, and the term* $t^L_{w_1,\ldots,w_m}$ *is injective.*

Proof. For an induction, it suffices to consider $m = 2$. Let I denote the intersection of the domains of LD_{w_1} and LD_{w_2}. Then I is the set of all terms that are substitutes both of $t^L_{w_1}$ and $t^L_{w_2}$. By Proposition 1.16 (unification), either $t^L_{w_1}$ and $t^L_{w_2}$ are not unifiable, and, in this case, I is empty, or they admit a most general unifier (f, g), and, if $t^L_{w_1,w_2}$ is the unique canonical term obtained by permuting the variables of $(t^L_{w_1})^f$, I is the set of all substitutes of $t^L_{w_1,w_2}$. If w_1 and w_2 are positive words, the terms $t^L_{w_1}$ and $t^L_{w_2}$ are injective, hence they are unifiable, and $t^L_{w_1,w_2}$ exists and is injective. ∎

Independence

Using the previous description of the domain and image of the operators LD_w, we can now prove that the monoids $\mathcal{G}_{LD,X}$ and $\mathcal{G}^+_{LD,X}$ do not depend on X:

Proposition 1.22. (**independence**) *Up to isomorphism, the geometry monoids* $\mathcal{G}_{LD,X}$ *and* $\mathcal{G}^+_{LD,X}$ *do not depend on* X.

The point is to prove that, if LD_w and $\mathrm{LD}_{w'}$ coincide on terms in one variable, then they coincide everywhere. This follows from the general properties of term substitution. We first prove a few auxiliary results.

Lemma 1.23. *Assume that* w *is a word on* $\mathbf{A} \cup \mathbf{A}^{-1}$, *and* LD_w *maps* t *to* t'. *Then, for every substitution* h, LD_w *maps* t^h *to* t'^h.

Proof. According to Proposition 1.10 (domain), if the operator LD_w maps t to t', there exists a substitution f satisfying $t = (t^L_w)^f$ and $t' = (t^R_w)^f$. Then we have $t^h = (t^L_w)^{fh}$, and $t'^h = (t^R_w)^{fh}$. By Proposition 1.10 (domain) again, this means that LD_w maps t^h to t'^h. ∎

Lemma 1.24. *Assume that* X *is a nonempty set, and that* t, t' *are terms in* T_∞ *that admit the same* X*-substitutes. Then there exists a permutation* f *of* $\{x_1, x_2, \ldots\}$ *such that* $t' = t^f$ *holds.*

Proof. Let h be an arbitrary mapping of $\{x_1, x_2, \ldots\}$ into X. The term t^h is an X-substitute of t, hence of t', so the skeleton of t^h, which is the skeleton of t, must include the skeleton of t'. By symmetry, t and t' have the same skeleton.

Assume now that α, β are addresses in the outline of t and that x_i occurs at α and β in t. Then we have $\mathrm{sub}(s, \alpha) = \mathrm{sub}(s, \beta)$ for every substitute s of t. Now, if distinct variables x_j, x_k occurred at α and β in t', defining a substitution h by $h(x_j) = x$, $h(x_k) = x{\cdot}x$, where x is an arbitrary element of X, would give a substitute s' of t' with $\mathrm{sub}(s', \alpha) = x$, $\mathrm{sub}(s', \beta) = x{\cdot}x$, hence one that cannot be a substitute of t. It follows that the family of all pairs of the form $(\mathrm{var}(t, \alpha), \mathrm{var}(t', \alpha))$ for α in the common outline of t and t' can be extended into a permutation of $\{x_1, x_2, \ldots\}$. ∎

Lemma 1.25. *Assume that X is a nonempty set and the operators $\mathrm{LD}_{w,X}$ and $\mathrm{LD}_{w',X}$ coincide. Then we have $t^L_w = t^L_{w'}$ and $t^R_w = t^R_{w'}$.*

Proof. Up to replacing X with one of its subsets, we may assume that X is a singleton. The domain of $\mathrm{LD}_{w,X}$ is the set of all X-substitutes of t^L_w, while the domain of $\mathrm{LD}_{w',X}$ is the set of all X-substitutes of $t^L_{w'}$. These sets coincide, hence, by Lemma 1.24, there exists a permutation f of $\{x_1, x_2, \ldots\}$ satisfying $t^L_{w'} = (t^L_w)^f$. Because t^L_w and $t^L_{w'}$ are canonical terms, f must be the identity mapping, *i.e.*, $t^L_w = t^L_{w'}$ holds.

Assume now that t' is an X-substitute of t^R_w, say $t' = (t^R_w)^h$. Let $t = (t^L_w)^h$. Then t belongs to the domain of $\mathrm{LD}_{w,X}$, hence of $\mathrm{LD}_{w',X}$, and we have $t' = t \bullet w = t \bullet w'$ by hypothesis. Hence t' is an X-substitute of $t^R_{w'}$. By a symmetric argument, every X-substitute of $t^R_{w'}$ is an X-substitute of t^R_w, and, as above, we deduce that $t^R_w = t^R_{w'}$ holds. ∎

We can now prove Proposition 1.22 (independence).

Proof. Let X be an arbitrary nonempty set. By Lemma 1.25, $\mathrm{LD}_{w,X} = \mathrm{LD}_{w',X}$ implies $t^L_w = t^L_{w'}$ and $t^R_w = t^R_{w'}$: these equalities do not involve X, and they imply $\mathrm{LD}_{w,X'} = \mathrm{LD}_{w',X'}$ for every set X'. ∎

Owing to the previous result, we shall drop the subscript X and write $\mathcal{G}_{LD}$ and $\mathcal{G}^+_{LD}$ henceforth.

The compatibility property

We have shown in Lemma 1.25 that a sufficient condition for two operators LD_w, $\mathrm{LD}_{w'}$ to be equal is that they take the same value on all one variable terms. In case of positive words, we can obtain the same conclusion under the weaker hypothesis that the operators LD_w and $\mathrm{LD}_{w'}$ take the same value on one term:

Proposition 1.26. (compatibility) *Assume that u and u' are words on* **A**, *and the operators LD_u and $\mathrm{LD}_{u'}$ take the same value on at least one term, i.e., $t \bullet u = t \bullet u'$ holds for at least one term t. Then $\mathrm{LD}_u = \mathrm{LD}_{u'}$ holds.*

The proof will resort to some specific geometric preservation properties of the operators LD_u. We recall from Chapter VI that the address α is said to cover the address β if there exists an address γ such that α is $\gamma 1^p$ for some $p \geq 1$ and β lies below $\gamma 0$.

Definition 1.27. **(covering)** Assume that t is a term in T_∞. We say that x_i *covers* x_j in t if there exist two addresses α, β in the outline of t such that x_i occurs at α in t, x_j occurs at β, and α covers β.

Lemma 1.28. *Assume $t \in T_\infty$, and $t' = t \bullet \gamma$. Let x_i be the rightmost variable of the subterm* $\mathrm{sub}(t, \gamma 10)$. *For $j \neq i$, the variables covered by x_j in t' are exactly those variables covered by x_j in t; the variables covered by x_i in t' are those variables covered by x_j in t, completed with those variables occurring in the subterm* $\mathrm{sub}(t, \gamma 0)$.

Proof. We assume that x_j occurs at $\alpha 1^p$ in t', and we consider the various possible cases. For $\alpha <_{LR} \gamma$, we have $\mathrm{sub}(t', \alpha) = \mathrm{sub}(t, \alpha)$, and the variables that are covered by the occurrence $\alpha 1^p$ of x_j both in t' and in t are those variables that occur at a non-final address in $\mathrm{sub}(t, \alpha)$. For $\alpha = \gamma 00\beta$ or $\alpha = \gamma 10\beta$, we have $\mathrm{sub}(t', \alpha) = \mathrm{sub}(t, \gamma 0\beta)$, and, again, the variables covered by the occurrence $\alpha 1^p$ of x_j in t' are the variables covered by the occurrence $\gamma 0\beta 1^p$ of x_j in t. For $\alpha = \gamma 01\beta$, we have $\mathrm{sub}(t', \alpha) = \mathrm{sub}(t, \gamma 10\beta)$. So, for β not of the form 1^q, *i.e.*, if α is not the rightmost address in t' under $\gamma 01$, the variables covered by the occurrence $\alpha 1^p$ of x_j in t' are those variables covered by the occurrence $\gamma 10\beta 1^p$ of x_j in t. For $\beta = 1^q$, the result is similar, but, in addition, all variables occurring in $\mathrm{sub}(t', \gamma 00)$, *i.e.*, all variables occurring in $\mathrm{sub}(t, \gamma 0)$ are covered. Finally, if $\gamma 11$ is a prefix of α, or if $\alpha >_{LR} \gamma$ holds, the variables covered by the occurrence $\alpha 1^p$ of x_j in t' and t coincide. Thus, all possible addresses in t' have been considered, and the reader can check that, conversely, all possible addresses in t have been considered as well. ∎

We are now able to establish Proposition 1.26 (compatibility).

Proof. Assume that both LD_u and $\mathrm{LD}_{u'}$ map t to t'. Without loss of generality, we may assume that t belongs to T_∞. Let us assume first that t is injective. Then t can be written as $(t_u^L)^h$, where h is an injective substitution. We claim that the term t_u^L belongs to the domain of the operator $\mathrm{LD}_{u'}$, *i.e.*, that the skeleton of $t_{u'}^L$ is included in the skeleton of t_u^L. Indeed, let us say that the variable x_i is (h, x_k)-*good* if x_i occurs at a non-final address in some (unique) term $h(x_k)$. Assume that x_i is (h, x_k)-good. Then, for every term s, the variables covered by x_i in a term of the form s^h are exactly those variables covered by x_i in $h(x_k)$, as shown by an induction on s. We have $t = (t_u^L)^h$ and $t' = (t_u^R)^h$, so we deduce that the variables covered by x_i both in t and t' are the variables covered by x_i in $h(x_k)$. Now assume that t_u^L does not belong to the domain of $\mathrm{LD}_{u'}$: this means that u' can be written as $u'_1 \cdot \gamma \cdot u'_2$ for some address γ, where t^L_u belongs to the domain of $\mathrm{LD}_{u'_1}$, and $t^L_u \bullet u'_1$ does not belong to the domain of LD_γ. Let s be the latter term. By hypothesis, the term t, which is $(t^L_u)^h$, belongs to the domain of $\mathrm{LD}_{u'}$, hence the term $t \bullet u'_1$, which is s^h, belongs to the domain of LD_γ. Hence the address $\gamma 10$ belongs to the skeleton of s^h, but not to the skeleton of s. Let γ' be the longest prefix of $\gamma 10$ that belongs to the skeleton of s: necessarily γ' belongs to the outline of s, *i.e.*, γ' is the address of a variable, say x_k, in t. (*cf.* Figure 1.1). Let x_i be the rightmost variable in $\mathrm{sub}(s^h, \gamma 10)$, and let x_j be any variable occurring in $\mathrm{sub}(s^h, \gamma 0)$. By definition, x_i is (h, x_k)-good, and x_i does not cover x_j in s^h—consider the two slightly different cases $\gamma' \sqsubseteq \gamma$ and $\gamma = \gamma' 1$. Now, by the previous lemma, x_i covers x_j in the LD-expansion $s^h \bullet \gamma$, and, therefore, in the LD-expansion t' of that term, a contradiction. As the words u and u' play symmetric roles, we conclude that the terms t^L_u and $t^L_{u'}$ have the same skeleton, and, therefore, that they are equal since they are canonical and injective. So we have $t = (t^L_u)^h = (t^L_{u'})^h$, which implies $t' = (t^R_u)^h = (t^R_{u'})^h$. Hence $t^R_u = t^R_{u'}$ holds, and so does $\mathrm{LD}_u = \mathrm{LD}_{u'}$.

It remains to get rid of the hypothesis that t is an injective term. So assume that t is arbitrary, and let s be an injective term with the same skeleton as t. Because u and u' are positive, the only constraint for belonging to the domain of the operators LD_u and $\mathrm{LD}_{u'}$ is to have a skeleton that is large enough, according to Proposition 1.10(ii) (domain). So s belongs to the domains of these operators. Let s_1 and s'_1 be the corresponding LD-expansions. Let x_k be an arbitrary fixed variable, and h be the substitution that maps every variable to x_k. Then the terms t^h and s^h coincide by hypothesis, so, by Lemma 1.23, the terms t'^h, $s_1{}^h$, $s'_1{}^h$ coincide as well: this means that the terms s_1 and s'_1 have the same skeleton. Now, they are LD-equivalent, since both of them are LD-expansions of s, so, by Proposition VI.1.17 (uniqueness), they must be equal. Hence there exists an injective term that admits the same image under LD_u and $\mathrm{LD}_{u'}$, and we are back to the case we have treated first. ■

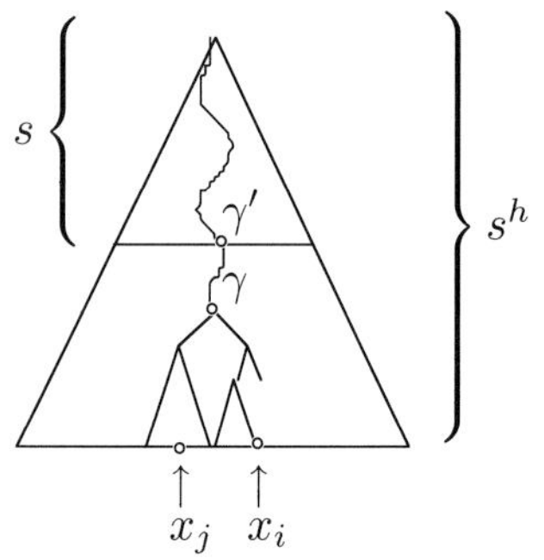

Figure 1.1. Proof of Proposition 1.26 (compatibility)

Exercise 1.29. *(**compatibility**) Assume that u and u' are words on $\mathbf{A}$, and that there exists a term t such that the LD-expansions $t \bullet u$ and $t \bullet u'$ exist and have the same skeleton. Prove* $\mathrm{LD}_u = \mathrm{LD}_{u'}$. *[Hint: Use Proposition VI.1.17 (uniqueness).]*

Exercise 1.30. *(**braidlike expansions**) (i) Say that a word on $\mathbf{A}$ is special if it has the form $1^{e_1} \cdot \cdots \cdot 1^{e_p}$ with $e_{i+1} \le e_i + 1$. Prove that, for every word u involving no address where 0 occurs, there exists a special word u' satisfying* $\mathrm{LD}_{u'} = \mathrm{LD}_u$ *[Hint:* LD_{1^e} *and* $\mathrm{LD}_{1^{e+2}}$ *commute.]*

(ii) Prove that, for u a special word and t in the domain of LD_u*, the pair $(\mathrm{Out}(t), \mathrm{Out}(t \bullet u))$ determines u, and so does $(\mathrm{Out}(t), \mathrm{Out}(\mathrm{sub}(t \bullet u, 0))$ when u ends with ϕ. [Hint: Use induction on $\mathrm{ht}_R(t_u^L)$. Writing $u = u_1 \cdot \phi \cdot u_2 \cdot \cdots \cdot \phi \cdot u_{p+1}$, where $u_1, \ldots, u_{p+1}$ are special and do not contain ϕ, and $u_1, \ldots, u_p$ finish with 1, set $t_0 = t$, $t_i = t \bullet u_0 \cdot \phi \cdot \ldots \cdot u_i \cdot \phi$, and show that $\mathrm{Out}(t_i)$ can be recovered from $\mathrm{Out}(t \bullet u)$.]*

---

2. Relations in the Geometry Monoid

We enter now the core of our study, namely the description of those relations that connect the LD-expansion operators LD_α: their existence expresses that, when t and t' are LD-equivalent terms, there exist in general several ways to transform t into t' by applying left self-distributivity, *i.e.*, there exist several words w such that LD_w maps t to t'. Actually, many relations of the type $\mathrm{LD}_w = \mathrm{LD}_{w'}$ hold in the geometry monoid $\mathcal{G}_{LD}$. In this section, we establish a list of such relations, but we do not prove that this list is complete in the sense that every relation between the operators LD_w is a consequence of the considered relations: this will be done in the next chapter.

Orthogonality relations

As a preliminary remark, let us observe that all relations in $\mathcal{G}_{LD}$ are compatible with left shift. The following notation will be used very often in the sequel:

Definition 2.1. (**shift**) For w a word on $\mathbf{A} \cup \mathbf{A}^{-1}$, and γ and address, we denote by $\mathrm{sh}_\gamma(w)$, or simply γw, the *γ-shift* of w, defined as the word obtained from w by replacing every address $\alpha^{\pm 1}$ with $\gamma\alpha^{\pm 1}$.

Thus, for $w = \alpha_1 \cdot \cdots \cdot \alpha_p$, we have $\gamma w = \gamma\alpha_1 \cdot \cdots \cdot \gamma\alpha_p$, not to be confused with $\gamma \cdot w$, the length $p+1$ word $\gamma \cdot \alpha_1 \cdot \cdots \cdot \alpha_p$ obtained by multiplying w by γ on the left (to this end we shall most often keep the operator '$\cdot$' for word concatenation).

Lemma 2.2. *For every α, $\mathrm{LD}_w = \mathrm{LD}_{w'}$ implies $\mathrm{LD}_{\alpha w} = \mathrm{LD}_{\alpha w'}$.*

Proof. An induction on $\lg(w)$ shows that t belongs to the domain of $\mathrm{LD}_{\alpha w}$ if and only if $\mathrm{sub}(t, \alpha)$ exists and belongs to the domain of LD_w. In this case, the image of t under $\mathrm{LD}_{\alpha w}$ is the term obtained from t by replacing the subterm $\mathrm{sub}(t, \alpha)$ by its image under LD_w. ∎

The first family of relations in $\mathcal{G}_{LD}$ expresses that operators associated with orthogonal addresses act independently.

Proposition 2.3. **(orthogonality)** *Assume that α and β are orthogonal addresses. Then the operators LD_α and LD_β commute.*

Proof. (Figure 2.1) The action of LD_α and LD_β involve disjoint subterms, and, therefore, their order has no influence on the final result. More precisely, t belongs to the domain both of $\mathrm{LD}_\alpha \cdot \mathrm{LD}_\beta$ and of $\mathrm{LD}_\beta \cdot \mathrm{LD}_\alpha$ if the addresses $\alpha 10$ and $\beta 10$ belong to its skeleton. In this case, the image of t under the previous operators is the term t' specified by the equalities $\mathrm{sub}(t', \alpha) = \mathrm{sub}(t, \alpha) \bullet \phi$, $\mathrm{sub}(t', \beta) = \mathrm{sub}(t, b) \bullet \phi$, $\mathrm{sub}(t', \gamma) = \mathrm{sub}(t, \gamma)$ for $\gamma \perp \alpha, \beta$. ∎

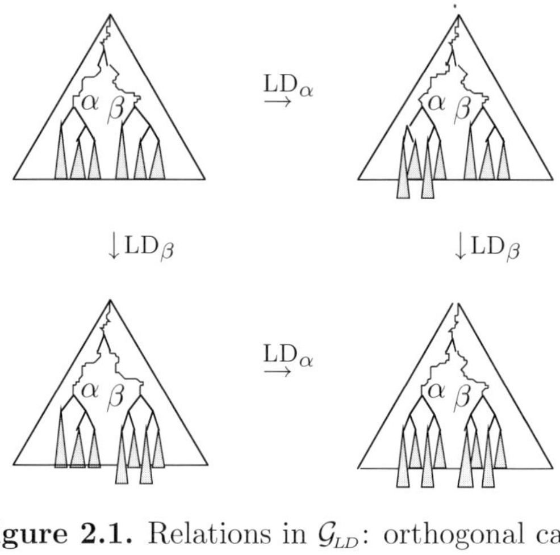

Figure 2.1. Relations in $\mathcal{G}_{LD}$: orthogonal case

Corollary 2.4. *Assume that w_1, w_2 are words on $\mathbf{A} \cup \mathbf{A}^{-1}$, and every address in w_1 is orthogonal to every address in w_2. Then LD_{w_1} and LD_{w_2} commute.*

Inheritance relations

The second family of relations originates in some inheritance relations that connect the addresses in t and in t' when t' is an LD-expansion of t. As no variable is repeated twice in the left-hand side of the identity (LD), every variable in an LD-expansion t' of t has a well defined origin in t. Equivalently, every address in the outline of t has a well defined set of *heirs* in the outline of t'. An inductive definition is easy.

Definition 2.5. (**heirs**) Assume that B is a set of addresses, and u is a word on $\mathbf{A}$. The set $\mathrm{Heir}(B,u)$ of all *heirs* of elements of B under the action of LD_u is inductively defined by the following clauses:

(i) $\mathrm{Heir}(B,u)$ exists if and only if $\mathrm{Heir}(\{\beta\},u)$ exists for every β in B, and, in this case, we have $\mathrm{Heir}(B,u) = \bigcup_{\beta\in B}\mathrm{Heir}(\{\beta\},u)$;

(ii) For every B, we have $\mathrm{Heir}(B,\varepsilon) = B$;

(iii) For α in $\mathbf{A}$, $\mathrm{Heir}(\{\beta\},\alpha)$ exists if and only if $\beta \not\sqsubseteq \alpha 1$ holds, and we have

$$\mathrm{Heir}(\{\beta\},\alpha) = \begin{cases} \{\beta\} & \text{for } \beta \perp \alpha \text{ or } \alpha 11 \sqsubseteq \beta,\\ \{\alpha 00\gamma, \alpha 10\gamma\} & \text{for } \beta = \alpha 0\gamma,\\ \{\alpha 01\gamma\} & \text{for } \beta = \alpha 10\gamma. \end{cases}$$

(iv) For $u = \alpha \cdot u_0$, we have $\mathrm{Heir}(B,u) = \mathrm{Heir}(\mathrm{Heir}(B,\alpha),u_0)$, when it exists.

The easy verification of the following results is left to the reader:

Lemma 2.6. *Assume that u is a word on $\mathbf{A}$, and β is an address.*

(i) The set $\mathrm{Heir}(\{\beta\},u)$ exists if and only if some address in the outline of t_u^L is a prefix of β.

(ii) If $\mathrm{Heir}(\{\beta\},u)$ is defined, so is $\mathrm{Heir}(\{\beta\gamma\},u)$ for every γ, and we have $\mathrm{Heir}(\{\beta\gamma\},u) = \{\beta'\gamma\ ;\ \beta' \in \mathrm{Heir}(\{\beta\},u)\}$.

(iii) The elements of every set $\mathrm{Heir}(\{\beta\},u)$ are pairwise orthogonal.

(iv) Assume $t' = t \bullet u$, and β belongs to the skeleton of t. If $\mathrm{Heir}(\{\beta\},u)$ is defined, then we have $\mathrm{sub}(t',\beta') = \mathrm{sub}(t,\beta)$ for every β' in $\mathrm{Heir}(\{\beta\},u)$.

Observe that Point (iv) always applies when the address β lies in the outline of t, *i.e.*, when β is the address of a variable in t; then $\mathrm{Heir}(\{\beta\},u)$ is the family of all occurrences in the outline of t' that come from β in t, in an obvious sense. In particular, if the variable x occurs at β and only there in t, then $\mathrm{Heir}(\{\beta\},u)$ is exactly the set of those addresses where x occurs in t'.

Example 2.7. *(**heirs**) Assume* $u = \phi \cdot 1$. *We have seen that* $t^L_{\phi\cdot 1}$ *is*

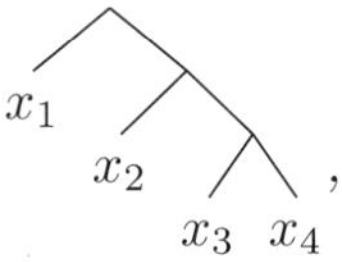

, which is mapped to

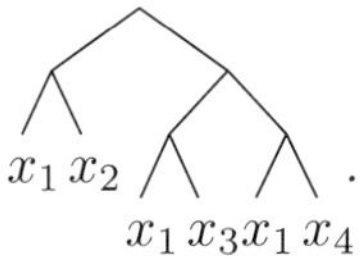

. Hence, those addresses β *for which* $\mathrm{Heir}(\{\beta\}, \phi\cdot 1)$ *is not defined are* ϕ, 1 *and* 11. *The reader can check* $\mathrm{Heir}(\{0\}, \phi\cdot 1) = \{00, 100, 110\}$, *which corresponds to the fact that the variable* x_1 *occurring at* 0 *in* $t^L_{\phi\cdot 1}$ *has three copies with addresses* 00, 100 *and* 110 *in* $t^R_{\phi\cdot 1}$. *Similarly we have* $\mathrm{Heir}(\{10\}, \phi\cdot 1) = \{01\}$, $\mathrm{Heir}(\{110\}, \phi\cdot 1) = \{101\}$, *and* $\mathrm{Heir}(\{111\}, \phi\cdot 1) = \{111\}$. *Lemma 2.6(ii) implies* $\mathrm{Heir}(\{0\gamma\}, \phi\cdot 1) = \{00\gamma, 100\gamma, 110\gamma\}$ *for every address* γ.

Besides the set $\mathrm{Heir}(B, u)$, which contains all heirs of addresses in B, we shall also consider the leftmost element of $\mathrm{Heir}(B, u)$ when B is a singleton.

Definition 2.8. (**left heir**) Assume that β is an address, and u is a positive word in $\mathbf{A}^*$. The address $\mathrm{heir}(\beta, u)$ is possibly defined by the following clauses:

(i) For $u = \varepsilon$, we have $\mathrm{heir}(\beta, u) = \beta$;

(ii) For α in $\mathbf{A}$, $\mathrm{heir}(\beta, u)$ exists if and only if $\beta \not\sqsubseteq \alpha 1$ holds, and we have

$$\mathrm{heir}(\beta, \alpha) = \begin{cases} \beta & \text{for } \beta \perp \alpha \text{ or } \alpha 11 \sqsubseteq \beta, \\ \alpha 01\beta' & \text{for } \beta = \alpha 10\beta', \\ \alpha 00\beta' & \text{for } \beta = \alpha 0\beta'. \end{cases}$$

(iii) For $u = \alpha \cdot u_0$, we have $\mathrm{heir}(\beta, u) = \mathrm{heir}(\mathrm{heir}(\beta, \alpha), u_0)$.

Again, we leave to the reader the easy verification of the following result:

Lemma 2.9. *Assume that* u *is a word on* $\mathbf{A}$, LD_u *maps* t *to* t', *and* β *belongs to the outline of* t. *Then* $\mathrm{heir}(\beta, u)$ *exists, it belongs to the outline of* t', *and it is the leftmost address in* $\mathrm{Heir}(\{\beta\}, u)$.

With the previous notion at hand, we can now state the expected relations in the monoid $\mathcal{G}_{LD}$ precisely.

Proposition 2.10. (**inheritance**) *Assume that* u *is a word on* $\mathbf{A}$, β *is an address, and* $\mathrm{Heir}(\{\beta\}, u)$ *is defined. Then we have*

$$\mathrm{LD}_\beta \bullet \mathrm{LD}_u = \mathrm{LD}_u \bullet \prod_{\beta' \in \mathrm{Heir}(\{\beta\}, u)} \mathrm{LD}_{\beta'}. \tag{2.1}$$

Proof. (Figure 2.2) The idea is obvious: if LD_u maps t to t', then there is a copy of $\mathrm{sub}(t, \beta)$ at each address of $\mathrm{Heir}(\{\beta\}, u)$ in t'. So first applying LD_β to t, and then LD_u, or first applying LD_u and, subsequently, applying the operator $\mathrm{LD}_\emptyset$ to each of the copies of $\mathrm{sub}(t, \beta)$ in t' lead to the same result.

For a formal proof, we argue inductively on the length of u. For u empty, the result is obvious. Assume $\lg(u) = 1$, *i.e.*, u is a single address, say α. The hypothesis that $\mathrm{Heir}(\{\beta\}, \alpha)$ is defined means that either β is orthogonal to α, or that either $\alpha 0$, or $\alpha 10$, or $\alpha 11$ is a prefix of β. The case $\beta \perp \alpha$ is covered by Proposition 2.3 (orthogonality). Assume $\beta = \alpha 0 \gamma$. We claim that $\mathrm{LD}_\beta \bullet \mathrm{LD}_\alpha = \mathrm{LD}_\alpha \bullet \mathrm{LD}_{\alpha 00\gamma} \bullet \mathrm{LD}_{\alpha 10\gamma}$ holds. Let t be an arbitrary term. Then t belongs to the domain of $\mathrm{LD}_\beta \bullet \mathrm{LD}_\alpha$ if and only if the address $\beta 10$ belongs to its skeleton. Then $\alpha 00\gamma 10$ and $\alpha 10\gamma 10$ belong to the skeleton of $t \bullet \alpha$, and, therefore, the latter belongs to the domain of $\mathrm{LD}_{\alpha 00\gamma} \bullet \mathrm{LD}_{\alpha 10\gamma}$, and t belongs to the domain of $\mathrm{LD}_\alpha \bullet \mathrm{LD}_{\alpha 00\gamma} \bullet \mathrm{LD}_{\alpha 10\gamma}$. The argument is similar if we start with the hypothesis that t belongs to the latter domain and wish to show that it belongs then to the domain of $\mathrm{LD}_\beta \bullet \mathrm{LD}_\alpha$. Then showing $t \bullet \beta \cdot \alpha = t \bullet \alpha \cdot \alpha 00\gamma \cdot \alpha 10\gamma$ is an easy computation. The argument is analogous for $\alpha 10 \sqsubseteq \beta$ and $\alpha 11 \sqsubseteq \beta$.

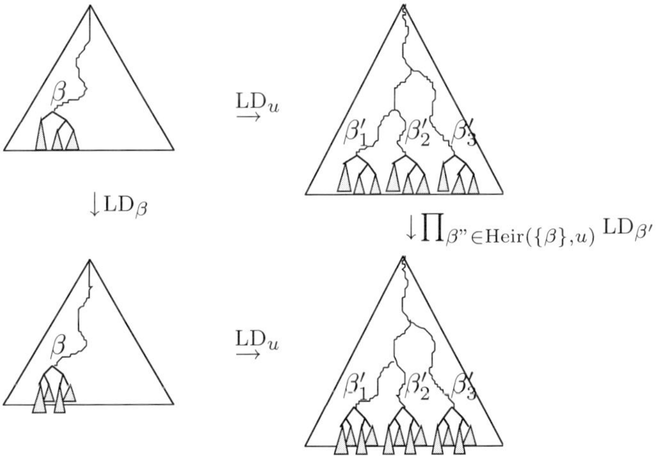

Figure 2.2. Relations in $\mathcal{G}_{LD}$: inheritance case

Assume now $\lg(u) \geq 2$, say $u = \alpha \cdot u_0$ with $\alpha \in \mathbf{A}$. By construction, the hypothesis that $\mathrm{Heir}(\{\beta\}, u)$ exists implies that $\mathrm{Heir}(\{\beta\}, \alpha)$ and $\mathrm{Heir}(\mathrm{Heir}(\{\beta\}, \alpha), u_0)$ exist, and that the latter is equal to $\mathrm{Heir}(\{\beta\}, u)$. By induction hypothesis, we have

$$\mathrm{LD}_\beta \bullet \mathrm{LD}_u = \mathrm{LD}_\alpha \bullet \prod_{\beta' \in \mathrm{Heir}(\{\beta\}, \alpha)} \mathrm{LD}_{\beta'} \bullet \mathrm{LD}_{u_0}.$$

By induction hypothesis again, we have, for each β' in $\mathrm{Heir}(\{\beta\},\alpha)$,

$$\mathrm{LD}_{\beta'} \bullet \mathrm{LD}_{u_0} = \mathrm{LD}_{u_0} \bullet \prod_{\beta''\in\mathrm{Heir}(\{\beta'\},u_0)} \mathrm{LD}_{\beta''},$$

and we obtain

$$\mathrm{LD}_{\beta} \bullet \mathrm{LD}_{u} = \mathrm{LD}_{\alpha} \bullet \mathrm{LD}_{u_0} \bullet \prod_{\beta'\in\mathrm{Heir}(\{\beta\},\alpha)} \prod_{\beta''\in\mathrm{Heir}(\{\beta'\},u_0)} \mathrm{LD}_{\beta''}. \tag{2.2}$$

Now, by Lemma 2.6(iii), the addresses β' in $\mathrm{Heir}(\{\beta\},\alpha)$ are pairwise orthogonal, so the associated operators $\mathrm{LD}_{\beta'}$ commute, and the double product in (2.2) is equal to the product $\prod_{\beta'\in\mathrm{Heir}(\{\beta\},u)} \mathrm{LD}_{\beta'}$ of (2.1). ■

Distribution relations

We describe now a third family of relations in $\mathcal{G}_{LD}$. Contrary to the previous families, which relie on uniform geometric considerations and do not really depend on the considered identity, this family is specific to the case of left self-distributivity. The idea is that distributing according to the operation $*$ of Section V.3 commutes with every operator LD_u up to a shift.

We recall that the term $t_0 * t$ is the LD-expansion of $t_0{\cdot}t$ obtained by distributing t_0 everywhere in t. Hence, there exists a positive word u, *a priori* depending on t_0 and t, such that LD_u maps $t_0{\cdot}t$ to $t_0 * t$. Such a word, which depends on t only, is computed easily.

Definition 2.11. (word $\boldsymbol{\delta}_t$) For t a term, the word $\boldsymbol{\delta}_t$ is defined by:

$$\boldsymbol{\delta}_t = \begin{cases} \varepsilon & \text{for } t \text{ a variable,}\\ \phi\cdot 1\boldsymbol{\delta}_{t_2}\cdot 0\boldsymbol{\delta}_{t_1} & \text{for } t = t_1{\cdot}t_2. \end{cases} \tag{2.3}$$

Proposition 2.12. (word $\boldsymbol{\delta}_t$) *Assume that t_0, t, t' are terms. Then $t_0{\cdot}t'$ lies in the domain of $\mathrm{LD}_{\boldsymbol{\delta}_t}$ if and only if the skeleton of t' includes the skeleton of t, and $\mathrm{LD}_{\boldsymbol{\delta}_t}$ maps $t_0{\cdot}t$ to $t_0 * t$ for every t_0.*

Proof. We use induction on t. The results are obvious if t is a variable. Otherwise, assume $t = t_1{\cdot}t_2$. The term $t_0{\cdot}t'$ belongs to the domain of $\mathrm{LD}_{\boldsymbol{\delta}_t}$ if and only if it belongs to the domain of LD_ϕ and its image belongs to the domain of $\mathrm{LD}_{1\boldsymbol{\delta}_{t_2}} \bullet \mathrm{LD}_{0\boldsymbol{\delta}_{t_1}}$. By definition of LD_ϕ, this happens if and only if t' has the form $t'_1{\cdot}t'_2$, and $t_0{\cdot}t'_e$ belongs to the domain of $\mathrm{LD}_{\boldsymbol{\delta}_{t_e}}$ for $e = 1, 2$. By induction hypothesis, this is true if and only if the skeleton of t'_e includes the skeleton of t_e for $e = 1, 2$, hence if and only if the skeleton of t' includes the skeleton of t. Then, using the induction hypothesis, we find

$$(t_0 \cdot t) \bullet \boldsymbol{\delta}_t = ((t_0 \cdot t_1) \cdot (t_0 \cdot t_2)) \bullet (0\boldsymbol{\delta}_{t_1} \cdot 1\boldsymbol{\delta}_{t_2})$$

$$= ((t_0 \cdot t_1) \bullet \delta_{t_1}) \cdot ((t_0 \cdot t_2) \bullet \delta_{t_2}) = (t_0 * t_1) \cdot (t_0 * t_2) = t_0 * t. \quad \blacksquare$$

The previous result implies that the skeleton of the term $t^L_{\delta_t}$ is the skeleton of $x{\cdot}t$ with x an arbitrary variable. An easy induction shows that δ_t is an $<$-decreasing enumeration of the inner part of the skeleton of the term t.

Example 2.13. (**word δ_t**) *Let $t = x_1{\cdot}(x_2{\cdot}x_3{\cdot}x_4){\cdot}x_5$. The skeleton of t (in $<$-decreasing order) is $\{\phi, 1, 11, 10, 101, 1011, 1010, 100, 0\}$, its inner part is $\{\phi, 1, 10, 101\}$, so we have $\delta_t = \phi \cdot 1 \cdot 10 \cdot 101$. Observe that, if t is an injective term, as here, the condition that the skeleton of t' includes the skeleton of t means that $t' = t^f$ holds for some substitution f. Then we have $(t_0{\cdot}t') \bullet \delta_t = t^g$ where $g(x) = t_0{\cdot}f(x)$ holds for every variable x: applying LD_{δ_t} means distributing until the outline of t is reached.*

The third type of relations we shall establish in the monoid $\mathcal{G}_{LD}$ involves the action of the words δ_t:

Proposition 2.14. (**distribution**) *Assume that u is a word on $\mathbf{A}$, and that LD_u maps t to t'. Then we have*

$$\mathrm{LD}_{\delta_t} \bullet \mathrm{LD}_u = \mathrm{LD}_{1u} \bullet \mathrm{LD}_{\delta_{t'}}. \tag{2.4}$$

Proof. (Figure 2.3) Again, the idea is simple. Applying LD_{δ_t} to the term $t_0{\cdot}t$ means distributing t_0 everywhere in t down to the level of the leaves, or, equivalently, replacing the term $t_0{\cdot}t$ with the term t^h where h is the substitution defined by $h(x) = t_0{\cdot}x$. Now, if LD_u maps t to t', then LD_{1u} maps $t_0{\cdot}t$ to $t_0{\cdot}t'$, and, by Lemma 1.23, LD_u maps t^h to t'^h. Now t'^h is the result of replacing every variable in t' by its product with t_0, *i.e.*, it is the term $t_0 * t'$, hence the result of applying $\mathrm{LD}_{\delta_{t'}}$ to $t_0{\cdot}t'$.

In order to make the proof precise—in particular, in order to verify that the domains of the operators coincide—we resort once more to an induction on the length of u. The result is obvious for $u = \varepsilon$. Assume now $\lg(u) = 1$, say $u = \alpha$ with $\alpha \in \mathbf{A}$. We use induction on the length of α as a word on $\{0, 1\}$. Assume first $\alpha = \phi$. We have to verify the equality

$$\mathrm{LD}_{\delta_t} \bullet \mathrm{LD}_{\phi} = \mathrm{LD}_1 \bullet \mathrm{LD}_{\delta_{t'}} \tag{2.5}$$

when LD_ϕ maps t to t'. The hypothesis that t belongs to the domain of LD_ϕ implies that 10 belongs to the skeleton of t, *i.e.*, that t can be expressed as $t_1{\cdot}(t_2{\cdot}t_3)$, and t' is then $(t_1{\cdot}t_2){\cdot}(t_1{\cdot}t_3)$. By definition, we have

$$\delta_t = \phi \cdot 1 \cdot 11\delta_{t_3} \cdot 10\delta_{t_2} \cdot 0\delta_{t_1},$$
$$\delta_{t'} = \phi \cdot 1 \cdot 11\delta_{t_3} \cdot 10\delta_{t_1} \cdot 0 \cdot 01\delta_{t_2} \cdot 00\delta_{t_1},$$

and, therefore, we have to compare the operators

$$\mathrm{LD}_{\emptyset} \bullet \mathrm{LD}_{1} \bullet \mathrm{LD}_{11\delta_{t_3}} \bullet \mathrm{LD}_{10\delta_{t_2}} \bullet \mathrm{LD}_{0\delta_{t_1}} \bullet \mathrm{LD}_{\emptyset}, \tag{2.6}$$

$$\mathrm{LD}_{1} \bullet \mathrm{LD}_{\emptyset} \bullet \mathrm{LD}_{1} \bullet \mathrm{LD}_{11\delta_{t_3}} \bullet \mathrm{LD}_{10\delta_{t_1}} \bullet \mathrm{LD}_{0} \bullet \mathrm{LD}_{01\delta_{t_2}} \bullet \mathrm{LD}_{00\delta_{t_1}}. \tag{2.7}$$

Using the commutation relations of Propositions 2.3 (orthogonality) and 2.10 (inheritance), we can transform the above expressions. First, in (2.6), we have $\mathrm{LD}_{0\delta_{t_1}} \bullet \mathrm{LD}_{\emptyset} = \mathrm{LD}_{\emptyset} \bullet \mathrm{LD}_{10\delta_{t_1}} \bullet \mathrm{LD}_{00\delta_{t_1}}$, and, similarly, $\mathrm{LD}_{10\delta_{t_2}} \bullet \mathrm{LD}_{\emptyset} = \mathrm{LD}_{\emptyset} \bullet \mathrm{LD}_{01\delta_{t_2}}$, while $\mathrm{LD}_{\emptyset}$ and $\mathrm{LD}_{11\delta_{t_3}}$ commute. Using orthogonality relations to move the factor $\mathrm{LD}_{10\delta_{t_1}}$ to the left, we find that the operator in (2.6) is equal to

$$\mathrm{LD}_{\emptyset} \bullet \mathrm{LD}_{1} \bullet \mathrm{LD}_{\emptyset} \bullet \mathrm{LD}_{11\delta_{t_3}} \bullet \mathrm{LD}_{10\delta_{t_1}} \bullet \mathrm{LD}_{01\delta_{t_2}} \bullet \mathrm{LD}_{00\delta_{t_1}}. \tag{2.8}$$

Similarly, by orthogonality, we can move the factor LD_{0} to the left in (2.7), and the latter becomes

$$\mathrm{LD}_{1} \bullet \mathrm{LD}_{\emptyset} \bullet \mathrm{LD}_{1} \bullet \mathrm{LD}_{0} \bullet \mathrm{LD}_{11\delta_{t_3}} \bullet \mathrm{LD}_{10\delta_{t_1}} \bullet \mathrm{LD}_{01\delta_{t_2}} \bullet \mathrm{LD}_{00\delta_{t_1}}. \tag{2.9}$$

So the operators in (2.6) and (2.7) are equal if and only if those in (2.8) and (2.9) are, and this in turn is true if and only if

$$\mathrm{LD}_{\emptyset} \bullet \mathrm{LD}_{1} \bullet \mathrm{LD}_{\emptyset} = \mathrm{LD}_{1} \bullet \mathrm{LD}_{\emptyset} \bullet \mathrm{LD}_{1} \bullet \mathrm{LD}_{0} \tag{2.10}$$

holds. The latter equality is checked directly: the canonical pair of terms associated with both sides is

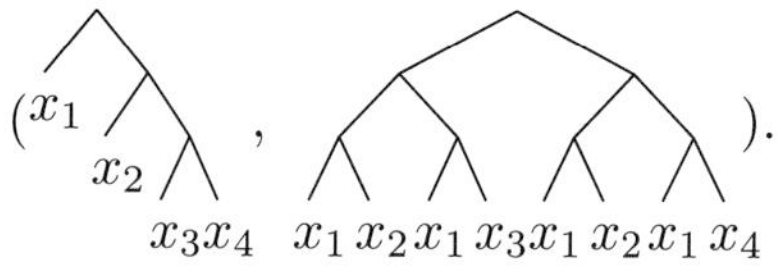

Assume now $\alpha = e\beta$, with $e = 0$ or $e = 1$. As $t \bullet \alpha$ is defined, t is not a variable. Write $t = t_1 \cdot t_2$, and $t' = t'_1 \cdot t'_2$. Assume first $e = 1$. Then $t' = t \bullet \alpha$ implies $t'_1 = t_1$ and $t'_2 = t_2 \bullet \beta$. By induction hypothesis, we have

$$\mathrm{LD}_{\delta_{t_2}} \bullet \mathrm{LD}_{\beta} = \mathrm{LD}_{1\beta} \bullet \mathrm{LD}_{\delta_{t'_2}},$$

which, by Lemma 2.2, implies $\mathrm{LD}_{1\delta_{t_2}} \bullet \mathrm{LD}_{1\beta} = \mathrm{LD}_{11\beta} \bullet \mathrm{LD}_{1\delta_{t'_2}}$, and, therefore,

$$\mathrm{LD}_{\emptyset} \bullet \mathrm{LD}_{1\delta_{t_2}} \bullet \mathrm{LD}_{1\beta} = \mathrm{LD}_{\emptyset} \bullet \mathrm{LD}_{11\beta} \bullet \mathrm{LD}_{1\delta_{t'_2}}.$$

Multiplying both sides by $\mathrm{LD}_{0\delta_{t_1}}$, which commutes with $\mathrm{LD}_{1\beta}$ by orthogonality, and using the fact that $\mathrm{LD}_{\emptyset}$ and $\mathrm{LD}_{11\beta}$ commute by Proposition 2.10 (inheritance), we obtain

$$\mathrm{LD}_{\delta_t} \bullet \mathrm{LD}_{1\beta} = \mathrm{LD}_{11\beta} \bullet \mathrm{LD}_{\delta_{t'}}.$$

The argument is similar for $\alpha = 0\beta$, using now $\mathrm{LD}_{\emptyset} \bullet \mathrm{LD}_{10\beta} = \mathrm{LD}_{01\beta} \bullet \mathrm{LD}_{\emptyset}$.

Assume now $\lg(u) \geq 2$, say $u = \alpha \cdot u_0$, with $\alpha \in \mathbf{A}$. Assuming that LD_{α}

maps t to t'' and using the induction hypothesis, we find

$$\mathrm{LD}_{\delta_t} \bullet \mathrm{LD}_u = \mathrm{LD}_{\delta_t} \bullet \mathrm{LD}_\alpha \bullet \mathrm{LD}_{u_0} = \mathrm{LD}_{1\alpha} \bullet \mathrm{LD}_{\delta_{t''}} \bullet \mathrm{LD}_{u_0}$$
$$= \mathrm{LD}_{1\alpha} \bullet \mathrm{LD}_{1u_0} \bullet \mathrm{LD}_{\delta_{t'}} = \mathrm{LD}_{1u} \bullet \mathrm{LD}_{\delta_{t'}}. \qquad \blacksquare$$

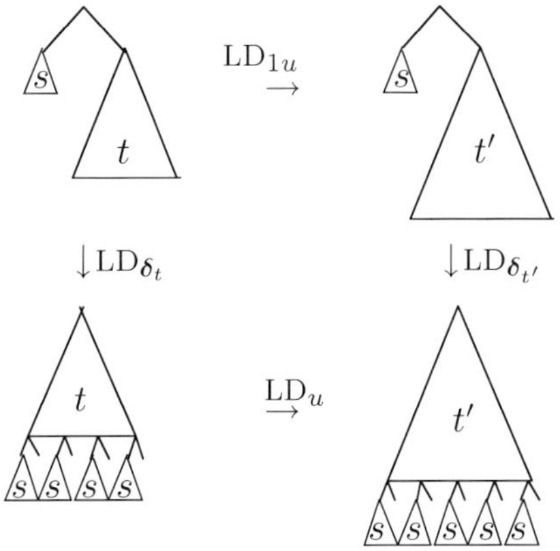

Figure 2.3. Relations in $\mathcal{G}_{LD}$: distribution case

---

Exercise 2.15. (associativity) *Adapt the previous results to the (easier) case where the associativity identity (A): $x(yz) = (xy)z$ replaces the left self-distributivity identity $x(yz) = (xy)(xz)$. In particular, write geometric relations between the associated operators A^+_α.*

Exercise 2.16. (inheritance) *Assume that α is an address which, as a word, contains p letters 0 and q letters 1. Prove that $\mathrm{heir}(\alpha, u) = 0^p1^q$ holds for some word u on* **A**.

Exercise 2.17. (origin) *(i) Assume that t_0 is a canonical term, and t is an LD-expansion of t_0. For α in the outline of t and α_0 in the outline of t_0, we say that α_0 is the* origin *of α in t_0 if α belongs to $\mathrm{Heir}(\{\alpha_0\}, u)$ for some positive word u such that LD_u maps t_0 to t. Prove that this definition does not depend on the choice of u.*

(ii) Prove that, for each address α in the outline of t, the last digit in the t_0-code of $\mathrm{cut}(t, \alpha)$ (as defined in Chapter VI) is the index of the variable that occurs at the origin of α in t_0.

---

3. Syntactic Proofs

We claim that the list of relations in the monoid $\mathcal{G}_{LD}$ established in Section 2 is complete, *i.e.*, any relation between the operators LD_w is a consequence of the previous relations. The proof of this result requires a rather tricky argument that will be completed in Chapter VIII. However, the general idea is simple: if every relation connecting the operators LD_w is a consequence of the relations we have listed, it should be possible to establish for each of the results obtained so far an abstract proof that resorts to our distinguished relations exclusively. This is the program we shall apply from now on. In this section, we prove what we call a syntactic version of the confluence property of Section V.3, namely a equivalence result for words in the monoid $(\mathbf{A} \cup \mathbf{A}^{-1})^*$ that directly implies the confluence property when words are replaced with the associated operators LD_w.

LD-relations

The list of relations obtained in Section 2 is not minimal. It will be convenient here to extract a proper sublist, in particular in order to obtain a family of relations eligible for the word reversing approach of Chapter II.

Definition 3.1. (LD-relation, relations $\equiv^+$ and $\equiv$) An *LD-relation* is defined to be a pair of words on $\mathbf{A}$ of one of the following five types:

$$(\gamma 0\alpha \cdot \gamma 1\beta\ ,\ \gamma 1\beta \cdot \gamma 0\alpha) \qquad \text{type } \perp$$
$$(\gamma 0\alpha \cdot \gamma\ ,\ \gamma \cdot \gamma 00\alpha \cdot \gamma 10\alpha) \qquad \text{type } 0$$
$$(\gamma 10\alpha \cdot \gamma\ ,\ \gamma \cdot \gamma 01\alpha) \qquad \text{type } 10$$
$$(\gamma 11\alpha \cdot \gamma\ ,\ \gamma \cdot \gamma 11\alpha) \qquad \text{type } 11$$
$$(\gamma 1 \cdot \gamma \cdot \gamma 1 \cdot \gamma 0\ ,\ \gamma \cdot \gamma 1 \cdot \gamma) \qquad \text{type } 1$$

Then $\equiv^+$ is defined to be the congruence on the monoid $\mathbf{A}^*$ generated by all LD-relations, and $\equiv$ is defined to be the congruence on the monoid $(\mathbf{A} \cup \mathbf{A}^{-1})^*$ generated by all LD-relations together with all pairs $(\gamma \cdot \gamma^{-1}, \varepsilon)$ and $(\gamma^{-1} \cdot \gamma, \varepsilon)$.

Example 3.2. ***(relations $\equiv^+$)*** *We have in $\mathbf{A}^*$*

$$\begin{aligned}
\phi \cdot 1 \cdot \phi \cdot 11 \cdot 1 &\equiv^+ 1 \cdot \phi \cdot 1 \cdot 0 \cdot 11 \cdot 1 && \textit{(type 1)}\\
&\equiv^+ 1 \cdot \phi \cdot 1 \cdot 11 \cdot 0 \cdot 1 && \textit{(type } \perp\textit{)}\\
&\equiv^+ 1 \cdot \phi \cdot 1 \cdot 11 \cdot 1 \cdot 0 && \textit{(type } \perp\textit{)}\\
&\equiv^+ 1 \cdot \phi \cdot 11 \cdot 1 \cdot 11 \cdot 10 \cdot 0 && \textit{(type 1)}\\
&\equiv^+ 1 \cdot 11 \cdot \phi \cdot 1 \cdot 11 \cdot 10 \cdot 0, && \textit{(type 11)}
\end{aligned}$$

so, for instance, $\phi \cdot 1 \cdot \phi \cdot 11 \cdot 1 \equiv^+ 1 \cdot 11 \cdot \phi \cdot 1 \cdot 11 \cdot 10 \cdot 0$ *holds:*

The following result follows from the verifications of Section 2:

Proposition 3.3. (operator) *For* u, u' *in* $\mathbf{A}^*$, $u \equiv^+ u'$ *implies* $\mathrm{LD}_u = \mathrm{LD}_{u'}$.

Proof. It suffices to prove the result when (u, u') is an LD-relation. Proposition 2.3 (orthogonality) gives the result for type $\perp$, Proposition 2.10 (inheritance) for types 0, 10 and 11, and Proposition 2.14 (distribution) for type 1. ■

In the case of words w, w' on $\mathbf{A} \cup \mathbf{A}^{-1}$, we shall see in Chapter VIII that $w \equiv w'$ implies a strong connection between the operators LD_w and $\mathrm{LD}_{w'}$, but the problem of possibly empty domains makes it difficult to state the connection directly.

Our first task is to check that the relations we have selected actually imply all relations of Section 2.

Lemma 3.4. *For* u, u' *words on* $\mathbf{A}$, $u \equiv^+ u'$ *implies* $\alpha u \equiv^+ \alpha u'$ *for every* α. *Similarly, for* w, w' *words on* $\mathbf{A} \cup \mathbf{A}^{-1}$, $w \equiv w'$ *implies* $\alpha w \equiv \alpha w'$ *for every* α.

Proof. The relation $u \equiv^+ u'$ holds if and only if there exists a finite sequence $u_0, \ldots, u_n$ satisfying $u = u_0$, $u' = u_n$, and, for each intermediate index i, the pair (u_{i-1}, u_i) has the form $(v_1 v v_2, v_1 v' v_2)$ where $\{v, v'\}$ is an LD-relation. By definition, for every LD-relation $\{v, v'\}$ and every address α, the pair $\{\alpha v, \alpha v'\}$ is still an LD-relation. Then $\alpha u_0, \ldots, \alpha u_n$ witnesses that $\alpha u \equiv^+ \alpha u'$ holds. The argument is similar in the case of $\equiv$. ■

Lemma 3.5. *Assume that* u_1 *and* u_2 *are words on* $\mathbf{A}$, *and every address in* u_1 *is orthogonal to every address in* u_2. *Then we have*

$$u_1 \cdot u_2 \equiv^+ u_2 \cdot u_1, \tag{3.1}$$

$$0u_1 \cdot 0u_2 \cdot \phi \equiv^+ \phi \cdot 00u_1 \cdot 00u_2 \cdot 10u_1 \cdot 10u_2, \tag{3.2}$$

$$10u_1 \cdot 10u_2 \cdot \phi \equiv^+ \phi \cdot 01u_1 \cdot 01u_2, \tag{3.3}$$

$$11u_1 \cdot 11u_2 \cdot \phi \equiv^+ \phi \cdot 11u_1 \cdot 11u_2. \tag{3.4}$$

Proof. Use an induction on $\lg(u_1) + \lg(u_2)$. The hypothesis implies that every address in $10u_1$ is orthogonal to every address in $00u_2$, and, therefore, that these addresses commute with respect to $\equiv^+$. ■

Proposition 3.6. (syntactic inheritance) *Assume that* u *is a word on* $\mathbf{A}$, β *is an address, and* $\mathrm{Heir}(\{\beta\}, u)$ *is defined. Then we have*

$$\beta \cdot u \equiv^+ u \cdot \prod_{\beta' \in \mathrm{Heir}(\{\beta\}, u)} \beta'. \tag{3.5}$$

Proof. Consider the proof of Proposition 2.10 (inheritance). It consists of an induction on the length of u. The case of length 1 corresponds to LD-relations of types ($\perp$), (0), (10) and (11) respectively. For the induction, the only requirement is for $\mathrm{LD}_{\beta'_1}$ and $\mathrm{LD}_{\beta'_2}$ to commute when the addresses β'_1 and β'_2 are orthogonal: but, in this case, the words $\beta'_1 \cdot \beta'_2$ and $\beta'_2 \cdot \beta'_1$ are $\equiv^+$-equivalent, and the argument remains the same. ■

Proposition 3.7. (syntactic distribution) *Assume that u is a word on* **A**, *and that LD_u maps t to t'. Then we have*

$$\boldsymbol{\delta}_t \cdot u \equiv^+ 1u \cdot \boldsymbol{\delta}_{t'}. \tag{3.6}$$

Proof. Again we come back to the proof of Proposition 2.14 (distribution). We use induction on $\lg(u)$. Assume $\lg(u) = 1$, say $u = \alpha$ with $\alpha \in \mathbf{A}$. We argue inductively on the length of α. Assume first $\alpha = \phi$. The hypothesis is that LD_ϕ maps t to t', and we wish to prove $\boldsymbol{\delta}_t \cdot \phi \equiv^+ 1 \cdot \boldsymbol{\delta}_{t'}$. The hypothesis that t lies in the domain of the operator LD_ϕ implies that t can be decomposed into $t_1 {\cdot} (t_2 {\cdot} t_3)$. Now we have

$$\begin{aligned}
\boldsymbol{\delta}_t \cdot \phi &= \phi \cdot 1 \cdot 11\boldsymbol{\delta}_{t_3} \cdot 10\boldsymbol{\delta}_{t_2} \cdot 0\boldsymbol{\delta}_{t_1} \cdot \phi \\
&\equiv^+ \phi \cdot 1 \cdot 11\boldsymbol{\delta}_{t_3} \cdot 10\boldsymbol{\delta}_{t_2} \cdot \phi \cdot 10\boldsymbol{\delta}_{t_1} \cdot 00\boldsymbol{\delta}_{t_1} && (0)\\
&\equiv^+ \phi \cdot 1 \cdot 11\boldsymbol{\delta}_{t_3} \cdot \phi \cdot 01\boldsymbol{\delta}_{t_2} \cdot 10\boldsymbol{\delta}_{t_1} \cdot 00\boldsymbol{\delta}_{t_1} && (10)\\
&\equiv^+ \phi \cdot 1 \cdot \phi \cdot 11\boldsymbol{\delta}_{t_3} \cdot 01\boldsymbol{\delta}_{t_2} \cdot 10\boldsymbol{\delta}_{t_1} \cdot 00\boldsymbol{\delta}_{t_1} && (11)\\
&\equiv^+ \phi \cdot 1 \cdot \phi \cdot 11\boldsymbol{\delta}_{t_3} \cdot 10\boldsymbol{\delta}_{t_1} \cdot 01\boldsymbol{\delta}_{t_2} \cdot 00\boldsymbol{\delta}_{t_1} && (\perp)\\
&\equiv^+ 1 \cdot \phi \cdot 1 \cdot 0 \cdot 11\boldsymbol{\delta}_{t_3} \cdot 10\boldsymbol{\delta}_{t_1} \cdot 01\boldsymbol{\delta}_{t_2} \cdot 00\boldsymbol{\delta}_{t_1} && (1)\\
&\equiv^+ 1 \cdot \phi \cdot 1 \cdot 11\boldsymbol{\delta}_{t_3} \cdot 10\boldsymbol{\delta}_{t_1} \cdot 0 \cdot 01\boldsymbol{\delta}_{t_2} \cdot 00\boldsymbol{\delta}_{t_1} = 1 \cdot \boldsymbol{\delta}_{t'}. && (\perp)
\end{aligned}$$

Assume now $\alpha = 0\beta$. Let us write $t = t_1 {\cdot} t_2$ and $t' = t'_1 {\cdot} t_2$. Then LD_β maps t_1 to t'_1, and the induction hypothesis gives $\boldsymbol{\delta}_{t_1} \cdot \beta \equiv^+ 1\beta \cdot \boldsymbol{\delta}_{t'_1}$. By Lemma 2.2, this implies $0\boldsymbol{\delta}_{t_1} \cdot 0\beta \equiv^+ 01\beta \cdot 0\boldsymbol{\delta}_{t'_1}$, and we deduce

$$\begin{aligned}
\boldsymbol{\delta}_t \cdot \alpha = \phi \cdot 1\boldsymbol{\delta}_{t_2} \cdot 0\boldsymbol{\delta}_{t_1} \cdot 0\beta &\equiv^+ \phi \cdot 0\boldsymbol{\delta}_{t_1} \cdot 0\beta \cdot 1\boldsymbol{\delta}_{t_2} && (\perp)\\
&\equiv^+ \phi \cdot 01\beta \cdot 0\boldsymbol{\delta}_{t'_1} \cdot 1\boldsymbol{\delta}_{t_2} && (\text{hyp.})\\
&\equiv^+ 10\beta \cdot \phi \cdot 0\boldsymbol{\delta}_{t'_1} \cdot 1\boldsymbol{\delta}_{t_2} = 1\alpha \cdot \boldsymbol{\delta}_{t'}. && (10)
\end{aligned}$$

The argument is similar for $\alpha = 1\beta$, and the induction on $\lg(u)$ is easy. ■

The word $\boldsymbol{\Delta}_t$

In Section V.3, we associated with every term t a distinguished LD-expansion ∂t of t. Here we explicitly determine a positive word $\boldsymbol{\Delta}_t$ such that the operator $\mathrm{LD}_{\boldsymbol{\Delta}_t}$ maps the term t to the term ∂t. Let us recall that the term ∂t is defined inductively by the rule

$$\partial t = \begin{cases} t & \text{for } t \text{ a variable,} \\ \partial t_1 * \partial t_2 & \text{for } t = t_1 \cdot t_2. \end{cases}$$

This leads to a possible choice for $\boldsymbol{\Delta}_t$. However, it will be more convenient to start with another equivalent approach.

Definition 3.8. **(word $\alpha^{(p)}$)** For α an address, we define $\alpha^{(0)} = \varepsilon$, and $\alpha^{(p)} = \alpha 1^{p-1} \cdot \alpha 1^{p-2} \cdot \cdots \cdot \alpha 1 \cdot \alpha$ for $p \geq 1$.

Example 3.9. **(word $\alpha^{(p)}$)** *By construction, the term t belongs to the domain of the operator $\mathrm{LD}_{\phi^{(r)}}$ if and only if $1^r 0$, or, equivalently, 1^{r+1}, belongs to its skeleton, i.e., if the right height of t is at least $r+1$. In this case, t can be written as*

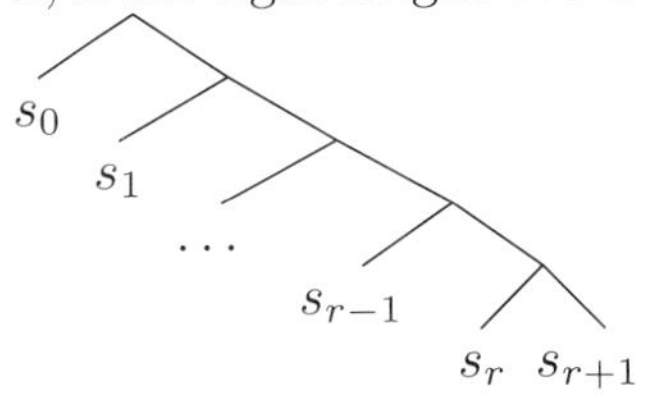

, and the image of t under $\mathrm{LD}_{\phi^{(r)}}$ is the term

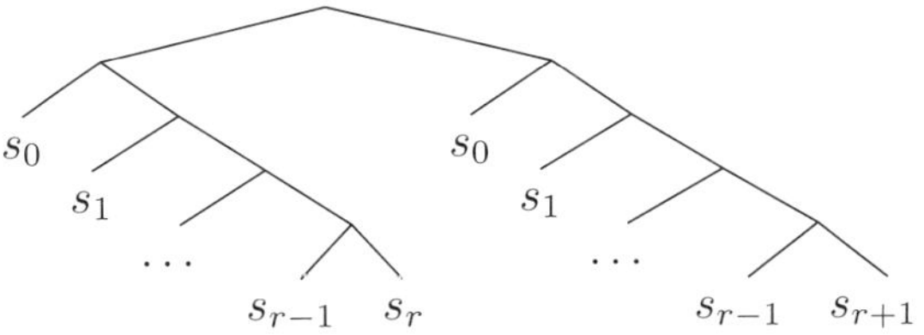

.

Lemma 3.10. *For $\mathrm{ht}_R(t) = r+1$, we have $\partial t = \partial s_1 \cdot \partial s_2$ with $s_1 \cdot s_2 = t \bullet \phi^{(r)}$.*

Proof. Assume $t = t_1 \cdot t_2$. We use induction on r. For $r = 0$, t_2 is a variable, say x, the result follows from $\partial(t_1 \cdot x) = \partial t_1 \cdot x$. Otherwise, let $s'_1 \cdot s'_2 = t_2 \bullet \phi^{(r-1)}$. By induction hypothesis, we have $\partial t_2 = \partial s'_1 \cdot \partial s'_2$, so we deduce

$$\partial t = \partial t_1 * (\partial s'_1 \cdot \partial s'_2) = (\partial t_1 * \partial s'_1) \cdot (\partial t_1 * \partial s'_2) = \partial(t_1 \cdot s'_1) \cdot \partial(t_1 \cdot s'_2).$$

Now, by construction, we have $s_e = t_1 \cdot s'_e$ for $e = 1, 2$. ∎

Definition 3.11. (**word $\boldsymbol{\Delta}_t$**) For t a term of right height $r+1$, the word $\boldsymbol{\Delta}_t$ is defined by:

$$\boldsymbol{\Delta}_t = \begin{cases} \varepsilon & \text{if } t \text{ is a variable,} \\ \phi^{(r)} \cdot 1\boldsymbol{\Delta}_{s_2} \cdot 0\boldsymbol{\Delta}_{s_1} & \text{otherwise, with } s_1{\cdot}s_2 = t \bullet \phi^{(r)}. \end{cases}$$

Example 3.12. *(***word $\boldsymbol{\Delta}_t$***) Let t be the term* 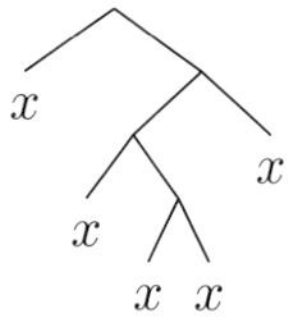 *x. The right height of t is 2, so the exponent of ϕ in $\boldsymbol{\Delta}_t$ is 1. The right subterm of the image of t under LD_ϕ is the term $s_2 = x^{[2]}$, while its left subterm is $s_1 = x^{[4]}$. We have $\partial s_2 = s_2$, hence $\boldsymbol{\Delta}_{s_2} = \varepsilon$. The right height of s_1 is 3, so the exponent of ϕ in $\boldsymbol{\Delta}_{s_1}$ is 2. The right and left subterms of the image of s_1 under $\mathrm{LD}_{\phi^{(2)}}$ are the terms $s_3 = s_4 = x^{[3]}$. The right height of s_3 is 2, so the exponent of ϕ in $\boldsymbol{\Delta}_{s_3}$ is 1. The right and left subterms of the image of s_3 under LD_ϕ are $x^{[2]}$, so we are done. By gathering the elements, we find*

$$\boldsymbol{\Delta}_t = \phi \cdot 0\boldsymbol{\Delta}_{s_1} = \phi \cdot 0^{(2)} \cdot 01\boldsymbol{\Delta}_{s_4} \cdot 00\boldsymbol{\Delta}_{s_3} = \phi \cdot 0^{(2)} \cdot 01 \cdot 00.$$

Proposition 3.13. (**word $\boldsymbol{\Delta}_t$**) *For every term t, we have $t \bullet \boldsymbol{\Delta}_t = \partial t$.*

The result is a straightforward consequence of Lemma 3.10. The role of the words $\boldsymbol{\Delta}_t$ will be crucial in our study of the geometry monoid $\mathcal{G}_{LD}$, as was the role of the terms ∂t in the study of LD-equivalence. Technically, the words $\boldsymbol{\Delta}_t$ share many properties with Garside's fundamental braid words $\boldsymbol{\Delta}_n$, to which they are deeply connected, as we shall see in Chapter VIII.

The confluence property

We have proved in Chapter V that any two LD-expansions of a given term admit a common LD-expansion. In the present framework, this means that, if t is a term and u, v are two positive words such that both $t \bullet u$ and $t \bullet v$ exist, then there exist words u' and v'—possibly depending on t—such that the LD-expansions $t \bullet uv'$ and $t \bullet vu'$ exist and are equal. This implies $\mathrm{LD}_{uv'} = \mathrm{LD}_{vu'}$, and, therefore, makes it plausible that $uv' \equiv^+ vu'$ holds. Here we shall establish a strong form of this result, namely that, if u, v are positive words on $\mathbf{A}$, there always exist positive words u', v' satisfying $uv' \equiv^+ vu'$, and, in addition, the domain of the operator $\mathrm{LD}_{uv'}$ is the intersection of the domains of LD_u

and LD_v. This result implies the confluence of LD-expansion as stated before, and it shows in addition that the complement words u', v' do not depend on t.

Our syntactic proof will follow the one of Chapter V, with the words $\boldsymbol{\Delta}_t$ playing the role of the terms ∂t. We begin with several inductive formulas expressing various ways of constructing the terms ∂t.

Lemma 3.14. *Assume $t = t_1{\cdot}t_2$. Then we have*

$$\boldsymbol{\Delta}_t \equiv^+ 1\boldsymbol{\Delta}_{t_2} \cdot 0\boldsymbol{\Delta}_{t_1} \cdot \boldsymbol{\delta}_{\partial t_2}, \tag{3.7}$$

$$\boldsymbol{\Delta}_t \equiv^+ 0\boldsymbol{\Delta}_{t_1} \cdot \boldsymbol{\delta}_{t_2} \cdot \boldsymbol{\Delta}_{t_2}, \tag{3.8}$$

$$\boldsymbol{\Delta}_t \equiv^+ \boldsymbol{\delta}_{t_2} \cdot \prod_{\alpha \in \mathrm{Out}(t_2)} \alpha 0\boldsymbol{\Delta}_{t_1} \cdot \boldsymbol{\Delta}_{t_2}. \tag{3.9}$$

Proof. We prove (3.7) using induction on t_2. Let $r + 1 = \mathrm{ht}_R(t)$, and $s_1{\cdot}s_2 = t \bullet \phi^{(r)}$. We recall that $\boldsymbol{\Delta}_t$ is defined to be $\phi^{(r)} \cdot 1\boldsymbol{\Delta}_{s_1} \cdot 0\boldsymbol{\Delta}_{s_0}$. For t_2 a variable, we have $\partial t_2 = t_2$, $r = 0$, hence $s_1 = t_1$, $s_2 = t_2$, and $\boldsymbol{\Delta}_t = 0\boldsymbol{\Delta}_{t_1}$, so (3.7) is an equality. Otherwise, let $s'_1{\cdot}s'_2 = t_2 \bullet \phi^{(r-1)}$. By construction, we have $s_e = t_1{\cdot}s'_e$ for $e = 1, 2$. The size of the right subterm of s_e, namely s'_e, is smaller than the size of the right subterm of t, namely t_2, so the induction hypothesis gives

$$\boldsymbol{\Delta}_{s_2} \equiv^+ 1\boldsymbol{\Delta}_{s'_2} \cdot 0\boldsymbol{\Delta}_{t_1} \cdot \boldsymbol{\delta}_{\partial s'_2}, \qquad \boldsymbol{\Delta}_{s_1} \equiv^+ 1\boldsymbol{\Delta}_{s'_1} \cdot 0\boldsymbol{\Delta}_{t_1} \cdot \boldsymbol{\delta}_{\partial s'_1},$$

and we deduce

$$\boldsymbol{\Delta}_t \equiv^+ \phi^{(r)} \cdot 11\boldsymbol{\Delta}_{s'_2} \cdot 10\boldsymbol{\Delta}_{t_1} \cdot 1\boldsymbol{\delta}_{\partial s'_2} \cdot 01\boldsymbol{\Delta}_{s'_1} \cdot 00\boldsymbol{\Delta}_{t_1} \cdot 0\boldsymbol{\delta}_{\partial s'_1}.$$

Using type $(\perp)$ relations, this can be rearranged into

$$\boldsymbol{\Delta}_t \equiv^+ \phi^{(r)} \cdot 11\boldsymbol{\Delta}_{s'_2} \cdot 01\boldsymbol{\Delta}_{s'_1} \cdot 10\boldsymbol{\Delta}_{t_1} \cdot 00\boldsymbol{\Delta}_{t_1} \cdot 1\boldsymbol{\delta}_{\partial s'_2} \cdot 0\boldsymbol{\delta}_{\partial s'_1}.$$

We have $\phi^{(r)} = 1^{(r-1)} \cdot \phi$, and using LD-relations of type (11), (10) and (0), we can push the factor ϕ to the right, thus obtaining

$$\boldsymbol{\Delta}_t \equiv^+ 1^{(r-1)} \cdot 11\boldsymbol{\Delta}_{s'_2} \cdot 10\boldsymbol{\Delta}_{s'_1} \cdot 0\boldsymbol{\Delta}_{t_1} \cdot \phi \cdot 1\boldsymbol{\delta}_{\partial s'_2} \cdot 0\boldsymbol{\delta}_{\partial s'_1}.$$

Now, by definition, we have $1^{(r-1)} \cdot 11\boldsymbol{\Delta}_{s'_2} \cdot 10\boldsymbol{\Delta}_{s'_1} = 1\boldsymbol{\Delta}_{s'_1 \cdot s'_2} = 1\boldsymbol{\Delta}_{t_2}$, and $\phi \cdot 1\boldsymbol{\delta}_{\partial s'_2} \cdot 0\boldsymbol{\delta}_{\partial s'_1} = \boldsymbol{\delta}_{\partial t_2}$. So we have obtained (3.7).

The other formulas follow easily. Indeed, we deduce (3.8) from (3.7) by using Proposition 3.7 (syntactic distribution), since, by construction, $\mathrm{LD}_{\boldsymbol{\Delta}_{t_1}}$ maps t_1 to ∂t_1. We deduce (3.9) from (3.8) by using Proposition 3.6 (syntactic inheritance), since, by construction again, the set $\mathrm{Heir}(\{0\}, \boldsymbol{\delta}_{t_1})$ exists and is equal to the set of all addresses $\beta 0$ for β in the outline of t_1. ∎

Remark 3.15. Let u be the word involved in the right hand side of (3.7). The diagram

$$t = t_1 \cdot t_2 \overset{1\boldsymbol{\Delta}_{t_2}}{\longmapsto} t_1 \cdot \partial t_2 \overset{0\boldsymbol{\Delta}_{t_1}}{\longmapsto} \partial t_1 \cdot \partial t_2 \overset{\boldsymbol{\delta}_{\partial t_2}}{\longmapsto} \partial t_1 * \partial t_2 = \partial t$$

makes it obvious that the operator LD_u maps the term t to the term ∂t. So, by Proposition 1.26 (compatibility), we know that the operators $\mathrm{LD}_{\boldsymbol{\Delta}_t}$ and LD_u coincide. However, the equivalence $\boldsymbol{\Delta}_t \equiv^+ u$ is a stronger result.

Now, we follow Section V.3. The first result is that the term ∂t is an LD-expansion of every basic LD-expansion of t. Its syntactic counterpart is:

Lemma 3.16. *Assume that t belongs to the domain of LD_α. Then there exists a positive word u satisfying $\alpha \cdot u \equiv^+ \boldsymbol{\Delta}_t$.*

Proof. We use induction on α. For $\alpha = \phi$, the result follows from Formula (3.9), which gives a word that explicitly begins with ϕ provided that the right subterm t_1 of t exists, *i.e.*, t is not a variable, and $\boldsymbol{\delta}_{t_1} \neq \varepsilon$ holds, *i.e.*, t_1 is not a variable, so for $\mathrm{ht}_R(t) \geq 2$, which is the case if t belongs to the domain of LD_ϕ. Assume now $\alpha = 0\beta$. Assume $t = t_1{\cdot}t_2$. Formula (3.7) shows that $\boldsymbol{\Delta}_t$ is $\equiv^+$-equivalent to a word that begins with $0\boldsymbol{\Delta}_{t_1}$. By construction, t_1 lies in the domain of LD_β, so, by induction hypothesis, $\boldsymbol{\Delta}_{t_1}$ is $\equiv^+$-equivalent to a positive word of the form $\beta \cdot u_0$, and we obtain

$$\boldsymbol{\Delta}_t \equiv^+ \alpha \cdot 0u_0 \cdot 1\boldsymbol{\Delta}_{t_2} \cdot \boldsymbol{\delta}_{\partial t_2}.$$

The argument is similar for $\alpha = 1\beta$, since, by type ($\perp$) relations, we have also

$$\boldsymbol{\Delta}_t \equiv^+ 1\boldsymbol{\Delta}_{t_2} \cdot 0\boldsymbol{\Delta}_{t_1} \cdot \boldsymbol{\delta}_{\partial t_2}.$$ ∎

The next step is the counterpart to the fact that the operator ∂ is increasing with respect to LD-expansion.

Lemma 3.17. *Assume that LD_α maps t to t'. Then there exists a positive word u satisfying $\alpha \cdot \boldsymbol{\Delta}_{t'} \equiv^+ \boldsymbol{\Delta}_t \cdot u$.*

Proof. We begin with the case $\alpha = \phi$. We argue inductively on the size of the subterm $\mathrm{sub}(t, 11)$, which exists since t is assumed to lie in the domain of LD_ϕ. Write $t = t_1{\cdot}(t_2{\cdot}t_3)$. Assume first that t_3 is a variable. Then we have $\mathrm{ht}_R(t) = 2$, and we have

$$\boldsymbol{\Delta}_t = \phi \cdot 1\boldsymbol{\Delta}_{s_2} \cdot 0\boldsymbol{\Delta}_{s_1}, \tag{3.10}$$

with $s_1 = t_1{\cdot}t_2$ and $s_2 = t_1{\cdot}t_3$. Then s_1 is the left subterm of t', and s_2 is its right subterm. So, by Formula (3.7), we have

$$\boldsymbol{\Delta}_t \equiv^+ 1\boldsymbol{\Delta}_{s_2} \cdot 0\boldsymbol{\Delta}_{s_1} \cdot \boldsymbol{\delta}_{\partial s_2}. \tag{3.11}$$

By comparing (3.10) and (3.11), we obtain $\phi \cdot \boldsymbol{\Delta}_{t'} \equiv^+ \boldsymbol{\Delta}_t \cdot \boldsymbol{\delta}_{\partial s_2}$, which has the desired form.

Assume now that t_3 is not a variable. Let $r+1 = \mathrm{ht}_R(t)$, $s_1{\cdot}s_2 = t \bullet \phi^{(r)}$, and $s'_1{\cdot}s'_2 = t' \bullet \phi^{(r)}$. By definition, and owing to the fact that the terms t and t' have the same right height, we have

$$\boldsymbol{\Delta}_t = \phi^{(r)} \cdot 1\boldsymbol{\Delta}_{s_2} \cdot 0\boldsymbol{\Delta}_{s_1}, \quad \text{and} \quad \boldsymbol{\Delta}_{t'} = \phi^{(r)} \cdot 1\boldsymbol{\Delta}_{s'_2} \cdot 0\boldsymbol{\Delta}_{s'_1}.$$

By construction, we have $s'_1 = s_1 \bullet \phi$, and $s'_2 = s_2 \bullet \phi$. Moreover, the size of the 11-subterms in s_1 and s_2 are smaller than the size of the 11-subterm in t. So, by induction hypothesis, there exist positive words u_1, u_2 satisfying $\phi \cdot \boldsymbol{\Delta}_{s'_e} \equiv^+ \boldsymbol{\Delta}_{s_e} \cdot u_e$ for $e = 1, 2$. Then, an induction gives the equivalence

$$\phi \cdot \phi^{(r)} \equiv^+ \phi^{(r)} \cdot 1 \cdot 0$$

for $r \geq 2$: the basic case is $r = 2$, where it is a type 1 relation. So we obtain

$$\begin{aligned}\phi \cdot \boldsymbol{\Delta}_{t'} = \phi \cdot \phi^{(r)}{\cdot}1\boldsymbol{\Delta}_{s'_2} \cdot 0\boldsymbol{\Delta}_{s'_1} &\equiv^+ \phi^{(r)} \cdot 1 \cdot 1\boldsymbol{\Delta}_{s'_2} \cdot 0 \cdot 0\boldsymbol{\Delta}_{s'_1} \\ &\equiv^+ \phi^{(r)} \cdot 1\boldsymbol{\Delta}_{s_2} \cdot 1u_2 \cdot 0\boldsymbol{\Delta}_{s_1} \cdot 0u_1 \\ &\equiv^+ \phi^{(r)} \cdot 1\boldsymbol{\Delta}_{s_2} \cdot 0\boldsymbol{\Delta}_{s_1} \cdot 1u_2 \cdot 0u_1 = \boldsymbol{\Delta}_t \cdot 1u_2 \cdot 0u_1.\end{aligned}$$

Assume now $\alpha = 0\beta$. For $t = t_1{\cdot}t_2$, we find $t' = t'_1{\cdot}t_2$, with $t'_1 = t_1 \bullet \beta$. By induction hypothesis, there exists a positive word u_1 satisfying $\beta \cdot \boldsymbol{\Delta}_{t'_1} \equiv^+ \boldsymbol{\Delta}_{t_1} \cdot u_1$. Starting from (3.7) and using Proposition 3.6 (syntactic inheritance), we obtain

$$\begin{aligned}\alpha \cdot \boldsymbol{\Delta}_{t'} &\equiv^+ \alpha \cdot 0\boldsymbol{\Delta}_{t'_1} \cdot 1\boldsymbol{\Delta}_{t_2} \cdot \boldsymbol{\delta}_{\partial t_2} \equiv^+ \mathrm{sh}_0(\beta \cdot \boldsymbol{\Delta}_{t'_1}) \cdot 1\boldsymbol{\Delta}_{t_2} \cdot \boldsymbol{\delta}_{\partial t_2} \\ &\equiv^+ \mathrm{sh}_0(\boldsymbol{\Delta}_{t_1} \cdot u_1) \cdot 1\boldsymbol{\Delta}_{t_2} \cdot \boldsymbol{\delta}_{\partial t_2} \equiv^+ 0\boldsymbol{\Delta}_{t_1} \cdot 1\boldsymbol{\Delta}_{t_2} \cdot 0u_1 \cdot \boldsymbol{\delta}_{\partial t_2} \\ &\equiv^+ 0\boldsymbol{\Delta}_{t_1} \cdot 1\boldsymbol{\Delta}_{t_2} \cdot \boldsymbol{\delta}_{\partial t_2} \cdot \prod_{\alpha \in \mathrm{Out}(t_2)} \alpha 0 u_1 \equiv^+ \boldsymbol{\Delta}_t \cdot \prod_{\alpha \in \mathrm{Out}(t_2)} \alpha 0 u_1.\end{aligned}$$

Assume finally $\alpha = 1\beta$. We have $t' = t_1{\cdot}t'_2$, with $t'_2 = t_1 \bullet \beta$. By induction hypothesis, we have $\beta \cdot \boldsymbol{\Delta}_{t'_2} \equiv^+ \boldsymbol{\Delta}_{t_2} \cdot u_2$ for some positive word u_2. We deduce

$$\begin{aligned}\alpha \cdot \boldsymbol{\Delta}_{t'} &\equiv^+ \alpha \cdot 0\boldsymbol{\Delta}_{t_1} \cdot 1\boldsymbol{\Delta}_{t'_2} \cdot \boldsymbol{\delta}_{\partial t'_2} \equiv^+ 0\boldsymbol{\Delta}_{t_1} \cdot \mathrm{sh}_1(\beta \cdot \boldsymbol{\Delta}_{t'_2}) \cdot \boldsymbol{\delta}_{\partial t'_2} \\ &\equiv^+ 0\boldsymbol{\Delta}_{t_1} \cdot \mathrm{sh}_1(\boldsymbol{\Delta}_{t_2} \cdot u_2) \cdot \boldsymbol{\delta}_{\partial t'_2} \equiv^+ 0\boldsymbol{\Delta}_{t_1} \cdot 1\boldsymbol{\Delta}_{t_2} \cdot 1u_2 \cdot \boldsymbol{\delta}_{\partial t'_2} \\ &\equiv^+ 0\boldsymbol{\Delta}_{t_1} \cdot 1\boldsymbol{\Delta}_{t_2} \cdot \boldsymbol{\delta}_{\partial t_2} \cdot u_2 \equiv^+ \boldsymbol{\Delta}_t \cdot u_2 :\end{aligned}$$

then $1u_2 \cdot \boldsymbol{\delta}_{\partial t'_2} \equiv^+ \boldsymbol{\delta}_{\partial t_2} \cdot u_2$ follows from Proposition 3.7 (syntactic distribution) as, by construction, we have $t_2 \bullet (\beta{\cdot}\boldsymbol{\Delta}_{t'_2}) = \partial t'_2 = t_2 \bullet \boldsymbol{\Delta}_{t_2} \cdot u_2$ and, therefore, $\partial t'_2 = \partial t_2 \bullet u_2$. ∎

Remark 3.18. The previous proof shows not only the existence of a positive word u satisfying $\phi \cdot \boldsymbol{\Delta}_{t'} \equiv^+ \boldsymbol{\Delta}_t \cdot u$, but it also gives an inductive formula for constructing such a word, namely

$$u = \prod_{\alpha \in \mathrm{Out}(\partial t_3)} (\alpha \cdot \alpha 0 \boldsymbol{\delta}_{\partial t_2})$$

for $t = t_1{\cdot}(t_2{\cdot}t_3)$ and $t' = t \bullet \phi$. This formula is easily understandable: ∂t is obtained from ∂t_3 by substituting every variable x with $\partial t_1 * (\partial t_2{\cdot}x)$, *i.e.*, $(\partial t_1{*}\partial t_2){\cdot}(\partial t_1{\cdot}x)$, while $\partial t'$ is obtained from ∂t_3 by substituting every variable x with $(\partial t_1 * \partial t_2) * (\partial t_1{\cdot}x)$, *i.e.*, $(\partial t_1 * \partial t_2) * (\partial t_1{\cdot}x)$. So $\partial t'$ is obtained from ∂t by applying $\mathrm{LD}_{\boldsymbol{\delta}_{\partial t_2 \cdot x}}$, *i.e.*, $\mathrm{LD}_{(\phi \cdot 0 \boldsymbol{\delta}_{\partial t_2})}$, at each address in the outline of ∂t_3.

Lemma 3.19. *Assume that u is a word on $\mathbf{A}$, and LD_u maps t to t'. Then there exists a positive word u' satisfying*

$$u \cdot \boldsymbol{\Delta}_{t'} \equiv^+ \boldsymbol{\Delta}_t \cdot u'. \tag{3.12}$$

Proof. We use induction on $\lg(u)$. For $u = \varepsilon$, the result is trivial. For $\lg(u) = 1$, the result is Lemma 3.17. Otherwise, assume $u = u_1 \cdot u_2$ where neither u_1 nor u_2 is empty. Let $t_1 = t \bullet u_1$. By induction hypothesis, there exist u'_1, u'_2 satisfying $u_1 \cdot \boldsymbol{\Delta}_{t_1} \equiv^+ \boldsymbol{\Delta}_t \cdot u'_1$ and $u_2 \cdot \boldsymbol{\Delta}_{t'} \equiv^+ \boldsymbol{\Delta}_{t_1} \cdot u'_2$. We deduce

$$u \cdot \boldsymbol{\Delta}_{t'} \equiv^+ u_1 \cdot \boldsymbol{\Delta}_{t_1} \cdot u'_2 \equiv^+ \boldsymbol{\Delta}_t \cdot u'_1 \cdot u'_2. \quad \blacksquare$$

Let us turn now to the general case. To this end, we iterate the words $\boldsymbol{\Delta}_t$.

Definition 3.20. (word $\boldsymbol{\Delta}_t^{(k)}$) For every term t, we put $\boldsymbol{\Delta}_t^{(0)} = \varepsilon$, and $\boldsymbol{\Delta}_t^{(k)} = \boldsymbol{\Delta}_t \cdot \boldsymbol{\Delta}_{\partial t} \cdot \cdots \cdot \boldsymbol{\Delta}_{\partial^{k-1} t}$ for $k \geq 1$.

Lemma 3.21. *Assume that u is a word on $\mathbf{A}$ of length k at most, and t lies in the domain of LD_u. Then there exists a word v' on $\mathbf{A}$ satisfying $u \cdot v' \equiv^+ \boldsymbol{\Delta}_t^{(k)}$.*

Proof. (Figure 3.1) We use induction on k. For $k = 0$, the result is trivial. Otherwise, write $u = u_0 \cdot \alpha$ with $\alpha \in \mathbf{A}$. By induction hypothesis, there exists a positive word v'_0 satisfying $u_0 \cdot v'_0 \equiv^+ \boldsymbol{\Delta}_t^{(k-1)}$. Let t' be the image of t under LD_{u_0}. By hypothesis, t' lies in the domain of LD_α, so, by Lemma 3.17, there exists a positive word v satisfying $\alpha \cdot v \equiv^+ \boldsymbol{\Delta}_{t'}$. Applying Lemma 3.19 to the terms t' and $\partial^{k-1} t$, we see that there exists a positive word v''_0 satisfying $v'_0 \cdot \boldsymbol{\Delta}_{\partial^{k-1} t} \equiv^+ \boldsymbol{\Delta}_{t'} \cdot v''_0$. We deduce

$$u{\cdot}v{\cdot}v''_0 = u_0{\cdot}\alpha{\cdot}v{\cdot}v''_0 \equiv^+ u_0{\cdot}\boldsymbol{\Delta}_{t'}{\cdot}v''_0 \equiv^+ u_0{\cdot}v'_0{\cdot}\boldsymbol{\Delta}_{\partial^{k-1} t} \equiv^+ \boldsymbol{\Delta}_t^{(k-1)}{\cdot}\boldsymbol{\Delta}_{\partial^{k-1} t} = \boldsymbol{\Delta}_t^{(k)}.$$

So taking $v' = v \cdot v''_0$ gives the result. $\blacksquare$

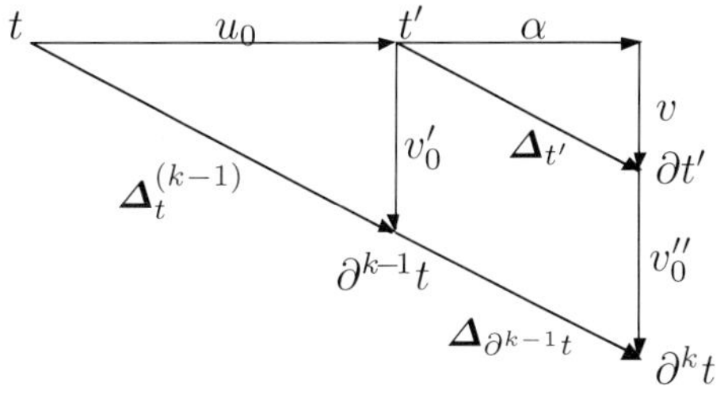

Figure 3.1. Divisors of $\boldsymbol{\Delta}_t^{(k)}$

Proposition 3.22. (**syntactic confluence**) *Assume that u, v are words on $\mathbf{A}$ of length k at most. Then there exist words u', v' on $\mathbf{A}$ satisfying*

$$u \cdot v' \equiv^+ v \cdot u' \equiv^+ \boldsymbol{\Delta}^{(k)}_{t^L_{u,v}}. \tag{3.13}$$

Proof. By Proposition 1.21 (intersection), the intersection of the domains of the operators LD_u and LD_v consists of the substitutes of the term $t^L_{u,v}$. Applying Lemma 3.21 to $t^L_{u,v}$ gives two positive words u', v' such that both $u \cdot v'$ and $v \cdot u'$ are $\equiv^+$-equivalent to $\boldsymbol{\Delta}^{(k)}_{t^L_{u,v}}$. ■

Observe that, in the situation of Proposition 3.22, the domain of the operators $\mathrm{LD}_{uv'}$ and $\mathrm{LD}_{vu'}$ is exactly the intersection of the domains of LD_u and LD_v, *i.e.*, we have found a common right multiple for u and v such that the associated operator has the largest possible domain.

We conclude with two corollaries involving words on $\mathbf{A} \cup \mathbf{A}^{-1}$. The first is the syntactic counterpart to the result that, if the LD-distance of t and t' is k at most, then $\partial^k t$ is an LD-expansion of t:

Proposition 3.23. (**quotient**) *If w is a word on $\mathbf{A} \cup \mathbf{A}^{-1}$ containing k positive addresses, and t lies in the domain of LD_w, then there exists a word u on $\mathbf{A}$ satisfying $w \cdot u \equiv \boldsymbol{\Delta}_t^{(k)}$.*

Proof. (Figure 3.2) The argument is similar to the one used for Lemma 3.21. We use induction on k. For $k = 0$, w consists of negative factors only, and defining u to be w^{-1} works, as $w \cdot w^{-1} \equiv \varepsilon$ holds. Assume now $k > 0$. Write $w = w_0 \cdot \alpha \cdot v^{-1}$, where α is an address and v is positive. By induction hypothesis, there exists a positive word u_0 satisfying $w_0 \cdot u_0 \equiv \boldsymbol{\Delta}_t^{(k-1)}$. Let t' be the image of t under LD_{w_0}. There exists a positive word v' satisfying $\alpha \cdot v' \equiv^+ \boldsymbol{\Delta}_{t'}$, and then a positive word u_0' satisfying $\boldsymbol{\Delta}_{t'} \cdot u_0' \equiv^+ u_0 \cdot \boldsymbol{\Delta}_{\partial^{k-1}t}$. Take $u = v \cdot v' \cdot u_0'$. ■

Finally, we deduce the following syntactic counterpart to the fact that any two LD-equivalent terms admit a common LD-expansion:

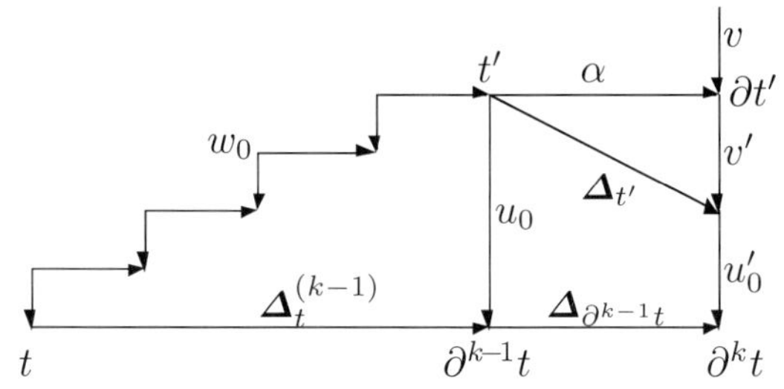

Figure 3.2. Quotients

Proposition 3.24. (**fraction**) *If w is a word on $\mathbf{A}\cup\mathbf{A}^{-1}$ such that the domain of LD_w is nonempty, then there exist words u, v on $\mathbf{A}$ satisfying $w \equiv u \cdot v^{-1}$*

(We shall see in the next chapter that the hypothesis about the domain of LD_w can be dropped.)

Exercise 3.25. *(**mass**) For every address α, define $\mu(\alpha) = 2^{-\lg(\alpha)}$, and extend μ to $\mathbf{A}^*$ additively. Prove that μ is invariant under $\equiv^+$.*

Exercise 3.26. *(**uniform distribution**) Assume that LD_{u_1} maps t_1 to t'_1 and LD_{u_2} maps t_2 to t'_2. Let $u = \prod_{\alpha\in\mathrm{Out}(t_2)} \alpha 0 u_1 \cdot u_2$. Prove that LD_u maps $t_1 * t_2$ to $t'_1 * t'_2$.*

Exercise 3.27. *(**word Δ_t**) For $t = t_1 \cdot \cdots \cdot t_p \cdot x$, with x a variable, prove*

$$\Delta_t \equiv^+ \Delta'_t \cdot 0\Delta_{s_1} \cdot \cdots \cdot 1^{p-1}0\Delta_{s_p},$$

with $\Delta'_t = (1^{p-2} \cdot \cdots \cdot \phi) \cdot (1^{p-2} \cdot \cdots \cdot 1) \cdot \cdots \cdot (1^{p-2})$, and $s_1{}^\wedge \cdots \cdot s_p \cdot x = t \bullet (1^{p-2} \cdot \cdots \cdot 1 \cdot \phi)$. Compare Δ'_t with Garside's braid word Δ_p.

Exercise 3.28. *(**substitution I**) Assume that h is a substitution. For t in T_1, define a positive word $u_{t,h}$ inductively by $u_{x,h} = \varepsilon$,*

$$u_{t_1\cdot t_2,h} = \prod_{\gamma\in\mathrm{Out}(t_2)} \gamma 0 u_{t_1,h} \cdot \prod_{\gamma\in\mathrm{Out}((\partial t_2)^h)\setminus\mathrm{Out}(\partial t_2)} \gamma \cdot u_{t_2,h}.$$

Prove that, for every term t, the operator $\mathrm{LD}_{u_{t,h}}$ maps $(\partial t)^h$ to $\partial(t^h)$.

Exercise 3.29. *(**substitution II**) We wish to prove that, if s is a substitute of t, there exists a positive word u satisfying $\Delta_s \equiv^+ \Delta_t \cdot u$ using induction on t. Assume $s = t^h$ and $t = t_1 \cdot t_2$. By induction hypothesis, there exist positive words u_1, u_2 such that $\Delta_{s_e} \equiv^+ \Delta_{t_e} \cdot u_e$ holds for $e = 1, 2$, where s_e is $(t_e)^h$. Prove $\Delta_s \equiv^+ 0\Delta_{t_1} \cdot 0u_1 \cdot \delta_{s_2} \cdot \Delta_{s_2}$. [Use Formula (3.8).] Deduce*

$$\boldsymbol{\Delta}_s \equiv^+ 0\boldsymbol{\Delta}_{t_1} \cdot \boldsymbol{\delta}_{s_2} \cdot \boldsymbol{\Delta}_{s_2} \cdot \prod_{\gamma \in \mathrm{Out}(\partial s_2)} \gamma 0 u_1.$$

[Use inheritance relations and the fact that the heir of the address 0 under $\boldsymbol{\delta}_{s_2} \cdot \boldsymbol{\Delta}_{s_2}$ is the set of all addresses $\gamma 0$ with γ in the outline of ∂s_2.] Establish

$$\boldsymbol{\delta}_{s_2} \equiv^+ \boldsymbol{\delta}_{t_2} \cdot \prod_{\gamma \in \mathrm{Out}(t_2)} \gamma \boldsymbol{\delta}_{\mathrm{sub}(s,1\gamma)}.$$

Prove that, for γ in the outline of t_2, the heirs of γ under $\boldsymbol{\Delta}_{t_2}$ exist, and we have for every positive word v

$$\gamma v \cdot \boldsymbol{\Delta}_{t_2} \equiv^+ \boldsymbol{\Delta}_{t_2} \cdot \prod_{\gamma' \in \mathrm{Heir}(\{\gamma\}, \boldsymbol{\Delta}_{t_2})} \gamma' v.$$

Deduce $\boldsymbol{\Delta}_s \equiv^+ 0\boldsymbol{\Delta}_{t_1} \cdot \boldsymbol{\delta}_{t_2} \cdot \boldsymbol{\Delta}_{t_2} \cdot u'$ with

$$u' = \prod_{\gamma \in \mathrm{Out}(t_2)} \prod_{\gamma' \in \mathrm{Heir}(\{\gamma\}, \boldsymbol{\Delta}_{t_2})} \gamma' \boldsymbol{\delta}_{\mathrm{sub}(s,1\gamma)} \cdot u_2 \cdot \prod_{\gamma \in \mathrm{Out}(\partial s_2)} \gamma 0 u_1.$$

[Apply the previous relation with $v = \boldsymbol{\delta}_{\mathrm{sub}(s,1\gamma)}$.] Conclude. Compare with Exercise 3.28 (substitution I).

Exercise 3.30. (operation $\partial^\bullet$) *(i) Let $t = (x_1 \cdot x_2 \cdot x_3) \cdot x_4$. Show that ∂t is not the least common LD-expansion of all basic LD-expansions of t.*

(ii) For $\alpha \in \mathbf{A}$, define $\alpha^\bullet$ to be ε if α finishes with 1, and to be $\beta^{(k)}$ for $\alpha = \beta 1^k 00^*$. For t a term, define $\boldsymbol{\Delta}_t^\bullet = \prod^{<_{LR}}_{\alpha \in \mathrm{Out}(t)} \alpha^\bullet$. Show that the term t lies in the domain of the operator $\mathrm{LD}_{\boldsymbol{\Delta}_t^\bullet}$, and that, if $\partial^\bullet t$ is defined to be $t \bullet \boldsymbol{\Delta}_t^\bullet$, then $\partial^\bullet t$ is the least common LD-expansion of all basic LD-expansions of t.

(iii) Let $t = x_1 \cdot x_2 \cdot x_3 \cdot x_4$, and $t' = t \bullet 1$. Check that $\partial^\bullet t'$ is not an LD-expansion of $\partial^\bullet t$. Similarly, let $t = (x_1 \cdot x_2 \cdot x_3) \cdot (x_4 \cdot x_5)$, and $t' = t \bullet \phi \cdot \phi \cdot 0$. Check that $(\partial^\bullet)^3 t$ is not an LD-expansion of t'. [Thus $\partial^\bullet$ is more minimal than ∂, but it does not share its compatibility with LD-expansions, and it is therefore of little interest.]

---

4. Cuts and Collapsing

In Chapter V, we established a number of LD-equivalence results. Now, each time $t =_{LD} t'$ holds, there must exist a word w on $\mathbf{A} \cup \mathbf{A}^{-1}$ such that the operator LD_w maps t to t'. We shall now revisit some of the results of Chapter V so as to obtain effective counterparts where each equivalence $t =_{LD} t'$ is replaced with a more precise equality $t' = t \bullet w$. Here, we revisit the results about cuts, and establish syntactic versions. We also consider left subterms, and another operation called collapsing not introduced so far. Although some technical verifications are tedious, the general scheme is simple and intuitive. However, it can be mentioned that most of the results in this section are not needed in the sequel of this text, excepted for one result in Section VIII.4.

Cuts

Let us begin with cuts. By Proposition VI.2.20 (cuts of expansion), if the term t' is an LD-expansion of the term t, and s is a cut of t, then some cut s' of t' is an LD-expansion of s. An effective version of this result consists in determining the cut s' and an operator that maps s to s' in terms of the initial cut s and of an operator that maps t to t'.

Definition 4.1. (cut of word) For β in $\mathbf{A}$, and u a word on $\mathbf{A}$, $\mathrm{cut}(u, \beta)$ is the word possibly defined by the inductive clauses:
(i) $\mathrm{cut}(\varepsilon, \beta) = \varepsilon$;
(ii) $\beta <_{LR} \alpha 11$ implies $\mathrm{cut}(\alpha, \beta) = \varepsilon$;
(iii) $\beta >_{LR} \alpha 10$ implies $\mathrm{cut}(\alpha, \beta) = \alpha'$, where α' is the address obtained from α by deleting all 0's in the maximal prefix of α that is also a prefix of β;
(iv) $\mathrm{cut}(\alpha \cdot v, \beta) = \mathrm{cut}(\alpha, \beta) \cdot \mathrm{cut}(v, \mathrm{heir}(\beta, \alpha))$, if the latter exists.

Example 4.2. ***(cut of word)*** *For $\beta = 1001011$ and $u = 100 \cdot 100010$, we find*

$$\mathrm{cut}(u, \beta) = \mathrm{cut}(100, \beta) \cdot \mathrm{cut}(100010, \mathrm{heir}(\beta, 100)),$$

if the involved addresses exist. Now, the greatest common prefix of β and 100 is 100. Hence $\mathrm{cut}(100, \beta)$ is obtained from 100 by replacing the prefix 100, i.e., 100 itself, with 1, i.e., it is 1. Then, $\mathrm{heir}(\beta, 100)$ is $\beta_1 = 1000111$. The greated common prefix of β_1 and 100010 is 10001. So $\mathrm{cut}(100010, \beta_1)$ is obtained from 100010 by replacing the prefix 10001 with 11, i.e., it is 110. Finally, $\mathrm{cut}(u, \alpha)$ exists, and it is $1 \cdot 110$.

Then the result is what we could expect:

Proposition 4.3. (effective cut) *Assume that u is a word on $\mathbf{A}$, t is a term such that $t \bullet u$ exists, and β is an address in the outline of t. Then $\mathrm{cut}(u, \beta)$ exists, and we have*

$$\mathrm{cut}(t \bullet u, \mathrm{heir}(\beta, u)) = \mathrm{cut}(t, \beta) \bullet \mathrm{cut}(u, \beta).$$

Proof. We use induction on the length of u. The result is trivial for $u = \varepsilon$. Assume that u has length 1, say $u = \alpha$. The hypothesis that $t \bullet \alpha$ exists and β belongs to the outline of t implies that $\mathrm{heir}(\beta, \alpha)$ exists. We consider the various possible cases. Assume $\beta <_{LR} \alpha 110$. Then we have $\mathrm{cut}(t', \mathrm{heir}(\beta, \alpha)) = \mathrm{cut}(t, \beta)$, so the result is true as $\mathrm{cut}(\alpha, \beta)$ is defined to be ε. Assume now $\alpha 101 <_{LR} \beta$. Then we make the proof using induction on the length of the greatest common prefix γ of β and α.

Assume first $\gamma = \phi$. The only possibility is that there exist β_1 and α_0 satisfying $\alpha = 0\alpha_0$ and $\beta = 1\beta_1$. Then we have $\mathrm{cut}(t, \beta) = \mathrm{sub}(t, 0) \cdot \mathrm{cut}(\mathrm{sub}(t, 1), \beta_1)$, $\mathrm{heir}(\beta, \alpha) = \beta$, and $\mathrm{cut}(t', \beta) = \mathrm{sub}(t', 0) \cdot \mathrm{cut}(\mathrm{sub}(t, 1), \beta_1)$. By construction, we have $\mathrm{sub}(t', 0) = \mathrm{sub}(t, 0) \bullet \alpha_0$, and, therefore,

$$\mathrm{cut}(t', \mathrm{heir}(\beta, \alpha)) = \mathrm{cut}(t, \beta) \bullet 0\alpha_0,$$

which is the desired result.

Assume now that γ begins with 0. Then there exist β_0 and α_0 satisfying $\beta = 0\beta_0$ and $\alpha = 0\alpha_0$. We have $\mathrm{cut}(t, \beta) = \mathrm{cut}(\mathrm{sub}(t, 0), \beta_0)$, $\mathrm{heir}(\beta, \alpha) = 0\mathrm{heir}(\beta_0, \alpha_0)$, and $\mathrm{cut}(t', \mathrm{heir}(\beta, \alpha)) = \mathrm{cut}(\mathrm{sub}(t', 0), \mathrm{heir}(\beta_0, \alpha_0))$. By induction hypothesis, the latter term is the image of the previous one under $\mathrm{LD}_{\mathrm{cut}(\alpha_0, \beta_0)}$, which is also $\mathrm{LD}_{\mathrm{cut}(\alpha, \beta)}$ by definition of $\mathrm{cut}(\alpha, \beta)$.

Finally, assume that γ begins with 1. We write now $\beta = 1\beta_1$ and $\alpha = 1\alpha_1$. We have $\mathrm{cut}(t, \beta) = \mathrm{sub}(t, 0) \cdot \mathrm{cut}(\mathrm{sub}(t, 1), \beta_0)$, $\mathrm{heir}(\beta, \alpha) = 1\mathrm{heir}(\beta_1, \alpha_1)$, and $\mathrm{cut}(t', \mathrm{heir}(\beta, \alpha)) = \mathrm{sub}(t, 0) \cdot \mathrm{cut}(\mathrm{sub}(t', 1), \mathrm{heir}(\beta_1, \alpha_1))$. By induction hypothesis, the latter term is the image of the previous one under $\mathrm{LD}_{1\mathrm{cut}(\alpha_0, \beta_0)}$, which is $\mathrm{LD}_{\mathrm{cut}(\alpha, \beta)}$ again by definition of $\mathrm{cut}(\alpha, \beta)$.

We then use induction on the length of u. Assume $u = \alpha \cdot u_0$. By the result above, the operator $\mathrm{LD}_{\mathrm{cut}(\alpha, \beta)}$ maps $\mathrm{cut}(t, \beta)$ to $\mathrm{cut}(t \bullet \alpha, \mathrm{heir}(\beta, \alpha))$. Then, by induction hypothesis, the operator $\mathrm{LD}_{\mathrm{cut}(u_0, \mathrm{heir}(\beta, \alpha))}$ maps $\mathrm{cut}(t \bullet \alpha, \mathrm{heir}(\beta, \alpha))$ to $\mathrm{cut}(t \bullet \alpha \cdot u_0, \mathrm{heir}(\beta, \alpha \cdot u_0))$, and, therefore, the product map, which is $\mathrm{LD}_{\mathrm{cut}(u, \beta)}$ by definition, maps $\mathrm{cut}(t, \beta)$ to $\mathrm{cut}(t', \mathrm{heir}(\beta, u))$. ∎

Left subterms

Iterated left subterms are special cases of cuts, and the general result about LD-expansion of cuts subsumes a weaker result involving only iterated left subterms. Here we state an effective version of this result. First we can compute the exponent of the iterated left subterm involved in Proposition V.4.6 (left subterm).

Definition 4.4. (dilatation) For $u \in \mathbf{A}^*$ and $p \geq 0$, the integer $\mathrm{dil}(p, u)$ is defined by the inductive clauses:

$$\mathrm{dil}(p, \varepsilon) = p, \quad \mathrm{dil}(p, \alpha) = \begin{cases} p+1 & \text{for } 0^p \sqsubset \alpha, \\ p & \text{otherwise,} \end{cases} \quad \mathrm{dil}(p, \alpha \cdot v) = \mathrm{dil}(\mathrm{dil}(p, \alpha), v).$$

Using an easy induction on the length of u, we obtain:

Lemma 4.5. *Assume that u is a word on $\mathbf{A}$.*

(i) If t is a term, and $t \bullet u$ and $\mathrm{left}^p(t)$ exist, then $\mathrm{left}^{\mathrm{dil}(p,u)}(t \bullet u)$ is an LD-expansion of $\mathrm{left}^p(t)$;

(ii) We have $\mathrm{heir}(0^p1^q, u) = 0^{\mathrm{dil}(p,u)}1^q$ for q large enough.

Then we can also compute a word describing the LD-expansion involved in Proposition V.4.6 (left subterm).

Definition 4.6. (left part) Assume $u \in \mathbf{A}^*$ and $p \geq 0$. Then $\mathrm{left}(u,p)$ is defined to be the word $\mathrm{cut}(u, 0^p1^q)$, where q is large enough to guarantee that $\mathrm{heir}(0^p1^q, u)$ exists.

Going back to the definition of the operator $\mathrm{cut}(u,\gamma)$, we see that the inductive rules for the construction of $\mathrm{left}(u,p)$ are

$$\mathrm{left}(\varepsilon,p) = \varepsilon, \qquad \mathrm{left}(\alpha,p) = \begin{cases} \alpha' & \text{for } \alpha = 0^p\alpha', \\ \varepsilon & \text{otherwise,} \end{cases}$$
$$\mathrm{left}(\alpha \cdot v, p) = \mathrm{left}(\alpha,p) \cdot \mathrm{left}(v, \mathrm{dil}(p,\alpha)).$$

Applying Proposition 4.3 (effective cut of expansion) gives the following effective form for Proposition V.4.6 (left subterm):

Proposition 4.7. (effective left subterm) *Assume that u is a word on $\mathbf{A}$, t is a term, and $t \bullet u$ and $\mathrm{left}^p(t)$ exist. Then we have*

$$\mathrm{left}^{\mathrm{dil}(p,u)}(t \bullet u) = \mathrm{left}^p(t) \bullet \mathrm{left}(u,p).$$

Collapsing

Cutting a term t at the address α amounts to ignoring the part of t that lies at the right of α. Collapsing addresses is a symmetric operation that enables us to ignore the part of t that lies at the left of a given address. In the sequel, we use $\emptyset$ to denote an empty term, hence not an element of the set T_X. The outline of $\emptyset$ is defined to be the empty set.

Definition 4.8. (collapsing) Assume that t is a term, and B is a subset of the outline of t. We define $\mathrm{coll}(t,B)$, the term obtained from t by *collapsing* the occurrences in B, by the following clauses:

(i) For t a variable, we put $\mathrm{coll}(t,\emptyset) = t$, and $\mathrm{coll}(t, \{\not{o}\}) = \emptyset$.

(ii) For $t = t_1 \cdot t_2$, and $B = \mathrm{sh}_0(B_1) \cup \mathrm{sh}_1(B_2)$, we put

$$\mathrm{coll}(t, B) = \begin{cases} \emptyset & \text{if } \mathrm{coll}(t_2, B_2) = \emptyset \text{ holds,} \\ \mathrm{coll}(t_2, B_2) & \text{if } \mathrm{coll}(t_1, B_1) = \emptyset \text{ holds,} \\ \mathrm{coll}(t_1, B_1) \cdot \mathrm{coll}(t_2, B_2) & \text{otherwise.} \end{cases}$$

Example 4.9. **(collapsing)** *The term* $\mathrm{coll}(t, B)$ *is obtained from* t *by replacing all variables with address in* B *by* $\emptyset$, *and then evaluating the result using the rules* $\emptyset \cdot a = a$ *and* $a \cdot \emptyset = \emptyset$ *for every* a. *For instance, for* $t = x_1 \cdot (x_2 \cdot x_3 \cdot x_4) \cdot x_5$ *and* $B = \{100, 1011\}$, *collapsing* B *amounts to evaluating* $x_1 \cdot (\emptyset \cdot x_3 \cdot \emptyset) \cdot x_5$, *yielding* $x_1 \cdot x_5$.

Due to the geometric definition of collapsing, we have the following commutativity result:

Proposition 4.10. **(collapsing)** *Assume that* u *is a word on* $\mathbf{A}$, *and* LD_u *maps* t *to* t'. *Then, for every subset* B *of* $\mathbf{A}$ *such that* $\mathrm{Heir}(B, u)$ *is defined,* $\mathrm{coll}(t', \mathrm{Heir}(B, u))$ *is an LD-expansion of the term* $\mathrm{coll}(t, B)$.

This qualitative statement will be refined below into an effective version which implies it, so we skip the proof, and directly turn to this refinement. To this end, it suffices to introduce the convenient notion of collapsing for words.

Definition 4.11. **(collapsing for words)** Assume that u is a word on $\mathbf{A}$, and B is a set of addresses such that $\mathrm{Heir}(B, u)$ is defined. Then $\mathrm{coll}(u, B)$ is the positive word possibly defined by the inductive clauses:

(i) $\mathrm{coll}(\varepsilon, B) = \varepsilon$;

$$\text{(ii) } \mathrm{coll}(\gamma, B) = \begin{cases} \varepsilon & \text{if } B \text{ contains an address of the form } \gamma 01^m, \\ & \quad \text{or an address of the form } \gamma 101^m, \text{ or an address} \\ & \quad \text{of the form } \gamma 111^m, \text{ or an address } \alpha \text{ that covers } \gamma, \\ \gamma' & \text{otherwise, where } \gamma' \text{ is obtained from } \gamma \text{ by deleting} \\ & \quad \text{the final 1 of every prefix of } \gamma \text{ of the form } \beta 1 \text{ such} \\ & \quad \text{that } B \text{ contains an address of the form } \beta 01^m; \end{cases}$$

(iii) $\mathrm{Fcoll}(\gamma \cdot v, B) = \mathrm{coll}(\gamma, B) \cdot \mathrm{coll}(v, \mathrm{Heir}(B, \gamma))$.

Example 4.12. **(collapsing)** *For* $u = 1 \cdot \phi$ *and* $B = \{0, 100\}$, *we find*

$$\begin{aligned} \mathrm{coll}(u, B) &= \mathrm{coll}(1, \{0, 100\}) \cdot \mathrm{coll}(\phi, \mathrm{heir}(B, 1)) \\ &= \mathrm{coll}(1, \{0, 100\}) \cdot \mathrm{coll}(\phi, \{0, 1000, 1100\}). \end{aligned}$$

The first factor is ϕ: *indeed* 1 *is a prefix of* 1 *such that* B *contains an address of the form* 01^m, *namely* 0 *itself; on the other hand,* B *contains no address of the form* 101^m, 1101^m, 111^m, *neither does it contain any address that covers* 1. *The second factor is empty: indeed, the set* $\{0, 1001, 1101\}$ *contains the address* 0, *which is of the form* 01^m. *So* $\mathrm{coll}(u, B)$ *is the length* 1 *word* ϕ.

The result takes the expected form:

Proposition 4.13. (**effective collapsing**) *Assume that* u *is a word on* $\mathbf{A}$, B *is a set of addresses such that* $\mathrm{Heir}(B, u)$ *is defined, and* t *is a term such that* $t \bullet u$ *exists. Then the word* $\mathrm{coll}(u, B)$ *exists and we have*

$$\mathrm{coll}(t \bullet u, \mathrm{Heir}(B, u)) = \mathrm{coll}(t, B) \bullet \mathrm{coll}(u, B).$$

Proof. The argument is the same as for cuts. The definition of $\mathrm{coll}(\alpha, B)$ has been made so as to obtain the result. ■

Example 4.14. (**effective collapsing**) *Assume* $u = 1 \cdot \phi$ *and* $B = \{0, 100\}$ *as in Example* 4.12. *Let* $t = x_1 \cdot (x_2 \cdot x_3) \cdot (x_4 \cdot x_5)$, *a minimal term in the domain of* LD_u *whose outline includes* B. *Then Proposition* 4.13 *states that the following diagram is commutative:*

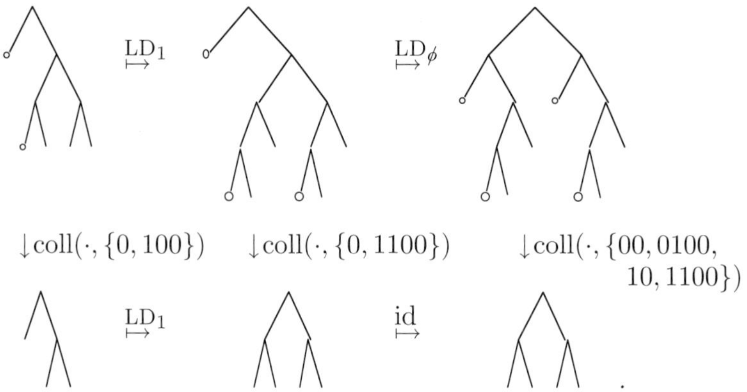

Here the variables to be collapsed have been marked with a circle.

Syntactic versions

We have associated with every word u on $\mathbf{A}$ derived objects depending on the operator LD_u only, namely the integers $\mathrm{dil}(p, u)$, the addresses $\mathrm{heir}(\beta, u)$, the sets $\mathrm{Heir}(B, u)$, and the various words $\mathrm{cut}(u, \beta)$, $\mathrm{left}(u, p)$ and $\mathrm{coll}(u, B)$. If u' is $\equiv^+$-equivalent to u, the operators LD_u and $\mathrm{LD}_{u'}$ coincide, so we directly obtain:

Lemma 4.15. *Assume that u, u' are $\equiv^+$-equivalent words on* **A**. *Then we have*

$$\mathrm{dil}(p,u) = \mathrm{dil}(p,u'), \quad \mathrm{heir}(\beta,u) = \mathrm{heir}(\beta,u'), \quad \text{and } \mathrm{Heir}(B,u) = \mathrm{Heir}(B,u')$$

whenever $\mathrm{heir}(\beta,u)$ *and* $\mathrm{Heir}(B,u)$ *respectively exists.*

When we consider the words $\mathrm{cut}(u,\beta)$, we only deduce that $u \equiv^+ u'$ implies $\mathrm{LD}_{\mathrm{cut}(u,\beta)} = \mathrm{LD}_{\mathrm{cut}(u',\beta)}$ when the latter are defined, and this is not enough to deduce the expected syntactic result $\mathrm{cut}(u,\beta) \equiv^+ \mathrm{cut}(u',\beta)$. The aim of this section is to establish the latter equivalence and its counterpart involving collapsing directly.

Proposition 4.16. **(syntactic cut)** *Assume that u, u' are $\equiv^+$-equivalent positive words, and $\mathrm{cut}(u,\beta)$ exists. Then $\mathrm{cut}(u',\beta)$ exists, and we have*

$$\mathrm{cut}(u,\beta) \equiv^+ \mathrm{cut}(u',\beta). \tag{4.1}$$

Proof. Due to the inductive definition of cut, it suffices to establish the result when (u,u') is an LD-relation. Indeed, assume that the result holds for the pair (u,u'), and let u_1, u_2 be arbitrary positive words. Assume that $\mathrm{cut}(u_1 \cdot u \cdot u_2, \beta)$ exists. By definition, we have

$$\mathrm{cut}(u_1 \cdot u \cdot u_2, \beta) = \mathrm{cut}(u_1,\beta) \cdot \mathrm{cut}(u,\beta_1) \cdot \mathrm{cut}(u_2,\beta_2),$$

with $\beta_1 = \mathrm{heir}(\beta,u_1)$ and $\beta_2 = \mathrm{heir}(\beta_1,u)$. The hypothesis $u' \equiv^+ u$ implies $\mathrm{LD}_u = \mathrm{LD}_{u'}$, and, therefore, $\beta_2 = \mathrm{heir}(\beta_1,u')$. Hence, $\mathrm{cut}(u,\beta_1) \equiv^+ \mathrm{cut}(u',\beta_1)$ implies $\mathrm{cut}(u_1 \cdot u \cdot u_2,\beta) \equiv^+ \mathrm{cut}(u_1 \cdot u' \cdot u_2,\beta)$.

We consider the various types of LD-relations, and, for each of them, the various possible positions of the address β. Consider first an LD-relation of the form $(\gamma u, \gamma u')$. For $\beta <_{LR} \gamma$, *i.e.*, if we cut before γ, the action of $\mathrm{LD}_{\gamma u}$ and $\mathrm{LD}_{\gamma u'}$ on the cut at β is trivial, and we have

$$\mathrm{cut}(\gamma u,\beta) = \mathrm{cut}(\gamma u',\beta) = \varepsilon.$$

For $\beta >_{LR} \gamma$, the action of $\mathrm{LD}_{\gamma u}$ and $\mathrm{LD}_{\gamma u'}$ on the cut at β is entirely inside this cut, and we have

$$\mathrm{cut}(\gamma u,\beta) = \gamma' u, \quad \text{and} \quad \mathrm{cut}(\gamma u',\beta) = \gamma' u',$$

where γ' is obtained from γ by deleting all 0's in the greatest prefix of γ that is also a prefix of β. Hence $\mathrm{cut}(\gamma u,\beta) \equiv^+ \mathrm{cut}(\gamma u',\beta)$ holds.

So it remains to consider the case where γ is a prefix of β. Up to using a shift, we may assume $\gamma = \phi$. Let us consider first the pair (u,u'), with $u = \phi \cdot 1 \cdot \phi$ and $u' = 1 \cdot \phi \cdot 1 \cdot 0$. For $0 \sqsubseteq \beta$, say $\beta = 0\beta_0$, we have

$$\mathrm{cut}(u,\beta) = \mathrm{cut}(\phi \cdot 1 \cdot \phi, 0\beta_0) = \mathrm{cut}(1 \cdot \phi, 00\beta_0) = \mathrm{cut}(\phi, 00\beta_0) = \varepsilon,$$

$$\begin{aligned}\mathrm{cut}(u', \beta) = \mathrm{cut}(1 \cdot \phi \cdot 1 \cdot 0, 0\beta_0) &= \mathrm{cut}(\phi \cdot 1 \cdot 0, 0\beta_0)\\ &= \mathrm{cut}(1 \cdot 0, 00\beta_0) = \mathrm{cut}(0, 00\beta_0) = \varepsilon,\end{aligned}$$

and the result is true. For $10 \sqsubseteq \beta$, say $\beta = 10\beta_0$, we find

$$\begin{aligned}\mathrm{cut}(u, \beta) = \mathrm{cut}(\phi \cdot 1 \cdot \phi, 10\beta_0) &= \mathrm{cut}(1 \cdot \phi, 01\beta_0) = \mathrm{cut}(\phi, 01\beta_0) = \varepsilon,\\ \mathrm{cut}(u', \beta) = \mathrm{cut}(1 \cdot \phi \cdot 1 \cdot 0, 10\beta_0) &= \mathrm{cut}(\phi \cdot 1 \cdot 0, 100\beta_0)\\ &= \mathrm{cut}(1 \cdot 0, 010\beta_0) = \mathrm{cut}(0, 010\beta_0) = \varepsilon,\end{aligned}$$

and, again, the result is true. For $110 \sqsubseteq \beta$, say $\beta = 110\beta_0$, we have now

$$\begin{aligned}\mathrm{cut}(u, \beta) = \mathrm{cut}(\phi \cdot 1 \cdot \phi, 110\beta_0) &= \phi \cdot \mathrm{cut}(1 \cdot \phi, 110\beta_0) = \phi \cdot \mathrm{cut}(\phi, 101\beta_0) = \phi,\\ \mathrm{cut}(u', \beta) = \mathrm{cut}(1 \cdot \phi \cdot 1 \cdot 0, 110\beta_0) &= \mathrm{cut}(\phi \cdot 1 \cdot 0, 101\beta_0)\\ &= \mathrm{cut}(1 \cdot 0, 011\beta_0) = \mathrm{cut}(0, 0110\beta_0) = \phi,\end{aligned}$$

as was required. Finally, for $111 \sqsubseteq \beta$, say $\beta = 111\beta_0$, we have

$$\begin{aligned}\mathrm{cut}(u, \beta) = \mathrm{cut}(\phi{\cdot}1{\cdot}\phi, 111\beta_0) &= \phi{\cdot}\mathrm{cut}(1{\cdot}\phi, 111\beta_0) = \phi{\cdot}1{\cdot}\mathrm{cut}(\phi, 111\beta_0) = \phi{\cdot}1{\cdot}\phi,\\ \mathrm{cut}(u', \beta) = \mathrm{cut}(1 \cdot \phi \cdot 1 \cdot 0, 111\beta_0) &= 1 \cdot \mathrm{cut}(\phi \cdot 1 \cdot 0, 111\beta_0)\\ &= 1{\cdot}\phi{\cdot}\mathrm{cut}(1{\cdot}0, 111\beta_0) = 1{\cdot}\phi{\cdot}1{\cdot}\mathrm{cut}(0, 0110\beta_0) = 1{\cdot}\phi{\cdot}1{\cdot}0,\end{aligned}$$

and the result holds. The proof is similar for the other LD-relations. ■

The scheme is the same when collapsing is involved.

Proposition 4.17. (**syntactic collapsing**) *Assume that u, u' are $\equiv^+$-equivalent words on $\mathbf{A}$, and B is a set of addresses such that* $\mathrm{Heir}(B, u)$ *exists. Then we have*

$$\mathrm{coll}(u, B) \equiv^+ \mathrm{coll}(u', B). \tag{4.2}$$

Proof. As for cuts, it suffices to prove the result when (u, u') is an LD-relation. Also, an easy induction shows that, if the set B is the disjoint union of two sets B_1, B_2, then we have

$$\mathrm{coll}(u, B) = \mathrm{coll}(\mathrm{coll}(u, B_1), \mathrm{Heir}(B_2, \mathrm{coll}(u, B_1))),$$

so that we can assume that B is a singleton, say $\{\beta\}$. Assume $u = \gamma \cdot \gamma 1 \cdot \gamma$ and $u' = \gamma 1 \cdot \gamma \cdot \gamma 1 \cdot \gamma 0$ for instance. For $\beta \perp \gamma$, we have $\mathrm{coll}(u, \{\beta\}) = \gamma' \cdot \gamma' 1 \cdot \gamma'$ for some γ' ($\gamma' = \gamma$ for $\beta >_{LR} \gamma$), and $\mathrm{coll}(u', \{\beta\}) = \gamma' 1 \cdot \gamma' \cdot \gamma' 1 \cdot \gamma' 0$, so the result is true. For $\gamma 0 \sqsubseteq \beta$, but β not of the form $\gamma 01^m$, we have $\mathrm{coll}(u, \{\beta\}) = u$, and $\mathrm{coll}(u', \{\beta\}) = u'$, so, again, the result is true. For $\beta = \gamma 01^m$, we have $\mathrm{coll}(u, \{\beta\}) = \gamma = \mathrm{coll}(u', \{\beta\})$: again the result is true. The argument is similar for $\gamma 10 \sqsubseteq \beta$ and $\gamma 110 \sqsubseteq \beta$ is a prefix of β: for β not of the form $\gamma 101^m$ or $\gamma 1101^m$, we find u and u', otherwise, we find twice γ. Finally, for $\gamma 111$ a prefix of β, we find u and u' for β not of the form $\gamma 1^m$, and twice the empty word ε otherwise. The case of the other LD-relations is similar. ■

Exercise 4.18. **(left subterm)** *Assume that u, u' are $\equiv^+$-equivalent words on* **A**. *Prove that, for every p, we have* $\mathrm{left}(u,p) \equiv^+ \mathrm{left}(u',p)$.

Exercise 4.19. **(left division)** *Assume that t_0, t are terms, and $t_0{\cdot}t =_{LD} t_0^{[2]}{\cdot}t$ holds. Let t' be a common LD-expansion for $t_0{\cdot}t$ and $t_0^{[2]}{\cdot}t$. Assume that LD_{u_1} maps $t_0{\cdot}t$ to t', and LD_{u_2} maps $t_0^{[2]}{\cdot}t$ to t'. Define $B_e = \mathrm{Heir}(\{01^*\}, u_e)$ for $e = 1, 2$, and $t_1 = \mathrm{coll}(B_1 \cup B_2, t')$. Prove $t =_{LD} t_0{\cdot}t_1$. [Thus every algorithm constructing a common LD-expansion for LD-equivalent terms leads to a left division algorithm.]*

---

5. Notes

Most of the results of Sections 1 and 2 have appeared, at least implicitly or in a different setting, in (Dehornoy 89*a*) and (Dehornoy 94*a*); the syntactic proofs of Sections 3 and 4 have not been published so far.

Introducing a geometry monoid similar to $\mathcal{G}_{LD}$ is possible for every given algebraic identity, or, more generally, every family of algebraic identities. This approach has not been developed systematically so far, and it seems that the first explicit description appears in (Dehornoy 89*a*), with the goal of investigating free LD-systems. The general case is mentioned in (Dehornoy 93). We think that this approach deserves further investigation, as it seems to capture deep properties of the considered identities. This is the case for left self-distributivity at least, as we shall see in the next chapters. We refer to the appendix of Chapter IX for a brief survey and, in particular, a comparison between the cases of left self-distributivity and of associativity.

Instead of considering the monoid $\mathcal{G}_{LD}$, with an everywhere defined multiplication but one with a possibly empty result, we could consider the set $\mathcal{G}_{LD} \setminus \{\emptyset\}$ equipped with the partial operation that maps (f, g) to $f \bullet g$ only if the result is nonempty. A *groupoid* is defined as a small category with inverses, or, equivalently, as a set equipped with a partial binary operation such that, if ab and bc exist, so do $(ab)c$ and $a(bc)$, and we have $(ab)c = a(bc)$, and, if ab exists, so do $a^{-1}(ab)$ and $(ab)b^{-1}$, and we have $a^{-1}(ab) = b$ and $(ab)b^{-1} = a$ (Paterson 99). The structure considered above on $\mathcal{G}_{LD} \setminus \{\emptyset\}$ resembles a groupoid, as every operator LD_w has an inverse $\mathrm{LD}_{w^{-1}}$. However, the groupoid conditions are not satisfied: for $a = \mathrm{LD}_\phi$, $b = \mathrm{LD}_0$, and $c = \mathrm{LD}_{\phi^{-1}}$, we have seen that ab and bc exist, *i.e.*, they are nonempty, while abc is empty. Similarly, the equality $a^{-1}(ab) = b$ is false unless the domain of b is included in the image of a. Therefore, using the notion of a groupoid here seems elusive.

Let us conclude with another negative remark. When we are given an inverse semigroup S, it is usual to project S to a group by using the congruence $\sim$ such that $a \sim a'$ holds if $ea = ea'$ holds for some idempotent element e. Here, the empty operator is an idempotent of $\mathcal{G}_{LD}$, so $\sim$ collapses all elements of $\mathcal{G}_{LD}$, and the associated group is trivial. Thus, it is definitely not easy to connect $\mathcal{G}_{LD}$ with a group directly. We shall see how to do it using LD-relations in the next chapter.

VIII
The Group of Left Self-Distributivity

The geometry monoid $\mathcal{G}_{LD}$ that describes the action of the left self-distributivity identity on terms is not a group, and, contrary to the case of associativity, we cannot obtain a group by merely using a projection. However, we know a family of relations holding in $\mathcal{G}_{LD}$, namely these LD-relations that define $\equiv^+$, and we have observed that many properties of $\mathcal{G}_{LD}$ can be established by using these relations exclusively. In this chapter, we investigate the abstract group G_{LD} for which LD-relations form a presentation. The hypothesis that G_{LD} must resemble $\mathcal{G}_{LD}$ is kept as a leading principle, and indeed we can show that all geometrical parameters defined in $\mathcal{G}_{LD}$ admit counterparts in G_{LD}. On the other hand, the group G_{LD} turns out to be an extension of the braid group B_∞—this being the precise content of our slogan: "The geometry of left self-distributivity is an extension of the geometry of braids." Many results about G_{LD} and B_∞ originate in this connection. In particular, braid exponentiation and braid ordering come from an operation and a relation on G_{LD} that somehow explain them and make their construction natural.

Technically, the group G_{LD} behaves like a sort of generalized Artin group. It shares several properties with such groups, yet a number of technical problems arise from G_{LD}, contrary to B_∞, not being a direct limit of finite type groups.

The divisions of the chapter are as follows. In Section 1, we introduce the group G_{LD} and the corresponding monoid M_{LD}. We observe that the braid group B_∞ is a quotient of G_{LD}, a result connected with the action of braids on left self-distributive systems. We also observe that the presentation of G_{LD} is associated with a complement, and verify that this complement satisfies all conditions of Chapter II. In Section 2, we embed T_1 into G_{LD} by using the absorption property of Chapter V to associate with every term a distinguished word that describes its construction. We deduce a complete description of the connection between $\mathcal{G}_{LD}$ and G_{LD}, and explain how braid exponentiation arises. By extending the approach to the case of several variables, we show how charged braids then appear naturally. In Section 3, we introduce two different (pre)-orders on the group G_{LD}. The first is a preorder connected with the braid ordering, and using it gives a purely syntactical proof for the acyclicity property of free LD-systems, one that does not use braid exponentiation. The second relation is a linear ordering on G_{LD} which is compatible with multiplication on both sides. In Section 4, we show that shifting the addresses gives a family of injective endomorphisms of G_{LD}, which amounts to determining certain parabolic subgroups of G_{LD}. Finally, we introduce in Section 5 the notion of a simple element in the monoid M_{LD}, which is an exact analog of the notion of a simple braid in B_∞^+. We establish in particular a normal form result in M_{LD} which directly extends the braid normal form of Chapter II.

1. The Group G_{LD} and the Monoid M_{LD}

Here we consider the group G_{LD} and the monoid M_{LD} for which the LD-relations of Chapter VII form a presentation. In this section, we investigate the connection between G_{LD} and Artin's braid group B_∞, we observe that the presentation of G_{LD} and M_{LD} is associated with a complement as defined in Chapter II, and we establish that this complement is coherent and convergent, which allows us to deduce several significant properties of M_{LD} and G_{LD}.

LD-relations vs. braid relations

We recall that $\mathbf{A}$ denotes the set of all binary addresses, *i.e.*, of all finite sequences of 0's and 1's; $\mathbf{A}^*$ denotes the free monoid generated by $\mathbf{A}$, *i.e.*, the set of all words on $\mathbf{A}$, while $(\mathbf{A} \cup \mathbf{A}^{-1})^*$ denotes the free monoid generated by $\mathbf{A}$ and a disjoint copy $\mathbf{A}^{-1}$ of $\mathbf{A}$. The words on $\mathbf{A}$ are said to be positive.

Definition 1.1. (group G_{LD}) We define G_{LD} to be the group generated by a family of elements $\{g_\alpha \,;\, \alpha \in \mathbf{A}\}$ subject to the relations

$$g_\alpha \, g_\beta = g_\beta \, g_\alpha \qquad \text{for } \alpha \perp \beta \qquad \text{(type } \perp)$$

$$g_{\alpha 0\beta} \, g_\alpha = g_\alpha \, g_{\alpha 10\beta} \, g_{\alpha 00\beta} \qquad \text{(type 0)}$$

$$g_{\alpha 10\beta} \, g_\alpha = g_\alpha \, g_{\alpha 01\beta} \qquad \text{(type 10)}$$

$$g_{\alpha 11\beta}\, g_{\alpha} = g_{\alpha}\, g_{\alpha 11\beta} \qquad \text{(type 11)}$$
$$g_{\alpha 1}\, g_{\alpha}\, g_{\alpha 1}\, g_{\alpha 0} = g_{\alpha}\, g_{\alpha 1}\, g_{\alpha} \qquad \text{(type 1)}$$

The submonoid of G_{LD} generated by all g_α's is denoted G^+_{LD}.

The relations that define G_{LD} are the LD-relations of Section VII.2. Hence, by construction, the mapping $\alpha \mapsto g_\alpha$ induces an isomorphism of $(\mathbf{A} \cup \mathbf{A}^{-1})^*/\equiv$ onto G_{LD}, where $\equiv$ is the congruence considered in Chapter VII.

Proposition 1.2. (projection) *The mapping* pr *defined by*

$$\mathrm{pr}(g_\alpha) = \begin{cases} 1 & \text{if the address } \alpha \text{ contains at least one } 0, \\ \sigma_{i+1} & \text{for } \alpha = 1^i, \text{ i.e., } 11\ldots1, \ i \text{ times } 1, \end{cases}$$

induces a surjective homomorphism of G_{LD} onto B_∞.

Proof. Let f be the homomorphism of the free monoid $(\mathbf{A} \cup \mathbf{A}^{-1})^*$ into B_∞ that maps α to $\mathrm{pr}(g_\alpha)$. Then $f(u) = f(u')$ holds for every LD-relation (u, u'). Let us consider for instance the LD-relation $(\alpha 1 \cdot \alpha \cdot \alpha 1 \cdot \alpha 0, \alpha \cdot \alpha 1 \cdot \alpha)$. If α contains one 0 at least, both sides are collapsed. For $\alpha = 1^i$, they are mapped to $\sigma_{i+2}\sigma_{i+1}\sigma_{i+2}$ and $\sigma_{i+1}\sigma_{i+2}\sigma_{i+1}$ respectively, hence equal in B_∞. ■

By definition, the kernel of pr is the normal subgroup of G_{LD} generated by those elements g_α where α contains at least one 0. The subgroup Ker(pr) is very large. No complete description is known, but we shall see in Section 4 that it includes in particular infinitely many copies of G_{LD} itself.

The existence of the projection of G_{LD} onto B_∞ is connected with the action of braids on LD-systems considered in Chapters I and III. Assume that S is a binary system. Then n-strand braid words act on S^n by

$$\sigma_1 : (a_1, a_2, a_3, \ldots, a_n) \longmapsto (a_1a_2, a_1, a_3, \ldots, a_n).$$

If we identify the finite sequence $(a_1, \ldots, a_n)$ with the binary tree 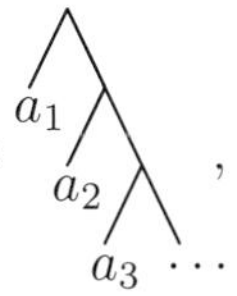

, the action becomes:

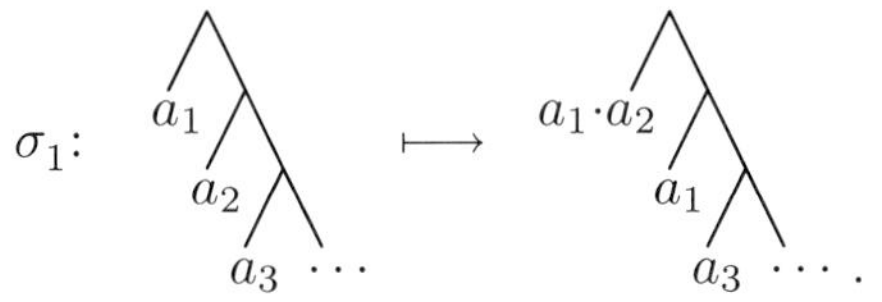

Now, the self-distributivity operators LD_α also act on binary trees. For instance, we have

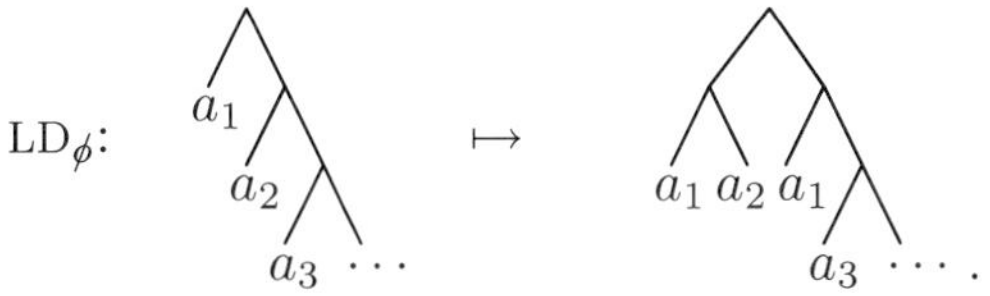

The only difference between the two actions is that, in the one case, we evaluate the expression a_1a_2 in S, while, in the other case, we keep it as a formal product. Now, let us consider the action of the equivalent braid words $\sigma_1\sigma_2\sigma_1$ and $\sigma_2\sigma_1\sigma_2$, and compare it with the action of the corresponding operators $\mathrm{LD}_{\phi\cdot 1\cdot\phi}$ and $\mathrm{LD}_{1\cdot\phi\cdot 1}$. As we see in Figure 1.1, in the case of braids, the actions of $\sigma_1\sigma_2\sigma_1$ and $\sigma_2\sigma_1\sigma_2$ coincide provided the operation on S is left self-distributive. In the case of self-distributivity operators, the actions of $\mathrm{LD}_{\phi\cdot 1\cdot\phi}$ and $\mathrm{LD}_{1\cdot\phi\cdot 1}$ do not coincide, but introducing the operator LD_0 is what is needed to obtain the equality. So g_0 measures the obstruction to an action of braids when the system S does not satisfy Identity (LD). Now, when S is an LD-system, we can collapse the additional term g_0, and what remains is the action of braids. So, the action of B_∞ is what remains from the action of G_{LD} when we collapse all operators but those acting on the rightmost branch of the tree.

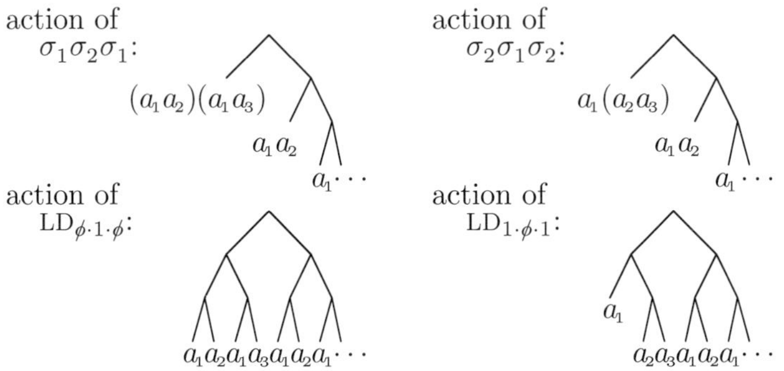

Figure 1.1. Action of B_∞ *vs.* action of G_{LD}

The monoid M_{LD}

As the positive geometry monoid $\mathcal{G}_{LD}^+$ plays a significant role beside $\mathcal{G}_{LD}$, it is natural to introduce the monoid for which LD-relations form a presentation.

Definition 1.3. (monoid M_{LD}) We define M_{LD} to be the monoid with the same presentation as G_{LD}; we use g_α^+ for the generator associated with α.

By construction, the mapping $\alpha \mapsto g_\alpha^+$ induces an isomorphism of $\mathbf{A}^*/\equiv^+$ onto M_{LD}, and the mapping $g_\alpha^+ \mapsto g_\alpha$ induces a surjective homomorphism of M_{LD} onto G_{LD}^+; we shall discuss in Section IX.5 the possible injectivity of this homomorphism, *i.e.*, the question of whether M_{LD} embeds in G_{LD}.

As in the case of G_{LD}, mapping g_α^+ to 1 if the address α contains at least one 0, and to σ_{i+1} for $\alpha = 1^i$ defines a surjective homomorphism of M_{LD} onto B_∞^+.

For each LD-relation (u, u'), and for each address γ, the shifted pair $(\gamma u, \gamma u')$ is still an LD-relation. We easily obtain.

Proposition 1.4. (**shift**) *For each address γ, the mapping $\mathrm{sh}_\gamma : g_\alpha \mapsto g_{\gamma\alpha}$ induces an endomorphism of G_{LD}, and the mapping $\mathrm{sh}^+_\gamma : g^+_\alpha \mapsto g^+_{\gamma\alpha}$ induces an injective endomorphism of M_{LD}.*

Proof. The existence follows from the invariance of LD-relations under shift. To prove injectivity in the case of M_{LD}, we observe that, if (u, u') is an LD-relation and all generators occurring in u belong to the image of sh_γ, so do all generators occurring in u'. Hence, if $\gamma v \equiv^+ \gamma v'$ holds, all intermediate words in a sequence of elementary transformations from γv to $\gamma v'$ belong to the image of sh_γ, hence $v \equiv^+ v'$ holds as well. ∎

We shall prove in Section 4 that the endomorphisms sh_γ of G_{LD} are injective as well, but the proof requires some care, and the result is not needed now.

The crucial point for our study of M_{LD} is that, for every pair of addresses (α, β), there exists exactly one LD-relation of the form $g^+_\alpha \cdots = g^+_\beta \cdots$ With the definitions of Chapter II, this means that our presentation of M_{LD} is associated with a complement on the right.

Definition 1.5. (**complement**) We denote by f the mapping of $\mathbf{A} \times \mathbf{A}$ into $\mathbf{A}^*$ defined by

$$f(\alpha, \beta) = \begin{cases} \alpha 10\gamma \cdot \alpha 00\gamma & \text{for } \beta = \alpha 0\gamma, \\ \alpha 01\gamma & \text{for } \beta = \alpha 10\gamma, \\ \beta \cdot \alpha & \text{for } \beta = \alpha 1, \\ \varepsilon & \text{for } \beta = \alpha, \\ \beta \cdot \alpha \cdot \beta 0 & \text{for } \alpha = \beta 1, \\ \beta & \text{for } \alpha \not\sqsubseteq \beta 1 \text{ or } \alpha 11 \sqsubseteq \beta. \end{cases}$$

By definition, the monoid M_{LD} is the monoid associated with f on the right, while G_{LD} is the corresponding group. As in Chapter II, we shall use , $\backslash$, N, R for the (right) reversing relation, the extension to positive words, the numerator and the denominator associated with f respectively.

We have observed that the braid group B_∞ is a projection of the group G_{LD}, and, similarly, the braid monoid B^+_∞ is a projection of the monoid M_{LD}. The existence of such surjective morphisms originates in the existence of a similar surjective map for the complements. We denote by pr the surjective homomorphism of $(\mathbf{A} \cup \mathbf{A}^{-1})^*$ onto BW_∞ that maps α to σ_{i+1} if the address α has the form 1^i, and to ε otherwise. This mapping induces the surjective homomorphism of G_{LD} onto B_∞ considered above, and there is no ambiguity in our notation.

Lemma 1.6. *(i) The projection* pr *of $(\mathbf{A} \cup \mathbf{A}^{-1})^*$ onto BW_∞ is a morphism with respect to the complements f and f_B, in the sense that*

$$\mathrm{pr}(f(\alpha, \beta)) = \mathrm{pr}(\alpha)\backslash\mathrm{pr}(\beta) \tag{1.1}$$

holds for all addresses α, β.

(ii) If u, v are words on $\mathbf{A}$ and $u\backslash v$ exists, we have

$$\mathrm{pr}(u\backslash v) = \mathrm{pr}(u)\backslash\mathrm{pr}(v). \tag{1.2}$$

(iii) If w is a word on $\mathbf{A} \cup \mathbf{A}^{-1}$ and $w \curvearrowright w'$ holds, i.e., w is reversible to w', then $\mathrm{pr}(w) \curvearrowright \mathrm{pr}(w')$ holds as well. In particular, the equalities

$$\mathrm{pr}(N(w)) = N_R(\mathrm{pr}(w)) \qquad \textit{and} \qquad \mathrm{pr}(D(w)) = D_R(\mathrm{pr}(w)) \tag{1.3}$$

hold whenever $N(w)$ and $D(w)$ exist.

Proof. Formula (1.1) follows from the definitions. The other results follow from an induction on the number of reversing steps (*cf.* Exercise II.2.21). ■

Another similarity between the complement for G_{LD} and the braid complement involves the partial actions on terms and on sequences from an LD-system previously defined. Lemma III.1.4 tells us that, if w is a braid word such that $\vec{a} \bullet w$ is defined, and w is right reversible to w', then $\vec{a} \bullet w'$ is defined as well, and we have $\vec{a} \bullet w' = \vec{a} \bullet w$. A similar phenomenon happens with the complement f and the partial action of $(\mathbf{A} \cup \mathbf{A}^{-1})^*$ on terms.

Lemma 1.7. *Assume that w, w' are words on $\mathbf{A} \cup \mathbf{A}^{-1}$ and w is reversible to w'. Assume that t is a term, and $t \bullet w$ exists. Then $t \bullet w'$ exists as well, and we have $t \bullet w' = t \bullet w$.*

Proof. It suffices to consider one step of reversing. Let us look at $w = 1^{-1} \cdot \phi$, $w' = \phi \cdot 1 \cdot 0 \cdot \phi^{-1} \cdot 1^{-1}$, actually the most complicated case. Assuming that t lies in the domain of $\mathrm{LD}_1^{-1} \bullet \mathrm{LD}_\phi$ implies that t has the form

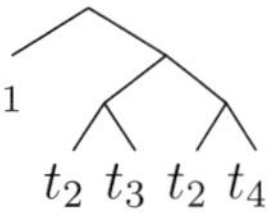

. Then, t lies in the domain of $\mathrm{LD}_\phi \bullet \mathrm{LD}_1 \bullet \mathrm{LD}_0$, and we find $t \bullet \phi \cdot 1 \cdot 0 =$

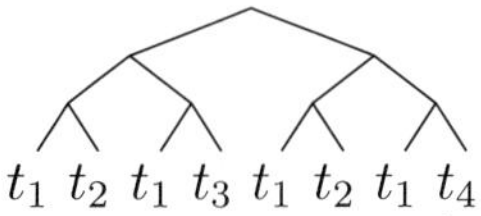

, which belongs to the image of $\mathrm{LD}_1 \bullet \mathrm{LD}_\phi$. Finally, we obtain $t \bullet w = t \bullet w' =$

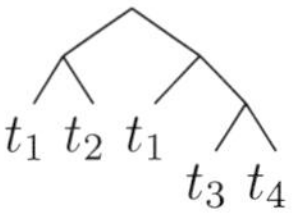

. ■

Completeness of word reversing

According to the scheme of Chapter II, we shall study the possible coherence and convergence of the complement f the monoid M_{LD} is associated with. The projection results of Lemma 1.6 are useless here: actually, deduction goes the other direction: proving results about f, like coherence or convergence, as will be done below, will re-prove the corresponding results about the braid complement.

Let us begin with norm. Contrary to the case of braids, the length of the words is not preserved under LD-relations, so the problem is nontrivial.

Proposition 1.8. (atomicity) *The complement f admits a norm, hence the monoid M_{LD} is atomic.*

Proof. Let us define, for u a word on $\mathbf{A}$,

$$\nu(u) = \mathrm{size}(t_u^R) - \mathrm{size}(t_u^L). \tag{1.4}$$

Firstly, $u' \equiv^+ u$ implies $\mathrm{LD}_u = \mathrm{LD}_{u'}$, hence $t_u^L = t_{u'}^L$ and $t_u^R = t_{u'}^R$, and, finally, $\nu(u') = \nu(u)$. Assume that α is an address, and u is a word on $\mathbf{A}$. By definition, we have $t_{\alpha \cdot u}^R = (t_{\alpha \cdot u}^L \bullet \alpha) \bullet u$, hence there exists a substitution f satisfying $t_{\alpha \cdot u}^L \bullet \alpha = (t_u^L)^f$ and $t_{\alpha \cdot u}^R = (t_u^R)^f$. We deduce

$$\begin{aligned}
\nu(\alpha \cdot u) &= \mathrm{size}(t_{\alpha \cdot u}^R) - \mathrm{size}(t_{\alpha \cdot u}^L)\\
&= \mathrm{size}(t_{\alpha \cdot u}^R) - \mathrm{size}(t_{\alpha \cdot u}^L \bullet \alpha) + \mathrm{size}(t_{\alpha \cdot u}^L \bullet \alpha) - \mathrm{size}(t_{\alpha \cdot u}^L)\\
&= \mathrm{size}((t_u^R)^f) - \mathrm{size}((t_u^L)^f) + \mathrm{size}(t_{\alpha \cdot u}^L \bullet \alpha) - \mathrm{size}(t_{\alpha \cdot u}^L)\\
&> \mathrm{size}((t_u^R)^f) - \mathrm{size}((t_u^L)^f) \geq \mathrm{size}(t_u^R) - \mathrm{size}(t_u^L) = \nu(u).
\end{aligned}$$

A similar argument gives $\nu(u \cdot \alpha) > \nu(u)$, and ν is a norm for f. ∎

It follows that the atoms of the monoid M_{LD} are the elements g_α^+, and, that, for every word u on $\mathbf{A}$, the length of every word u' satisfying $u' \equiv^+ u$ is bounded above by the integer $\nu(u)$ defined in (1.4).

Let us turn to the completeness of word reversing. Owing to Propositions II.2.5 (complete) and II.2.9 (locally coherent), we are left with the question of establishing that f is coherent on $\mathbf{A}$.

Proposition 1.9. (coherence) *The complement f is coherent on $\mathbf{A}$.*

Proof. We consider all triples (α, β, γ) in $\mathbf{A}^3$ and, for each of them, we prove the three relations

$$\begin{aligned}
&(\alpha\backslash\beta)\backslash(\alpha\backslash\gamma) \equiv^{++} (\beta\backslash\alpha)\backslash(\beta\backslash\gamma),\\
&(\beta\backslash\gamma)\backslash(\beta\backslash\alpha) \equiv^{++} (\gamma\backslash\beta)\backslash(\gamma\backslash\alpha),\\
&(\gamma\backslash\alpha)\backslash(\gamma\backslash\beta) \equiv^{++} (\alpha\backslash\gamma)\backslash(\alpha\backslash\beta),
\end{aligned}$$

where we recall $u' \equiv^{++} u$ means that $u^{-1}u'$ is reversible to ε. As LD-relations are closed under shift, we may assume that the greatest common prefix of α, β, γ is empty, and, because we consider the three cyclic permutations of the addresses simultaneously, we may choose the ordering of α, β, γ.

Case 1. *Two addresses are equal.* Assume $\alpha = \beta$. Then we have

$$(\alpha\backslash\beta)\backslash(\alpha\backslash\gamma) = \varepsilon\backslash(\alpha\backslash\gamma) = \alpha\backslash\gamma,$$
$$(\beta\backslash\alpha)\backslash(\beta\backslash\gamma) = \varepsilon\backslash(\beta\backslash\gamma) = \beta\backslash\gamma = \alpha\backslash\beta;$$
$$(\beta\backslash\gamma)\backslash(\beta\backslash\alpha) = (\beta\backslash\gamma)\backslash\varepsilon = \varepsilon,$$
$$(\gamma\backslash\beta)\backslash(\gamma\backslash\alpha) = (\gamma\backslash\alpha)\backslash(\gamma\backslash\alpha) = \varepsilon,$$

and this is enough, as α and β play symmetric roles.

Case 2. *One address is orthogonal to the greatest common prefix of the two other ones.* We assume that γ is orthogonal to the greatest common prefix of α and β. Then γ is orthogonal to every address occurring in $\alpha\backslash\beta$ and $\beta\backslash\alpha$, and we have

$$(\alpha\backslash\beta)\backslash(\alpha\backslash\gamma) = (\alpha\backslash\beta)\backslash\gamma = \gamma,$$
$$(\beta\backslash\alpha)\backslash(\beta\backslash\gamma) = (\beta\backslash\alpha)\backslash\gamma = \gamma;$$
$$(\beta\backslash\gamma)\backslash(\beta\backslash\alpha) = \gamma\backslash(\beta\backslash\alpha) = \beta\backslash a,$$
$$(\gamma\backslash\beta)\backslash(\gamma\backslash\alpha) = \beta\backslash\alpha.$$

Again this is enough, since α and β play symmetric roles.

Case 3. *One address is a prefix of the greatest common prefix of the two others.* We assume that γ is a prefix of the greatest common prefix δ of α and β. As we assumed that α, β and γ have no common prefix, we have $\gamma = \phi$.

Case 3.1. *The address* 0 *is a prefix of α and β.* Write $\alpha = 0\alpha_0$, $\beta = 0\beta_0$, and assume $\alpha_0\backslash\beta_0 = \beta_1 \cdot \cdots \cdot \beta_q$ and $\beta_0\backslash\alpha_0 = \alpha_1 \cdot \cdots \cdot \alpha_p$. Then we have $\alpha\backslash\beta = 0\beta_1 \cdot \cdots \cdot 0\beta_q$, and $\beta\backslash\alpha = 0\alpha_1 \cdot \cdots \cdot 0\alpha_p$, hence

$$(\alpha\backslash\beta)\backslash(\alpha\backslash\gamma) = (0\beta_1 \cdot \cdots \cdot 0\beta_q)\backslash\phi = \phi,$$
$$(\beta\backslash\alpha)\backslash(\beta\backslash\gamma) = (0\alpha_1 \cdot \cdots \cdot 0\alpha_p)\backslash\phi = \phi;$$
$$(\beta\backslash\gamma)\backslash(\beta\backslash\alpha) = \phi\backslash(0\alpha_1 \cdot \cdots \cdot 0\alpha_p) = 10\alpha_1 \cdot 00\alpha_1 \cdot \cdots \cdot 10\alpha_p \cdot 00\alpha_p,$$
$$(\gamma\backslash\beta)\backslash(\gamma\backslash\alpha) = (10\beta_0 \cdot \cdots \cdot 00\beta_q)\backslash(10\alpha_1 \cdot \cdots \cdot 00\alpha_p)$$
$$= 10\alpha_1 \cdot \cdots \cdot 10\alpha_p \cdot 00\alpha_1 \cdot \cdots \cdot 00\alpha_p.$$

It remains to compute

$$(10\alpha_1 \cdot 00\alpha_1 \cdot \cdots \cdot 10\alpha_p \cdot 00\alpha_p)\backslash(10\alpha_1 \cdot \cdots \cdot 10\alpha_p \cdot 00\alpha_1 \cdot \cdots \cdot 00\alpha_p),$$

and the symmetric expression. Using type $\perp$ relations between $10\alpha_i$ and $00\alpha_j$, we see that the previous words are empty, and we are done.

Case 3.2. *The address* 1 *is a proper prefix of α and β.* Write $\alpha = 1e\alpha_0$, $\beta = 1e\beta_0$, with $e = 0$ or $e = 1$. As every factor in $\alpha\backslash\beta$ and $\beta\backslash\alpha$ begins with $1e$, and, for every δ, we have $1e\delta\backslash\phi = \phi$ and $\phi\backslash 1e\delta = e1\delta$, we obtain

$$(\alpha\backslash\beta)\backslash(\alpha\backslash\gamma) = (\alpha\backslash\beta)\backslash\phi = \phi,$$
$$(\beta\backslash\alpha)\backslash(\beta\backslash\gamma) = (\beta\backslash\alpha)\backslash\phi = \phi;$$
$$(\beta\backslash\gamma)\backslash(\beta\backslash\alpha) = \phi\backslash\mathrm{sh}_{1e}(\beta_0\backslash\alpha_0) = \mathrm{sh}_{e1}(\beta_0\backslash\alpha_0),$$
$$(\gamma\backslash\beta)\backslash(\gamma\backslash\alpha) = e1\beta_0\backslash e1\alpha_0 = \mathrm{sh}_{e1}(\beta_0\backslash\alpha_0),$$

which is enough for this case.

Case 3.3. *The address 1 is the greatest common prefix of α and β.*

Case 3.3.1. *The addresses α and β are orthogonal.* We may assume $\alpha = 10\alpha_0$ and $\beta = 11\beta_0$. Then we have

$$(\alpha\backslash\beta)\backslash(\alpha\backslash\gamma) = 11\beta_1\backslash\phi = \phi,$$
$$(\beta\backslash\alpha)\backslash(\beta\backslash\gamma) = 10\alpha_0\backslash\phi = \phi;$$
$$(\beta\backslash\gamma)\backslash(\beta\backslash\alpha) = \phi\backslash 10\alpha_0 = 01\alpha_0,$$
$$(\gamma\backslash\beta)\backslash(\gamma\backslash\alpha) = 11\beta_1\backslash 01\alpha_0 = 01\alpha_0;$$
$$(\gamma\backslash\alpha)\backslash(\gamma\backslash\beta) = 10\alpha_0\backslash 11\beta_1 = 11\beta_1,$$
$$(\alpha\backslash\gamma)\backslash(\alpha\backslash\beta) = \phi\backslash 11\beta_1 = 11\beta_1,$$

and we are done.

Case 3.3.2. *The addresses α and β are comparable.* We may assume that β is a strict prefix of α, hence $\beta = 1$.

Case 3.3.2.1. *The address 10 is a prefix of α,* say $\alpha = 10\alpha_0$. We have

$$(\alpha\backslash\beta)\backslash(\alpha\backslash\gamma) = 1\backslash\phi = \phi\cdot 1\cdot 0,$$
$$(\beta\backslash\alpha)\backslash(\beta\backslash\gamma) = (110\alpha_0\cdot 100\alpha_0)\backslash(\phi\cdot 1\cdot 0) = \phi\cdot 1\cdot 0;$$
$$(\beta\backslash\gamma)\backslash(\beta\backslash\alpha) = (\phi\cdot 1\cdot 0)\backslash(110\alpha_0\cdot 100\alpha_0)$$
$$= (1\cdot 0)\backslash(110\alpha_0\cdot 010\alpha_0) = 0\backslash(101\alpha_0\cdot 010\alpha_0) = 101\alpha_0\cdot 001\alpha_0,$$
$$(\gamma\backslash\beta)\backslash(\gamma\backslash\alpha) = (1\cdot\phi)\backslash 01\alpha_0 = \phi\backslash 01\alpha_0 = 101\alpha_0\cdot 001\alpha_0;$$
$$(\gamma\backslash\alpha)\backslash(\gamma\backslash\beta) = 01\alpha_0\backslash(1\cdot\phi) = 1\cdot\phi,$$
$$(\alpha\backslash\gamma)\backslash(\alpha\backslash\beta) = \phi\backslash 1 = 1\cdot\phi,$$

and we are done for this case.

Case 3.3.2.2. *The address 11 is a strict prefix of α.* Write $\alpha = 11e\alpha_0$ with $e = 0$ or $e = 1$. We have

$$(\alpha\backslash\beta)\backslash(\alpha\backslash\gamma) = 1\backslash\phi = \phi\cdot 1\cdot 0,$$
$$(\beta\backslash\alpha)\backslash(\beta\backslash\gamma) = 1e1\alpha_0\backslash(\phi\cdot 1\cdot 0) = \phi\cdot 1\cdot 0;$$
$$(\beta\backslash\gamma)\backslash(\beta\backslash\alpha) = (\phi\cdot 1\cdot 0)\backslash 1e1\alpha_0 = (1\cdot 0)\backslash e11\alpha_0 = e11\alpha_0,$$
$$(\gamma\backslash\beta)\backslash(\gamma\backslash\alpha) = (1\cdot\phi)\backslash 11e\alpha_0 = \phi\backslash 1e1\alpha_0 = e11\alpha_0;$$
$$(\gamma\backslash\alpha)\backslash(\gamma\backslash\beta) = 11e\alpha_0\backslash(1\cdot\phi) = 1\cdot\phi,$$
$$(\alpha\backslash\gamma)\backslash(\alpha\backslash\beta) = \phi\backslash 1 = 1\cdot\phi,$$

and we are done.

Case 3.3.2.3. *The address α is* 11. This case is critical. We find (see Figure 2.1)

$$(\alpha\backslash\beta)\backslash(\alpha\backslash\gamma) = (1\cdot 11\cdot 10)\backslash\phi = \phi\cdot 1\cdot 11\cdot 10\cdot 0\cdot 01\cdot 00,$$
$$(\beta\backslash\alpha)\backslash(\beta\backslash\gamma) = (11\cdot 1)\backslash(\phi\cdot 1\cdot 0) = \phi\cdot 1\cdot 0\cdot 11\cdot 01\cdot 10\cdot 00,$$

and we then verify $\phi\cdot 1\cdot 11\cdot 10\cdot 0\cdot 01\cdot 00 \equiv^{++} \phi\cdot 1\cdot 0\cdot 11\cdot 01\cdot 10\cdot 00$. Similarly, we have

$$(\beta\backslash\gamma)\backslash(\beta\backslash\alpha) = (\phi\cdot 1\cdot 0)\backslash(11\cdot 1) = 11\cdot 1\cdot\phi,$$
$$(\gamma\backslash\beta)\backslash(\gamma\backslash\alpha) = (1\cdot\phi)\backslash 11 = 11\cdot 1\cdot\phi,$$

and we have equality. Finally, we have

$$(\gamma\backslash\alpha)\backslash(\gamma\backslash\beta) = 11\backslash(1\cdot\phi) = 1\cdot 11\cdot 10\cdot\phi\cdot 1\cdot 0,$$
$$(\alpha\backslash\gamma)\backslash(\alpha\backslash\beta) = \phi\backslash(1\cdot 11\cdot 10) = 1\cdot\phi\cdot 11\cdot 1\cdot 01\cdot 0,$$

and we verify $1\cdot 11\cdot 10\cdot\phi\cdot 1\cdot 0 \equiv^{++} 1\cdot\phi\cdot 11\cdot 1\cdot 01\cdot 0$.

Case 3.4. *The address ϕ is the greatest common prefix of α and β.* We may assume $\alpha \perp \beta$, for, otherwise, one of α, β should be ϕ, and we are in case 1. So we assume $\alpha = 0\alpha_0$, while 1 is a prefix of β.

Case 3.4.1. *The address 1 is a strict prefix of β.* Write $\beta = 1e\beta_0$, with $\varepsilon = 0$ or $e = 1$. We find

$$(\alpha\backslash\beta)\backslash(\alpha\backslash\gamma) = 1e\beta_0\backslash\phi = \phi,$$
$$(\beta\backslash\alpha)\backslash(\beta\backslash\gamma) = 0\alpha_0\backslash\phi = \phi;$$
$$(\beta\backslash\gamma)\backslash(\beta\backslash\alpha) = \phi\backslash 0\alpha_0 = 10\alpha_0 \cdot 00\alpha_0,$$
$$(\gamma\backslash\beta)\backslash(\gamma\backslash\alpha) = e1\beta_0\backslash(10\alpha_0 \cdot 00\alpha_0) = 10\alpha_0 \cdot 00\alpha_0;$$
$$(\gamma\backslash\alpha)\backslash(\gamma\backslash\beta) = (10\alpha_0 \cdot 00\alpha_0)\backslash e1\beta_0 = e1\beta_0,$$
$$(\alpha\backslash\gamma)\backslash(\alpha\backslash\beta) = \phi\backslash 1e\beta_0 = e1\beta_0,$$

and we are done.

Case 3.4.2. *The address β is equal to* 1. Then we find

$$(\alpha\backslash\beta)\backslash(\alpha\backslash\gamma) = 1\backslash\phi = \phi \cdot 1 \cdot 0,$$
$$(\beta\backslash\alpha)\backslash(\beta\backslash\gamma) = 0\alpha_0\backslash(\phi \cdot 1 \cdot 0) = \phi \cdot 1 \cdot 0;$$
$$(\beta\backslash\gamma)\backslash(\beta\backslash\alpha) = (\phi \cdot 1 \cdot 0)\backslash 0\alpha_0 = 110\alpha_0 \cdot 100\alpha_0 \cdot 010\alpha_0 \cdot 000\alpha_0,$$
$$(\gamma\backslash\beta)\backslash(\gamma\backslash\alpha) = (1 \cdot \phi)\backslash(10\alpha_0 \cdot 00\alpha_0) = 110\alpha_0 \cdot 010\alpha_0 \cdot 100\alpha_0 \cdot 000\alpha_0,$$

and we find $110\alpha_0 \cdot 100\alpha_0 \cdot 010\alpha_0 \cdot 000\alpha_0 \equiv^{++} 110\alpha_0 \cdot 010\alpha_0 \cdot 100\alpha_0 \cdot 000\alpha_0$. Finally, we have

$$(\gamma\backslash\alpha)\backslash(\gamma\backslash\beta) = (10\alpha_0 \cdot 00\alpha_0)\backslash(1 \cdot \phi) = 1 \cdot \phi,$$
$$(\alpha\backslash\gamma)\backslash(\alpha\backslash\beta) = \phi\backslash 1 = 1 \cdot \phi,$$

and we are done for this case.

Case 4. *The greatest common prefix of two addresses is a prefix of the third address.* We assume that the greatest common prefix of α and β is a prefix of γ. If $\alpha \perp \beta$ holds, there exists δ satisfying $\delta 0 \sqsubseteq \alpha$ and $\delta 1 \sqsubseteq \beta$, or conversely. Then we have either $\delta 0 \sqsubseteq \gamma$, and, then, β is orthogonal to the greatest common prefix of α and γ, or $\delta 1 \sqsubseteq \gamma$, and α is orthogonal to the greatest common prefix of β and γ. In both cases, we are in case 2 above. Finally, if $\alpha \perp \beta$ fails, we may assume $\beta \sqsubset \alpha$. Then $\beta \sqsubseteq \gamma$ necessarily holds, and, therefore, β is a prefix of the greatest common prefix of α and γ, and we are in case 3 above. So the proof is complete. ■

By Proposition II.2.1 (coherent), we deduce:

Proposition 1.10. (completeness) *Word reversing is complete for f, i.e., for all words u, u' on* **A**, *$u \equiv^+ u'$ is equivalent to $u^{-1}u' \curvearrowright \varepsilon$.*

The next step is to prove that word reversing using f always converges.

Proposition 1.11. (convergence) *The complement f is convergent. If u, v are words on* **A** *and their cumulated lengths in 0's and 1's is ℓ, reversing of $u^{-1}v$ requires at most $\exp^*(O(\ell))$ steps*

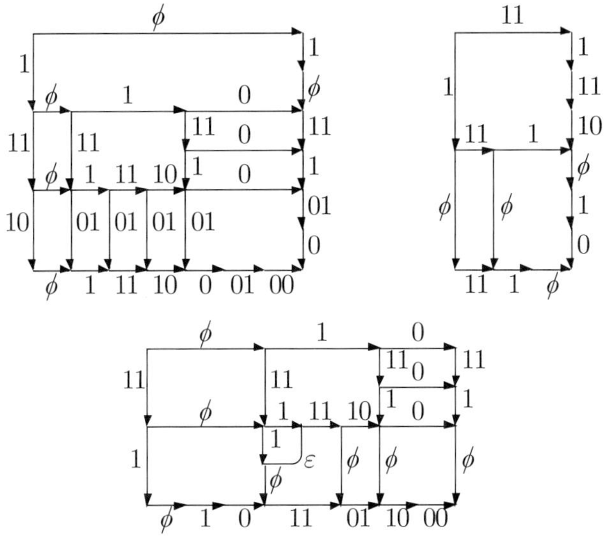

Figure 2.1. Coherence of the complement f

Proof. (We recall that $\exp^*(n)$ is a base 2 tower of exponentials of height n.) For convergence, by Proposition II.2.16 (lcm), it suffices that we show that, for all positive words u, v, there exist positive words u', v' satisfying $u\,v' \equiv^+ v\,u'$. This follows from Proposition VII.3.22 (syntactic confluence).

For complexity, Proposition VII.3.22 tells us that $\partial^\ell t^L_{u,v}$ represents a common right multiple of the classes of u and v in M_{LD}. An induction shows that, when the cumulated length of u and v in terms of 0's and 1's is ℓ, then the size of the term $t^L_{u,v}$ lies in $\mathcal{O}(\ell)$. Hence the size of $\partial^\ell t^L_{u,v}$ is bounded above by $\exp^*(\mathcal{O}(\ell))$. We conclude using Exercise II.3.23 (complexity). ■

It follows that the results of Chapter II apply to the monoid M_{LD}. We obtain:

Proposition 1.12. **(left cancellative)** *The monoid M_{LD} is left cancellative.*

Proposition 1.13. **(lattice)** *Every finite subset of M_{LD} admits a right lcm and a left gcd, and M_{LD} equipped with the operations $\vee_R$ and $\wedge_L$ is a lattice.*

Proposition 1.14. **(fraction)** *(i) We have $G_{LD} = (G^+_{LD})(G^+_{LD})^{-1}$, i.e., every element of G_{LD} admits an expression of the form uv^{-1} with u, v words on $\mathbf{A}$.*
(ii) For w, w' words on $\mathbf{A} \cup \mathbf{A}^{-1}$, the relation $w \equiv w'$ holds, i.e., w and w' represent the same element of G_{LD}, if and only if there exist positive words v, v' satisfying

$$N(w)\,v \equiv^+ N(w')\,v' \quad \text{and} \quad D(w)\,v \equiv^+ D(w')\,v'. \tag{1.5}$$

(iii) In particular, for u, u' words on $\mathbf{A}$, $u \equiv u'$ holds if and only if $uv \equiv^+ u'v$ holds for some word v on $\mathbf{A}$.

A natural question is whether M_{LD} is also associated with a complement on the left. The answer is positive, but the involved complement fails to be left coherent. Actually, left lcm's do not always exist in M_{LD}, and it is false that every element of G_{LD} can be expressed as a left fraction (see Exercise 1.20).

Exercise 1.15. (kernel) *Prove that the kernel of the projection of G_{LD} onto B_∞ is the normal subgroup of G_{LD} generated by the elements of the form $g_{0\alpha}$. [Hint: Every generator $g_{1^i0\beta}$ is conjugated in G_{LD} with some generator $g_{0\alpha}$, as we have $g_{1^i0\beta} = g_{1^{i-1}} \cdots g_\phi g_{01^i\beta} g_\phi^{-1} \cdots g_{1^{i-1}}^{-1}$.]*

Exercise 1.16. (other presentation) *(i) Write $a_i = g_{1^{i-1}}$. Prove that G_{LD} is generated by the sequence $a_1, a_2, \ldots$, and that $a_i a_j = a_j a_i$ holds for $|i-j| \geq 2$. [Hint: For $\alpha = \beta 01^j$, prove, using relations of type 10,*

$$g_\alpha = g_\beta^{-1}\, g_{\beta 1}^{-1} \cdots g_{\beta 1^{j-1}}^{-1}\, g_{\beta 1^j 0}\, g_{\beta 1^{j-1}} \cdots g_{\beta 1}\, g_\beta,$$

then, using a relation of type 1,

$$g_\alpha = g_\beta^{-1} g_{\beta 1}^{-1} \cdots g_{\beta 1^{j-1}}^{-1} g_{\beta 1^{j+1}}^{-1} g_{\beta 1^j}^{-1} g_{\beta 1^{j+1}}^{-1} g_{\beta 1^j}\, g_{\beta 1^{j+1}}\, g_{\beta 1^j}\, g_{\beta 1^{j-1}} \cdots g_{\beta 1}\, g_\beta.$$

Deduce using induction on the number of 0's in α that g_α belongs to the subgroup of G_{LD} generated by the a_i's.]

(ii) Prove that G_{LD} is not of finite type. [Hint: B_∞ is not of finite type.]

Exercise 1.17. (abelianization) *Prove that the abelianization G_{LD}^{ab} of G_{LD} is obtained by adding the relations $g_{\alpha 0} = g_\alpha - g_{\alpha 1}$, $g_{\alpha 10} = g_{\alpha 01}$. Deduce that G_{LD}^{ab} is isomorphic to $\mathbf{Z}[z]$, the image of g_{1^i} in G_{LD}^{ab} being z^i.*

Exercise 1.18. (braidlike) *Say that a word on $\mathbf{A}$ is braidlike if it only contains addresses of the form 1^m. Prove that, for every word u on $\mathbf{A}$, there exist words $v, u_0, u_1, \ldots$ on $\mathbf{A}$ such that v is braidlike and $u \equiv^+ v \cdot \prod_i 1^i 0 u_i$ holds.*

Exercise 1.19. (no minimal section) *Prove that there exists no minimal section for the projection of M_{LD} onto B_∞^+ in the sense that there is no section s of pr such that every element a of M_{LD} is a right multiple of $s(\mathrm{pr}(a))$. [Hint: Consider $a = g_\phi g_1 g_\phi g_1 g_\phi$, which projects onto $\sigma_1\sigma_2\sigma_1\sigma_2\sigma_1$. Show that $s(\sigma_1\sigma_2\sigma_1\sigma_2\sigma_1)$ should be one of a, $g_\phi g_1 g_1 g_\phi g_1$, or $g_1 g_\phi g_1 g_1 g_\phi$, and that none of these elements is convenient.]*

Exercise 1.20. (**left complement**) *(i) Prove there exists a unique complement f_L on $\mathbf{A}$ such that M_{LD} is associated with f_L on the left.*

(ii) Prove that the domain of the operator $\mathrm{LD}_\phi \cdot \mathrm{LD}_0 \bullet \mathrm{LD}_\phi^{-1}$ is empty. Deduce that no equality $u_1 \cdot \phi \cdot 0 \equiv^+ v_1 \cdot \phi$ with u_1, v_1 positive words may hold. Conclude that the left reversing of the word $1 \cdot 10 \cdot 1^{-1}$ (using the complement f_L) does not terminate.

(iii) Consider $u = 1 \cdot \phi \cdot 01$, $u' = 1 \cdot 10 \cdot \phi$, $v = 1$. Check that the left reversing of uv^{-1} succeeds, leading to the word $\phi^{-1} \cdot 1 \cdot \phi \cdot 0 \cdot 01$, while the left reversing of $u'v^{-1}$ leads to $w = 1 \cdot 10 \cdot 1^{-1} \cdot \phi^{-1} \cdot 1 \cdot \phi \cdot 0$. Conclude that the complement f_L is not left coherent.

(iv) Check the equivalence $\phi \cdot 1 \cdot 10 \cdot \phi \equiv^+ 1 \cdot \phi \cdot 0 \cdot 01 \cdot 1$. Show that the left reversing of $(\phi \cdot 1 \cdot 10 \cdot \phi)(1 \cdot \phi \cdot 0 \cdot 01 \cdot 1)^{-1}$ does not lead to the empty word. Conclude that left word reversing is not complete for f_L.

(v) Use the previous results to prove the existence of a pair (a, b) in M_{LD} such that a and b have no left lcm, and the existence of an element in G_{LD} that cannot be expressed as a left fraction $a^{-1}b$ with a, b in G_{LD}^+.

Exercise 1.21. (**right cancellation**) *Assume that u, u' are words on $\mathbf{A}$ that satisfy $u \cdot 1^i \equiv^+ u' \cdot 1^i$. Prove that the braid words $\mathrm{pr}(u)$ and $\mathrm{pr}(u')$ are equivalent, and conclude that we must have $\mathrm{pr}(u\backslash u') = \mathrm{pr}(u'\backslash u) = \varepsilon$. [Hint: Use Lemma 1.6.] Deduce that there exist positive words u_0, u_1, ... and v satisfying $u\backslash u' = \prod_i 1^i 0 u_i$ and $1^i \equiv^+ \prod_i 1^i 0 u_i \cdot v$. Conclude that $u \equiv^+ u'$ holds.*

---

2. The Blueprint of a Term

Among the specific properties of left self-distributivity, the phenomenon called absorption by right powers in Chapter V plays a significant role. As every result stating that two terms t, t' are LD-equivalent, it admits a syntactic counterpart consisting in explicitly computing a sequence of addresses, *i.e.*, a word on $\mathbf{A} \cup \mathbf{A}^{-1}$, say $w(t, t')$, such that t' is obtained from t by applying left self-distributivity successively at the addresses indicated in $w(t, t')$. In the current case, we consider the equivalence

$$x^{[p+1]} =_{LD} t \cdot x^{[p]} \tag{2.1}$$

which we know holds for p large enough for every term t in T_1. Here we show that some corresponding word $w(x^{[p+1]}, t{\cdot}x^{[p]})$ can be chosen so as to depend on t only, and that this word can be used as a syntactic copy of t on which we can simulate many properties. This word—or its class in G_{LD}—will be called the *blueprint* of t. Studying such blueprints will lead us to the braid exponentiation of Chapter I, as well as to a complete description of the connection between the monoid $\mathcal{G}_{LD}$ and the group G_{LD}.

The syntactic content of absorption by right powers

In order to find a definition for the blueprint of a term informally introduced above, it suffices to consider the inductive proof of Formula (2.1) given in Section V.2. For $t = x$, (2.1) is an equality. Otherwise, for $t = t_1 \cdot t_2$, and p large enough, we have written

$$x^{[p+1]} =_{LD} t_1 \cdot x^{[p]} =_{LD} t_1 \cdot (t_2 \cdot x^{[p-1]}) =_{LD} (t_1 \cdot t_2) \cdot (t_1 \cdot x^{[p-1]}) =_{LD} t \cdot x^{[p]}. \quad (2.2)$$

Assume that f_1 and f_2 are operators in $\mathcal{G}_{LD}$ such that, for p large enough, f_1 maps $x^{[p+1]}$ to $t_1 \cdot x^{[p]}$ and, similarly, f_2 maps $x^{[p+1]}$ to $t_2 \cdot x^{[p]}$. Then (with an obvious meaning) $\mathrm{sh}_1(f_2)$ maps $t_1 \cdot x^{[p]}$ to $t_1 \cdot (t_2 \cdot x^{[p-1]})$, and $\mathrm{sh}_1(f_1^{-1})$ maps $t \cdot (t_1 \cdot x^{[p-1]})$ to $t \cdot x^{[p]}$. Now, by definition, LD_ϕ maps $t_1 \cdot (t_2 \cdot x^{[p-1]})$ to $(t_1 \cdot t_2) \cdot (t_1 \cdot x^{[p-1]})$, *i.e.*, $t \cdot (t_1 \cdot x^{[p-1]})$. By composing the previous operators, we conclude that

$$f_1 \bullet \mathrm{sh}_1(f_2) \bullet \mathrm{LD}_\phi \bullet \mathrm{sh}_1(f_1^{-1}) \quad (2.3)$$

maps $x^{[p+1]}$ to $t \cdot x^{[p]}$. Formula (2.3), which is nothing more than a syntactic translation of (2.2), tells us how to define the blueprint of a term inductively:

Definition 2.1. (**blueprint**) For t a term in T_1, the *blueprint* of t is the word χ_t inductively defined by $\chi_x = \varepsilon$ and

$$\chi_{t_1 \cdot t_2} = \chi_{t_1} \cdot 1\chi_{t_2} \cdot \phi \cdot 1\chi_{t_1}^{-1}. \quad (2.4)$$

Example 2.2. (**blueprint**) *Let $t = (x \cdot x) \cdot x$. We have $\chi_x = \varepsilon$, hence $\chi_{x \cdot x} = \varepsilon \cdot \varepsilon \cdot \phi \cdot \varepsilon^{-1} = \phi$, and, finally, we find $\chi_t = \phi \cdot \varepsilon \cdot \phi \cdot \phi^{-1} = \phi \cdot \phi \cdot 1^{-1}$.*

By construction, we have:

Proposition 2.3. (**effective absorption**) *For every term t in T_1, the operator LD_{χ_t} maps $x^{[p+1]}$ to $t \cdot x^{[p]}$ for $p \geq \mathrm{ht}(t)$.*

Here we arrive at the key point. Assume that t and t' are LD-equivalent terms. Then there exists a word w on $\mathbf{A} \cup \mathbf{A}^{-1}$ such that LD_w maps t to t'. Now, by Proposition 2.3 (effective absorption), the operator $\mathrm{LD}_{\chi_{t'}}$ maps $x^{[p+1]}$ to $t' \cdot x^{[p]}$

for n large enough. Similarly, the operator LD_{χ_t} maps $x^{[p+1]}$ to $t{\cdot}x^{[p]}$. As LD_w maps t to t', the operator LD_{0w} maps $t{\cdot}x^{[p]}$ to $t'{\cdot}x^{[p]}$, and we have found two operators that map $x^{[p+1]}$ to $t'{\cdot}x^{[p]}$, namely $\mathrm{LD}_{\chi_{t'}}$ and $\mathrm{LD}_{\chi_t} \bullet \mathrm{LD}_{0w}$. We have seen in Chapter VII that, if u and u' are positive words and the operators LD_u and $\mathrm{LD}_{u'}$ take the same value on at least one term, *i.e.*, if $t \bullet u = t \bullet u'$ holds for at least one term t, then LD_u and $\mathrm{LD}_{u'}$ coincide. We did not prove that this implies $u \equiv^+ u'$, neither did we prove a similar result for arbitrary words, and we have so far no way for deducing the relation $w \equiv w'$ from a possible equality $\mathrm{LD}_w = \mathrm{LD}_{w'}$. However, the definition of $\equiv$ has been made by considering those relations that we know are satisfied by the operators LD_w. So, if our analysis is correct, we can expect that an equality of the form $t \bullet w = t \bullet w'$ implies $w \equiv w'$, and, therefore, we can expect that

$$\chi_{t \bullet w} \equiv \chi_t \cdot 0w. \tag{2.5}$$

holds. The point is that, if this equivalence is true, then, necessarily, it can be established by a direct verification, and this is what we are going to do now.

To this end, it is convenient that we introduce the binary operation $\wedge$ on $(\mathbf{A} \cup \mathbf{A}^{-1})^*$ such that $\chi_{t_1 \cdot t_2}$ is $\chi_{t_1} {}^\wedge \chi_{t_2}$.

Definition 2.4. (exponentiation) For u, v words on $\mathbf{A} \cup \mathbf{A}^{-1}$, we define

$$u \wedge v = u \cdot 1v \cdot \phi \cdot 1u^{-1}. \tag{2.6}$$

With this notation, the definition of the blueprint can be restated as:

Lemma 2.5. *The mapping χ is the homomorphism of T_1 into $((\mathbf{A} \cup \mathbf{A}^{-1})^*, \wedge)$ that maps x to ε, i.e., χ_t is $t(\varepsilon)$ evaluated in $((\mathbf{A} \cup \mathbf{A}^{-1})^*, \wedge)$.*

Let us consider (2.5) in the case $w = \phi$. Assuming $t' = t \bullet \phi$ means that there exist three terms t_1, t_2, t_3 satisfying $t = t_1{\cdot}(t_2{\cdot}t_3)$ and $t' = (t_1{\cdot}t_2){\cdot}(t_1{\cdot}t_3)$. So, proving (2.5) here amounts to proving the equivalence

$$(u \wedge v) \wedge (u \wedge w) \equiv u \wedge (v \wedge w) \cdot 0 \tag{2.7}$$

with $u = \chi_{t_1}$, $v = \chi_{t_2}$, and $w = \chi_{t_3}$. Now, this is the matter of a simple verification:

Lemma 2.6. *Let u, v, w be arbitrary words on $\mathbf{A} \cup \mathbf{A}^{-1}$. Then (2.7) holds.*

Proof. Applying the definition and LD-relations, we find

$$\begin{aligned}
(u \wedge v) \wedge (u \wedge w) &= u \cdot 1v \cdot \phi \cdot 1u^{-1} \cdot 1u \cdot 11w \cdot 1 \cdot 11u^{-1} \cdot \phi \cdot 11u \cdot 1^{-1} \cdot 11v^{-1} \cdot 1u^{-1} \\
&\equiv u \cdot 1v \cdot 11w \cdot \phi \cdot 1 \cdot \phi \cdot 11u^{-1} \cdot 11u \cdot 1^{-1} \cdot 11v^{-1} \cdot 1u^{-1} \\
&\equiv u \cdot 1v \cdot 11w \cdot \phi \cdot 1 \cdot \phi \cdot 1^{-1} \cdot 11v^{-1} \cdot 1u^{-1}
\end{aligned}$$

$$\equiv u \cdot 1v \cdot 11w \cdot 1 \cdot \phi \cdot 1 \cdot 0 \cdot 1^{-1} \cdot 11v^{-1} \cdot 1u^{-1}$$
$$\equiv u \cdot 1v \cdot 11w \cdot 1 \cdot \phi \cdot 0 \cdot 11v^{-1} \cdot 1u^{-1}$$
$$\equiv u \cdot 1v \cdot 11w \cdot 1 \cdot 11v^{-1} \cdot \phi \cdot 1u^{-1} \cdot 0 = u \wedge (v \wedge w) \cdot 0. \quad \blacksquare$$

Let us consider now (2.5) when w consists of a single address of the form $\alpha = 0\beta$. The hypothesis $t' = t \bullet \alpha$ yields $t = t_1 \cdot t_2$, $t' = t_1' \cdot t_2$ with $t_1' = t_1 \bullet \beta$. If we assume, by an induction hypothesis, that $\chi_{t_1'} \equiv \chi_{t_1} \cdot \beta$ holds, then proving (2.5) amounts to proving the equivalence

$$(u \wedge 0w) \wedge v \equiv u \wedge v \cdot 00w \tag{2.8}$$

with $u = \chi_{t_1}$, $v = \chi_{t_2}$, and $w = \beta$. Similarly, proving (2.5) in the case $\alpha = 1\beta$ amounts to proving a special case of

$$u \wedge (v \cdot 0w) \equiv u \wedge v \cdot 01w. \tag{2.9}$$

As above, such equivalences can be established by a simple verification.

Lemma 2.7. *Let u, v, w be arbitrary words on $\mathbf{A} \cup \mathbf{A}^{-1}$. Then (2.8) and (2.9) hold.*

Proof. We use induction on the length of w. The result is obvious for $w = \varepsilon$. Assume $\lg(w) = 1$, say $w = \alpha^e$ with α an address and $e = \pm 1$. For $e = +1$, we obtain

$$\begin{aligned}(u \cdot 0\alpha) \wedge v &= u \cdot 0\alpha \cdot 1v \cdot \phi \cdot 10\alpha^{-1} \cdot 1u^{-1}\\ &\equiv u \cdot 1v \cdot 0\alpha \cdot \phi \cdot 10\alpha^{-1} \cdot 1u^{-1}\\ &\equiv u \cdot 1v \cdot \phi \cdot 10\alpha \cdot 00\alpha \cdot 10\alpha^{-1} \cdot 1u^{-1}\\ &\equiv u \cdot 1v \cdot \phi \cdot 00\alpha \cdot 1u^{-1} \equiv u \cdot 1v \cdot \phi \cdot 1u^{-1} \cdot 00\alpha \equiv (u \wedge v) \cdot 00\alpha\end{aligned}$$

as was expected. For $e = -1$, applying the previous result to $u \cdot 0\alpha^{-1}$ gives $((u \cdot 0\alpha^{-1}) \cdot 0\alpha) \wedge v \equiv (u \cdot 0\alpha^{-1}) \wedge v \cdot 00\alpha$, hence $(u \cdot 0\alpha^{-1}) \wedge v \equiv (u \wedge v) \cdot 00\alpha^{-1}$. The case $\lg(w) \geq 2$ follows from an easy induction.

We prove now (2.9) using a similar induction. Again, the only nontrivial case is when w consists of a single address, say $w = \alpha$. Then we obtain

$$\begin{aligned}u \wedge (v \cdot 0\alpha) &= u \cdot 1v \cdot 10\alpha \cdot \phi \cdot 1u^{-1}\\ &\equiv u \cdot 1v \cdot \phi \cdot 01\alpha \cdot 1u^{-1} \equiv u \cdot 1v \cdot \phi \cdot 1u^{-1} \cdot 01\alpha = (u \wedge v) \cdot 01\alpha. \quad \blacksquare\end{aligned}$$

We easily deduce a proof of the conjectured equivalence (2.5):

Proposition 2.8. (**action** I) *Assume that t is a term in T_1, and w is a word on $\mathbf{A} \cup \mathbf{A}^{-1}$ such that $t \bullet w$ exists. Then we have*

$$\chi_{t \bullet w} \equiv \chi_t \cdot 0w. \tag{2.10}$$

Proof. We use induction on the length of w. For $w = \varepsilon$, the result is obvious. Assume $\lg(w) = 1$, say $w = \alpha^e$ with α an address and $e = \pm 1$. Assume first $e = +1$. We use induction on the length of α. For $\alpha = \phi$, the result follows from Lemma 2.6, as was noted above. For $\alpha = 0\beta$, it follows from the induction hypothesis and from (2.8) in Lemma 2.7; for $\alpha = 1\beta$, it follows similarly from the induction hypothesis and from (2.9) in Lemma 2.7. The argument is the same for $e = -1$. Assume now $\lg(w) \geq 2$. Write $w = w_1 w_2$. Applying the induction hypothesis, we find $\chi_{t \bullet w} \equiv \chi_{t \bullet w_1} \cdot 0w_2 \equiv \chi_t \cdot 0w_1 \cdot 0w_2 = \chi_t \cdot 0w$. ∎

The exponentiation on G_{LD}

As a first application of the previous results, we show how to construct a left self-distributivity operation on a quotient of G_{LD}. The latter turns out to be closely connected with braid exponentiation.

As the shift mapping sh_1 on $(\mathbf{A} \cup \mathbf{A}^{-1})^*$ is compatible with the congruence relation $\equiv$, the exponentiation of (2.6) induces a well defined operation on the group G_{LD}, which we shall denote by ^ as well. Thus, for a, b in G_{LD}, we have

$$a \mathbin{^\wedge} b = a \cdot \mathrm{sh}_1(b) \cdot g_\phi \cdot \mathrm{sh}_1(a^{-1}). \tag{2.11}$$

The exponentiation on G_{LD} is not a left self-distributive operation. However, the main interest of our approach is that we can measure the lack of self-distributivity exactly. Indeed, translating Lemma 2.6 immediately gives:

Lemma 2.9. *For all a, b, $c \in G_{LD}$, we have*

$$(a \mathbin{^\wedge} b) \mathbin{^\wedge} (a \mathbin{^\wedge} c) = (a \mathbin{^\wedge} (b \mathbin{^\wedge} c)) \cdot g_0. \tag{2.12}$$

So, in order to construct a possible left self-distributive operation from the exponentiation of G_{LD}, and owing to (2.12), we have to collapse g_0. The first idea could be to use the normal subgroup of G_{LD} generated by g_0, but exponentiation need not be compatible with the corresponding congruence. Now, Lemma 2.7 gives:

Lemma 2.10. *For all a, b, c, d in G_{LD}, we have*

$$(a \cdot \mathrm{sh}_0(c)) \mathbin{^\wedge} (b \cdot \mathrm{sh}_0(d)) = (a \mathbin{^\wedge} b) \cdot \mathrm{sh}_{00}(c) \cdot \mathrm{sh}_{01}(d). \tag{2.13}$$

We are thus led to consider right cosets associated with the subgroup $\mathrm{sh}_0(G_{LD})$ of G_{LD}.

Proposition 2.11. **(self-distributive operation)** *Let G_0 be the subgroup of G_{LD} generated by all elements $g_{0\alpha}$. Then the exponentiation of G_{LD} induces a left self-distributive operation on the homogeneous set G_{LD}/G_0.*

Proof. Write $a \cong a'$ for $a^{-1}a' \in G_0$. Formula (2.13) shows that exponentiation on G_{LD} is compatible with $\cong$, and (2.12) gives $(a\hat{}b)\hat{}(a\hat{}c) \cong a\hat{}(b\hat{}c)$ for all elements a, b, c of G_{LD}. ■

We obtain in this way a natural explanation for the existence of the left self-distributive exponentiation of braids. Indeed, we have $B_\infty = G_{LD}/\widehat{G_0}$, where $\widehat{G_0}$ is the normal subgroup of G_{LD} generated by G_0 (Exercise 1.15). So braid exponentiation is the operation induced on $G_{LD}/\widehat{G_0}$ by the above left self-distributive operation on G_{LD}/G_0, which itself is nothing but the translation of the absorption property of monogenic LD-systems.

The connection between $\mathcal{G}_{LD}$ and G_{LD}

The next application is a complete description of the connection between the monoid $\mathcal{G}_{LD}$ and the group G_{LD}.

To this end, Formula (2.10) can be used directly (see Exercise 2.28). However, a more simple proof can be obtained by using an alternative formula where the shift mapping sh_0 does not appear. This new formula is reminiscent of the braid equality (III.1.3)

$$\prod^{sh}(\vec{a} \bullet b) = (\prod^{sh}\vec{a}) \cdot b \tag{2.14}$$

which connects the action of braids on sequences of braids with a right multiplication—(2.10) also establishes such a connection, but with the endomorphism sh_0 added.

Definition 2.12. (half-blueprint) For t in T_1, the *half-blueprint* of t is the word χ_t^* inductively defined by $\chi_x^* = \varepsilon$ and

$$\chi_{t_1 \cdot t_2}^* = \chi_{t_1} \cdot 1\chi_{t_2}^*. \tag{2.15}$$

Exponentiation in $(\mathbf{A} \cup \mathbf{A}^{-1})^*$ is reminiscent of a twisted conjugacy, and replacing χ_t with χ_t^* amounts to replacing conjugacy with half-conjugacy as defined in Chapter I. Observe for instance that $\chi_{x^{[n]}}^*$ is the empty word for every n. In every case, the word χ_t^* is a prefix of the word χ_t, essentially its first half; the precise connection between χ_t and χ_t^* is as follows:

Lemma 2.13. *For t a term in T_1 of right height r, we have*

$$\chi_t \equiv \chi_t^* \cdot \phi^{(r)} \cdot 1\chi_t^{*-1}. \tag{2.16}$$

Proof. We use induction on t. For $t = x$, the result is obvious. Assume $t = t_1 \cdot t_2$. Using the induction hypothesis and a LD-relation of type 11, we find

$$\chi_t = \chi_{t_1} \cdot 1\chi_{t_2} \cdot \phi \cdot 1\chi_{t_1}^{-1}$$

$$\begin{aligned}&\equiv \chi_{t_1}\cdot \mathrm{sh}_1(\,\chi^*_{t_2}\cdot \phi^{(r-1)}\cdot 1\chi^{*\;-1}_{t_2}\,)\cdot\phi\cdot 1\chi^{-1}_{t_1}\\&\equiv \chi_{t_1}\cdot 1\chi^*_{t_2}\cdot 1^{(r-1)}\cdot 11\chi^{*\;-1}_{t_2}\cdot\phi\cdot 1\chi^{-1}_{t_1}\\&\equiv \chi_{t_1}\cdot 1\chi^*_{t_2}\cdot\phi^{(r)}\cdot 11\chi^{*\;-1}_{t_2}\cdot 1\chi^{-1}_{t_1}=\chi^*_t\cdot\phi^{(r)}\cdot 1\chi^{*-1}_t.\end{aligned}$$ ■

The counterpart to Formula (2.10) involving half-blueprints is as follows:

Proposition 2.14. (action II) *Assume that t is a term in T_1 and w is a word on $\mathbf{A}\cup\mathbf{A}^{-1}$ such that $t\bullet w$ exists. Then we have*

$$\chi^*_{t\bullet w}\equiv \chi^*_t\cdot w. \tag{2.17}$$

Proof. We use induction on the length of w. As above, the point is the case when w consists of a single positive address, say $w=\alpha$. Let us write $t=t_1\cdot t_2\cdots$, and let $w_i=\chi_{t_i}$. There are two cases, according to whether α has the form 1^i or not. Assume first $\alpha=1^{i-1}0\beta$. Applying the definition, Proposition 2.8 (action I), and LD-relations of type $\perp$, we find

$$\begin{aligned}\chi^*_{t\bullet w}&=w_1\cdot 1w_2\cdot\dots\cdot 1^{i-1}\chi_{t_i\bullet\beta}\cdot 1^iw_{i+1}\cdots\\&\equiv w_1\cdot 1w_2\cdot\dots\cdot 1^{i-1}w_i\cdot 1^{i-1}0\beta\cdot 1^iw_{i+1}\cdots\\&\equiv w_1\cdot 1w_2\cdot\dots\cdot 1^{i-1}w_i\cdot 1^iw_{i+1}\cdot\dots\cdot 1^{i-1}0\beta=\chi^*_t\cdot 1^{i-1}0\beta.\end{aligned}$$

Assume now $\alpha=1^{i-1}$. We find

$$\begin{aligned}\chi^*_{t\bullet w}&=w_1\cdot 1w_2\cdot\dots\cdot 1^{i-1}\chi_{t_i\cdot t_{i+1}}\cdot 1^iw_i\cdot 1^{i+1}w_{i+2}\cdots\\&\equiv w_1\cdot 1w_2\cdot\dots\cdot 1^{i-1}w_i\cdot 1^iw_{i+1}\cdot 1^{i-1}\cdot 1^iw_i^{-1}\cdot 1^iw_i\cdot 1^{i+1}w_{i+2}\cdots\\&\equiv w_1\cdot 1w_2\cdot\dots\cdot 1^{i-1}w_i\cdot 1^iw_{i+1}\cdot 1^{i-1}\cdot 1^{i+1}w_{i+2}\cdots\\&\equiv w_1\cdot 1w_2\cdot\dots\cdot 1^{i-1}w_i\cdot 1^iw_{i+1}\cdot 1^{i+1}w_{i+2}\cdot\dots\cdot 1^{i-1}.\end{aligned}$$ ■

Describing the connection between $\mathcal{G}_{LD}$ and G_{LD} is now easy.

Proposition 2.15. (connection) *Assume that w, w' are words on $\mathbf{A}\cup\mathbf{A}^{-1}$, and the domains of LD_w and $\mathrm{LD}_{w'}$ are not disjoint. Then the following are equivalent:*

(i) There exists at least one term t in T_1 satisfying $t\bullet w=t\bullet w'$;

(ii) For every t, we have $t\bullet w=t\bullet w'$ when the latter terms exist;

(iii) The relation $w\equiv w'$ holds.

In particular, for u, u' positive words, $\mathrm{LD}_u=\mathrm{LD}_{u'}$ is equivalent to $u\equiv u'$.

Proof. Assume that both LD_w and $\mathrm{LD}_{w'}$ both map t to t'. By Proposition 2.14 (action II), we have

$$\chi^*_t\cdot w\equiv\chi^*_{t'}\equiv\chi^*_t\cdot w',$$

hence $w\equiv\chi^{*-1}_t\cdot\chi^*_{t'}\equiv w'$.

Conversely, assume that $w \equiv w'$ holds, and both $t \bullet w$ and $t \bullet w'$ exist. By Lemma 2.8, the term t also belongs to the domain of the operators $\mathrm{LD}_{N(w)} \bullet \mathrm{LD}_{D(w)}^{-1}$ and $\mathrm{LD}_{N(w')} \bullet \mathrm{LD}_{D(w')}^{-1}$. By Proposition 1.14 (fraction), there exists two positive words v, v' satisfying

$$N(w) \cdot v \equiv^{+} N(w') \cdot v' \quad \text{and} \quad D(w) \cdot v \equiv^{+} D(w') \cdot v'. \tag{2.18}$$

Then we have

$$\begin{aligned} t \bullet w &= t \bullet (N(w) \cdot D(w)^{-1}) = t \bullet (N(w) \cdot v \cdot v^{-1} \cdot D(w)^{-1}) \\ &= t \bullet (N(w') \cdot v' \cdot v'^{-1} \cdot D(w')^{-1}) = t \bullet (N(w') \cdot D(w')^{-1}) = t \bullet w'. \end{aligned}$$

In the case of positive words u, u', the intersection of the domains of LD_u and $\mathrm{LD}_{u'}$ is never empty, and, by Proposition VII.1.26 (compatibility), the conditions (i) and (ii) are equivalent to $\mathrm{LD}_u = \mathrm{LD}_{u'}$. ■

It would be tempting to introduce the relation $\sim$ on $\mathcal{G}_{LD} \setminus \{\emptyset\}$ such that $f \sim f'$ holds if there exists at least one term t satisfying $f(t) = f'(t)$, and to try to prove that $\mathcal{G}_{LD} \setminus \{\emptyset\}$ quotiented by $\sim$ embeds in G_{LD}. Unfortunately, this does not make sense, as $\sim$ is not a transitive relation: consider for instance $f = \mathrm{LD}_{0^{-1} \cdot \phi^{-1} \cdot \phi \cdot 0}$, $f' = \mathrm{LD}_{\varepsilon}$, and $f'' = \mathrm{LD}_{\phi^{-1} \cdot \phi}$; then f, f', and f'' are the identity mappings of their respective domains, the domains of f and f' intersect, and so do the domains of f' and f'', but the domains of f and f'' are disjoint, as $\mathrm{LD}_{\phi \cdot 0 \cdot \phi^{-1}}$ is empty. Such problems however cannot occur when we restrict to positive transformations, as the intersections of domains are never empty. Thus, we deduce from Proposition 2.15:

Proposition 2.16. (embedding) *The monoid $\mathcal{G}_{LD}^{+}$ is isomorphic to the submonoid G_{LD}^{+} of G_{LD}.*

We conclude this section with an application. As the complement f is coherent and convergent, the word problem of the monoid M_{LD} is decidable: two positive words u, u' represent the same element of M_{LD} if and only if reversing $u^{-1}u'$ ends with an empty word. As we have not proved that the monoid M_{LD} is right cancellative, we cannot resort to the results of Chapter II to solve the word problem for the group G_{LD}. However, the previous result gives a solution.

Proposition 2.17. (word problem) *The group G_{LD} has a solvable word problem.*

Proof. Assume that w is a word on $\mathbf{A} \cup \mathbf{A}^{-1}$. Then $w \equiv \varepsilon$ is equivalent to $N(w) \equiv D(w)$. By Proposition 2.15 (connection), the latter condition is equivalent to $\mathrm{LD}_{N(w)} = \mathrm{LD}_{D(w)}$. This equality can be decided effectively: indeed, we have seen in Chapter VII how to determine the terms t_u^L and t_u^R for every word u, and $\mathrm{LD}_{N(w)} = \mathrm{LD}_{D(w)}$ is equivalent to the conjunction of $t_{N(w)}^L = t_{D(w)}^L$ and $t_{N(w)}^R = t_{D(w)}^R$. ■

The action of G_{LD} on terms

Associating with every word w on $\mathbf{A} \cup \mathbf{A}^{-1}$ the operator LD_w defines a partial action of $(\mathbf{A} \cup \mathbf{A}^{-1})^*$ on terms. It follows from Proposition 2.15 (equivalence) that this action induces a well defined partial action of the group G_{LD}. We extend our notation in the natural way: for t a term and a an element G_{LD}, we define $t \bullet a$ to be the term $t \bullet w$ where w is an expression of a such that $t \bullet w$ exists, if such an expression exists. By Lemma 2.8, if $t \bullet w$ exists, so does $t \bullet N(w)D(w)^{-1}$, and , if the latter exists, so does $t \bullet N(w')D(w')^{-1}$ for every word w' satisfying $w' \equiv w$. Thus $t \bullet a$ exists if and only if $t \bullet w$ does, where w is an arbitrary expression of a as a right fraction.

In this framework, Proposition 2.14 (action II) gives, for every term t in T_1 and every a in G_{LD} such that $t \bullet a$ is defined, the equality

$$\boldsymbol{\chi}^*_{t \bullet a} = \boldsymbol{\chi}^*_t \; a \tag{2.19}$$

which is an exact counterpart of (2.14). In particular, if $x^{[n]} \bullet a$ exists, we have

$$\boldsymbol{\chi}^*_{x^{[n]} \bullet a} = a. \tag{2.20}$$

An application of the previous results is that a term of T_1 is completely determined by the class of its blueprint in G_{LD}.

Definition 2.18. (functions χ and χ^*) For t in T_1, the classes of $\boldsymbol{\chi}_t$ and $\boldsymbol{\chi}^*_t$ in G_{LD} are denoted χ_t and χ^*_t respectively.

Proposition 2.19. (injectivity) *The mapping χ is injective on T_1.*

Proof. Assume $\boldsymbol{\chi}_t \equiv \boldsymbol{\chi}_{t'}$. Then, by Proposition 2.15 (equivalence), the operators $\mathrm{LD}_{\boldsymbol{\chi}_t}$ and $\mathrm{LD}_{\boldsymbol{\chi}_{t'}}$ agree on every term in the intersection of their domains. By construction, the term $x^{[p+1]}$ belongs to this intersection for p large enough. For such a p, we have

$$x^{[p+1]} \bullet \boldsymbol{\chi}_t = t \cdot x^{[p]}, \quad \text{and} \quad x^{[p+1]} \bullet \boldsymbol{\chi}_{t'} = t' \cdot x^{[p]}.$$

The hypothesis implies $t \cdot x^{[p]} = t' \cdot x^{[p]}$, hence $t = t'$. ∎

Definition 2.20. (special) Assume that a belongs to G_{LD}. We say that a is *special* if it belongs to the image of the mapping χ, *i.e.*, if a is the image of a (necessarily unique) term in T_1. The set of all special elements in G_{LD} is denoted by G^{sp}_{LD}.

The group G_{LD} plays with respect to the braid group B_∞ the same role as the absolutely free system T_1 plays with respect to the free LD-system FLD_1: B_∞

includes a copy B_{sp} of FLD_1, while G_{LD} includes a copy G_{LD}^{sp} of T_1, as shown in the commutative diagram

$$\begin{array}{ccccc} T_1 & \xrightarrow{\chi} & G_{LD}^{sp} & \subseteq & G_{LD} \\ \downarrow & & \downarrow & & \downarrow \\ \mathrm{FLD}_1 & \xrightarrow{\cong} & B_{sp} & \subseteq & B_\infty . \end{array}$$

We then have counterparts for the special decompositions of braids. Let us say that the element a of G_{LD} admits a special decomposition if it admits at least one decomposition of the form

$$a = a_1 \cdot \mathrm{sh}_1(a_2) \cdot \mathrm{sh}_{11}(a_3) \cdot \cdots \tag{2.21}$$

where a_1, a_2, ... are special. Observe that, if $a_i = \chi_{t_i}$ holds for every i, then the product involved in (2.21) is χ_t^*, where t is the term $t_1 \cdot t_2 \cdot \cdots$. Thus admitting a special decomposition means lying in the image of χ^*.

Proposition 2.21. (special decomposition) *Assume that a is an element of G_{LD} such that LD_a is nonempty. Then a admits in G_{LD} the decomposition*

$$a = \chi_t^{*-1} \chi_{t \bullet a}^* \tag{2.22}$$

where t is an arbitrary term in the domain of LD_a. In particular, if $x^{[n]}$ belongs to the domain of LD_a, then a admits the special decomposition $a = \chi_{x^{[n]} \bullet a}^*$.

Proof. The results follow from (2.19) and (2.20) directly. ∎

The previous result does not apply to those elements a of G_{LD} such that LD_a is empty. However, even such an element admits a decomposition involving special elements, as a can be expressed as bc^{-1} with $b, c \in G_{LD}^+$, hence as $b_1^{-1} b_2 c_2^{-1} c_1$ with b_1, b_2, c_1, c_2 admitting a special decomposition.

The case of several variables

In this section, we show how extending the previous construction to arbitrary terms naturally leads to the group CB_∞ of charged braids introduced in Section V.6.

For t in T_1, the blueprint χ_t of t has been defined so that LD_{χ_t} maps $x^{[p+1]}$ to $t \cdot x^{[p]}$ for p large enough. Thus χ_t describes a construction of t from $x^{[p+1]}$—we could use the extended term x^∞ to have a uniform starting point. As left self-distributivity preserves the set of variables of the terms it is applied to, we cannot extend the definition so as to construct terms with more than one variable. However, we can do it at the expense of introducing new operators that change the variables.

Definition 2.22. (**variable shift**) Assume that α is an address. We define the partial *variable shift* operator VS_α on T_∞ as follows: The term t belongs to the domain of VS_α if and only if α belongs to the skeleton of t, and, in this case, the image of t under VS_α is obtained from t by shifting by one unit the indices of all variables occurring below α in t.

Example 2.23. *(**operator** VS_α) Let t be the term* 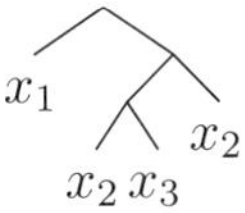

. Then t belongs to the domain of the operators VS_{10} and VS_{10}^{-1}. The images are obtained by shifting by $+1$ and -1 respectively the variables in the 10-subterm of t, so they are respectively

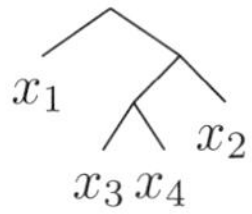

and
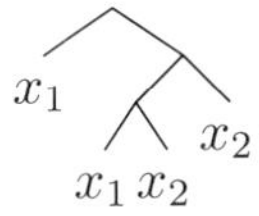

.

Observe that every term belongs to the domain of VS_ϕ, while the domain of VS_ϕ^{-1} consists of those terms where x_1 does not occur.

Now, the idea is obvious: using the additional operators VS_α, we can construct from $x^{[p]}$ not only terms in T_1, but also arbitrary terms in T_∞.

The first step is to introduce the *completed geometry monoid* $\widetilde{\mathcal{G}}_{LD}$ generated by all operators LD_α and VS_β and their inverses. The elements of these monoids are represented by words on $\mathbf{A}\cup\mathbf{A}^{-1}$, augmented with a new alphabet that represents the VS_β operators and their inverses. We use $\widetilde{\beta}$ for the letter that corresponds to the operator VS_β. The shift mapping sh_γ is extended by $\mathrm{sh}_\gamma(\widetilde{\beta}) = \widetilde{\gamma\beta}$. Then, as in Section VII.2, we look for the geometric relations satisfied by the operators. It is easy to obtain the following list:

$$
\begin{gathered}
\mathrm{VS}_\alpha \approx \mathrm{VS}_{\alpha 1} \cdot \mathrm{VS}_{\alpha 0}, \qquad \mathrm{VS}_\alpha \cdot \mathrm{VS}_\beta = \mathrm{VS}_\beta \cdot \mathrm{VS}_\alpha \\
\mathrm{VS}_{\gamma 0\alpha} \cdot \mathrm{LD}_{\gamma 1\beta} = \mathrm{LD}_{\gamma 1\beta} \cdot \mathrm{VS}_{\gamma 0\alpha}, \qquad \mathrm{VS}_\gamma \cdot \mathrm{LD}_{\gamma\alpha} = \mathrm{LD}_{\gamma\alpha} \cdot \mathrm{VS}_\gamma, \\
\mathrm{VS}_{\gamma 0\alpha} \cdot \mathrm{LD}_\gamma = \mathrm{LD}_\gamma \cdot \mathrm{VS}_{\gamma 10\alpha} \cdot \mathrm{VS}_{\gamma 00\alpha}, \quad \mathrm{VS}_{\gamma 10\alpha} \cdot \mathrm{LD}_\gamma = \mathrm{LD}_\gamma \cdot \mathrm{VS}_{\gamma 01\alpha}, \\
\mathrm{VS}_{\gamma 11\alpha} \cdot \mathrm{LD}_\gamma = \mathrm{LD}_\gamma \cdot \mathrm{VS}_{\gamma 11\alpha}
\end{gathered} \tag{2.23}
$$

in addition to the already known relations between the operators LD_α—we use $f \approx f'$ means that the operator f and f' agree on the intersection of their domains. Then, we extend the construction of the blueprint.

Definition 2.24. (**blueprint**) For t a term in T_∞, the *blueprint* of t is the word χ_t inductively defined by $\chi_{x_i} = \widetilde{0}^{i-1}$ and $\chi_{t_1 \cdot t_2} = \chi_{t_1} \cdot 1\chi_{t_2} \cdot \phi \cdot 1\chi_{t_1}^{-1}$.

For instance, for $t = (x_1 \cdot x_2) \cdot x_3$, we find $\chi_t = \widetilde{10} \cdot \phi \cdot \widetilde{10} \cdot \widetilde{10} \cdot \phi \cdot 1^{-1} \cdot \widetilde{110}^{-1}$.

An immediate induction gives the expected result:

Lemma 2.25. *Assume that t is a term in T_∞. Then, for p large enough, the operator LD_{χ_t} maps $x^{[p+1]}$ to $t \cdot x^{[p]}$.*

Like $\mathcal{G}_{LD}$, the completed geometry monoid $\widetilde{\mathcal{G}}_{LD}$ is not a group, so we consider instead the group $\widetilde{G}_{LD}$ for which the relations in (2.23) constitute a presentation. We use $\widetilde{g}_\beta$ for the class of $\widetilde{\beta}$ in $\widetilde{G}_{LD}$. We introduce on $\widetilde{G}_{LD}$ the exponentiation defined by

$$a \wedge b = a \cdot \mathrm{sh}_1(b) \cdot g_\phi \cdot \mathrm{sh}_1(a^{-1}),$$

and the homomorphism χ of T_∞ into $(\widetilde{G}_{LD}, \wedge)$ that maps x_i to $\widetilde{g}_0^{i-1}$. Lemmas 2.9 and 2.10 extend to $\widetilde{G}_{LD}$. Hence, it suffices to collapse the generators $g_{0\alpha}$ to obtain a self-distributive operation on the quotient. Owing to the case of rank 1, we can expect the image of T_∞ in the quotient group to be a free LD-system. Thus, we have to determine what remains from the relations of $\widetilde{G}_{LD}$ under the collapsing.

Actually, we first observe that all operators VS_β are not needed to construct terms. Considering the operators VS_{1^j0} is sufficient, as applying the relations of (2.23) lets no other operators appear. We write $\widetilde{G}'_{LD}$ for the subgroup of $\widetilde{G}_{LD}$ generated by all g_α and those generators $\widetilde{g}_\beta$ of the above type.

Collapsing the generators $g_{0\alpha}$ of G_{LD} amounts to projecting G_{LD} onto the braid group B_∞ by mapping g_α to σ_{i+1} for $\alpha = 1^i$, and to 1 otherwise. When we consider a similar process in $\widetilde{G}'_{LD}$, the generators $\widetilde{g}_{1^j0}$ are not collapsed. Let us denote by ρ_{j+1} the image of $\widetilde{g}_{1^j0}$ in the projection. What remains from (2.23) is the list of all braid relations, augmented with the relations

$$\rho_j \rho_k = \rho_k \rho_j \qquad \text{for all } j,\ k,$$

$$\rho_j \sigma_i = \sigma_i \rho_j \qquad \text{for } j < i \text{ or } j > i+1,$$

$$\rho_i \rho_{i+1} \sigma_i = \sigma_i \rho_i \rho_{i+1} \qquad \text{for all } i.$$

We recognize the defining relations of CB_∞. So we can state:

Proposition 2.26. **(charged braids)** *The image of $\widetilde{G}'_{LD}$ under the projection that collapses every generator $g_{0\alpha}$ is the group CB_∞.*

It follows that the exponentiation of $\widetilde{G}_{LD}$ induces a left self-distributive operation on CB_∞—as was already checked in Chapter V—and that the mapping χ induces a well defined homomorphism of the free LD-system FLD_∞ into $(CB_\infty, \wedge)$ that maps x_i to ρ_1^{i-1}. The example of rank 1 and ordinary braids suggests that this homomorphism is likely to be an embedding. This result is not obvious, as new relations could have been added when collapsing, but, precisely, it is the content of Proposition V.6.14 (charged braids free).

Thus, the results have been extended from rank 1 to arbitrary rank completely, and, in particular, we have obtained the following commutative diagram

$$\begin{array}{ccc} T_\infty & \xrightarrow{\chi} & \widetilde{G}'_{LD} \\ \downarrow & & \downarrow \\ \mathrm{FLD}_\infty & \longrightarrow & CB_\infty \end{array}$$

where the bottom horizontal arrow is an embedding.

Exercise 2.27. **(abelianization)** *Show that using the abelianized group $G_{LD}^{ab} = \mathbf{Z}[z]$ instead of G_{LD} leads to the blueprint χ^{ab} inductively defined by $\chi_x^{ab} = 1$, $\chi_{t_1 \cdot t_2}^{ab} = (1-z)\chi_{t_1}^{ab} + z\chi_{t_2}^{ab} + 1$. Show that collapsing the image of g_0 amounts to putting $z = 1$, and conclude that the left self-distributive operation obtained is $\mathbf{Z}$ equipped with $a^\wedge b = b + 1$.*

Exercise 2.28. **(left cancellation)** *Prove that exponentiation in G_{LD} is left cancellative.*

Exercise 2.29. **(mapping χ^*)** *Assume $t \in T_1$ and $\mathrm{ht}_R(t) = r$. Prove that, for n large enough, $\mathrm{LD}_{\chi_t^*}$ maps $x^{[p]}$ to the term obtained from t by replacing the rightmost variable with $x^{[p-r]}$.*

Exercise 2.30. **(evaluation)** *Let us consider terms in $T_{(\mathbf{A}\cup\mathbf{A}^{-1})^*}$ and $T_{G_{LD}}$, i.e., terms whose variables are themselves words on $\mathbf{A}\cup\mathbf{A}^{-1}$ or elements of G_{LD}. For t in $T_{(\mathbf{A}\cup\mathbf{A}^{-1})^*}$, we write $\mathrm{eval}(t)$ for the evaluation of t in $((\mathbf{A}\cup\mathbf{A}^{-1})^*, {}^\wedge)$, i.e., for the element of $(\mathbf{A}\cup\mathbf{A}^{-1})^*$ iteratively defined by $\mathrm{eval}(t) = t$ if t is a variable, and $\mathrm{eval}(t_1 \cdot t_2) = \mathrm{eval}(t_1)^\wedge\mathrm{eval}(t_2)$. The definition is similar for $t \in T_{G_{LD}}$.*

(i) For t in T_1, prove $\chi_t = \mathrm{eval}(t(\varepsilon))$, where $t(\varepsilon)$ denotes the term in $T_{(\mathbf{A}\cup\mathbf{A}^{-1})^*}$ obtained by substituting x with ε.

(ii) For t in $T_{(\mathbf{A}\cup\mathbf{A}^{-1})^*}$ and w in $\mathbf{A}\cup\mathbf{A}^{-1}$ such that $t \bullet w$ exists, prove $\mathrm{eval}(t \bullet w) \equiv \mathrm{eval}(t) \cdot 0w$. [Hint: Use induction on the length of w, as in the proof of Proposition 2.8 (action I).] Re-deduce Proposition 2.8.

(iii) For t in $T_{(\mathbf{A}\cup\mathbf{A}^{-1})^*}$, say $t = t_1 \cdot \cdots \cdot t_p \cdot x$ with x a variable, define $\mathrm{eval}^*(t) = \mathrm{eval}(t_1) \cdot 1\mathrm{eval}(t_2) \cdot 11\mathrm{eval}(t_3) \cdot \cdots \cdot 1^p\mathrm{eval}(x)$. Prove $\chi_t^* = \mathrm{eval}^*(t(\varepsilon))$ for t in T_1.

(iv) For t in $T_{(\mathbf{A}\cup\mathbf{A}^{-1})^*}$ and w in $(\mathbf{A}\cup\mathbf{A}^{-1})^*$ such that $t \bullet w$ exists, prove $\mathrm{eval}^*(t \bullet w) \equiv \mathrm{eval}^*(t) \cdot w$. Re-deduce Proposition 2.14 (action II).

(v) For t in $T_{G_{LD}}$ and a in G_{LD} such that $t \bullet a$ exists, prove $\mathrm{eval}^*(t \bullet a) = \mathrm{eval}^*(t) \cdot a$, where $\mathrm{eval}^*(t)$ is defined as in (iii).

Exercise 2.31. **(embedding)** *Prove that χ is injective on T_∞.*

---

3. Order Properties in G_{LD}

We shall see now how the braid order of Chapter III originates in some natural preorder on G_{LD}. Actually, we introduce below two different (pre)orders on G_{LD}, both relying on its action on terms by left self-distributivity. One consists in comparing a and b by analysing the action of the (non-uniquely defined) numerators and denominators of ab^{-1} on iterated left subterms, and it results in a right invariant preorder connected with the braid order, which gives a direct proof that left division in free LD-systems has no cycle, and a new solution for the word problem of Identity (LD).

The second order on G_{LD} resorts to a linear ordering of terms, and consists in comparing a and b by using their action on terms, as was done for braids in Chapter III; it results in a two-sided invariant linear order on G_{LD}. This order can be interpreted in terms of an action of G_{LD} on the Cantor line.

A preordering on G_{LD}

We have established in Section 3 that G_{LD} includes a copy G_{LD}^{sp} of T_1. The latter set is equipped with a preordering $\sqsubset_{LD}$. A natural idea is to define on G_{LD} a relation that includes this copy of $\sqsubset_{LD}$.

Let us start from the proof of the confluence property as given in Chapter V. Assume that t, t' are terms in T_1. Let $w = \chi_t^{-1} \cdot \chi_{t'}$. By construction, LD_w maps $t{\cdot}x^{[p]}$ to $t'{\cdot}x^{[p]}$ for p large enough. Then, we have

$$(t \cdot x^{[p]}) \bullet N(w) = (t' \cdot x^{[p]}) \bullet D(w),$$

and the latter term, say t_0, is a common LD-expansion of $t{\cdot}x^{[p]}$ and $t'{\cdot}x^{[p]}$. By Lemma VII.4.5, we have

$$t \bullet N(w) = \mathrm{left}(t \cdot x^{[p]}) \bullet N(w) = \mathrm{left}^{\mathrm{dil}(1,N(w))}(t_0),$$
$$t' \bullet D(w) = \mathrm{left}(t' \cdot x^{[p]}) \bullet D(w) = \mathrm{left}^{\mathrm{dil}(1,D(w))}(t_0).$$

In particular, the term t is LD-equivalent to $\mathrm{left}^{\mathrm{dil}(1,N(w))}(t_0)$ and t' is LD-equivalent to $\mathrm{left}^{\mathrm{dil}(1,D(w))}(t_0)$. Hence $t \sqsubset_{LD} t'$, $t =_{LD} t'$, or $t \sqsupset_{LD} t'$ holds according to whether $\mathrm{dil}(1, D(w)) < \mathrm{dil}(1, N(w))$, $\mathrm{dil}(1, D(w)) = \mathrm{dil}(1, N(w))$, or $\text{dil}(1, D(w)) > \text{dil}(1, N(w))$ does. This leads us to introduce the following partition on $(\mathbf{A} \cup \mathbf{A}^{-1})^*$:

Definition 3.1. (sets P_+, P_0, P_-) Assume that w is a word on $\mathbf{A} \cup \mathbf{A}^{-1}$. We say that w belongs to P_+ (*resp.* P_0, *resp.* P_-) if we have

$$\text{dil}(1, D(w)) < \text{dil}(1, N(w)) \quad (\textit{resp.} =, \textit{resp.} >). \tag{3.1}$$

Lemma 3.2. *The sets P_+, P_0, and P_- form a partition of $(\mathbf{A} \cup \mathbf{A}^{-1})^*$; each of them is saturated under $\equiv$ and closed under product. Moreover, we have $P_0\, P_+ \subseteq P_+$, $P_+\, P_0 \subseteq P_+$, $P_0\, P_- \subseteq P_-$, and $P_-\, P_0 \subseteq P_-$.*

Proof. Assume $w \equiv w'$. Then there exist positive words v, v' satisfying $N(w)\, v \equiv^+ N(w')\, v'$ and $D(w)\, v \equiv^+ D(w')\, v'$. By definition of the 'dil' mapping and Lemma VII.4.5, we have

$$\text{dil}(\text{dil}(1, D(w)), v) = \text{dil}(1, D(w)v) = \text{dil}(1, D(w')v') = \text{dil}(\text{dil}(1, D(w')), v'),$$
$$\text{dil}(\text{dil}(1, N(w)), v) = \text{dil}(1, N(w)v) = \text{dil}(1, N(w')v') = \text{dil}(\text{dil}(1, N(w')), v').$$

By construction, the mappings $\text{dil}(\cdot, v)$ and $\text{dil}(\cdot, v')$ are increasing, so we deduce that $\text{dil}(1, D(w)) < \text{dil}(1, N(w))$ is equivalent to $\text{dil}(1, D(w')) < \text{dil}(1, N(w'))$, and the same for $=$ and $>$.

Assume now w_1, $w_2 \in P_+$. We find

$$\begin{aligned}
\text{dil}(1, D(w_1 w_2)) &= \text{dil}(1, D(w_2)\, (N(w_2)\backslash D(w_1))) \\
&= \text{dil}(\text{dil}(1, D(w_2)), N(w_2)\backslash D(w_1)) \\
&< \text{dil}(\text{dil}(1, N(w_2)), N(w_2)\backslash D(w_1)) \\
&= \text{dil}(1, N(w_2)\, (N(w_2)\backslash D(w_1))) \\
&= \text{dil}(1, D(w_1)\, (D(w_1)\backslash N(w_2))) \\
&= \text{dil}(\text{dil}(1, D(w_1)), D(w_1)\backslash N(w_2)) \\
&< \text{dil}(\text{dil}(1, N(w_1)), D(w_1)\backslash N(w_2)) \\
&= \text{dil}(1, N(w_1)\, (D(w_1)\backslash N(w_2))) = \text{dil}(1, N(w_1 w_2)).
\end{aligned}$$

The computation is similar in the other cases. ∎

Definition 3.3. (relations $\prec$ and $\simeq$) For a, b in G_{LD}, we say that $a \prec b$ (*resp.* $a \simeq b$) holds if $a^{-1}b$ admits an expression in P_+ (*resp.* in P_0).

Applying Lemma 3.2, we obtain directly:

Proposition 3.4. (preorder) *The relation $\prec$ is a preorder on G_{LD}, and $\simeq$ is the associated equivalence relation; both are invariant under left multiplication.*

We deduce a new solution of the word problem for (LD) in the case of one variable terms:

Proposition 3.5. (**word problem**) *For all terms t_1, t_2 in T_1, $t_1 \sqsubset_{LD} t_2$ is equivalent to $\chi_{t_1} \prec \chi_{t_2}$, and $t_1 =_{LD} t_2$ is equivalent to $\chi_{t_1} \simeq \chi_{t_2}$*

Proof. The computation at the beginning of this section shows that $\chi_{t_1}^{-1} \cdot \chi_{t_2}$ belonging to P_+, P_0, or P_- implies $t_1 \sqsubset_{LD} t_2$, $t_1 =_{LD} t_2$, or $t_1 \sqsupset_{LD} t_2$ respectively. The latter cases exclude each other, so the implication is an equivalence. ∎

Example 3.6. (**word problem**) *Every factor in the above expressions is effectively computable. For instance, let $t_1 = (x{\cdot}x){\cdot}(x{\cdot}(x{\cdot}x))$ and $t_2 = x{\cdot}((x{\cdot}x){\cdot}(x{\cdot}x))$. We compute*

$$\chi_{t_1} = \phi \cdot 11 \cdot 1 \cdot \phi \cdot 1^{-1}, \quad \chi_{t_2} = 1 \cdot 11 \cdot 1 \cdot 11^{-1} \cdot \phi,$$
$$D(\chi_{t_1}^{-1} \cdot \chi_{t_2}) = 11 \cdot 1 \cdot 0 \cdot 11 \cdot 01 \cdot 10 \cdot 00, \quad N(\chi_{t_1}^{-1} \cdot \chi_{t_2}) = 1 \cdot 11 \cdot 01 \cdot 1 \cdot 0,$$

and, finally, $\mathrm{dil}(1, D(\chi_{t_1}^{-1} \cdot \chi_{t_2})) = 1 = \mathrm{dil}(1, N(\chi_{t_1}^{-1} \cdot \chi_{t_2}))$. *So we have $\chi_{t_1}^{-1} \cdot \chi_{t_2} \in P_0$, hence $t_1 =_{LD} t_2$.*

Another consequence is:

Corollary 3.7. *Assume that a, b are special elements of G_{LD}. Then*
(i) $a \prec b$ holds in G_{LD} if and only if $\mathrm{pr}(a) <_L \mathrm{pr}(b)$ holds in B_∞;
(ii) $a \simeq b$ holds in G_{LD} if and only if $\mathrm{pr}(a) = \mathrm{pr}(b)$ holds in B_∞.

Proof. Assume $a = \chi_{t_1}$, $b = \chi_{t_2}$. By Proposition 3.4, $a \prec b$ is equivalent to $t_1 \sqsubset_{LD} t_2$, which, by Proposition III.2.14(ii) (order), is equivalent to $t_1(1)^{B_\infty} <_L t_2(1)^{B_\infty}$. Now, by construction, $t(1)$ evaluated in $(B_\infty, {}^\wedge)$ is $\mathrm{pr}(\chi_t)$ for every term t in T_1. The argument is similar for $\simeq$. ∎

A natural question is whether the equivalences of Corollary 3.7, which connect the order on special braids with the preorder $\prec$ on G_{LD}, extend to the whole of G_{LD}. They do not, for the definition of $\prec$ takes into account the image of the left subterm only. For instance, we have $\mathrm{dil}(1, \phi \cdot 0) = \mathrm{dil}(1, \phi \cdot \phi) = 3$, hence $g_\phi g_0 \simeq g_\phi^2$ holds in G_{LD}, while $\mathrm{pr}(g_\phi g_0) = \sigma_1 \neq \mathrm{pr}(g_\phi^2) = \sigma_1^2$ holds in B_∞.

The last point in the proof of Proposition 3.4 (preorder), namely the relations $t_1 \sqsubset_{LD} t_2$, $t_1 =_{LD} t_2$, or $t_1 \sqsupset_{LD} t_2$ excluding each other, is equivalent to the statement that $t \sqsubset_{LD} t$ holds for no term t in T_1, *i.e.*, to the property that left division in the free LD-system FLD_1 has no cycle. This property has been deduced in Section I.3 from the similar property for exponentiation on $\mathrm{Aut}(\mathrm{FG}_\infty)$. With the current tools, we can give a purely syntactic—and very simple—proof.

Proposition 3.8. (**syntactic acyclicity**) *No term t in T_1 satisfies $t \sqsubset_{LD} t$.*

Proof. It suffices to prove directly that $t_1 \sqsubset_{LD} t_2$ implies $\chi_{t_1}^{-1} \cdot \chi_{t_2} \in P_+$. Then $t \sqsubset_{LD} t$ would imply $\varepsilon \in P_+$, which is impossible, as P_+ and P_0 are disjoint. So, assume $t_1 \sqsubset_{LD} t_2$. By definition, there exist terms t'_1, t'_2 satisfying $t'_1 =_{LD} t_1$, $t'_2 =_{LD} t_2$ and t'_1 is an iterated left subterm of t'_2, say $t'_1 = \mathrm{left}^p(t'_2)$. By Proposition 2.8 (action I), there exist two words w_1, w_2 such that $\chi_{t_e} \equiv \chi_{t'_e} \cdot 0w_e$ holds for $e = 1, 2$, and, therefore, we have

$$\chi_{t_1}^{-1} \cdot \chi_{t_2} \equiv 0w_1^{-1} \cdot \chi_{t'_1}^{-1} \cdot \chi_{t'_2} \cdot 0w_2. \tag{3.2}$$

Assume $t'_2 = (\cdots((t'_1{\cdot}s_1){\cdot}s_2)\cdots){\cdot}s_p$. Applying the definition of the blueprint gives a decomposition of the form

$$\chi_{t'_2} = \chi_{t'_1} \cdot 1u_0 \cdot \phi \cdot 1u_1 \cdot \phi \cdot \cdots \cdot \phi \cdot 1u_p,$$

and we deduce from (3.2) that $\chi_{t_1}^{-1} \cdot \chi_{t_2}$ is equivalent to

$$0w_1^{-1} \cdot 1u_0 \cdot \phi \cdot 1u_1 \cdot \phi \cdot \cdots \cdot \phi \cdot 1u_p \cdot 0w_2. \tag{3.3}$$

Now, ϕ belongs to P_+, as we have $\mathrm{dil}(1, \phi) = 2$, and $\mathrm{dil}(1, \varepsilon) = 1$. On the other hand, every word of the form $0u$ or $1u$ belongs to P_0 by definition. Hence, the word in (3.3) belongs to P_+, and so does $\chi_{t_1}^{-1} \cdot \chi_{t_2}$. ■

The left ordering of terms

Terms can be equipped with several orderings in connection with their tree structure. Here, we introduce a linear order on T_∞ that uses the left height as a discrimimant.

Terms have been constructed as words, and we have used the right Polish form where the product of t_1 and t_2 is defined to be the word $t_1t_2\bullet$. Here we shall resort for a while to the left Polish form, where the product of t_1 and t_2 is defined to be $\bullet t_1t_2$.

Definition 3.9. (left Polish form) For t a term, the *left Polish form* of t is the word $(t)_L$ inductively defined by $(t)_L = t$ for t a variable, and $(t_1{\cdot}t_2)_L = \bullet(t_1)_L(t_2)_L$.

For instance, we have

$$(x_1 \cdot (x_2 \cdot (x_3 \cdot x_4)) \cdot x_5)_L = (x_1x_2x_3x_4{\bullet}{\bullet}x_5{\bullet}{\bullet})_L = {\bullet}x_1{\bullet}{\bullet}x_2{\bullet}x_3x_4x_5.$$

When the term t is viewed as a tree, the word $(t)_L$ is obtained by enumerating the variables of t from left to right and writing before the variable occurring at α as many letters $\bullet$ as there are final 0's in α, a counterpart to right Polish form, where the variable occurring at α is followed by as many letters $\bullet$ as there are final 1's in α.

Definition 3.10. (left order) For t_1, $t_2 \in T_\infty$, we say that $t_1 <_L t_2$ holds if the word $(t_1)_L$ precedes the word $(t_2)_L$ in the lexicographical extension of $x_1 < x_2 < \ldots < \bullet$.

By construction, the relation $<_L$ is a linear order on T_∞, and x_1 is minimal for $<_L$. Letting $\mathrm{ht}_L(t)$ denote the left height of the term t, *i.e.*, the length of the leftmost branch of t, we see that the word $(t)_L$ begins with $\mathrm{ht}_L(t)$ letters $\bullet$ followed by a variable. So, $\mathrm{ht}_L(t_1) < \mathrm{ht}_L(t_2)$ implies $t_1 <_L t_2$. Observe that the linear order $<_L$ does not include the partial orders connected with the right Polish form considered in Chapter V: for instance, we have $x_1{\cdot}x_2 \ll x_2$, hence $x_1{\cdot}x_2 < x_2$, but $x_1{\cdot}x_2 >_L x_2$ holds, as the word $(x_1{\cdot}x_2)_L$, namely $\bullet x_1 x_2$, lies after x_2.

Our aim is to prove that the action of G_{LD} on T_∞ preserves $<_L$:

Proposition 3.11. (invariance) *Assume t, $t' \in T_\infty$. Then $t <_L t'$ is equivalent to $t \bullet a <_L t' \bullet a$ for every a in G_{LD} such that $t \bullet a$ and $t' \bullet a$ exist (if any).*

The proof begins with auxiliary results about the left Polish notation and the left ordering of terms. First, we have the following counterpart to Lemma V.1.7 (we recall that $\#_X(w)$ and $\#_\bullet(w)$ denote the number of letters from X and of $\bullet$ in w respectively).

Lemma 3.12. *A word w on the alphabet $X \cup \{\bullet\}$ is the left Polish form of a term if and only if we have $\#_X(w) = \#_\bullet(w) + 1$, and $\#_X(u) \le \#_\bullet(u)$ for every proper prefix u of w.*

Lemma 3.13. *The inequality $t_1 <_L t_2$ implies $t_1{\cdot}t_3 <_L t_2{\cdot}t_4$ for all t_3, t_4.*

Proof. Lemma 3.12 implies that a proper prefix of the left Polish form of a term is never the left Polish form of a term. Hence $t_1 <_L t_2$ holds if and only if the words $(t_1)_L$ and $(t_2)_L$ have a clash of the type "variable *vs.* $\bullet$". Then the words $(t_1{\cdot}t_3)_L$ and $(t_2{\cdot}t_4)_L$, *i.e.*, $\bullet(t_1)_L(t_3)_L$ and $\bullet(t_2)_L(t_4)_L$, have a similar clash. ∎

Lemma 3.14. *Assume t_1, $t_2 \in T_\infty$. If t_1 is a variable, say $t_1 = x_i$, $t_1 <_L t_2$ holds unless we have $t_2 = x_j$ with $j < i$. If t_1 is not a variable, then $t_1 <_L t_2$ holds if and only if we have either $\mathrm{sub}(t_1, 0) <_L \mathrm{sub}(t_2, 0)$, or $\mathrm{sub}(t_1, 0) = \mathrm{sub}(t_2, 0)$ and $\mathrm{sub}(t_1, 1) <_L \mathrm{sub}(t_2, 1)$.*

Proof. Assume $t_1 <_L t_2$, and neither t_1 nor t_2 are variables. Three cases are possible. If $\mathrm{sub}(t_1, 0) <_L \mathrm{sub}(t_2, 0)$ holds, then, by Lemma 3.13, $t_1 <_L t_2$ holds. Symmetrically, if $\mathrm{sub}(t_1, 0) >_L \mathrm{sub}(t_2, 0)$ holds, then $t_1 >_L t_2$ holds. Finally, if $\mathrm{sub}(t_1, 0)$ and $\mathrm{sub}(t_2, 0)$ are equal, then, by definition, $t_1 <_L t_2$ holds if and only if $\mathrm{sub}(t_1, 1) <_L \mathrm{sub}(t_2, 1)$ does. ∎

Lemma 3.15. *Assume that t_1, t_2 are terms in T_∞. Then the following are equivalent:*

(i) The relation $t_1 <_L t_2$ holds;

(ii) There exists an address α in $\mathrm{Skel}(t_1) \cap \mathrm{Skel}(t_2)$ such that $\mathrm{sub}(t_1, \beta) = \mathrm{sub}(t_2, \beta)$ holds for every β in the left edge of α, and $\mathrm{sub}(t_1, \alpha) <_L \mathrm{sub}(t_2, \alpha)$ holds;

(iii) There exists an address α in $\mathrm{Out}(t_1) \cap \mathrm{Skel}(t_2)$ such that $\mathrm{sub}(t_1, \beta) = \mathrm{sub}(t_2, \beta)$ holds for every β in the left edge of α, and either $\mathrm{sub}(t_2, \alpha)$ is a variable larger than $\mathrm{var}(t_1, \alpha)$, or it is not a variable.

Proof. An induction on α shows that, if α belongs to the skeleton of the term t and $(\alpha_1, \ldots, \alpha_p)$ is the left edge of α, then the word $(t)_L$ begins with

$$\bullet^{k_1}(\mathrm{sub}(t, \alpha_1))_L \bullet^{k_2}(\mathrm{sub}(t, \alpha_2))_L \ldots \bullet^{k_p}(\mathrm{sub}(t, \alpha_p))_L \bullet^{k}(\mathrm{sub}(t, \alpha))_L, \tag{3.4}$$

where k_i is the number of final 0's in α_i and k is the number of final 0's in α: the result is obvious for $\alpha = \phi$, and, otherwise, we apply the inductive definition of the left edge as given in Chapter VI. Then, by definition of a lexicographical ordering, it follows from (3.4) and from the fact that a proper prefix of a left Polish form is never the left Polish form of a term that (ii) implies (i).

By construction, (iii) implies (ii). Finally, assuming (i), and letting α be the address of the first position where the words $(t_1)_L$ and $(t_2)_L$ disagree, we obtain (iii) using the explicit expansion of (3.4). ■

Lemma 3.16. *Assume that t_1, t_2 are $\sqsubset_{LD}$-comparable terms in T_∞. Then the following are equivalent:*

(i) The relation $t_1 <_L t_2$ holds;

(ii) There exists an address α in $\mathrm{Out}(t_1) \cap \mathrm{Skel}(t_2)$ such that $\mathrm{sub}(t_1, \beta) = \mathrm{sub}(t_2, \beta)$ holds for every β in the left edge of α, and $\mathrm{sub}(t_2, \alpha)$ is a term of left height at least 1 whose leftmost variable is $\mathrm{var}(t_1, \alpha)$.

Proof. Assume (i). Then there exists an address α satisfying the conditions of Lemma 3.15(iii). We claim that $\mathrm{sub}(t_2, \alpha)$ cannot be a variable. Indeed, assume $\mathrm{sub}(t_1, \alpha) = x_i$ and $\mathrm{sub}(t_2, \alpha) = x_j$ with $j > i$. Then, letting $(\alpha_1, \ldots, \alpha_p)$ denote the left edge of α, we see that (the right Polish form of) t_1 begins with $\mathrm{sub}(t_1, \alpha_1) \ldots \mathrm{sub}(t_1, \alpha_p) x_i$, while (the right Polish form of) t_2 begins with $\mathrm{sub}(t_1, \alpha_1) \ldots \mathrm{sub}(t_1, \alpha_p) x_j$. Thus $t_1 \ll t_2$ holds, contradicting the hypothesis that t_1 and t_2 are $\sqsubset_{LD}$-comparable. So the only possibility is $\mathrm{sub}(t_2, \alpha)$ not to be a variable, and its leftmost variable to be x_i. This gives (ii). Moreover, the conditions of Lemma 3.15(iii) hold for the terms $t_1^\dagger$ and $t_2^\dagger$ as well, so we have $t_1^\dagger <_L t_2^\dagger$, and (iii) holds as well. That (ii) implies (i) follows from Lemma 3.15. ■

We deduce that the left order of terms is invariant under substitution:

Proposition 3.17. (**substitution**) *Assume that t_1, t_2 belong to T_∞, f is a substitution of T_∞, and at least one of the following conditions holds:*

(i) We have $f(x_i) <_L f(x_{i+1})$ and $\mathrm{ht}_L(f(x_i)) = \mathrm{ht}_L(f(x_{i+1}))$ for every i;

(ii) The terms t_1 and t_2 are $\sqsubseteq_{LD}$-comparable.

Then $t_1 <_L t_2$ is equivalent to $t_1^f <_L t_2^f$.

Proof. As $<_L$ is a linear order, it suffices to prove that $t_1 <_L t_2$ implies $t_1^f <_L t_2^f$. So assume $t_1 <_L t_2$. By Lemma 3.15, there exists an address α such that $\mathrm{sub}(t_1, \beta) = \mathrm{sub}(t_2, \beta)$ holds for every β in the left edge of α, $\mathrm{sub}(t_1, \alpha)$ is a variable say x_i, and $\mathrm{sub}(t_2, \alpha)$ is either a variable x_j with $j > i$, or it is a term that is not a variable. When Condition (ii) holds, by Lemma 3.16, we can assume that $\mathrm{sub}(t_2, \alpha)$ is a term with leftmost variable x_i and left height at least 1. Applying f, we deduce $\mathrm{sub}(t_1^f, \beta) = \mathrm{sub}(t_2^f, \beta)$ for every β in the left edge of α. Then we have $\mathrm{sub}(t_1^f, \alpha) = f(x_i)$. We claim that $\mathrm{sub}(t_1^f, \alpha) <_L \mathrm{sub}(t_2^f, \alpha)$ always holds, which, by Lemma 3.15, implies $t_1^f <_L t_2^f$. Three cases are to be considered.

If Condition (i) holds and $\mathrm{sub}(t_2, \alpha)$ is x_j with $j > i$, we obtain $\mathrm{sub}(t_1^f, \alpha) = f(x_i) <_L f(x_j) = \mathrm{sub}(t_2^f, \alpha)$. If Condition (i) holds and $\mathrm{sub}(t_2, \alpha)$ is not a variable, the hypothesis on f implies $\mathrm{ht}_L(\mathrm{sub}(t_2^f, \alpha)) > \mathrm{ht}_L(\mathrm{sub}(t_1^f, \alpha))$, hence $\mathrm{sub}(t_1^f, \alpha) <_L \mathrm{sub}(t_2^f, \alpha)$. If Condition (ii) holds and $\mathrm{sub}(t_2, \alpha)$ is a term with leftmost variable x_i and left height $k \geq 1$, we find $\mathrm{ht}_L(\mathrm{sub}(t_1^f, \alpha)) = \mathrm{ht}_L(f(x_i))$, and $\mathrm{ht}_L(\mathrm{sub}(t_2^f, \alpha)) = \mathrm{ht}_L(f(x_i)) + k$, hence $\mathrm{sub}(t_1^f, \alpha) <_L \mathrm{sub}(t_2^f, \alpha)$. ■

The previous result implies that every substitution of T_1 preserves $<_L$, and that, if t_1 and t_2 are $\sqsubseteq_{LD}$-comparable terms in T_∞, then $t_1 <_L t_2$ is equivalent to $t_1^\dagger <_L t_2^\dagger$, where we recall $t^\dagger$ denotes the projection of t on T_1.

We are now ready to prove Proposition 3.11 (invariance).

Proof. As G_{LD} is generated by the elements g_α with $\alpha \in \mathbf{A}$, it suffices to prove the result for the latter elements, *i.e.*, to prove that, if α is an address, and t, t' are terms in the domain of the operator LD_α, then $t <_L t'$ is equivalent to $t \bullet \alpha <_L t' \bullet \alpha$. As the action of α is injective, it suffices to prove that $t <_L t'$ implies $t \bullet \alpha <_L t' \bullet \alpha$. We use induction on α. Assume first $\alpha = \phi$. The hypothesis that t and t' belong to the domain of LD_ϕ implies that we can write $t = t_1 \cdot (t_2 \cdot t_3)$, $t' = t'_1 \cdot (t'_2 \cdot t'_3)$, and we have the explicit decompositions

$$(t)_L = \bullet(t_1)_L \bullet(t_2)_L(t_3)_L \quad \text{and} \quad (t \bullet \phi)_L = \bullet\bullet(t_1)_L(t_2)_L \bullet(t_1)_L(t_3)_L,$$

and similar dashed counterparts. By Lemma 3.14, three cases only are possible, namely $t_1 <_L t'_1$, or $t_1 = t'_1$ and $t_2 <_L t'_2$, or $t_1 = t'_1$, $t_2 = t'_2$ and $t_3 <_L t'_3$: the result is clear in each case.

Assume now $\alpha = 0\beta$. Write $t = t_1 \cdot t_2$ and $t' = t'_1 \cdot t'_2$. Then we have $t \bullet \alpha = (t_1 \bullet \beta) \cdot t_2$ and $t' \bullet \alpha = (t'_1 \bullet \beta) \cdot t'_2$. Two cases are possible. For $t_1 <_L t'_1$, the induction hypothesis implies $t_1 \bullet \beta <_L t'_1 \bullet \beta$, and, therefore, $t \bullet \alpha <_L t' \bullet \alpha$. For $t_1 = t'_1$ and $t_2 <_L t'_2$, we have $t_1 \bullet \beta = t'_1 \bullet \beta$, and, again, $t \bullet \alpha <_L t' \bullet \alpha$.

Assume finally $\alpha = 1\beta$. Then we have $t \bullet \alpha = t_1 \cdot (t_2 \bullet \beta)$ and $t' \bullet \alpha = t'_1 \cdot (t'_2 \bullet \beta)$. Two cases are possible again. For $t_1 <_L t'_1$, we obtain $t \bullet \alpha <_L t' \bullet \alpha$ directly. For $t_1 = t'_1$ and $t_2 <_L t'_2$, the induction hypothesis implies $t_2 \bullet \beta <_L t'_2 \bullet \beta$, and we deduce $t \bullet \alpha <_L t' \bullet \alpha$ again. ■

A linear order on G_{LD}

In Chapter III, we used the action of braids on the linearly ordered set $(B_{sp}, \sqsubset)^{\mathbf{N}}$ to define a linear order on B_∞. Here we use the action of G_{LD} on the linearly ordered set $(T_\infty, <_L)$ to similarly define a linear order on G_{LD}. The order so defined is compatible with multiplication on both sides, so G_{LD} is a bi-orderable group.

We begin with an ordering on the submonoid G^+_{LD} of G_{LD}. We recall from Chapter VII that, for all words u, v on $\mathbf{A}$, there exists a unique canonical term $t^L_{u,v}$ such that the intersection of the domains of LD_u and LD_v is the set of substitutes of $t^L_{u,v}$. As $u \equiv^+ u'$ and $v \equiv^+ v'$ imply $t^L_{u,v} = t^L_{u',v'}$, t^L induces a well defined mapping, still denoted t^L, on $G^+_{LD} \times G^+_{LD}$.

Lemma 3.18. *For a, b in G^+_{LD}, the following are equivalent:*
(i) There exists a term t in T_∞ such that $t \bullet a <_L t \bullet b$ holds;
(ii) The inequality $t^L_{a,b} \bullet a <_L t^L_{a,b} \bullet b$ holds;
(iii) For every term t such that $t \bullet a$ and $t \bullet b$ exist, $t \bullet a <_L t \bullet b$ holds.

Proof. That (ii) implies (i) and (iii) implies (ii) is clear. So assume (i). By construction, there exists a substitution f satisfying $t = (t^L_{a,b})^f$, and our hypothesis is the inequality $(t^L_{a,b})^f \bullet a <_L (t^L_{a,b})^f \bullet b$, *i.e.*, $(t^L_{a,b} \bullet a)^f <_L (t^L_{a,b} \bullet b)^f$. The terms $t^L_{a,b} \bullet a$ and $t^L_{a,b} \bullet b$ are LD-equivalent, hence, by Proposition 3.17 (substitution), the previous inequality is equivalent to $t^L_{a,b} \bullet a <_L t^L_{a,b} \bullet b$, which gives (ii), and, then, to $(t^L_{a,b} \bullet a)^g <_L (t^L_{a,b} \bullet b)^g$ for every substitution g, which gives (iii). ■

Definition 3.19. (order on G^+_{LD}) For a, b in G^+_{LD}, we say that $a < b$ holds if the equivalent conditions of Lemma 3.18 are satisfied.

Proposition 3.20. (order on G^+_{LD}) *(i) The relation $<$ is a linear order on the monoid G^+_{LD} that is both left and right compatible with multiplication, and admits 1 as a minimal element.*

(ii) If a, b, a', b' are elements of G^+_{LD} satisfying $ab^{-1} = a'b'^{-1}$, then $a < b$ is equivalent to $a' < b'$

Proof. (i) The relation $<$ is irreflexive as $<_L$ is. Assume $a < b < c$. There exist two substitutions f, g satisfying

$$t^L_{a,b,c} = (t^L_{a,b})^f = (t^L_{b,c})^g. \tag{3.5}$$

By Lemma 3.18, $a < b$ implies $t^L_{a,b} \bullet a <_L t^L_{a,b} \bullet b$, hence, by (3.5), $t^L_{a,b,c} \bullet a <_L t^L_{a,b,c} \bullet b$. Similarly, $b < c$ implies $t^L_{b,c} \bullet b <_L t^L_{b,c} \bullet c$, hence, by (3.5), $t^L_{a,b,c} \bullet b <_L t^L_{a,b,c} \bullet c$. By transitivity of $<_L$, we deduce $t^L_{a,b,c} \bullet a <_L t^L_{a,b,c} \bullet c$, which in turn gives $a < c$ by Lemma 3.18.

Assume now $a < b$, and let c be an arbitrary element of G^+_{LD}. Let t be a term in the domains of LD_{ca} and LD_{cb}. By construction, the term $t \bullet c$ belongs to the domain of LD_a and LD_b, so the hypothesis $a < b$ implies $(t \bullet c) \bullet a <_L (t \bullet c) \bullet b$, which in turn implies $ca < cb$ by definition. With the same hypotheses, assume that t belongs to the domains of LD_{ac} and LD_{bc}. Then $t \bullet a < t \bullet b$ holds by hypothesis, and, by Proposition 3.11 (invariance under G_{LD}), this implies $(t \bullet a) \bullet c <_L (t \bullet b) \bullet c$, hence $ac < bc$ by definition.

Finally, assume $a \neq 1$. We claim that $t <_L t \bullet a$ holds whenever $t \bullet a$ is defined. It suffices to consider the case of a single address α. For $\alpha = \phi$, the result follows from the equality $\mathrm{ht}_L(t \bullet \phi) = \mathrm{ht}_L(t) + 1$. Otherwise, we use induction on α, or we resort to Lemma 3.15: by the previous argument, we have $\mathrm{sub}(t, \alpha) <_L \mathrm{sub}(t \bullet \alpha, \alpha)$, and, by construction, $\mathrm{sub}(t, \beta) = \mathrm{sub}(t \bullet \alpha, \beta)$ holds for every β in the left edge of α.

(ii) By Proposition 1.14 (fraction), there exist c, c' in G^+_{LD} satisfying $ac = a'c'$ and $bc = b'c'$. Assume $a < b$. Using the compatibility of the order with multiplication, we deduce $ac < bc$, *i.e.*, $a'c' < b'c'$, hence $a' \leq b'$. ∎

Definition 3.21. (**order on G_{LD}**) For c, d in G_{LD}, we say that $c < d$ holds if $cd^{-1} = ab^{-1}$ holds for some elements a, b of G^+_{LD} satisfying $a < b$.

Proposition 3.22. (**order on G_{LD}**) *(i) The relation $<$ is a linear order on G_{LD} that extends the order $<$ on G^+_{LD}. This order is compatible with multiplication on both sides, and, therefore, it is compatible with conjugacy.*

(ii) The group G_{LD} is a bi-orderable group. Hence G_{LD} is torsion free, and the group algebra $\mathbf{C}G_{LD}$ admits no zero divisor.

Proof. (i) For a, b in G^+_{LD}, $1 = ab^{-1}$ implies $a = b$, hence $a \not< b$, so, for every c in G_{LD}, $c < c$ is impossible. Assume now $c_1 < c_2 < c_3$ in G_{LD}. There exist a_1, b_1, a_2, b_2 in G^+_{LD} satisfying $c_1c_2^{-1} = a_1b_1^{-1}$, $c_2c_3^{-1} = a_2b_2^{-1}$, $a_1 < b_1$, and $a_2 < b_2$. Let a_3, b_3 be elements of G^+_{LD} satisfying $a_2b_3 = b_1a_3$. We find

$$c_1c_3^{-1} = a_1b_1^{-1}a_2b_2^{-1} = (a_1a_3)(b_2b_3)^{-1}.$$

Then $a_1 < b_1$ implies $a_1a_3 < b_1a_3$, and $a_2 < b_2$ implies $a_2b_3 < b_2b_3$. By hypothesis, we have $b_1a_3 = a_2b_3$, so we deduce $a_1a_3 < b_2b_3$, and, therefore, $c_1 < c_3$. Hence $<$ is an ordering on G_{LD}.

Assume that a, b belong to G^+_{LD} and $a < b$ holds in the sense of G^+_{LD}. Then ab^{-1} is an expression of ab^{-1} with a, b in G^+_{LD} and $a < b$, and $a < b$ holds in the sense of G_{LD}. Thus $<$ on G_{LD} extends $<$ on G^+_{LD}.

Assume now c_1, c_2, $c \in G_{LD}$ and $c_1 < c_2$. By definition, there exist a, b in G^+_{LD} satisfying $c_1c_2^{-1} = ab^{-1}$ and $a < b$. Then we have $(c_1c)(c_2c)^{-1} = ab^{-1}$, so $c_1c < c_2c$ holds as well. On the other hand, let us express c as $a_0b_0^{-1}$ with a_0, b_0 in G^+_{LD}. There exist a_1, b_1, a_2, b_2, a_3, b_3 in G^+_{LD} satisfying $b_0a_1 = aa_2$, $b_0b_1 = bb_2$, $a_2a_3 = b_2b_3$ (Figure 4.1), and we find

$$(cc_1)(cc_2)^{-1} = a_0b_0^{-1}ab^{-1}b_0a_0^{-1} = (a_0a_1a_3)(a_0b_1b_3)^{-1}. \tag{3.6}$$

Then $a < b$ implies $b_0a_1a_3 = aa_2a_3 < ba_2a_3 = bb_2b_3 = b_0b_1b_3$, from which we deduce $a_1a_3 < b_1b_3$, and, therefore, $a_0a_1a_3 < a_0b_1b_3$ using compatibility with multiplication on the left twice. By (3.6), this gives $cc_1 < cc_2$.

Point (ii) is a straightforward consequence of (i). ■

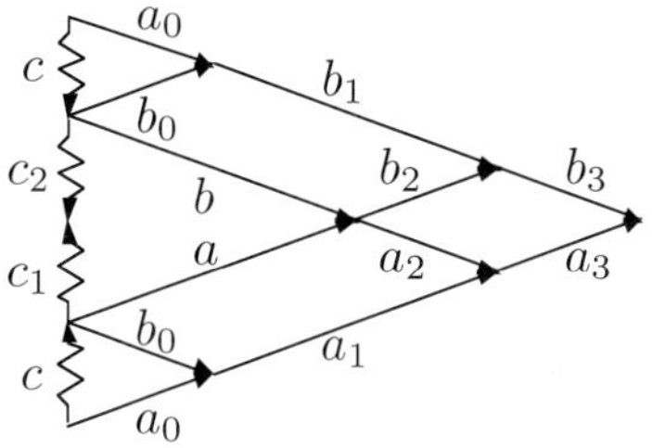

Figure 4.1. Compatibility with left multiplication

The action of the group G_{LD} on terms is a partial action. In particular, some elements c of G_{LD} do not act, *i.e.*, the operator LD_c is empty. However, when the action exists, the connection between the order in G_{LD} and the left order of terms is what we can expect:

Proposition 3.23. (action) *(i) For c, d in G_{LD} and t in T_∞ such that $t \bullet c$ and $t \bullet d$ exist, $c < d$ is equivalent to $t \bullet c <_L t \bullet d$.*

(ii) For c in G_{LD} and t in T_∞ such that $t \bullet c$ exists, $c > 1$ is equivalent to $t \bullet c >_L t$.

Proof. (i) Assume that w_1, w_2 are words on $\mathbf{A} \cup \mathbf{A}^{-1}$ that represent c and d respectively, and the terms $t \bullet w_1$ and $t \bullet w_2$ are defined. Let $t' = t \bullet w_1$. Then $t' \bullet w_1^{-1}w_2$ is defined. By Proposition 1.11 (convergence), there exist two words u, v on $\mathbf{A}$ such that $w_1^{-1}w_2$ is reversible to uv^{-1}. Then, by Lemma 3.23, the term $t' \bullet uv^{-1}$ exists, and we have $t' \bullet uv^{-1} = t' \bullet w_1^{-1}w_2 = t \bullet w_2 = t \bullet d$. Let a and b be the classes of u and v in G^+_{LD}. By construction, we have $c^{-1}d = ab^{-1}$.

Assume $c < d$. By construction, we have $ca = db$, hence $a > b$. The term $t \bullet d$ need not belong to the domain of the operator LD_u in general. However, as u

is a positive word, the only possible obstruction is the skeleton of t being too small. Hence, at the expense of possibly replacing t with a sufficiently large substitute, we may assume that $t \bullet da$ is defined. Now the hypothesis $a > b$ implies $t \bullet da >_L t \bullet db = t \bullet ca$, and, therefore, $t \bullet d >_L t \bullet c$. The argument is symmetric for $c > d$.

Point (ii) follows from (i) by taking $d = 1$. ■

Corollary 3.24. *For t_1, t_2 in T_1, we have $t_1 <_L t_2$ in T_1 if and only if $\chi_{t_1} < \chi_{t_2}$ holds in G_{LD}.*

Proof. By construction, we have

$$x^{[p+1]} \bullet \chi_{t_1} = t_1 \cdot x^{[p]}, \quad x^{[p+1]} \bullet \chi_{t_2} = t_2 \cdot x^{[p]}$$

for p large enough. By Proposition 3.23 (action), the inequality $\chi_{t_1} < \chi_{t_2}$ holds in G_{LD} if and only if the inequality $t_1 \cdot x^{[p]} <_L t_2 \cdot x^{[p]}$ holds in T_∞, hence in T_1. The latter is equivalent to $t_1 <_L t_2$. ■

Exercise 3.25. **(term order)** *(i) Assume that t, t_1, t_2 are terms in T_∞ and the outline of t is included in the skeleton of t_1 and t_2. Let $(\alpha_1, \ldots, \alpha_p)$ be the left–right enumeration of $\mathrm{Out}(t)$. Prove that $t_1 <_L t_2$ holds if and only if the sequence $(\mathrm{sub}(t_1, \alpha_1), \ldots, \mathrm{sub}(t_1, \alpha_p))$ precedes the sequence $(\mathrm{sub}(t_2, \alpha_1), \ldots, \mathrm{sub}(t_2, \alpha_p))$ in the lexicographical extension of $<_L$ to T_∞^*.*

(ii) Assume that t_1 and t_2 belong to T_1. Prove that $t_1 <_L t_2$ holds if and only if $\overrightarrow{\mathrm{Out}}(t_1) \sqsubset^* \overrightarrow{\mathrm{Out}}(t_2)$ does, where $\overrightarrow{\mathrm{Out}}(t)$ denotes the left–right increasing enumeration of the outline of t, and $\sqsubset^*$ denotes the lexicographical extension to $\mathbf{A}^*$ of the prefix ordering $\sqsubset$ on $\mathbf{A}$.

Exercise 3.26. **(right ordering)** *Denote by $<_R$ the linear order on T_∞ obtained by lexicographically extending $\bullet < x_1 < x_2 < \ldots$ to (right Polish) terms. Investigate the compatibility of $<_R$ with substitutions and action of G_{LD}. Deduce a new linear order on G_{LD}^+ which is compatible with left multiplication. Is it compatible with right multiplication?*

Exercise 3.27. **(counter-examples)** *(i) Find two terms t_1, t_2 satisfying $t_1 \sqsubset_{LD} t_2$ and $t_1 >_L t_2$. Deduce two elements c, d in G_{LD} satisfying $c \prec d$ and $c > d$. [Hint: Take $c = \chi_{t_1}$ and $d = \chi_{t_2}$.]*

(ii) Find two elements c, d in G_{LD} such that $c < d$ holds in G_{LD}, while $\mathrm{pr}(c) >_L \mathrm{pr}(d)$ holds in B_∞. [Hint: Take $c = g_\phi g_1 g_\phi$, $d = g_1 g_\phi g_1 g_{11}$.]

Exercise 3.28. **(action on the Cantor space and the real line)** *(i) Let $\widehat{\mathbf{A}}$ denote the Cantor space consisting of all $\mathbf{N}$-indexed sequences of 0's and 1's. For $s \in \widehat{\mathbf{A}}$ and $a \in G^+_{LD}$, define $a \bullet s$ in $\widehat{\mathbf{A}}$ as follows: for $s = \beta s_0$, where β is the unique prefix of s that belongs to the outline of t^L_a, we put $a(s) = \alpha s_0$, where α is the origin of β under a, i.e., the (unique) address satisfying $\beta \in \mathrm{Heir}(\{\alpha\}, a)$. Prove that the mapping $s \mapsto a(s)$ is surjective, and it gives a left action on $\widehat{\mathbf{A}}$, i.e., $1{\bullet}s = s$ and $(ab){\bullet}s = a{\bullet}(b{\bullet}s)$ hold for every s in $\widehat{\mathbf{A}}$.*

(ii) Prove that this action preserves the order in the sense that $a < b$ holds in G^+_{LD} if and only if there exists s in $\widehat{\mathbf{A}}$ such that $a \bullet s = b \bullet s$ holds for $s \le s_0$, but $a \bullet s < b \bullet s$ holds for $s > s_0$, s close enough to s_0 (the order on $\widehat{\mathbf{A}}$ is the lexicographical order).

(iii) Show that carrying the previous action of G^+_{LD} on $\widehat{\mathbf{A}}$ to the interval $[0, 1)$ of $\mathbf{R}$ using the binary expansion of the reals amounts to associating with every element a of G^+_{LD} a continuous, piecewise affine mapping f_a of $[0, 1)$ into itself. Check that f_ϕ is defined by

$$f_\phi(x) = \begin{cases} 2x & \text{for } 0 \le x < 1/4, \\ x + 1/4 & \text{for } 1/4 \le x < 1/2, \\ 2x - 1 & \text{for } 1/2 \le x < 3/4, \\ x & \text{for } 3/4 \le x < 1. \end{cases}$$

Prove that $a < b$ holds in G^+_{LD} if and only if there exists a real x_0 satisfying $a(x) = b(x)$ for $x \le x_0$ and $a(x_0 + \varepsilon) < b(x_0 + \varepsilon)$ for ε small enough.

(iv) Develop a similar study for the group G_A corresponding to associativity. [In this case, the mappings f_a are bijections, and the action is defined on the group, and not only on the monoid.]

---

4. Parabolic Subgroups

We have observed in Section 1 that, for every address γ, the shift mapping sh_γ defines an endomorphism of G_{LD}. The aim of this short section is to prove:

Proposition 4.1. **(shift)** *For every address γ, the shift endomorphism sh_γ of G_{LD} is injective.*

For B a subset of $\mathbf{A}$, let us denote by G_B the subgroup of G_{LD} generated by the elements g_α with $\alpha \in B$—such a subgroup is usually called a *parabolic* subgroup in the framework of Artin groups. By construction, the image of sh_γ is the parabolic subgroup of G_{LD} associated with the subset $\mathrm{sh}_\gamma(\mathbf{A})$. Then Proposition 4.1 (shift) is a counterpart to the result that a parabolic subgroup of an Artin group is the Artin group associated with the induced graph. In the current case, we have proved that the parabolic subgroup associated with the subset $\mathrm{sh}_\gamma(\mathbf{A})$ of $\mathbf{A}$ is isomorphic to G_{LD}, *i.e.*, to the "Artin" group associated with the graph $\mathbf{A}$, which, as a graph, is isomorphic to $\mathrm{sh}_\gamma(\mathbf{A})$.

The method we shall use is the same as for proving the injectivity of the shift mapping on braids (Lemma I.3.3). In the case of braids, the idea was to remove the leftmost strand; in the current case, we shall use a convenient combination of cutting and collapsing to get rid of all addresses that do not begin with γ.

Geometric operators on M_{LD}

In Chapter VII, we established that a number of geometric results about the action of the operators LD_w on terms admit a purely syntactic counterpart. This enables us to define analogs in M_{LD}.

For instance, we proved that, if u and u' are words on $\mathbf{A}$ satisfying $u' \equiv^+ u$, and B is a subset of $\mathbf{A}$ such that $\mathrm{Heir}(B, u)$ exists, then $\mathrm{Heir}(B, u')$ exists as well, and we have $\mathrm{Heir}(B, u') = \mathrm{Heir}(B, u)$. We can therefore define, for a in M_{LD}, $\mathrm{Heir}(B, a)$ to be the common value of $\mathrm{Heir}(B, u)$ for u an arbitrary (positive) expression of a, when it exists. Similarly, for α an address, we define $\mathrm{heir}(\alpha, a)$ to be the common value of $\mathrm{heir}(\alpha, u)$ for u a positive expression of a, when it exists. Then, by construction, we have the induction formulas

$$\mathrm{Heir}(B, ab) = \mathrm{Heir}(\mathrm{Heir}(B, a), b), \tag{4.1}$$
$$\mathrm{heir}(\alpha, ab) = \mathrm{heir}(\mathrm{heir}(\alpha, a), b) \tag{4.2}$$

when they are defined.

Another example is the cut application of Chapter VII. Applying Proposition VII.4.16 (syntactic cut), we see that the mapping $u \mapsto \mathrm{cut}(u, \beta)$ induces a well defined mapping on the subset of M_{LD} made of those elements a for which $\mathrm{heir}(\beta, a)$ exists. For a in M_{LD}, we naturally write $\mathrm{cut}(a, \beta)$ for the value of $\mathrm{cut}(u, \beta)$ where u is an arbitrary positive expression of a. Then we have:

$$\mathrm{cut}(ab, \beta) = \mathrm{cut}(a, \beta)\ \mathrm{cut}(b, \mathrm{heir}(\beta, a)), \tag{4.3}$$

and Proposition VII.4.3 (effective cut) then gives:

Proposition 4.2. (cut) *Assume that a belongs to M_{LD}, t is a term such that $t \bullet a$ exists, and β is an address in the outline of t. Then we have*

$$\mathrm{cut}(t \bullet a, \mathrm{heir}(\beta, a)) = \mathrm{cut}(t, \beta) \bullet \mathrm{cut}(a, \beta).$$

Using similarly Proposition VII.4.17 (syntactic collapsing), we see that, for a in M_{LD}, and B a set of addresses such that $\mathrm{Heir}(B,a)$ exists, an element $\mathrm{coll}(a,B)$ of M_{LD} can be defined to be the common value of $\mathrm{coll}(u,B)$ for u an arbitrary expression of a. Then we have

$$\mathrm{coll}(ab,B)=\mathrm{coll}(a,B)\ \mathrm{coll}(b,\mathrm{Heir}(B,a)), \tag{4.4}$$

and Proposition VII.4.13 (effective collapsing) gives:

Proposition 4.3. (collapsing) *Assume that a belongs to M_{LD}, B is a set of addresses such that $\mathrm{Heir}(B,a)$ exists, and t is a term such that $t \bullet a$ exists. Then we have*

$$\mathrm{coll}(t\bullet a,\mathrm{Heir}(B,a))=\mathrm{coll}(t,B)\bullet\mathrm{coll}(a,B).$$

Remark 4.4. Let us define, for w a (positive) braid word, $\mathrm{coll}(w,i)$ to be the braid word coding the braid diagram obtained from that of w by removing the strand initially at position i. Then $\mathrm{coll}(\cdot,i)$ induces a well defined mapping on B_∞, and we have for all a, b in B_∞

$$\mathrm{coll}(ab,i)=\mathrm{coll}(a,i)\ \mathrm{coll}(b,\mathrm{perm}(a)^{-1}(i)), \tag{4.5}$$

a formula which is directly comparable with (4.4) above. Using (4.5), it is easy to give a purely algebraic proof of the injectivity of sh on B_∞, similar to the proof of the injectivity of sh_γ on G_{LD} that will be given below. Formulas (4.3) and (4.4)—as well as (4.5) in the case of braids—show that the mappings cut and coll are sort of skew endomorphisms on M_{LD}.

Injectivity of shift

We are now ready to establish Proposition 4.1 (shift).

Proof. Assume that w, w' are words on $\mathbf{A}\cup\mathbf{A}^{-1}$, and $\gamma w\equiv\gamma w'$ holds. By Proposition 1.14 (fraction), there exist positive words v, v' satisfying

$$N(\gamma w)\cdot v\equiv^+ N(\gamma w')\cdot v' \quad\text{and}\quad D(\gamma w)\cdot v\equiv^+ D(\gamma w')\cdot v'.$$

By construction, the latter relations are also

$$\gamma N(w)\cdot v\equiv^+\gamma N(w')\cdot v' \quad\text{and}\quad \gamma D(w)\cdot v\equiv^+\gamma D(w')\cdot v'. \tag{4.6}$$

Our aim is to eliminate the initial γ's in (4.6). To this end, we use the mappings cut and coll. First, there exists an integer p such that both

$$\mathrm{heir}(\gamma 1^p,\gamma N(w)\cdot v) \quad\text{and}\quad \mathrm{heir}(\gamma 1^p,\gamma D(w)\cdot v)$$

exist: indeed, the only possible obstruction for such addresses to exist is $\gamma 1^p$ being too short. Let us then apply the mapping $\mathrm{cut}(\cdot, \gamma 1^p)$ to (4.6). By definition, we have

$$\mathrm{cut}(\gamma N(w) \cdot v, \gamma 1^p) = \mathrm{cut}(\gamma N(w), \gamma 1^p) \cdot \mathrm{cut}(v, \mathrm{heir}(\gamma 1^p, \gamma N(w))).$$

Now, by construction, we have $\mathrm{cut}(\gamma N(w), \gamma 1^p) = 1^m(N(w)$, where m is the number of 1's in γ, and $\mathrm{heir}(\gamma 1^p, \gamma N(w)) = 1^m$. Thus we deduce

$$\gamma_1 N(w) \cdot v_1 \equiv^+ \gamma_1 N(w') \cdot v_1', \quad \text{and} \quad \gamma_1 D(w) \cdot v_1 \equiv^+ \gamma_1 D(w') \cdot v_1', \tag{4.7}$$

with $\gamma_1 = 1^m$, $v_1 = \mathrm{cut}(v, 1^m)$ and $v_1' = \mathrm{cut}(v', 1^m)$. So, at this point, we have eliminated all 0's in the address γ.

We now resort to collapsing. As above, we can find an integer q large enough to make sure that, letting $B = \{01^q, 101^q, 1101^q, \ldots, 1^{m-1}01^q\}$, the sets

$$\mathrm{Heir}(B, 1^m N(w) \cdot v_1) \quad \text{and} \quad \mathrm{Heir}(B, 1^m D(w) \cdot v_1)$$

exist. Then, we have

$$\mathrm{coll}(\gamma_1 N(w) \cdot v_1, B) = \mathrm{coll}(\gamma_1 N(w), B) \cdot \mathrm{coll}(v_1, \mathrm{Heir}(B, \gamma_1 N(w)).$$

We have $\mathrm{coll}(\gamma_1 N(w), B) = N(w)$, and $\mathrm{Heir}(B, \gamma_1 N(w)) = B$ by definition. We deduce from (4.7)

$$N(w) \cdot v_2 \equiv^+ N(w') \cdot v_2' \quad \text{and} \quad D(w) \cdot v_2 \equiv^+ D(w') \cdot v_2' \tag{4.8}$$

with $v_2 = \mathrm{coll}(v_1, B)$ and $v_2' = \mathrm{coll}(v_1', B)$. Now, (4.8) implies $w \equiv w'$. ∎

Exercise 4.5. (kernel) *(i) Prove that the kernel of the projection of G_{LD} onto B_∞ includes the direct product of infinitely many copies of G_{LD}. [Hint: $\mathrm{Ker}(\mathrm{pr})$ includes $\prod_i \mathrm{sh}_{1^i0}(G_{LD})$; apply Proposition 4.1.]*

(ii) Define $\mathrm{pr}^{-1}(\sigma_i) = 1^{i-1}$, and extend pr^{-1} into an alphabetical homomorphism of BW_∞^+ into $\mathbf{A}^*$. Prove that, for every u in $\mathbf{A}^*$, we have $u \equiv^+ \mathrm{pr}^{-1}(\mathrm{pr}(u)) \cdot \prod_i 1^i0u_i$ for some u_1, u_2, ... in $\mathbf{A}^*$.

(iii) Assume u, $v \in BW_\infty^+$ with $u \equiv v$. Show that

$$\mathrm{pr}^{-1}(u) \cdot \textstyle\prod_i 1^i0u_i \equiv^+ \mathrm{pr}^{-1}(v) \cdot \prod_i 1^i0v_i$$

holds for some u_1, v_1, u_2, v_2, ... in $\mathbf{A}^*$.

(iv) Assume $\mathrm{pr}(w) \equiv \varepsilon$, where w is a word on $\mathbf{A} \cup \mathbf{A}^{-1}$. Show that $w \equiv u \cdot \prod_i 1^i0w_i \cdot u^{-1}$ holds for some words u, w_1, w_2, ... with u positive. [Hint: Apply (ii) and (iii) to $N(w)$ and $D(w)$.] Deduce that every element of G_{LD} in $\mathrm{Ker}(\mathrm{pr})$ can be expressed as aba^{-1} where a belongs to the submonoid generated by the elements g_{1^i} and b belongs to $\prod_i \mathrm{sh}_{1^i0}(G_{LD})$.

Exercise 4.6. (**left subterm**) *For a in M_{LD} and $p \geq 0$, define* $\mathrm{dil}(a,p)$ *in* $\mathbf{N}$ *and* $\mathrm{left}(a,p)$ *in M_{LD} to be the common values of* $\mathrm{dil}(u,p)$ *and* $\mathrm{left}(u,p)$ *respectively for u representing a. Prove the equality* $\mathrm{left}(ab,p) = \mathrm{left}(a,p)\,\mathrm{left}(b,\mathrm{dil}(p,a))$ *Prove that, for $a \in M_{LD}$, and t a term such that $t \bullet a$ and* $\mathrm{left}^p(t)$ *exist, we have* $\mathrm{left}^{\mathrm{dil}(p,a)}(t \bullet a) = \mathrm{left}^p(t) \bullet \mathrm{left}(a,p)$.

Exercise 4.7. (**alternative proof**) *Prove that* LD_w *and* $\mathrm{LD}_{w'}$ *agreeing on one term implies $w \equiv w'$, using Proposition 3.23 (action I) instead of Proposition 2.14 (action II). [Use the injectivity of* sh_0*.]*

Exercise 4.8. (**decomposition**) *Assume that B is a finite set of pairwise orthogonal addresses. Assume $\prod_{\alpha \in B} \alpha u_\alpha \equiv^+ \prod_{\alpha \in B} \alpha v_\alpha$, where all u_α, v_α are words on* $\mathbf{A}$. *Prove $u_\alpha \equiv^+ v_\alpha$ for each α in B.*

---

5. Simple Elements in M_{LD}

Simple braids play a significant role in the study of the monoid B_∞^+ in Chapter II. In this section, we develop the notion of a simple element in the monoid M_{LD}. The intuition is similar in both cases, and so are the results. Actually, the simple elements in M_{LD} are mapped to simple braids under the projection of M_{LD} onto B_∞^+, and all results we shall establish in M_{LD} re-prove their braid counterpart. The main distinction is that B_∞^+ is the union of the finitely generated monoids B_n^+, and, in each of them, there exists a Garside element, namely Δ_n. This is not true in M_{LD}, but we shall see that the class Δ_t of the word $\boldsymbol{\Delta}_t$ can be used as a counterpart to the braid Δ_n.

In the case of B_n^+, we have introduced a special family of positive braid words, called permutation braid words, and the main result is that a braid b can be represented by a permutation braid word if and only if it is a divisor of Δ_n, and if and only if it can be represented by a braid diagram where any two strands cross at most once. Our approach in the case of M_{LD} will be similar: we shall introduce the notion of a *permutation-like* word on $\mathbf{A}$ by an explicit definition, and the notion of a *simple* element of M_{LD} by a geometrical property of the corresponding LD-operators. Then we prove the following result, which is directly reminiscent of Proposition II.4.14 (simple)—which it implies:

Proposition 5.1. (**simple**) *For a in M_{LD}, the following are equivalent:*

(i) The element a is permutation-like;

(ii) The element a is simple;
(iii) There exists a term t such that a is a left divisor of Δ_t.
Moreover, these conditions are consequences of
(iv) There exists a term t such that a is a right divisor of Δ_t.

As an application, we shall deduce a unique normal form in M_{LD}.

Permutation-like elements

Let us recall from Chapter VII that, for α in $\mathbf{A}$ and $p \geq 0$, $\alpha^{(p)}$ is defined to be the word $\alpha 1^{p-1} \cdot \alpha 1^{p-2} \cdot \dots \cdot \alpha 1 \cdot \alpha$ for $p \geq 1$, and to be ε for $p = 0$. We also recall that, for α, β in $\mathbf{A}$, $\alpha > \beta$ means that α is a proper prefix of β, or α lies on the right of β: thus, for instance, $\phi > 1 > 0$ holds.

Definition 5.2. (permutation-like, exponent) We say that a word u on $\mathbf{A}$ is a *permutation-like* word if u can be factorized as $\alpha_1{}^{(p_1)} \cdot \dots \cdot \alpha_\ell{}^{(p_\ell)}$ with $\alpha_1 > \dots > \alpha_\ell$; then, for $\alpha \in \mathbf{A}$, the *exponent* $\mathrm{e}(\alpha, u)$ of α in u is defined to be the integer p such that $\alpha^{(p)}$ appears in u, if it exists, and to be 0 otherwise. An element of M_{LD} is said to be a *permutation-like* element if it can be represented by a permutation-like word.

As $\alpha^{(0)}$ has been defined to be the empty word, a permutation-like word can be written as $\prod^{>}_{\alpha \in \mathbf{A}} \alpha^{(p_\alpha)}$, where $(p_\alpha \; ; \; \alpha \in \mathbf{A})$ is a sequence of nonnegative integers with finitely many positive entries only. A length 1 word, *i.e.*, a single address, is a permutation-like word. By definition, the projection of a permutation-like element of M_{LD} on B_∞^+ is a permutation braid.

Example 5.3. (permutation-like) *Let $u = 11 \cdot 1 \cdot \phi \cdot 1 \cdot 001 \cdot 00$. We have $u = (11 \cdot 1 \cdot \phi) \cdot (1) \cdot (001 \cdot 00) = \phi^{(3)} \cdot 1^{(1)} \cdot 00^{(2)}$, and $\phi > 1 > 00$. So u is a permutation-like word, with $\mathrm{e}(\phi, u) = 3$, $\mathrm{e}(0, u) = 0$, and $\mathrm{e}(1, u) = 1$.*

By definition, every permutation-like word has the form $\phi^{(p)} \cdot 1u_2 \cdot 0u_1$, where u_1 and u_0 are permutation-like words: this will enable us to develop inductive arguments.

Lemma 5.4. *A permutation-like element in M_{LD} admits a unique representation by a permutation-like word: if a is a permutation-like element, the unique permutation-like word representing a depends on the operator LD_a only.*

Proof. Assume that u is a permutation-like word. We show that the exponents $\mathrm{e}(\alpha, u)$ are determined by the operator LD_u using induction on the size of t_u^L. If t_u^L is a variable, we have $\mathrm{LD}_u = \mathrm{id}$, so $u = \varepsilon$, and the result is true. Otherwise, assume $u = \phi^{(p)} \cdot 1u_2 \cdot 0u_1$. Let $t = t_u^L$. Since t_u^L belongs to the domain of $\mathrm{LD}_{\phi^{(p)}}$, its right height is at least $p+1$. Let $t_1{\cdot}t_2 = t_u^L \bullet \phi^{(p)}$, and $x_{f(i)} = \mathrm{var}_R(\mathrm{sub}(t_u^L, 1^i0))$. By construction, we have $\mathrm{var}_R(t_1) = x_{f(p)}$. Now we have $t_u^R = t_u^L \bullet u = (t_1 \bullet u_1){\cdot}(t_2 \bullet u_2)$, we deduce $x_{f(p)} = \mathrm{var}_R(\mathrm{sub}(t_u^R, 0))$. This shows that t_u^R determines p, and, therefore, so does u. Then, for $e = 1, 2$, the term t_e belongs to the domain of LD_{u_e}, it is injective, and it is smaller than t_u^L. As t_e is a substitute of $t_{u_e}^L$, the latter term is smaller than t_u^L as well. Hence, by induction hypothesis, $t_e \bullet u_e$ determines the exponents in u_e, and so does $t \bullet u$, as $t_e \bullet u_e = \mathrm{sub}(t \bullet u, e)$ holds. ∎

It follows that, for every permutation-like element a and every address α, we can define without ambiguity the exponent of α in a as the exponent of α in the unique permutation-like word that represents a.

Simple elements

We recall from Chapter VII that, if t is a term, the variable x_i is said to cover the variable x_j in t if there exist two addresses α, β in the outline of t such that x_i occurs at α in t, x_j occurs at β in t, and α covers β.

Definition 5.5. (**semi-injective**) The term t is said to be *semi-injective* if no variable covers itself in t.

For a term t to be semi-injective means that, for every subterm s of t, the rightmost variable of s occurs only once in s. Thus every injective term is semi-injective, but the converse is not true: for instance, the term $(x_1{\cdot}x_2){\cdot}(x_1{\cdot}x_3)$, which is is not injective since x_1 occurs twice, is semi-injective.

Non-semi-injective terms have good closure properties.

Lemma 5.6. *Non semi-injective terms are closed under substitution and LD-expansion.*

Proof. Assume that t is non-semi-injective. Then some variable x_i occurs both at $\alpha 1^r$ and $\alpha 0 \beta$ in t. Let h be an arbitrary substitution, and let $x_k = \mathrm{var}_R(h(x_i))$, $q = \mathrm{ht}_R(h(x_i))$. Then x_k occurs at $\alpha 1^{r+q}$ and $\alpha 0 \beta 1^q$ in t^h. Hence t^h is not semi-injective. On the other hand, we have seen in Lemma VII.1.28 that LD-expansions never delete coverings: if x_i covers x_j in t, it covers x_j in every LD-expansion of t. This applies in particular when x_i covers itself. ∎

We introduce a semantic notion of simplicity that is analogous to the condition that any two strands cross at most once in a braid diagram.

Definition 5.7. (**simple**) For a in M_{LD}, we say that a is *simple* if the image of LD_a contains at least one semi-injective term. For u in $\mathbf{A}^*$, we say that u is *simple* if the class of u in M_{LD} is simple.

Lemma 5.8. *Assume that a is an element of M_{LD}. Then, the following are equivalent:*
(i) The element a is simple;
(ii) The term t_a^R is semi-injective;
(iii) The operator LD_a maps every injective term to a semi-injective term.

Proof. The term t_a^L is injective, and the operator LD_a maps t_a^L to t_a^R, so (iii) implies (ii), and (ii) implies (i). Assume (i). Let t be a term in the domain of LD_a such that $t \bullet a$ exists and is semi-injective. By Proposition VII.1.10 (domain), there exists a substitution h satisfying $t = (t_a^L)^h$ and $t \bullet a = (t_a^R)^h$. By Lemma 5.6, $(t_a^R)^h$ being semi-injective implies t_a^R being semi-injective, so (ii) holds. Assume now (ii), and let t be an injective term in the domain of LD_a. Then there exists a substitution h satisfying $t = (t_a^L)^h$, and t being injective means that we can assume that h is an injective substitution. Now we have $t \bullet a = (t_a^R)^h$, and such a term being not semi-injective would imply t_a^R itself being not semi-injective. ∎

Using the closure properties of non-semi-injective terms, we obtain the following closure property for simple elements of M_{LD}. Observe that the corresponding result for permutation-like elements is not clear—a situation parallel to what happened in the case of simple braids and permutation braids.

Lemma 5.9. *Every divisor of a simple element of M_{LD} is simple.*

Proof. Assume that a is not simple, and let b, c be arbitrary elements of M_{LD}. By Proposition VII.1.10 (domain), the term t_{ba}^R is a substitute of t_a^R, and the term t_{bac}^R is an LD-expansion of the previous term. By hypothesis, t_a^R is not semi-injective, hence, by Lemma 5.6, t_{ba}^R and t_{bac}^R are not semi-injective either. Hence bac is not simple. ∎

We shall prove eventually that permutation-like elements and simple elements in M_{LD} coincide. For the moment, let us observe that one direction is easy.

Lemma 5.10. *Every permutation-like element of M_{LD} is simple.*

Proof. We show that using induction on the size of the term t_u^L that, if u is a permutation-like word, then u is simple. By definition, we have $u = \phi^{(p)} \cdot 1u_2 \cdot 0u_1$ where u_1 and u_2 are permutation-like words. Let $t = t_u^L$. Then

$t \bullet \phi^{(p)}$ exists, and, therefore, the right height of t is at least $p+1$, *i.e.*, t can be written as $t_1 \cdot \cdots \cdot t_{p+2}$. Then we have $t \bullet \phi^{(p)} = t'_1 \cdot t'_2$, with

$$t'_1 = t_1 \cdot \cdots \cdot t_p \cdot t_{p+1} \qquad \text{and} \qquad t'_2 = t_1 \cdot \cdots \cdot t_p \cdot t_{p+2}.$$

By hypothesis, for $e = 1, 2$, t'_e is an injective term that lies in the domain of the operator LD_{u_e}. The terms t'_1 and t'_2 are smaller than t, so, $t^{L}_{u_1}$ and $t^{L}_{u_2}$ are smaller than t^{L}_{u}. By induction hypothesis, the LD-expansions $t'_1 \bullet u_1$ and $t'_2 \bullet u_2$ are semi-injective terms. Hence $t \bullet u$, which is $(t'_1 \bullet u_1) \cdot (t'_2 \bullet u_2)$, is semi-injective as well, for the rightmost variable of $t'_2 \bullet u_2$, which is $\mathrm{var}_R(t'_2)$, occurs neither in t'_1 nor in $t'_1 \bullet u_1$. ∎

Example 5.11. (**not permutation-like**) *We have obtained a criterion for proving that an element of M_{LD} is not permutation-like. For instance, $g^+_\phi\, g^+_\phi$ is not simple, and, therefore, it is not permutation-like: indeed we have $(x_1 \cdot x_2 \cdot x_3) \bullet \phi^2 = ((x_1 \cdot x_2) \cdot x_1) \cdot ((x_1 \cdot x_2) \cdot x_3))$, a non-semi-injective term since the variable x_1 occurs both at 01 and 000.*

Product of permutation-like elements

Our goal is now to establish the converse of Lemma 5.10. We begin with a series of computation formulas. The point is to determine the permutation-like decomposition of the products $\alpha^{(p)} \cdot \phi^{(q)}$, when it exists. Two cases arise, according to whether α contains at least one 0 or not.

Lemma 5.12. *Assume $\alpha = 1^m 0 \beta$. Then $\alpha^{(p)} \cdot \phi^{(q)}$ is simple for all p, q, and we have*

$$\alpha^{(p)} \cdot \phi^{(q)} \equiv^+ \begin{cases} \phi^{(q)} \cdot \alpha^{(p)} & \text{for } q < m, \\ \phi^{(q)} \cdot (01^m\beta)^{(p)} & \text{for } q = m, \\ \phi^{(q)} \cdot (1\alpha)^{(p)} \cdot (0\alpha)^{(p)} & \text{for } q > m. \end{cases}$$

Proof. Assume $p = 1$. For $m \geq q+1$, $1^m 0 \beta$ commutes with every factor in $\phi^{(q)}$ by type 11 relations, so α commutes with $\phi^{(q)}$. For $m = q$, using q successive type 10 relations, we obtain

$$\begin{aligned} \alpha \cdot \phi^{(m)} &= 1^m 0 \beta \cdot 1^{m-1} \cdot \cdots \cdot \phi \\ &\equiv^+ 1^{m-1} \cdot 1^{m-1} 01 \beta \cdot 1^{m-2} \cdot \cdots \cdot \phi \\ &\cdots \\ &\equiv^+ 1^{m-1} \cdot \cdots \cdot 1 \cdot 101^{m-1} \beta \cdot \phi \\ &\equiv^+ 1^{m-1} \cdot \cdots \cdot 1 \cdot \phi \cdot 01^m \beta = \phi^{(m)} \cdot 01^m \beta. \end{aligned}$$

For $m < q$, we find

$$\begin{aligned}
\alpha \cdot \phi^{(q)} &= 1^m 0\beta \cdot (1^{m+1})^{(q-m-1)} \cdot 1^m \cdot \phi^{(m)} \\
&\equiv^+ (1^{m+1})^{(q-m-1)} \cdot 1^m 0\beta \cdot 1^m \cdot \phi^{(m)} && (\perp)\\
&\equiv^+ (1^{m+1})^{(q-m-1)} \cdot 1^m \cdot 1^m 10\beta \cdot 1^m 00\beta \cdot \phi^{(m)} && (0)\\
&\equiv^+ (1^{m+1})^{(q-m-1)} \cdot 1^m \cdot 1^m 10\beta \cdot \phi^{(m)} \cdot 01^m 0\beta \\
&\equiv^+ (1^{m+1})^{(q-m-1)} \cdot 1^m \cdot \phi^{(m)} \cdot 1^m 10\beta \cdot 01^m 0\beta = \phi^{(q)} \cdot 1\alpha \cdot 0\alpha. && (11)
\end{aligned}$$

Extending the result to the case $p > 1$ is easy in the first two cases. In the last case, we observe that $1\alpha 1^{p-1} \cdot 0\alpha 1^{p-1} \cdot \cdots \cdot 1\alpha \cdot 0\alpha$ is equivalent to $(1\alpha)^{(p)} \cdot (0\alpha)^{(p)}$ using type $\perp$ relations. ■

Lemma 5.13. *Assume $\alpha = 1^m$. Then $\alpha^{(p)} \cdot \phi^{(q)}$ is simple unless $m < q \leq m+p$ holds, and we have*

$$\alpha^{(p)} \cdot \phi^{(q)} \equiv^+ \begin{cases} \phi^{(q)} \cdot \alpha^{(p)} & \text{for } q < m, \\ \phi^{(p+q)} & \text{for } q = m, \\ \phi^{(q)} \cdot (1^{m+1})^{(p)} \cdot (01^m)^{(p)} & \text{for } q > m+p. \end{cases}$$

Proof. For $q < m$, every factor in $\phi^{(q)}$ commutes with every factor in $\alpha^{(p)}$ by type 11 relations, so $\alpha^{(p)}$ and $\phi^{(q)}$ commute. For $q = m$, we have $\alpha^{(p)} \cdot \phi^{(q)} = \phi^{(p+q)}$ by definition. For $q > m+p$, we use induction on p. For $p = 1$, hence $q \geq m+2$, we find

$$\begin{aligned}
1^m \cdot \phi^{(q)} &= 1^m \cdot (1^{m+2})^{(q-m-2)} \cdot 1^{m+1} \cdot 1^m \cdot \phi^{(m)} \\
&\equiv^+ (1^{m+2})^{(q-m-2)} \cdot 1^m \cdot 1^{m+1} \cdot 1^m \cdot \phi^{(m)} && (11)\\
&\equiv^+ (1^{m+2})^{(q-m-2)} \cdot 1^{m+1} \cdot 1^m \cdot 1^{m+1} \cdot 1^m 0 \cdot \phi^{(m)} && (1)\\
&= (1^m)^{(q-m)} \cdot 1^{m+1} \cdot 1^m 0 \cdot \phi^{(m)} \\
&\equiv^+ (1^m)^{(q-m)} \cdot 1^{m+1} \cdot \phi^{(m)} \cdot 01^m && (\text{Lemma 5.12})\\
&\equiv^+ (1^m)^{(q-m)} \cdot \phi^{(m)} \cdot 1^{m+1} \cdot 01^m = \phi^{(q)} \cdot 1^{m+1} \cdot 01^m. && (11)
\end{aligned}$$

Assume now $p > 1$. We have

$$\begin{aligned}
(1^m)^{(p)} \cdot \phi^{(q)} &= (1^{m+1})^{(p-1)} \cdot 1^m \cdot \phi^{(q)} \\
&\equiv^+ (1^{m+1})^{(p-1)} \cdot \phi^{(q)} \cdot 1^{m+1} \cdot 01^m && (\text{ind. hyp.})\\
&\equiv^+ \phi^{(q)} \cdot (1^{m+2})^{(p-1)} \cdot (01^{m+1})^{(p-1)} \cdot 1^{m+1} \cdot 01^m && (\text{ind. hyp.})\\
&\equiv^+ \phi^{(q)} \cdot (1^{m+2})^{(p-1)} \cdot 1^{m+1} \cdot (01^{m+1})^{(p-1)} \cdot 01^m && (\perp)\\
&= \phi^{(q)} \cdot (1^{m+1})^{(p)} \cdot (01^m)^{(p)}.
\end{aligned}$$

The previous explicit formulas show that, in the three previous cases, $\alpha^{(p)} \cdot \phi^{(q)}$ is a permutation-like element. So it only remains to prove that the product is not simple in the case $m < q \leq m+p$. Owing to Lemma 5.10, it suffices to exhibit an injective term whose image under the operator $\mathrm{LD}_{\alpha^{(p)}} \bullet \mathrm{LD}_{\phi^{(q)}}$ is not semi-injective. Let $t = x_1 \cdot \cdots \cdot x_{m+p+2}$. Then $t \bullet \alpha^{(p)}$ exists, and it is

$$x_1 \cdot \cdots \cdot x_m \cdot (x_{m+1} \cdot \cdots \cdot x_{m+p} \cdot x_{m+p+1}) \cdot x_{m+1} \cdot \cdots \cdot x_{m+p} \cdot x_{m+p+2}.$$

Applying $\mathrm{LD}_{\phi^{(q)}}$ to this term gives a term whose 01^m-subterm is

$$(x_{m+1} \cdot \cdots \cdot x_{m+p+1}) \cdot x_{m+1} \cdot \cdots \cdot x_{q+1},$$

and the rightmost variable of this subterm, namely x_{q+1}, also occurs in its left subterm, so it is not semi-injective—see an example in Figure 5.1. ■

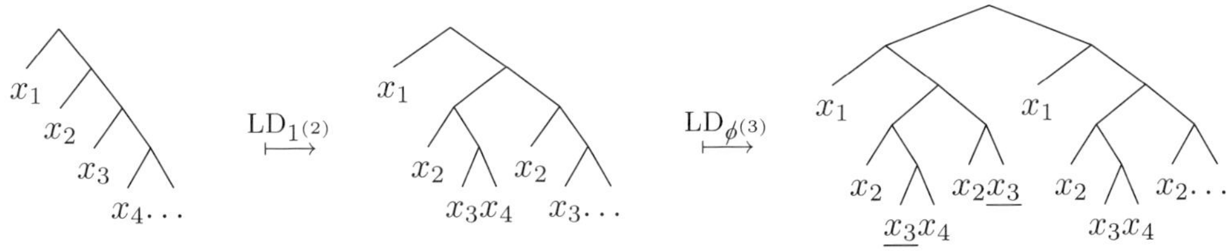

Figure 5.1: A non-simple case: $m = 1$, $p = 2$, $q = 3$.

We can now determine whether a permutation-like element remains a permutation-like element when an additional factor $\phi^{(q)}$ is added.

Lemma 5.14. *Assume that u is a permutation-like word, and $q \geq 0$ holds. Let $r = q + \mathrm{e}(1^q, u)$. Then $u \cdot \phi^{(q)}$ is simple if and only if $m + \mathrm{e}(1^m, u) < r$ holds for $0 \leq m < q$; in this case, $u \cdot \phi^{(q)}$ is a permutation-like element, and we have $\mathrm{e}(\phi, u \cdot \phi^{(q)}) = r$.*

Proof. Decompose u as

$$\prod_{m=0}^{\infty} (1^m)^{(p_m)} \cdot \prod_{m=\infty}^{0} 1^m 0 u_m,$$

where all u_m are permutation-like words. We add the factor $\phi^{(q)}$ on the right, and try to push this factor to the left and integrate it in the decomposition. By Lemma 5.12, we cross the right product: $u \cdot \phi^{(q)}$ is $\equiv^+$-equivalent to

$$\prod_{m=0}^{\infty} (1^m)^{(p_m)} \cdot \phi^{(q)} \cdot \prod_{m=\infty}^{q+1} 1^m 0 u_m \cdot 01^q u_q \cdot \prod_{m=q-1}^{0} (1^{m+1} 0 u_m \cdot 01^m 0 u_m),$$

hence, using type $\perp$ relations, to

$$\prod_{m=0}^{\infty} (1^m)^{(p_m)} \cdot \phi^{(q)} \cdot \prod_{m=\infty}^{q+1} 1^m 0 u_m \cdot \prod_{m=q-1}^{0} 1^{m+1} 0 u_m \cdot 01^q u_q \cdot \prod_{m=q-1}^{0} 01^m 0 u_m.$$

It remains to study the expression $\prod_{m=0}^{\infty}(1^m)^{(p_m)} \cdot \phi^{(q)}$. We use now Lemma 5.13 to push $\phi^{(q)}$ to the left. First, we have

$$\prod_{m=q}^{\infty}(1^m)^{(p_m)} \cdot \phi^{(q)} \equiv^+ \phi^{(r)} \cdot \prod_{m=q+1}^{\infty}(1^m)^{(p_m)},$$

with $r = q+p_q$, *i.e.*, $r = q + \mathrm{e}(1^q, a)$, and we are left with $\prod_{m=0}^{q-1}(1^m)^{(p_m)} \cdot \phi^{(r)}$. By Lemma 5.13, two cases are possible. Either the condition $q-1+p_{q-1} \geq r$ holds, and then $(1^{q-1})^{(p_{q-1})} \cdot \phi^{(r)}$ is not simple, and, therefore, by Lemma 5.9, $u \cdot \phi^{(q)}$ is not simple either. Or $q-1+p_{q-1} < r$ holds, and $(1^{q-1})^{(p_{q-1})} \cdot \phi^{(r)}$ is $\equiv^+$-equivalent to $\phi^{(r)} \cdot (1^q)^{(p_{q-1})} \cdot (01^{q-1})^{(p_{q-1})}$. We continue with the product $(1^{q-2})^{(p_{q-2})} \cdot \phi^{(r)}$. Again two cases are possible: in the one case, $u \cdot \phi^{(q)}$ is not simple, in the other, it is, we can push the factor $\phi^{(q)}$ to the left, and the process continues. Finally, if the condition $m+p_m < r$ fails for some m, $u \cdot \phi^{(q)}$ is not simple; if the condition holds for every m, the factor $\phi^{(q)}$ migrates to the leftmost position, and we obtain that $u \cdot \phi^{(p)}$ is $\equiv^+$-equivalent to

$$\phi^{(r)} \cdot \prod_{m=0}^{q-1}(1^{m+1})^{(p_m)} \cdot \prod_{m=0}^{q-1}(01^m)^{(p_m)} \cdot \prod_{m=\infty}^{q+1} 1^m 0 u_m \cdot \prod_{m=q-1}^{0} 1^{m+1} 0 u_m \cdot 01^q u_q \cdot \prod_{m=q-1}^{0} 01^m 0 u_m,$$

which can be rearranged using relations of type $\perp$ and renumbering into

$$\phi^{(r)} \cdot \prod_{m=1}^{q}(1^m)^{(p_{m-1})} \cdot \prod_{m=\infty}^{q+1} 1^m 0 u_m \cdot \prod_{m=q}^{1} 1^m 0 u_{m-1} \cdot \prod_{m=0}^{q-1}(01^m)^{(p_m)} \cdot 01^q u_q \cdot \prod_{m=q-1}^{0} 01^m 0 u_m,$$

an explicit permutation-like element of M_{LD}. ∎

Proposition 5.15. (equivalence) *An element of M_{LD} is a permutation-like element if and only if it is simple.*

Proof. We have already seen that every permutation-like element is simple. We establish now, using induction on the size of t_u^L, that u being a simple word implies u being $\equiv^+$-equivalent to a permutation-like word. If t_u^L is a variable, we have $u = \varepsilon$, a permutation-like word. Otherwise, u can be decomposed as $v \cdot \alpha^{(q)}$ with $q \geq 1$. By Lemma 5.9, v is simple, so, by induction hypothesis, it is $\equiv^+$-equivalent to some permutation-like word v'. We show inductively on the length of α that $v' \cdot \alpha^{(q)}$ is $\equiv^+$-equivalent to some permutation-like word. For $\alpha = \phi$, the previous lemma gives the result. Otherwise, assume $\alpha = e\beta$, with $e = 0$ or $e = 1$. There exist an integer p and permutation-like words v_1, v_0 satisfying $v' \equiv^+ \phi^{(p)} \cdot 0v_1 \cdot 0v_0$. By Lemma 5.9 again, $ev_e \cdot \alpha^{(q)}$ is simple, which implies that $v_e \cdot \beta^{(q)}$ is simple too, since a subterm of a semi-injective term is semi-injective. By induction hypothesis, $v_e \cdot \beta^{(q)}$ is $\equiv^+$-equivalent to some permutation-like word. Hence, so are $ev_e \cdot \alpha^{(q)}$, and $\phi^{(p)} \cdot (1-e)v_{1-e} \cdot ev_e \cdot \alpha^{(q)}$, hence u. ∎

Simple LD-expansions

In the study of LD-expansions, and, in particular, in the proof of confluence of Chapter V, the key step is to introduce for each term t the distinguished LD-expansion ∂t of t, which is a common LD-expansion of all basic LD-expansions of t. Here we introduce the natural notion of a simple LD-expansion, and we show that ∂t is a maximal simple LD-expansion of t, this corresponding to the element of M_{LD} represented by $\boldsymbol{\Delta}_t$ being a maximal simple element in M_{LD}—as Δ_n is a maximal simple braid in B_n^+.

Definition 5.16. (element Δ_t) For t a term, the class of the word $\boldsymbol{\Delta}_t$ in M_{LD} is denoted Δ_t. Similarly, the class of $\boldsymbol{\Delta}_t^{(k)}$ is denoted $\Delta_t^{(k)}$.

Definition 5.17. (simple expansion) We say that t' is a *simple* LD-expansion of t if $t' = t \bullet u$ holds for some simple word u.

By Lemma 5.4, there exists a one-to-one correspondence between the simple LD-expansions of a term t and the permutation-like words u such that t belongs to the domain of LD_u.

Proposition 5.18. (maximal simple) *For every term t, the term ∂t is the maximal simple LD-expansion of t, and $\boldsymbol{\Delta}_t$ is the (unique) permutation-like word u such that LD_u maps t to ∂t.*

Proof. We already know that $\mathrm{LD}_{\boldsymbol{\Delta}_t}$ maps t to ∂t. That $\boldsymbol{\Delta}_t$ is a permutation-like word follows from its explicit definition. We prove using induction on t that no LD-expansion of ∂t may be a semi-injective term. The result is vacuously true for t a variable. Assume $t = t_1{\cdot}t_2$. We consider first LD-expansion at ϕ. The equality $\partial t = \partial t_1 * \partial t_2$ shows that every variable occurring in t except possibly the rightmost one occurs both in the left and the right subterm of ∂t. So the rightmost variable of $\mathrm{sub}(\partial t, 10)$, say x_i, occurs in $\mathrm{sub}(\partial t, 0)$ also, hence, when LD_ϕ is applied to ∂t, x_i covers itself in the resulting LD-expansion, which therefore is not semi-injective. Consider now LD-expansion at α, where α is a nonempty address, say $\alpha = e\beta$ with $e = 0$ or $e = 1$. We have $\partial t = \partial s_1{\cdot}\partial s_2$, with $s_1{\cdot}s_2 = t{\bullet}\phi^{(r)}$ for some r. By construction, we have $\mathrm{sub}(\partial t{\bullet}\alpha, e) = \partial s_e{\bullet}\beta$, and, s_e is smaller than t, the induction hypothesis implies that $\partial s_e \bullet \beta$ is not semi-injective. So $t \bullet \alpha$ is not semi-injective either. ∎

Corollary 5.19. *For every term t, the element Δ_t is simple, and it is maximal in the sense that $\Delta_t a$ is simple for no element a such that $t \bullet \Delta_t a$ exists.*

Proposition 5.20. (divisors of Δ) *For every a of M_{LD}, the following are equivalent:*

(i) The element a is simple;

(ii) There exists a term t such that a is a left divisor of Δ_t in M_{LD};

(iii) For every term t such that $t \bullet a$ exists, a is a left divisor of Δ_t in M_{LD}.

Proof. By definition, (iii) implies (ii), and, by Lemma 5.9, (ii) implies (i) as Δ_t is simple. So the point is to prove that (i) implies (iii). We prove using induction on t that, if a is a permutation-like element of M_{LD} and $t \bullet a$ is defined, then a is a left divisor of Δ_t. The result is true when t is a variable. Otherwise, let $r+1$ be the right height of t. By definition, t belongs to the domain of LD_a, so the inequality $m + \mathrm{e}(1^m, a) \leq r$ holds for every m between 0 and r, and there exists a least q satisfying $q + \mathrm{e}(1^q, a) = r$. By Lemma 5.14, we deduce that $a \cdot \phi^{(q)}$ is simple, with $\mathrm{e}(\phi, a \cdot \phi^{(q)}) = r$. This means that there exist simple elements a_1, a_2 satisfying

$$a \cdot \lambda^{(q)} = \lambda^{(r)} \cdot 1a_2 \cdot 0a_1.$$

By construction, the term t belongs to the domain of $\mathrm{LD}_{\phi^{(r)}}$. Let $s_1 \cdot s_2$ be its image. By definition, the term $s_e \bullet a_e$ is defined for $e = 1, 2$, and, by construction, s_e is smaller than t. Hence, by induction hypothesis, there exists an element b_e of M_{LD} satisfying $a_e\, b_e = \Delta_{s_e}$. We deduce

$$a \cdot \phi^{(q)} \cdot 1b_2 \cdot 0b_1 = \phi^{(r)} \cdot 1a_2 \cdot 0a_1 \cdot 1b_2 \cdot 0b_1 = \phi^{(r)} \cdot 1\Delta_{s_2} \cdot 0\Delta_{s_1} = \Delta_t. \quad \blacksquare$$

We can now complete the proof of Proposition 5.1 (simple).

Proof. First, (i) implying (ii) is Lemma 5.10; (ii) implying (i) is Proposition 5.15 (equivalence). Then Δ_t is simple, so, by Lemma 5.9, (iii) and (iv) imply (ii). Finally (ii) implies (iii) by Proposition 5.20 (divisors of Δ). ■

We do not claim that every simple element of M_{LD} is a right divisor of an element Δ_t, a question we shall address in Chapter IX.

Another consequence of Proposition 5.20 (divisors of Δ) is:

Proposition 5.21. (lcm) *Simple elements of M_{LD} are closed under right lcm and operation $\backslash$.*

Proof. Assume that a, b are simple elements of M_{LD}. Let t be a term both in the domain of LD_a and in domain of LD_b. Then Δ_t is a common right multiple of a and b, hence it is a right multiple of $a \vee_R b$. By Proposition 5.20 (divisors of Δ), the latter element, which divides an element of the form Δ_t, is simple. Then $a\backslash b$ is a right divisor of Δ_t, so, by Lemma 5.9, it is simple as well. ■

As an application, we obtain that, if the positive words u, v can be decomposed into the product of p and q simple words respectively, then the determination of the words $u\backslash v$ and $v\backslash u$ can be done by computing at most pq lcm's of simple words (see Exercise 5.25).

In terms of LD-expansions, we deduce:

Proposition 5.22. (simple expansion) *The term ∂t is an LD-expansion of every simple LD-expansion of t—thus, the simple LD-expansions of a term t are exactly those LD-expansions of t of which ∂t is an LD-expansion.*

Proof. Assume that t' is a simple LD-expansion of t. By definition, there is a simple element a of M_{LD} satisfying $t' = t \bullet a$. Now $a \vee_R \Delta_t$ is simple. By Proposition VII.3.22 (syntactic confluence), there exists a common right multiple b of a and Δ_t such that the domain of LD_b is the intersection of the domains of LD_a and LD_{Δ_t}, hence t belongs to the domain of LD_b, and, therefore, of $\mathrm{LD}_{a\vee_R\Delta_t}$. Then Corollary 5.19 implies $a \vee_R \Delta_t = \Delta_t$, *i.e.*, a is a left divisor of Δ_t, and, therefore, ∂t is an LD-expansion of $t \bullet a$. ∎

Another application is existence of a normal form for M_{LD}.

Proposition 5.23. (normal form) *Every element of M_{LD} admits a unique expression of the form $u_1 \dots u_p$, where $u_1, \dots, u_p$ are permutation-like words and, for $k \geq 2$ and $\alpha \in \mathbf{A}$, if g_α^+ is a left divisor of $\overline{u_k}$ in M_{LD}, then $\overline{u_{k-1}\cdot g_\alpha^+}$ is not simple.*

Proof. By Proposition II.3.4 (lcm monoid), M_{LD} is an atomic right lcm monoid. For every address α, the generator g_α^+ is simple. By Proposition 5.21 (lcm), simple elements make a saturated family in M_{LD}. We then apply Proposition II.3.11 (decomposition). ∎

Exercise 5.24. (projection) *(i) Show that the projection of a permutation-like word on $\mathbf{A}$ is a permutation braid word, and that the projection of a simple word on $\mathbf{A}$ is a positive braid word u with the property that, in the braid diagram associated with u, any two strands cross at most once.*

(i)) Prove that, for every term t, the projection of Δ_t is Δ_n, where n is the right height of t. Compare their properties.

Exercise 5.25. (**simple complement**) *Let $\mathbf{A}^s$ denote the set of all permutation-like words on $\mathbf{A}$. We say that a word on $\mathbf{A}$ is simple if it represents a simple element of M_{LD}. Our aim is to directly construct a complement f^s on $\mathbf{A}^s$ as was done in Section II.5 for braids.*

(i) Assume that u, v are simple words on $\mathbf{A}$. Prove that there exist two integers p', q' such that the words $u \cdot \phi^{(q')}$ and $v \cdot \phi^{(p')}$ are simple and ϕ has the same exponent in both; in addition, every term that lies both in the domain of LD_u and of LD_v lies in the domain of $\mathrm{LD}_u \bullet \mathrm{LD}_{\phi^{(q')}}$ and $\mathrm{LD}_v \bullet \mathrm{LD}_{\phi^{(p')}}$. [Hint: For w a simple word and m an integer, let $\widehat{w}(m) = m + \mathrm{e}(1^m, w)$; say that r is accessible for w if there exists q satisfying $r = \widehat{w}(q)$ and $\widehat{w}(m) < \widehat{w}(q)$ for $0 \le m < q$: by Lemma 5.14, the accessible integers are the possible exponents of ϕ in those simple words one can obtain from w by multiplying it on the right by some word of the form $\phi^{(q)}$. Observe that $\widehat{w}$ eventually coincides with the identity, so an accessible r exists. Let r be the least accessible, and let p', q' be the smallest integers satisfying $r = \widehat{u}(q') = \widehat{v}(p')$. Show that at least one of $p' < r$, $q' < r$ holds. Assume $\widehat{u}(q') < r$: by definition, the exponent of $1^{q'}$ in u is the positive integer $r - q'$. Let t be any term in the domain of LD_u. Then $t \bullet (1^{q'})^{(r-q')}$ exists, i.e., the right height of t is at most $q' + (r - q') + 1$, i.e., $r + 1$. Conclude that t lies in the domain of $\mathrm{LD}_{\phi^{(r)}}$, and, a fortiori, both in the domains of $\mathrm{LD}_{\phi^{(q')}}$ and of $\mathrm{LD}_{\phi^{(p')}}$, hence in the domains of $\mathrm{LD}_u \bullet \mathrm{LD}_{\phi^{(q')}}$ and $\mathrm{LD}_v \bullet \mathrm{LD}_{\phi^{(p')}}$.] Apply the previous construction to $u = \phi \cdot 11^{(2)}$ and $v = 1^{(3)} \cdot 11$. [One finds $u \cdot \phi^{(2)} \equiv^+ \phi^{(4)} \cdot 1 \cdot 0$, and $v \cdot \phi \equiv^+ \phi^{(4)} \cdot 11$.]

(ii) Prove that there exists an explicit mapping f^s of $\mathbf{A}^s \times \mathbf{A}^s$ into $\mathbf{A}^s$ such that, for all permutation-like words u, v, we have $u \cdot f^s(v, u) \equiv^+ v \cdot f^s(u, v)$; in addition, the latter words are simple, and the domain of the associated operator is the intersection of the domains of LD_u and LD_v. [Hint: Use (i) and an induction on the size of $t^L_{u,v}$.] Apply the construction to the words u, v of (ii). [One finds $f^s(v, u) = \phi^{(2)} \cdot 1^{(2)}$, $f^s(u, v) = \phi \cdot 1 \cdot 11 \cdot 10 \cdot 0$, $u \vee_R v = \phi^{(4)} \cdot 1^{(2)} \cdot 11 \cdot 10 \cdot 0$.]

(iii) Using (ii), re-prove that the simple elements of M_{LD} are closed under right lcm and operation \, and deduce a new proof of the confluence property.

Exercise 5.26. (**degree** 2) *Say that the term t' is a degree 2 expansion of the term t if $\partial^2 t$ is an LD-expansion of t'. Prove that a degree 2 LD-expansion of t need not be a simple LD-expansion of a simple LD-expansion of t. [Hint: Consider $t = x^{[4]}$ and $t' = t \bullet u$, with $u = 1 \cdot 1 \cdot \phi \cdot 0 \cdot 00 \cdot 00$, and prove that $\partial t''$ is an LD-expansion of t' for no t'' of which t' is an LD-expansion.] Compare with the case of braids.*

[All divisors of Δ_n^2 can be expressed as the product of two simple braids.]

Exercise 5.27. **(operator ∂)** *(i) For u a simple word on* **A** *define $f(u)$ to be $u\backslash\boldsymbol{\Delta}_{t_u^L}$. Prove that f induces a well defined mapping on M_{LD}, and that $\mathrm{LD}_{f(u)}$ maps t_u^R to ∂t_u^L.*

(ii) Define ∂u to be $f(u)\backslash\boldsymbol{\Delta}_{t_u^R}$. Prove that ∂ induces a well defined mapping on M_{LD}, and that $\mathrm{LD}_{\partial u}$ maps ∂t_u^L to ∂t_u^R. Deduce that, if t' is a simple LD-expansion of the term t, then $\partial t'$ is a simple LD-expansion of ∂t. What is the counterpart of ∂ on B_n^+? [Answer: The flip automorphism.]

Exercise 5.28. **(no right normal form)** *Let $a = g_\gamma^+ g_1^+ g_\phi^+ g_1^+$. Prove that a admits no unique maximal simple right divisor in M_{LD}. [Hint: Show that $g_1^+ g_\phi^+ g_1^+$ and $g_0^+ g_1^+$ are maximal simple right divisors of a.]*

---

6. Notes

The results of Sections 1, 2, some of the results in Sections 3 and 4, and Exercise 5.25 (simple complement) appear, sometimes implicitly, in (Dehornoy 94*a*). The results about charged braids are from (Dehornoy *94c*). The results about the linear order on G_{LD} and about simple elements have not yet appeared in print.

We think that considering the group G_{LD} gives a really deep insight into the left self-distributivity identity. Although probably more complicated as it requires verifying the coherence and the convergence of the complement involved in M_{LD} first, the syntactic proof of the acyclicity of left division in free LD-systems given in Section 3 may be preferred to the proof given in Chapter I, because it resorts to geometrical properties of left self-distributivity exclusively, rather than to the specific properties of a particular LD-system. Using this proof and the argument of Proposition III.2.18 (σ_i-positive), we obtain a complete proof of the acyclicity of division in $(B_\infty, {}^\wedge)$, of existence of the braid order, and of decidability of the word problem of Identity (LD): this is the original scheme of (Dehornoy 94*a*)—and we shall mention in the appendix to Chapter IX that this scheme is the only one working in the case of other identities.

One of the main questions left open is whether $\mathrm{LD}_u = \mathrm{LD}'_u$ implies $u \equiv^+ u'$ when u, u' are words on **A**. We shall discuss the problem in the next chapter. Let us mention here that the methods of Section 2 do not apply directly as the blueprints are not positive words in general. Now, we can associate with every term t in T_1 positive words, namely the numerator χ_t^+ and the denominator χ_t^- of χ_t. Experiments suggest that $t' = t \bullet u$ with u a positive word could imply $\chi_{t'}^+ \equiv^+ \chi_t^+ \cdot v$ and $0u \cdot \chi_{t'}^- \equiv^+ \chi_t^- \cdot v$ for some positive v, but, even if this is true, it is not clear that these formulas determine u up to $\equiv^+$-equivalence.

No upper bound about the complexity of word reversing in $\mathbf{A}^*$ is known, except the coarse bound of Proposition 1.11 (convergence). In the case of B_∞, we have seen that replacing the atomic complement f_B with the simple complement f_B^s improves the complexity bound, which decreases from exponential to polynomial.

Question 6.1. *Which upper bound for the complexity of the word problem in M_{LD} can be deduced from using the simple complement f^s of Exercise 5.25?*

The injectivity of the mapping χ on T_1 implies that, if a is a special element of G_{LD}, then there exists exactly one decomposition $a = b$^c with b, c special.

Question 6.2. *Assume that a is an element of G_{LD} that can be expressed as b^c; Are b and c unique?*

About order properties on terms and on G_{LD}, the existence of a possible connection between the braid order and the linear order $<$ on G_{LD} is open.

Finally, we conjecture a positive answer to the following question, which could be connected to Laver's conjectures of Chapter IV and V:

Question 6.3. *Is the LD-equivalence class of every term well ordered by $<_L$?*

IX
Progressive Expansions

In this chapter, we deepen our study of left self-distributivity and, describing the geometry of the terms $\partial^k t$ more closely, we prove partial results about the convergence of the Polish Algorithm and the Embedding Conjecture. The Polish Algorithm is a natural syntactic method for deciding LD-equivalence of terms, and the question of whether it always converges is one of the most puzzling open questions about left self-distributivity. The Embedding Conjecture claims that the monoid M_{LD} embeds in the group G_{LD}. The main technical notion in the chapter is the notion of a progressive LD-expansion, a particular kind of LD-expansion where self-distributivity is applied to positions that move from from left to right only. Its interest lies in the uniqueness properties it entails.

The organization of the chapter is as follows. In Section 1, we introduce the Polish Algorithm and the notion of a progressive word. In Section 2, we define the notion of an element of FLD_∞ *appearing* in a term, and we study an associated covering relation. In Section 3, we continue the analysis of the terms $\partial^k t$ initiated in Chapter VI. We prove that the latter terms satisfy a strong self-similarity property called perfectness, and we deduce partial results about the practical determination of the t_0-normal form of a term. In Section 4, we deduce convergence results for the Polish Algorithm; we establish in particular that the algorithm always converges when running on terms that are $\sqsubset_{LD}$-comparable to injective terms. We also extend the results established in

Section VIII.5 for simple elements of M_{LD} to a more general notion of degree k elements where $\partial^k t$ replaces ∂t. In Section 5, we establish explicit decompositions for the elements Δ_t, which we recall are the counterparts to Garside's braids Δ_n. In Section 6, we prove a number of particular instances of the Embedding Conjecture by establishing the existence of certain sets of terms called confluent families. Finally, we conclude this chapter and Part B with a brief appendix describing what remains from the current approach when the left self-distributivity identity is replaced with another algebraic identity, in particular associativity.

1. The Polish Algorithm

After Chapter V, we know that the word problem of (LD) is decidable, *i.e.*, there exists an algorithm which, starting with arbitrary terms t_1, t_2 in T_∞, decides whether $t_1 =_{LD} t_2$ holds. The algorithms of Chapter V are semantic in nature in that the decision method consists in comparing the values obtained by evaluating t_1 and t_2 in some concrete LD-system, the braid group B_∞, or Larue's group LB_∞. In Chapter VI, we described other algorithms which are more syntactic in nature and consist in computing some normal forms of t_1 and t_2. The computation is always possible, and we have established a primitive recursive upper bound on its complexity. However, the process is intractable in general. In this section, we introduce a new syntactic algorithm called the Polish Algorithm, which is both natural and simple.

Hard and soft clashes

Let us come back here to the right Polish form of terms.

Definition 1.1. (clash) Assume that t_1, t_2 are terms in T_∞. We say that t_1 and t_2 have a *hard clash* at position p if the first $p-1$ letters of t_1 and t_2 coincide, and either distinct variables occur at p in t_1 and t_2, or a variable occurs at p in t_1, while t_2 has length less than p, or *vice versa.* We say that t_1 and t_2 have a *soft clash* at position p if the first $p-1$ letters of t_1 and t_2 coincide, a variable occurs at p in t_1, while $\bullet$ occurs at p in t_2, or *vice versa.*

Example 1.2. (clash) *Consider for instance*

$$\begin{cases} t_1 = x_1x_2\|\bullet x_1x_3x_4\bullet\bullet\bullet, \\ t_2 = x_1x_2\|x_3\bullet x_2x_4\bullet\bullet\bullet. \end{cases}$$

Then t_1 and t_2 have a soft clash at position 3 (as indicated by the $\|$ sign): the two first letters of t_1 and t_2 coincide, but the third letters do not.

We recall that, for t a term, and α an address in the skeleton of t, $\alpha 0^*$ and $\alpha 1^*$ respectively denote the unique addresses of the form $\alpha 0^p$ and $\alpha 1^q$ in the outline of t, a convention that avoids introducing the exponents p, q explicitly.

Lemma 1.3. *(i) Assume that t_1, t_2 are non-equal LD-equivalent terms. Then t_1 and t_2 have a soft clash.*

(ii) More precisely, assume that t_2 is the LD-expansion of t_1 at α. Then t_1 and t_2 have a soft clash at p, where p is the position of $\alpha 110^*$ in t_1 and the position of $\alpha 0^*$ in t_2.

Proof. (i) By definition, the terms t_1 and t_2 have a hard clash if and only if one of $t_1 \ll t_2$, $t_2 \ll t_1$, $t_1 \sqsubset t_2$, $t_2 \sqsubset t_1$ holds. By the results of Section V.5, t_1 and t_2 are LD-equivalent if and only if none of the previous relations hold.

(ii) Let s_1, s_2 and s_3 be the subterms of t_1 at $\alpha 0$, $\alpha 10$, and $\alpha 11$ respectively. Then we have explicit decompositions

$$\begin{cases} t_1 = & \cdots s_1\, s_2\, s_3 \bullet \bullet \cdots, \\ t_2 = & \cdots s_1\, s_2 \bullet s_1\, s_3 \bullet \bullet \cdots, \end{cases}$$

so t_1 and t_2 have a soft clash at the position p which corresponds to the first letter of the subterm s_3 in t_1, hence to address $\alpha 110^*$ in t_1, and to the first letter $\bullet$ in t_2, hence to address $\alpha 0^*$ in t_2. ∎

Lemma 1.4. *Assume that t_1, t_2 have a soft clash. Then, for every substitution f, the terms t_1^f and t_2^f have a soft clash.*

Proof. As a word, the term t_1^f is obtained from t_1 by substituting every variable x_i with its image under f, and keeping every letter $\bullet$. Assume that t_1 and t_2 have a soft clash at p, say that the variable x occurs at p in t_1, while $\bullet$ occurs in t_2. Letting w be the common prefix of t_1 and t_2 of length $p-1$, the word t_1^f begins with $w^f x^f$, while t_2^f begins with $w^f \bullet$. The first letter of the term x^f is necessarily a variable, hence t_1^f and t_2^f have a soft clash at position $\lg(w^f)+1$. ∎

Let us address the question of deciding whether two given terms t_1, t_2 are LD-equivalent. After Lemma 1.3(i), three cases are possible:

- either t_1 and t_2 are equal: in this case, they are obviously LD-equivalent;
- or they have a hard clash: in this case, they are not LD-equivalent;
- or they have a soft clash.

In the latter case, we can make no conclusion. Now, if t_1 and t_2 have a soft clash at position p with, say, $\bullet$ occurring at p in t_1, expanding t_2 gives a new term t_2' having a soft clash with t_2: if t_2' is chosen so that the clash between t_2 and t_2' is at p, then the possible clash between t_1 and t_2' is at $p+1$ at least.

Example 1.5. (pushing a clash) *Let t_1 and t_2 be as in Example 1.2; they have a soft clash at position 3. Choose $t_2' = t_2 \bullet \phi \cdot 0$. We find*

$$\begin{cases} t_1 = x_1x_2{\bullet}x_1x_3\|x_4{\bullet}{\bullet}{\bullet}, \\ t_2' = x_1x_2{\bullet}x_1x_3\|{\bullet}{\bullet}x_1x_2x_4{\bullet}{\bullet}{\bullet}, \end{cases}$$

so t_1 and t_2' have a soft clash at position 6.

The point is that there exists a canonical way for choosing an LD-expansion as above, *i.e.*, for fixing a soft clash without changing the prefix before the clash.

Definition 1.6. (word $\alpha_{(k)}$) For $\alpha \in \mathbf{A}$, we define $\alpha_{(k)} = \alpha \cdot \alpha 0 \cdot \cdots \cdot \alpha 0^{k-1}$ for $k \geq 1$ and $\alpha_{(0)} = \varepsilon$.

For instance, the word $\phi \cdot 0$ used in Example 1.5 above is the word $\phi_{(2)}$.

Lemma 1.7. *Assume that t is a term. Then t belongs to the domain of $\mathrm{LD}_{\alpha_{(k)}}$ if and only if $\alpha 10^k$ belongs to the skeleton of t. In this case, the terms t and $t \bullet \alpha_{(k)}$ have a soft clash at the position of $\alpha 10^{k-1}10^*$ in t, which is also the position of $\alpha 0^k$ in $t \bullet \alpha_{(k)}$.*

Proof. Assume that t belongs to the domain of $\mathrm{LD}_{\alpha_{(k)}}$. Let $t' = t \bullet \alpha_{(k)}$. An induction on k shows that the $\alpha 10^k$-subterm of t is defined, and that, letting $s_0 = \mathrm{sub}(t, \alpha 0)$, $s_1 = \mathrm{sub}(t, \alpha 10^k)$, and $s_j = \mathrm{sub}(t, \alpha 10^{k-j+1}1)$ for $2 \leq j \leq k+1$, we have

$$\begin{cases} \mathrm{sub}(t, \alpha) = s_0s_1s_2{\bullet} \cdots s_k{\bullet}s_{k+1}{\bullet}{\bullet}, \\ \mathrm{sub}(t', \alpha) = s_0s_1{\bullet}s_0s_2{\bullet}{\bullet} \cdots s_0s_k{\bullet}{\bullet}s_0s_{k+1}{\bullet}{\bullet}, \end{cases}$$

i.e., t' is obtained from t by distributing s_0 to s_1, ..., s_{k+1} (Figure 1.1). Let w be the greater common prefix of the words t and t'. In the word t, w is the prefix that ends with s_0s_1, thus w is followed by the leftmost variable of s_2, while, in t', w is followed by the letter $\bullet$ whose address is $\alpha 0^k$. Let $p-1$ be the length of w. By construction, t and t' coincide up to the $p-1$-th letter and they have a soft clash at position p (see Figure 1.1). ■

Lemma 1.8. *Assume that the terms t_1 and t_2 have a soft clash at position p, with $\bullet$ at p in t_1. Then the address of p in t_2 has the form $\alpha 10^{k-1}10^*$, the first p letters of the terms t_1 and $t_2 \bullet \alpha_{(k)}$ coincide, and $\alpha_{(k)}$ is the only word of the form $\beta_{(j)}$ with this property.*

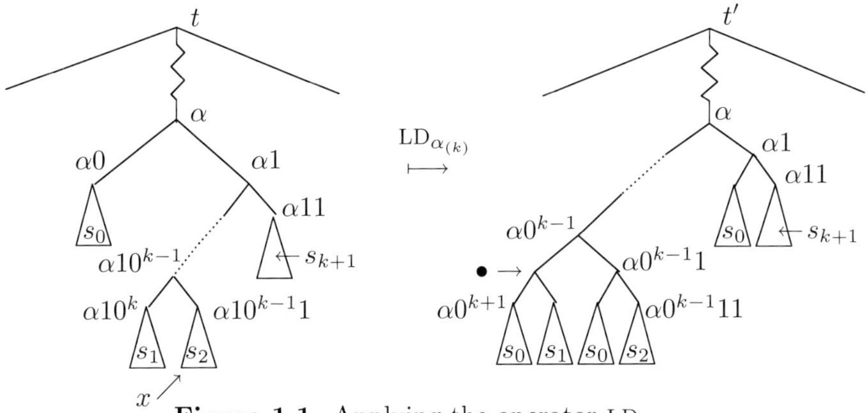

Figure 1.1. Applying the operator $\mathrm{LD}_{\alpha_{(k)}}$

Proof. By hypothesis, the first $p-1$ letters in t_1 and t_2 coincide, and the p-th letter in t_1 is $\bullet$, while the p-th letter in t_2 is a variable. Let γ_e be the address of p in t_e, for $e = 1, 2$. Proving that γ_2 has the form $\alpha 10^{k-1}10^*$ means proving that it contains at least two 1's. Let w be the length $p-1$ common prefix of t_1 and t_2. By construction, we have

$$\#_1(\gamma_2) = \#_X(wx) - \#_\bullet(wx) - 1 = \#_X(w) - \#_\bullet(w),$$

where we recall $\#_X(u)$ and $\#_\bullet(u)$ respectively denote the number of variables and of $\bullet$'s in u, and $\#_1(\gamma)$ denotes the number of 1's in γ. Similarly, we have

$$0 \leq \#_1(\gamma_1) = \#_X(w\bullet) - \#_\bullet(w\bullet) - 1 = \#_X(w) - \#_\bullet(w) - 2,$$

hence $\#_1(\gamma_2) = \#_X(w) - \#_\bullet(w) \geq 2$. By Lemma 1.7, we deduce that t_2 belongs to the domain of $\mathrm{LD}_{\alpha_{(k)}}$, that t_2 and $t_2 \bullet \alpha_{(k)}$ have a soft clash at p, and that, for $\beta \neq \alpha$ or $j \neq k$, the clash between t_2 and $t_2 \bullet \beta_{(j)}$ is not at p. ∎

The Polish Algorithm

Now, a natural idea is to iterate the previous process.

Definition 1.9. (**Polish Algorithm**) Assume that the terms t_1, t_2 have a soft clash at p, with $\bullet$ occurring at p in t_1. We say that the Polish Algorithm running on (t_1, t_2) *returns* (t_1', t_2') *in one step*, with *output* $(\varepsilon, \alpha_{(k)})$, if we have $t_1' = t_1$ and $t_2' = t_2 \bullet \alpha_{(k)}$, where $\alpha 10^{k-1}10^*$ is the address of p in t_2. The definition for $\bullet$ occurring at p in t_2 is symmetric.

For $m \geq 0$, we say that the Polish Algorithm running on (t_1, t_2) *returns* (t_1', t_2') *in m steps*, with *output* (u_1, u_2), if there exist two sequences of terms $t_e^{(0)} = t_e$, $t_e^{(1)}$, ..., $t_e^{(m)} = t_e'$, $e = 1, 2$ such that the Polish Algorithm running on $(t_1^{(i)}, t_2^{(i)})$ returns $(t_1^{(i+1)}, t_2^{(i+1)})$ in one step for every i, and (u_1, u_2) is obtained by concatenating the outputs of each step.

Finally, we say that the Polish Algorithm running on (t_1, t_2) *converges* in m steps to (t_1', t_2') if it returns (t_1', t_2') in m steps and t_1' and t_2' do not have a soft clash.

Example 1.10. (Polish Algorithm) *Let us consider the pair of Example 1.2, namely*

$$\begin{cases} t_1 = x_1x_2\|\bullet x_1x_3x_4\bullet\bullet\bullet, \\ t_2 = x_1x_2\|x_3\bullet x_2x_4\bullet\bullet\bullet. \end{cases}$$

We have seen above that t_1 and t_2 have a soft clash at position 3 (we mark the longest common prefix), and that this clash can be repaired by expanding t_2 using $\mathrm{LD}_{\phi_{(2)}}$. Thus, the Polish Algorithm running on (t_1, t_2) returns in one step the pair

$$\begin{cases} t_1' = x_1x_2\bullet x_1x_3\|x_4\bullet\bullet\bullet, \\ t_2' = x_1x_2\bullet x_1x_3\|\bullet\bullet x_1x_2x_4\bullet\bullet\bullet, \end{cases}$$

with output $(\varepsilon, \phi_{(2)})$. Now, the terms t_1' and t_2' have a soft clash at position 6. As $\bullet$ occurs at 6 in t_2', we expand t_1'. The address of 6 in t_1' is 111, so we apply $\mathrm{LD}_{1_{(1)}}$, and get the pair

$$\begin{cases} t_1'' = x_1x_2\bullet x_1x_3\bullet\|x_1x_4\bullet\bullet\bullet, \\ t_2'' = x_1x_2\bullet x_1x_3\bullet\|\bullet x_1x_2x_4\bullet\bullet\bullet. \end{cases}$$

The terms t_1'' and t_2'' have a soft clash at position 7, whose address in t_1'' is 110. We expand t_1'' using $\mathrm{LD}_{\phi_{(1)}}$, and find

$$\begin{cases} t_1''' = x_1x_2\bullet x_1x_3\bullet\bullet x_1x_2\|\bullet x_1x_4\bullet\bullet\bullet, \\ t_2''' = x_1x_2\bullet x_1x_3\bullet\bullet x_1x_2\|x_4\bullet\bullet\bullet. \end{cases}$$

The terms t_1''' and t_2''' have a soft clash at position 10, whose address in t_2''' is 111. We expand t_2''' using $\mathrm{LD}_{1_{(1)}}$, and find

$$\begin{cases} t_1'''' = x_1x_2\bullet x_1x_3\bullet\bullet x_1x_2\bullet x_1x_4\bullet\bullet\bullet\|, \\ t_2'''' = x_1x_2\bullet x_1x_3\bullet\bullet x_1x_2\bullet x_1x_4\bullet\bullet\bullet\|. \end{cases}$$

The terms t_1'''' and t_2'''' are equal. So the Polish Algorithm running on the pair (t_1, t_2) converges to (t_1'''', t_1'''') in 4 steps. The successive outputs are $(\varepsilon, \phi_{(2)})$, $(1_{(1)}, \varepsilon)$, $(\phi_{(1)}, \varepsilon)$, and $(\varepsilon, 1_{(1)})$, hence the global output is the pair $(1_{(1)} \cdot \phi_{(1)}, \phi_{(2)} \cdot 1_{(1)})$.

By definition, if the Polish Algorithm running on (t_1, t_2) returns (t_1', t_2') with output (u_1, u_2), then we have $t_1' = t_1 \bullet u_1$ and $t_2' = t_2 \bullet u_2$, and, in particular, $t_1' =_{LD} t_1$ and $t_2' =_{LD} t_2$. This implies that, when it converges, the Polish Algorithm solves the word problem of (LD) and decides the canonical ordering in the following strong sense:

Proposition 1.11. **(correctness)** *Assume that the Polish Algorithm running on the pair (t_1, t_2) converges to the pair (t'_1, t'_2). Then*
(i) $t_1 =_{LD} t_2$ is equivalent to $t'_1 = t'_2$;
(ii) $t_1 \sqsubset_{LD} t_2$ is equivalent to $t'_1 \sqsubset t'_2$;
(iii) $t_1 \ll_{LD} t_2$ is equivalent to $t'_1 \ll t'_2$.

Proof. By construction, we have $t'_e =_{LD} t_e$ for $e = 1, 2$, and the conditions of the proposition are sufficient. They are also necessary, for the hypothesis that (t'_1, t'_2) have no soft clash implies that exactly one of $t'_1 = t'_2$, $t'_1 \sqsubset t'_2$, $t'_1 \sqsupset t'_2$, $t'_1 \ll t'_2$, $t'_1 \gg t'_2$ holds. ■

At this point, the remaining question is the convergence of the Polish Algorithm. Here we mention some preliminary observations. We shall return to the question in Section 4 below.

Lemma 1.12. *Assume that the Polish Algorithm running on (t_1, t_2) converges, and that s_1 and s_2 are cuts of t_1 and t_2 respectively. Then the Polish Algorithm running on (s_1, s_2) converges as well.*

Proof. Let q_e be the largest position of a variable in s_e. Assume first that t_1 and t_2 have no clash before $\inf(q_1, q_2)$. Then either s_1 and s_2 are equal, in which case the result is true, or one is a cut of the other, say s_1 is a cut of s_2. Let w be the longest common prefix of the words s_1, s_2. By construction, we have $s_1 = w\bullet^k$ for some k, and $s_2 = wxw_1$ for some variable x. For $k = 0$, s_1 and s_2 have a hard clash, and the algorithm halts. Otherwise, applying the algorithm amounts to replacing s_2 with some term s'_2 that begins with $w\bullet$. Now s_1 is a cut of s'_2, and, letting w' be the longest common prefix of s_1 and s'_2, we have now $s_1 = w'\bullet^{k-1}$. Using induction on k, we conclude that the Polish Algorithm running on (s_1, s'_2) converges in $k - 1$ steps, and, therefore, that the Polish Algorithm running on (s_1, s_2) converges in k steps.

Assume now that t_1 and t_2 have a clash at position p with $p \leq \min(q_1, q_2)$. Then s_1 and s_2 have a clash of the same kind at p. If the clash is hard, the algorithm halts for (s_1, s_2) as it does for (t_1, t_2). If the clash is soft, it results in expanding say t_1 into t'_1, and s_1 into s'_1. Using the explicit decomposition of a cut of Lemma VI.2.7 and the definition of the operator $\alpha_{(k)}$, we see that s'_1 is a cut of t'_1, and we conclude using induction on the number of steps needed for the algorithm to converge when running on (t_1, t_2). ■

Definition 1.13. **(lifting)** Assume that $(t_1, t_2, \ldots)$ is a sequence of terms in T_∞. We say that the sequence $(t'_1, t'_2, \ldots)$ is a *lifting* of $(t_1, t_2, \ldots)$ if there exists a mapping f of $\{x_1, x_2, \ldots\}$ into itself such that $t_i = (t'_i)^f$ holds for each i.

Thus a lifting of a term t is obtained from t by changing the names of the variables with the only constraint that distinct variables cannot be replaced with the same variable. We do not require the replacement to be functional: different occurrences of the same variable may be replaced with different variables, so that, for instance, every term is a lifting of a one variable term.

Lemma 1.14. *Assume that (t'_1, t'_2) is a lifting of (t_1, t_2).*

(i) If the Polish Algorithm running on (t_1, t_2) converges, then it converges as well when running on (t'_1, t'_2).

(ii) Conversely, if t'_1 and t'_2 are $\sqsubseteq_{LD}$-comparable, and the Polish Algorithm running on (t'_1, t'_2) converges, then it converges as well when running on (t_1, t_2).

Proof. Let us compare how the Polish Algorithm runs on (t_1, t_2) and on (t'_1, t'_2). Assume that the algorithm converges for (t_1, t_2) in m steps. We claim that it converges for (t'_1, t'_2) in at most m steps. For an induction, it suffices to consider the first step. Assume that t_1 and t_2 have a clash at position p. Then, either t'_1 and t'_2 have a variable clash at some position $< p$, in which case the Polish Algorithm halts immediately, or they coincide at least until position $p - 1$. If the clash of t_1 and t_2 at p is strong, then t'_1 and t'_2 have a strong clash of the same type at p: indeed, different variables in t_1 and t_2 must come from distinct variables in t'_1 and t'_2. If the clash of t_1 and t_2 at p is soft, t'_1 and t'_2 also have a soft clash at p. In this case, the pair $((t'_1)^{(1)}, (t'_2)^{(1)})$ obtained when treating the clash of t'_1 and t'_2 at p is a lifting of the pair $(t_1^{(1)}, t_2^{(1)})$ obtained when treating the clash of t_1 and t_2, and the induction goes on. This proves (i).

For (ii), we prove similarly that, if the algorithm converges for (t'_1, t'_2) in m steps, then it converges for (t_1, t_2) in m steps too. The argument is the same as above. The only problem is that t'_1 and t'_2 may have a variable clash that t_1 and t_2 do not have, as the mapping involved in the lifting need not be injective. Now the hypothesis of t'_1 and t'_2 being $\sqsubseteq_{LD}$-comparable discards this possibility. ∎

Proposition 1.15. (sufficient) *Assume that the Polish Algorithm converges when running on every pair of LD-equivalent one variable terms. Then it converges always.*

Proof. Let t_1, t_2 be arbitrary terms. Let $t_1^\dagger$ and $t_2^\dagger$ be the projections of t_1 and t_2 obtained by replacing every variable with x. By Proposition V.2.11 (absorption property), there exist p and q such that $t_1^\dagger \cdot x^{[p]}$ and $t_2^\dagger \cdot x^{[q]}$ are LD-equivalent. If we suppose that the Polish Algorithm converges when running on $(t_1^\dagger \cdot x^{[p]}, t_2^\dagger \cdot x^{[q]})$, then, by Lemma 1.12, it converges when running on $(t_1^\dagger, t_2^\dagger)$. By Lemma 1.14, it converges when running on (t_1, t_2) as well. ∎

Remark 1.16. The Polish Algorithm may be extended easily so as to run on an arbitrary sequence of terms $(t_1, t_2, \ldots)$ rather than on a pair (t_1, t_2) only.

Indeed, let us say that $(t_1, t_2, \ldots)$ has a soft clash at p if all terms t_i have the same length $p-1$ prefix, and $\bullet$ occurs at p in at least one of the t_i's, while variables occur at p in the other terms. In this case, we can apply to the latter terms convenient operators $\mathrm{LD}_{\alpha(k)}$ (depending on the terms) so as to push the soft clash further to the right. Then we iterate the process, removing those terms that are shorter than the current clash position. If the process converges, *i.e.*, if we eventually obtain a sequence of terms with no soft clash, the final sequence consists of pairwise $\sqsubset$- or $\ll$-comparable terms. In particular, ending with a constant sequence $(t', t', \ldots)$ means having found a common LD-expansion for all terms t_i. The easy details are left to the reader.

Progressive LD-expansions

We know that, if t_1 and t_2 are LD-equivalent terms, then they admit a common LD-expansion. The principle of the Polish Algorithm running on (t_1, t_2) is to try to construct such a common LD-expansion, *i.e.*, to find positive words u_1, u_2 satisfying $t_1 \bullet u_1 = t_2 \bullet u_2$. Here, we study those words arising as the output of the Polish algorithm, which will be called progressive words.

Definition 1.17. (progressive) For u a word on $\mathbf{A}$, we say that u is *progressive* if it admits a decomposition

$$u = \alpha_{1(k_1)} \cdot \cdots \cdot \alpha_{m(k_m)}$$

such that $\alpha_{i-1} 0^{k_{i-1}} <_{LR} \alpha_i 1 0^{k_i - 1} 1$ holds for $1 < i \le m$. We say that the term t' is a *progressive LD-expansion* of the term t if we have $t' = t \bullet u$ for some progressive word u.

Lemma 1.18. *The output of the Polish Algorithm consists of progressive words.*

Proof. Assume that $\alpha_{(k)}$ and $\beta_{(j)}$ are successive factors in an output word of the Polish Algorithm. This means that there exist three terms t, t', t'' satisfying $t' = t \bullet \alpha_{(k)}$ and $t'' = t' \bullet \beta_{(j)}$. By hypothesis, the terms t and t' have a soft clash at some position p, and the terms t' and t'' have a soft clash at some position q. Now, the fact that $\beta_{(j)}$ occurs after $\alpha_{(k)}$ in an output word implies $p < q$: the clashes always go right, and never back. By Lemma 1.7, the address of p in the middle term t' is $\alpha 0^k$, while the address of q in t' is $\beta 1 0^{j-1} 1 0^*$. As the address function is increasing with respect to the linear order $<$ of addresses, the hypothesis $p < q$ translates into $\alpha 0^k < \beta 1 0^{j-1} 1 0^*$, hence to $\alpha 0^k <_{LR} \beta 1 0^{j-1} 1$. ∎

Proposition 1.19. (progressive) *The following are equivalent:*

(i) The term t_2 is a progressive LD-expansion of the term t_1;

(ii) The Polish Algorithm running on the pair (t_1, t_2) converges to (t_2, t_2).

Proof. We prove using induction on m that, if u is a word on $\mathbf{A}$ that admits a progressive decomposition of length m, then the Polish Algorithm running on every pair $(t, t \bullet u)$ returns $(t \bullet u, t \bullet u)$ in m steps with output (u, ε). For $m = 0$, u must be empty, and the result is true. Assume $m > 0$. Let us write $t_e^{(i)}$ for the term obtained after i steps of the algorithm. Then the clash between $t_1^{(0)}$, *i.e.*, t_1, and $t_1^{(m)}$, *i.e.*, $t_1 \bullet u$, must be the clash between $t_1^{(0)}$ and $t_1^{(1)}$, as the subsequent clashes occur farther to the right. Hence the Polish Algorithm must guess the first factor of u at the first step, and the induction is obvious.

Conversely, if the Polish Algorithm running on (t_1, t_2) converges to some pair (t, t), then, by Lemma 1.18, t is a progressive LD-expansion both of t_1 and t_2. ∎

Corollary 1.20. *Assume that u is a word on $\mathbf{A}$ that is $\equiv$-equivalent to some progressive word. Then u is $\equiv$-equivalent to a unique progressive word, namely the word u' such that the Polish Algorithm running on an arbitrary pair $(t, t \bullet u)$ converges with output (u', ε).*

Example 1.21. (**not progressive**) *The previous criterion allows us to prove that some words are not $\equiv$-equivalent to a progressive word. For instance, let u be the word $\phi \cdot 00$. First u is not progressive, for the only possible decomposition is $\phi_{(1)} \cdot 00_{(1)}$, and $\phi 0 = 0 <_{LR} 0011$ fails We prove that u is not $\equiv$-equivalent to any progressive word as follows: let t be the term*

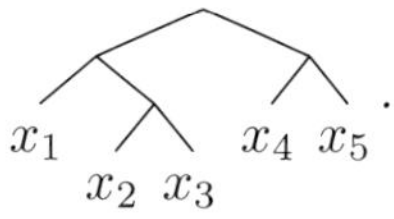

. Then $t \bullet \phi \cdot 00$ exists, and it is equal to $t' =$

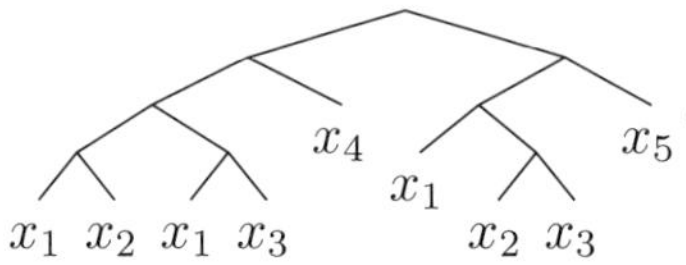

. When we run the Polish Algorithm on (t, t'), it first expands t to $t'' =$

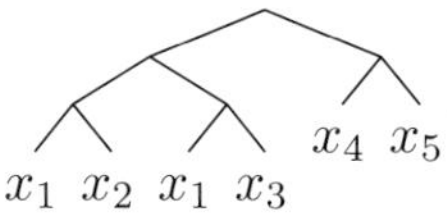

, and then it expands t'' into $t''' =$

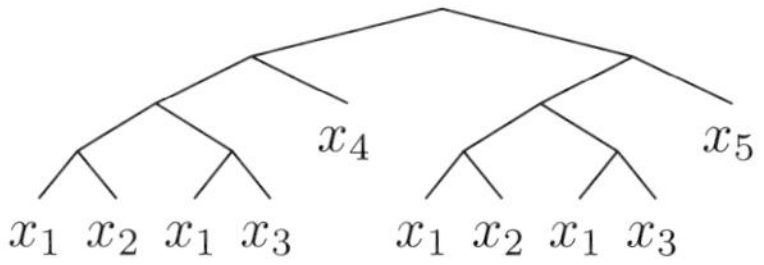

. Thus, it converges in three steps, but to the pair (t''', t''') and not to the pair (t', t'). The reason is clear: t' is an LD-expansion of t, but when we try to determine a common LD-expansion for t and t' starting from the left, we first meet the factor $x_1x_2{\bullet}x_1x_3{\bullet\bullet}$ of t' and conclude that the factor $x_1x_2x_3{\bullet\bullet}$ in t has to be expanded. Now, this factor has a second copy in t', which

was not expanded. So, when $x_1x_2x_3\bullet\bullet$ has been expanded before being duplicated, we cannot go back and can only obtain a second factor $x_1x_2\bullet x_1x_3\bullet\bullet$ which is not in t'.

Exercise 1.22. (progressive) *Assume that the term $t_1{\cdot}t_2$ is a progressive LD-expansion of the term t. Prove that $t_1 * t_2$ is a progressive LD-expansion of t as well. [Hint: Use induction on the left height of t_2.]*

---

2. The Content of a Term

The central notion in the construction of the normal forms of Chapter VI is the notion of a cut of a term. This notion will play a significant role in the sequel as well. However, it will be convenient to consider here LD-equivalence classes of cuts rather than cuts themselves. The idea is to associate with every term t the family of all LD-classes of cuts of t. This family will be called the content of t. If t' is an LD-expansion of t, the content of t' includes the content of t, and the key point is to compare the positions where the elements of the content of t appears in t and in t'.

Appearing in a term

We recall that, for t a term and α an address in the outline of t, the cut of t at α is the term, denoted $\mathrm{cut}(t, \alpha)$, obtained by ignoring the part of t lying on the right of α; as a word, this amounts to completing the prefix of t that ends at α with enough letters $\bullet$ to obtain a well formed term.

Definition 2.1. (appearing, content) Assume that t is a term in T_∞, a is an element of the free LD-system FLD_∞, and α is an address not of the form 1^*. We say that *a appears in t at α* if a is the LD-class of $\mathrm{cut}(t, \alpha)$. The *content* of t is defined to be the set of all elements of FLD_∞ that appear in t, completed with $\bar{t}$.

We shall say that the element a appears in t *below* α if a appears in t at some address lying below α, *i.e.*, of the form $\alpha\beta$. It follows from Proposition VI.2.8 (cut order) that a given element of FLD_∞ appears at most once in a given term, so we can speak without ambiguity of *the* address where a appears in t. The following results are restatements or easy consequences of Proposition VI.2.8 (cut order), Lemma VI.3.2, and Lemma VI.2.19.

Lemma 2.2. *Assume that a and b appear in t at α and β respectively. Then $\alpha >_{LR} \beta$ implies $a \sqsupset b$.*

Proposition 2.3. (appears *vs.* $\sqsubseteq$) *For t a term in T_∞, and a in FLD_∞, the following are equivalent:*
(i) The inequality $a \sqsubset \bar{t}$ holds;
(ii) The element a appears in some term t' that is LD-equivalent to t;
(iii) The element a appears in some term $\partial^k t$.

Proposition 2.4. (LD-expansion) *(i) Assume that t' is an LD-expansion of t. Then every element appearing in t appears in t' as well, hence the content of t is included in the content of t'.*

(ii) More precisely, assume $t' = t \bullet \alpha$. Then the elements appearing in t' are those elements appearing in t, augmented with the elements ab such that a appears at $\alpha 101^*$ in t and b appears below $\alpha 0$ in t.

The covering graph of a term

The left–right ordering of addresses has been partitioned into two suborderings using the notion of covering: the address α lies on the right of the address β, written $\alpha >_{LR} \beta$, if there exists γ such that α lies below $\gamma 1$ and β lies below $\gamma 0$, and, if this is the case, we say that α covers β if, in addition, α has the form $\gamma 1^*$, *i.e.*, α is $\gamma 1^q$ for some q; otherwise, we say that α uncovers β. As every element appearing in a term t appears at a well defined address, we deduce a covering relation for the elements that appear in t.

Definition 2.5. (covering, uncovering) Assume that t is a term in T_∞. We say that a *covers* (*resp. uncovers*) b in t if a and b appear in t, and the address where a appears in t covers (*resp. uncovers*) the address where b appears in t.

Example 2.6. *(**covering**) Let t be the term $x_1{\cdot}(x_2{\cdot}x_3{\cdot}x_4){\cdot}x_5$, i.e.,*

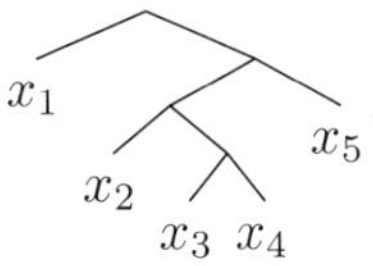

. Four elements appear in t, namely x_1, x_1x_2, $x_1x_2x_3$ and $x_1x_2x_3x_4$, which appear at 0, 100, 1010, and 1011 respectively. By definition, the address 100 uncovers 0, hence x_1x_2 uncovers x_1 in t. Similarly, $x_1x_2x_3$ uncovers x_1 and x_1x_2 in t, and $x_1x_2x_3x_4$ uncovers x_1, but covers x_1x_2 and $x_1x_2x_3$ in t.

Using the explicit values of the cuts given in Lemma VI.2.19, we see that expanding a term t at α causes the cut of t at $\alpha 10$ to cover all cuts of t at addresses below $\alpha 0$:

Proposition 2.7. (covering) *(i) Assume that t' is an LD-expansion of t, and that a covers b in t. Then a covers b in t'.*

(ii) More precisely, assume $t' = t \bullet \alpha$. Let a be the element that appears at $\alpha 101^*$ in t. Then the covering graph of t' is the covering graph of t, augmented with the following pairs:

- *all pairs (a, b) with b appearing below $\alpha 0$ in t;*
- *all pairs (ab, ac) with b, c appearing below $\alpha 0$ in t and b covering c in t;*
- *all pairs (c, ab) with b appearing below $\alpha 0$ in t and c covering a in t.*

For every term t, the elements that appear in t make a chain with respect to $\sqsubset$, and this chain is isomorphic to the outline of t ordered by the left–right ordering. For every address α in the outline of t, the addresses β that are covered by α are those addresses that satisfy $\alpha >_{LR} \beta \geq_{LR} \mu_t(\alpha)$. Thus, we have the following property:

Lemma 2.8. *Assume that t is a term, and a appears at α in t. Then the elements covered by a in t are those elements b that appear in t and satisfy $a \sqsupset b \sqsupseteq a_0$, where a_0 is the element that appears at $\mu_t(\alpha)$ in t.*

We have established in Chapter VI a relation between each non-final cut of a term t and the next cut of t. For a appearing in t, this results in a similar relation between a and the $\sqsubset$-next element that appears in t.

Definition 2.9. (almost equal) For a, a' elements of FLD_∞, we say that a and a' are *almost equal* if they can be represented by almost LD-equivalent terms, *i.e.*, terms that are LD-equivalent up to a possible change of the rightmost variable.

As the rightmost variable of a term is preserved by left self-distributivity, the previous definition is non-ambiguous. We deduce from Proposition VI.2.14 (next cut) the following result:

Lemma 2.10. *Assume that t is a term, and a appears in t at α. Let a_0 be the element that appears in t at $\mu_t(\alpha)$. Then the successor of a in the content of t is almost equal to aa_0.*

Steep terms

We shall consider now several families of terms defined by order constraints. The interest of such terms lies in the possibility of describing the content of their LD-expansions precisely. The first notion is the notion of an increasing term, and its weakening, the notion of a quasi-increasing term.

Definition 2.11. (increasing, quasi-increasing) Assume that t is a term in T_∞. We say that t is *increasing* if the variables of t enumerated from left to right make a strictly increasing sequence. We say that t is *quasi-increasing* if, for every internal address α in the skeleton of t, the inequality $\mathrm{var}(t, \alpha 10^*) > \mathrm{var}(t, \alpha 0^*)$ holds.

By definition, every increasing term is injective, *i.e.*, no variable occurs twice, and it is quasi-increasing, as $\alpha 10^* >_{LR} \alpha 0^*$ holds.

Example 2.12. (quasi-increasing) *A quasi-increasing term need not be increasing. For instance, the term* $(x_1{\cdot}x_2){\cdot}x_2$ *is not increasing, but it is quasi-increasing: there are two internal addresses, namely* ϕ *and* 0, *and, in both cases, we have* $\mathrm{var}(t, \alpha 10^*) = x_2$, *and* $\mathrm{var}(t, \alpha 0^*) = x_1$.

Lemma 2.13. *Assume that t is a quasi-increasing term, and a covers b in t. Then $\bar{t} \gg ab$ holds, and, therefore, $ab \sqsubset \bar{t}$ does not.*

Proof. Assume that a appears at α in t, and b appears at β. Let α_0 be the least address covered by α in the outline of t, and α^+ be the immediate successor of α in this set. Let a_0 denote the LD-class of $\mathrm{cut}(t, \alpha_0)$, and a^+ denote the LD-class of $\mathrm{cut}(t, \alpha^+)$. By Proposition VI.2.14 (next cut), $\mathrm{cut}(t, \alpha^+)$ is LD-equivalent to the term obtained from $\mathrm{cut}(t, \alpha){\cdot}\mathrm{cut}(t, \alpha_0)$ by replacing the right variable $\mathrm{var}(t, \alpha_0)$ with the variable $\mathrm{var}(t, \alpha^+)$. Let γ be the greatest common prefix of the addresses α and α^+. Then t being quasi-increasing at γ implies $\mathrm{var}(t, \alpha^+) > \mathrm{var}(t, \alpha_0)$, and we deduce $a^+ \gg aa_0$. By construction, we have $\bar{t} \sqsupseteq a^+$, and $b \sqsupseteq a_0$, hence $ab \sqsupseteq aa_0$, so we deduce $\bar{t} \gg ab$. ∎

We introduce now properties of (quasi)-increasing terms that are preserved under LD-equivalence, which is not the case for (quasi)-increasingness itself.

Definition 2.14. (steep) For a in FLD_∞, we say that a is *steep* if $a > b$ implies $a \gg b^{[2]}$ for every b. For t in T_∞, we say that t is *steep* if the LD-class of t is steep.

As $a \gg b$ always implies $a \gg b^{[2]}$, the nontrivial part in the definition involves those elements b that satisfy $a \sqsupset b$. Observe that no element of FLD_1 but the generator x_1 is steep. The following criterion characterizes steep terms by a purely local condition, one that involves only t and not the LD-expansions of t.

Proposition 2.15. (**steep**) *(i) Assume that t is a term. Then t is steep if and only if $\overline{t} \gg a^{[2]}$ holds for every element a appearing in t.*

(ii) Assume $t = t_1 \cdot t_2$. A sufficient condition for t to be steep is t_1 and t_2 being steep and satisfying $t_2 \gg_{LD} t_1$.

Proof. (i) Say that t is *presteep* if it satisfies the condition of (i). By definition, if a appears in t, then $\overline{t} \sqsupset a$ holds, and, therefore, being steep implies being presteep. On the other hand, if t' is an LD-expansion of t, and t' is presteep, then t is presteep as well, since every element appearing in t appears in t'. So, in order to prove that presteepness implies steepness, it suffices to prove that every LD-expansion of a presteep term is presteep, and, for an induction, it suffices to consider the case of a basic LD-expansion.

So, assume that t is a presteep term, and $t' = t \bullet \alpha$ holds. By Proposition 2.4 (LD-expansion), those elements that appear in t' are the elements that appear in t, augmented with new elements of the form ab, where a and b appear in t. As $\overline{t'} = \overline{t}$ holds, it suffices to verify that $\overline{t} \gg (ab)^{[2]}$ holds for the latter elements. First, we observe that $\overline{t} \sqsupset a$ implies $a\overline{t} \sqsupset a^{[2]}$, hence the hypothesis $\overline{t} \gg a^{[2]}$ implies $\overline{t} \gg a\overline{t}$. Then, applying the hypothesis $\overline{t} \gg b^{[2]}$, we find

$$\overline{t} \gg a\overline{t} \gg ab^{[2]} = (ab)^{[2]}.$$

(ii) Assume that t satisfies the conditions and a appears in t. Three cases are possible. For a appearing in t_1, the hypothesis $\overline{t_1} \gg a^{[2]}$ implies $\overline{t_1}\,\overline{t_2} \gg a^{[2]}$. For $a = \overline{t_1}$, the hypothesis $t_2 \gg_{LD} t_1$ implies $\overline{t_1}\,\overline{t_2} \gg \overline{t_1}\,\overline{t_1} = a^{[2]}$. Finally, for $a = \overline{t_1}b$ with b appearing in t_2, the hypothesis $\overline{t_2} \gg b^{[2]}$ implies $\overline{t_1}\,\overline{t_2} \gg \overline{t_1}b^{[2]} = a^{[2]}$. By (i), we conclude that t is steep. ∎

Lemma 2.16. *Every quasi increasing term is steep.*

Proof. We use induction on t. If t is a variable, the result is vacuously true. Assume $t = t_1 \cdot t_2$. By definition, every subterm of a quasi-increasing term is quasi-increasing. Hence, by induction hypothesis, t_1 and t_2 are steep. Moreover, the quasi-increasing condition at ϕ in t gives $\mathrm{var}(t, 10^*) > \mathrm{var}(t, 0^*)$. Hence $t_2 \gg t_1$ holds, and Proposition 2.15(ii) (steep) applies. ∎

Steep elements enjoy the following order constraints that will be used several times in the sequel.

Lemma 2.17. *Assume that a is steep and $a > a_1, \ldots, a > a_p$ holds in FLD_∞. Then we have $a > a_1 \cdots a_p$, and $a \sqsupset a_1 \cdots a_p$ is possible only if $a \sqsupset a_i \sqsupset a_{i+1} \cdots a_p$ holds for every i.*

Proof. We use induction on $p \geq 1$. For $p = 1$, the result is obvious. Assume $p \geq 2$. If $a \gg a_1$ holds, then $a \gg a_1 a_2 \cdots a_p$ does. Assume now $a \sqsupset a_1$. By definition, $a \gg a_1^{[2]}$ holds. Now, $a \sqsupset a_1$ implies $a_1 a \sqsupset a_1^{[2]}$, and, therefore, $a \gg a_1 a$. By induction hypothesis, we have $a > a_2 \cdots a_p$, so we deduce

$$a \gg a_1 a > a_1 a_2 \cdots a_p,$$

which implies $a > a_1 a_2 \cdots a_p$. Moreover, if $a \gg a_2 \cdots a_p$ holds, we find

$$a \gg a_1 a \gg a_1 a_2 \cdots a_p,$$

hence $a \gg a_1 a_2 \cdots a_p$.

Assume now $a \sqsupset a_1 a_2 \cdots a_p$. The previous argument shows that the only possible case is $a \sqsupset a_1$ and $a \sqsupset a_2 \cdots a_p$. The relation $a_2 \cdots a_p \sqsupseteq a_1$ is impossible, as it would imply $a \sqsupset a_1 a_2 \cdots a_p \sqsupseteq a_1^{[2]}$, contradicting the steepness of a. Hence, we have $a_1 \sqsupset a_2 \cdots a_p$, and, therefore, $a \sqsupset a_2 \cdots a_p$. By induction hypothesis, this implies $a_i \sqsupset a_{i+1} \cdots a_p$ for $2 \leq i < p$. ∎

Exercise 2.18. (content) *Prove that a term (and not only its LD-equivalence class) is determined by its content. [Hint: Write $t = x_{i_1} \bullet^{p_1} x_{i_2} \bullet^{p_2} \ldots x_{i_m} \bullet^{p_m}$, and show that i_ℓ and p_ℓ can be deduced from the right variable and the right height of the ℓ-th element of the content of t inductively.]*

Exercise 2.19. (function μ_t) *(i) For t a term, and a appearing in t at α, define $\mu_t(a)$ to be the element that appears in t at $\mu_t(\alpha)$. Assume that t' is an LD-expansion of t. Prove that $\mu_{t'}(a) \sqsubseteq \mu_t(a)$ holds for every a appearing in t.*

(ii) Assume $t' = t \bullet \gamma$, c appears at $\gamma 101^*$ in t, and c_0 appears at $\gamma 0^*$ in t.

Prove $\mu_{t'}(a) = \begin{cases} \mu_t(a) & \text{for } a \text{ appearing in } t, \\ c_0 & \text{for } a = c, \\ c\mu_t(a) & \text{for } a = cb \text{ with } b \text{ appearing in } t \text{ below } \gamma 0. \end{cases}$

Exercise 2.20. (semisteep) *Say that the term t is semisteep if $\bar{t} > a^{[2]}$ holds for every a that appears in t. Prove that semisteep terms are closed under LD-equivalence.*

Exercise 2.21. (steep) *Prove the following refinement of Lemma 2.17: If a is steep, and $a > a_1$, ..., $a > a_p$ hold, then $a \sqsupset a_1 \cdots a_p$ implies $a \sqsupset a_{i_1} \cdots a_{i_k}$ for every increasing subsequence $(i_1, \ldots, i_k)$ of $(1, \ldots, p)$.*

Exercise 2.22. **(subterm)** *Assume that t is a quasi-increasing term, and that a covers b in t. Assume that s is a subterm of a term t' that is LD-equivalent to t. Prove $\overline{s} \neq ab$. [Hint: Use Lemma 2.17.]*

Exercise 2.23. **(origin)** *Assume that t is an LD-expansion of t_0, and a uncovers b in t. Prove that the last digit in the t_0-code of ab (i.e., the rightmost variable in the t_0-normal form of ab) is the same as the last digit in the t_0-code of b. [Hint: The cuts of $\partial^k t_0$ associated with a and ab have the same origin in t_0.]*

Exercise 2.24. **(square)** *(i) Say that the term t contains a square if the inequality $\overline{t} \sqsupseteq a^{[2]}$ holds for at least one a. Prove that every substitute of a term containing a square contains a square, and deduce that no steep term may be a substitute of a term containing a square.*

(ii) Assume that t, t' are LD-inequivalent terms and at least one of them is steep. Assume that (I) is an algebraic identity consisting of terms that contain a square. Prove that adding (I) to (LD) cannot make t and t' equivalent. [Hint: Assume that t is steep, and consider a possible sequence of $(LD) + (I)$-equivalent terms starting from t: each term in the sequence is steep, hence (I) can never be applied.] Deduce that the identity $x(yz) = xx$ cannot be deduced from (LD) with the additional identities $(xx)y = y(xx) = xx$.

3. Perfect Terms

In this section, we address the following question: assuming that t is a term, and a, b appear in t, does ab appear in t, or, more generally, in some LD-expansion of t? The problem is very difficult in general, but we shall establish here a complete answer for a family of particular terms called perfect terms. The general idea is that a term t is perfect if t already contains almost all products of the form ab that appear in some LD-expansion of t. The main result is that every term of the form $\partial^k t$ with t an injective term is perfect. It leads in particular to a simple algorithmic method for determining the t-normal form when t is injective.

Products of cuts

Assume that t is a term and a appears in t. A first, trivial observation is that, for a given a, the set of those b's such that ab appears in an LD-expansion of t make an initial segment in the content of t.

Lemma 3.1. *Assume that t is a term, and ab appears in some LD-expansion of t. Then $b' \sqsubseteq b$ implies that ab' appears in some LD-expansion of t.*

Proof. Assume $b' \sqsubseteq b$. Then we have $ab' \sqsubseteq ab \sqsubset \bar{t}$, hence, by Proposition 2.3 (appears *vs.* $\sqsubseteq$), ab' appears in some LD-expansion of t. ■

Let us address the above question of ab appearing in an LD-expansion of t when a and b appear in t. We shall consider three cases according to the position of b in the order of FLD_∞, and prove the following results: If b is very small, then ab always appears in some LD-expansion of t, actually in ∂t; If b is large, and t is steep, then ab never appears in any LD-expansion of t; In the third case, if t is what we shall call full, then ab appears in some LD-expansion of t if and only if it appears in t, and there exists an effective algorithm to determine if this is the case. Here we begin with the first two cases, which are easy.

Lemma 3.2. *Assume that t is a term, and a uncovers b in t. Then ab does not appear in t, but it appears in ∂t.*

Proof. Assume that a appears at α in t, and b appears at β. By hypothesis, (α, β) is a descent in t. Let γ be the address $\pi_t((\alpha, \beta))$. By Proposition VI.4.4 (decomposition), we have

$$\mathrm{cut}(\partial t, \gamma) =_{LD} \mathrm{cut}(t, \alpha) \cdot \mathrm{cut}(t, \beta),$$

and ab appears at γ in ∂t. The injectivity of the mapping π_t implies that γ cannot be written as $\pi_t((\gamma_0))$ for any address γ_0 in t. This shows that the LD-class of $\mathrm{cut}(\partial t, \gamma)$ does not appear in t. ■

Lemma 3.3. *Assume that t is a steep term, a, b appear in t, and $a \leq b$ holds. Then ab appears in no LD-expansion of t.*

Proof. As a and b appear in t, $a \leq b$ implies $a \sqsubseteq b$, and, therefore, $a^{[2]} \sqsubseteq ab$. As t is steep, we have $a^{[2]} \ll \bar{t}$. We deduce $ab \ll \bar{t}$, which forbids $ab \sqsubset \bar{t}$. ■

Observe that the previous argument works under the weaker hypothesis that a and b appear in some LD-expansion of t only. At this point, we know that the situation is as depicted in Figure 3.1.

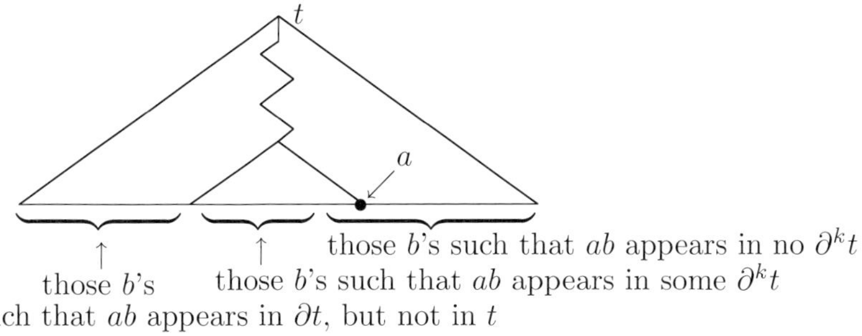

Figure 3.1. Appearing in $\partial^k t$

Full terms

We turn to the difficult case, namely when a covers b in t. Examples show that it may happen that ab does not appear in t, but appears in some LD-expansion $\partial^k t$ with $k \geq 1$ (*cf.* Exercise 3.34), and we have no control of the situation. So we shall restrict ourselves to particular terms. An optimal hypothesis is to require that ab, if it appears in some LD-expansion of t, already appears in t.

Definition 3.4. (**full**) Assume that t is a term. We say that t is *full* if ab appears in t whenever a covers b in t and $ab \sqsubset \overline{t}$ holds.

Example 3.5. (**full**) *The terms $x^{[m]}$ are full: no non-final address in the outline of $x^{[m]}$ covers any address, so the condition is vacuously true.*

Let us consider now the term $\partial x^{[m]}$. The skeleton of $\partial x^{[m]}$ is a complete binary tree of height m. By the results of Chapter VI, the elements that appear in $\partial x^{[m]}$ have the form $x^{[i_1]} \cdots x^{[i_{p+1}]}$ with $i_1 > \ldots > i_{p+1}$. Let a, b be two elements appearing in $\partial x^{[m]}$, say $a = x^{[i_1]} \cdots x^{[i_{p+1}]}$, $b = x^{[j_1]} \cdots x^{[j_{q+1}]}$. Then a covers b in $\partial x^{[m]}$ if and only if $p \leq q$ and $i_1 = j_1$, ..., $i_p = j_p$ hold. In this case, we have $ab = x^{[i_1]} \cdots x^{[i_p]} x^{[j_{p+1}]} \cdots x^{[j_{q+1}]}$, which appears in $\partial x^{[m]}$. Hence every term $\partial x^{[m]}$ is full.

Lemma 3.6. *Every cut of a full term is full.*

Proof. Assume that t is full, and t' is a cut of t. Assume that a covers b in t' and $ab \sqsubset \overline{t'}$ holds. By definition of covering, a covers b in t, and $ab \sqsubset \overline{t}$ holds by transitivity of $\sqsubset$. As t is full, ab appears in t. Now $ab \sqsubset \overline{t'}$ implies that ab appears in t on the left of $\overline{t'}$, hence it appears in t' as well. ∎

Proposition 3.7. (**substitution**) *Assume that some substitute of t is full. Then t is full as well.*

Proof. Assume that f is a substitution, and the term t^f is full. As observed in Chapter V, f induces a well defined endomorphism of FLD_∞, and we use a^f for the image of a under this endomorphism. Assume that a covers b in t, and $ab \sqsubset \bar{t}$ holds. Then a^f appears in t^f. More precisely, if a appears in t at α, then a^f appears in t^f at $\alpha 1^p$, with $p = \mathrm{ht}_R(\mathrm{var}(t,\alpha)^f)$. Similarly, b^f appears in t^f, and the previous explicit formula shows that a^f covers b^f in t^f. Moreover, the hypothesis $ab \sqsubset \bar{t}$ implies $a^f b^f \sqsubset \overline{t^f}$, as f is an endomorphism and $\sqsubset$ is definable from the product and, therefore, preserved under f.

On the one hand, $ab \sqsubset \bar{t}$ holds, hence ab appears in some term $\partial^k t$ at some address say α. Let $x_i = \mathrm{var}(\partial^k t, \alpha)$, and $p = \mathrm{ht}_R(f(x_i))$. Then, by construction, $a^f b^f$ occurs in $(\partial^k t)^f$ at $\alpha 1^p$.

On the other hand, the term t^f is full, so we deduce that $a^f b^f$ appears in t^f, at some address $\beta\gamma$ with β in the outline of t. Let $x_j = \mathrm{var}(t,\beta)$. Then $\partial^k t$ is an LD-expansion of t, hence $a^f b^f$ appears in $(\partial^k t)^f$ at the address $\beta'\gamma$, where β' is the heir of β in $\partial^k t$. The comparison of the above results implies $\alpha 1^p = \beta'\gamma$, $j = i$, $\beta' = \alpha$ and $\gamma = 1^p$. Then $\mathrm{cut}(\partial^k t, \alpha)$ is an LD-expansion of $\mathrm{cut}(t,\beta)$, *i.e.*, ab appears at β in t. Hence t is full. ■

Corollary 3.8. *Every lifting of a full term is full.*

Perfect terms

We are interested in proving fullness results for large families of terms. Example 3.5 could suggest that ∂t is full whenever t is. As shown in Exercise 3.35 below, this is not the case. An optimal technical hypothesis is to require steepness and fullness simultaneously. In some sense, these hypotheses are complementary: when $abc \sqsubset \bar{t}$ is assumed, steepness implies $a \sqsupset bc$, *i.e.*, it says that bc is small, while fullness implies (in good cases) that $a(bc)$ appears in t, hence bc is not too small.

Definition 3.9. (**perfect**) Assume that t is a term. We say that t is *perfect* if it is both steep and full.

Lemma 3.10. *Every quasi-increasing term is perfect.*

Proof. Assume that t is quasi-increasing. By Lemma 2.16, t is steep. Assume that a covers b in t. By Lemma 2.13, $ab \sqsubset \bar{t}$ does not hold, and the fullness condition is vacuously true in t. ■

Our main result here is that perfect terms are closed under operation ∂:

Proposition 3.11. (**closure**) *If t is perfect, then $\partial^k t$ is perfect for every k.*

The proof builds on several auxiliary results.

Lemma 3.12. *Assume that t is a perfect term, and we have $a \sqsubset \bar{t}$, $b \sqsubset \bar{t}$, $c \sqsubset \bar{t}$, and $abc \sqsubset \bar{t}$. Assume moreover that ab and bc appear in t. Then ac and abc appear in t as well.*

Proof. The term t is steep, hence, by Lemma 2.13, the hypothesis $abc \sqsubset \bar{t}$ implies $bc \sqsubset a$. As ab appears in t, a covers b in t, and, therefore, it covers bc as well, since the latter appears on the right of b in t. As t is full, we deduce that $a(bc)$ appears in t.

As for ac, we know that $c \sqsubset b$ holds, since bc appears in t, hence we have $ac \sqsubset ab \sqsubset \bar{t}$. Moreover, a covers b in t since ab appears in t, and, similarly, b covers c in t. By transitivity of covering, a covers c in t. As t is full, this implies that ac appears in t. ∎

The next step is to determine the content of ∂t from the content of t.

Definition 3.13. (**decomposition**) Assume that t is a term and a appears in ∂t. By Proposition VI.4.4 (decomposition), there exists a unique decreasing sequence $(a_1, \ldots, a_{p+1})$ in $\mathrm{FLD}_\infty^{p+1}$ such that $a = a_1 \cdots a_{p+1}$ holds and a_i uncovers a_{i+1} in t for every i. This unique sequence $(a_1, \ldots, a_{p+1})$ will be called the *t-decomposition of a*.

Lemma 3.14. *Assume that a covers b in ∂t, and $(a_1, \ldots, a_{p+1})$ is the t-decomposition of a. Then the t-decomposition of b has the form $(a_1, \ldots, a_p, b_{p+1}, \ldots, b_{q+1})$ with $a_{p+1} > b_{p+1}$.*

Proof. Assume that a appears at α in t, and that b appears at β. There exist two descents in t, say $(\alpha_1, \ldots, \alpha_{p+1})$ and $(\beta_1, \ldots, \beta_{q+1})$, of which α and β are the images under the mapping π_t. The explicit definition of π_t (Definition VI.4.3) shows that α covers β if and only if $p \leq q$ holds, the first p terms of $(\beta_1, \ldots, \beta_{q+1})$ coincide with the first p terms of $(\alpha_1, \ldots, \alpha_{p+1})$, and $\alpha_{p+1} >_{LR} \beta_{p+1}$ holds. Introducing a_i to be the class of $\mathrm{cut}(t, \alpha_i)$ and b_j to be the class of $\mathrm{cut}(t, \beta_j)$ gives the result. ∎

Now, the problem is as follows: We assume that t is a perfect term, and we try to prove that ∂t is perfect as well. So, we assume that a covers b in ∂t, and study whether ab appears in ∂t. By the previous lemma, we know that the t-decompositions of a and b have the form $(a_1, \ldots, a_{p+1})$ and $(a_1, \ldots, a_p, b_{p+1}, \ldots, b_{q+1})$ with $a_{p+1} > b_{p+1}$. Then we find

$$\begin{aligned} ab &= (a_1 \cdots a_p a_{p+1})(a_1 \ldots a_p b_{p+1} \ldots b_{q+1}) \\ &= a_1 \cdots a_p a_{p+1} b_{p+1} \ldots b_{q+1}. \end{aligned} \tag{3.1}$$

The latter sequence is almost the t-decomposition of an element appearing in t: if a_{p+1} uncovers b_{p+1} in t, *i.e.*, if the corresponding sequence of addresses

$(\alpha_1, \ldots, \alpha_p, \alpha_{p+1}, \beta_{p+1}, \ldots, \beta_{q+1})$ is a descent in t, then, letting γ be the image of this descent under π_t, we see that ab appears at γ in ∂t. Now, if a_{p+1} covers b_{p+1} in t, (3.1) is no longer the t-decomposition of an element appearing in ∂t. But, in this case, the hypothesis that t is perfect implies that the element $a_{p+1}b_{p+1}$ appears in t, and we can replace the sequence of (3.1) with a new sequence. This leads us to the notion of t-reduction of a sequence.

Definition 3.15. **(reduction)** Assume that $\vec{a}$, $\vec{a}'$ are finite nonempty sequences of elements of FLD_∞ appearing in the term t. We say that $\vec{a}$ is t-*reducible* to $\vec{a}'$ in one step if $\vec{a}$ and $\vec{a}'$ either have the form

$$\begin{cases} \vec{a} = (a_1, \ldots, a_{i-1}, a_i, a_{i+1}, a_{i+2}, \ldots, a_{r+1}), \\ \vec{a}' = (a_1, \ldots, a_{i-1}, a_i a_{i+1}, a_i, a_{i+2}, \ldots, a_{r+1}), \end{cases}$$

for some $i < r$, or they have the form

$$\begin{cases} \vec{a} = (a_1, \ldots, a_{i-1}, a_r, a_{r+1}), \\ \vec{a}' = (a_1, \ldots, a_{i-1}, a_r a_{r+1}). \end{cases}$$

We say that $\vec{a}$ is t-reducible to $\vec{a}'$ in n steps if there exists a length $n+1$ sequence from $\vec{a}$ to $\vec{a}'$ such that every component is t-reducible to the next one in one step, and that $\vec{a}$ is t-irreducible if it is t-reducible in one step to no sequence.

The idea of reduction is to use the product of FLD_∞ to replace the covering patterns, and to iterate the process until no more covering remains, if this is possible.

Example 3.16. **(reduction)** *Let t be the term $\partial(x_1{\cdot}x_2{\cdot}x_3{\cdot}x_4)$. Then t has 7 cuts, corresponding to the scheme*

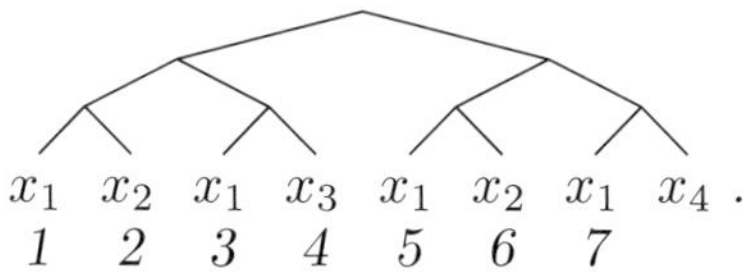

Consider $\vec{a} = (7, 4, 2, 1)$, where the digits stand for the classes of the corresponding cuts. Then $\vec{a}$ is not t-irreducible: $\mathrm{cut}_7(t){\cdot}\mathrm{cut}_4(t)$ does not appear in t, but we have $\mathrm{cut}_4(t){\cdot}\mathrm{cut}_2(t) =_{LD} \mathrm{cut}_6(t)$. Thus, $\vec{a}$ is t-reducible in one step to $(7, 6, 4, 1)$. The latter sequence in turn is not t-irreducible, as we have $\mathrm{cut}_4(t){\cdot}\mathrm{cut}_1(t) =_{LD} \mathrm{cut}_5(t)$. Thus, $\vec{a}$ is t-reducible in two steps to $(7, 6, 5)$, and, similarly, it is t-reducible in 3 steps to $(7, 7)$. The latter sequence is t-irreducible.

Several verifications are in order. For $\vec{a}$ a sequence as above, say $\vec{a} = (a_1, \ldots, a_r)$, we write $\prod \vec{a}$ for $a_1 \cdots a_r$.

Lemma 3.17. *Let t be an arbitrary term.*

(i) Assume that $\vec{a}$ is t-reducible to $\vec{a}'$. Then we have $\prod \vec{a}' = \prod \vec{a}$.

(ii) Every sequence of t-reductions converges to a t-irreducible sequence in a finite number of steps.

Proof. Point (i) follows from the definition of reduction. For (ii), we observe that, if $\vec{a}$ is t-reducible to $\vec{a}'$, then $\vec{a}'$ is strictly larger than $\vec{a}$ in the lexicographic extension of $\sqsubset$ to sequences from FLD_∞. The length of the sequences is preserved or it decreases under t-reduction, and only finitely many elements appear in the term t, hence any sequence of t-reductions must result in a t-irreducible sequence after a finite number of steps (with an explicit bound). ■

With our definition, t-reduction is not a functional process: a given sequence may be t-reducible to several distinct sequences. However, in good cases, the choice of the t-reduction steps does not matter.

Lemma 3.18. *Assume that t is a perfect term, and $\vec{a}$ is a sequence of elements appearing in t and satisfying $\prod \vec{a} \sqsubset \bar{t}$. Then t-reduction from $\vec{a}$ is confluent, i.e., if the sequence $\vec{a}$ is t-reducible both to $\vec{a}_1$ and $\vec{a}_2$, there exists a sequence $\vec{a}'$ such that both $\vec{a}_1$ and $\vec{a}_2$ are t-reducible to $\vec{a}'$.*

Proof. As t-reduction is Noetherian, *i.e.*, it has no infinite descending sequence, it suffices to prove that it is locally confluent, *i.e.*, to show that, if $\vec{a}$ is t-reducible to $\vec{a}_0$, and $\vec{a}_0$ is t-reducible in one step to $\vec{a}_1$ and $\vec{a}_2$, then there exists a sequence $\vec{a}'$ such that both $\vec{a}_1$ and $\vec{a}_2$ are t-reducible to $\vec{a}'$. Assume that $\vec{a}_0$ is $(a_1, \ldots, a_{r+1})$, and $\vec{a}_1$ and $\vec{a}_2$ are obtained from $\vec{a}$ using reduction at positions i and j respectively. For $|i-j| \geq 2$, the reductions involve disjoint fragments of the sequences, and we can define $\vec{a}'$ to be the sequence obtained from $\vec{a}_0$ by reducing both at i and at j. Thus the point is the case $|i-j| = 1$. Assume for instance $i+1 = j < r$. Then we have

$$\vec{a}_1 = (\ldots\ ,\ a_{i-1}\ ,\ a_i a_{i+1}\ ,\ a_i\ ,\ a_{i+2}\ ,\ a_{i+3}\ ,\ \ldots),$$
$$\vec{a}_2 = (\ldots\ ,\ a_{i-1}\ ,\ a_i\ ,\ a_{i+1} a_{i+2}\ ,\ a_{i+1}\ ,\ a_{i+3}\ ,\ \ldots),$$

By Lemma 3.17, we have $\prod \vec{a}_0 = \prod \vec{a}$, and, therefore, $\prod \vec{a}_0 \sqsubset \bar{t}$. By Lemma 2.17, we deduce $a_i a_{i+1} a_{i+2} \cdots \sqsubset \bar{t}$, and, therefore, $a_i a_{i+1} a_{i+2} \sqsubset \bar{t}$. Applying Lemma 3.12, we see that $a_i a_{i+2}$ and $a_i a_{i+1} a_{i+2}$ appear in t as well. This implies that the sequences $\vec{a}_1$ and $\vec{a}_2$ both are t-reducible to the sequence

$$\vec{a}' = (\ldots\ ,\ a_{i-1}\ ,\ a_i a_{i+1} a_{i+2}\ ,\ a_i a_{i+1}\ ,\ a_i\ ,\ a_{i+3}\ ,\ \ldots).$$

The case $i+1 = j = r$ is similar, with the terminal sequence

$$\vec{a}' = (\ldots\ ,\ a_{i-1}\ ,\ a_i a_{i+1} a_{i+2}).$$ ■

Proposition 3.19. **(reduction)** *Assume that t is a perfect term, and $\vec{a}$ is a sequence of elements appearing in t and satisfying $\prod \vec{a} \sqsubset \bar{t}$. Then t-reduction from $\vec{a}$ leads to a unique t-irreducible sequence, which is the t-decomposition of an element of ∂t. In particular, $\prod \vec{a}$ appears in ∂t.*

Proof. Consider an arbitrary sequence of t-reductions from $\vec{a}$. By Lemma 3.17, it must have a finite length and end with some t-irreducible sequence say $\vec{a}_*$. Assume that $\vec{a}_0$ is an arbitrary sequence that $\vec{a}$ is t-reducible to. By Lemma 3.18, $\vec{a}_*$ and $\vec{a}_0$ have to be t-reducible to some sequence $\vec{a}'_*$, and the hypothesis that $\vec{a}_*$ is t-irreducible implies that $\vec{a}'_*$ coincides with $\vec{a}_*$, *i.e.*, $\vec{a}_0$ is t-reducible to $\vec{a}_*$. Hence $\vec{a}_*$ is the unique t-irreducible sequence accessible from $\vec{a}$. Assume $\vec{a}_* = (a_1, \dots, a_{r+1})$. By Lemma 2.17, the hypothesis $\prod \vec{a} \sqsubset \bar{t}$ forces $\vec{a}$ to be decreasing. Assume that a_i covers a_{i+1} for some i. Then, the hypothesis that $a_1 \cdots a_i a_{i+1} \sqsubset \bar{t}$ holds and that t is perfect implies that $a_i a_{i+1}$ appears in t, which contradicts the hypothesis that $\vec{a}_*$ is t-irreducible. Hence we conclude that a_i uncovers a_{i+1} in t for every i: this means that $\vec{a}_*$ is the t-decomposition of an element of ∂t. ∎

In the above framework, we shall denote by $\mathrm{red}_t(\vec{a})$ the unique t-irreducible sequence obtained from $\vec{a}$ using t-reduction. We can now describe the elements ab that appear in the term ∂t completely.

Lemma 3.20. *Assume that t is a perfect term, that a covers b in ∂t, and $ab \sqsubset \bar{t}$ holds. Then the element ab appears in ∂t, and, letting $(a_1, \dots, a_{p+1})$ and $(b_1, \dots, b_{q+1})$ be respectively the t-decompositions of a and b, the t-decomposition of ab is* $\mathrm{red}_t(a_1, \dots, a_{p+1}, b_{p+1}, \dots, b_{q+1})$.

Proof. By Lemma 3.14, we know that the t-decompositions of a and b have the form $(a_1, \dots, a_p, a_{p+1})$ and $(a_1, \dots, a_p, b_{p+1}, \dots, b_{q+1})$ respectively. Let $\vec{c}$ be the sequence $(a_1, \dots, a_{p+1}, b_{p+1}, \dots, b_{q+1})$. By construction, the elements of $\vec{c}$ appear in t, and we have $\prod \vec{c} = ab \sqsubset \bar{t}$. By Proposition 3.19 (reduction), the sequence $\mathrm{red}_t(\vec{c})$ is the t-decomposition of an element of ∂t. This means that ab appears in ∂t, and $\mathrm{red}_t(\vec{c})$ is its t-decomposition. ∎

We can now easily complete the proof of Proposition 3.11 (closure):

Proof. For an induction, it suffices to assume that t is perfect, and to prove that ∂t is perfect as well. Now, the term ∂t is steep, as it is LD-equivalent to t, and Lemma 3.20 tells us that ∂t is full. ∎

Clearly, the order of the variables involved in the definition of increasing and steep terms is unessential, and we can restate the results in a slightly more general form by allowing permutations of the variables.

Definition 3.21. (**quasi-injective**) For t in T_∞, we say that t is *quasi-injective* if $t = t_0^f$ holds for some quasi-increasing term t_0 and some permutation f of $\{x_1, x_2, \ldots\}$.

Thus a quasi-injective term is a quasi-increasing term up to a permutation of the variables. In particular, every injective term is quasi-injective.

Proposition 3.22. (**full**) *If t is quasi-injective, then $\partial^k t$ is full for every k.*

Proof. Assume $t = t_0^f$, where t_0 is quasi-increasing and f is a permutation of the variables. By Lemma 2.16, t_0 is steep, and, by Lemma 3.10, it is perfect. Then Proposition 3.11 (closure) implies that the same holds for every term $\partial^k t_0$, which is $(\partial^k t)^f$. Now, by construction, fullness is preserved under a permutation of the variables. Indeed, assume that s is perfect, and f is a permutation of the variables. Still using f to denote the automorphism of FLD_∞ induced by f, we see that a appears in s if and only if $f(a)$ appears in s^f, and, similarly, a covers b in s if and only if $f(a)$ covers $f(b)$ in s^f. ∎

The table of a term

If t is a term in T_∞, then, by the results of Chapter VI, the elements of FLD_∞ appearing in some LD-expansion of t are specified by t-codes. Describing the pairs (a, b) of elements appearing in some LD-expansion of t such that ab still appears in some LD-expansion of t is equivalent to finding, for every pair $(\boldsymbol{c}_1, \boldsymbol{c}_2)$ of t-codes, the unique t-code $\boldsymbol{c}$ satisfying $\mathrm{cut}_{\boldsymbol{c}}(t) =_{LD} \mathrm{cut}_{\boldsymbol{c}_1}(t)\cdot\mathrm{cut}_{\boldsymbol{c}_2}(t)$, if such a code exists. This leads us to the following notion:

Definition 3.23. (**table**) Assume that t is a term. The *k-table* of t is defined to be the pair $(C_t^k, *_t^k)$, where C_t^k is the set of all t-codes of degree $\leq k$ except the last one, and $*_t^k$ is the partial binary operation on C_t^k defined by

$$\boldsymbol{c}_1 *_t^k \boldsymbol{c}_2 = \begin{cases} \boldsymbol{c} & \text{if } \mathrm{cut}_{\boldsymbol{c}_1}(t)\cdot\mathrm{cut}_{\boldsymbol{c}_2}(t) =_{LD} \mathrm{cut}_{\boldsymbol{c}}(t) \text{ holds,} \\ \bot & \text{if } \mathrm{cut}_{\boldsymbol{c}_1}(t)\cdot\mathrm{cut}_{\boldsymbol{c}_2}(t) \text{ appears in no LD-expansion of } t. \end{cases}$$

The table of t is the union of all k-tables of t.

Thus $\boldsymbol{c}_1 *_t^k \boldsymbol{c}_2$ is undefined if $\mathrm{cut}_{\boldsymbol{c}_1}(t)\cdot\mathrm{cut}_{\boldsymbol{c}_2}(t)$ appears in some LD-expansion of t, but not in t: the value is not yet known. On the other hand, $\bot$ indicates a permanent obstruction. Observe that, by definition, every k-table is a partial left self-distributive table, in the sense that, if the values of $a * (b * c)$ and $(a * b) * (a * c)$ both exist, then they are equal.

Example 3.24. (**table**) *Let $t = x^{[4]}$. Three elements appear in t, namely x, $x^{[2]}$, and $x^{[3]}$. Thus C_t^0 consists of the three codes 1, 2, 3. Similarly, seven ($= 2^3 - 1$) elements appear in ∂t, and we have $C_t^1 = \{1, 2, 2.1, 3, 3.1, 3.2, 3.21\}$. The reader is invited to check that the operations $*_t^0$ and $*_t^1$ are given by the arrays*

$*_t^0$	1	2	3
1	2	3	$\bot$
2	.	3	$\bot$
3	.	.	$\bot$

$*_t^1$	1	2	2.1	3	3.1	3.2	3.21
1	2	3	3.2	$\bot$	$\bot$	$\bot$	$\bot$
2	2.1	3	$\bot$	$\bot$	$\bot$	$\bot$	$\bot$
2.1	.	.	3	$\bot$	$\bot$	$\bot$	$\bot$
3	3.1	3.2	3.21	$\bot$	$\bot$	$\bot$	$\bot$
3.1	.	.	.	.	3.2	$\bot$	$\bot$
3.2	.	.	.	.	3.21	$\bot$	$\bot$
3.21	.	.	.	.	.	.	$\bot$

The above results enable us to completely describe the table of a quasi-injective term. In the sequel, we do not distinguish between t-normal terms and their codes. So we can speak of the product of two codes, or of the decomposition of a degree k code as a sequence of degree $k-1$ codes. Also, for $*$ a partial operation on C, we define $*$-reduction for sequences of elements of C using the same scheme as above for terms: here the basic step consists in replacing $(\boldsymbol{c}_1, \ldots, \boldsymbol{c}_r)$ with $(\boldsymbol{c}_1, \ldots, \boldsymbol{c}_{i-1}, \boldsymbol{c}_i * \boldsymbol{c}_{i+1}, \boldsymbol{c}_i, \boldsymbol{c}_{i+2}, \ldots, \boldsymbol{c}_r)$ when $\boldsymbol{c}_i * \boldsymbol{c}_{i+1}$ is defined.

Proposition 3.25. (**table**) *Assume that t is a quasi-injective term of size n.*

(i) The set C_t^0 is $\{1, \ldots, n-1\}$; the operation $*_t^0$ is defined by $\boldsymbol{c}_1 *_t^0 \boldsymbol{c}_2 = \bot$ for $\mathrm{cut}_{\boldsymbol{c}_1}(t)$ covering $\mathrm{cut}_{\boldsymbol{c}_2}(t)$ in t, and undefined otherwise.

(ii) For $k > 0$, the k-table of t is obtained from the $(k-1)$-table as follows:
(#) The set C_t^k is the set of all codes $\boldsymbol{c}_1 \cdots \boldsymbol{c}_{p+1}$ such that $\boldsymbol{c}_i$ belongs to C_t^{k-1} and $\boldsymbol{c}_i *_t^{k-1} \boldsymbol{c}_{i+1}$ is undefined for every i;
(##) For $\boldsymbol{c}_1$, $\boldsymbol{c}_2$ in C_t^k satisfying $\boldsymbol{c}_1 \le \boldsymbol{c}_2$, we have $\boldsymbol{c}_1 *_t^k \boldsymbol{c}_2 = \bot$;
(###) For $\boldsymbol{c}_1$, $\boldsymbol{c}_2$ in C_t^k satisfying $\boldsymbol{c}_1 > \boldsymbol{c}_2$ and of the form

$$\boldsymbol{c}_1 = \boldsymbol{c}'_1 \cdots \boldsymbol{c}'_p \boldsymbol{c}'_{p+1}, \qquad \boldsymbol{c}_2 = \boldsymbol{c}'_1 \cdots \boldsymbol{c}'_p \boldsymbol{c}''_{p+1} \cdots \boldsymbol{c}''_{q+1}$$

with $\boldsymbol{c}'_{p+1} > \boldsymbol{c}''_{p+1}$, we have $\boldsymbol{c}_1 *_t^k \boldsymbol{c}_2 = \boldsymbol{c}'''_1 \cdots \boldsymbol{c}'''_r$ where $(\boldsymbol{c}'''_1, \ldots, \boldsymbol{c}'''_r)$ is the irreducible sequence obtained from $(\boldsymbol{c}'_1, \ldots, \boldsymbol{c}'_{p+1}, \boldsymbol{c}''_{p+1}, \ldots, \boldsymbol{c}''_{q+1})$ by using $*_t^{k-1}$-reduction, if this code belongs to C_t^k, and we have $\boldsymbol{c}_1 *_t^k \boldsymbol{c}_2 = \bot$ otherwise;
(####) For $\boldsymbol{c}_1$, $\boldsymbol{c}_2$ in C_t^k satisfying $\boldsymbol{c}_1 > \boldsymbol{c}_2$ but not of the form (###), $\boldsymbol{c}_1 *_t^k \boldsymbol{c}_2$ is undefined.

Proof. Without loss of generality, we may assume that t is quasi-increasing. Using induction on k, we prove the result of the proposition together with the additional fact that $\boldsymbol{c}_1 *_t^k \boldsymbol{c}_2$ being undefined means that $\mathrm{cut}_{\boldsymbol{c}_1}(t)$ uncovers $\mathrm{cut}_{\boldsymbol{c}_2}(t)$ in $\partial^k t$. For $k = 0$, everything is true as, by definition, no cut of a quasi-injective term may be LD-equivalent to the product of two other cuts. The induction then follows from Lemma 2.13, which, in addition to proving that $\partial^k t$ is perfect, also gives the effective computation method by a reduction. ∎

Example 3.26. **(table)** *Let t be $x_1{\cdot}x_2{\cdot}x_3{\cdot}x_4$. Then t is increasing, hence eligible for the previous result—while its substitute $x^{[4]}$ is not. The 0-table and the 1-table of t are*

$*_t^0$	1	2	3
1	$\bot$	$\bot$	$\bot$
2	.	$\bot$	$\bot$
3	.	.	$\bot$

$*_t^1$	1	2	2.1	3	3.1	3.2	3.21
1	$\bot$	$\bot$	$\bot$	$\bot$	$\bot$	$\bot$	$\bot$
2	2.1	$\bot$	$\bot$	$\bot$	$\bot$	$\bot$	$\bot$
2.1	.	.	$\bot$	$\bot$	$\bot$	$\bot$	$\bot$
3	3.1	3.2	3.21	$\bot$	$\bot$	$\bot$	$\bot$
3.1	.	.	.	.	$\bot$	$\bot$	$\bot$
3.2	.	.	.	.	3.21	$\bot$	$\bot$
3.21	.	.	.	.	.	.	$\bot$

(No 0-reduction is possible.) Forty-one elements appear in $\partial^2 t$, and writing the complete 2-table of t is tedious. We give below the trace of the 2-table of t on C_t^1, i.e., we fill the empty entries of the 1-table:

$*_t^2$ *(excerpt)*	1	2	2.1	3	3.1	3.2	3.21
1	$\bot$	$\bot$	$\bot$	$\bot$	$\bot$	$\bot$	$\bot$
2	2.1	$\bot$	$\bot$	$\bot$	$\bot$	$\bot$	$\bot$
2.1	2.101	2.102	$\bot$	$\bot$	$\bot$	$\bot$	$\bot$
3	3.1	3.2	3.21	$\bot$	$\bot$	$\bot$	$\bot$
3.1	3.101	3.102	3.1021	3.103	$\bot$	$\bot$	$\bot$
3.2	3.201	3.202	3.2021	3.203	3.21	$\bot$	$\bot$
3.21	3.2101	3.2102	3.21021	3.2103	3.21031	3.21032	$\bot$

Actually, the part above is the trivial part of the 2-table. More interesting features appear when 1-reduction is possible. Let us consider one example. As $3.2 *_t^1 3$ is undefined, 3.203 belongs to C_t^2, and so does 3.201 for the same reason. Let us compute $3.203 *_t^2 3.201$. We have $3.203 > 3.201$, so we are in case (###) or (####). The 1-decompositions are $(3.2, 3)$ and $(3.2, 1)$ respectively, so we are in case (###), and we have to reduce the sequence $(3.2, 3, 1)$. We see in the table that $3.2 *_t^1 3$ is not defined, but we find $3 *_t^1 1 = 3.1$. Hence $(3.2, 3, 1)$ reduces to $(3.2, 3.1)$. Now, we find $3.2 *_t^1 3.1 = 3.21$. So $(3.2, 3.1)$ reduces to (3.21). The code 3.21 belongs to C_t^2, and we deduce $3.203 *_t^2 3.201 = 3.21$.

The reader can check other cases, for instance $3.203 *_t^2 3.2021 = \bot$.

Computation of normal form

Computing the table of a term t_0 is connected with determining the t_0-normal form. Indeed, assume that t_1 and t_2 are terms and the t_0-normal form of t_1 and t_2 exists, *i.e.*, we have $t_1 \sqsubseteq_{LD} t_0$ and $t_2 \sqsubseteq_{LD} t_0$. Let $\boldsymbol{c}_i$ be the t_0-code of t_e for $e = 1, 2$. Let us consider $t = t_1 \cdot t_2$. Finding the t_0-normal form of t, if it exists, amounts to computing $\boldsymbol{c}_1 *_{t_0} \boldsymbol{c}_2$. If t_0 is quasi-injective, this can be done using Proposition 3.25 (table). The following observation allows us to possibly use the method even if the base term t_0 is not quasi-injective.

Lemma 3.27. *Assume that t'_0 is a lifting of t_0, say $t_0 = (t'_0)^f$.*

(i) Assume that the t'_0-normal form of t exists. Then the t_0-normal form of t^f exists as well, and the t_0-code of t^f is the t'_0-code of t.

(ii) The table of t'_0 is included in the table of t_0.

Proof. (i) For each k, we have $\partial^k t = (\partial^k t'_0)^f$, and the outlines of the term $\partial^k t$ and $\partial^k t'_0$ are equal, so, in particular, they have the same descents. The construction of the normal form involves the geometry of the terms $\partial^k t$ only, whatever the names of the variables are, so an induction shows that, for every k and every cut t of $\partial^k t'_0$, the t_0-normal form of the cut t^f of $\partial^k t_0$ is the image under f of the t'_0-normal form of s, *i.e.*, it is obtained from the t'_0-normal form of t by replacing $\mathrm{cut}_{\boldsymbol{c}}(t'_0)$ with $\mathrm{cut}_{\boldsymbol{c}}(t_0)$ everywhere. Hence the codes are the same.

(ii) The relations $\boldsymbol{c}_1 *_{t'_0} \boldsymbol{c}_2 = \boldsymbol{c}$ and $\mathrm{cut}_{\boldsymbol{c}_1}(t'_0) \cdot \mathrm{cut}_{\boldsymbol{c}_2}(t'_0) =_{LD} \mathrm{cut}_{\boldsymbol{c}}(t'_0)$ are equivalent. The latter implies $\mathrm{cut}_{\boldsymbol{c}_1}(t_0) \cdot \mathrm{cut}_{\boldsymbol{c}_2}(t_0) =_{LD} \mathrm{cut}_{\boldsymbol{c}}(t_0)$, which is in turn equivalent to $\boldsymbol{c}_1 \ *_{t_0} \ \boldsymbol{c}_2 = \boldsymbol{c}$. ∎

Definition 3.28. (precoding) Assume that t_0 is an extended term, and t is a term. The *t_0-precode of* t is the precode $\boldsymbol{c}$ possibly defined as follows: for t a cut of t_0, say $t = \mathrm{cut}_i(t)$, then $\boldsymbol{c}$ is defined to be i; otherwise, $\boldsymbol{c}$ is defined to be the product of the precodes of the left and the right subterms of t, if they exist.

Because of possible variable clashes, the t_0-precode of a term need not exist. On the other hand, for $t \in T_1$, the x^∞-precode of t always exists.

By definition, a t_0-precode $\boldsymbol{c}$ is a term (whose variables are integers). If $*$ is a partial binary operation defined on precodes, we define the evaluation of $\boldsymbol{c}$ in the table of $*$ inductively in the usual way: for $\boldsymbol{c}$ an integer, the evaluation of $\boldsymbol{c}$ is $\boldsymbol{c}$ itself; for $\boldsymbol{c} = \boldsymbol{c}_1\boldsymbol{c}_2$, the evaluation of $\boldsymbol{c}$ is $\boldsymbol{c}'_1 * \boldsymbol{c}'_2$, where $\boldsymbol{c}'_e$ is the evaluation of $\boldsymbol{c}_e$, if it exists.

Proposition 3.29. (computation of normal form) *Assume that t_0 is an extended term, and t is a term satisfying $t \sqsubseteq_{LD} t_0$. Let t'_0 be a quasi-injective lifting of t_0. Then the t_0-code of t is the evaluation of the t_0-precode of t in the table of t'_0, when this evaluation is defined.*

Proof. Let $\boldsymbol{c}'$ be the t_0-precode of t, and t_0' be a quasi-injective lifting of t_0 such that $\boldsymbol{c}'$ evaluates to $\boldsymbol{c}$ in the table of t_0'. By definition, we have $\mathrm{cut}_{\boldsymbol{c}}(t_0') =_{LD} \mathrm{cut}_{\boldsymbol{c}'}(t_0')$, hence $\mathrm{cut}_{\boldsymbol{c}}(t_0) =_{LD} \mathrm{cut}_{\boldsymbol{c}'}(t_0) = t$. By construction, $\boldsymbol{c}$ is a t_0'-code, and, by Lemma 3.27, it is also a t_0-code. By uniqueness, this code must be the t_0-code of t. ■

Example 3.30. (computation of normal form) *Let t be the term*

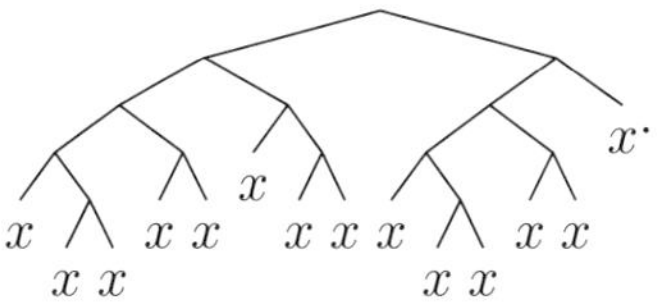

The $x^{[4]}$-precode of t is $((3{\cdot}2){\cdot}3){\cdot}((3{\cdot}2){\cdot}1))$—also denoted 3.20303201. The term $x_1{\cdot}x_2{\cdot}x_3{\cdot}x_4$ is an injective lifting of $x^{[4]}$. Let us look for the evaluation of the above precode in the table of $x_1{\cdot}x_2{\cdot}x_3{\cdot}x_4$. Using the values computed in Example 3.26, we find

$$((3 * 2) * 3) * ((3 * 2) * 1)) = (3.2 * 3) * (3.2 * 1) = 3.203 * 3.201 = 3.21.$$

The computation has succeeded, so we deduce that 3.21 is the $x^{[4]}$-code of t, i.e., that $(x{\cdot}x{\cdot}x){\cdot}(x{\cdot}x){\cdot}x$ is its $x^{[4]}$-normal form.

The previous result holds for every lifting of the base term t_0, including t_0 itself: if the t_0-precode of t can be evaluabled in the t_0-table, then we obtain the t_0-normal form of t directly. The point is that, if t_0 is not quasi-injective, we have no way for computing the t_0-table practically—although, theoretically, all this is doable by computing the terms $\partial^k t_0$ explicitly. For instance, no complete description of the 2-table of x^{∞} is known, and Exercise 3.35 suggests that such a description may be very complicated.

Exercise 3.31. (reduction) *Let $t = \partial(x_1{\cdot}x_2{\cdot}x_3{\cdot}x_4)$. Prove that 7.4 covers 7.201 in ∂t, and check the value $7.4 *_t^1 7.201 = 7.64007601$.*

Exercise 3.32. (not full) *Let $u = 0 \cdot \phi \cdot 1000$. Show that u is progressive. Check that t_u^L is $(x_1{\cdot}x_2{\cdot}x_3){\cdot}((x_4{\cdot}x_5){\cdot}(x_6{\cdot}x_7))$, and that t_u^R is not full. [So a progressive LD-expansion of an increasing term need not be full.]*

Exercise 3.33. **(covering)** *Assume that t' is a progressive LD-expansion of t, and that a covers b and b covers c in t', while a uncovers b and b uncovers c in t. Prove that, in t', ab, ac, bc and abc appear, a covers bc, and ab covers ac. Is the result true if t' is an LD-expansion, but not a progressive one?*

Exercise 3.34. **(counter-examples)** *(i) Let $t = x_1^{[2]}{\cdot}x_2{\cdot}x_1{\cdot}x_3$. Prove that t is full. Let $a = x_1^{[2]}$ and $b = x_1$. Check that a covers b in t, ab does not appear in t, but $(x_1^{[2]}x_2)ab$ appears in the LD-expansion $t \bullet \phi$ of t.*

(ii) Let $t = ((x_1{\cdot}x_2){\cdot}x_1{\cdot}x_3){\cdot}x_1{\cdot}x_2{\cdot}x_4$. Check that t is steep. Let a be the class of $\mathrm{cut}_4(t)$, and b be the class of $\mathrm{cut}_3(t)$. Prove that a covers b in t, and ab does not appear in t, but it appears in ∂t.

Exercise 3.35. **(not full)** *(i) Let $t = \partial x^{[4]}$. Draw a picture of ∂t, and, for each address in the outline of ∂t, write the descent of t it comes from.*

(ii) Show that 7.54 covers 7.531 in ∂t, and therefore, in $\partial x^{[6]}$ as well.

(iii) Prove the following equivalences between the cuts of $\partial x^{[6]}$: $4.31 =_{LD} 7.5$, $5.75 =_{LD} [14].6$ (we write $[14]$ for the code of the 14-th cut of $\partial x^{[6]}$), $7.[14]6 =_{LD} 7.[14]076$, $7.[14] =_{LD} [23].[22]$, and, finally, $(7.54){\cdot}(7.531) =_{LD} [23].[22]076$. Observe that the latter term is $\partial x^{[6]}$-normal of degree 2, and conclude that $\partial^2 x^{[6]}$ is not full.

Exercise 3.36. **(partial LD-system)** *(i) Define on the set S of all finite nonempty decreasing sequences of positive integers a partial binary operation as follows: $(a_1, \ldots, a_{p+1})(b_1, \ldots, b_{q+1})$ exists if and only if we have $q \geq p$ and $b_i = a_i$ for $i \leq p$, and, in this case, it is equal to $(a_1, \ldots, a_p, a_{p+1}, b_{q+1}, \ldots, b_{q+1})$. Prove that S is a partial LD-system in the sense that either abc and $(ab)(ac)$ both exist and they are equal, or neither exists. Establish a connection with the partial LD-system associated with the table of the term $x_1x_2x_3\cdots$.*

4. Convergence Results

Here we use the fullness results of the previous section to prove partial convergence results for the Polish Algorithm. We deduce various properties of progressive LD-expansions, and generalize some of the results obtained in Section VIII.5 for simple elements to the case of arbitrary elements of M_{LD}.

The Polish Algorithm below a full term

The main convergence result we are going to prove here is as follows:

Proposition 4.1. (convergence) *(i) Assume that the terms t_1 and t_2 are LD-expansions of $\sqsubseteq$-comparable terms. Then the Polish Algorithm running on (t_1, t_2) converges.*

(ii) Let t_1, t_2 be arbitrary terms. Then, for k large enough, the Polish Algorithm running on $(\partial^k t_1, t_2)$ and on $(\partial^k t_1, \partial^k t_2)$ converges.

The main ingredient of the proof is the convergence of the Polish Algorithm when the initial terms are included in some full term, in the following sense:

Definition 4.2. (included) For t, t_0 terms, we say that t is *included* in t_0 if the content of t is included in the content of t_0, *i.e.*, if every element that appears in t also appears in t_0, and $\overline{t}$ itself appears in t_0, or $t =_{LD} t_0$ holds.

Then we have:

Proposition 4.3. (Polish for full) *Assume that t_0 is a full term of size N, and $t_1, \ldots, t_m$ are included in t_0. Then the Polish Algorithm running on $(t_1, \ldots, t_m)$ converges in at most N steps, and the final terms still are included in t_0.*

Let us begin with preliminary results.

Lemma 4.4. *Assume that t is included in t_0, and that a covers b in t. Then a covers b in t_0 as well.*

Proof. Let a_0 be the least element covered by a in t. Let a^+ be the least element that appears on the right of a in t, or $\overline{t}$ if a is the largest element appearing in t. By hypothesis, a_0 appears in t_0, and a^+ appears in t_0 or it is $\overline{t_0}$. By Lemma 2.10, the elements a^+ and aa_0 are almost equal, *i.e.*, they may differ only by the rightmost variable. In particular, their projections $(a^+)^\dagger$ and $(aa_0)^\dagger$ on FLD_1 are equal. Let b_0 be least element covered by a in t_0, and b^+ be the least element that appears on the right of a in t_0, or $\overline{t_0}$ if a is the largest element appearing in t_0. We have similarly $(b^+)^\dagger = (ab_0)^\dagger$, and, by construction, $b^+ \sqsubseteq a^+$ holds. Hence, we find,

$$a^\dagger b_0^\dagger = (b^+)^\dagger \sqsubseteq (a^+)^\dagger = a^\dagger a_0^\dagger,$$

which implies $b_0^\dagger \sqsubseteq a_0^\dagger$, and, therefore, $b_0 \sqsubseteq a_0$, since, otherwise, we would get $a_0 \sqsubset b_0$ and $a_0^\dagger \sqsubset b_0^\dagger$. Now b satisfies $a_0 \leq b < a$, hence it satisfies $b_0 \leq b < a$ as well, and, therefore, it is covered by a in t_0. ■

Lemma 4.5. *Assume that t_0 is a full term, t_1, t_2 are included in t_0, and the Polish Algorithm running on (t_1, t_2) returns (t'_1, t'_2) in one step. Then t'_1 and t'_2 are included in t_0.*

Proof. (Figure 4.1) By hypothesis, t_1 and t_2 have a soft clash. Let us assume that the clash occurs at position p, with a variable in t_1 and a letter $\bullet$ in t_2. Let $\gamma 10^m 1^*$ be the address of p in t_1, and let $a_1, \ldots, a_m$ be the elements that appear in t_1 respectively at the addresses $\gamma 10^m 1^*$, $\gamma 10^{m-1} 1^*$, $\ldots$, $\gamma 101^*$. Let a_0 be the element that appears at $\gamma 10^*$, and, finally, let $b_1, \ldots, b_q$ be the increasing enumeration of those elements that appear in t_1 below $\gamma 0$. With our hypothesis, we have $t'_2 = t_2$ and $t'_1 = t_1 \bullet \gamma_{(m)}$. Hence, the elements that appear in t'_1 are those elements that appear in t_1, plus new elements, namely the elements $a_i b_j$ with $1 \leq i \leq m$ and $1 \leq j \leq q$. Our aim is to prove that these elements appear in t_0.

To this end, we use the covering relations in t_0, and the hypothesis that t_0 is full. First, for $j = 1, \ldots, q-1$, b_q covers b_j in t_1, so, by Lemma 4.4, b_q covers b_j in t_0 as well. Similarly, for $i = 2, \ldots, m$, a_i covers a_1 in t_1, so a_i covers a_1 in t_0 as well. Now, a_1 covers a_0 in t_1, and, by hypothesis, the terms t_1 and t_2 coincide up to a_1, with the exception that a_1 covers more elements in t_2 than in t_1. As the immediate predecessor of a_0 in t_1 and t_2 is b_q, a_1 must cover b_q in t_2. This implies that a_1 covers b_q in t_0. The covering relation being transitive, we conclude that, for every $i = 1, \ldots, m$ and every $j = 1, \ldots, q$, a_i covers b_j in t_0. Now, the elements $a_i b_j$ appear in t'_1, which is an LD-expansion of t_1, so we must have $a_i b_j \sqsubset \overline{t_1}$, and, therefore $a_i b_j \sqsubset \overline{t_0}$. Hence $a_i b_j$ appears in some LD-expansion of t_0. As a_i covers b_j in t_0, and t_0 is full, the hypothesis of $a_i b_j$ appearing in some LD-expansion of t_0 implies $a_i b_j$ already appearing in t_0, and we are done. ■

We can now prove Proposition 4.3 (Polish for full).

Proof. By the previous lemma (extended from pairs to arbitrary sequences of terms for $n > 2$, which is straightforward), all terms involved in the algorithm are included in t_0, and, therefore, their size is bounded by N. As the positions of the clashes involved in the successive steps make an increasing sequence, there are at most N steps. ■

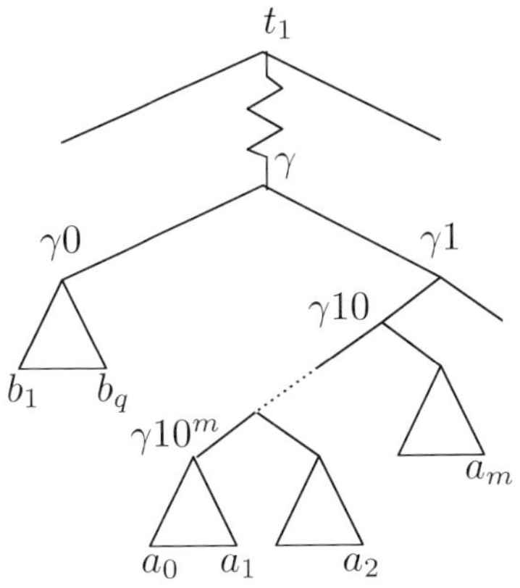

Figure 4.1. Convergence of the Polish Algorithm

The first application is a criterion for recognizing those terms of which a full term is an LD-expansion.

Proposition 4.6. (criterion) *Assume that t_0, t are terms and t_0 is full. Then the following are equivalent:*

(i) The term t is included in the term t_0;

(ii) Some cut of t_0 is an LD-expansion of t;

(iii) Some cut of t_0 is a progressive LD-expansion of t.

Proof. For every term t_0, (ii) implies (i), and (iii) implies (ii). Now, assume that t_0 is full and t is included in t_0. Proposition 4.3 (Polish for full) tells us that the Polish Algorithm running on (t, t_0) converges to a pair (t'_1, t'_2) with t'_1, t'_2 included in t_0. By construction, t'_2 is an LD-expansion of t_0, so the only possibility is $t'_2 = t_0$. Finally, t'_1 and t'_2 having no soft clash implies that t'_1, which is a progressive LD-expansion of t, is a cut of t'_2, *i.e.*, of t_0. ∎

Lemma 4.7. *Assume that the pair (t_1, t_2) admits a lifting (t'_1, t'_2) such that $t'_1 \sqsubseteq_{LD} t'_0$ and $t'_2 \sqsubseteq_{LD} t'_0$ hold for some (quasi)injective term t'_0. Then the Polish Algorithm running on (t_1, t_2) converges.*

Proof. For k large enough, the terms t'_1 and t'_2 are included in $\partial^k t'_0$, which is full. Hence the Polish Algorithm for (t'_1, t'_2) converges. By Lemma 1.14, this implies that it converges for (t_1, t_2) as well. ∎

We can now establish Proposition 4.1 (convergence) easily.

Proof. (i) Assume $t_e = s_e \bullet u_e$ for $e = 1, 2$, where u_e is a positive word, and $s_1 \sqsubseteq s_2$ holds. Let s'_2 be an injective lifting of s_2, and s'_1 be the induced lifting of s_1. Put $t'_e = s'_e \bullet u_e$, which exists as u_e is positive. Then (t'_1, t'_2) is a lifting of (t_1, t_2), and Lemma 4.7 applies.

(ii) Let $t_1^\dagger$ and $t_2^\dagger$ be the projections of t_1 and t_2 on T_1. Choose m satisfying $t_2^\dagger \sqsubseteq_{LD} t_1^\dagger \cdot x^{[m]}$. For k large enough, some cut of $\partial^k(t_1^\dagger \cdot x^{[m]})$ is an LD-expansion of $t_2^\dagger$. By (i), the Polish Algorithm converges when running on $(\partial^k(t_1^\dagger \cdot x^{[m]}), t_2^\dagger)$, hence when running on $(\partial^k t_1^\dagger, t_2^\dagger)$ since $\partial^k t_1$ is a cut of $\partial^k(t_1^\dagger \cdot x^{[m]})$. By construction, $(\partial^k t_1, t_2)$ is a lifting of $(\partial^k t_1^\dagger, t_2^\dagger)$, so Lemma 1.14 implies that the Polish Algorithm converges for $(\partial^k t_1, t_2)$ as well.

On the other hand, by Proposition V.4.12 (distinguished expansion), some term admits two iterated left subterms s_1, s_2 that are LD-expansions of $t_1^\dagger$ and $t_2^\dagger$ respectively. For k large enough, the term $\partial^k t_e^\dagger$ is an LD-expansion of s_e, and, always by (i), the Polish algorithm converges when running on $(\partial^k t_1^\dagger, \partial^k t_2^\dagger)$, hence on $(\partial^k t_1, \partial^k t_2)$. ■

Example 4.8. **(limitations)** *Let us consider one-variable terms. The previous criterion applies to all pairs (t_1, t_2) for which we can find a lifting, i.e., a renaming of the variables, consisting of terms $\sqsubseteq_{LD}$-comparable to some common quasi-injective term. If only one term were involved, we could consider an arbitrary injective lifting, which always exists. The point is that, if we rename the variables of t_1 say x_1, x_2, ..., there need not exist a lifting of t_2 giving a $\sqsubseteq_{LD}$-comparable term. The smallest pair for which the obstruction occurs is the pair*

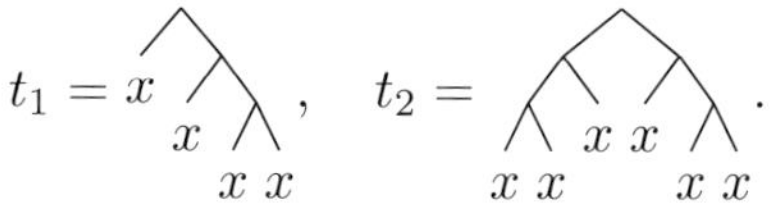

Up to a permutation, the only (quasi)injective lifting of t_1 is $x_1 \cdot x_2 \cdot x_3 \cdot x_4$, but no lifting of t_2 is $\sqsubseteq_{LD}$-comparable with $x_1 \cdot x_2 \cdot x_3 \cdot x_4$. The LD-expansion $t_2 \bullet 1$ admits one such lifting, namely

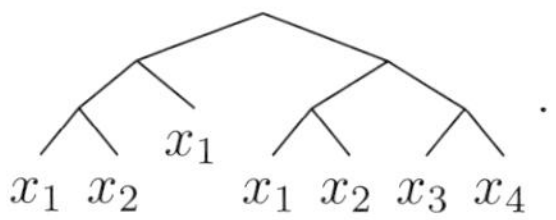

Now this lifting cannot be carried to t_2 because of the clash between the variables x_1 and x_3 at 100 and 110.

Degree k expansions

We have seen in Section VIII.5 that, for each term t, the family of all LD-expansions of t of which ∂t is an LD-expansion, *i.e.*, the family of all simple LD-expansions of t, enjoys good closure properties. In particular, any two simple LD-expansions of t admit a common LD-expansion, which is still a simple LD-expansion of t. Using the previous results, we can introduce the more general notion of a degree k LD-expansion of t, and prove similar closure properties.

Definition 4.9. (degree k) We say that t' is a *degree* k LD-expansion of t if t' is an LD-expansion of t and $\partial^k t$ is an LD-expansion of t', but $\partial^{k-1} t$ is not.

We know that, if t' is an LD-expansion of t, then $\partial^k t$ is an LD-expansion of t' for every k large enough, and every LD-expansion admits a well defined degree. By the results of Section VIII.5, the degree 1 LD-expansions of t are the proper simple LD-expansions of t.

Proposition 4.10. (characterization) *Assume that t' is an LD-expansion of t. Then the following are equivalent:*
(i) The term t' is included in $\partial^k t$;
(ii) The term t' is an LD-expansion of t of degree at most k;
(iii) The term $\partial^k t$ is a progressive LD-expansion of t'.

Proof. Let t_0 be an injective lifting of t, say $t = (t_0)^f$. Then the term $\partial^k t_0$ is full, and Proposition 4.7 (criterion) shows that (i), (ii), and (iii) are equivalent for LD-expansions of t_0. Now, for each LD-expansion t' of t, there exists a unique LD-expansion t'_0 of t_0 that has the same outline as t'. Then we have $t' = (t'_0)^f$, and each of (i), (ii), (iii) for t' is equivalent to its counterpart for t'_0. ■

Proposition 4.11. (common expansion) *Assume that $t_1, \ldots, t_m$ are LD-expansions of t of degree at most k. Then the Polish Algorithm running on $(t_1, \ldots, t_m)$ returns a common LD-expansion of $t_1, \ldots, t_m$, which is an LD-expansion of t of degree at most k.*

Proof. As above, we use an injective lifting t_0 of t. Proposition 4.3 (Polish for full) gives the result for the LD-expansions of t_0. We then deduce the result for LD-expansions of t using a projection. ■

The next results are a little surprising. We know that some LD-expansions are not progressive. Now the result tells us that the lack of progressivity can always be repaired by applying the operator ∂^k with k large enough.

Proposition 4.12. (operator ∂) *Assume that t' is an LD-expansion of t of degree at most k. Then, for every positive n, $\partial^{nk} t$ is a progressive LD-expansion of $\partial^{(n-1)k} t'$ of degree at most k, and $\partial^{nk} t'$ is a progressive LD-expansion of $\partial^{nk} t$ of degree at most k.*

Proof. We use induction on $n \geq 1$. Assume first $n = 1$. The hypothesis is that $\partial^k t$ is an LD-expansion of t'. By Proposition 4.10 (characterization), $\partial^k t$ is a progressive LD-expansion of t'. Now, as t' is an LD-expansion of t, $\partial^k t'$ is an LD-expansion of $\partial^k t$, hence the LD-class of every cut of $\partial^k t$ appears in $\partial^k t'$, *i.e.*, $\partial^k t$ is included in $\partial^k t'$. By Proposition 4.10 again, $\partial^k t$ is an LD-expansion of t' of degree k at most. Applying the previous argument to the pair $(t', \partial^k t)$ instead of (t, t') shows that $\partial^k t'$ is a progressive LD-expansion of $\partial^k t$ of degree k at most. Assume now $n \geq 2$. By induction hypothesis, the term $\partial^{(n-1)k} t'$ is an LD-expansion of $\partial^{(n-1)k} t$ of degree k at most. Hence, by the above argument, the term $\partial^k(\partial^{(n-1)k} t)$, *i.e.*, $\partial^{nk} t$, is a progressive LD-expansion of $\partial^{(n-1)k} t'$ of degree k at most. Always by the same argument, the term $\partial^k(\partial^{(n-1)k} t')$, *i.e.*, $\partial^{nk} t'$, is a progressive LD-expansion of $\partial^{nk} t$ of degree k at most. ■

Corollary 4.13. *(i) Assume that t' is an LD-expansion of t. Then $\partial^k t'$ is a progressive LD-expansion of $\partial^k t$ for k large enough.*

(ii) Assume that t and t' are LD-equivalent. Then, for k large enough, $\partial^k t$ and $\partial^k t'$ admit a common progressive LD-expansion.

Proof. (i) For k large enough, t' is an LD-expansion of t of degree k at most. Then we can apply Proposition 4.12 (operator ∂).

(ii) The terms t and t' admit a common LD-expansion t''. For k large enough, t'' is an LD-expansion of t and t' of degree k at most. Hence $\partial^k t''$ is a progressive LD-expansion of $\partial^k t$ and $\partial^k t'$. ■

Exercise 4.14. (included) *Assume that t_0, t are LD-equivalent terms and t_0 is full. Prove that the following are equivalent:*

(i) The term t is included in the term t_0;

(ii) The term t_0 is an LD-expansion of t;

(iii) The term t_0 is a progressive LD-expansion of t.

Exercise 4.15. (acyclicity) *Deduce from Corollary 4.13 a new proof of the acyclicity property for left division in free LD-systems. [Hint: Observe that, if t' is a common LD-expansion for t and $\mathrm{left}^p(t)$, then, for every k, there exists p_k such that $\partial^k t'$ is a common LD-expansion for ∂t and $\mathrm{left}^{p_k}(\partial^k t)$. For the latter expansions to be progressive is impossible.]*

---

5. Effective Decompositions

The previous results show that, for each term t and each nonnegative integer k, the term $\partial^k t$ is a progressive LD-expansion of the term t, and, therefore, the element $\Delta_t^{(k)}$ admits a progressive decomposition. In this short (unessential) section, we investigate the case $k = 1$ more closely, and, in particular, we establish the progressive decomposition of Δ_t.

Markings

We begin with a simple interpretation for the progressive expansion of t to ∂t in terms of markings of letters.

Definition 5.1. (weight) Assume that t is a term, and p is a position in t where a variable occurs. We define the *weight* $\mathrm{wt}(p, t)$ of p in t to be the number of $\bullet$ that immediately follow the p-th letter in t, and the *maximal weight* $\mathrm{wt}_{\max}(p, t)$ to be $\#_X(w) - \#_\bullet(w)$ where w is the length $p - 1$ prefix of t.

Example 5.2. (weight) *Let t be the term $x_1 \cdot (x_2 \cdot x_3 \cdot x_4) \cdot x_5$. The term t is the word $x_1 x_2 x_3 x_4 \bullet\bullet x_5 \bullet\bullet$, and the weights and maximal weights of the five occurrences of variables in t are respectively*

t:	x_1	x_2	x_3	x_4	$\bullet$	$\bullet$	x_5	$\bullet$	$\bullet$
weight:	0	0	0	2	–	–	2	–	–
maximal weight:	0	1	2	3	–	–	2	–	– .

The characterization of right Polish terms in Lemma V.1.7 implies that, if α is the address of p in t, then $\mathrm{wt}(p, t)$ is the number of final 1's in α, while $\mathrm{wt}_{\max}(p, t)$ is the total number of 1's in α. Hence, $\mathrm{wt}(p, t) \le \mathrm{wt}_{\max}(p, t)$ always holds. The following lemma is a restatement of Lemma 1.7. It shows that LD-expansion always increases the weight.

Lemma 5.3. *Assume that $t \bullet \alpha_{(k)}$ is defined. Let p be the unique position of a variable in t with an address of the form $\alpha 1 0^k 1^*$. Then applying $\mathrm{LD}_{\alpha_{(k)}}$ to t preserves the weights of all positions before p, while it increases the weight of p by one unit.*

Lemma 5.4. *Assume that p is a position in t, and t' is an LD-expansion of t that begins with the same p letters as t. Then $\mathrm{wt}(p, t) \le \mathrm{wt}(p, t') \le \mathrm{wt}_{\max}(p, t)$ holds.*

Proof. The first inequality follows from Lemma 5.3. The second one follows from the previous remark that $\mathrm{wt}(p, t') \leq \mathrm{wt}_{\max}(p, t')$ always holds, and from the fact that t and t' coinciding up to p implies $\mathrm{wt}_{\max}(p, t') = \mathrm{wt}_{\max}(p, t)$. ∎

Definition 5.5. (marking, realization) Assume that t is a term. A *marking* of t is an $\mathbf{N}$-valued function f defined on all positions in t where a variable occurs and satisfying $\mathrm{wt}(p, t) \leq f(p) \leq \mathrm{wt}_{\max}(p, t)$ for each p. For f a marking of t, the *realization* $\mathrm{real}(f, t)$ of f is the word defined as follows: if $f(p) = \mathrm{wt}(p, t)$ holds for every p, then we put $\mathrm{real}(f, t) = \varepsilon$; otherwise, letting p be the smallest integer satisfying $f(p) > \mathrm{wt}(p, t)$, and $\alpha 10^k 1^*$ be the address of p in t, we put $\mathrm{real}(f, t) = \alpha_{(k)} \cdot \mathrm{real}(f', t')$, where t' is the image of t under $\mathrm{LD}_{\alpha_{(k)}}$, and f' is the marking of t' defined by $f'(p') = f(p)$ for p the origin of p' in t.

Example 5.6. (**marking**) *Let t be the term $x_1{\cdot}(x_2{\cdot}x_3{\cdot}x_4){\cdot}x_5$. A typical marking of t is*

t:	x_1	x_2	x_3	x_4	•	•	x_5	•	•
weight:	0	0	0	2	–	0	2	–	–
f:	0	1	1	2	–	–	2	–	– .

Realizing the marking f consists in applying left self-distributivity so as to increase the weights until the values prescribed by f are obtained. In the current case, position 1 (x_1) has current weight 0, and prescribed weight 0, so we do nothing. Then position 2 (x_2) has current weight 0, while f prescribes weight 1. Then we apply the unique operator $\mathrm{LD}_{\alpha_{(k)}}$ that preserves position 1 and increases the weight of position 2, here $\mathrm{LD}_{\phi_{(2)}}$. We obtain in this way a new term t', and propagate the marking f of t into a marking f' of t' by copying the values:

t':	x_1	x_2	•	x_1	x_3	x_4	•	•	•	x_1	x_5	•	•
weight:	0	1	–	0	0	2	–	–	–	0	2	–	–
f' :	0	1	–	0	1	2	–	–	–	0	2	–	– .

We now iterate the process. By construction, positions 1 and 2 have the desired weights. So does 4. Now position 5 has weight 0, while f' prescribes 1. We then apply the operator that increases the weight of 5 by one unit, namely $\mathrm{LD}_{01_{(1)}}$, and obtain

t":	x_1	x_2	•	x_1	x_3	•	x_1	x_4	•	•	•	x_1	x_5	•	•
weight:	0	1	–	0	1	–	0	2	–	–		–0	2	–	–
f":	0	1	–	0	1	–	0	2	–	–	–	0	2	–	– .

All prescribed weights have been obtained, so $\mathrm{real}(f, t) = \phi_{(2)} \cdot 01_{(1)}$ holds.

By constuction, every word that is the realization of a marking is progressive. The converse implication is not true, because, with marking, we cannot go very far in LD-expansions.

Lemma 5.7. *Assume that f is a marking for t, and t' is the result of realizing f on t. Let g be the marking of $x{\cdot}t$ defined by $g(1) = 0$, $g(p+1) = f(p) + 1$. Then the result of realizing g on $x{\cdot}t$ is $x * t'$.*

Proof. We use induction on t. For t a variable, we have $t' = t$, and the result is true. Otherwise, assume $t = (\ldots(y{\cdot}t_1){\cdot}\ldots){\cdot}t_k$, where y is a variable. Thus, as a word, t is $yt_1\bullet\ldots t_k\bullet$. By construction, the realization of f on t is the term $yt'_1\bullet\ldots t'_k\bullet$, where t'_i is the realization of the restriction of f to t_i. Indeed, the last position in each t_i already has in t the maximal possible weight, and, therefore, it cannot be involved in an LD-expansion.

Let us now consider the realization of g on $x{\cdot}t$. We have $g(1) = 0$, so there is nothing to do for the initial position. Then, we have $f(1) = 0$, hence $g(2) = 1$, while, in t, position 2 has weight 0. Hence the first step of realization consists in applying $\mathrm{LD}_{\phi_{(k)}}$ to $x{\cdot}t$, obtaining $xy\bullet xt_1\bullet\ldots xt_k\bullet$. The resulting marking assigns 0 to each letter x, while the letters in the terms t_i receive their values for f augmented by 1. Using the induction hypothesis and the observation above, we deduce that the realization of the marking on the term $xt_i\bullet$ yields the term $x * t'_i$, and, therefore, the realization of the whole marking ends with the term $xy\bullet(x * t_1)\bullet\ldots(x * t_k)\bullet$, which is $x * (yt'_1\bullet\ldots t'_k\bullet)$, *i.e.*, $x * t'$. ∎

Proposition 5.8. **(marking)** *For every term t, those LD-expansions of t that are obtained by means of markings on t are the simple progressive expansions of t. In particular, the term ∂t is obtained by realizing the maximal marking of t, namely f such that $f(p) = \mathrm{wt}_{\max}(p, t)$ holds for every p.*

Proof. We first establish the last point using induction on t. For t a variable, the realization is trivial, and the result is true. Otherwise, assume $t = t_1{\cdot}t_2$. The maximal marking of t consists of the maximal marking f_1 of t_1, followed by "$f_2 + 1$", where f_2 is the maximal marking of t_2. The realization of f on t first consists in realizing f_1 on $t_1{\cdot}t_2$, which, by induction hypothesis, leads to $\partial t_1{\cdot}t_2$. Then, we realize "$f_2 + 1$" on $\partial t_1{\cdot}t_2$. By induction hypothesis, realizing f_2 on t_2 leads to ∂t_2, hence, by Lemma 5.7, realizing "$f_2 + 1$" on $x{\cdot}t_2$ yields $x * \partial t_2$. Substituting x with ∂t_1 shows that realizing "$f_2 + 1$" on $\partial t_1{\cdot}t_2$ yields $\partial t_1 * \partial t_2$, which is ∂t.

Assume now that f is an arbitrary marking for t, and let t' be the term obtained by realizing f on t. Let f' be the marking of t' obtained by assigning with every position p' the maximal weight of the origin of p' in t. The above argument shows that the realization of f' on t' is the term ∂t. Thus ∂t is an LD-expansion of t', and t' is a simple LD-expansion of t. ∎

Example 5.9. ***(marking for Δ_t)*** *Let t be the term $x_1{\cdot}(x_2{\cdot}(x_3{\cdot}x_4)){\cdot}x_5$ once more. The maximal weights are as follows:*

$$\begin{array}{cccccccccc} x_1 & x_2 & x_3 & x_4 & \bullet & \bullet & x_5 & \bullet & \bullet \\ 0 & 1 & 2 & 3 & - & - & 2 & - & - & . \end{array}$$

Giving weight 1 to position 2 leads to

$$\begin{array}{cccccccccccccc} x_1 & x_2 & \bullet & x_1 & x_3 & x_4 & \bullet & \bullet & \bullet & x_1 & x_5 & \bullet & \bullet \\ 0 & 1 & - & 0 & 2 & 3 & - & - & - & 0 & 2 & - & - & . \end{array}$$

Giving weight 2 to position 5 leads (in two steps) to

$$\begin{array}{cccccccccccccccccccccc} x_1 & x_2 & \bullet & x_1 & x_3 & \bullet & \bullet & x_1 & x_2 & \bullet & x_1 & x_4 & \bullet & \bullet & \bullet & x_1 & x_5 & \bullet & \bullet \\ 0 & 1 & - & 0 & 2 & - & - & 0 & 1 & - & 0 & 3 & - & - & - & 0 & 2 & - & - & . \end{array}$$

Attention! When a letter is duplicated, all copies inherit the marking of the letter they come from: for the leftmost copy, this is still the maximal weight in the current word, but, for the other copies, this is not the maximal weight, as there are more 1's in the new addresses than in the address they come from. If we systematically assign the current maximal weight, the process never comes to an end.

The progressive expression of Δ_t

The previous results have given a complete description of the progressive way of transforming t into ∂t. However, this description does not give an explicit progressive expression for Δ_t. Here we give such an expression. We recall that, for α an address, $\#_e(\alpha)$ denotes the number of letters e in α. We use $\#_0^{fin}(\alpha)$ to be the number of final 0's in α.

Definition 5.10. (words $\widehat{\alpha}$ and $\boldsymbol{\Delta}_t^{prog}$) For α an address, we define $\widehat{\alpha} = \varepsilon$ if there is no 1 in α, and $\widehat{\alpha} = (0^p1^{q-1})_{(r)} \cdot (0^p1^{q-2})_{(r)} \cdot \cdots \cdot (0^p)_{(r)}$ otherwise, with $r = \#_0^{fin}(\alpha)$, $p = \#_0(\alpha) - r$, and $q = \#_1(\alpha)$. For t a term, we put

$$\boldsymbol{\Delta}_t^{prog} = \prod_{\alpha \in \mathrm{Out}(t)}^{<_{LR}} \widehat{\alpha}.$$

Lemma 5.11. *For every address α, the word $\widehat{\alpha}$ is progressive. For every term t, the word $\boldsymbol{\Delta}_t^{prog}$ is progressive.*

Proof. Assume $p \geq 0$, $r \geq 1$. For every positive j, we have $0^p1^j0^r <_{LR} 0^p1^{j-1}10^r$, hence every word $\widehat{\alpha}$ is progressive.

We claim that the following property holds: *Assume that t is a term, α is an address in $\mathrm{Out}(t)$ that finishes with 0, and β is the next address in $\mathrm{Out}(t)$ that finishes with 0; then the word $\widehat{\alpha} \cdot \widehat{\beta}$ is progressive.* Indeed, the hypothesis implies that there exists an address γ satisfying $\gamma 0 \sqsubseteq \alpha$ and $\beta = \gamma 10^*$. Let $p = \#_0(\alpha) - \#_0^{fin}(\alpha)$, $r = \#_0^{fin}(\alpha)$, $p' = \#_0(\beta) - \#_0^{fin}(\beta) = \#_0(\gamma)$, $q' = \#_1(\beta)$, $r' = \#_0^{fin}(\beta)$. The last factor in $\widehat{\alpha}$ is $0^p_{(r)}$, while the first factor in $\widehat{\beta}$ is $(0^{p'}1^{q'-1})_{(r')}$.

By construction, we have $p' = \#_0(\beta) - \#_0^{fin}(\beta) < \#_0(\alpha) = p + r$, hence $0^{p+r} <_{LR} 0^{p'}1^{q'-1}10^{r'-1}1$. So $\widehat{\alpha} \cdot \widehat{\beta}$ is progressive. The result of the lemma follows, for an address α gives a nonempty factor in $\boldsymbol{\Delta}_t^{prog}$ if and only if α ends with 0. ∎

Lemma 5.12. *Assume that α is an address. Let $m = \#_0(\alpha)$, $q = \#_1(\alpha)$. Then we have*

$$\widehat{\alpha 0} \equiv^+ \widehat{\alpha} \cdot 0^m 1^{q-1} \cdot 0^m 1^{q-2} \cdot \dots \cdot 0^m. \tag{5.1}$$

Proof. Let $p = \#_0(\alpha) - \#_0^{fin}(\alpha)$, $r = \#_0^{fin}(\alpha)$. Then we have $\#_0^{fin}(\alpha 0) = r + 1$, hence, by definition,

$$\begin{aligned}\widehat{\alpha 0} &= (0^p 1^{q-1})_{(r+1)} \cdot (0^p 1^{q-2})_{(r+1)} \cdot \dots \cdot (0^p)_{(r+1)} \\ &= (0^p 1^{q-1})_{(r)} \cdot 0^p 1^{q-1} 0^r \cdot (0^p 1^{q-2})_{(r)} \cdot 0^p 1^{q-2} 0^r \cdot \dots \cdot (0^p)_{(r)} \cdot 0^{p+r}. \end{aligned} \tag{5.2}$$

By Proposition VII.3.6 (syntactic inheritance), we have

$$0^p 10^r \cdot (0^p)_{(r)} \equiv^+ (0^p)_{(r)} \cdot 0^{p+r} 1 = (0^p)_{(r)} \cdot 0^m 1,$$

$$\begin{aligned} 0^p 110^r \cdot (0^p 1)_{(r)} \cdot (0^p)_{(r)} &\equiv^+ (0^p 1)_{(r)} \cdot 0^p 10^r 1 \cdot (0^p)_{(r)} \\ &\equiv^+ (0^p 1)_{(r)} \cdot (0^p)_{(r)} \cdot 0^{p+r} 11 = (0^p 1)_{(r)} \cdot (0^p)_{(r)} \cdot 0^m 11. \end{aligned}$$

Similar formulas hold for each factor. For the last one, we find

$$\begin{aligned} & 0^p 1^{q-1} 0^r \cdot (0^p 1^{q-2})_{(r)} \cdot \dots \cdot (0^p)_{(r)} \\ & \quad \equiv^+ (0^p 1^{q-2})_{(r)} \cdot 0^p 101^{q-3} 0^r \cdot (0^p 1^{q-3})_{(r)} \cdot \dots \cdot (0^p)_{(r)} \\ & \dots \equiv^+ (0^p 1^{q-2})_{(r)} \cdot \dots \cdot (0^p)_{(r)} \cdot 0^{p+r} 1^{q-1} = (0^p 1^{q-2})_{(r)} \cdot \dots \cdot (0^p)_{(r)} \cdot 0^m 1^{q-1}. \end{aligned}$$

Inserting the values in (5.2) gives the result. ∎

We recall from Definition VII.2.11 that, for t a term, $\boldsymbol{\delta}_t$ denotes the positive word obtained by enumerating the internal part of the skeleton of t starting with the root.

Lemma 5.13. *Assume that t is a term, and γ is an address. Let $m = \#_0(\gamma)$, $q = \#_1(\gamma)$. Then we have*

$$\prod_{\alpha \in \mathrm{Out}(t)}^{<_{LR}} \widehat{\gamma\alpha} \equiv^+ \widehat{\gamma} \cdot 0^m 1^{q-1} \boldsymbol{\delta}_t \cdot 0^m 1^{q-2} \boldsymbol{\delta}_t \cdot \dots \cdot 0^m \boldsymbol{\delta}_t \cdot 0^m \boldsymbol{\Delta}_t^{prog}. \tag{5.3}$$

Proof. We prove the formula simultaneously for every γ using induction on t. If t is a variable, then both factors in (5.3) are $\widehat{\gamma}$. Otherwise, assume $t = t_1 \cdot t_2$.

Let u_γ be the word $\prod_{\alpha\in\mathrm{Out}(t)} \widehat{\gamma\alpha}$. The induction hypothesis gives

$$\begin{aligned}
u_\gamma &= \prod_{\alpha\in\mathrm{Out}(t_1)}^{<_{LR}} \widehat{\gamma 0\alpha} \cdot \prod_{\alpha\in\mathrm{Out}(t_2)}^{<_{LR}} \widehat{\gamma 1\alpha} \\
&\equiv^+ \widehat{\gamma 0} \cdot 0^{m+1}1^{q-1}\boldsymbol{\delta}_{t_1} \cdot \cdots \cdot 0^{m+1}\boldsymbol{\delta}_{t_1} \cdot 0^{m+1}\boldsymbol{\Delta}^{prog}_{t_1} \\
&\qquad\qquad \cdot \widehat{\gamma 1} \cdot 0^m 1^q \boldsymbol{\delta}_{t_2} \cdot \cdots \cdot 0^m \boldsymbol{\delta}_{t_2} \cdot 0^m \boldsymbol{\Delta}^{prog}_{t_2}.
\end{aligned}$$

By Lemma 5.12, we have $\widehat{\gamma 0} \equiv^+ \widehat{\alpha} \cdot 0^m 1^{q-1} \cdot \, \cdots \, \cdot 0^m$, while, by definition, $\widehat{\gamma 1}$ is empty. By repeateadly applying formulas of the type $\gamma \cdot \gamma 10u \equiv^+ \gamma 01u \cdot \gamma$, which are special cases of Proposition VII.3.6 (syntactic inheritance), we push the factors $0^{m+1}1^j\boldsymbol{\delta}_{t_1}$ to the left

$$\begin{aligned}
u_\gamma &\equiv^+ \widehat{\gamma} \cdot 0^m 1^{q-1} \cdot \cdots \cdot 0^m \cdot 0^{m+1}1^{q-1}\boldsymbol{\delta}_{t_1} \cdot \cdots \cdot 0^{m+1}\boldsymbol{\delta}_{t_1} \cdot 0^{m+1}\boldsymbol{\Delta}^{prog}_{t_1} \\
&\qquad\qquad \cdot 0^m 1^q \boldsymbol{\delta}_{t_2} \cdot \cdots \cdot 0^m \boldsymbol{\delta}_{t_2} \cdot 0^m \boldsymbol{\Delta}^{prog}_{t_2} \\
&\equiv^+ \widehat{\gamma} \cdot 0^m 1^{q-1} \cdot 0^m 1^{q-1} 0 \boldsymbol{\delta}_{t_1} \cdot 0^m 1^{q-2} \cdot \cdots \cdot 0^m \\
&\qquad \cdot 0^{m+1}1^{q-2}\boldsymbol{\delta}_{t_1} \cdot \cdots \cdot 0^{m+1}\boldsymbol{\delta}_{t_1} \cdot 0^{m+1}\boldsymbol{\Delta}^{prog}_{t_1} \cdot 0^m 1^q \boldsymbol{\delta}_{t_2} \cdot \cdots \cdot 0^m \boldsymbol{\delta}_{t_2} \cdot 0^m \boldsymbol{\Delta}^{prog}_{t_2} \\
\ldots &\equiv^+ \widehat{\gamma} \cdot 0^m 1^{q-1} \cdot 0^m 1^{q-1} 0 \boldsymbol{\delta}_{t_1} \cdot 0^m 1^{q-2} \cdot 0^m 1^{q-2} 0 \boldsymbol{\delta}_{t_1} \\
&\qquad \cdot \ldots \cdot 0^m \cdot 0^{m+1}\boldsymbol{\delta}_{t_1} \cdot 0^{m+1}\boldsymbol{\Delta}^{prog}_{t_1} \cdot 0^m 1^q \boldsymbol{\delta}_{t_2} \cdot \cdots \cdot 0^m \boldsymbol{\delta}_{t_2} \cdot 0^m \boldsymbol{\Delta}^{prog}_{t_2}.
\end{aligned}$$

Similarly, using type $\perp$ and 11 relations, we push the factors $0^m 1^j \boldsymbol{\delta}_{t_2}$ to the left, and obtain

$$\begin{aligned}
u_\gamma &\equiv^+ \widehat{\gamma} \cdot 0^m 1^{q-1} \cdot 0^m 1^{q-1} 0 \boldsymbol{\delta}_{t_1} \cdot 0^m 1^q \boldsymbol{\delta}_{t_2} \cdot 0^m 1^{q-2} \cdot 0^m 1^{q-2} 0 \boldsymbol{\delta}_{t_1} \cdot 0^m 1^{q-1} \boldsymbol{\delta}_{t_2} \\
&\qquad \cdot \ldots \cdot 0^m \cdot 0^{m+1}\boldsymbol{\delta}_{t_1} \cdot 0^m 1 \boldsymbol{\delta}_{t_2} \cdot 0^{m+1}\boldsymbol{\Delta}^{prog}_{t_1} \cdot 0^m \boldsymbol{\delta}_{t_2} \cdot 0^m \boldsymbol{\Delta}^{prog}_{t_2} \\
&= \widehat{\gamma} \cdot 0^m 1^{q-1} \boldsymbol{\delta}_t \cdot 0^m 1^{q-2} \boldsymbol{\delta}_t \cdot \ldots \cdot 0^m \boldsymbol{\delta}_t \cdot 0^{m+1}\boldsymbol{\Delta}^{prog}_{t_1} \cdot 0^m \boldsymbol{\delta}_{t_2} \cdot 0^m \boldsymbol{\Delta}^{prog}_{t_2} \\
&= \widehat{\gamma} \cdot 0^m 1^{q-1} \boldsymbol{\delta}_t \cdot 0^m 1^{q-2} \boldsymbol{\delta}_t \cdot \ldots \cdot 0^m \boldsymbol{\delta}_t \cdot 0^m (0\boldsymbol{\Delta}^{prog}_{t_1} \cdot \boldsymbol{\delta}_{t_2} \cdot \boldsymbol{\Delta}^{prog}_{t_2}).
\end{aligned} \tag{5.4}$$

Writing (5.4) for $\gamma = \phi$, hence $m = q = 0$, yields $u_\phi \equiv^+ 0\boldsymbol{\Delta}^{prog}_{t_1} \cdot \boldsymbol{\delta}_{t_2} \cdot \boldsymbol{\Delta}^{prog}_{t_2}$, and we finally obtain

$$u_\gamma \equiv^+ \widehat{\gamma} \cdot 0^m 1^{q-1} \boldsymbol{\delta}_t \cdot 0^m 1^{q-2} \boldsymbol{\delta}_t \cdot \ldots \cdot 0^m \boldsymbol{\delta}_t \cdot 0^m u_\phi,$$

the expected relation, as, by definition, $u_\phi = \boldsymbol{\Delta}^{prog}_t$ holds. ∎

Proposition 5.14. (progressive form of Δ_t) *For every term t, we have $\boldsymbol{\Delta}^{prog}_t \equiv^+ \boldsymbol{\Delta}_t$. Hence $\boldsymbol{\Delta}^{prog}_t$ is the (unique) progressive expression of Δ_t. More generally, for every address γ with $m = \#_0(\gamma)$, $q = \#_1(\gamma)$, the word $\prod^{<_{LR}}_{\alpha\in\mathrm{Out}(t)} \widehat{\gamma\alpha}$ is the progressive expression of $\overline{\widehat{\gamma}} \cdot 0^m 1^{q-1} \delta_t \cdot 0^m 1^{q-2} \delta_t \cdot \, \cdots \, \cdot 0^m \delta_t \cdot 0^m \Delta_t$, where δ_t is the class of $\boldsymbol{\delta}_t$ in M_{LD}.*

Proof. First, we prove $\boldsymbol{\Delta}^{prog}_t \equiv^+ \boldsymbol{\Delta}_t$ using induction on t. For t a variable, we have $\boldsymbol{\Delta}_t = \boldsymbol{\Delta}^{prog}_t = \varepsilon$. Otherwise, assume $t = t_1{\cdot}t_2$. Using Lemma VII.3.14, the induction hypothesis, and Lemma 5.13, we find

$$\boldsymbol{\Delta}_t \equiv^+ 0\boldsymbol{\Delta}_{t_1} \cdot \boldsymbol{\delta}_{t_2} \cdot \boldsymbol{\Delta}_{t_2} \equiv^+ 0\boldsymbol{\Delta}^{\mathit{prog}}_{t_1} \cdot \boldsymbol{\delta}_{t_2} \cdot \boldsymbol{\Delta}^{\mathit{prog}}_{t_2} \equiv^+ \boldsymbol{\Delta}^{\mathit{prog}}_{t}.$$

As the word $\boldsymbol{\Delta}^{\mathit{prog}}_{t}$ is progressive, the first part of the result is proved.

Applying the previous result, we obtain

$$\widehat{\gamma}{\cdot}0^m1^{q-1}\boldsymbol{\delta}_t{\cdot}0^m1^{q-2}\boldsymbol{\delta}_t{\cdot}\cdots{\cdot}0^m\boldsymbol{\delta}_t{\cdot}0^m\boldsymbol{\Delta}_t \equiv^+ \widehat{\gamma}{\cdot}0^m1^{q-1}\boldsymbol{\delta}_t{\cdot}0^m1^{q-2}\boldsymbol{\delta}_t{\cdot}\cdots{\cdot}0^m\boldsymbol{\delta}_t{\cdot}0^m\boldsymbol{\Delta}^{\mathit{prog}}_t.$$

By Lemma 5.13, the latter word is $\equiv^+$-equivalent to $\prod_{\alpha\in\mathrm{Out}(t)}\widehat{\gamma\alpha}$, which is progressive. Indeed, it is a subword of the word $\boldsymbol{\Delta}^{\mathit{prog}}_{s}$ where s is any term satisfying $\mathrm{sub}(s,\gamma) = t$. ∎

In particular, considering the case when γ has the form 1^q yields for every term t and every positive integer q the relation

$$1^{q-1}\boldsymbol{\delta}_t \cdot \cdots \cdot \boldsymbol{\delta}_t \cdot \boldsymbol{\Delta}_t \equiv^+ \prod_{\alpha\in\mathrm{Out}(t)} \widehat{1^q\alpha}, \tag{5.5}$$

where the latter word is progressive.

Exercise 5.15. **(dynamic marking)** *Extend the notion of marking as follows: starting with the term t, we assign to every position in t its maximal weight; then, when the marking is realized, the value assigned to a new position is either its maximal weight if the corresponding cut has t-degree at most $k-1$, or the weight of the position it comes from if the corresponding cut has degree k or more. Prove that the realization of this modified marking ends with the term $\partial^k t$.*

Exercise 5.16. **(progressive form of Δ_t)** *(i) For t a term and $q \geq 0$, define $\boldsymbol{\Delta}_{t,0} = \boldsymbol{\Delta}_t$ and $\boldsymbol{\Delta}_{t,q} = 1^{q-1}\boldsymbol{\delta}_t \cdot \cdots \cdot 1\boldsymbol{\delta}_t \cdot \boldsymbol{\delta}_t \cdot \boldsymbol{\Delta}_t$ for $q > 0$. Prove that $\mathrm{LD}_{\boldsymbol{\Delta}_{t,q}}$ maps $\partial x^{[q]}{\cdot}t$ to $\partial(x^{[q]}{\cdot}t)$.*

(ii) Assume $t = xt_1\bullet\ldots t_h\bullet$, i.e., $t = (\ldots(x{\cdot}t_1){\cdot}\ldots){\cdot}t_h$, with x a variable. Prove the relations

$$\boldsymbol{\Delta}_{t,0} \equiv^+ 0^{h-1}\boldsymbol{\Delta}_{t_1,1} \cdot 0^{h-2}\boldsymbol{\Delta}_{t_2,1} \cdot \cdots \cdot \boldsymbol{\Delta}_{t_h,1},$$

$$\boldsymbol{\Delta}_{t,q} \equiv^+ 1^{q-1}_{(h)} \cdot \cdots \cdot 1_{(h)} \cdot \phi_{(h)} \cdot 0^{h-1}\boldsymbol{\Delta}_{t_1,q+1} \cdot 0^{h-2}\boldsymbol{\Delta}_{t_2,q+1} \cdot \cdots \cdot \boldsymbol{\Delta}_{t_h,q+1}.$$

Deduce (without using Proposition 5.14) that $\boldsymbol{\Delta}_{t,q}$ is $\equiv^+$-equivalent to a progressive word which begins with $1^{q-1}_{(h)} \cdot 1^{q-2}_{(h)} \cdot \ldots \cdot \phi_{(h)}$ if $q \geq 1$ holds and h is the left height of t, and with $0^{h_1}_{(h_2)}$ if $q = 0$ holds and $0^{h_1}10^{h_2}$ is the leftmost address in $\mathrm{Out}(t)$ that contains the pattern 10.

Exercise 5.17. **(operation $\partial^\bullet$)** *Show that, for every term t, the word $\mathbf{\Delta}^\bullet_t$ of Exercise VII.3.30 is progressive.*

---

6. The Embedding Conjecture

Finally, we come back to one of the problems left open in Chapter VIII, namely the question of whether the monoid M_{LD} embeds in the group G_{LD}. In this section, we deduce several partial positive results from the properties of full terms.

Equivalent forms of the conjecture

Here we address the following conjecture:

Conjecture 6.1. (**Embedding Conjecture**) *The monoid M_{LD} embeds in the group G_{LD}, i.e., for all words u, u' on $\mathbf{A}$, $u' \equiv u$ implies (and, therefore, is equivalent to) $u' \equiv^+ u$.*

Several equivalent forms can be stated.

Proposition 6.2. (**equivalence** I) *The Embedding Conjecture is equivalent to each of the following statements:*

(i) The monoid M_{LD} admits right cancellation;

(ii) The monoid $\mathcal{G}^+_{LD}$ is isomorphic to the monoid M_{LD}, i.e., for all words u, u' on $\mathbf{A}$, $\mathrm{LD}_{u'} = \mathrm{LD}_u$ implies (and, therefore, is equivalent to) $u' \equiv^+ u$.

Proof. Equivalence with (i) follows from Proposition II.3.17 (embedding), as we know that M_{LD} is associated on the right with a coherent and convergent complement. Equivalence with (ii) follows from Proposition VIII.2.16 (embedding), which tells us that $\mathrm{LD}_{u'} = \mathrm{LD}_u$ is equivalent to $u' \equiv u$. ∎

If the monoid $\mathcal{G}^+_{LD}$ is isomorphic to the monoid M_{LD}, the lattice structure associated with left division in M_{LD} gives an isomorphic lattice structure for $\mathcal{G}^+_{LD}$ equipped with the relation of being an LD-expansion.

Proposition 6.3. (**lattice**) *If the Embedding Conjecture is true, then, for each term t, the family of all LD-expansions of t is a lattice with respect to the relation of being an LD-expansion.*

Proof. Assume that $t \bullet a$ and $t \bullet b$ are LD-expansions of t. Then $t \bullet (a \vee_R b)$ is a common LD-expansion of $t \bullet a$ and $t \bullet b$. Let $t \bullet c$ be an arbitrary common LD-expansion of $t \bullet a$ and $t \bullet b$. The Embedding Conjecture implies that a and b are left divisors of c in M_{LD}, hence $a \vee_R b$ is a left divisor of c, and $t \bullet c$ is an LD-expansion of $t \bullet (a \vee_R b)$. ■

In order to study the question more closely, we pose a definition.

Definition 6.4. (**Embedding Conjecture**) Assume that a is an element of M_{LD}. We say that the Embedding Conjecture is true *for* a if the canonical projection of M_{LD} onto G^+_{LD} is injective on a, *i.e.*, if $\mathrm{LD}_{a'} = \mathrm{LD}_a$ implies $a' = a$.

Thus the Embedding Conjecture is true if and only if it is true for every element of M_{LD}.

Lemma 6.5. *Assume that the Embedding Conjecture is true for a.*

(i) For all a', b in M_{LD}, $ab = a'b$ implies $a' = a$.

(ii) The Embedding Conjecture is true for every right divisor of a.

Proof. (i) If $a'b = ab$ holds, we have $\mathrm{LD}_{a'b} = \mathrm{LD}_{ab}$, hence $\mathrm{LD}_{a'} = \mathrm{LD}_a$ since LD_b is injective. The Embedding Conjecture being true for a then implies $a' = a$.

(ii) Assume $a = bc$, and $\mathrm{LD}_{c'} = \mathrm{LD}_c$. Then we have $\mathrm{LD}_{bc'} = \mathrm{LD}_{bc} = \mathrm{LD}_a$, hence $bc' = bc$, and $c' = c$ as M_{LD} is left cancellative. ■

The rest of this section is devoted to proving the Embedding Conjecture for particular elements of M_{LD}.

Proposition 6.6. (**simple**) *The Embedding Conjecture is true for every simple element of M_{LD}.*

Proof. Assume that a is a simple element of M_{LD}, and $\mathrm{LD}_a = \mathrm{LD}_{a'}$ holds. By definition, a' is simple as well, and, by Proposition VIII.5.15 (equivalence), a and a' are permutation-like elements. By Lemma VIII.5.4, they are equal, since they are both represented by the same permutation-like word determined by LD_a. ■

Confluent families

Here we shall prove the following result:

Proposition 6.7. (**right divisors**) *The Embedding Conjecture is true for every element of M_{LD} that is a right divisor of some element $\Delta_t^{(k)}$.*

The point in the proof is the notion of a confluent family of terms. For u, v words on $\mathbf{A}$, we recall that $u\backslash v$ and $v\backslash u$ denote the (unique) words such that $u^{-1}\cdot v$ is reversible to $(u\backslash v)\cdot(v\backslash u)^{-1}$. We use $u\vee_R v$ for $u\cdot(u\backslash v)$—which is natural as, if u represents a and v represents b, then $u\vee_R v$ represents the right lcm $a\vee_R b$ of a and b in M_{LD}.

Definition 6.8. (**confluent**) Assume that F is a family of terms. We say that F is *confluent* if, for all positive words u, v, F contains $t\bullet(u\vee_R v)$ whenever it contains t, $t\bullet u$, and $t\bullet v$. We say that F is *locally confluent* if, for all addresses α, β, F contains $t\bullet(\alpha\vee_R\beta)$ whenever it contains t, $t\bullet\alpha$, and $t\bullet\beta$.

The interest of confluent families for the Embedding Conjecture lies in the following criterion.

Lemma 6.9. *Let a be an element of M_{LD}. Then the following are equivalent:*

(i) The Embedding Conjecture is true for a;

(ii) For every term t in the domain of LD_a, the family of all LD-expansions of t of which $t\bullet a$ is an LD-expansion is confluent;

(iii) There exist a term t in the domain of LD_a and a confluent family F containing t and $t\bullet a$ such that $t\bullet a$ is terminal in F, i.e., no proper LD-expansion of $t\bullet a$ belongs to F.

Proof. Assume that the Embedding Conjecture is true for a, and t is a term in the domain of LD_a. Let F be the family of all LD-expansions of t of which $t\bullet a$ is an LD-expansion. Let w be a positive expression of a. Assume that t', $t'\bullet u$, and $t'\bullet v$ belong to F. By definition, there exist positive words w_0, u_1, v_1 satisfying $t'=t\bullet w_0$, $t\bullet a=t'\bullet(u\cdot v_1)=t'\bullet(v\cdot u_1)$ (Figure 6.1). Hence we have $\mathrm{LD}_w=\mathrm{LD}_{w_0\cdot u\cdot v_1}=\mathrm{LD}_{w_0\cdot v\cdot u_1}$, and, therefore, $w\equiv w_0\cdot u\cdot v_1\equiv w_0\cdot v\cdot u_1$. The Embedding Conjecture for a implies $w\equiv^+ w_0\cdot u\cdot v_1\equiv^+ w_0\cdot v\cdot u_1$, hence $u\cdot v_1\equiv^+ v\cdot u_1$. There must exist a positive word w_1 satisfying $v_1=(u\backslash v)\cdot w_1$ and $u_1=(v\backslash u)\cdot w_1$. Hence we have $t\bullet a=(t'\bullet(u\vee_R v))\bullet w_1$, and $t'\bullet(u\vee_R v)\in F$. Hence F is confluent, and (i) implies (ii).

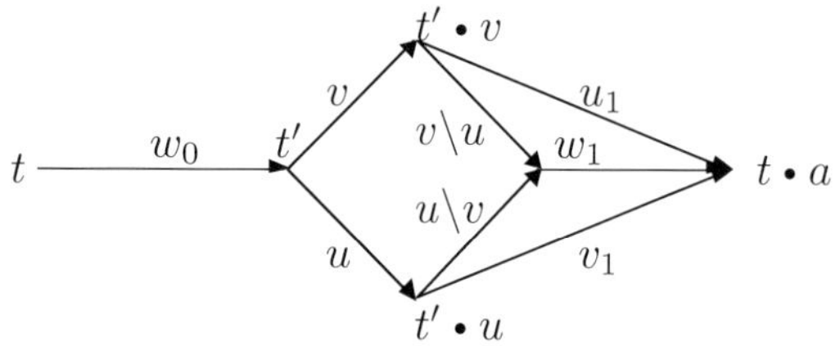

Figure 6.1: Confluent family

It is clear that (ii) implies (iii), as, by construction, the term $t \bullet a$ is terminal in every family satisfying the conditions of (ii).

Finally, assume that F is a confluent family, and $t{\bullet}a$ is terminal in F. Assume that u represents a, and $u' \equiv u$, hence $\mathrm{LD}_{u'} = \mathrm{LD}_u$, holds. By construction, we have $t{\bullet}u' = t{\bullet}u$. As $t{\bullet}u$ is an LD-expansion of $t{\bullet}u$ and $t{\bullet}u'$ (!), the definition of a confluent family implies $t \bullet (u{\vee_R}u') \in F$. The hypothesis that $t \bullet u$ is terminal in F implies $t \bullet (u \vee_R u') = t \bullet u$, and the only possibility is $u \backslash u' = \varepsilon$, and, symmetrically, $u' \backslash u = \varepsilon$. Hence $u' \equiv^+ u$ holds, and the Embedding Conjecture is true for a. ■

The termination of word reversing in $\mathbf{A}^*$ implies that confluent families admit a local characterization.

Lemma 6.10. *Assume that F is a family of terms. Then F is confluent if and only if it is locally confluent.*

Proof. By definition, (i) implies (ii). Assume (ii). We establish that F contains $t \bullet (u \vee_R v)$ whenever it contains $t \bullet u$ and $t \bullet v$ using induction on the complexity of the pair (u, v), *i.e.*, on the number m of elementary steps needed to reverse the word $u^{-1}v$, as in the proof of Proposition II.1.15 (positive equivalence). For $m = 0$, at least one of u, v is empty, and the result is true. Assume $m \geq 1$. Then u and v are not empty. Write $u = \alpha \cdot u_0$, $v = \beta \cdot v_0$, with $\alpha, \beta \in \mathbf{A}$. There exist positive words u_1, v_1, ..., u_3, v_3 as represented in Figure 6.2. By definition, we have $t \bullet (\alpha{\vee_R}\beta) \in F$. By construction, the complexity of $(u_0, \alpha\backslash\beta)$ is less than m, hence the induction hypothesis implies $t \bullet (\alpha \cdot u_0 \cdot v_1) \in F$. Similarly, the complexity of $(v_0, \beta\backslash\alpha)$ is less than m, hence the induction hypothesis implies $t \bullet (\beta \cdot v_0 \cdot u_1) \in F$. Finally, the complexity of (u_2, v_2) is less than m, and the induction hypothesis implies $t \bullet (\alpha \cdot u_0 \cdot v_1 \cdot v_3) = t \bullet (u{\vee_R}v) \in F$. ■

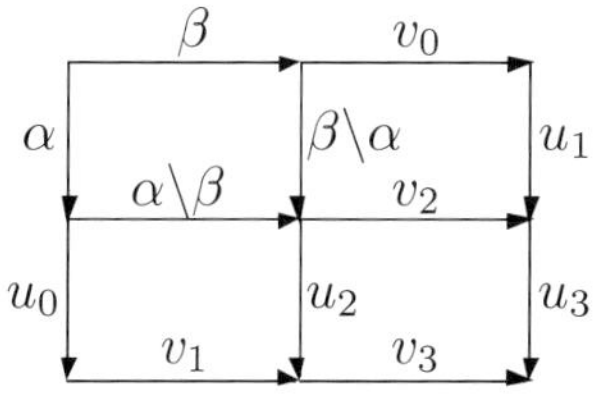

Figure 6.2. Induction for confluence

By gathering the previous results, we deduce the following criterion:

Proposition 6.11. (criterion) *Assume that a is an element of M_{LD}. Then the following are equivalent:*

(i) The Embedding Conjecture is true for a;

(ii) There exists a locally confluent family of terms with a terminal term in the image of LD_a.

This gives new forms for the Embedding Conjecture.

Proposition 6.12. (**equivalence** II) *The Embedding Conjecture is equivalent to each of the following statements:*

(i) For all terms t, t', and all addresses α, β, if t' is an LD-expansion of $t \bullet \alpha$ and $t \bullet \beta$, then it is an LD-expansion of $t \bullet (\alpha \vee_R \beta)$ as well.

(ii) For all terms t, t', and all positive words u, v, if t' is an LD-expansion of $t \bullet u$ and $t \bullet v$, then it is an LD-expansion of $t \bullet (u \vee_R v)$ as well.

Proof. Assume that the Embedding Conjecture is true, and assume that t' is an LD-expansion of $t \bullet u$ and $t \bullet v$. Then t' is an LD-expansion of t. By Lemma 6.9, the family F of all LD-expansions of t of which t' is an LD-expansion is confluent. By hypothesis, F contains $t \bullet u$ and $t \bullet v$, hence it contains $t \bullet (u \vee_R v)$, which means that t' is an LD-expansion of $t \bullet (u \vee_R v)$. So (ii) holds.

Conversely, (ii) implies (i) by definition. Assume now (i), and let a be an arbitrary element of M_{LD}. Let t be a term in the domain of LD_a, and F be the family of all LD-expansions of t of which $t \bullet a$ is an LD-expansion. We claim that F is locally confluent. Indeed, assume that t', $t' \bullet \alpha$ and $t' \bullet \beta$ belong to F. By construction, $t' \bullet (\alpha \vee_R \beta)$ is an LD-expansion of t, and, as $t \bullet a$ is an LD-expansion both of $t' \bullet \alpha$ and $t' \bullet \beta$, then, by hypothesis, it must be an LD-expansion of $t' \bullet (\alpha \vee_R \beta)$ as well. Hence $t' \bullet (\alpha \vee_R \beta)$ belongs to F, which is locally confluent, and the Embedding Conjecture is true for a. ■

The problem for establishing the conditions of Proposition 6.12 (equivalence II) is that no intrinsic characterization of the LD-expansions of a term is known in general. Now, in the case of perfect terms, such a characterization is known, and a confluent family of the desired type may be constructed.

Lemma 6.13. *Assume that t_0 is a perfect term. Then the family consisting of all terms included in t_0 is confluent.*

Proof. By Lemma 6.10, it suffices to show that F is locally confluent. Assume that t, $t \bullet \alpha$ and $t \bullet \beta$ are included in t_0. We consider the various possible cases. If the addresses α and β are orthogonal, then we have $\alpha \vee_R \beta = \alpha \cdot \beta$, and the content of $t \bullet \alpha{\cdot}\beta$ is the union of the contents of $t \bullet \alpha$ and $t \bullet \beta$, so the result is clear.

Assume now that α and β are $\sqsubseteq$-comparable, say $\beta \sqsupseteq \alpha$ for instance. The case $\beta = \alpha$ is trivial. For $\beta \sqsupseteq \alpha 10$ and $\beta \sqsupseteq \alpha 11$, the content of $t \bullet \alpha{\cdot}\beta$ is the union of the contents of $t \bullet \alpha$ and $t \bullet \beta$, and the result is clear again. There remain the cases $\beta = \alpha 1$ and $\beta \sqsupseteq \alpha 0$. Assume first $\beta = \alpha 1$. We have $\alpha \backslash \beta = \alpha 1 \cdot \alpha$. The content of the term $t \bullet (\alpha \vee_R \beta)$ is the union of the contents of $t \bullet \alpha$, $t \bullet \beta$, and of additional elements of the form abc where a appears in t at $\alpha 1101^*$, b appears at $\alpha 101^*$, and c appears below $\alpha 0$ (*cf.* Figure 6.3). Consider such an element. By construction, abc appears in an LD-expansion of t, hence we have

$abc \sqsubset \overline{t} \sqsubseteq \overline{t_0}$. Moreover, ab appears in $t \bullet \alpha$, bc appears in $t \bullet \beta$, hence both appear in t_0, and Lemma 3.12 implies that abc appears in t_0. So the result is true.

The final case is $\beta \sqsupseteq \alpha 0$, say $\beta = \alpha 0 \gamma$. Then we have $\alpha \backslash \beta = \alpha 10\gamma \cdot \alpha 00\gamma$. The same argument as above shows that the content of $t \bullet (\alpha \vee_R \beta)$ is the union of the contents of $t \bullet \alpha$, $t \bullet \beta$, and of additional elements of the form abc where a appears in t at $\alpha 101^*$, b appears at $\alpha 0 \gamma 101^*$, and c appears below $\alpha 0 \gamma 0$. As in the case $\beta = \alpha 1$, we have $abc \sqsubset \overline{t_0}$, ab and bc appear in s, and Lemma 3.12 implies that abc appears in t_0. ■

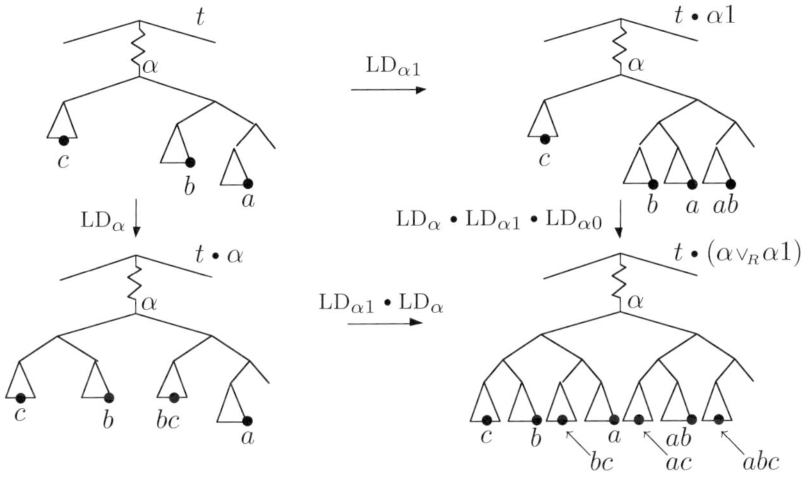

Figure 6.3. Confluence

We deduce the following criterion.

Proposition 6.14. (embedding) *Assume that a is an element of M_{LD}. Then the following are equivalent:*

(i) The term t_a^R is full (hence perfect);

(ii) The image of the operator LD_a contains at least one full term.

If the above conditions are satisfied, the Embedding Conjecture is true for a.

Proof. The equivalence of (i) and (ii) follows from Proposition 3.7 (substitution), as the terms in the image of LD_a are the substitutes of t_a^R. Assume that the term t_a^R is full. By construction, the term t_a^L is increasing, hence steep. Hence t_a^R, which is LD-equivalent to t_a^L, is steep as well, and, therefore, it is perfect. By Lemma 6.13, the family of all terms included in t_a^R is confluent, and, by construction, the term t_a^R is terminal in this family. By Proposition 6.14 (criterion), we conclude that the Embedding Conjecture is true for a. ■

Using the results of Section 3, we obtain the Embedding Conjecture for a large family of elements. First, we can complete the proof of Proposition 6.7 (right divisors):

Proof. The element $\Delta_t^{(k)}$ depends on the skeleton of the term t only, so we may assume without loss of generality that t is increasing. Then the term $\partial^k t$ is the image of t under $\mathrm{LD}_{\Delta_t^{(k)}}$, and it is perfect. Hence $\Delta_t^{(k)}$ and, by Lemma 6.5, every right divisor of $\Delta_t^{(k)}$ in M_{LD}, is eligible for the criterion of Proposition 6.14 (embedding). ∎

Another application of the criterion involves those elements of M_{LD} that admit an expression containing addresses of the form 1^i only.

Definition 6.15. (braidlike) An element a of M_{LD} is said to be *braidlike* if it belongs to the submonoid generated by the elements $g_{1^i}^+$.

The mapping $\sigma_i \mapsto g_{1^{i-1}}^+$ induces a section s for the canonical projection of M_{LD} onto B_∞^+, and its image is the set of all braidlike elements. Note that s is not a homomorphism, and that every special element of G_{LD} is braidlike.

Lemma 6.16. *Assume that a is a braidlike element of M_{LD}. Then, for every perfect term t in the domain of LD_a, the term $t \bullet a$ is perfect. So, in particular, the term t_a^R is perfect.*

Proof. Assume that t is perfect, and t' is the image of t under the operator LD_{1^i}. The term t' is steep since it is LD-equivalent to t, and the only problem is fullness. We look at those pairs (a', b') such that a' covers b' in t' and $a'b' \sqsubset \overline{t'}$, *i.e.*, $a'b' \sqsubset \overline{t}$, holds. If a' and b' already appear in t, and a' covers b' in t, then the hypothesis that t is full implies that $a'b'$ appears in t, hence in t'. Let a be the element that appears at $1^{i+1}01^*$ in t. By Proposition 2.7 (covering), three cases remain to be considered. The first case is $a' = a$ and b' appearing at $1^i0\beta$ in t. In this case, $a'b'$ appears in t' at $1^{i+1}0\beta$.

The second case is $a' = ab$, $b' = ac$ with b appearing at $1^i0\beta$ and c at $1^i0\gamma$ and β covering γ. Assume $a'b' \sqsubset \overline{t}$, *i.e.*, $a(bc) \sqsubset \overline{t}$. Then t being steep implies $bc \sqsubset \overline{t}$, and its being full implies that bc appears in t. If a covers bc in t, then $a(bc)$, which is $a'b'$, appears in t' as well, and we are done. Otherwise, bc appears below 1^i0 at some address, say $1^i0\beta'$, and, in this case, $a'b'$ appears in t' at $1^{i+1}0\beta'$. The study of this case is therefore complete.

The last case is a' covering a in t. Now, by definition, this case is impossible, as the only addresses covering an address of the form 1^i01^* are those addresses of the form 1^{i+j}. ∎

Applying Proposition 6.14 (embedding), we deduce:

Proposition 6.17. (**braidlike**) *The Embedding Conjecture is true for every braidlike element of M_{LD}.*

Let us conclude with a syntactic result. For u a word on **A**, the Embedding Conjecture is true for the element $\overline{u}$ of M_{LD} if, for every word u' on **A**, $u' \equiv u$ implies $u' \equiv^+ u$. Thus, a consequence of Proposition 6.7 (right divisors) is that, for u being a word on **A**, $u \equiv \boldsymbol{\Delta}_t^{(k)}$ implies $u \equiv^+ \boldsymbol{\Delta}_t^{(k)}$. Then, we can refine Proposition VII.3.22 (syntactic confluence) into

Proposition 6.18. (**progressive confluence**) *Assume that u, v are words on* **A** *of length k at most. Then there exist progressive words u', v' satisfying*

$$u \cdot v' \equiv^+ v \cdot u' \equiv^+ \boldsymbol{\Delta}_{t^L_{u,v}}^{(k)}. \tag{6.1}$$

Proof. Let $t = t^L_{u,v}$. Then the term $\partial^k t$ is an LD-expansion of $t \bullet u$ and $t \bullet v$, and, by Proposition 4.11 (common expansion), it is a progressive LD-expansion of these terms. This means that there exist words u', v' satisfying $u \cdot v' \equiv v \cdot u' \equiv \boldsymbol{\Delta}_t^{(k)}$. By the above remark, the latter are $\equiv^+$-equivalences. ∎

Proposition 6.19. (**degree** k) *Assume that a belongs to M_{LD} and $t \bullet a$ exists. Then the following are equivalent:*

(i) The element a is a left divisor of $\Delta_t^{(k)}$ in M_{LD};

(ii) The term $\partial^k t$ is an LD-expansion of $t \bullet a$, i.e., the latter is an LD-expansion of t of degree at most k.

Proof. That (i) implies (ii) is clear. Conversely, (ii) implies that $u \cdot v \equiv \boldsymbol{\Delta}_t^{(k)}$ holds for some positive words u, v such that u represents a. As the Embedding Conjecture is true for $\Delta_t^{(k)}$, this implies $u \cdot v \equiv^+ \boldsymbol{\Delta}_t^{(k)}$. ∎

Exercise 6.20. (**connection**) *(i) Assume that a is an element of M_{LD} and the normal form of a (as in Proposition VIII.5.23) has length k at most. Prove that, for every term t in the domain of LD_a, the term $t \bullet a$ is an LD-expansion of t of degree at most k.*

(ii) Conversely, assume that $t \bullet a$ is an LD-expansion of t of degree k at most. Prove that a is a left divisor of $\Delta_t^{(k)}$ in M_{LD}. Can we deduce that the normal form of a has length k at most? [No, cf. Exercise VIII.5.26.]

Exercise 6.21. **(conditional distributivity)** *(i) Assume that $\succ$ is a binary relation on FLD_∞. For every term t, we define the family $F_\succ(t)$ of all LD-expansions of t conditioned by $\succ$ to be the closure of $\{t\}$ under the following operation: for $t' \in F_\succ(t)$, we have $t' \bullet \alpha \in F_\succ(t)$ if and only if the condition $\mathrm{cut}(t', \alpha 101^*) \succ \mathrm{cut}(t', \beta)$ holds for every address below $\alpha 0$ in the outline of t'. Prove that a sufficient condition for the family $F_\succ(t_0)$ to be confluent is that the conjunction of $a \succ b$ and $b \succ c$ implies $a \succ c$ and $a \succ bc$.*

(ii) Assume that $a \succ b$ is the condition $\mathrm{var}_R(a) > \mathrm{var}_R(b)$. Prove that $F_\succ(t)$ is a finite confluent family. [Hint: Observe that $F_\succ(t)$ consists only of simple LD-expansions of t.] What is the terminal element of $F_\succ(t)$ if t is supposed to be increasing?

(iii) Assume that t_0 is a fixed perfect term and $a \succ b$ is the condition "$ab \ll \overline{t_0}$ or ab appears in t_0". Prove that $\succ$ satisfies the hypotheses of (i). Interpret the case of degree k LD-expansions in this framework.

---

7. Notes

The Polish Algorithm and the notion of a progressive LD-expansion were considered at the end of the 1980s by the author as a possible way for proving that left division in free LD-systems has no cycle using the scheme of Exercise 4.15. Actually, this algorithm appears naturally when one tries to solve the word problem of Identity (LD) using words, and it was probably considered by several authors independently, in particular by R. Laver. The first reference seems to be (Dehornoy 92*a*), where the correctness of the algorithm is established assuming that left division has no cycle in free LD-systems. (Dehornoy 93*a*) contains partial results about simple LD-expansions, and (Dehornoy 97*a*) introduces the notion of a full term, but, without the stronger notions considered in Sections 2 and 3, it is impossible to state any strong convergence statement. The other results in the current chapter have not yet appeared. The closure of perfect terms under ∂ established in Section 3 was conjectured for a long time.

Exercise 2.24 (square) is from (Dehornoy 98*a*), answering a question of (Ježek *97*) about those LD-systems that admit a zero element, *i.e.*, an element 0 such that $x0 = 0x = 0$ holds for every x, and $xx = 0$ holds for every x. Call such systems zeropotent. Ježek shows that $x(yz) = 0$ holds in every finite zeropotent LD-system. Exercise 2.24 shows that the identity $x(yz) = 0$ is not true in a free zeropotent LD-system, for this would mean that the identity $x(yz) = xx$ can be deduced from (LD) with the additional identities $(xx)y = y(xx) = xx$.

The main open question here remains the algorithm:

Conjecture 7.1. *The Polish Algorithm converges for every initial pair of terms.*

A proof of this result would imply the following strengthening of the confluence property (Proposition V.3.14).

Conjecture 7.2. *Any two LD-equivalent terms admit a common progressive LD-expansion.*

By Proposition 4.1 (convergence), the result is true if the terms we consider are themselves LD-expansions of a common term, but the general case is open.

The fact that every term of the form $\partial x^{[n]}$, as well as the terms $\partial^2 x^{[n]}$ for $n \leq 5$ are full implies that any possible counter-example to the convergence of the Polish Algorithm has to involve large terms, and practical experiments seem to be of little use here: as anyone can check, the algorithm succeeds on all terms with a reasonable size, and the complexity increases very quickly.

The argument of Section 4, *i.e.*, the principle of proving convergence by showing that all terms appearing in the process remain in some sense below some fixed large term is the only argument currently known. Theoretically, one could also hope to prove termination using a tricky, possibly even simple, way of assigning a natural number, or an ordinal, to each pair of terms in such a way that this parameter decreases when the algorithm is run. Unfortunately, no idea in this direction has been proposed so far, and the problem seems deep enough to make a naive approach unlikely to succeed.

Deciding the Embedding Conjecture could be easier. The conjecture being true for several unrelated families of elements of M_{LD}, namely simple elements, braidlike elements, and elements $\Delta_t^{(k)}$ (and their right divisors) might suggest that some uniform argument should unify the partial results and perhaps prove the conjecture. However, the proofs of Section 6 seem to be optimal in some sense, and there is no evidence why the conjecture should be true, except for esthetic reasons: there could exist a large term t and two complicated words u, v of degree 2 or 3 such that u and v are not $\equiv^+$-equivalent but the LD-expansions $t \bullet u$, $t \bullet v$ have the same skeleton.

The Embedding Conjecture implies that the family of all LD-expansions of a term is a lattice. It is not true that any two LD-equivalent terms need to be LD-expansions of a common term, *i.e.*, an LD-equivalence class need not be a lower semilattice, but the question for the upper bound is open.

Question 7.3. *Is every LD-equivalence class equipped with the relation of being an LD-expansion an upper semilattice, i.e., do any two LD-equivalent terms admit a smaller common LD-expansion?*

The question of whether a possible proof of convergence for the Polish Algorithm would provide any progress with the Embedding Conjecture is open, and so are many questions about progressive LD-expansions in general.

Conjecture 7.4. *Assume that the term t' is a progressive LD-expansion of the term t. Then the term $\partial t'$ is a progressive LD-expansion of every simple LD-expansion of t.*

Due to the lack of transitivity of the relation "being a progressive LD-expansion", even the natural fact that, if t' is a progressive LD-expansion of t, so is $\partial t'$ remains open. Only weak partial results are known, such as the fact that, if t' is a basic LD-expansion of t, then $\partial t'$ is a progressive LD-expansion of t (with a rather complicated proof). Technical statements such as Conjecture 7.4 were considered as lemmas toward an inductive proof of $\partial^k t$ being a progressive LD-expansion of t for every k. The latter result is now established using the techniques of Section 4, but the statement above, as well as various analogs, remains open.

Confluent families provide powerful tools for constructing progressive words even outside the framework of perfect terms. However, we know no general method for associating with every confluent family F in which t is terminal a new confluent family ∂F in which ∂t would be terminal. Simple LD-expansions of elements of F are not sufficient, as a degree 2 expansion need not be a simple expansion of a simple expansion.

Let us conclude with a statement of a completely different flavour:

Conjecture 7.5. (Pospichal) *Assume that t, t' are LD-equivalent terms of height at most n. Then there exists a sequence of terms $t_0 = t, t_1, \ldots, t_p = t'$ such that, for each i, t_i is a basic LD-expansion of t_{i-1} or vice versa and each term t_i has height at most n.*

In the previous framework, a natural idea is to start with a word w on $\mathbf{A} \cup \mathbf{A}^{-1}$ such that the operator LD_w maps t to t', and to try to transform w into an $\equiv$-equivalent word w' with the desired property, *i.e.*, one such that $t \bullet w'$ exists and, for every prefix w_0 of w', the term $t \bullet w_0$ has height at most n. In particular, we can systematically explore all words obtained from w using left and right reversing: left reversing need not terminate, but no problem arises here provided we restrict ourselves to words w' such that $t \bullet w'$ exists. Now, even if solutions exist, it is not clear that the previous process leads to them, for it could happen that introducing new temporary pairs xx^{-1} or $x^{-1}x$ is needed.

Appendix. Other Algebraic Identities

The study developed in the previous chapters for the specific case of left self-distributivity can be developed for other algebraic identities as well. Here we briefly compare the results obtained in various cases.

Assume that $\vec{I}$ is a family of algebraic identities involving a single binary operation (extending to more general cases is easy). For each identity I_k, and each address α in **A**, we can introduce the (partial) operator $\mathrm{I}_{k,\alpha}$ on T_∞ that corresponds to applying I_k (considered as an oriented pair) to the α-subterm—actually, we obtain a well defined operator, *i.e.*, a functional correspondence, only for an identity such that the same variables occur on both sides, as is the case for all natural examples. Then, we consider the monoid $\mathcal{G}_{\vec{I}}$ generated by all operators $\mathrm{I}_{k,\alpha}^{\pm 1}$. The argument of Proposition VII.1.22 (independence) shows that, in every case, the geometry monoid $\mathcal{G}_{\vec{I}}$ does not depend on the set of variables. It also does not depend on the choice of the identities, but only on the equational variety they define: if we replace the initial family $\vec{I}$ with a new family $\vec{J}$ defining the same equational variety, *i.e.*, such that every identity in $\vec{J}$ is a consequence of $\vec{I}$, and conversely, then the geometry monoids $\mathcal{G}_{\vec{I}}$ and $\mathcal{G}_{\vec{J}}$ are isomorphic.

The case of associativity

The case of the associativity identity (A): $x(yz) = (xy)z$ is one of the simplest cases—and it seems to be the only one that has been investigated in the literature along the lines considered here, yet if in a slightly different framework. We refer to (Dehornoy 96) for some details.

From a syntactic point of view, the associativity and the left self-distributivity identities are very similar. The main difference, which explains why associativity is much simpler, is that the latter is symmetric.

Thus, for every address α, we consider the operator A_α on T_∞ that consists in replacing the α-subterm, supposed to be of the form $t_1{\cdot}(t_2{\cdot}t_3)$, with $(t_1{\cdot}t_2){\cdot}t_3$, and we let $\mathcal{G}_A$ be the monoid generated by all operators $\mathrm{A}_\alpha^{\pm 1}$. As in Section VII.2, we look at those relations connecting the operators A_α. With obvious notations, we find now

$$\gamma 0\alpha \cdot \gamma 1\beta \equiv^+ \gamma 1\beta \cdot \gamma 0\alpha, \tag{7.1}$$

$$\gamma 0\alpha \cdot \gamma \equiv^+ \gamma \cdot 00\alpha, \tag{7.2}$$

$$\gamma 10\alpha \cdot \gamma \equiv^+ \gamma \cdot \gamma 01\alpha, \tag{7.3}$$

$$\gamma 11\alpha \cdot \gamma \equiv^+ \gamma \cdot \gamma 1\alpha, \tag{7.4}$$

$$\gamma \cdot \gamma \equiv^+ \gamma 1 \cdot \gamma \cdot \gamma 1. \tag{7.5}$$

Relation (7.1) is a trivial orthogonality relation, Relations (7.2) to (7.4) are inheritance relations. The most interesting one is (7.5), which is the counterpart to the LD-relation of type (1)

$$\gamma \cdot \gamma 1 \cdot \gamma \equiv^{+} \gamma 1 \cdot \gamma \cdot \gamma 1 \cdot \gamma 0.$$

Relation (7.5) is a formulation in terms of operators and addresses of the well known MacLane–Stasheff pentagon relation for associativity (MacLane 63), (Stasheff 63).

The confluence property, *i.e.*, the result that any two A-equivalent term admit a common A-expansion, is true, and, actually, trivial: in this case, we can define ∂t to be the term obtained from t by pushing all parentheses to the left using associativity, and, then, t being A-equivalent to t' corresponds to $\partial t = \partial t'$—which makes constructing normal forms trivial. The previous property relies on the fact that no variable is duplicated when associativity is applied, so, in some sense, associativity cannot be repeated: all A-expansions are simple A-expansions.

The sequel is easy. We consider the group G_A and the monoid M_A for which Relations (7.1) to (7.5) form a presentation. As in the case of (LD), the presentation is associated with a complement (on the right). The complement admits a norm: here we cannot use the size of the terms, which is preserved by the operators A_α, but we can use $\nu(u) = \mu_0(t_u^R) - \mu_0(t_u^L)$, where $\mu_0(t)$ is defined to be the total number of 0's involved in all addresses of the outline of t. Then the complement is coherent on $\mathbf{A}$, and, as in Chapter VIII, we deduce that word reversing is complete, and, therefore, the monoid M_A is left cancellative. For the complement to be convergent easily follows from the fact that, for every term t and every a in M_A such that $t \bullet a$ exists, the term ∂t is an A-expansion of $t \bullet a$. Here, the word $\boldsymbol{\Delta}_t$ can be defined by the inductive formula

$$\boldsymbol{\Delta}_t = 0\boldsymbol{\Delta}_{t_1} \cdot 1\boldsymbol{\Delta}_{t_2} \cdot \phi \cdot 0 \cdot \dots \cdot 0^{\mathrm{size}(t_2)-2}$$

for $t = t_1{\cdot}t_2$. So we conclude that right lcm's exist in M_A, and, as Identity (A) is symmetric, the monoid M_A being left cancellative implies its being right cancellative as well. Using the results of Chapter II, we conclude that M_A embeds in G_A (so the Embedding Conjecture is true in this case), and that G_A is the corresponding group of fractions.

We can now describe the connection between the monoid $\mathcal{G}_A$ and the group G_A completely using the method of Section VIII.2. Here, we can use the blueprint χ_t defined by

$$\chi_t = \chi_{t_1} \cdot 1\chi_{t_2} \cdot \phi$$

for $t = t_1{\cdot}t_2$: the property leading to this definition is that, for p large enough, $t{\cdot}x^{[p]}$ is A-equivalent to $x^{[p+\mathrm{size}(t)]}$ for p large enough—this is not the absorption property used in the case of (LD). Using χ_t, one shows that, for all word w, w' on $\mathbf{A} \cup \mathbf{A}^{-1}$, $w \equiv w'$ holds if and only if $t \bullet w = t \bullet w'$ holds for every term t whenever the latter terms exists, if and only if $t \bullet w = t \bullet w'$ holds for at least one such term t. The point is that, in the current case, all terms t_w^L and t_w^R are injective, and, therefore, the operator A_w is never empty, so G_A is simply the quotient of $\mathcal{G}_A$ under the congruence "to agree on at least one term". Also, the submonoid $\mathcal{G}_A^+$ of $\mathcal{G}_A$ generated by the operators A_α alone is isomorphic to M_A.

All results of the current chapter admit trivial counterparts when associativity is considered. Every element of the monoid M_A admits a unique progressive expression, and the counterpart of the Polish Algorithm trivially converges, as every A-equivalence class is finished. A linear order on G_A can be deduced from the linear ordering $<_L$ of terms of Section VIII.3. This order can also be constructed by associating with every element of G_A a piecewise affine homeomorphism of the real line as in Exercise VIII.3.28 (Ghys & Sergiescu 87).

The group G_A has been considered in (Mc Kenzie & Thomson 73), it is known as Richard Thompson's group F, and it enjoys many interesting properties: see (Cannon, Floyd, Parry 96). Let us mention here that a significant difference between G_{LD} and G_A is that the latter is of finite type: with our current notations, G_A is generated by g_ϕ and g_1.

The case of central duplication

Let us conclude with the case of the identity

$$x(yz) = (xy)(yz), \tag{CD}$$

which will be called *central duplication* as the central factor y is duplicated. It turns out that this identity is very similar to (LD), though the technical verifications are quite different (Dehornoy *00a*). A confluence property is still true, and it is connected with the existence of a nontrivial operator ∂, as for (LD). The crucial relation in the monoid $\mathcal{G}_{CD}$ now takes the form

$$\gamma \cdot 1 \cdot \gamma \equiv^+ \gamma 1 \cdot \gamma \cdot \gamma 0,$$

and one can still check that the associated complement is normed, coherent, and convergent on the right. Then there exists a convenient notion of blueprint; the inductive formula is now

$$\chi_t = \chi_{t_1} \cdot 1\chi_{t_2} \cdot \phi \cdot 1\chi_{t_2}^{-1} \tag{7.6}$$

for $t = t_1{\cdot}t_2$. One deduces a connection between the monoid $\mathcal{G}_{CD}$ and the group G_{CD} associated with the CD-relations which is exactly similar to the connection between $\mathcal{G}_{LD}$ and G_{LD}. This leads to a solution for the word problem of Identity (CD), and to the construction of a realization of the free CD-system of rank 1 involving the right cosets associated with the subgroup $G_0 = \mathrm{sh}_0(G_{CD})$ generated by all elements $g_{\alpha 0}$. As in Proposition VIII.3.5 (word problem), we can solve the word problem for one-variable terms by showing that $t_1 =_{CD} t_2$ is equivalent to

$$\mathrm{dil}(1, D(\chi_{t_1}^{-1} \cdot \chi_{t_2})) = \mathrm{dil}(1, N(\chi_{t_1}^{-1} \cdot \chi_{t_2})),$$

and the operation on G_0-cosets is induced by the operation

$$a * b = a \cdot \mathrm{sh}_1(b) \cdot g_\phi \cdot \mathrm{sh}_1(b^{-1})$$

on G_{CD}, a direct translation of (7.6). As in the case of (LD), a crucial point is that left division in free CD-system has no cycle—and, therefore, iterated left divisibility gives an ordering on free CD-systems. A noticeable difference is as follows: quotienting G_{LD} by the normal subgroup generated by $\mathrm{sh}_0(G_{LD})$ yields the braid group B_∞, and an LD-system with domain B_∞ for which we have a direct proof that left division has no cycle; in the case of (CD), the normal subgroup generated by $\mathrm{sh}_0(G_{CD})$ is all of G_{CD}: so the previous approach fails, and the only known proof of the acyclicity of left division is the syntactic proof using the argument of Proposition VIII.3.8 (syntactic acyclicity).

Many questions about (CD) remain open, in particular the Embedding Conjecture and the convergence of the counterpart to the Polish Algorithm. Probably the most interesting question is to understand why the results for (LD) and (CD) are so similar, while the presentations of the associated monoids and the technical computations look unrelated.

Other identities

Other examples directly eligible for the current approach are for instance the identities $x(yz) = (xz)(yz)$—here everything works, but the relations lead to a monoid which is not right cancellative—or $x(yz) = (xy)(x(xz))$—here, a blueprint exists, but confluence seems to fail.

Different results (and presumably different tools) are to be expected for those identities that do not preserve the ordering of the variables, like commutativity $xy = yx$, or entropicity $(xy)(zu) = (xz)(yu)$. It seems that almost nothing is known so far.

Finally, let us recall that introducing a geometry monoid is also possible when several identities are considered simultaneously. We could therefore consider the variety of all monoids, or of all groups. However, as describing the free objects seems to be the most natural application of the method, its interest in such cases is not clear. A more promising case could be that of (LD) together with the idempotency identity $x = xx$: as we shall mention in Section X.4, almost nothing is known in this case, and studying the geometry monoid could be helpful.

Part C

Other LD-Systems

X More LD-Systems

So far we have described several examples of LD-systems: conjugacy of a group, braid exponentiation, injection bracket. In this chapter, we give new examples, and mention some general results about LD-systems, and about those special families obtained by prescribing additional identities.

The organization of the chapter is as follows. In Section 1, we introduce an infinite family of finite monogenic LD-systems that we call the Laver tables. For each n, there exists one Laver table with 2^n elements, and these tables organize into a projective system. Here we establish their basic properties. In Section 2, we sketch Drápal's result that every finite monogenic LD-system can be constructed from a Laver table using some uniform scheme. In Section 3, we introduce multi-LD-systems, which are sets equipped with several mutually left distributive operations, and we describe free multi-LD-systems in terms of free LD-systems. In Section 4, we consider idempotents in LD-systems, and LDI-systems, which are those LD-systems where every element is idempotent. We establish Joyce's result that group conjugacy is axiomatized by the axioms of LD-quasigroups plus idempotency when both "aba^{-1}" and "$a^{-1}ba$" are considered, and Larue's result that group conjugacy is not axiomatized by (LD) plus idempotency when "aba^{-1}" only is used. Finally, in Section 5, we consider LRD-systems, which are those LD-systems that also satisfy the right self-distributivity law. The main result relates divisible LRD-systems with commutative Moufang loops. In an appendix, we explicitly display the Laver tables A_n for $n \leq 6$, and, partially, for $n \leq 10$.

1. The Laver Tables

Here we introduce an infinite family of monogenic LD-systems. These remarkable LD-systems were discovered by R. Laver while studying elementary embeddings (see Chapter XII), and we call them Laver tables. Here, we construct the Laver tables directly, and establish some basic properties. These tables turn out to be extremely complicated from a combinatorial point of view. We shall come back to them in Chapter XIII.

Left monogenic LD-systems

Let us address the question of constructing a finite monogenic LD-system. We can start with an incomplete table, and try to complete it by using Identity (LD). Here, we consider the case when the first column is given. So, let us try to construct a monogenic LD-system $(\{1, \dots, N\}, *)$ where every element is a left power of 1, *i.e.*, is $1_{[p]}$ for some p—we recall that the left powers $x_{[p]}$ are inductively defined by $x_{[1]} = x$ and $x_{[p+1]} = x_{[p]} \cdot x$. Without loss of generality, we may assume $1_{[p]} = p$ for $1 \leq p \leq N$, and the first column of our table takes the form

	1	2	...	N
1	2			
$\vdots$				
a	$a+1$			
$\vdots$				
$N-1$	N			
N				

Only the last value $N * 1$, *i.e.*, $1_{[N+1]}$, remains to be fixed. Here we consider the case $1_{[N+1]} = 1$, *i.e.*, we assume that the values in the first column form a cycle. We shall see below that everything is then determined, *i.e.*, there is at most one way to complete the table.

Notation 1.1. (projection) For a an integer, we denote by $(a)_N$ the unique integer x in $\{1, \dots, N\}$ satisfying $x = a \bmod N$.

With this notation, the condition for the values in the first column of our (possible) table is that, for every a in $\{1, \dots, N\}$, we have

$$a * 1 = (a+1)_N. \tag{1.1}$$

Lemma 1.2. *(i) For every N, there exists a unique operation $*$ on $\{1, \dots, N\}$ satisfying* (1.1) *and, for all a, b,*

$$a * (b * 1) = (a * b) * (a * 1). \tag{1.2}$$

(ii) The following relations hold in the resulting system:

$$a * b \quad \begin{cases} = b & \text{for } a = N, \\ = a + 1 & \text{for } b = 1, \text{ and for } a * (b-1) = N, \\ > a * (b-1) & \text{otherwise.} \end{cases} \tag{1.3}$$

For $a < N$, there exists $p \leq N - a$ and $c_1 = a + 1 < c_2 < \ldots < c_p = N$ such that, for every b, we have $a * b = c_i$ with $i = (b)_p$, hence, in particular, $a * b > a$.

Proof. The table is constructed using a double induction: $a * b$ is computed using descending induction on a, and, for each value of a, using ascending induction on b. The induction hypothesis is that the values are well defined and (1.3) holds. By (1.1), this is true for $a = N$ and $b = 1$. For $a = N$ and $b > 1$, say $b = c + 1$, (1.2) prescribes $a * b = (a * c) * (a * 1)$, hence, by induction hypothesis, $a * b = c * 1 = c + 1 = b$. Assume now $a < N$. For $b = 1$, we have $a * b = a + 1$ by (1.1). Otherwise, assume $b = c + 1$. Again, (1.2) requires $a * b = (a * c) * (a * 1)$. By induction hypothesis, $a * c > a$ holds, all values $(a * c) * d$ have already been determined, and there is only one possibility for the value of $a * b$. Two cases are possible. Either we have $a * c = N$, and then $(a * c) * (a * 1) = a * 1 = a + 1$, or we have $a * c < N$, and then, by induction hypothesis, $(a * c) * (a * 1) > a * c$. Again, the induction hypothesis is maintained. So the table is completely determined, and (1.2) holds since all pairs (a, b) have been considered during the constructing process.

For the last statement, (1.3) implies that, for $a < N$, the successive values of $a * 1$, $a * 2$, $\ldots$ increase as long as the value N is not reached. Since the first value is $a + 1$, this happens after at most $N - a$ steps. Now, $a * p = N$ implies $a * (p + b) = a * b$ for every positive b, as shown by an induction. Indeed, for $b = 1$, we have $a * (p + b) = a + 1 = a * b$ by (1.3). Otherwise, assume $b = c + 1$. By (1.2), we have

$$a * (p + b) = (a * (p + c)) * (a * 1) = (a * c) * (a * 1) = a * (c + 1) = a * b. \quad \blacksquare$$

Definition 1.3. (system S_N, period) The binary system involved in Lemma 1.2 is called S_N; for $1 \leq a \leq N$, the least integer p such that $a * p = N$ holds in S_N is called the *period* of a in S_N.

Lemma 1.2 implies strong constraints on the lower part of the table of S_N. Indeed, the last row is the identity. Then the period of $N - 1$ must be 1, hence we have $(N - 1) * b = N$ for every b. Similarly, the period of $N - 2$ is 2: it cannot be 1 as $(N - 2) * 1 = N - 1$ holds. So we have

$$(N-2) * b = \begin{cases} N - 1 & \text{if } b \text{ is odd,} \\ N & \text{if } b \text{ is even.} \end{cases} \tag{1.4}$$

Computing the other rows quickly turns out to be difficult.

Example 1.4. **(system** S_N) *The tables of* S_4 *and* S_5 *are respectively:*

	1	2	3	4
1	2	4	2	4
2	3	4	3	4
3	4	4	4	4
4	1	2	3	4

	1	2	3	4	5
1	2	4	5	2	4
2	3	4	5	3	4
3	4	5	4	5	4
4	5	5	5	5	5
5	1	2	3	4	5

Indeed, assuming for instance that the rows of 5, 4 *and* 3 *in* S_5 *have been constructed, the row of* 2 *is determined by:*

$$\begin{aligned}&2 * 1 = 3,\\ &2 * 2 = (2 * 1) * (2 * 1) = 3 * 3 = 4,\\ &2 * 3 = (2 * 2) * (2 * 1) = 4 * 3 = 5,\end{aligned}$$

and the values 3, 4 *are then repeated. The periods of* 1, ..., 5 *in* S_5 *are* 3, 3, 2, 1, *and* 5 *respectively.*

Lemma 1.5. *Every LD-system admitting a single generator* g *satisfying* $g_{[N+1]} = g$ *and* $g_{[a]} \neq g$ *for* $a \leq N$ *is isomorphic to* S_N.

Proof. Assume that S is an LD-system generated by an element g satisfying the above conditions. A double induction gives, for a, $b \leq N$, the equality $g_{[a]}g_{[b]} = g_{[a*b]}$, where $a * b$ refers to the product in S_N. So the set of all left powers of g is closed under product, and S, which has exactly N elements, is isomorphic to S_N. ∎

At this point, the natural question is whether the system S_N defined above is left self-distributive: by construction, certain occurrences of the left distributivity identity hold in the table, but this does not guarantee that it holds for all possible triples. Actually, it need not. In the case of S_5, we find

$$2 * (2 * 2) = 3 \neq (2 * 2) * (2 * 2) = 5,$$

so $(S_5, *)$ is *not* an LD-system. The result is as follows:

Proposition 1.6. **(LD-systems)** *(i) If* N *is not a power of* 2, *there exists no LD-system satisfying* (1.1).

(ii) For each n, *there exists a unique LD-system with domain* $\{1, \dots, 2^n\}$ *that satisfies* (1.1), *namely the system* S_{2^n} *of Lemma* 1.2.

We begin with auxiliary lemmas.

Lemma 1.7. *The binary system S_N is an LD-system if and only if $a * N = N$ holds for every a, and, therefore, if and only if the period of each row divides N.*

Proof. Since $a * b = N$ holds if and only if b is a multiple of the period of the a-th row, the two conditions of the lemma are equivalent. Let us first assume that the operation $*$ is left distributive. Then, in particular, we have for every a

$$a * N = a * (N * N) = (a * N) * (a * N),$$

hence the element $a * N$ is idempotent. Now (1.3) shows that no element but N is idempotent in S_N, and, therefore, the conditions of the lemma hold.

Conversely, assume that these conditions hold. We show $a * (b * c) = (a * b) * (a * c)$ using descending induction on a, and, for a given a, descending induction on b and ascending induction on c. For $a = N$, the equality amounts to

$$N * (b * c) = b * c = (N * b) * (N * c),$$

which is true. Assume $a < N$. For $b = N$, we have, by using the assumption $a * N = N$,

$$a * (N * c) = a * c = N * (a * c) = (a * N) * (a * c).$$

Assume now $b < N$. For $c = 1$, the equality is (1.2); for $c = d + 1$, we find

$$a * (b * c) = a * (b * (d * 1)) = a * ((b * d) * (b * 1)). \tag{1.5}$$

By (1.3), we have $b * d > b$, hence, by induction hypothesis,

$$a * ((b * d) * (b * 1)) = (a * (b * d)) * (a * (b * 1)). \tag{1.6}$$

As $d < c$ and $1 < c$ hold, we have, always by induction hypothesis,

$$a * (b * d) = (a * b) * (a * d), \quad a * (b * 1) = (a * b) * (a * 1),$$

hence, using (1.6),

$$a * (b * c) = ((a * b) * (a * d)) * ((a * b) * (a * 1)). \tag{1.7}$$

Now $a * b > a$ holds, and we may apply the induction hypothesis to conclude that the second member of (1.7) is $(a * b) * ((a * d) * (a * 1))$, which is, by (1.2), $(a * b) * (a * (d * 1))$, *i.e.*, $(a * b) * (a * c)$. So $*$ is left self-distributive. ■

For instance, we see that the last column in the table of S_4 is constant with value 4: the previous lemma implies that S_4, contrary to S_5, is an LD-system.

The next step is to compare various systems S_N: we investigate whether projecting *modulo* N induces a homomorphism of $S_{N'}$ onto S_N for $N' > N$.

Lemma 1.8. *(i) Assume that S is an LD-system and $g_{[N'+1]} = g$ holds in S. Then mapping a to $g_{[a]}$ defines a homomorphism of $S_{N'}$ into S.*

(ii) In particular, if S_N is an LD-system and N divides N', then mapping a to $(a)_N$ defines a homomorphism of $S_{N'}$ onto S_N.

Proof. (i) Let $\pi(a) = g_{[a]}$. We prove $\pi(a * b) = \pi(a) \cdot \pi(b)$ using a double induction, first descending on a, then ascending on b. For $a = N'$ and $b = 1$, we have

$$\pi(a * b) = \pi(N' * 1) = \pi(1) = g = g_{[N'+1]} = g_{[N']} \cdot g = \pi(a) \cdot \pi(b).$$

For $b = c + 1$, we find

$$\pi(a * b) = \pi(N' * b) = \pi(b) = g_{[b]} = (g_{[N'+1]})_{[b]} = g_{[N']} \cdot g_{[b]} = \pi(a) \cdot \pi(b).$$

Assume now $a < N'$. For $b = 1$, we have

$$\pi(a * b) = g_{[a+1]} = g_{[a]} \cdot g = \pi(a) \cdot \pi(1).$$

For $b = c + 1$, we find, using (1.2), $\pi(a * b) = \pi((a * c) * (a * 1))$. As $a * c > a$ holds, the induction hypothesis implies

$$\begin{aligned}\pi(a * b) &= \pi(a * c) \cdot \pi(a * 1) = (\pi(a) \cdot \pi(c)) \cdot (\pi(a) \cdot \pi(1)) \\ &= (g_{[a]} \cdot g_{[c]}) \cdot (g_{[a]} \cdot g) = g_{[a]} \cdot (g_{[c]} \cdot g) = g_{[a]} \cdot g_{[b]} = \pi(a) \cdot \pi(b).\end{aligned}$$

(ii) We apply (i) with $S = S_N$ and $g = 1$. This is legal as, by construction, $1_{[a]} = (a)_N$ holds in S_N. Hence $1_{[N'+1]} = 1$ holds when N divides N'. ■

We deduce the fundamental relation between S_N and S_{kN}:

Lemma 1.9. *Assume that S_N is an LD-system, and N is a proper divisor of N'. Write $*'$ for the operation of $S_{N'}$. Then, for $1 \le b \le N'$, the following equalities hold:*

(i) for $N' - N < a \le N'$,

$$a *' b = (a)_N * (b)_N + N' - N,$$

so, in particular, the period of a in $S_{N'}$ is that of $(a)_N$ in S_N;

(ii) for $a = N' - N$,

$$a *' b = (b)_N + N' - N,$$

so, in particular, the period of a in $S_{N'}$ is N;

(iii) for $N' - 2N \le a < N' - N$,

$$a *' b = (a)_N * (b)_N + N' - 2N + \delta(a, b)N,$$

with $\delta(a, b) \in \{0, 1\}$, so, in particular, the period of a in $S_{N'}$ is either that of $(a)_N$ in S_N, or its double.

Proof. (i), (ii) Assume $N' = kN$. Let $M = N' - N = (k - 1)N$. By (1.3), the possible values for $(M + a) *' b$ are above M, hence they belong to $\{N' - N + 1, \ldots, N'\}$. By Lemma 1.8, projection *modulo* N is a homomorphism π of $S_{N'}$ onto S_N, hence we have $\pi(a *' b) = \pi(a) * \pi(b)$. The formulas correspond to the only remaining possibility.

(iii) The argument is similar: the permitted values are the $2N$ integers $N' - 2N + 1, \ldots, N'$, which gives two possibilities once the value *modulo* N is given by the projection. Let p be the period of $(a)_N$ in S_N. Then $a *' p$ is either N', or $N' - N$. In the first case, the period of a in $S_{N'}$ is p. If the second case, it is greater. For $p < b < 2p$, the value of $a *' b$ cannot be N', since *modulo* N it equals $(a)_N * (b - p)$. On the other hand, we have $a *' (2p) > N' - N$ and $a *' (2p) = N \pmod N$: the only possibility is $a *' (2p) = N'$, and $2p$ is the period of a in $S_{N'}$. ■

Example 1.10. *($S_{N'}$ vs. S_N) Consider $N = 4$ and $N' = 12$. The table of S_4 determines the rows of 8 to 11 in the table of S_{12}, and it gives partial indications about the rows 4 to 7:*

	1	2	3	4	5	6	7	8	9	10	11	12
4	5	6/10	7/11	8/12	5/9	6/10	7/11	8/12	5/9	6/10	7/11	8/12
5	6	8/12	6/10	8/12	6/10	8/12	6/10	8/12	6/10	8/12	6/10	8/12
6	7	8/12	7/11	8/12	7/11	8/12	7/11	8/12	7/11	8/12	7/11	8/12
7	8	8/12	8/12	8/12	8/12	8/12	8/12	8/12	8/12	8/12	8/12	8/12
8	9	10	11	12	9	10	11	12	9	10	11	12
9	10	12	10	12	10	12	10	12	10	12	10	12
10	11	12	11	12	11	12	11	12	11	12	11	12
11	12	12	12	12	12	12	12	12	12	12	12	12
12	1	2	3	4	5	6	7	8	9	10	11	12

By Lemma 1.2(ii), the values on each row increase as long as N' is not reached. This implies that the values of $\delta(a, 1)$, $\delta(a, 2)$, etc. must increase as well: so $\delta(a, b)$ is continuously 0 until it becomes 1, and it then remains 1. For instance, we deduce from the fact that the row of 4 in S_4 is $(1, 2, 3, 4)$ that the row of 4 in S_{12} is one of the sequences $(5, 10, 11, 12)$, $(5, 6, 11, 12)$, $(5, 6, 7, 12)$, and $(5, 6, 7, 8, 9, 10, 11, 12)$. Actually, the correct sequence is the latter, as the reader can check

	1	2	3	4	5	6	7	8	9	...
4	5	6	7	8	9	10	11	12	5	...
5	6	8	10	12	6	...				
6	7	8	11	12	7	...				
7	8	12	8	12	8	...				

Lemma 1.11. *Assume that the periods of $a + 1$, $a + 2$, $\ldots$, $a + (d - 1)$ in S_N*

divide a. Then, for $1 \le b \le d$, we have $a * b = a + b$ in S_N.

Proof. Use induction on b. For $b = 1$, the result is true by (1.1). For $b = c+1$, (1.2) and the induction hypothesis imply

$$a * b = (a * c) * (a * 1) = (a + c) * (a + 1).$$

Our assumption is that the period of $a + c$ divides a. We deduce

$$(a + c) * (a + 1) = (a + c) * 1 = (a + c) + 1 = a + b. \qquad \blacksquare$$

For instance, in the table of S_{12}, the periods of the rows 5 to 11 all divide 8, because all periods divide 4 in the table of S_4: the previous lemma then implies $4 * b = 4 + b$ for $1 \le b \le 8$.

We are now able to prove Proposition 1.6 (LD-system).

Proof. By Lemma 1.2, if a binary system S contains $1, \ldots, N$ and satisfies (1.1) and (1.2), then the subsystem of S generated by 1 is S_N. We show inductively on $n \ge 0$ that all systems S_{2^n} are LD-systems. The result is true for $n = 0$. For $n \ge 1$, the induction hypothesis tells us that $S_{2^{n-1}}$ is an LD-system, so, by Lemma 1.7, every period in $S_{2^{n-1}}$ divides 2^{n-1}, hence, by Lemma 1.9, every period in S_{2^n} divides 2^n. So, by Lemma 1.7 again, S_{2^n} is an LD-system.

Assume now $N = p2^n$, with p an odd number, $p \ge 3$. Since S_{2^n} is an LD-system, the periods of $1, \ldots, 2^n - 1$ in S_{2^n} divide 2^n, and even 2^{n-1} since no element except 2^n may have period 2^n. So, by Lemma 1.9, the periods of the $2^{n+1} - 1$ elements $N - 2^{n+1} + 1, \ldots, N - 1$ in S_N all divide 2^n. Hence Lemma 1.11 implies

$$(N - 2^{n+1}) * b = N - 2^{n+1} + b$$

in S_N for $1 \le b \le 2^{n+1}$. In particular, the period of $N - 2^{n+1}$ in S_N is 2^{n+1}, and it does not divide N. Hence S_N cannot be an LD-system. $\blacksquare$

The Laver tables

Building on Proposition 1.6 (LD-system), we introduce our final notations:

Definition 1.12. (**Laver table**) For $n \ge 0$, the n-th Laver table $(A_n, *_n)$ is defined to be the LD-system S_{2^n}, *i.e.*, the unique LD-system with domain $\{1, 2, \ldots, 2^n\}$ that satisfies (1.1).

By Lemma 1.2, every row in the table A_n is periodic and, by Lemma 1.7, the corresponding period is a power of 2.

Definition 1.13. **(period)** For $1 \leq a \leq 2^n$, the *period* of a in A_n is the number of distinct values in the row of a in A_n; $o_n(a)$ is defined to be the unique number such that $2^{o_n(a)}$ is the period of a in A_n.

Example 1.14. **(Laver tables)** *The first Laver tables are*

A_0	1
1	1

A_1	1	2
1	2	2
2	1	2

A_2	1	2	3	4
1	2	4	2	4
2	3	4	3	4
3	4	4	4	4
4	1	2	3	4

A_3	1	2	3	4	5	6	7	8
1	2	4	6	8	2	4	6	8
2	3	4	7	8	3	4	7	8
3	4	8	4	8	4	8	4	8
4	5	6	7	8	5	6	7	8
5	6	8	6	8	6	8	6	8
6	7	8	7	8	7	8	7	8
7	8	8	8	8	8	8	8	8
8	1	2	3	4	5	6	7	8

A_4	1	2	3	4	5	6	7	8	9	10	11	12	13	14	15	16
1	2	12	14	16	2	12	14	16	2	12	14	16	2	12	14	16
2	3	12	15	16	3	12	15	16	3	12	15	16	3	12	15	16
3	4	8	12	16	4	8	12	16	4	8	12	16	4	8	12	16
4	5	6	7	8	13	14	15	16	5	6	7	8	13	14	15	16
5	6	8	14	16	6	8	14	16	6	8	14	16	6	8	14	16
6	7	8	15	16	7	8	15	16	7	8	15	16	7	8	15	16
7	8	16	8	16	8	16	8	16	8	16	8	16	8	16	8	16
8	9	10	11	12	13	14	15	16	9	10	11	12	13	14	15	16
9	10	12	14	16	10	12	14	16	10	12	14	16	10	12	14	16
10	11	12	15	16	11	12	15	16	11	12	15	16	11	12	15	16
11	12	16	12	16	12	16	12	16	12	16	12	16	12	16	12	16
12	13	14	15	16	13	14	15	16	13	14	15	16	13	14	15	16
13	14	16	14	16	14	16	14	16	14	16	14	16	14	16	14	16
14	15	16	15	16	15	16	15	16	15	16	15	16	15	16	15	16
15	16	16	16	16	16	16	16	16	16	16	16	16	16	16	16	16
16	1	2	3	4	5	6	7	8	9	10	11	12	13	14	15	16

(see Appendix for further examples.) The periods of 1 *in* $A_0, \ldots, A_4$ *are respectively* 1, 1, 2, 4 *and* 4, *corresponding to the equalities*

$$o_0(1) = 0, \quad o_1(1) = 0, \quad o_2(1) = 1, \quad o_3(1) = 2, \quad o_4(1) = 2.$$

Observe that the above values are non-decreasing.

We can summarize the previous results as follows:

Proposition 1.15. (**basics**) *(i) For every n, $(A_n, *_n)$ is an LD-system; in A_n, 1 is the unique generator, 2^n is the unique idempotent, and we have $2^n *_n a = a$ and $a *_n 2^n = 2^n$ for every a.*

(ii) For $n' \geq n$, $a \mapsto (a)_{2^n}$ is a surjective morphism of $A_{n'}$ onto A_n.

(iii) All possible morphisms of A_m into A_n have the form $b \mapsto a *_n (b)_{2^n}$ for some fixed a. They all map 2^m to 2^n.

Proof. (i) That 1 is the only element that generates $(A_n, *_n)$ comes from the fact that $2^n *_n 1 = 1$ is the only occurrence of 1 in the table of A_n: so 1 cannot belong to the sub-LD-system of A_n generated by any other element. That 2^n is the only idempotent element follows from (1.3).

Point (ii) follows from Lemma 1.8.

(iii) Assume that f is a homomorphism of A_m into A_n. There exists a unique integer a satisfying $f(1) = a *_n 1$. An induction gives $f(b) = a_{[b]} = a *_n (b)_{2^n}$ for every b. Now, the equality $2^m *_m 2^m = 2^m$ implies $f(2^n) *_n f(2^n) = f(2^n)$, and therefore $f(2^n) = 2^m$, since 2^n is the only idempotent element of A_n. ∎

Proposition 1.16. (**projection**) *For every n, and every a with $1 \leq a \leq 2^n$, there exists a number $\theta_{n+1}(a)$ with $0 \leq \theta_{n+1}(a) \leq 2^{o_n(a)}$ and $\theta_{n+1}(2^n) = 0$ such that, for every b with $1 \leq b \leq 2^n$, we have*

$$a *_{n+1} b = a *_{n+1} (2^n + b) = \begin{cases} a *_n b & \text{for } b \leq \theta_{n+1}(a), \\ a *_n b + 2^n & \text{for } b > \theta_{n+1}(a), \end{cases} \tag{1.8}$$

$$(2^n + a) *_{n+1} b = (2^n + a) *_{n+1} (2^n + b) = a *_n b + 2^n. \tag{1.9}$$

Proof. Assume $a \leq 2^n$. Let $p = 2^{o_n(a)}$, and let $c_1 = a + 1, c_2, \ldots, c_p = 2^n$ be the values of $a *_n 1, \ldots, a *_n p$. By Lemma 1.9, for $1 \leq b \leq 2p$, we have $a *_{n+1} b = a *_n b + \delta(a,b)2^n$ for some $\delta(a,b)$ that is 0 or 1. Moreover, $a *_{n+1} b = 2^{n+1}$ is possible only for $b = p$ or $b = 2p$. So, we have

$$a *_{n+1} 1 < \ldots < a *_{n+1} p,$$

and $\delta(a,b) = 1$ implies $\delta(a,c) = 1$ for $b \leq c \leq p$. Two cases may occur. For $\delta(a,p) = 0$, *i.e.*, $a *_{n+1} p = 2^n$, the period of a in A_{n+1} is $2p$, $\delta(a,b) = 1$ holds for $p < b \leq 2p$, and we have $\theta_{n+1}(a) = p$. For $\delta(a,p) = 1$, we have $\theta_{n+1}(a) < p$, and the period of a in A_{n+1} is p. The rest is clear. ∎

Definition 1.17. (**threshold**) The integer $\theta_{n+1}(a)$ involved in Proposition 1.16 is called the *threshold* of a in A_{n+1}.

The previous proposition gives the following inductive formula for the periods: for $1 \leq a \leq 2^{n+1}$, we have

$$o_{n+1}(a) = \begin{cases} o_n(a) + 1 & \text{for } a \leq 2^n, \text{ if } \theta_{n+1}(a) = 2^{o_n(a)} \text{ holds}, \\ o_n(a) & \text{for } a \leq 2^n, \text{ if } \theta_{n+1}(a) < 2^{o_n(a)} \text{ holds}, \\ o_n(a - 2^n) & \text{for } 2^n < a < 2^{n+1}, \\ n + 1 & \text{for } a = 2^{n+1}. \end{cases}$$

As shown in Figure 2.1, the complete table of A_{n+1} is obtained from the table of A_n once the thresholds $\theta_{n+1}(1), \ldots, \theta_{n+1}(2^n)$ are known—the last one is always 0.

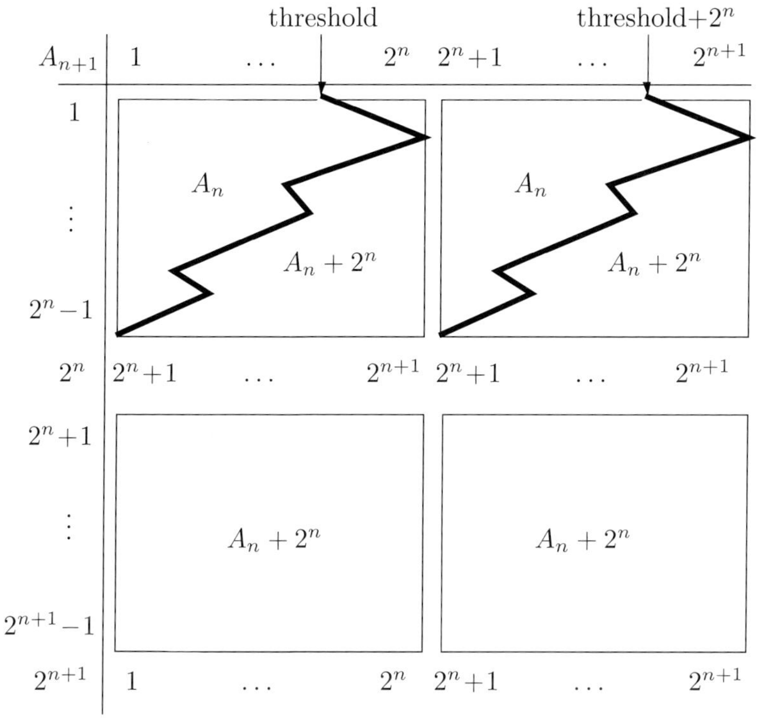

Figure 2.1: From A_n to A_{n+1} (Attention! The top part is not really correct: the periodic pattern in each row repeats more than twice in general.)

Example 1.18. (threshold) *From the tables of A_2, A_3, and A_4, we find*

$$\begin{array}{llll}\theta_3(1) = 2, & \theta_3(2) = 2, & \theta_3(3) = 1, & \theta_3(4) = 0,\\ \theta_4(1) = 1, & \theta_4(2) = 1, & \theta_4(3) = 2, & \theta_4(4) = 4,\\ \theta_4(5) = 2, & \theta_4(6) = 2, & \theta_4(7) = 1, & \theta_4(8) = 0.\end{array}$$

We obtain in this way a short description of A_n: the above 12 values contain all information needed for constructing A_4 (16 elements) from A_2.

Remark 1.19. By renaming the elements of A_n so that a becomes $2^n - a$, we obtain an LD-system A'_n, in which $2^n - 1$ is a generator and 0 is an idempotent. Then the table of A'_n is a subtable of the table of $A'_{n'}$ for $n' > n$. This makes the definition of an inductive limit of the tables A'_n natural. Similarly, we could use 0 rather than 2^n in the case of A_n. The optimal choice could be to identify 0 and 2^n and consider that the elements of A_n live in $\mathbf{Z}/2^n\mathbf{Z}$. However, it is convenient for the inductions to use the ordering $1 < 2 < \ldots < 2^n$, which explains our choice.

Presentation of A_n

By construction, 1 is a generator of A_n, and we have $1_{[2^n+1]} = 1$, which implies that $1_{[m+1]} = 1$ holds in A_n if and only if m is a multiple of 2^n. This property characterizes A_n.

Proposition 1.20. (presentation) *As an LD-system, A_n admits the presentation $\langle g \ ; \ g_{[m+1]} = g \rangle$ for every number m of the form $2^n(2p+1)$.*

Proof. Assume $m = 2^n(2p+1)$. Then A_n is an LD-system generated by an element, namely 1, that satisfies $1_{[m+1]} = 1$. In order to prove (i), it suffices that we prove that every LD-system S generated by an element g that satisfies $g_{[m+1]} = g$ is a quotient of A_n. For such a system S, by Lemma 1.8, the mapping $\pi : a \mapsto g_{[a]}$ defines a homomorphism of S_m into S. Since g generates S, this homomorphism is surjective. So S is isomorphic to the quotient S' of S_m under the congruence $\pi(a) = \pi(a')$, *i.e.*, $g_{[a]} = g_{[a']}$. Let m' be the least integer with $m' > 1$ satisfying $g_{[m'+1]} = g$: then the values $g_{[1]}, \ldots, g_{[m']}$ are distinct, so the congruence $\pi(a) = \pi(a')$ is the relation $(a)_{m'} = (a')_{m'}$, S' has m' elements, and m' divides m. Moreover S' satisfies the relations (1.1) and (1.2), hence S' coincides with $S_{m'}$. Since it is an LD-system, we have $m' = 2^{n'}$ for some n', and m' dividing m implies $n' \leq n$. We deduce $S' = A_{n'}$, and S is isomorphic to $A_{n'}$, a quotient of A_n. ∎

Let us conclude the section with the subsystems of A_n.

Proposition 1.21. (subsystems) *Every element of A_n generates a subsystem that is isomorphic to A_m for some $m \leq n$.*

Proof. Let a be an arbitrary element of A_n. Consider the left powers of a. Then $a_{[m]} \neq 2^n$ implies $a_{[m+1]} > a_{[m]}$. Hence there must exist a smallest integer N satisfying $a_{[N]} = 2^n$, and we have then

$$a < a_{[2]} < \ldots < a_{[N]} = 2^n,$$

and, therefore, $a_{[N+1]} = a$. By Lemma 1.5, the subsystem of A_n generated by a is isomorphic to the system S_N. Moreover, this system is an LD-system, so, by Proposition 1.6 (LD-system), N must be a power of 2, and $N \leq 2^n$ implies $N = 2^m$ for some $m \leq n$. ∎

The previous result involves *monogenic* subsystems of A_n only. Besides, there exist subsystems with more than one generator: for instance, the subset $\{2, 3, \dots, 2^n\}$ is closed under $*_n$, and it provides a subsystem of A_n with $2^n - 1$ elements.

Exercise 1.22. (**not closed**) *Are the LD-systems A_n closed, i.e., does $a *_n c = a^{[2]} *_n c$ imply $(\exists b)(a *_n b = c)$? [Answer: No: A_0 to A_3 are closed, but A_4 is not, as we have $1 *_4 4 = 2 *_4 4 = 16$, while 4 is not left divisible by 1, since the multiples of 1 are 2, 12, 14, and 16.]*

Exercise 1.23. (**equation**) *Show that, in A_n, the equation $x * x = 2^n$ has $n + 1$ solutions, namely the elements $2^n - 2^p$ with $0 \leq p \leq n$.*

Exercise 1.24. (**identity**) *Show that, for m of the form $2^n(2p + 1)$, the identity $x_{[m+1]} = x$ holds in A_n, and that A_n is free in the variety defined by (LD) and $x_{[m+1]} = x$. [Hint: If $a_{[m+1]} = a$ holds in an LD-system, $b_{[m+1]} = b$ holds for every b in the sub-LD-system generated by a.]*

---

2. Finite Monogenic LD-systems

In the case of associative systems, *i.e.*, of semigroups, monogenic systems can be described easily. The situation for LD-systems is more complicated. In this section, we sketch the description of all finite monogenic LD-systems given by A. Drápal. The main result is that every such system is constructed from the Laver tables using a uniform scheme. The construction involves several parameters and some free choices, resulting in a large and rather intricate family. However, the Laver tables definitely appear as the fundamental objects here, and their role may be compared with the role of the groups $\mathbf{Z}/n\mathbf{Z}$ in the description of finite (semi)groups.

Most of the proofs in this section will be sketched, or omitted.

Right monogenic LD-systems

A natural way for constructing finite monogenic LD-systems is to investigate the finite quotients of the known LD-systems. Now all examples involving conjugacy are idempotent, thus giving rise to trivial monogenic subsystems only, and almost nothing is known about the finite quotients of $(B_\infty, \wedge)$ and $(\mathfrak{I}_\infty, [\,])$. There remains the trivial operation $a^\wedge b = b + 1$ on the integers, which leads to the following finite LD-systems.

Definition 2.1. (table $C_{m,p}$) For $m \geq 0$ and $p \geq 1$, the LD-system $C_{m,r}$ is the set $\{1, \ldots, m+r\}$ equipped with the operation

$$a * b = \begin{cases} b+1 & \text{for } b < m+r, \\ m+1 & \text{for } b = m+r. \end{cases}$$

The LD-system $C_{m,r}$ is trivial in the sense that the product of a and b depends on b only. For $m = 0$, *i.e.*, in the case of a group $\mathbf{Z}/p\mathbf{Z}$, every element is a generator, while, for $m \geq 1$, the element 1 is the only generator, and 1 does not belong to $C_{m,r} * C_{m,r}$.

The system $C_{m,r}$ has the property that every element is a right power of the generator. Let us call such monogenic LD-systems *right* systems. It turns out that every right LD-system is closely connected with some system $C_{m,r}$.

Proposition 2.2. (right LD-systems) *Assume that S is a finite LD-system where every element is a right power of some fixed element. Then S is isomorphic to some system $C_{m,r}$, or to some system $V_A(C_{2,r})$ with $A \subseteq \{2, \ldots, 2+r\}$, where $V_A(C_{2,r})$ is the system obtained from $C_{2,r}$ by replacing $a * 1 = 2$ with $a * 1 = 2 + r$ for $a \in A$.*

Proof. (Sketch) Let g be an element of S such that every element is a right power of g. In particular g generates S. There must exist m, r such that the right powers of g are g, $g^{[2]}$, $\ldots$, $g^{[m+r]}$, and $g^{[m+r+1]} = g^{[m+1]}$ holds. An induction on a shows that the mapping $a \mapsto g^{[a]}$ is an isomorphism of $C_{m,r}$ onto S, *i.e.*, we have $g^{[a]}g^{[b]} = g^{[c]}$ where c is $a * b$ computed in $C_{m,r}$, except possibly in the cases $m = 2$, $a \geq 2$, $b = 1$ where the argument implies only that the value is either $g^{[2]}$ or $g^{[2+r]}$. One then checks that none of the latter values contradicts left self-distributivity. ■

Corollary 2.3. *Up to isomorphism, the finite monogenic LD-systems that are right divisible are the systems $C_{0,r}$ with $r \geq 1$.*

Proof. Let S be such an LD-system, and g be a generator of S. By hypothesis, every element of S appears in every row in the table of S, and, in particular, in the row of g. This shows inductively that every element of S is a right power of g. We apply the previous proposition, and observe that $C_{m,r}$ is right divisible for $m = 2$, while $V_A(C_{2,r})$ is never right divisible. ■

Systems of type I

The Laver tables A_n and the systems $C_{m,r}$ are finite, monogenic LD-systems, and so are all finite products of such systems. We shall see here that the systems A_n and the systems $C_{m,r}$ are, in some sense, the two ends in the classification of finite monogenic LD-systems, and that every such system can be described in terms of quotients of products $A_n \times C_{m,r}$.

The LD-systems A_n have been defined as those monogenic LD-systems where the first column is a cyclic permutation. By Proposition 1.20 (presentation), A_n is also characterized by the presentation $\langle g\ ;\ g_{[2^n+1]} = g\rangle$. Here, we shall use still another characterization. We know that the left translation of A_n associated with 2^n is the identity mapping. The existence of such an element actually characterizes A_n:

Lemma 2.4. *Assume that S is a finite, monogenic LD-system, and there exists e in S satisfying $ea = a$ for every a. Then S is isomorphic to some A_n.*

(The proof is omitted.) Finite monogenic LD-systems can be divided into two families.

Definition 2.5. (type) Assume that S is a finite monogenic LD-system. We say that S is of *type I* if $S = SS$ holds, *i.e.*, every element of S can be expressed as the product of two elements; otherwise, we say that S is of *type II.*

Lemma 2.6. *Assume that S is a finite monogenic LD-system.*

(i) The system S is of type I if and only if some translation of S is bijective;

(ii) If S is of type II, then S has a unique generator, namely the (necessarily unique) element of $S \setminus SS$.

Proof. (i) Assume that g generates S, and $g = ab$ holds. The set aS is a sub-LD-system of S that contains g, hence it is all of S. So the left translation associated with a is surjective, and it is therefore injective since S is finite.

(ii) The sub-LD-system of S generated by g is always included in the set $\{g\} \cup SS$. If the latter is S, the difference $S \setminus SS$ is included in $\{g\}$. ∎

Saying that S is of type I means that all elements of S appear in the table of S (viewed as a two-dimensional array): thus, the distinction is between the cases where the generator appears in the table and those where it does not.

Example 2.7. (type) *The system A_n is of type I, since the generator 1 appears in the table: $1 = 2^n *_n 1$ holds. The system $C_{0,r}$ is of type I, as we have $1 = r * 1$, but $C_{m,r}$ is of type II for $m \geq 1$.*

The product of two monogenic LD-systems is an LD-system, but it is not monogenic in general. However, the product $A_n \times C_{0,r}$ is generated by $(1,1)$, and we obtain in this way new examples of type I LD-systems. Actually, we can obtain more by letting the index "r" vary, *i.e.*, by replacing the product $A_n \times C_{0,r}$ by a more flexible construction of the form $\prod_{a \in A_n} C_{0,\rho(a)}$.

Definition 2.8. (tables $D_{n,\rho}$) For $n \geq 1$ and ρ a mapping of $\{1, \ldots, 2^n\}$ into $\mathbf{N}_+$, $D_{n,\rho}$ is defined to be the set

$$\{(a,i)\ ;\ 1 \leq a \leq 2^n\ ,\ 1 \leq i \leq \rho(a)\}$$

equipped with the operation $(a,i) * (b,j) = (a *_{A_n} b, i *_{C_{0,\rho(a)}} j)$.

The direct product $A_n \times C_{0,r}$ corresponds to ρ being the constant function with value r. Because the rows of (the table of) $C_{0,r}$ all are equal, the rows of $D_{n,\rho}$ that correspond to values (a,i) and (a,j) have to coincide. On the other hand, the rows in A_n are pairwise distinct, so the rows of (a,i) and (b,j) with $i \neq j$ cannot coincide in $D_{n,\rho}$. The main idea here is that the table of $D_{n,\rho}$ is obtained from the table of A_n by *splitting* the element a into $\rho(a)$ many elements, and that the operation of $D_{n,\rho}$ on these values is cyclic.

If the values of the integers $\rho(a)$ are chosen randomly, the system $D_{n,\rho}$ need not be left self-distributive. However, it is easy to describe those constraints on ρ that guarantee $D_{n,\rho}$ to be an LD-system.

Lemma 2.9. *(i) Assume that ρ satisfies the condition:*

$$A_n *_n a \subseteq A_n *_n b \quad \textit{implies} \quad \rho(a) \mid \rho(b). \tag{2.1}$$

Then $D_{n,\rho}$ is an LD-system generated by $(1,1)$.

(ii) Conversely, every quotient of $A_n \times C_{0,r}$ is isomorphic to $D_{n,\rho}$ for some function ρ that satisfies (2.1) *and is such that $\rho(1)$ divides r.*

Observe that Condition (2.1) above implies that $\rho(a)$ divides $\rho(1)$ for every a: $D_{n,\rho}$ is obtained from $A_n \times C_{0,\rho(1)}$ by collapsing some of the cycles, this being done in a coherent way, *i.e.*, in such a way that values that have been identified at level b remain identified at all levels a such that a divides b in A_n.

The following notions were considered in Exercise I.2.21:

Definition 2.10. (left and right equivalence) Assume that S is a binary system. For a, a' in S, we say that $a \equiv_L^S a'$ (or, simply, $a \equiv_L a'$) holds if $ax = a'x$ holds for every x in S. Similarly, we say that $a \equiv_R^S a'$ holds if $xa = xa'$ holds for every x in S.

If we consider the table of S, $a \equiv_L a'$ means that the rows of a and a' coincide, while $a \equiv_R a'$ means that the columns of a and a' do.

Lemma 2.11. (**left and right equivalence**) If S is an LD-system, then $\equiv_R^S$ is a congruence, while $\equiv_L^S$ is an equivalence relation that is compatible with multiplication on the left (but not necessarily with multiplication on the right).

The main technical point for completing the description of type I systems is:

Lemma 2.12. *A finite monogenic LD-system S is of type I if and only if the congruence $\equiv_R^S$ is trivial, i.e., if and only if any two columns in the table of S are distinct. In this case, the equivalence relation $\equiv_L^S$ is a congruence.*

From there, one obtains:

Proposition 2.13. (**type I**) *Up to isomorphism, the finite monogenic LD-systems of type I are the LD-systems $D_{n,\rho}$.*

Proof. (Sketch) Assume that S is an LD-system of type I. Then the quotient $S/\equiv_L$ is a finite, monogenic LD-system, and some row in its table is the identity. By Lemma 2.4, this quotient $S/\equiv_L$ is isomorphic to A_n for some n. Let π_S denote the associated projection

$$\pi_S : S \twoheadrightarrow A_n, \tag{2.2}$$

and r be the cardinality of $\pi_S^{-1}(1)$. One constructs a surjective morphism of $A_n \times C_{0,r}$ onto S. So S is isomorphic to a quotient of the latter product, and, by Lemma 2.9, this quotient is isomorphic to $D_{n,\rho}$ for some function ρ. ■

Systems of type II

The only examples of systems of type II we have met so far are the systems $C_{m,r}$ with $m > 1$. Our first step will be to restart with the systems $D_{n,\rho}$ and to construct more systems of type II. The principle is that any system of type I can be transformed into a system of type II by adding a new, external generator.

Lemma 2.14. *Assume that S is an LD-system of type I generated by g. Let $S^\#$ be $S \cup \{g^\#\}$, where $g^\#$ is a new element. Extend the operation of S so that $g^\# \equiv_L g$ and $g^\# \equiv_R g$ hold. Then $S^\#$ is a system of type II generated by $g^\#$.*

Proof. The elements g and $g^\#$ behave in the same way in all evaluations. Formally, for every term $t(x_0, x_1, \dots, x_n)$ with $t \neq x_0$, and for all $a_1, \dots, a_n$ in S, we have $t^{S^\#}(g^\#, a_1, \dots, a_n) = t^S(g, a_1, \dots, a_n)$, as shown by an induction on t. This implies for instance $g^\#(ab) = g(ab) = (ga)(gb) = (g^\# a)(g^\# b)$. By considering all possible positions of $g^\#$ in an expression $x(yz)$ (there are seven of them since repetitions are allowed), we prove that $S^\#$ is an LD-system.

Next, there exist a and b in S satisfying $g = ab$, since S has type I. Since g generates S, there exist terms $t_0(x)$, $t_1(x)$ such that $a = t_0(g)$ and $b = t_1(g)$ hold. By the above result, we have $t_0(g^\#)t_1(g^\#) = ab = g$: in other words, the sub-LD-system of $S^\#$ generated by $g^\#$ contains g, and, therefore, it includes S. So $g^\#$ generates $S^\#$. Now, $g^\#$ does not belong to $S^\# S^\#$, *i.e.*, $S^\#$ has type II. ■

Example 2.15. **(extension)** *Starting with A_2, we obtain an LD-system $A_2^\#$ of type II with 5 elements:*

	$1^\#$	1	2	3	4
$1^\#$	2	2	4	2	4
1	2	2	4	2	4
2	3	3	4	3	4
3	4	4	4	4	4
4	1	1	2	3	4

The idea now is to resort to the type II systems $D_{n,\rho}^\#$. Actually, a still larger family will be considered. Indeed, in the case of type I, there was no reason to consider the systems $C_{m,r}$ with $m > 1$, since the latter systems have type II and all systems of type I have been obtained. For type II, the restriction is no longer possible, and the basic systems we shall consider are those obtained as above, but starting with $C_{m,r}$ rather than with $C_{0,r}$ only.

Definition 2.16. **(tables $E_{n,\mu,\rho}$)** For $n \geq 1$, μ a mapping of $\{1, \ldots, 2^n\}$ into $\mathbf{N}$ and ρ a mapping of $\{1, \ldots, 2^n\}$ into $\mathbf{N}_+$, $E_{n,\mu,\rho}$ is defined to be the set

$$\{1\} \cup \{(a, i) \ ; \ 1 \leq a \leq 2^n \ , \ 1 \leq i \leq \mu(a) + \rho(a)\}$$

equipped with the operation $(a, i) * (b, j) = (a *_{A_n} b, i *_{C_{\mu(a),\rho(a)}} j)$, extended by $1 \equiv_L (1, 1)$ and $1 \equiv_R (1, 1)$.

Example 2.17. **(tables $E_{n,\mu,\rho}$)** *When μ is constant with value 0, $E_{n,\mu,\rho}$ is, up to renaming $(1, 1)^\#$ by 1, the system $D_{n,\rho}^\#$. For instance, the five element LD-system $A_2^\#$ of the previous example is (isomorphic to) $E_{2,\mu,\rho}$, with $\mu(1) = \ldots = \mu(4) = 0$ and $\rho(1) = \ldots = \rho(4) = 1$.*

As in the case of the systems $D_{n,\rho}$, we can describe those constraints the mappings μ and ρ must satisfy for $E_{n,\mu,\rho}$ to be an LD-system.

Lemma 2.18. *Assume that the mapping ρ satisfies Condition* (2.1), *and that*

$$A_n *_n a \subseteq A_n *_n b \quad \textit{implies} \quad \mu(a) \leq \mu(b) + 1. \tag{2.3}$$

Then $E_{n,\mu,\rho}$ is an LD-system of type II generated by 1.

We could expect a natural counterpart to Proposition 2.13 (type I) stating that every system of type II is isomorphic to some system $E_{n,\mu,\rho}$. Such a result is *not* true. Proposition 2.2 (right tables) shows the problem in a simple case: besides the basic systems $C_{m,r}$, we find in addition some variants $V_A(C_{2,r})$ obtained from $C_{2,r}$ by randomly changing some values in the first column. The phenomenon is general: in the case of a type II system S with generator g, the values in Sg, and possibly in $S(Sg)$, are more or less random, so that the resemblance with the systems $E_{n,\mu,\rho}$ only becomes evident for $S(S(Sg))$ and beyond. For S a binary system, we define inductively $S^{[p]}$ for $p \geq 1$ by $S^{[1]} = S$ and $S^{[p]} = SS^{[p-1]} = \{ab \; ; \; a \in S, b \in S^{[p-1]}\}$ for $p \geq 2$.

Definition 2.19. (**variant**) We say that the LD-system S' is a *p-variant* of the system S if there exists a surjective homomorphism θ: $S' \twoheadrightarrow S$ such that θ is injective on $S^{[p]}$.

For instance, every system $V_A(C_{2,r})$ mentioned above is a 2-variant of $C_{2,r}$. With this notion, the result is as follows:

Proposition 2.20. (**type II**) *Up to isomorphism, all finite monogenic LD-systems of type II are 3-variants of LD-systems $E_{n,\mu,\rho}$.*

Proof. (Sketch) Assume that S is a finite, monogenic LD-system of type II. By Lemma 2.12, the congruence $\equiv_R^S$ is not trivial, *i.e.*, at least two columns in the table of S are equal. Define a sequence of monogenic LD-systems by $S_0 = S$ and

$$S_{i+1} = S_i / \equiv_R^{S_i} .$$

As long as S_i has type II, the congruence $\equiv_R^{S_i}$ is not trivial, and the size of S_{i+1} is less than the size of S_i. It follows that there must exist some finite index ℓ such that S_ℓ has type I. By Proposition 2.13 (type I), S_ℓ is isomorphic to $D_{n,\rho}$ for some n and ρ, and we have obtained a surjective morphism

$$\pi'_S : S \twoheadrightarrow D_{n,\rho}. \tag{2.4}$$

Now some left ideal H_S of S is isomorphic to $D_{n,\rho}$, and every long enough composition of left translations maps every element of S into H_S. This property is clear in the case of the systems $E_{n,\mu,\rho}$: if ℓ is the supremum of the numbers $\mu(a)$ for $1 \leq a \leq 2^n$, *i.e.*, of the lengths of the tails, then applying ℓ left translations to an element gives an element in the cyclic part of $E_{n,\mu,\rho}$, *i.e.*, one of the form (a, i) with $\mu(a) < i \leq \mu(a) + \rho(a)$, and no further left translation can push us out of H_S. The point is that the property is general, and that the operation of S can be reconstructed from that of H_S and from the lengths of the tails, *i.e.*, from the distances to H_S (the index ℓ involved in the iterative construction of the quotients S_i is the supremum of these lengths)—except for those values in Sg and $S(Sg)$, which amounts to introducing a 3-variant in the construction. ■

We can say a little more. First, as was the case for the variants $V_A(C_{2,r})$, the free values in the 3-variants of the systems $E_{n,\mu,\rho}$ have to obey some constraints: for instance, the values in $S(Sg)$ have to be chosen in such a way that $c = ab$ implies $a(bg) = c(ag)$. Let us simply mention here that listing these constraints effectively is possible.

Finally, we can describe exactly those systems of type II that are isomorphic to some system $E_{n,\mu,\rho}$, *i.e.*, those for which the basic construction works. For S a finite monogenic LD-system, let us denote by π_S^* the surjective homomorphism of S onto some system A_n obtained by composing the morphism π'_S of (2.4) and the corresponding morphism $\pi_{D_{n,\rho}}$ of (2.2). By construction, π_S^* consists in iteratively identifying those elements with equal associated columns until a table with distinct columns is obtained, and, then, identifying those elements with equal associated rows. Now $a \equiv_L^S b$ implies $\pi_S^*(a) \equiv_L^{A_n} \pi_S^*(b)$, and $\equiv_L^{A_n}$ is trivial, so $a \equiv_L^S b$ implies $\pi_S^*(a) = \pi_S^*(b)$. Let us say that a type II system is *slender* when the converse implication is true. The meaning of this notion is clear: in every case, saying that A_n is the image of S under π_S^* means that S is obtained from A_n by splitting some values and adding a new, external generator. This splitting process can correspond both to introducing cycles, possibly with tails, as in the case of the systems $E_{n,\mu,\rho}$, and to introducing new values in Sg or $S(Sg)$. In the latter case, the rows are distinct although they correspond to the same element of A_n. Slenderness is the hypothesis that forbids such a phenomenon.

Proposition 2.21. (slender type II) *Up to isomorphism, the slender finite monogenic LD-systems of type II are the LD-systems $E_{n,\mu,\rho}$.*

The conclusion of this study is that every finite monogenic LD-system is obtained in an essentially trivial way from some Laver table. So, definitely, the whole complexity of the subject lies in the description of the latter systems.

Exercise 2.22. (constant column) *Assume that S is a finite monogenic LD-system. Show that there exists an element c in S such that the value of xc does not depend on x, i.e., the table of S admits a constant column. [Hint: Assuming that g generates S, consider a function f such that, for every a in S, the equality $ag^{[p]} = g^{[p+1]}$ holds for $p \geq f(a)$, and put $c = g^{[N]}$ where N is the supremum of all values $f(a)$ for a in S.]*

Exercise 2.23. **(tables T_m)** *Write the table $D_{1,\rho}$ for $\rho(1) = m$, $\rho(2) = 1$. Show that $D_{1,\rho}$ is isomorphic to the table T_m of Exercise V.2.15.*

Exercise 2.24. **(cycles)** *Assume that S is an LD-system with n elements. Show that, for every a in S, $a_{[p]} = a_{[q]}$ holds for some p, q satisfying $p < q \leq n+1$. Deduce that left division in S admits cycles.*

Exercise 2.25. **(special injections)** *(i) Let $\mathfrak{I}_{sp}$ be the closure of sh under bracket in $\mathfrak{I}_\infty$ (cf. Section I.2). Show that all elements of $\mathfrak{I}_{sp}$ coincide with sh eventually, i.e., for every f in $\mathfrak{I}_{sp}$, $f(p) = p+1$ holds for p large enough.*

(ii) Prove that, for $f \in \mathfrak{I}_{sp}$, either $f(0) = 0$ or $0 \notin \mathrm{Im}(f)$ holds.

(iii) For $f \in \mathfrak{I}_\infty$, define $\widetilde{f}$ by $\widetilde{f}(0) = 0$ and $\widetilde{f}(k) = f(k)$ for $k \geq 1$. Prove that $f \in \mathfrak{I}_{sp}$ implies $\widetilde{f} \in \mathfrak{I}_{sp}$. [Hint: Prove $\widetilde{f[g]} = f[\widetilde{g}]$ using (iii).]

(iv) Describe all monogenic sub-LD-systems of $\mathfrak{I}_\infty$ in terms of $\mathfrak{I}_{sp}$ and groups $\mathbf{Z}$ or $\mathbf{Z}/m\mathbf{Z}$ equipped with conjugacy.

Exercise 2.26. **(associative operation)** *(i) Assume that S is a monogenic LD-system, and g generates S. Show that every equality $a_1 \ldots a_p g = b_1 \ldots b_q g$ implies $a_1 \ldots a_p x = b_1 \ldots b_q x$ for every x in S.*

(ii) Show that the formula $a \times b = a_1 \ldots a_p b$ for $a = a_1 \ldots a_p g$ gives on S a well defined operation. Prove that $(S, \times, g)$ is a monoid, that the identities $x \times g = x$, $x \times (yz) = (x \times y)(x \times z)$, $g \times x = x$, $(xy) \times z = x(y \times z)$ hold in S, and that they completely define the operation $\times$.

(iii) Show that the previous construction applies to the case of the LD-system $(\mathfrak{I}_{sp}, [\,])$ of Exercise 2.25 (special injections), and that the equality $f[\mathrm{sh}] \times g[\mathrm{sh}] = f[g[\mathrm{sh}]] = (f \circ g)[\mathrm{sh}]$ holds for all f, g in $\mathfrak{I}_{sp}$.

Exercise 2.27. **(infinite monogenic)** *Define ^ on $\mathbf{Z}$ by a^$1 = a+1$ for $a \geq 1$, a^$1 = a-1$ for $a \leq 0$, and a^$b = 0$ for $b \neq 1$. Prove that $(\mathbf{Z},$ ^$)$ is an LD-system and 1 is a generator. [Hint: a^b^c is always 0.]*

Exercise 2.28. **(system K)** *(i) Let K be the set $\{(p,n) \in \mathbf{N}^2; p \neq 0, 2\}$ equipped with*

$$\begin{cases} (p,n)\cdot(1,0) = (p,n+1), \\ (p,n)\cdot(q,0) = (q+1,0) & \text{for } q \geq 3, \\ (p,n)\cdot(q,m) = (3,0) & \text{for } m \geq 1. \end{cases}$$

Prove that K is an LD-system generated by $(1,0)$, and we have $(1,0)_{[n]} = (1,n)$, and $(1,1)^{[p]} = (3,0)$ for $p \geq 3$.

(ii) Consider on K the ordering $\sqsubset$

$$\ldots \sqsubset (4,0) \sqsubset (3,0) \sqsubset \ldots \sqsubset (4,1) \sqsubset (3,1) \sqsubset \ldots \sqsubset (4,2) \sqsubset (3,2) \sqsubset \ldots \sqsubset (1,2) \sqsubset (1,1) \sqsubset (1,0),$$

i.e., for p, $q \geq 3$, $(p,n) \sqsubset (q,m)$ holds for $n < m$ and for $n = m$ with $p \geq q$, while $(1,n) \sqsubset (1,m)$ holds for $n > m$, and $(p,n) \sqsubset (1,m)$ always holds for $p \geq 3$. Show that $\sqsubset$ is compatible with the product in the sense that $x \sqsubseteq x'$ and $y \sqsubseteq y'$ imply $xy \sqsubseteq x'y'$. In particular, the left translations of K are non-decreasing [but not increasing: we have $(3,1) = (3,1)\cdot(3,2) = (3,1)\cdot(4,2)$].

Exercise 2.29. (extension to subsets) *(i) Assume that S is an LD-system. For A, $B \subseteq S$, let $AB = \{ab\,;\ a \in A, b \in B\}$. Define inductively $A^{[1]} = A_{[1]} = A$, $A^{[p]} = AA^{[p-1]}$, $A_{[p]} = A_{[p-1]}A$ for $p \geq 2$. Show that the powerset $\mathfrak{P}(S)$ equipped with this product is an LD-system, and we have $S^{[p+1]} \subseteq S^{[p]}$ and $S_{[n+1]} \subseteq S_{[n]}$.*

(ii) For $c \in S$, let $\lambda_R(c) = \sup\{r\,;\ c \in S^{[r]}\}$ (possibly ∞). Assume $q \geq 2$, and $a \in S^{[q+1]}$, say $a = a_0a_1\ldots a_q$. Show that there exists a decomposition $a'_0a'_1\ldots a'_q$ of a with $\lambda_R(a'_0) \geq q$ and deduce $S^{[p]}S^{[q]} = S^{[q+1]}$ for $q \geq 2$. [Hint: Use $a_0a_1\ldots a_q = (a_0a_1)a_0a_2\ldots a_q = ((a_0a_1)a_0)(a_0a_1)a_2\ldots a_q$, and $\lambda_R((a_0a_1)a_0) \geq \lambda_R(a_0) + 1$.]

(iii) Denote by $S^{[p,n]}$ the family of all elements in S that admit a decomposition of the form $(\ldots((a_0a_1)\ldots)a_n$ with $a_0 \in S^{[p]}$. Prove $S^{[p,n]}S^{[q]} = S^{[q+1]}$ for $p \geq 1$, $q \geq 2$ and $n \geq 0$. [Hint: Argue as in (ii), using $\lambda_L(c) = \sup\{k\,;\ c \in S^{[p,k]}\}$.]

(iv) Prove $S^{[p,n]}S^{[q,m]} = S^{[3]}$ for $p, q \geq 1$, $n \geq 0$ and $m \geq 1$. [Hint: Prove that, for all p, n, every element of $S^{[3]}$ admits a decomposition $a_0a_1a_2$ with $a_0 \in S^{[p,n]}$. The argument used to increase the values of the parameters λ_R and λ_L on a_0 works for a_1 too, so, after a sufficient number of applications of (LD), we obtain a decomposition $a'_0a'_1a'_2$ with $a'_1 \in S^{[q,m-1]}$, and, therefore, we have $a'_1a'_2 \in S^{[q,m]}$, while the parameters for a'_0 are at least equal to those for a_0, i.e., a'_0 belongs to $S^{[p,n]}$.]

(v) Assume that S is an LD-system. Show that the sub-LD-system of $\mathfrak{P}(S)$ generated by S is a quotient of the LD-system K of Exercise 2.28, and that the sub-LD-system of $\mathfrak{P}(K)$ generated by K is isomorphic to K.

---

3. Multi-LD-Systems

A number of examples of LD-systems actually involve several left self-distributive operations, which in addition are mutually left distributive. This leads us to the notion of a multi-LD-system. Here we describe a uniform operation that associates with every multi-LD-system an LD-system with an extended domain, and prove the correspondence to be bijective in some sense. So multi-LD-systems are not more general than LD-systems. In particular, a free p-LD-system of rank n is essentially a free LD-system of rank pn.

Examples

Definition 3.1. (multi-LD-system) A *multi-LD-system* is defined to be a set equipped with a family of binary operations $(*_i)_{i \in I}$ such that the left distributivity identity

$$x *_i (y *_j z) = (x *_i y) *_j (x *_i z)$$

is satisfied for every i, j in I, and, in particular, for $i = j$.

An LD-system is a multi-LD-system with a single operation. Observe that, if we denote by ad^i_a the left translation $x \mapsto a *_i x$, then the distributivity identity for $*_i$ and $*_j$ can be rewritten as

$$\mathrm{ad}^i_x(y *_j z) = \mathrm{ad}^i_x(y) *_j \mathrm{ad}^i_x(z). \tag{3.1}$$

Thus $(S, (*_i)_{i \in I})$ is a multi-LD-system if and only if each left translation ad^i_a is an endomorphism.

Many LD-systems turn out to be parts of multi-LD-systems.

Example 3.2. (trivial) *Let S be an arbitrary set, and $(f_i)_{i \in I}$ be a family of functions of S into itself. Define $*_i$ by $x *_i y = f_i(y)$. Then $(S, (*_i)_{i \in I})$ is a multi-LD-system.*

Example 3.3. (lattice) *A lattice, and, in particular, a Boolean algebra, is a bi-LD-system with respect to the upper and lower bound operations.*

Example 3.4. (LD-quasigroup) *By Proposition I.2.8 (second operation), an LD-quasigroup is a bi-LD-system $(Q, {}^\wedge, {}_\vee)$ satisfying the additional identities $x{}^\wedge(x{}_\vee y) = x{}_\vee(x{}^\wedge y) = y$.*

Example 3.5. (**braids**) *We constructed in Chapter I a left self-distributive operation on Artin's braid group B_∞, namely the exponentiation defined by $a^\wedge b = a\ \mathrm{sh}(b)\ \sigma_1\ \mathrm{sh}(a^{-1})$. A straightforward verification shows that the alternative operation*

$$a \vee b = a\ \mathrm{sh}(b)\ \sigma_1^{-1}\ \mathrm{sh}(a^{-1})$$

is also left self-distributive, and that $(B_\infty, {}^\wedge, {}_\vee)$ is a bi-LD-system, i.e., $^\wedge$ and $_\vee$ are mutually left distributive.

Example 3.6. (**mean**) *Let V be a left module over a ring R. For λ in R define the binary operation $*_\lambda$ on V by*

$$x *_\lambda y = \lambda x + (1 - \lambda) y.$$

The operation $*_\lambda$ is both left and right distributive, but, in general, it is neither commutative (except for $2\lambda = 1$) nor associative (except for $\lambda^2 = \lambda$). For fixed λ, $(V, *_\lambda)$ is an LD-system. Now, for all λ and μ, we have

$$x *_\lambda (y *_\mu z) = (1 - \lambda)x + \lambda(1 - \mu)y + \lambda\mu z = (x *_\lambda y) *_\mu (x *_\lambda z),$$

so $(V, (*_\lambda)_{\lambda \in R})$ is a multi-LD-system.

The hull of a multi-LD-system

We observe now that the operations of a multi-LD-system can always be gathered into a unique left distributive operation on a larger set.

Proposition 3.7. (**hull**) *(i) Assume that $(S\,, (*_i)_{i \in I})$ is a multi-LD-system. Then the set $S \times I$ equipped with*

$$(x, i) \cdot (y, j) = (x *_i y, j)$$

is an LD-system.

(ii) Conversely, assume that $S \times I$ is an LD-system such that $(x, i){\cdot}(y, j)$ has the form $(f(x, y, i), j)$ for some mapping f of $S^2 \times I$ to S. Define $*_i$ on S by $x *_i y = f(x, y, i)$. Then $(S\,, (*_i)_{i \in I})$ is a multi-LD-system, and the LD-system given by (i) is the initial LD-system.

Proof. The verification of (i) is straightforward. For (ii), we note that, for every k in I, $x *_i (y *_j z)$ is the first component of $(x, i){\cdot}((y, j){\cdot}(z, k))$, and $(x *_i y) *_j (x *_i z)$ is the first component of $((x, i){\cdot}(y, j)){\cdot}((x, i){\cdot}(z, k))$. ∎

Definition 3.8. (**hull**) The LD-system $S \times I$ of Proposition 3.7(i) is called the *hull* of the multi-LD-system $(S\,,\,(*_i)_{i\in I})$.

Example 3.9. (**hull**) *The hull of the multi-LD-system* $(V, (*_\lambda)_{\lambda\in R})$ *of Example* 3.6 *(mean) consists of* $V \times R$ *equipped with*

$$(x, \lambda) \cdot (y, \mu) = ((1 - \lambda)x + \lambda y, \mu).$$

Free multi-LD-systems

For the uniform reasons explained in Section V.1, for each fixed family of operators indexed by I, and each nonempty set X, there exists a *free* multi-LD-system based on X. Here we describe free multi-LD-systems completely. The result is:

Proposition 3.10. (**free multi-LD-systems**) *(i) The hull of a free multi-LD-system* $(S\,,\,(*_i)_{i\in I})$ *based on* X *is a free LD-system based on* $X \times I$.

(ii) Conversely, assume that S *is a free LD-system based on* $X \times I$. *Let* i_0 *be an arbitrary fixed element of* I, *let* S_0 *be the left ideal of* S *generated by the elements* (x, i_0) *with* $x \in X$. *For* i *in* I, *let* f_i *denote the function on* S_0 *that maps* $a_1 \ldots a_p(x, i_0)$ *to* $a_1 \ldots a_p(x, i)$ *for all* $a_1, \ldots, a_p$. *Finally, for* i *in* I, *define a binary operation* $*_i$ *on* S_0 *by*

$$x *_i y = f_i(x)\, y.$$

Then $(S_0, (*_i)_{i\in I})$ *is a free multi-LD-system based on* $X \times \{i_0\}$.

In order to prove this result, we introduce the set $T_{X,I}$ of all terms constructed using variables in X and operators $*_i$ indexed by I. Then the free multi-LD-system (with I-indexed operations) based on X is $T_{X,I}/{=^I_{LD}}$, where $=^I_{LD}$ is the congruence generated by all pairs of the form

$$(t_1 *_{i_1} (t_2 *_{i_2} t_3)\ ,\ (t_1 *_{i_1} t_2) *_{i_2} (t_1 *_{i_1} t_3))$$

with $t_1, t_2, t_3 \in T_{X,I}$ and $i_1, i_2 \in I$. The study of the relation $=^I_{LD}$ on $T_{X,I}$ turns out to be easy, for it reduces to the study of $=_{LD}$ on the extended set of terms $T_{X\times I}$. Indeed, let us introduce two mappings f, g as follows: For t in $T_{X,I}$ and k in I, define the term $f(t, k)$ in $T_{X\times I}$ by

$$f(x, k) = (x, k), \qquad f(t_1 *_i t_2, k) = f(t_1, i) \cdot f(t_2, k).$$

Conversely, for t in $T_{X\times I}$, define the pair $g(t) = (h(t), i(t))$ in $T_{X,I} \times I$ by

$$g(x, k) = (x, k), \qquad g(t_1 \cdot t_2) = (h(t_1) *_{i(t_1)} h(t_2), i(t_2)).$$

Lemma 3.11. *The mappings f and g are bijections, g is the inverse of f, and $f(t,k) =_{LD} f(t',k')$ holds in $T_{X\times I}$ if and only if we have $t =^I_{LD} t'$ and $k = k'$.*

Proof. Firstly, we verify inductively that $f\circ g$ and $g\circ f$ are the identity mappings of their domains. Then, by construction, the rightmost variable of every term of the form $f(t,k)$ has the form (x,k) for some x in X. As the rightmost variable is invariant under LD-equivalence, $f(t,k) =_{LD} f(t',k')$ implies $k = k'$. So we are left with the question of proving that $f(t,k) =_{LD} f(t',k)$ holds in $T_{X\times I}$ if and only if $t =^I_{LD} t'$ holds in T^I_X.

Let us prove first that $t =^I_{LD} t'$ implies $f(t,k) =_{LD} f(t',k)$ for every k. As the congruence $=^I_{LD}$ is generated by the pairs (t,t') with t' a basic LD-expansion of t, it is enough to prove the result for such pairs (t,t'). Assume $t' = t\bullet\alpha$. We use induction on α. Assume $\alpha = \phi$. Write $t = t_1 *_i (t_2 *_j t_3)$ and $t' = (t_1 *_i t_2) *_j (t_1 *_j t_3)$. Then we find

$$f(t,k) = f(t_1,i)\cdot(f(t_2,j)\cdot f(t_3,k)),$$
$$f(t',k) = (f(t_1,i)\cdot f(t_2,j))\cdot(f(t_1,i)\cdot f(t_3,k)).$$

Hence $f(t',k)$ is a basic LD-expansion of $f(t,k)$, and, in particular, $f(t,k) =_{LD} f(t',k)$ holds.

Assume now $\alpha = 0\beta$. Write $t = t_1 *_i t_2$. Then we have $t' = t'_1 *_i t_2$, with $t'_1 = t_1\bullet\beta$. By construction, we have

$$f(t,k) = f(t_1,i)\cdot f(t_2,k), \qquad f(t',k) = f(t'_1,i)\cdot f(t_2,k),$$

and the induction hypothesis implies $f(t_1,i) =_{LD} f(t'_1,i)$. We deduce $f(t,k) =_{LD} f(t',k)$. The argument is similar for $\alpha = 1\beta$.

The proof that $f(t,k) =_{LD} f(t',k)$ implies $t =^I_{LD} t'$ is similar: as every term in $T_{X\times I}$ belongs to the image of f, it suffices that we consider the case when $f(t',k)$ is a basic LD-expansion of $f(t,k)$, and we do it as above. ∎

Remark 3.12. When we associate a binary tree with every term, considering several operators amounts to adding labels to the internal nodes of the trees. The geometrical meaning of the previous lemma is that, as far as left self-distributivity is concerned, this additional information can be carried to the leaves. More precisely, applying the isomorphism of the lemma amounts to transferring the information from each internal node to the leaf that lies on the rightmost possible position below the left successor of that node. This makes sense for every internal node, and, conversely, every leaf except the rightmost one receives in this way an additional information from a unique internal node. For instance, the correspondence exchanges the terms

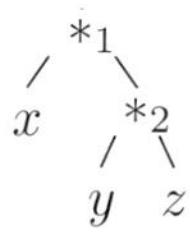

and

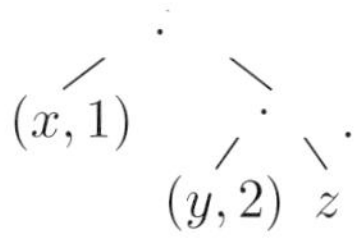

We can now easily complete the proof of Proposition 3.10 (free multi-LD-systems).

Proof. (i) Assume that $(S\,,(*_i)_{i\in I})$ is a free multi-LD-system with I-indexed operations based on X. Let $\overline{f}$ map every term in $T_X^I \times I$ to the LD-class of $f(t,k)$ in $\mathrm{FLD}_{X\times I}$. By Lemma 3.11, $\overline{f}$ is compatible with the congruence $=^I_{LD} \times =$ on $T_X^I \times I$, and, therefore, it induces a well defined mapping $\overline{\overline{f}}$ of $S \times I$ onto $\mathrm{FLD}_{X\times I}$. Moreover, $\overline{\overline{f}}$ preserves the operation induced on $S \times I$ by the operation

$$(t_1, i_1) \cdot (t_2, i_2) = (t_1 *_{i_1} t_2, i_2),$$

on $T_{X,I} \times I$, *i.e.*, the operation of the hull of S. Conversely, a similar argument shows that the function of $T_{X\times I}$ onto $S \times I$ that maps every term t in $T_{X,I}$ to the pair $(\overline{h(t)}, i(t))$ factors through $=^I_{LD}$, and the induced mapping is an inverse of $\overline{\overline{f}}$. Hence the latter is an isomorphism.

(ii) The invariance of the rightmost variable under LD-equivalence guarantees that the functions f_i are well defined. Then, $(S_0, (*_i)_{i\in I})$ is a multi-LD-system generated by $X \times \{i_0\}$ by definition of $*_i$. Lemma 3.11 implies that this multi-LD-system is free based on $X \times \{i_0\}$ as, by construction, the evaluation in S_0 of a term t of T_X^I is equal to the LD-class of $f(t, i_0)$. So, two such terms t, t' have the same evaluation if and only if $f(t, i_0) =_{LD} f(t', i_0)$ holds, hence if and only if $t =^I_{LD} t'$ does. ■

Thus, we obtain a realization of the free p-LD-system (p operations) of rank 1 by considering the left ideal of FLD_p generated by x_1, *i.e.*, the set of all elements of the form $a_1 \ldots a_k x_1$, equipped with the operations $*_1, \ldots, *_p$ defined by

$$(a_1 \ldots a_k x_1) *_i (b_1 \ldots b_j x_1) = (a_1 \ldots a_k x_i) b_1 \ldots b_j x_1.$$

More generally, a free p-LD-system of rank n embeds in a free LD-system of rank pn.

Exercise 3.13. **(disjoint sum)** *Assume that $(S_i, *_i)_{i\in I}$ is a family of disjoint LD-systems. Let S be the union of all S_i's. Show that defining $a\hat{\ }b = a *_i b$ if both a and b belong to S_i, and $a\hat{\ }b = b$ otherwise, gives S the structure of an LD-system.*

Exercise 3.14. **(skew operation)** *(i) Assume that S is an LD-system and $(f_i)_{i\in I}$ is a family of mappings of S into itself such that $f_i(xy) = xf_i(y)$ holds for all i, x, y. Define $*_i$ by $x *_i y = f_i(x)y$. Prove that $(S\,, (*_i)_{i\in I})$ is a multi-LD-system.*

(ii) Deduce that, if S is an LD-system, so is $(S, *)$, where $*$ is defined by $x * y = (xx)y$.

Exercise 3.15. **(half-conjugacy)** *Define a multi-LD-quasigroup to be a multi-LD-system where all left translations are bijective. Let G be a group. For a in G, define the operation $*_a$ by $x *_a y = xax^{-1}y$. Show that $(G\,,\,(*_a)_{a\in A})$ is a multi-LD-quasigroup, and that the inverse of the operation $*_a$ is the operation $*_{a^{-1}}$. Check that the hull of $(G\,,\,(*_a)_{a\in A})$ is $(G \times A, {}^\wedge)$ equipped with half-conjugacy.*

Exercise 3.16. **(special power)** *Assume that $(S\,,\,(*_i)_{i\in I})$ is a multi-LD-system. For $\vec{\imath} = (i_1, \ldots, i_n)$ in I^n, define $*_{\vec{\imath}}$ on S^n by*

$$\vec{x} *_{\vec{\imath}} \vec{y} = (x_1 *_{i_1} \cdots *_{i_{n-1}} x_n *_{i_n} y_1, \ldots, x_1 *_{i_1} \cdots *_{i_{n-1}} x_n *_{i_n} y_n).$$

Prove that $(S^n, (*_{\vec{\imath}})_{\vec{\imath}\in I^n})$ is a multi-LD-system. [In the case of an LD-system and of $n = 2$, the previous construction gives $(x_1, x_2) * (y_1, y_2) = (x_1x_2y_1, x_1x_2y_2)$, which is different from the standard direct power $(x_1, x_2)\,(y_1, y_2) = (x_1y_1, x_2y_2)$.]

Exercise 3.17. **(adjoint monoid)** *Define the adjoint monoid of a multi-LD-system $(S\,,\,(*_i)_{i\in I})$ to be the submonoid of $\mathrm{End}(S)$ generated by all left translations ad^i_a, where ad^i_a is defined to map x to $a *_i x$ for every x. Show that mapping ad^i_a to $\mathrm{ad}_{(a,i)}$ defines an isomorphism of the adjoint monoid of $(S\,,\,(*_i)_{i\in I})$ onto the adjoint monoid of the hull of $(S\,,\,(*_i)_{i\in I})$.*

---

4. Idempotent LD-systems

Here we gather a few remarks about idempotents in LD-systems, and about those LD-systems in which every element is idempotent, which we call LDI-systems. Although many classical examples of LD-systems are LDI-systems, it turns out that our knowledge of LDI-systems is far from complete, and, in particular, very little is known about free LDI-systems.

Idempotent elements and LDI-systems

Definition 4.1. (idempotent, LDI-system) Assume that S is a binary system. An element a of S is said to be *idempotent* if $aa = a$ holds. The system S is said to *idempotent* if every element of S is idempotent, *i.e.*, if S satisfies the identity

$$xx = x. \tag{I}$$

An idempotent LD-system is called an *LDI-system.*

> **Example 4.2.** (**mean**) *Assume that R is a ring, and E is a left R-module. Then, for each λ in R, E equipped with the operation $*_\lambda$ defined by $a *_\lambda b = (1-\lambda)a + \lambda b$ is an LDI-system. More generally, $(E, (*_\lambda)_{\lambda \in R})$ is, in an obvious sense, a multi-LDI-system.*

The example of $\mathbf{Z}$ equipped with the operation $a * b = b + 1$ shows that there exist LD-systems, and even LD-quasigroups, with no idempotent. However, rather weak hypotheses imply that idempotents exist in an LD-system, and, in this case, they are many in number.

Lemma 4.3. *Assume that S is an LD-system. If nonempty, the set of all idempotents in S is a left ideal of S i.e., if e is idempotent, so is ae for every a.*

Proof. Applying Identity (LD) gives $(ae)(ae) = a(ee) = ae$ if $ee = e$ holds. ∎

Proposition 4.4. (right cancellative) *Every right cancellative LD-system is an LDI-system.*

Proof. Let a be an arbitrary element of S. By (LD), we have $a(aa) = (aa)(aa)$: this implies $a = aa$ if right cancellation of aa is allowed. ∎

Quandles

As recalled in Example 3.4, an LD-quasigroup is a bi-LD-system $(Q, {}^\wedge, {}_\vee)$ satisfying the additional identities $x^\wedge(x_\vee y) = x_\vee(x^\wedge y) = y$. We have seen in Section I.2 that each operation in an LD-quasigroup determines the other one, so that, if we define an implicit LD-quasigroup to be an LD-system with bijective left translations, there exists exactly one way to complete an implicit LD-quasigroup into an (explicit) LD-quasigroup.

Lemma 4.5. *Assume that $(Q, {}^\wedge, {}_\vee)$ is an LD-quasigroup. Then the operation ${}^\wedge$ is idempotent if and only if the operation ${}_\vee$ is.*

Proof. Assume that $\wedge$ is idempotent. For a in Q, we find $a \vee a = a \vee (a \wedge a) = a$. ■

Thus there is only one possible notion of an idempotent LD-quasigroup:

Definition 4.6. (quandle) Assume that $(Q, \wedge, \vee)$ is an LD-quasigroup. We say that Q is an *LDI-quasigroup*, or a *quandle*, if $\wedge$ and $\vee$ are idempotent.

Example 4.7. (group conjugacy) *Every group G equipped with*

$$a \wedge b = aba^{-1}, \qquad a \vee b = a^{-1}ba \tag{4.1}$$

is a quandle. Variants are $a \wedge b = a^n ba^{-n}$, $a \vee b = a^{-n}ba^n$ with n a fixed integer, or $a \wedge b = af(ba^{-1})$, $a \vee b = f^{-1}(a^{-1}b)a$ with $f \in \mathrm{Aut}(G)$.

In the sequel, we denote by $(LDQI)$ the family of eight identities that define quandles, namely the six identities (LDQ) defining LD-quasigroups completed with the two identities $x \wedge x =$ and $x \vee x = x$.

We have shown in Section V.1 that, if FG_X is a free group based on X, and $\wedge$ denotes the half-conjugacy operation on $\mathrm{FG}_X \times X$, then the subsystem of $\mathrm{FG}_X \times X$ generated by $\{1\} \times X$ is a free LD-quasigroup based on $\{1\} \times X$. The LD-quasigroup $(\mathrm{FG}_X \times X, \wedge)$ is not a quandle—actually it contains no idempotent—but we can easily deduce a description of free quandles.

Proposition 4.8. (free quandle) *Assume that X is a nonempty set. Let FG_X be a free group based on X, and $\wedge$, $\vee$ be the conjugacy operations defined in (4.1). Then the subsystem of $(\mathrm{FG}_X, \wedge, \vee)$ generated by X is a free quandle based on X.*

Proof. We follow the scheme of Section V.1, and use the same notations. Let NT'_X be the subset of NT_X consisting of those terms $x_1 *^{e_1} x_2 *^{e_2} \ldots x_p *^{e_p} x$ that satisfy the additional requirement $x_p \neq x$. We claim that every term in $T_{(X,\wedge,\vee)}$ is LDQI-equivalent to a term in NT'_X. By Lemma V.1.16, every term in $T_{(X,\wedge,\vee)}$ is LDQ-equivalent, hence LDQI-equivalent to some term in NT_X. Then idempotency implies that every term in NT_X is LDQI-equivalent to some term in NT'_X, as, for $x_p = x$, the term $x_1 *^{e_1} \ldots *^{e_{p-1}} x_p *^{e_p} x$ is LDQI-equivalent to $x_1 *^{e_1} \ldots *^{e_{p-1}} x$.

Owing to Lemma V.1.15, it suffices now to verify that different terms in NT'_X receive different evaluations in $(\mathrm{FG}_X, \wedge, \vee)$. This is clear as, by definition, the evaluation of $x_1 *^{e_1} \ldots x_p *^{e_p} x$ is the reduced word $x_1^{e_1} \ldots x_p^{e_p} x x_p^{-e_p} \ldots x_1^{-e_1}$. ■

Corollary 4.9. *The word problem for $(LDQI)$ is decidable.*

Proof. We decide whether two terms t, t' in $T_{(X,\wedge,\vee)}$ are LDQI-equivalent by evaluating them in $(\mathrm{FG}_X, {}^\wedge, {}_\vee)$: the terms are equivalent if and only if the evaluations are equal. Alternatively, we can directly determine two terms t_0, t'_0 in NT'_X LDQI-equivalent to t and t' respectively by using the inductive proof of Lemma V.1.16. Then t and t' are equivalent if and only if $t_0 = t'_0$ holds. ∎

Corollary 4.10. *The eight identities* $(LDQI)$ *axiomatize double conjugacy in groups, i.e., those identities that are satisfied in every structure* $(G, {}^\wedge, {}_\vee)$, *where* G *is a group and* ${}^\wedge$ *and* ${}_\vee$ *are the conjugacy operations of* (4.1), *are exactly the consequences of* $(LDQI)$.

Proof. If an identity is satisfied in every system $(G, {}^\wedge, {}_\vee)$, it is also satisfied in the subsystem of $(\mathrm{FG}_X, {}^\wedge, {}_\vee)$ generated by X, hence in the free quandle based on X. Therefore, it is satisfied in every quandle, *i.e.*, it is a consequence of $(LDQI)$. ∎

In some sense, the previous results complete the study both of free quandles and of conjugacy in a free group when we consider the two operations ${}^\wedge$ and ${}_\vee$ simultaneously.

Let us consider now similar questions in the case of one operation. Letting FG_X always denote a free group based on X, and ${}^\wedge$ denote the conjugacy operation on FG_X defined by $a^\wedge b = aba^{-1}$, we could expect from the previous results the subsystem of $(\mathrm{FG}_X, {}^\wedge)$ generated by X to be a free LDI-system. As in Corollaries 4.9 and 4.10, this would solve the word problem of (LDI), and show that (LDI) axiomatizes conjugacy in groups. This however is *not* true.

Lemma 4.11. *In every group, the conjugacy operation satisfies the identities*

$$((x \wedge y) \wedge y) \wedge x = (x \wedge y) \wedge ((y \wedge x) \wedge x), \tag{4.2}$$

$$((x \wedge y) \wedge y) \wedge (x \wedge z) = (x \wedge y) \wedge ((y \wedge x) \wedge z). \tag{4.3}$$

Proof. Both sides in (4.3) expand into $xyx^{-1}yxy^{-1}zyx^{-1}y^{-1}xy^{-1}x^{-1}$. Identity (4.2) follows from (4.3) by putting $z = x$ and applying idempotency. ∎

On the other hand, we have:

Lemma 4.12. *Identity* (4.2) *fails in the hull of the lattice* $(\{0,1\}, \vee, \wedge\}$.

Proof. Here $\vee$ and $\wedge$ denote the standard supremum and infimum operations on $\{0,1\}$. Then $(\{0,1\}, \vee, \wedge\}$ is an idempotent bi-LDI-system. Its hull consists of four elements, namely $(0,\vee)$, $(0,\wedge)$, $(1,\vee)$, and $(1,\wedge)$, and it is an LDI-system. Put $a = (0,\wedge)$, $b = (1,\vee)$. Computing $ab = (0 \wedge 1, \vee) = (0,\vee)$, *etc.* we find $((ab)b)a = (1,\wedge)$ and $(ab)((ba)a) = (0,\wedge)$, so (4.2) fails for $x = a$ and $y = b$. ∎

Hence, it is not true that (LDI) axiomatizes conjugacy in a group: (4.2) and (4.3) are examples of identities satisfied by group conjugacy that are not consequences of (LDI).

Exercise 4.13. (**injection bracket**) *Show that the idempotents of the LD-system $(\mathfrak{I}_\infty, [\,])$ are the bijections of $\mathbf{N}_+$.*

Exercise 4.14. (**left cancellative**) *Assume that S is an LDI-system. For $a, b \in S$, say that $a \sim b$ holds if there exist p and $c_1, \ldots, c_p$ satisfying $c_1 \ldots c_p a = c_1 \ldots c_p b$. Prove that $\sim$ is a congruence on S, that $S/\sim$ is a left cancellative LDI-system, and that it is the largest left cancellative quotient of S.*

Exercise 4.15. (**identities**) *(i) Prove that Identities (4.2) and (4.3) hold in every left cancellative LDI-system. [Hint: Multiply both sides of (4.3) by y.]*

(ii) Prove that (4.2) and (4.3) hold in every left divisible LDI-system (we say that $(S, {}^\wedge)$ is left divisible if $(\forall a, c)(\exists b)(a{}^\wedge b = c)$ holds). [Hint: Write $y{}^\wedge t = z$.]

Exercise 4.16. (**not left cancellative**) *Define $t_1 = ((x{\cdot}y){\cdot}y){\cdot}x$, $t_2 = (x{\cdot}y){\cdot}((y{\cdot}x){\cdot}x)$ [so Identity (4.2) is $t_1 = t_2$], $t_3 = (t_2{\cdot}y){\cdot}t_2$, and $t_4 = (t_2{\cdot}y){\cdot}x$. Prove $y{\cdot}t_1 =_{LDI} y{\cdot}t_2$, and deduce $t_3{\cdot}t_4 =_{LDI} t_3$. Evaluate $t_3((0,\vee),(1,\vee))$ and $t_4((0,\vee),(1,\vee))$ in the hull of $(\{0,1\},\vee,\wedge)$, and deduce that $t_3 =_{LDI} t_4$ fails. Conclude that the free LDI-systems of rank 2 or more are not left cancellative.*

---

5. Two-Sided Self-Distributivity

Let us define a right self-distributive system, or RD-system, to be a set equipped with a binary operation that satisfies the right self-distributivity identity

$$(xy)z = (xz)(yz). \tag{RD}$$

Of course, the properties of RD-systems are similar to those of LD-systems *mutatis mutandis*. Things become different when both left and right distributivities are considered simultaneously.

Definition 5.1. (**LRD-system**) An *LRD-system* is a set equipped with a binary operation that satisfies both (LD) and (RD).

Example 5.2. (**lattice**) *Assume that* $(L, \leq)$ *is a lattice, i.e., an ordered set where each pair* $\{a, b\}$ *of elements of* L *has a lower and an upper bound, respectively denoted* $a \wedge b$ *and* $a \vee b$*. Then both* $(L, \wedge)$ *and* $(L, \vee)$ *are LRD-systems, and, in an obvious sense,* $(L, \wedge, \vee)$ *is a bi-LRD-system.*

Lattices are LRD-systems that satisfy a number of additional identities, as the operations are idempotent, commutative and associative. Actually many LRD-systems are idempotent.

Lemma 5.3. *Assume that* S *is an LRD-system. Then every element in* $S(SS)$ *and* $(SS)S$ *is idempotent.*

Proof. Let I be the set of all idempotents in S. Use $=_{RD}$ for RD-equivalence. For every x, in S, we have $(x(xx))(x(xx)) =_{RD} (xx)(xx) =_{LD} x \cdot xx$, hence $x(xx)$ is idempotent, and so is $(xx)x$ by a symmetric argument. Now, for $x, y, z \in S$, we find $x(yz) =_{LD} (xy)(xz) =_{LD} ((xy)x)((xy)z) =_{RD} ((xx)(yx))((xy)z) =_{LD} ((xx)y)((xx)x))((xy)z)$. By Lemma 4.3, I is both a left and a right ideal, hence $(xx)x \in I$ implies $x(yz) \in I$, *i.e.*, we have $S(SS) \subseteq I$. The proof is similar for $(SS)S$. ∎

Proposition 5.4. (**idempotent**) *Assume that* S *is an LRD-system such that at least one left translation of* S*, or at least one right translation of* S*, is surjective. Then* S *is idempotent.*

Proof. Assume that the right translation $x \mapsto xa$ is surjective. Then we have $a = ba$ for some b, and therefore $a = b(ba)$. By Lemma 5.3, the latter element is idempotent. Now, for every element x, there exists y satisfying $x = ya$, and $a = aa$ implies $xx = (ya)(ya) = y(aa) = ya = x$. ∎

So, in particular, every left or right divisible LRD-system is idempotent. On the other hand, there exist non-idempotent LRD-systems, as shows Exercise 5.12.

Definition 5.5. (**n-nilpotent**) Assume that S is a semigroup. We say that S is *n-nilpotent* if the set S^n has cardinal 1, *i.e.*, there exists an element 0 such that $a_1 \ldots a_n = 0$ holds for all $a_1, \ldots, a_n$ in S.

Lemma 5.6. *Every 3-nilpotent semigroup is an LRD-system.*

Proof. Let S be a 3-nilpotent semigroup. For all a, b, c in S, we have $a(bc) = (ab)(ac) = (ab)c = (ac)(bc) = 0$, where 0 is the unique element of S^3. ■

Idempotent LRD-systems and 3-nilpotent semigroups appear as the two ends in the variety of all LRD-systems, as we shall prove now that every LRD-system can be decomposed in terms of systems of these two types. For S a binary system, and I an ideal of S, we denote by $\cong_I$ the equivalence relation on S generated by $I \times I$, *i.e.*, the set of all pairs (x, x') with x, x' in I augmented with the set of all diagonal pairs (x, x) with $x \notin I$. The relation $\cong_I$ is a congruence, so the quotient set $S/\!\cong_I$ has a natural induced product. We denote by $x \bmod I$ the equivalence class of x with respect to $\cong_I$.

Proposition 5.7. (decomposition) *Assume that S is an LRD-system. Let I be the set of all idempotents in S. Then the mapping*

$$f : x \mapsto (\,(xx)x\ ,\ x \bmod I\,)$$

gives a decomposition of S into a subsystem of the direct product of an idempotent LRD-system, namely I, and a 3-nilpotent semigroup, namely $S/\!\cong_I$.

Proof. By Lemma 4.3, the set I is a left ideal of S, and a right ideal as well since S is right self-distributive. Let $\varepsilon(x) = (xx)x$. We have seen above that $\varepsilon(x) \in I$ always holds. As $x \in I$ implies $\varepsilon(x) = x$, ε maps S onto I. For x, $y \in S$, we have $\varepsilon(x) \in I$, hence $\varepsilon(x)y \in I$, and, therefore,

$$\varepsilon(x)\varepsilon(y) = \varepsilon(\varepsilon(x)y) = \varepsilon(x)y = \varepsilon(xy),$$

i.e., ε is a homomorphism of S onto I. Now assume $f(x) = f(x')$. Either we have $x \notin I$, hence necessarily $x' = x$. Or we have $x \in I$, hence $\varepsilon(x) = x$, and, necessarily, $x' \in I$ as well, and we find $x' = \varepsilon(x') = \varepsilon(x) = x$. Hence f is injective. Finally Lemma 5.3 implies that $S/\!\cong_I$ is a 3-nilpotent semigroup. ■

Divisible LRD-systems

A typical example of a divisible LRD-system is a module equipped with the centroid operation $x * y = (1-\lambda)x + \lambda y$, where λ is a fixed scalar. This example can be generalized by replacing the scalar multiplications $x \mapsto (1-\lambda)x$, $y \mapsto \lambda y$ by more general endomorphisms. Moreover, the underlying structure of an abelian group can be weakened by keeping the hypothesis that inverses exist, but removing associativity. This leads to the notion of a commutative Moufang loop.

Definition 5.8. (commutative Moufang loop, nucleus) A *loop* is defined to be a binary system with a unit such that all left and right translations are bijections. A *commutative Moufang loop*, or *CML*, is defined to be a loop $(L, +)$ satisfying in addition the Moufang identity

$$(x + y) + (x + z) = (x + x) + (y + z)$$

(which forces $+$ to be commutative). The *nucleus* of $(L, +)$ is defined to be the set of all a in L satisfying $(a + y) + z = a + (y + z)$ for all y, z.

Every abelian group G is a commutative Moufang loop, and is its own nucleus. The following result describes all divisible LRD-systems in terms of commutative Moufang loops.

Proposition 5.9. (divisible LRD) *(i) Assume that $(L, +)$ is a commutative Moufang loop, and f, g are surjective endomorphisms of $(L, +)$ such that, for every x in L, $f(x) + g(x) = x$ holds and $x + f(x)$ belongs to the nucleus of L. Define $x * y = f(x) + g(y)$. Then $(L, *)$ is a divisible LRD-system.*

(ii) Conversely, assume that $(S, *)$ is a divisible LRD-system. Then there exists an operation $+$ on S and two surjective endomorphisms f, g of $(S, +)$ such that $(S, +)$ is a commutative Moufang loop and $*$ is obtained from $+$ as in (i).

Proof. (Sketch) (i) Let 0 be the unit of L. For every x in L, there exists a unique element denoted $-x$ that satisfies $(-x) + x = 0$, and, then, $x + ((-x) + y) = y$ holds for every y. The point is that the equality $x + (f(y) + f(z)) = (f(x) + f(y)) + (g(x) + f(z))$ holds for all x, y, z when f, g satisfy the conditions of the proposition. This implies $x + (y + z) = (f(x) + y) + (g(x) + z)$ since f and g are supposed to be surjective, and one deduces that the operation $*$ is left and right self-distributive.

(ii) We know that $(S, *)$ is idempotent. Let e be *any* element of S. Write ad_L and ad_R for the left and the right translations associated with e respectively. Then ad_L and ad_R are surjective by hypothesis. Choose sections, say h and k, of ad_L and ad_R, and define $x + y = h(x) * k(y)$. One successively shows that h and k commute, that $x * y = \mathrm{ad}_R(x) + \mathrm{ad}_L(y)$ holds for every x and y, that $(S, +)$ is a loop with unit e, that the mappings ad_L and ad_R are endomorphisms of $(S, +)$, that the operation $+$ is commutative and $(S, +)$ is a Moufang loop, and, finally, that $x + \mathrm{ad}_R(x)$ belongs to the nucleus of the CML $(S, +)$ for every x. ■

A particular case of a commutative Moufang loop is an abelian group. Those LRD-systems that are associated with abelian groups using the above scheme can be characterized by an additional identity.

Definition 5.10. (medial) A binary system S is said to be *medial*—or *entropic*—if it satisfies the identity

$$(xy)(uv) = (xu)(yv). \qquad (E)$$

It is easy to deduce from Proposition 5.9 the following result:

Proposition 5.11. (divisible medial LRD) *(i) Assume that $(G, +)$ is an abelian group, and f, g are surjective endomorphisms of $(G, +)$ such that $f(x)+g(x) = x$ holds for every x. Define $*$ by $x * y = f(x) + g(y)$. Then $(G, *)$ is a divisible medial LRD-system.*

(ii) Conversely, assume that $(S, *)$ is a divisible medial LRD-system. Then there exists an operation $+$ on S and two surjective endomorphisms f, g of $(S, +)$ such that $(S, +)$ is an abelian group and $*$ is obtained from $+$ as in (i).

Exercise 5.12. (non-idempotent) *Prove that $\mathbf{R}^3$ equipped with the operation $(a, b, c) * (a', b', c') = (0, 0, ab')$ is a 3-nilpotent, not 2-nilpotent, semigroup. [Observe that S is isomorphic to 3×3 upper triangular matrices]. Check that $(S, *)$ is not idempotent.*

Exercise 5.13. (Moufang loop) *(i) Assume that G is a (non-abelian) group and $(xy)^3 = x^3y^3$ holds for all x, y in G. Prove that $x+y = x^{-1}yx^2$ turns G into a commutative Moufang loop.*

(ii) Prove that $(\mathbf{Z}/3\mathbf{Z})^3$ equipped with $(x_1, x_2, x_3) * (y_1, y_2, y_3) = (x_1 + y_1, x_2 + y_2, x_3 + y_3 + (y_1 - x_1)(x_2y_3 - x_3y_2)$ is a commutative Moufang loop of order 81 [A non-associative CML of minimal size.]

Exercise 5.14. (identities) *Prove that Identities (4.2) and (4.3) hold in every LDI-system that satisfies the medial identity (E).*

---

6. Notes

Self-distributive operations have been considered for a long time. A systematic study was developed in the 1970s mostly by Czech algebraists at Prague, and the bibliography includes many references by J. Ježek, T. Kepka and P. Němec. We are indebted to Tomáš Kepka for his having prepared the following historical outline.

It seems that the first reference is (Peirce 1880). In the paper, the author explicitly formulates the (right) self-distributive law

$$(a + b) + c = (a + c) + (b + c), \qquad (a \times b) \times c = (a \times b) \times (b \times c).$$

The first explicit example of a non-associative self-distributive system seems to be (Schröder 1887), where the table

	0	1	2
0	0	2	1
1	2	1	0
2	1	0	2

is considered. The first paper fully devoted to self-distributive systems seems to be (Burstin & Mayer 29); it is concerned with two-sided self-distributive quasigroups. The first book mentioning results about self-distributuve systems (namely the results of the previous paper) is (Suškevič 37). The first example of a non-medial two-sided distributive system is (Bol 37): in the paper, an 81-element non-associative commutative Moufang loop is constructed: by Proposition 3.9 (medial), this is equivalent to a non-medial divisible LRD-system. According to a footnote in the paper, the construction is based on a hint communicated to Bol by Zassenhaus.

The first paper on two-sided distributive systems that are not quasigroups is (Ruedin 66), where Proposition 5.7 (decomposition) can be found. The connections between two-sided distributive systems, the medial (or entropic) law, and commutative Moufang loops were first studied in (Belousov 60). About commutative Moufang loops, probably the most interesting generalization known so far of groups when associativity is relaxed, general references are (Bruck 66) and (Chein, Pflugfelder & Smith 90). See also (Manin 72), in particular for the connection between Moufang loops and symmetric quasigroups. A comprehensive survey for the study of two-sided self-distributive systems is (Ježek, Kepka & Němec 81)—which is the first monograph devoted only to self-distributive structures.

Early references about the equational theories, *i.e.*, about those identities that follow from the self-distributivity law, are (Fajtlowicz & Mycielski 74) and (Ježek & Kepka 83) for the case of idempotent medial systems, and (Ježek & Kepka 84) for the case of two-sided self-distributivity.

The specific interest in one-sided self-distributivity is rather recent. The first reference could be (Takasaki 43), where idempotent left symmetric LD-systems are considered. Such systems are LD-quasigroups, and they were studied later by the pupils of Takasaki in Japan, and also in (Pierce 78, 79). General (non-idempotent) LD-systems were first studied for themselves in (Kepka 81); the first reference about right powers in a monogenic LD-system could be (Birkenmeier 85). The question of whether cancellative self-distributive systems can be embedded into quasigroups, which was studied in (Adhémar 70) in the two-sided case, is considered in (Kepka & Němec 77) in the one-sided case. Early references for construction of self-distributive operations starting from groups are (Hosszú 61), (Stein 59), (Belouzov 63).

Let us now mention specific sources for the material of this chapter.

The LD-systems A_n were introduced and investigated by R. Laver in the 1980s in his study of elementary embeddings in set theory, initially as a tool for computing critical ordinals (see Chapters XII and XIII). The main reference here is (Laver 95). Some basic results about periods appear in (Wehrung 91)—these results were also known to Laver and Knill—and a more complete description of the systems S_N and A_n appears in (Drápal 94).

The description of monogenic LD-systems where every element is a right power of the generator appears in (Sesboüé 96). The classification of all monogenic LD-systems, was completed in (Drápal 97) and (Drápal 97*a*). The example of Exercise 2.27 is due to R. Dougherty. Exercise 2.29 (extension to powerset) and the LD-system K are from (Birkenmeier 86) and (Kepka 81).

Multi-LD-systems seem to have been considered first in (Larue 99), where the hull is introduced. Free multi-LD-systems are described in (Dehornoy 97*c*). Most of the techniques developed in Part B of these notes extend to multi-LD-systems, in particular, those involving the operator ∂. However, the construction of canonical orders is not so natural because the right Polish notation is not optimal for multi-systems.

The description of free quandles appears in (Joyce 82). The other results about LDI-systems come from (Larue 99), in particular the result that group conjugacy satisfies Identities (4.2) and (4.3), and that the latter are not consequences of (LDI). These identities are also mentioned in (Drápal, Kepka & Musílek 94). Exercise 4.14 (left cancellative) is due to T. Kepka.

Proposition 5.9 (divisible LRD-systems) appears in (Kepka 79), which builds on (Murdoch 41), (Belousov 60) and (Soublin 71). Exercise 5.13 is from (Chein, Pflugfelder & Smith 90). Let us mention here a beautiful result about LRD-systems, namely Fischer's theorem (Fischer 64) that the group generated by the left translations of a finite LRD-quasigroup is a soluble group. Using the correspondence between LRD-quasigroups and commutative Moufang loops, Smith was able to deduce Fisher's theorem from the Bruck–Slaby result that gives a decomposition of every finite CML into the direct product of an abelian group and a nilpotent commutative loop whose order is a power of 3 (Smith 76).

As far as terminology is concerned, we have not always followed earlier traditions. In particular, the name "LD-groupoid" is used in the Czech literature, but we preferred here "LD-system" to avoid confusion with the usual definition of a groupoid as a small category with inverses. Our choice is consistent with the general name of "binary system" used after (Bruck 66) for a set equipped with a binary operation—one could also think of using "LD-magma" after Bourbaki, but the word "magma" is not widely used. The name "LD-algebra" has also been used, but, again, it could interfere with the usual notion of an algebra as a module or a vector space equipped with an additional structure. LD-quasigroups are called "racks" in (Fenn & Rourke 92) and (Devine *92*), but it seemed natural to use a name that is more understandable in itself.

Let us conclude with a few additional remarks and open questions about LDI-systems. In (Larue 99), an infinite family of identities I_1, I_2, ... is constructed so that the following holds: for each n, I_n is satisfied by group conjugacy, but it is not a consequence of (LDI) augmented with all I_j with $j \neq i$—the latter result is proved by constructing a specific LDI-system S_i in which I_i holds, but none of the I_j's, $j \neq i$; S_i is defined from a convenient operation $x^\wedge y = (1-\lambda_i)x + \lambda_i y$. There is no reason to believe that the previous family axiomatizes conjugacy in groups completely.

Question 6.1. Is group conjugacy finitely axiomatizable, *i.e.*, does there exist a finite family of identities $\vec{I}$ such that those identities satisfied by conjugacy in any group are the consequences of $\vec{I}$?

A result of (Larue *94a*) is that the identities satisfied by group conjugacy, which, by Proposition 4.8 (free quandle), coincide with those identities (involving operation $^\wedge$ only) satisfied in free quandles, also coincide with those identities satisfied in every left cancellative LDI-system.

Puzzling questions connect group conjugacy with the normal terms of Chapter VI.

Conjecture 6.2. (Dehornoy 99) *Define a special term to be a term in* T_∞ *obtained from a code (in the sense of Chapter VI) by replacing* i *with* x_i *for every* i. *Let* FG_∞ *be a free group based on an infinite sequence* x_1, x_2, ... *Then evaluation of special terms in* FG_∞ *equipped with conjugacy (resp. in* $\mathrm{FG}_\infty \times X$ *equipped with half-conjugacy) is injective.*

Conjecture 6.2 has been tested on hundreds of special terms (see Appendix of Chapter VI), and proved for degree 0, 1, and 2.

It is connected with the properties of the square function on free LD-systems. Let us say that an element a of FLD_∞ is a *tower* if we have $a = a_1 \ldots a_p$ with $a_p \sqsubset \cdots \sqsubset a_1 \sqsubseteq x_1 \cdots x_n$ for some n. Then Conjecture 6.2 is true if and only if the square function $a \mapsto a^{[2]}$ of FLD_∞ is injective on towers (Dehornoy 99). The idea is that, if the square function is injective on towers, we obtain an effective way for inversing the evaluation mapping on special terms, *i.e.*, for recovering a special term from its image under conjugacy or half-conjugacy. The problem is to control how free reductions occur in conjugate words, defined as those words obtained when evaluating a term using conjugacy. This can be done by associating with every letter in such a word a level that lives in the free LD-monoid FLDM_∞ (see Chapter XI) and in proving that, if the square function is injective, then reductions occur only between occurrences that lie on the same level.

Conjecture 6.2 is related to Question 6.1. Indeed, if Conjecture 6.2 is true, then group conjugacy satisfies no identity $t_1 = t_2$ with t_1, $t_2 \sqsubseteq_{LD} x_1 \cdot \cdots \cdot x_n$ except those that are consequence of (LD). Observe that the terms involved in (4.2) and (4.3) do not satisfy the above condition. Still another consequence of Conjecture 6.2 is that, for every injective term t_0 in T_∞, the evaluation mapping of FLD_∞ into FG_∞ equipped with conjugacy is injective on those elements a that satisfy $a \sqsubset \bar{t}_0$. Such a result could perhaps be proved directly using the methods of Chapter IX.

Some of the results established in Chapter V extend to LDI-systems. In particular, a confluence property holds, namely any two LDI-equivalent terms admit a common LDI-expansion (Larue 99). However, no order can be defined in free LDI-systems from left divisibility, for idempotency forbids antisymmetry, and it is not clear how to use the above result. Even the following problem is open:

Question 6.3. *Is the word problem for (LDI) decidable, i.e., does there exist an algorithm recognizing whether two terms are LDI-equivalent or not?*

We have seen that free LDI-systems are not left cancellative. A result of (Kepka 81) is that every left cancellative LDI-system embeds in a quandle of the same cardinality.

Question 6.4. (Drápal) *Are free LDI-systems right cancellative?*

A positive answer is conjectured. We think that investigating the geometry monoid $\mathcal{G}_{LDI}$ associated with Identities (LD) and (I) could be useful here.

Appendix. The First Laver Tables

No explicit formula is known for the value of the product $a *_n b$ of a and b in the table A_n, and we shall see in Chapter XIII that it is unlikely that such a formula may exist.

Here we display completely the tables A_n for $n \leq 6$ and give partial information about the tables A_n with $7 \leq n \leq 10$. It is neither necessary nor useful to give all values, since we know that the rows are periodic. So we enumerate the values in each row, and stop when the maximal value 2^n is reached: the next values are the previous ones repeating periodically. For $n \geq 5$, we skip the second half of the table: in order to compute $(2^{n-1} + a) *_n b$, it suffices to compute $a *_n b$, and either to keep the value if it lies above 2^{n-1}, or to add 2^{n-1} otherwise. Also, we omit the row of 2^{n-1}, which we know is a shifted identity.

Table of A_0:

1: 1 ||

Table of A_1:

1: 2 || **2**: 1, 2 ||

Table of A_2:

1: 2, 4 || **2**: 3, 4 || **3**: 4 || **4**: 1, 2, 3, 4 ||

Table of A_3:

1: 2, 4, 6, 8 || **2**: 3, 4, 7, 8 || **3**: 4, 8 || **4**: 5, 6, 7, 8 || **5**: 6, 8 || **6**: 7, 8 || **7**: 8 || **8**: 1, 2, 3, 4, 5, 6, 7, 8 ||

Table of A_4:

1: 2, 12, 14, 16 || **2**: 3, 12, 15, 16 || **3**: 4, 8, 12, 16 || **4**: 5, 6, 7, 8, 13, 14, 15, 16 || **5**: 6, 8, 14, 16 || **6**: 7, 8, 15, 16 || **7**: 8, 16 || **8**: 9, 10, 11, 12, 13, 14, 15, 16 || **9**: 10, 12, 14, 16 || **10**: 11, 12, 15, 16 || **11**: 12, 16 || **12**: 13, 14, 15, 16 || **13**: 14, 16 || **14**: 15, 16 || **15**: 16 || **16**: 1, 2, 3, 4, 5, 6, 7, 8, 9, 10, 11, 12, 13, 14, 15, 16 ||

Table of A_5:

1: 2, 12, 14, 16, 18, 28, 30, 32 || **2**: 3, 12, 15, 16, 19, 28, 31, 32 || **3**: 4, 8, 12, 16, 20, 24, 28, 32 || **4**: 5, 6, 7, 8, 13, 14, 15, 16, 21, 22, 23, 24, 29, 30, 31, 32 || **5**: 6, 24, 30, 32 || **6**: 7, 24, 31, 32 || **7**: 8, 16, 24, 32 || **8**: 9, 10, 11, 12, 13, 14, 15, 16, 25, 26, 27, 28, 29, 30, 31, 32 || **9**: 10, 28, 30, 32 || **10**: 11, 28, 31, 32 || **11**: 12, 16, 28, 32 || **12**: 13, 14, 15, 16, 29, 30, 31, 32 || **13**: 14, 16, 30, 32 || **14**: 15, 16, 31, 32 || **15**: 16, 32 ||

Table of A_6:

1: 2, 12, 14, 48, 50, 60, 62, 64 ‖ **2**: 3, 12, 15, 48, 51, 60, 63, 64 ‖ **3**: 4, 8, 12, 16, 20, 24, 28, 32, 36, 40, 44, 48, 52, 56, 60, 64 ‖ **4**: 5, 6, 7, 8, 45, 46, 47, 48, 53, 54, 55, 56, 61, 62, 63, 64 ‖ **5**: 6, 56, 62, 64 ‖ **6**: 7, 56, 63, 64 ‖ **7**: 8, 48, 56, 64 ‖ **8**: 9, 10, 11, 12, 45, 46, 47, 48, 57, 58, 59, 60, 61, 62, 63, 64 ‖ **9**: 10, 60, 62, 64 ‖ **10**: 11, 60, 63, 64 ‖ **11**: 12, 48, 60, 64 ‖ **12**: 13, 14, 15, 16, 29, 30, 31, 32, 45, 46, 47, 48, 61, 62, 63, 64 ‖ **13**: 14, 48, 62, 64 ‖ **14**: 15, 48, 63, 64 ‖ **15**: 16, 32, 48, 64 ‖ **16**: 17, 18, 19, 20, 21, 22, 23, 24, 25, 26, 27, 28, 29, 30, 31, 32, 49, 50, 51, 52, 53, 54, 55, 56, 57, 58, 59, 60, 61, 62, 63, 64 ‖ **17**: 18, 28, 30, 32, 50, 60, 62, 64 ‖ **18**: 19, 28, 31, 32, 51, 60, 63, 64 ‖ **19**: 20, 24, 28, 32, 52, 56, 60, 64 ‖ **20**: 21, 22, 23, 24, 29, 30, 31, 32, 53, 54, 55, 56, 61, 62, 63, 64 ‖ **21**: 22, 56, 62, 64 ‖ **22**: 23, 56, 63, 64 ‖ **23**: 24, 32, 56, 64 ‖ **24**: 25, 26, 27, 28, 29, 30, 31, 32, 57, 58, 59, 60, 61, 62, 63, 64 ‖ **25**: 26, 60, 62, 64 ‖ **26**: 27, 60, 63, 64 ‖ **27**: 28, 32, 60, 64 ‖ **28**: 29, 30, 31, 32, 61, 62, 63, 64 ‖ **29**: 30, 32, 62, 64 ‖ **30**: 31, 32, 63, 64 ‖ **31**: 32, 64 ‖ ...

Table of A_7:

1: 2, 12, 14, 112, 114, 124, 126, 128 ‖ **2**: 3, 12, 15, 48, 51, 60, 63, 64, 67, 76, 79, 112, 115, 124, 127, 128 ‖ **3**: 4, 8, 12, 80, 84, 88, 92, 96, 100, 104, 108, 112, 116, 120, 124, 128 ‖ **4**: 5, 6, 7, 8, 109, 110, 111, 112, 117, 118, 119, 120, 125, 126, 127, 128 ‖ **5**: 6, 120, 126, 128 ‖ **6**: 7, 120, 127, 128 ‖ **7**: 8, 112, 120, 128 ‖ **8**: 9, 10, 11, 12, 109, 110, 111, 112, 121, 122, 123, 124, 125, 126, 127, 128 ‖ **9**: 10, 124, 126, 128 ‖ **10**: 11, 124, 127, 128 ‖ **11**: 12, 112, 124, 128 ‖ **12**: 13, 14, 15, 80, 93, 94, 95, 96, 109, 110, 111, 112, 125, 126, 127, 128 ‖ **13**: 14, 112, 126, 128 ‖ **14**: 15, 112, 127, 128 ‖ **15**: 16, 32, 48, 64, 80, 96, 112, 128 ‖ **16**: 17, 18, 19, 20, 21, 22, 23, 24, 25, 26, 27, 28, 29, 30, 31, 32, 49, 50, 51, 52, 53, 54, 55, 56, 57, 58, 59, 60, 61, 62, 63, 64, 81, 82, 83, 84, 85, 86, 87, 88, 89, 90, 91, 92, 93, 94, 95, 96, 113, 114, 115, 116, 117, 118, 119, 120, 121, 122, 123, 124, 125, 126, 127, 128 ‖ **17**: 18, 28, 30, 96, 114, 124, 126, 128 ‖ **18**: 19, 28, 31, 96, 115, 124, 127, 128 ‖ **19**: 20, 24, 28, 32, 52, 56, 60, 64, 84, 88, 92, 96, 116, 120, 124, 128 ‖ **20**: 21, 22, 23, 24, 93, 94, 95, 96, 117, 118, 119, 120, 125, 126, 127, 128 ‖ **21**: 22, 120, 126, 128 ‖ **22**: 23, 120, 127, 128 ‖ **23**: 24, 96, 120, 128 ‖ **24**: 25, 26, 27, 28, 93, 94, 95, 96, 121, 122, 123, 124, 125, 126, 127, 128 ‖ **25**: 26, 124, 126, 128 ‖ **26**: 27, 124, 127, 128 ‖ **27**: 28, 96, 124, 128 ‖ **28**: 29, 30, 31, 32, 61, 62, 63, 64, 93, 94, 95, 96, 125, 126, 127, 128 ‖ **29**: 30, 96, 126, 128 ‖ **30**: 31, 96, 127, 128 ‖ **31**: 32, 64, 96, 128 ‖ ...

Table of A_8:

1: 2, 12, 14, 240, 242, 252, 254, 256 ‖ **2**: 3, 12, 15, 48, 51, 60, 63, 192, 195, 204, 207, 240, 243, 252, 255, 256 ‖ **3**: 4, 8, 12, 208, 212, 216,

220, 224, 228, 232, 236, 240, 244, 248, 252, 256 ‖ **4**: 5, 6, 7, 8, 237, 238, 239, 240, 245, 246, 247, 248, 253, 254, 255, 256 ‖ **5**: 6, 248, 254, 256 ‖ **6**: 7, 248, 255, 256 ‖ **7**: 8, 240, 248, 256 ‖ **8**: 9, 10, 11, 12, 237, 238, 239, 240, 249, 250, 251, 252, 253, 254, 255, 256 ‖ **9**: 10, 252, 254, 256 ‖ **10**: 11, 252, 255, 256 ‖ **11**: 12, 240, 252, 256 ‖ **12**: 13, 14, 15, 208, 221, 222, 223, 224, 237, 238, 239, 240, 253, 254, 255, 256 ‖ **13**: 14, 240, 254, 256 ‖ **14**: 15, 240, 255, 256 ‖ **15**: 16, 32, 48, 64, 80, 96, 112, 128, 144, 160, 176, 192, 208, 224, 240, 256 ‖ ...

Table of A_9:

1: 2, 12, 14, 240, 242, 252, 254, 256, 258, 268, 270, 496, 498, 508, 510, 512 ‖ **2**: 3, 12, 271, 304, 307, 316, 319, 448, 451, 460, 463, 496, 499, 508, 511, 512 ‖ **3**: 4, 8, 12, 464, 468, 472, 476, 480, 484, 488, 492, 496, 500, 504, 508, 512 ‖ **4**: 5, 6, 7, 8, 493, 494, 495, 496, 501, 502, 503, 504, 509, 510, 511, 512 ‖ **5**: 6, 504, 510, 512 ‖ **6**: 7, 504, 511, 512 ‖ **7**: 8, 496, 504, 512 ‖ **8**: 9, 10, 11, 12, 493, 494, 495, 496, 505, 506, 507, 508, 509, 510, 511, 512 ‖ **9**: 10, 508, 510, 512 ‖ **10**: 11, 508, 511, 512 ‖ **11**: 12, 496, 508, 512 ‖ **12**: 13, 14, 271, 464, 477, 478, 479, 480, 493, 494, 495, 496, 509, 510, 511, 512 ‖ **13**: 14, 496, 510, 512 ‖ **14**: 15, 240, 255, 256, 271, 496, 511, 512 ‖ **15**: 16, 32, 48, 64, 80, 96, 112, 128, 144, 160, 176, 192, 208, 224, 240, 256, 272, 288, 304, 320, 336, 352, 368, 384, 400, 416, 432, 448, 464, 480, 496, 512 ‖ ...

Table of A_{10}:

1: 2, 12, 14, 240, 242, 252, 254, 768, 770, 780, 782, 1008, 1010, 1020, 1022, 1024 ‖ **2**: 3, 12, 783, 816, 819, 828, 831, 960, 963, 972, 975, 1008, 1011, 1020, 1023, 1024 ‖ **3**: 4, 8, 12, 976, 980, 984, 988, 992, 996, 1000, 1004, 1008, 1012, 1016, 1020, 1024 ‖ **4**: 5, 6, 7, 8, 1005, 1006, 1007, 1008, 1013, 1014, 1015, 1016, 1021, 1022, 1023, 1024 ‖ **5**: 6, 1016, 1022, 1024 ‖ **6**: 7, 1016, 1023, 1024 ‖ **7**: 8, 1008, 1016, 1024 ‖ **8**: 9, 10, 11, 12, 1005, 1006, 1007, 1008, 1017, 1018, 1019, 1020, 1021, 1022, 1023, 1024 ‖ **9**: 10, 1020, 1022, 1024 ‖ **10**: 11, 1020, 1023, 1024 ‖ **11**: 12, 1008, 1020, 1024 ‖ **12**: 13, 14, 783, 976, 989, 990, 991, 992, 1005, 1006, 1007, 1008, 1021, 1022, 1023, 1024 ‖ **13**: 14, 1008, 1022, 1024 ‖ **14**: 15, 240, 255, 768, 783, 1008, 1023, 1024 ‖ **15**: 16, 32, 48, 64, 80, 96, 112, 128, 144, 160, 176, 192, 208, 224, 240, 256, 272, 288, 304, 320, 336, 352, 368, 384, 400, 416, 432, 448, 464, 480, 496, 512, 528, 544, 560, 576, 592, 608, 624, 640, 656, 672, 688, 704, 720, 736, 752, 768, 784, 800, 816, 832, 848, 864, 880, 896, 912, 928, 944, 960, 976, 992, 1008, 1024 ‖ ...

XI
LD-Monoids

A number of LD-systems can be equipped with a second, associative operation connected to the self-distributive operation by several mixed identities. Here we study such double structures, called LD-monoids. In particular, we discuss the question of completing a given LD-system into an LD-monoid. We give two solutions, and deduce a complete description of free LD-monoids. The global conclusion is that the self-distributive operation is the core of the structure, and that adding an associative product is essentially trivial. However, the case of braid exponentiation is not so simple, and applying the above mentioned completion scheme requires considering the extended braids of Section I.4.

The chapter is organized as follows. In Section 1, we give examples of LD-monoids, and discuss the problem of completing a given monoid into an LD-monoid. In Section 2, we address the problem of embedding a given LD-system into an LD-monoid; we describe a universal solution called the free completion, as well as another completion defined using composition of left translations. The construction applies in particular to the Laver tables. In Section 3, we show that the free LD-system FLD_X embeds in the free LD-monoid based on X, and that the latter can be constructed inside FLD_X. This enables us in particular to solve the word problem of free LD-monoids. In Section 4, we deduce that every free LD-monoid admits canonical linear orderings. As for free LD-systems, the existence of such orderings leads to a freeness criterion and to various algebraic properties like left cancellativity. Finally, in Section 5, we come back to the LD-monoid of extended braids as defined in Chapter I, and show that it includes many free LD-monoids of rank 1.

1. LD-Monoids

In this introductory section, we recall the definition of an LD-monoid, review some examples, and consider the question of defining the structure of an LD-monoid on a given monoid.

LD-monoids

Let us recall the definition of an LD-monoid as given in Section I.4—actually the definition is slightly different, but we shall verify the equivalence below:

Definition 1.1. (LD-monoid) We say that $(M, \cdot, 1, ^\wedge)$ is an *LD-monoid* if $(M, \cdot, 1)$ is a monoid, and the following identities are obeyed:

$$x \cdot y = (x \mathbin{^\wedge} y) \cdot x, \qquad (LDM_1)$$

$$(x \cdot y) \mathbin{^\wedge} z = x \mathbin{^\wedge} (y \mathbin{^\wedge} z), \qquad (LDM_2)$$

$$x \mathbin{^\wedge} (y \cdot z) = (x \mathbin{^\wedge} y) \cdot (x \mathbin{^\wedge} z). \qquad (LDM_3)$$

$$1 \mathbin{^\wedge} x = x, \qquad x \mathbin{^\wedge} 1 = 1. \qquad (LDM_4)$$

In general, the associative operation of an LD-monoid will be called its product, while the other operation will be called its exponentiation. Unless otherwise specified, we use 1 for the unit of the product. So, usually, we shall speak of "the LD-monoid $(M, \cdot, ^\wedge)$", or, more simply, of "the LD-monoid M".

The connection between LD-monoids and LD-systems is straightforward—and, therefore, the above definition is equivalent to that of Section I.4:

Proposition 1.2. (LD-system) *Exponentiation is left self-distributive in every LD-monoid.*

Proof. Using (LDM_2) and (LDM_1), we deduce

$$x \mathbin{^\wedge} (y \mathbin{^\wedge} z) = (x \cdot y) \mathbin{^\wedge} z = ((x \mathbin{^\wedge} y) \cdot x) \mathbin{^\wedge} z = (x \mathbin{^\wedge} y) \mathbin{^\wedge} (x \mathbin{^\wedge} z). \qquad \blacksquare$$

Observe that the identity $1 ^\wedge x = x$ in the definition of an LD-monoid is a consequence of the other identities. Indeed, $1 ^\wedge x = (1 ^\wedge x) \cdot 1 = 1 \cdot x = x$ follows from (LDM_1).

We observed in Chapter I that, if G is a group, then G equipped with the conjugacy operation $x ^\wedge y = xyx^{-1}$ is an LD-monoid. There exist a number of other examples.

Example 1.3. (abelian) *If M is an abelian monoid, then M equipped with $x ^\wedge y = y$ is an LD-monoid.*

Example 1.4. **(injection bracket)** *Let $\mathfrak{I}_\infty$ denote the set of all injections of $\mathbf{N}_+$ into itself. Then $(\mathfrak{I}_\infty, \circ, \mathrm{id}, [\,])$ is an LD-monoid, where $[\,]$ is the bracket operation defined in Section I.2. Here Identities (LDM_1) to (LDM_4) take the form*

$$f \circ g = f[g] \circ f, \tag{LDM_1}$$

$$(f \circ g)[h] = f[g[h]], \tag{LDM_2}$$

$$f[g \circ h] = f[g] \circ f[h], \tag{LDM_3}$$

$$\mathrm{id}[f] = f, \qquad f[\mathrm{id}] = \mathrm{id}. \tag{LDM_4}$$

Example 1.5. **(implication)** *Boolean implication is a left self-distributive operation: for all Boolean formulas A, B, C, the formulas $A \Rightarrow (B \Rightarrow C)$ and $(A \Rightarrow B) \Rightarrow (A \Rightarrow C)$ are equivalent. Moreover, conjunction is associative, "True" is a unit for conjunction, and conjunction together with implication satisfy the mixed identities of LD-monoids, i.e., $(\{0,1\}, \wedge, 1, \Rightarrow)$ is an LD-monoid, where $\wedge$ and $\Rightarrow$ are defined by*

$\wedge$	0	1
0	0	0
1	0	1

$\Rightarrow$	0	1
0	1	1
1	0	1

More generally, if $(B, \wedge, \vee, \neg, 0, 1)$ is a Boolean algebra, then $(B, \wedge, 1, \rightarrow)$ is an LD-monoid, where $\rightarrow$ is defined by $a \rightarrow b = \neg a \vee b$. Symmetrically, $(B, \vee, 0, \rightarrow')$ is an LD-monoid, where $a \rightarrow' b = \neg a \wedge b$.

The previous construction extends to every Heyting algebra. A Heyting algebra is a lattice $(H, \vee, \wedge, 0, 1)$ with an additional binary operation $\rightarrow$ such that, writing $x \le y$ for $x \vee y = y$, $c \le (a \rightarrow b)$ is equivalent to $(a \wedge c) \le b$. Every Boolean algebra becomes a Heyting algebra when $a \rightarrow b$ is defined, as above, to be $\neg a \vee b$, but, conversely, a Heyting algebra need not be a Boolean algebra as defining $\neg a$ to be $a \rightarrow 0$ does not give in general an involutive negation. An example of a Heyting algebra that is not a Boolean algebra in general is the family of all open sets in a topological space equipped with union, intersection, and $U \rightarrow V$ defined to be the interior of $U^c \cup V$. Then $\rightarrow$ is left self-distributive: $x \le (a \rightarrow (b \rightarrow c))$ is equivalent to $(x \wedge a \wedge b) \le c$, $x \le ((a \rightarrow b) \rightarrow (a \rightarrow c))$ is equivalent to $(x \wedge (a \rightarrow b) \wedge a) \le c$, and we have $a \wedge b = a \wedge (a \rightarrow b)$. The verification of (LDM_1), ..., (LDM_4) is similar.

LD-monoids can be defined in terms of an action of a monoid onto itself. The following notion, which has already been used occasionally, will be central here:

Definition 1.6. (left translation) For $(S, ^\wedge)$ a binary system, and a in S, we denote by $\mathrm{ad}_a^\wedge$ (or ad_a) the left translation of $(S, ^\wedge)$ associated with a, *i.e.*, the mapping $x \mapsto a^\wedge x$.

A binary system $(S, ^\wedge)$ is an LD-system if and only if , for every a in S, the left translation ad_a is an endomorphism of $(S, ^\wedge)$. More generally, a multiple binary system $(S\,, (*_i)_{i\in I})$ is a multi-LD-system if and only if, for every a in S and every i in I, the left translation $\mathrm{ad}_a^{*_i}$ is an endomorphism of $(S\,, (*_i)_{i\in I})$. In the case of LD-monoids, we obtain:

Proposition 1.7. (characterization) *(i) Assume that $(M, \cdot\,, ^\wedge)$ is an LD-monoid. Then the left translation mapping* $\mathrm{ad} : x \mapsto \mathrm{ad}_x^\wedge$ *is a homomorphism of $(M, \cdot\,)$ into* $\mathrm{End}(M, \cdot\,)$.

(ii) Conversely, assume that $(M, \cdot\,)$ is a monoid, and f is a homomorphism of $(M, \cdot\,)$ into $\mathrm{End}(M, \cdot\,)$ such that $x \cdot y = f(x)(y) \cdot x$ holds for all x, y. Define $x^\wedge y = f(x)(y)$. Then $(M, \cdot\,, ^\wedge)$ is an LD-monoid.

Proof. Identity (LDM_2) together with $1^\wedge x = x$ means that exponentiation defines an action of M on itself, and (LDM_3) together with $x^\wedge 1 = 1$ means that this action is compatible with the monoid structure on M. ■

Companion multi LD-systems

There exists a uniform way of defining new left self-distributive operations on every LD-monoid.

Proposition 1.8. (companion) *Assume that $(M, \cdot\,, ^\wedge)$ is an LD-monoid, and X is a subset of M. For x in X, define $*_x$ by $a *_x b = (a^\wedge x) \cdot b$. Then $(M, (*_x)_{x\in X})$ is a multi-LD-system.*

Proof. For all x, y, a, b, c, we find

$$\begin{aligned}(a{*_x}b) *_y (a *_x c) &= (((a \wedge x) \cdot b) \wedge y) \cdot (a \wedge x) \cdot c\\ &= ((a \wedge x) \wedge (b \wedge y)) \cdot (a \wedge x) \cdot c = (a \wedge x) \cdot (b \wedge y) \cdot c = a *_x (b *_y c).\end{aligned}$$ ■

Definition 1.9. (companion) In the previous situation, the multi-LD-system $(M, (*_x)_{x\in X})$ is called the X-*companion* of the LD-monoid M.

Example 1.10. (group conjugacy) *Assume that G is a group, and $^\wedge$ is the conjugacy operation $a^\wedge b = aba^{-1}$. The X-companion of the LD-monoid $(G, \cdot\,, ^\wedge)$ is $(G, (*_x)_{x\in G})$, with $a *_x b = axa^{-1}b$. The hull of the latter multi-LD-system is then $G \times X$ equipped with the half-conjugacy operation $(a, x)^\wedge(b, y) = (axa^{-1}b, y)$.*

We shall see in Section 3 below how the correspondence from an LD-monoid to its companions can be reversed in good cases.

Invertible elements

The exponentiation of an LD-monoid always acts like a conjugacy on invertible elements. For M a monoid, we denote by $U(M)$ the group of all invertible elements of M.

Proposition 1.11. (invertibles) *Assume that M is an LD-monoid.*

(i) For c in M, if c admits a left inverse, or a right inverse (with respect to product), then the latter is a two-sided inverse. In that case, $a^\wedge c$ is invertible for every a in M, and $c^\wedge a = cac^{-1}$ holds. So, $U(M)$ is a sub-LD-monoid of M, and it is the LD-monoid associated with conjugacy in the group $U(M)$.

(ii) The left coset equivalence relation associated with $U(M)$ is a congruence on the LD-monoid M, and the quotient LD-monoid $U(M)\backslash M$ has no invertible element except 1.

Proof. (i) Assume $cd = 1$. We have $dc = cddc = (c^\wedge d)cdc = (c^\wedge d)c = cd = 1$, and d is a two-sided inverse of c. Now, for every a, we have $(a^\wedge c)(a^\wedge c^{-1}) = a^\wedge(cc^{-1}) = a^\wedge 1 = 1$, hence $a^\wedge c^{-1}$ is an inverse of $a^\wedge c$. Moreover, we have $c^\wedge a = (c^\wedge a)cc^{-1} = cac^{-1}$.

(ii) For a, a' in M, write $a \equiv a'$ for $U(M)a = U(M)a'$. Assume $a \equiv a'$. Since $U(M)$ is a group, there exists c in $U(M)$ satisfying $a' = ca$. Clearly $ab \equiv a'b$ holds for every b. On the other hand, we have $ba' = bca = (b^\wedge c)ba$. We have seen that $b^\wedge c$ is invertible, so $ba \equiv ba'$ holds, and $\equiv$ is a congruence of the monoid M. Then, we have $b^\wedge a' = b^\wedge(ca) = (b^\wedge c)(b^\wedge a)$. Again, $b^\wedge c$ is invertible, hence $b^\wedge a \equiv b^\wedge a'$ holds. Finally, we have

$$a' \wedge b = (ca) \wedge b = c \wedge (a \wedge b) = c \cdot (a \wedge b) \cdot c^{-1} = c \cdot ((a \wedge b) \wedge (c^{-1})) \cdot (a \wedge b),$$

which gives $a^\wedge b \equiv a'^\wedge b$, as c and $(a^\wedge b)^\wedge(c^{-1})$ are invertible. Hence the equivalence relation $\equiv$ is compatible both with product and exponentiation in M, and the quotient $U(M)\backslash M$ inherits the structure of an LD-monoid. Assume finally that the class $U(M)a$ is invertible in $U(M)\backslash M$. This means that there exists b satisfying $ab \equiv 1$, *i.e.*, that ab is invertible in M. In other words, there exists c satisfying $abc = 1$, which implies that a is invertible on the right, and therefore invertible. Hence we have $U(M)a = U(M)$, the unit of $U(M)\backslash M$. ∎

It follows that, for every LD-monoid M, there exists an exact sequence of LD-monoids

$$1 \to U(M) \to M \to U(M)\backslash M \to 1, \tag{1.1}$$

where $U(M)$ is a group equipped with conjugacy and $U(M)\backslash M$ has no invertible element except its unit.

Example 1.12. (**injections**) *In the case of the LD-monoid $\mathfrak{I}_\infty$, the invertible elements are the bijections of $\mathbf{N}_+$, so the LD-monoid $U(\mathfrak{I}_\infty)$ is the conjugacy LD-monoid of the symmetric group $\mathfrak{S}_\infty$. Two injections f, g are equivalent modulo $\mathfrak{S}_\infty$ if and only if $f = hg$ holds for some bijection h: a necessary and sufficient condition is that $\mathrm{Im}(f)$ and $\mathrm{Im}(g)$ on the one hand, and $\mathrm{coIm}(f)$ and $\mathrm{coIm}(g)$ on the other hand, have equal cardinalities. Since f and g are injections, the cardinality of $\mathrm{Im}(f)$ and $\mathrm{Im}(g)$ is always that of $\mathbf{N}_+$. So the class of f in $\mathfrak{S}_\infty\backslash\mathfrak{I}_\infty$ is characterized by $\mathrm{card}(\mathrm{coIm}(f))$, which lives in $\mathbf{N}_+ \cup \{\infty\}$. The quotient operations on $\mathbf{N}_+ \cup \{\infty\}$ induced by the operations of $\mathfrak{I}_\infty$ are defined respectively by $p \circ q = p + q$ and $p[q] = q$, with $p \circ \infty = \infty \circ p = \infty$ for every p.*

The sub-LD-monoid M of $\mathfrak{I}_\infty$ generated by the shift injection sh is infinite, while $U(M)$ and $U(M)\backslash M$ are singletons. So the two components $U(M)$ and $U(M)\backslash M$ in (1.1) need not capture the structure of M.

The previous construction extends to injections on an arbitrary set A. Then the quotient LD-monoid $\mathfrak{S}_A\backslash\mathfrak{I}_A$ is the set of those (finite or infinite) cardinal numbers κ that satisfy $\mathrm{card}(A) = \kappa + \mathrm{card}(A)$, equipped with cardinal addition and the trivial exponentiation $\alpha\hat{\ }\beta = \beta$.

Completion of a monoid into an LD-monoid

Every LD-monoid is both a monoid and an LD-system, so two natural completion problems arise, namely completing a given monoid into an LD-monoid, and completing a given LD-system into an LD-monoid. Here we briefly consider the first problem.

By (LDM_1), every LD-monoid satisfies the weak commutativity condition

$$(\forall x, y)(\exists z)(zx = xy), \tag{1.2}$$

and, therefore, some monoids cannot be completed into LD-monoids. Now, if a monoid satisfies (1.2), defining the second operation so as to obtain an LD-monoid amounts to selecting one element z as in (1.2) in such a way that the compatibility identities (LDM_2) and (LDM_3) are satisfied, which amounts to selecting the element z in a functorial way. Such a construction is not always possible (*cf.* Exercise 1.20). A trivial case for a completion to exist is M being abelian. A more interesting sufficient condition is requiring uniqueness in (1.2).

Proposition 1.13. (**LD-completion**) *Assume that M is a monoid satisfying*

$$(\forall x, y)(\exists! z)(zx = xy). \tag{1.3}$$

Then there exists a unique way to complete M into an LD-monoid. The corresponding exponentiation is idempotent.

Proof. In order to obey Identity (LDM_1), we must define $a{\wedge}b$ to be the unique c satisfying $ca = ab$. Then $a1 = 1a = a$ implies $a{\wedge}1 = 1$ and $1{\wedge}a = a$ for every a. For all a, b, c in M, $(ab){\wedge}c$ is the unique d satisfying $dab = abc$. Now we have $(a{\wedge}(b{\wedge}c))ab = a(b{\wedge}c)b = abc$, and, therefore, $(a{\wedge}(b{\wedge}c))ab = (ab){\wedge}c$. Similarly $a{\wedge}(bc)$ is the unique d satisfying $da = abc$. We have $(a{\wedge}b)(a{\wedge}c)a = (a{\wedge}b)ac = abc$, hence $a{\wedge}(bc) = (a{\wedge}b)(a{\wedge}c)$. Finally, $(a{\wedge}a)a = aa$, which holds in every LD-monoid, implies $a{\wedge}a = a$ when (1.3) holds. ∎

If G is a group, (1.3) holds, and the left self-distributive operation so obtained is the conjugacy $x{\wedge}y = xyx^{-1}$. Proposition 1.13 (LD-completion) tells us that this is the only way to complete the group operation into an LD-monoid.

Example 1.14. (ordinals) *Here is a monoid satisfying Condition (1.3) without being a group. Let κ be an (infinite) ordinal that is closed under ordinal addition, i.e., one of the form ω^α. If $+^{opp}$ denotes the opposite of addition: $\alpha +^{opp} \beta = \beta + \alpha$, then the monoid $(\kappa, +^{opp})$ satisfies (1.3). Indeed, we have always $\alpha + \beta \geq \beta$, and, therefore, there exists a unique γ satisfying $\alpha + \beta = \beta + \gamma$. For $\kappa = \omega$, the monoid $(\kappa, +^{opp})$ is abelian, and exponentiation is trivial. For $\kappa > \omega$, we obtain an LD-monoid that is not an LD-quasigroup. Indeed the left translations of $(\kappa, +^{opp})$, i.e., the right translations of $(\kappa, +)$, are not injective, since $0 + \omega = 1 + \omega$ holds (these translations are not surjective either).*

Using the explicit definition of ordinal addition in Cantor's normal decomposition, we can avoid ordinals. For instance, the case $\kappa = \omega^2$ amounts to considering the monoid $\langle a, b \,;\, ab = a\rangle$. Every element can be written as $a^p b^q$ in a unique way, and exponentation is defined by

$$(a^p b^q) \wedge (a^r b^s) = a^r b^{[\![r \neq 0]\!]q + [\![p=0]\!]s}$$

where $[\![C]\!]$ is 1 or 0 according to whether C is true or false. Equivalently, this amounts to considering $(p,q){\wedge}(r,s) = (r, [\![r \neq 0]\!]q + [\![p = 0]\!]s)$ on $\mathbf{N}^2$.

Condition (1.3) does not hold in every LD-monoid. A typical example where Condition (1.2) holds, but Condition (1.3) does not, is the LD-monoid $\mathfrak{I}_\infty$. More generally, by Proposition 1.13 (LD-completion), Condition (1.3) fails in every LD-monoid where exponentiation is not idempotent.

Exercise 1.15. (identity element) *For S a binary system, and e in S, say that e is an identity element in S if $ea = a$ and $ae = e$ hold for every a.*

(i) Prove that, if S is an LD-system, and $e \notin S$ holds, then there exists exactly one way to extend the operation of S to $S \cup \{e\}$ so that the resulting system is an LD-system and e is an identity element.

(ii) Prove that the identity mapping is the only identity element in $(\mathfrak{I}_\infty, [\,])$.

Exercise 1.16. **(companion)** *Assume that M is an LD-monoid, and X is a subset of M. Show that the hull of the X-companion of M is the LD-system consisting of $M \times X$ equipped with the left distributive operation $(a, x)^\wedge(b, y) = ((a^\wedge x)\, b, y)$.*

Exercise 1.17. **(generators)** *Assume that M is an LD-monoid, and X generates M. Let S be the sub-LD-system of M generated by X. Prove that M is generated by S as a monoid, i.e., every element of M is a finite (possibly empty) product of elements of S. [Hint: Show that the submonoid of M generated by S is closed under exponentiation.]*

Exercise 1.18. **(graded LD-monoid)** *Say that M is a graded LD-monoid if there exists a partition $M = \coprod M_i$ satisifying $M_i \cdot M_j = M_{i+j}$ and $M_i{}^\wedge M_j \subseteq M_j$ for all i, j. The elements of M_i are called the degree i elements of M.*

(i) Let $\mathfrak{I}_\infty^{\rm fin}$ denote the sub-LD-monoid of $\mathfrak{I}_\infty$ consisting of those injections whose coimage is finite. Show that $\mathfrak{I}_\infty^{\rm fin}$ is a graded LD-monoid, the degree being the cardinality of the coimage.

(ii) Assume that M is a graded monoid. Show that the set S of all elements in M with degree 1 at most is a sub-LD-system of M, which generates M as a monoid.

Exercise 1.19. **(strong idempotent)** *Say that an element a of an LD-monoid M is a strong idempotent if $a^\wedge a = 1$ holds. Observe that 1 is the only strong idempotent in a group with conjugacy, while every element is a strong idempotent in a Boolean algebra.*

(i) Show that the identity mapping is the only strong idempotent in the LD-monoid $(\mathfrak{I}_\infty, \circ, [\,])$.

(ii) Assume that M is an LD-monoid. Show that every strong idempotent of M is idempotent with respect to the product, and that the set of all strong idempotents in M is a sub-LD-monoid of M, on which the relation $a^\wedge b = 1$ defines a preordering $\preceq$.

(iii) Show that a sufficient condition for $\preceq$ to be an ordering is the product being commutative.

Exercise 1.20. **(no LD-completion)** *(i) Let M be the monoid with presentation $\langle a, b\ ;\ a^3 = a, b^2 = 1, ab = a^2\rangle$. Show that M has six elements, namely 1, a, a^2, b, ba and ba^2, and that M satisfies Condition (1.2).*

(ii) Prove that there exists no exponentiation on M giving M the structure of an LD-monoid. [Hint: Show that $a^\wedge b$ must be a, and obtain a contradiction by comparing a^2 and $a^\wedge b^2$.]

Exercise 1.21. **(LRD-monoids)** *Define an LRD-monoid to be a monoid M equipped with two operations $^\wedge$ and $_\vee$ satisfying*

$$1^\wedge x = 1_\vee x = x, \qquad x^\wedge 1 = x_\vee 1 = 1,$$
$$x \cdot y = (x^\wedge y) \cdot x = y \cdot (y_\vee x),$$
$$(x \cdot y)^\wedge z = x^\wedge (y^\wedge z), \qquad (x \cdot y)_\vee z = y_\vee (x_\vee z),$$
$$x^\wedge (y \cdot z) = (x^\wedge y) \cdot (x^\wedge z), \qquad x_\vee (y \cdot z) = (x_\vee y) \cdot (x_\vee z).$$

We say that M is an LRDQ-monoid if M is an LRD-monoid satisfying $x^\wedge (x_\vee y) = x_\vee (x^\wedge y) = y$ in addition.

(i) Assume that G is a group. Show that G equipped with the operations $x^\wedge y = xyx^{-1}$, $x_\vee y = x^{-1}yx$ is an LRDQ-monoid.

(ii) Assume that $(M, \cdot, 1, ^\wedge, _\vee)$ is an LDRQ-monoid. Show that $(M, ^\wedge, _\vee)$ is an LD-quasigroup.

(iii) Conversely, assume that $(M, \cdot, 1, ^\wedge)$ is an LD-monoid, and that $(M, ^\wedge)$ is an (implicit) LD-quasigroup. Let $_\vee$ be the inverse of $^\wedge$ as given by Proposition I.2.8 (second operation). Show that $(M, \cdot, 1, ^\wedge, _\vee)$ is an LDRQ-monoid.

Exercise 1.22. **(LRD-completion)** *Let us consider the problem of completing a given monoid into an LRD-monoid (Exercise 1.21). A necessary condition for such a completion to exist is*

$$(\forall x, y)(\exists z_L, z_R)(x\, y = z_L\, x = y\, z_R), \tag{$*$}$$

and, by Proposition 1.13 (LD-completion), a sufficient condition is the stronger condition

$$(\forall x, y)(\exists! z_L, z_R)(x\, y = z_L\, x = y\, z_R). \tag{$**$}$$

(i) Assume that M is a monoid satisfying Condition $(*)$. Prove that M satisfies Condition $(**)$ if and only if it admits cancellation, and that, in this case, there exists a unique way to define on M the structure of an LRD-monoid. Moreover, this LRD-monoid is an LRDQ-monoid.

(ii) Assume that M is an abelian monoid. Verify that M equipped with $x^\wedge y = x_\vee y = y$ is an LRDQ-monoid. Deduce that the sufficient conditions of (i) are not necessary for M to be an LRDQ-monoid.

(iii) Assume that M is a monoid satisfying Condition $(*)$. Write $a \equiv a'$ if there exist x, y satisfying $xay = xa'y$. Show that $\equiv$ is a congruence

on M, and that $M/\equiv$ satisfies Condition $(**)$.

Exercise 1.23. **(shuffle product)** *Assume that M is an RD-monoid, i.e., a monoid equipped with a right exponentiation satisfying $x \cdot y = y \cdot x^y$ and the right counterparts of the identities defining LD-monoids. On the set $M^{\mathbf{N}}$ of all $\mathbf{N}$-indexed sequences of elements of M, let $\equiv$ denote the equivalence relation generated by the pairs $(s, s/f)$ where f is an injection of $\mathbf{N}$ into itself completed with $f(-1) = -1$ and s/f is the sequence defined by $s/f(i) = s(f(i-1)+1) \cdot \cdots \cdot s(f(i))$, i.e., s/f is obtained from s by grouping the factors as prescribed by f. For s_1, s_2 in $M^{\mathbf{N}}$, define $s_1 ⊔⊔ s_2$ to be the sequence s satisfying $s(2p) = s_1(p)^{s_2(0) \cdots s_2(p-1)}$, $s(2p+1) = s_2(p)$. Show that ⊔⊔ induces a well-defined associative operation on $M^{\mathbf{N}}/\equiv$, and the latter set is an RD-monoid. Show that M embeds in this RD-monoid provided M has no invertible element.*

---

2. Completion of an LD-System

We consider now the question of completing a given LD-system into an LD-monoid. We shall describe two solutions. The first works for every LD-system, and it enjoys a natural universal condition, but at the expense of enlarging the domain of the original system. The second one works only when some conditions are met, but, in good cases, it enables one to define the associative operation on the initial domain.

The free completion

Let $(S, ^\wedge)$ be an arbitrary LD-system. We extend the exponentiation of S to the free monoid S^* of all words on S by defining

$$(a_1 \cdots a_p) ^\wedge b = a_1 ^\wedge \cdots ^\wedge a_p ^\wedge b, \tag{2.1}$$

$$w ^\wedge (b_1 \cdots b_q) = (w ^\wedge b_1) \ \cdots \ (w ^\wedge b_q). \tag{2.2}$$

Here and in the sequel of this section, we use a, b, c for elements of S, and u, v, w for words on S. We complete the above definition with

$$u ^\wedge \varepsilon = \varepsilon, \qquad \varepsilon ^\wedge v = v, \tag{2.3}$$

where we recall ε denotes the empty word.

Lemma 2.1. *Assume that $(S, \wedge)$ is an LD-monoid. Then S^* equipped with the exponentiation defined in (2.1), ..., (2.3) is an LD-system. Identities (LDM_2), (LDM_3), and (LDM_4) hold in $(S^*, \cdot, \varepsilon, \wedge)$.*

Proof. We prove $u^\wedge(v^\wedge w) = (u^\wedge v)^\wedge(u^\wedge w)$ using induction on the length of w. For $w = \varepsilon$, the result follows from (2.3). Assume $w \in S$, say $w = c$. Assume $u = a_1 \cdots a_p$, $v = b_1 \cdots b_q$. Let $b'_j = a_1{}^\wedge \cdots {}^\wedge a_p{}^\wedge b_j$. Then we have $u^\wedge v = b'_1 \cdots b'_q$, and we find

$$\begin{aligned} u \wedge (v \wedge w) &= a_1 \wedge \cdots \wedge a_p \wedge b_1 \wedge \cdots \wedge b_q \wedge c \\ &= b'_1 \wedge a_1 \wedge \cdots \wedge a_p \wedge b_2 \wedge \cdots \wedge b_q \wedge c = \ldots \\ &= b'_1 \wedge \cdots \wedge b'_q \wedge a_1 \wedge \cdots \wedge a_p \wedge c = (u \wedge v) \wedge (u \wedge w). \end{aligned}$$

Finally, for $w = c_1 \cdots c_r$ with $c_1, \ldots, c_r \in S$, we have

$$\begin{aligned} u \wedge (v \wedge w) &= (u \wedge (v \wedge c_1)) \;\cdots\; (u \wedge (v \wedge c_r)), \\ (u \wedge v) \wedge (u \wedge w) &= (u \wedge v) \wedge (u \wedge c_1)) \;\cdots\; (u \wedge v) \wedge (u \wedge c_r)), \end{aligned}$$

and the equality follows from the induction hypothesis. Identities (LDM_2), (LDM_3), and (LDM_4) follow from the definition of exponentiation. ∎

Observe that the same argument shows that, for every n, (2.1), (2.2) and (2.3) define a left self-distributive operation on S^n.

At this point, we are close to having obtained the structure of an LD-monoid on S^*: only Identity (LDM_1) may fail. Now it suffices to go to a convenient quotient to force the latter identity.

Proposition 2.2. (**free completion**) *Assume that $(S, \wedge)$ is an LD-system. Let $\sim$ denote the equivalence relation on S^* generated by all pairs of the form $(uabv\,,\, u(a^\wedge b)av)$. Then $\sim$ is compatible both with concatenation and exponentiation, and the quotient structure $\widehat{S}$ is an LD-monoid. Moreover, $\widehat{S}$ is characterized up to isomorphism by the following universal property:*

> *The LD-system $(S, \wedge)$ embeds in $(\widehat{S}, \wedge)$, the monoid $(\widehat{S}, \cdot)$ is generated by (the image of) S, and, for every LD-monoid $(M, \cdot, \wedge)$ and every homomorphism f of $(S, \wedge)$ into $(M, \wedge)$, there exists a unique extension of f into a homomorphism of $(\widehat{S}, \cdot, \wedge)$ into $(M, \cdot, \wedge)$.*

Proof. In order to show that $\sim$ is compatible with concatenation and exponentiation on S^*, it suffices to verify the property for the pairs $(uabv\,,\, u(a^\wedge b)av)$, since these pairs generate $\sim$. The case of concatenation is trivial. In the case of exponentiation, using left self-distributivity, we have, for every a, b in S, and every u, v, w in S^*,

$$(uabv) \wedge w = u \wedge a \wedge b \wedge v \wedge w = u \wedge (a \wedge b) \wedge a \wedge v \wedge w = (u(a \wedge b)av) \wedge w.$$

Hence $u \sim u'$ implies u^$v = u'$^v for all u, u', v in S^*. On the other hand, for every c in S, we have

$$\begin{aligned} c \wedge (uabv) &= (c \wedge u)(c \wedge a)(c \wedge b)(c \wedge v) \\ &\sim (c \wedge u)((c \wedge a) \wedge (c \wedge b))(c \wedge a)(c \wedge v) \\ &\sim (c \wedge u)(c \wedge a \wedge b)(c \wedge a)(c \wedge v) = c \wedge (u(a \wedge b)av). \end{aligned}$$

By an induction on the length of v, we deduce that $u \sim u'$ implies v^$u = v$^u'. The quotient structure $\widehat{S}$ satisfies all identities that hold in S^*, in particular (LD), (LDM_2), (LDM_3), and (LDM_4). The $\sim$-class of a length 1 word, *i.e.*, of an element of S, contains this element only, as $\sim$ preserves the length, and no defining pair involves a length 1 word. So S embeds in $\widehat{S}$, and there is no danger in identifying every element of S with its image in $\widehat{S}$. Now S generates S^* under product, so it generates $\widehat{S}$. In order to show that $\widehat{S}$ is an LD-monoid, it remains to verify that (LDM_1) holds in $\widehat{S}$. By construction, we have $ab \sim (a$^$b)a$ in S^*, so (LDM_1) holds in the case of length 1 words. For the general case, assume $u = a_1 \cdots a_p$, $v = b_1 \cdots b_q$. We find:

$$\begin{aligned} uv = a_1 \cdots a_p b_1 \cdots b_q &\sim a_1 \cdots a_{p-1}(a_p \wedge b_1)a_p b_2 \cdots b_q \\ &\sim a_1 \cdots a_{p-2}(a_{p-1} \wedge a_p \wedge b_1)a_{p-1}a_p b_2 \cdots b_q \\ &\ldots \sim (u \wedge b_1)a_1 \cdots a_p b_2 \cdots b_q \\ &\ldots \sim (u \wedge b_1)(u \wedge b_2)a_1 \cdots a_p b_3 \cdots b_q \\ &\ldots \sim (u \wedge b_1)(u \wedge b_2) \ \cdots \ (u \wedge b_q)a_1 \cdots a_p = (u \wedge v)u. \end{aligned}$$

So $\widehat{S}$ is an LD-monoid.

Assume now that $(M, \cdot, ^\wedge)$ is an LD-monoid and f is a homomorphism of $(S, ^\wedge)$ into $(M, ^\wedge)$. Let f^* be the unique homomorphism of $(S^*, \cdot)$ into $(M, \cdot)$ that extends f. Then f^* preserves exponentiation. Indeed, as f^* preserves the product, it suffices to consider the case of u^b where u is a word on S and b is an element of S. Assume $u = a_1 \cdots a_p$. We find

$$\begin{aligned} f^*(u \wedge b) &= f^*(a_1 \wedge \cdots \wedge a_p \wedge b) \\ &= f(a_1) \wedge \cdots \wedge f(a_p) \wedge f(b) = (f(a_1) \cdots f(a_p)) \wedge f(b) = f^*(u) \wedge f^*(b). \end{aligned}$$

Next, we observe that f^* is compatible with the congruence $\sim$:

$$f^*(ab) = f(a) \cdot f(b) = (f(a) \wedge f(b)) \cdot f(a) = f(a \wedge b) \cdot f(a) = f^*((a \wedge b)a).$$

Hence f^* induces a well-defined homomorphism $\widehat{f}$ of $(\widehat{S}, \cdot, ^\wedge)$ into $(M, \cdot, ^\wedge)$, and $\widehat{f}$ extends f—up to identifying the elements of S with their image in $\widehat{S}$. Now, if f' is an arbitrary homomorphism of $\widehat{S}$ into M that extends f, then f' has to map the class of the word $a_1 \ldots a_p$ to the product of the images of a_1, $\ldots$, a_p, *i.e.*, to $\widehat{f}(a_1 \ldots a_p)$. So f' coincides with $\widehat{f}$, which is unique.

Finally, assume that M is an LD-monoid that includes S and satisfies the property of the proposition (we need not assume uniqueness of the extension). By hypothesis, the identity mapping of S into $\widehat{S}$ extends to a homomorphism f' of the LD-monoid $\widehat{S}$ into the LD-monoid M, and, similarly, it extends into a homomorphism f'' of the LD-monoid M into the LD-monoid $\widehat{S}$. Now $f'' \circ f'$ is an endomorphism of the LD-monoid $\widehat{S}$ that is the identity on S. The identity of $\widehat{S}$ is another endomorphism of $\widehat{S}$ with the same property, and these homomorphisms coincide on S, hence on the sub-LD-monoid generated by S, which is $\widehat{S}$. So, $f'' \circ f'$ is the identity. Similarly $f' \circ f''$ is the identity of M, and f is an isomorphism. ■

Applying Proposition 2.2 (free completion) to the identity embedding of S into M, we obtain the following result:

Corollary 2.3. *Assume that the LD-monoid M is generated, as a monoid, by the set X. Let S be the sub-LD-system of M generated by X. Then M is a quotient of $\widehat{S}$.*

Observe that the free completion of an LD-system is a graded LD-monoid (Exercise 1.18). Indeed, the length on S^* is invariant under relation $\sim$, so it induces a well-defined mapping of $\widehat{S}$ onto $\mathbf{N}$, which is the desired degree. This implies in particular that the LD-monoid $\widehat{S}$ never coincides with the initial LD-system S, which is the set of all degree 1 elements in $\widehat{S}$.

The adjoint completion

The free completion $\widehat{S}$ of an LD-system S is a big monoid, which properly includes the initial system. A smaller completion can be obtained, in good cases, by defining the structure of an LD-monoid on the monoid generated by the left translations of S and then carrying this structure back to S.

Definition 2.4. (adjoint monoid) Assume that S is a multi-LD-system. The *(left) adjoint monoid of S* is defined to be the submonoid $\mathrm{Ad}(S)$ of $\mathrm{End}(S)$ generated by all left translations ad_a.

By construction, the elements of $\mathrm{Ad}(S)$ are finite products of left translations, and $\mathrm{Ad}(S)$ equipped with composition is a monoid. A natural idea for defining a left self-distributive operation on $\mathrm{Ad}(S)$ is to use the mapping $\mathrm{ad} : a \mapsto \mathrm{ad}_a$ to carry the operation ^ to $\mathrm{Ad}(S)$ by defining

$$\mathrm{ad}_a \wedge \mathrm{ad}_b = \mathrm{ad}_{a^\wedge b}. \tag{2.4}$$

Such a definition makes sense only if the equivalence relation $\mathrm{ad}_a = \mathrm{ad}_{a'}$ (written $a \equiv_L a'$ in Section X.2) is compatible with exponentiation. Such a condition however is not sufficient to complete the construction, for arbitrary elements of $\mathrm{Ad}(S)$ are products of several translations. We are thus led to introducing a stronger condition.

Definition 2.5. (**nice**) Assume that $(S, {}^\wedge)$ is an LD-system. We say that $(S, {}^\wedge)$ is *nice* if each equality $\mathrm{ad}_{a_1} \circ \cdots \circ \mathrm{ad}_{a_p} = \mathrm{ad}_{a'_1} \circ \cdots \circ \mathrm{ad}_{a'_{p'}}$ in $\mathrm{Ad}(S)$ implies $\mathrm{ad}_{c^\wedge a_1} \circ \cdots \circ \mathrm{ad}_{c^\wedge a_p} = \mathrm{ad}_{c^\wedge a'_1} \circ \cdots \circ \mathrm{ad}_{c^\wedge a'_{p'}}$ for every c in S.

Example 2.6. (**half-conjugacy**) *Let us consider a free LD-quasigroup* $(\mathrm{FG}_X \times X, {}^\wedge)$ *with* $(a, x)^\wedge(b, y) = (axa^{-1}b, y)$. *The action of the left translation* $\mathrm{ad}_{(a,x)}$ *is left multiplication by* axa^{-1} *on the first entry. Thus* $\mathrm{ad}_{(a_1,x_1)}{}^\wedge \cdots {}^\wedge \mathrm{ad}_{(a_p,x_p)} = \mathrm{ad}_{(a'_1,x'_1)}{}^\wedge \cdots {}^\wedge \mathrm{ad}_{(a'_{p'},x'_{p'})}$ *is equivalent to*

$$a_1 x_1 a_1^{-1} \cdots a_p x_p a_p^{-1} = a'_1 x'_1 a'^{-1}_1 \cdots a'_{p'} x'_{p'} a'^{-1}_{p'}. \tag{2.5}$$

Replacing $(a_i; x_i)$ *with* $(a, x)^\wedge(a_i, x_i)$ *amounts to conjugating under* axa^{-1} *everywhere, so* (2.5) *implies*

$$\mathrm{ad}_{(a,x)^\wedge(a_1,x_1)}{}^\wedge \cdots {}^\wedge \mathrm{ad}_{(a,x)^\wedge(a_p,x_p)} = \mathrm{ad}_{(a,x)^\wedge(a'_1,x'_1)}{}^\wedge \cdots {}^\wedge \mathrm{ad}_{(a,x)^\wedge(a'_{p'},x'_{p'})}$$

for all a, x, *and* $(\mathrm{FG}_X \times X, {}^\wedge)$ *is nice.*

Being nice is what is needed for the adjoint monoid to be endowed with a natural structure of LD-monoid:

Proposition 2.7. (**adjoint**) *Assume that* S *is a nice LD-system. There exists on* $\mathrm{Ad}(S)$ *a unique operation* ${}^\wedge$ *such that* $\mathrm{ad}_a{}^\wedge\mathrm{ad}_b = \mathrm{ad}_{a^\wedge b}$ *holds for all* a, b *in* S. *Then* $(\mathrm{Ad}(S), \circ, {}^\wedge)$ *is an LD-monoid.*

Proof. By construction, the elements of $\mathrm{Ad}(S)$ have the form $\mathrm{ad}_{a_1} \circ \cdots \circ \mathrm{ad}_{a_p}$. If Identities (LDM_2) and (LDM_3) are to be verified, as well as Formula (2.4), the operation ${}^\wedge$ on $\mathrm{Ad}(S)$ must be defined by

$$(\mathrm{ad}_{a_1} \circ \cdots \circ \mathrm{ad}_{a_p}) {}^\wedge (\mathrm{ad}_{b_1} \circ \cdots \circ \mathrm{ad}_{b_q}) = \mathrm{ad}_{c_1} \circ \cdots \circ \mathrm{ad}_{c_q}, \tag{2.6}$$

with c_ℓ is $a_1{}^\wedge \cdots {}^\wedge a_p{}^\wedge b_\ell$ for $\ell = 1, \ldots, q$. Several (easy) verifications are in order. First we claim that (2.6) gives rise to a well defined operation. Assume $\mathrm{ad}_{a_1} \circ \cdots \circ \mathrm{ad}_{a_p} = \mathrm{ad}_{a'_1} \circ \cdots \circ \mathrm{ad}_{a'_{p'}}$ and $\mathrm{ad}_{b_1} \circ \cdots \circ \mathrm{ad}_{b_q} = \mathrm{ad}_{b'_1} \circ \cdots \circ \mathrm{ad}_{b'_{q'}}$. Then, for every j, we have $a_1{}^\wedge \cdots {}^\wedge a_p{}^\wedge b_j = a'_1{}^\wedge \cdots {}^\wedge a'_{p'}{}^\wedge b_j$ by definition, and, therefore,

$$(\mathrm{ad}_{a_1} \circ \cdots \circ \mathrm{ad}_{a_p}) {}^\wedge (\mathrm{ad}_{b_1} \circ \cdots \circ \mathrm{ad}_{b_q}) = (\mathrm{ad}_{a'_1} \circ \cdots \circ \mathrm{ad}_{a'_{p'}}) {}^\wedge (\mathrm{ad}_{b_1} \circ \cdots \circ \mathrm{ad}_{b_q}).$$

Assume now $\mathrm{ad}_{b_1} \circ \cdots \circ \mathrm{ad}_{b_q} = \mathrm{ad}_{b'_1} \circ \cdots \circ \mathrm{ad}_{b'_{q'}}$. As S is nice, we deduce

$$\mathrm{ad}_{a_p{}^\wedge b_1} \circ \cdots \circ \mathrm{ad}_{a_p{}^\wedge b_q} = \mathrm{ad}_{a_p{}^\wedge b'_1} \circ \cdots \circ \mathrm{ad}_{a_p{}^\wedge b'_{q'}},$$
$$\mathrm{ad}_{a_{p-1}{}^\wedge a_p{}^\wedge b_1} \circ \cdots \circ \mathrm{ad}_{a_{p-1}{}^\wedge a_p{}^\wedge b_q} = \mathrm{ad}_{a_{p-1}{}^\wedge a_p{}^\wedge b'_1} \circ \cdots \circ \mathrm{ad}_{a_{p-1}{}^\wedge a_p{}^\wedge b'_{q'}}, \textit{ etc.},$$
$$\mathrm{ad}_{a_1{}^\wedge \ldots {}^\wedge a_p{}^\wedge b_1} \circ \cdots \circ \mathrm{ad}_{a_1{}^\wedge \ldots {}^\wedge a_p{}^\wedge b_q} = \mathrm{ad}_{a_1{}^\wedge \ldots {}^\wedge a_p{}^\wedge b'_1} \circ \cdots \circ \mathrm{ad}_{a_1{}^\wedge \ldots {}^\wedge a_p{}^\wedge b'_{q'}}, \textit{ i.e.},$$
$$(\mathrm{ad}_{a_1} \circ \cdots \circ \mathrm{ad}_{a_p})^\wedge(\mathrm{ad}_{b_1} \circ \cdots \circ \mathrm{ad}_{b_q}) = (\mathrm{ad}_{a'_1} \circ \cdots \circ \mathrm{ad}_{a'_{p'}})^\wedge(\mathrm{ad}_{b'_1} \circ \cdots \circ \mathrm{ad}_{b'_{q'}}).$$

This completes the proof that (2.6) gives rise to a well defined operation $^\wedge$.

We claim that $(\mathrm{Ad}(S), \circ, {}^\wedge)$ is an LD-monoid. Identities (LDM_2) to (LDM_4) follow from the definition of $^\wedge$, so only (LDM_1) is to be verified. We have already observed that, for a, b in S, we have $\mathrm{ad}_a \circ \mathrm{ad}_b = \mathrm{ad}_{a^\wedge b} \circ \mathrm{ad}_a$, *i.e.*, $\mathrm{ad}_a \circ \mathrm{ad}_b = (\mathrm{ad}_a{}^\wedge \mathrm{ad}_b) \circ \mathrm{ad}_a$, which is (LDM_1) in the case of single translations. The extension to products of translations is easy:

$$\begin{aligned}
&(\mathrm{ad}_{a_1} \circ \ldots \circ \mathrm{ad}_{a_p}) \circ (\mathrm{ad}_{b_1} \circ \cdots \circ \mathrm{ad}_{b_q})\\
&\qquad = \mathrm{ad}_{a_1} \circ \cdots \circ \mathrm{ad}_{a_p{}^\wedge b_1} \circ \mathrm{ad}_{a_p} \circ \cdots \circ \mathrm{ad}_{b_q}\\
&\quad \ldots = \mathrm{ad}_{a_1{}^\wedge \ldots {}^\wedge a_p{}^\wedge b_1} \circ \mathrm{ad}_{a_1} \circ \cdots \circ \mathrm{ad}_{a_p} \circ \mathrm{ad}_{b_2} \circ \cdots \circ \mathrm{ad}_{b_q}\\
&\quad \ldots = \mathrm{ad}_{a_1{}^\wedge \ldots {}^\wedge a_p{}^\wedge b_1} \circ \cdots \circ \mathrm{ad}_{a_1{}^\wedge \ldots {}^\wedge a_p{}^\wedge b_q} \circ \mathrm{ad}_{a_1} \circ \cdots \circ \mathrm{ad}_{a_p}\\
&\qquad = ((\mathrm{ad}_{a_1} \circ \cdots \circ \mathrm{ad}_{a_p}) {}^\wedge (\mathrm{ad}_{b_1} \circ \cdots \circ \mathrm{ad}_{b_q})) \circ (\mathrm{ad}_{a_1} \circ \cdots \circ \mathrm{ad}_{a_p}). \qquad \blacksquare
\end{aligned}$$

Definition 2.8. (adjoint) In the above situation, the monoid $\mathrm{Ad}(S)$ equipped with the exponentiation defined in (2.6) is called the *adjoint LD-monoid of* S.

Example 2.9. **(half-conjugacy)** *We have seen that the free LD-quasigroup consisting of* $\mathrm{FG}_X \times X$ *equipped with half-conjugacy is nice. By construction,* $(a, x) \mapsto axa^{-1}$ *induces an embedding of* $\mathrm{Ad}(\mathrm{FG}_X \times X, {}^\wedge)$ *into the free group* FG_X. *Under this embedding, composition of translations becomes the product of* FG_X, *while operation* $^\wedge$ *becomes conjugacy. Hence the adjoint LD-monoid of* $(\mathrm{FG}_X \times X, {}^\wedge)$ *is the sub-LD-monoid of* $(\mathrm{FG}_X, \cdot, {}^\wedge)$ *generated by* X*—since* x *is the image of the translation* $\mathrm{ad}_{(1,x)}$. *Observe that the adjoint and the free completions are not isomorphic here, since the former is idempotent, while the latter is not.*

We give now several sufficient conditions for an LD-system to be nice.

Definition 2.10. (generic) Assume that S is an LD-system, and g belongs to S. We say that g is *generic* if the evaluation mapping $\mathrm{eval}_g : f \mapsto f(g)$ is injective on $\mathrm{Ad}(S)$.

Lemma 2.11. *Assume that* S *is a monogenic LD-system, and* g *generates* S. *Then* g *is generic in* S.

Proof. Every element of the monoid $\mathrm{Ad}(S)$ is an endomorphism of S. Hence, if $f(g) = f'(g)$ holds, so does $f(x) = f'(x)$ for every element x in the sub-LD-system of S generated by g. ∎

Proposition 2.12. (**nice**) *Assume that $(S, {}^\wedge)$ is an LD-system. Then each of the following conditions is sufficient for $(S, {}^\wedge)$ to be nice:*

(i) The LD-system $(S, {}^\wedge)$ is left divisible (hence every LD-quasigroup is nice);

(ii) There exists a generic element g in S such that $a_1{}^\wedge \cdots {}^\wedge a_p{}^\wedge g = a'_1{}^\wedge \cdots {}^\wedge a'_{p'}{}^\wedge g$ implies $(c^\wedge a_1)^\wedge \cdots {}^\wedge(c^\wedge a_p)^\wedge g = (c^\wedge a'_1)^\wedge \cdots {}^\wedge(c^\wedge a'_{p'})^\wedge g$ for every c;

(iii) The LD-system $(S, {}^\wedge)$ can be completed into an LD-monoid, and $\mathrm{ad}_a = \mathrm{ad}_{a'}$ implies $\mathrm{ad}_{c^\wedge a} = \mathrm{ad}_{c^\wedge a'}$ for every c.

Proof. (i) Assume $\mathrm{ad}_{a_1} \circ \cdots \circ \mathrm{ad}_{a_p} = \mathrm{ad}_{a'_1} \circ \cdots \circ \mathrm{ad}_{a'_{p'}}$, and let c be an arbitrary element of S. For each x in S, there exists y satisfying $x = c^\wedge y$, so we have

$$\begin{aligned}(c \wedge a_1) \wedge \ldots \wedge (c \wedge a_p) \wedge x &= (c \wedge a_1) \wedge \cdots \wedge (c \wedge a_p) \wedge (c \wedge y)\\ &= c \wedge a_1 \wedge \cdots \wedge a_p \wedge y = c \wedge a'_1 \wedge \cdots \wedge a'_{p'} \wedge y\\ &= (c \wedge a'_1) \wedge \cdots \wedge (c \wedge a'_{p'}) \wedge (c \wedge y) = (c \wedge a'_1) \wedge \cdots \wedge (c \wedge a'_{p'}) \wedge x.\end{aligned}$$

(ii) By definition of genericity, $\mathrm{ad}_{a_1} \circ \cdots \circ \mathrm{ad}_{a_p} = \mathrm{ad}_{a'_1} \circ \cdots \circ \mathrm{ad}_{a'_{p'}}$ is equivalent to $a_1{}^\wedge \cdots {}^\wedge a_p{}^\wedge g = a'_1{}^\wedge \cdots {}^\wedge a'_{p'}{}^\wedge g$.

(iii) Identity (LDM_2) implies $\mathrm{ad}_a \circ \mathrm{ad}_b = \mathrm{ad}_{ab}$, so the image of the mapping ad is closed under composition. Assume $\mathrm{ad}_{a_1} \circ \cdots \circ \mathrm{ad}_{a_p} = \mathrm{ad}_{a'_1} \circ \cdots \circ \mathrm{ad}_{a'_{p'}}$: this implies $\mathrm{ad}_{a_1 \cdots a_p} = \mathrm{ad}_{a'_1 \cdots a'_{p'}}$, hence $\mathrm{ad}_{c^\wedge(a_1 \cdots a_p)} = \mathrm{ad}_{c^\wedge(a'_1 \cdots a'_{p'})}$, then $\mathrm{ad}_{(c^\wedge a_1) \cdots (c^\wedge a_p)} = \mathrm{ad}_{(c^\wedge a'_1) \cdots (c^\wedge a'_{p'})}$, and, finally, $\mathrm{ad}_{c^\wedge a_1} \circ \cdots \circ \mathrm{ad}_{c^\wedge a_p} = \mathrm{ad}_{c^\wedge a'_1} \circ \cdots \circ \mathrm{ad}_{c^\wedge a'_{p'}}$. ∎

Lifting to the initial domain, first solution

Assume that S is a nice LD-system. We now try to lift the operations of the adjoint LD-monoid of S to S. If the mapping ad is injective, ad^{-1} can be used to carry the operations provided that the image of ad is closed under composition. The construction is then minimal, in the following sense:

Proposition 2.13. (**minimal completion**) *(i) Assume that $(S, {}^\wedge)$ is a nice LD-system such that the mapping* ad *is injective and its image is closed under composition. Then there exists a unique product on S that completes $(S, {}^\wedge)$ into an LD-monoid, namely that obtained by carrying the composition of* $\mathrm{Ad}(S)$ *using* ad^{-1}.

(ii) If M is an LD-monoid and the mapping ad *associated with exponentiation is injective, the conditions of (i) are satisfied, and the product so defined is the original product of M.*

Proof. (i) Existence is clear. And, by (LDM_2), there is no other choice for defining the product on S.

(ii) By (LDM_2), we have $\mathrm{ad}_{ab} = \mathrm{ad}_a \circ \mathrm{ad}_b$, hence the image of ad is closed under composition. As ad is supposed to be injective, $\mathrm{ad}_a = \mathrm{ad}_{a'}$ implies $a = a'$, hence $c\hat{}a = c\hat{}a'$ and $\mathrm{ad}_{c\hat{}a} = \mathrm{ad}_{c\hat{}a'}$ for every c. By Proposition 2.12(iii) (nice), this implies that $(M, \hat{})$ is nice. The rest is easy. ∎

Example 2.14. (injection bracket) *Let* sh *be the shift injection of* $\mathbf{N}_+$. *Then* sh *is right cancellable in* $(\mathfrak{I}_\infty, [\,])$. *Hence the mapping* $\mathrm{ad} : f \mapsto ad_f$ *is injective, and* $(\mathfrak{I}_\infty, [\,])$ *is trivially nice. We know that there exists a product that gives* $\mathfrak{I}_\infty$ *the structure of an LD-monoid, so we deduce from Proposition 2.13 (minimal completion) that composition is the only product that completes the bracket so as to obtain an LD-monoid.*

The associative product on A_n

Another example is given by the Laver tables of Chapter X:

Proposition 2.15. (Laver tables) *For each* n, *there exists a unique product* $\circ_n$ *on* A_n *such that Identity* (LDM_2) *is satisfied, namely the product defined by*

$$a \circ_n b = \begin{cases} a *_n (b+1) - 1 & \text{for } b < 2^n, \\ a & \text{for } b = 2^n. \end{cases} \tag{2.7}$$

Then $(A_n, \circ_n, *_n)$ *is an LD-monoid with unit* 2^n.

Proof. By definition of $\circ_n$, we have $\mathrm{ad}_a \circ \mathrm{ad}_b(1) = \mathrm{ad}_{a \circ_n b}(1)$ for all a, b, which implies $\mathrm{ad}_a \circ \mathrm{ad}_b = \mathrm{ad}_{a \circ_n b}$ since 1 generates A_n. It follows that the image of ad in $\mathrm{End}(A_n)$ is closed under composition. So, in order to prove that A_n is nice, it suffices to show that each equality of the form $\mathrm{ad}_a = \mathrm{ad}_{a_1} \circ \cdots \circ \mathrm{ad}_{a_p}$ implies

$$\mathrm{ad}_{c *_n a} = \mathrm{ad}_{c *_n a_1} \circ \cdots \circ \mathrm{ad}_{c *_n a_p} \tag{2.8}$$

for every c. We use induction on $p \geq 1$. For $p = 1$, the hypothesis is $a *_n 1 = a_1 *_n 1$, which implies $a = a_1$, whence $c *_n a = c *_n a_1$ and (2.8). Otherwise, we use descending induction on a_1. For $a_1 = 2^n$, ad_{a_1} is the identity, and the result follows from the induction hypothesis. Assume $a_1 < 2^n$. By using left self-distributivity, we find $\mathrm{ad}_{a_1} \circ \cdots \circ \mathrm{ad}_{a_p} = \mathrm{ad}_{a_1 *_n a_2} \circ \cdots \circ \mathrm{ad}_{a_1 *_n a_p} \circ \mathrm{ad}_{a_1}$, and the induction hypothesis implies $\mathrm{ad}_{c *_n a} = \mathrm{ad}_{c *_n (a_1 *_n a_2)} \circ \cdots \circ \mathrm{ad}_{c *_n (a_1 *_n a_p)} \circ \mathrm{ad}_{c *_n a_1}$, which gives (2.8) by applying (LD) again. Hence the LD-system $(A_n, *_n)$ is nice, and the adjoint LD-monoid exists. Now the mapping ad is injective on A_n, and it is surjective on the adjoint LD-monoid A_n, since its image is closed under composition. We then apply Proposition 2.13 (minimal completion).

Finally, we observe that (2.7) is the only possible operation on A_n satisfying (LDM_2), as, in this case, we must have $(a \circ_n b) *_n 1 = a *_n (b *_n 1) = a *_n (b+1)$, which gives (2.7). ∎

Example 2.16. ***(product in A_n)*** *Except for the bottom row, the columns in the table of $(A_n, \circ_n)$ are obtained by shifting the columns of $(A_n, *_n)$ by one unit to the left, and substracting 1. In particular, the next to last column is constant with value $2^n - 1$.*

$\circ_0$	1
1	1

$\circ_1$	1	2
1	1	1
2	1	2

$\circ_2$	1	2	3	4
1	3	1	3	1
2	3	2	3	2
3	3	3	3	3
4	1	2	3	4

$\circ_3$	1	2	3	4	5	6	7	8
1	3	5	7	1	3	5	7	1
2	3	6	7	2	3	6	7	2
3	7	3	7	3	7	3	7	3
4	5	6	7	4	5	6	7	4
5	7	5	7	5	7	5	7	5
6	7	6	7	6	7	6	7	6
7	7	7	7	7	7	7	7	7
8	1	2	3	4	5	6	7	8

The periodicity phenomena in $(A_n, {}^\wedge{}_n)$ appear also in $(A_n, \circ_n)$. Observe that the monoid $(A_n, \circ_n)$ is not monogenic for $n \geq 2$.

Because the product of the LD-monoid A_n can be defined from the left distributive operation, it adds nothing really new to the structure.

Lifting to the initial domain, second solution

When the previous construction does not work, we can consider a different method for lifting the operations of the adjoint LD-monoid back to the initial domain by defining *two* new operations.

Assume that S is an LD-system. For g in S, the image of the evaluation mapping $\mathrm{eval}_g : f \mapsto f(g)$ of $\mathrm{Ad}(S)$ into S is the left ideal $I_L(S, g)$ of S generated by g. If g is generic in S, eval_g is injective, and we can use eval_g^{-1} to carry the structure of $\mathrm{Ad}(S)$ to $I_L(S, g)$.

For future use, it is convenient to state the result for multi-LD-systems. Defining the adjoint monoid of a multi-LD-system $(S, (*_i)_{i \in I})$ as the monoid generated by all left translations ad_a^i, one easily extends the construction of the adjoint LD-monoid to nice multi-LD-systems (*cf.* Exercise 2.26). Then, using the bijection eval_g^{-1}, we can translate the formulas that define the product and the exponentiation of $\mathrm{Ad}(S)$, and we easily obtain:

Proposition 2.17. (**new operations**) *Assume that $(S, (*_i)_{i\in I})$ is a nice multi-LD-system and g is generic in S. Let $I_L(S,g)$ be the left ideal of S generated by g. Define on $I_L(S,g)$ new operations $\cdot$ and $^\wedge$ by*

$$(a_1 *_{i_1} \cdots *_{i_{p-1}} a_p *_{i_p} g) \cdot (b_1 *_{j_1} \cdots *_{j_{q-1}} *_{j_q} g) \\ = a_1 *_{i_1} \cdots *_{i_{p-1}} a_p *_{i_p} b_1 *_{j_1} \cdots *_{j_{q-1}} *_{j_q} g, \qquad (2.9)$$

$$(a_1 *_{i_1} \cdots *_{i_{p-1}} a_p *_{i_p} g) \,^\wedge\, (b_1 *_{j_1} \cdots *_{j_{q-1}} *_{j_q} g) = c_1 *_{j_1} \cdots *_{j_{q-1}} c_q *_{j_q} g, \quad (2.10)$$

with $c_\ell = a_1 *_{i_1} \cdots *_{i_{p-1}} a_p *_{i_p} b_\ell$. Then $(I_L(S,g), \cdot, g, ^\wedge)$ is an LD-monoid, which is isomorphic to the adjoint LD-monoid of $(S, (*_i)_{i\in I})$.

Definition 2.18. (**coadjoint**) In the situation of Proposition 2.17, the LD-monoid $(I_L(S,g), \cdot, g, ^\wedge)$ is called the *g-coadjoint* of S.

Example 2.19. ***(injection bracket)*** *We have already observed that the LD-systems $(\mathfrak{I}_\infty, [\,])$ is nice. The shift mapping* sh *is generic in $\mathfrak{I}_\infty$, since it is right cancellable. Hence, the* sh*-coadjoint LD-monoid exists: the left ideal of $\mathfrak{I}_\infty$ generated by* sh *is the set of those injections of the form $f[\mathrm{sh}]$ augmented with* sh. *The new operations $\cdot$ and $^\wedge$ on this ideal are*

$$f[\mathrm{sh}] \cdot g[\mathrm{sh}] = (f \circ g)[\mathrm{sh}], \qquad f[\mathrm{sh}] \,^\wedge\, g[\mathrm{sh}] = f[g][\mathrm{sh}].$$

Observe that, up to a right translation by sh, *we re-obtain the operations of the original LD-monoid $(\mathfrak{I}_\infty, \circ, [\,])$.*

Companion and adjoint

With the coadjoint LD-monoids, we have constructed (in good cases) LD-monoids on the domain of a given (multi)-LD-system. On the other hand, we have defined in Section 1 the companion multi-LD-systems of an LD-monoid. We prove now that the two constructions are inverses of one another in the following sense:

Proposition 2.20. (**companion** *vs.* **coadjoint**) *(i) Assume that M is an LD-monoid, and X is a subset of M. Then the operations of M coincide with those of the 1-coadjoint of its X-companion on the left ideal generated by 1. In particular, M is the 1-coadjoint of its M-companion.*

(ii) Assume that $(S, (*_i)_{i\in I})$ is a nice multi-LD-system and g is generic in S. Then the subsystem of S induced on the left ideal generated by g is the X-companion of its g-coadjoint, where X is $\{g *_i g; i \in I\}$.

We begin with an auxiliary result.

Lemma 2.21. *Assume that M is an LD-monoid. Then the M-companion multi-LD-system of M is nice, and 1 is generic in this multi-LD-system.*

Proof. Assume that $\mathrm{ad}_{a_1}^{x_1} \circ \cdots \circ \mathrm{ad}_{a_p}^{x_p} = \mathrm{ad}_{b_1}^{y_1} \circ \cdots \circ \mathrm{ad}_{b_q}^{y_q}$ holds in the multi-LD-system $(M, (*_z)_{z \in M})$. By construction, we have $(a_1{}^\wedge x_1) \cdots (a_p{}^\wedge x_p)\ x = (b_1{}^\wedge y_1) \cdots (b_q{}^\wedge y_q)\ x$ for every x, and, in particular, for $x = 1$. So we have $(a_1{}^\wedge x_1) \cdots (a_p{}^\wedge x_p) = (b_1{}^\wedge y_1) \cdots (b_q{}^\wedge y_q)$. Applying exponentiation by $c^\wedge z$ and expanding, we obtain

$$((c\,{}^\wedge z)\,{}^\wedge (a_1\,{}^\wedge x_1)) \cdots ((c\,{}^\wedge z)\,{}^\wedge (a_p\,{}^\wedge x_p)) = ((c\,{}^\wedge z)\,{}^\wedge (b_1\,{}^\wedge y_1) \cdots ((c\,{}^\wedge z)\,{}^\wedge (b_q\,{}^\wedge y_q)),$$

hence $((c *_z a_1)^\wedge x_1) \cdots ((c *_z a_p)^\wedge x_p) = ((c *_z b_1)^\wedge y_1) \cdots ((c *_z b_q)^\wedge y_q)$. By multiplying both sides by x on the right, we deduce

$$\mathrm{ad}_{c *_z a_1}^{x_1} \circ \cdots \circ \mathrm{ad}_{c *_z a_p}^{x_p} = \mathrm{ad}_{c *_z b_1}^{y_1} \circ \cdots \circ \mathrm{ad}_{c *_z b_q}^{y_q},$$

hence M is nice. As we have $\mathrm{ad}_{a_1}^{x_1} \circ \cdots \circ \mathrm{ad}_{a_p}^{x_p}(x) = \mathrm{ad}_{a_1}^{x_1} \circ \cdots \circ \mathrm{ad}_{a_p}^{x_p}(1) \cdot x$ for every x, 1 is generic. ∎

We can now establish Proposition 2.20 (companion *vs.* coadjoint).

Proof. (i) By Proposition 2.21 (companion), the 1-coadjoint of the X-companion of M exists. Let I be the left ideal of $(M, \{*_x; x \in X\})$ generated by 1. A typical element of I has the form $a_1 *_{x_1} \cdots *_{i_{p-1}} a_p *_{x_p} 1$, which, by definition, evaluates into $(a_1{}^\wedge x_1) \cdots (a_p{}^\wedge x_p)$. Assume $a = a_1 *_{x_1} \cdots *_{i_{p-1}} a_p *_{x_p} 1$ and $b = b_1 *_{y_1} \cdots *_{y_{q-1}} b_q *_{y_q} 1$. Then we obtain

$$ab = (a_1{}^\wedge x_1) \cdots (a_p{}^\wedge x_p)(b_1{}^\wedge y_1) \cdots (b_q{}^\wedge y_q) = a_1 *_{x_1} \cdots *_{i_{p-1}} a_p *_{x_p} b_1 *_{y_1} \cdots *_{y_{q-1}} b_q *_{y_q} 1.$$

Similarly, using Identities (LDM_2) and (LDM_3), $a^\wedge b$ expands into $c_1 \cdots c_q$, with

$$c_j = (a_1\,{}^\wedge x_1)\,{}^\wedge \cdots {}^\wedge\,(a_p\,{}^\wedge x_p)\,{}^\wedge\,(b_j\,{}^\wedge y_j) = a_1 *_{x_1} \cdots *_{i_{p-1}} a_p *_{x_p} b_j *_{y_j} 1.$$

These explicit formulas are those of the 1-coadjoint LD-monoid of the multi-LD-system $(M, \{*_x; x \in X\})$. In the case $X = M$, the left ideal of $(M, \{*_x; x \in X\})$ generated by 1 is all of M, as we have $x = (1^\wedge x)\ 1 = 1 *_x 1$ for every x in M.

(ii) Let $\cdot$ and ${}^\wedge$ denote the operations of the coadjoint LD-monoid of S at g. We claim that the equality

$$a *_i b = (a\,{}^\wedge (g *_i g)) \cdot b \tag{2.11}$$

holds for every a, b in the ideal $I_L(S, g)$ of S generated by g. Indeed, assuming $a = a_1 *_{i_1} \cdots *_{i_{p-1}} a_p *_{i_p} g$ and $b = b_1 *_{j_1} \cdots *_{j_{q-1}} *_{j_q} g$, we find

$$\begin{aligned}
(a\,{}^\wedge (g *_i g)) \cdot b &= ((a_1 *_{i_1} \cdots *_{i_{p-1}} a_p *_{i_p} g)\,{}^\wedge (g *_i g)) \cdot b \\
&= ((a_1 *_{i_1} \cdots *_{i_{p-1}} a_p *_{i_p} g) *_i g) \cdot b \\
&= (a *_i g) \cdot b = (a *_i g) \cdot (b_1 *_{j_1} \cdots *_{j_{q-1}} *_{j_q} g)
\end{aligned}$$

$$= a *_i b_1 *_{j_1} \cdots *_{j_{q-1}} *_{j_q} g = a *_i b.$$

Now (2.11) means that $(I_L(S,g); (*_i)_{i \in I})$ is the X-companion of the LD-monoid $(I_L(S,g), \cdot, g, ^\wedge)$, where X is the set of all $g *_i g$ with $i \in I$. ∎

Lemma 2.21 shows that the sufficient conditions for the coadjoint LD-monoid to exist also are necessary, and the previous construction is optimal.

To conclude, let us return to the case of an LD-system, *i.e.*, the case of a single self-distributive operation. Assume that $(S, *)$ is a nice LD-system and g is generic in S. Then the g-coadjoint LD-monoid of $(S, *)$ exists, which means that two new operations $\cdot$ and $^\wedge$ have been defined on the left ideal $I_L(S,g)$ of $(S, *)$ so that $(I_L(S,g), \cdot, g, ^\wedge)$ is an LD-monoid. In particular, $(I_L(S,g), ^\wedge)$ is an LD-system, and a natural question is the connection between this LD-system and the original LD-system $(S, *)$. The answer is given by Formula (2.10), of which a particular case is the equality

$$(x * g) ^\wedge (y * g) = (x * y) * g. \tag{2.12}$$

This expresses that the right translation $\widetilde{\mathrm{ad}}_g : x \mapsto x * g$ is a homomorphism of $(S, *)$ into $(I_L(S,g), ^\wedge)$. In particular, when $\widetilde{\mathrm{ad}}_g$ is injective, *i.e.*, when g is right cancellable in $(S, *)$, the homomorphism is injective. In this case, we have constructed LD-monoid operations on (a left ideal of) S at the expense of replacing the original self-distributive operation with a translated copy:

Proposition 2.22. (coadjoint completion) *Assume that $(S, *)$ is a nice LD-system, and that g is generic and right cancellable in S. Then the right translation $\widetilde{\mathrm{ad}}_g : x \mapsto x * g$ embeds $(S, *)$ into its g-coadjoint LD-monoid.*

Example 2.23. (injections) *The LD-system $\mathfrak{I}_{sp}$ is nice, and the injection* sh *is both generic and right cancellative. So the above result applies. The mapping $R_{\mathrm{sh}} : f \mapsto f[\mathrm{sh}]$ embeds the LD-system $(\mathfrak{I}_{sp}, [\ ])$ into the LD-monoid $(I_{\mathrm{sh}}, \cdot, ^\wedge)$—as was already observed above.*

In Exercise X.2.26, an associative operation $\times$ was constructed on every monogenic LD-system. In the case of $\mathfrak{I}_{sp}$, the operation is defined by $(f \times g)[\mathrm{sh}] = f[\mathrm{sh}] \circ g[\mathrm{sh}]$. So it coincides with the product of the coadjoint LD-monoid, and both constructions are equivalent here.

Exercise 2.24. (height) *For S an LD-system, define a height function on S to be a homomorphism of S into $(\mathbf{N}, ^\wedge)$, with $a^\wedge b = b + 1$.*

(i) For f in $\mathfrak{I}_\infty$, let $\mathrm{Fix}(f)$ denote the set of all fixed points of f (if any). Show that the mapping $f \mapsto \mathrm{card}(\mathrm{Fix}(f))$ is a height function on $(\mathfrak{I}_{sp}, [\,])$. [Hint: Prove $\mathrm{Fix}(g[h]) = g(\mathrm{Fix}(h)) \cup \mathrm{coIm}(g)$.]

(ii) Prove that no LD-system S admitting a height function can be completed into an LD-monoid. [Hint: (LDM_2) cannot hold.]

Exercise 2.25. (**free commutative**) *Assume that S is a trivial LD-system, i.e., $a^\wedge b = b$ always holds. Show that the free completion of S is the free commutative monoid generated by S.*

Exercise 2.26. (**multi-LD-system**) *Extend the construction of the adjoint LD-monoid to the case of multi-LD-systems. [Hint: Here being nice means that each equality $\mathrm{ad}_{a_1}^{i_1} \circ \cdots \circ \mathrm{ad}_{a_p}^{i_p} = \mathrm{ad}_{b_1}^{j_1} \circ \cdots \circ \mathrm{ad}_{b_q}^{j_q}$ implies the corresponding equality $\mathrm{ad}_{c*_k a_1}^{i_1} \circ \cdots \circ \mathrm{ad}_{c*_k a_p}^{i_p} = \mathrm{ad}_{c*_k b_1}^{j_1} \circ \cdots \circ \mathrm{ad}_{c*_k b_q}^{j_q}$.] Prove that a multi-LD-system is nice if and only if its hull is nice, and show that, in this case, the adjoint LD-monoids are isomorphic.*

Exercise 2.27. (**adjoint**) *(i) Assume that V is a vector space over a field k, and, for $\lambda \neq 0$ and x, y in V, define $x *_\lambda y = (1-\lambda)x + \lambda y$. Show that $(V, (*_\lambda)_{\lambda \in k \setminus \{0\}})$ is a nice multi-LD-system.*

(ii) Prove that the adjoint monoid $\mathrm{Ad}(V)$ is the set of all affine functions of the form $x \mapsto a + \lambda x$ with $\lambda \neq 0$ [Hint: Show that every function of this form is the product of at most two translations.]

(iii) Show that the adjoint LD-monoid of V is isomorphic to the set of all pairs (a, λ) in $V \times (k \setminus \{0\})$, equipped with the operations

$(a, \lambda) \circ (b, \mu) = (a + \lambda b, \lambda\mu)$, $(a, \lambda)^\wedge(b, \mu) = ((1-\mu)a + \lambda b, \mu)$.

[Map the translation ad_a^λ to the pair $((1-\lambda)a, \lambda)$.] Describe the sub-LD-monoid obtained when restricting to a fixed value λ_0.

Exercise 2.28. (**adjoint**) *Assume that M is an LD-monoid that is nice as an LD-system. Show that the mapping $\mathrm{ad} : a \mapsto \mathrm{ad}_a$ defines a surjective morphism of M onto the adjoint LD-monoid of the LD-system M.*

Exercise 2.29. (**homomorphisms**) *Assume that f is a homomorphism of $(A_m, *_m)$ into $(A_n, *_n)$. Prove that f also preserves the structures of LD-monoids. [Hint: First prove $f(2^m) = 2^n$, and $f(1) = 1$, and use the fact that 1 is generic in $(A_n, *_n)$.]*

---

3. Free LD-Monoids

By definition, being an LD-monoid means satisfying the identities:

$$\begin{cases} x \cdot (y \cdot z) = (x \cdot y) \cdot z, \\ x \cdot 1 = 1 \cdot x = 1\hat{}x = x, \quad x\hat{}1 = 1, \\ x \cdot y = (x\hat{}y) \cdot x, \\ (x \cdot y)\hat{}z = x\hat{}(y\hat{}z), \\ x\hat{}(y \cdot z) = (x\hat{}y) \cdot (x\hat{}z). \end{cases} \tag{LDM}$$

Hence LD-monoids form an equational variety, and free LD-monoids exist. Here we describe them. The main result is that free LD-monoids can be constructed inside free LD-systems as coadjoint LD-monoids. This implies that free LD-monoids are not more complicated than free LD-systems.

Notation 3.1. (free LD-systems) Free LD-systems will appear in this and the next section. In order to avoid confusion with the associative operations involved in LD-monoids, and contrary to what has been done in Part B, we shall use ^ for the self-distributive operation of free LD-systems. Accordingly, we use terms equipped with ^, and denote by $T_{(X,\hat{})}$ the corresponding set. Thus x, $x\hat{}x$, $(x\hat{}x)\hat{}x$ are typical elements of FLD_1.

Free LD-monoids as free completions

Definition 3.2. (free LD-monoid) For X a nonempty set, we denote by FLDM_X the free LD-monoid based on X; for $1 \le n \le \infty$, we write FLDM_n for the free LD-monoid based on $\{x_1, \ldots, x_n\}$.

Free LD-monoids can be constructed as free completions of free LD-systems:

Proposition 3.3. (free LD-monoid as free completion) *For every X, the free completion of FLD_X is a free LD-monoid based on X.*

Proof. By construction, the free completion $\widehat{\mathrm{FLD}}_X$ is an LD-monoid generated by X. Let f be a mapping of X into an LD-monoid M. By definition, there exists a (unique) extension f' of f into a homomorphism of the LD-system FLD_X into the LD-system M. By Proposition 2.2 (free completion), there exists a (unique) extension of f' into a homomorphism of the LD-monoid $\widehat{\mathrm{FLD}}_X$ into the LD-monoid M, so $\widehat{\mathrm{FLD}}_X$ satisfies the universal property that characterizes free LD-monoids. ∎

Proposition 3.4. (embedding) *For every X, the free LD-system FLD_X embeds in the free LD-monoid FLDM_X.*

Proof. For every LD-system S, the function that maps each element a of S to the class of a (viewed as a length 1 word) under $\sim$ is an embedding of S into its free completion $\widehat{S}$. ■

By construction, every element in $\widehat{\mathrm{FLD}}_X$ is the $\sim$-equivalence class of some word on FLD_X, and each element of FLD_X is the LD-equivalence class of some term in $T_{(X,\wedge)}$. This gives a natural representation for the elements of FLDM_X, *i.e.*, of $\widehat{\mathrm{FLD}}_X$, by finite sequences of terms.

Definition 3.5. (multiterm) Let X be a nonempty set. A *multiterm* on X is defined to be a finite sequence of terms in $T_{(X,\wedge)}$. For $t_1, \ldots, t_p$ in $T_{(X,\wedge)}$, the multiterm $(t_1, \ldots, t_p)$ is denoted $t_1{\wedge} \cdots {\wedge} t_p{\wedge}$. The set of all multiterms on X is denoted by $T^*_{(X,\wedge)}$, and the empty multiterm is denoted by ε.

So, for instance, the sequences (x_1), (x_1, x_2, x_3), $(x_1{\wedge}x_2, x_1)$ are multiterms. According to our convention, they will be denoted by $x_1{\wedge}$, $x_1{\wedge}x_2{\wedge}x_3{\wedge}$ and $(x_1{\wedge}x_2){\wedge}x_1{\wedge}$ respectively. If $\vec{t}$ is a multiterm, say $\vec{t} = t_1{\wedge} \cdots {\wedge}t_p{\wedge}$, and x is a variable, $\vec{t}x$ denotes the term $t_1{\wedge} \cdots {\wedge}t_p{\wedge}x$. The above apparently weird notation corresponds to the intuition that a multiterm is a term where the last variable has been removed, which we shall see is relevant here. As every term admits a unique decomposition of the form $t_1{\wedge} \cdots {\wedge}t_p{\wedge}x$ with x a variable, the mapping $i_x : \vec{t} \mapsto \vec{t}x$ defines a bijection between $T^*_{(X,\wedge)}$ and the subset of $T_{(X,\wedge)}$ made of those terms with rightmost variable x, x itself being the image of ε. In Section V.2, we associated with every term on X a binary tree with X-labelled leaves. We shall associate with multiterms similar trees where all leaves except the rightmost one have labels in X. For instance, the trees associated with the multiterms $x_1{\wedge}$, $x_1{\wedge}x_2{\wedge}$ and $(x_1{\wedge}x_2){\wedge}x_1{\wedge}$ are

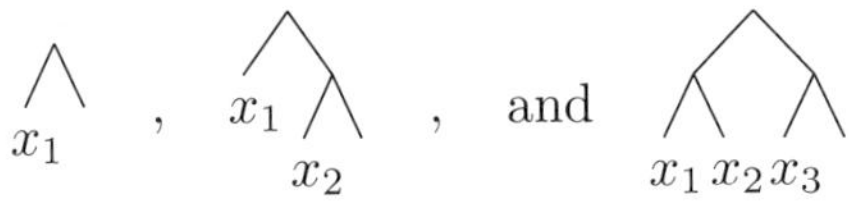

respectively.

Definition 3.6. (LD-equivalence) We let $=_{LD}$ denote the equivalence relation on $T^*_{(X,\wedge)}$ generated by the pairs:

type (i): $(t_1{\wedge} \cdots {\wedge}t_k{\wedge} \cdots {\wedge}t_p{\wedge},\ t_1{\wedge} \cdots {\wedge}t'_k{\wedge} \cdots {\wedge}t_p{\wedge})$ with $t'_k =_{LD} t_k$;

type (ii): $(t_1{\wedge} \cdots {\wedge}t_p{\wedge},\ t_1{\wedge} \cdots {\wedge}(t_i{\wedge}t_{i+1}){\wedge}t_i{\wedge} \cdots {\wedge}t_p{\wedge})$.

Two multiterms that are equivalent under $=_{LD}$ are said to be *LD-equivalent*.

Considering multiterms as trees, we recognize in the definition of $=_{LD}$ on $T^*_{(X,\wedge)}$ the usual notion of LD-equivalence: indeed, applying relations of both types (i) and (ii) amounts to replacing an incomplete subtree of the form $s_1{\wedge}s_2{\wedge}$ with the corresponding tree $(s_1{\wedge}s_2){\wedge}s_1{\wedge}$. This explains our notations.

Proposition 3.7. (**LD-equivalence**) *Two multiterms in $T^*_{(X,^\wedge)}$ represent the same element of* FLDM_X *if and only if they are LD-equivalent.*

Proof. Type (ii) relations mimic the definition of the relation $\sim$, hence multiterms up to LD-equivalence form the free completion of $T_{(X,^\wedge)}/=_{LD}$. We then apply Proposition 3.3 (free LD-monoid as free competion). ∎

Free LD-monoids as coadjoints

We are thus left with the question of describing LD-equivalence of multiterms. This turns out to be easy, as LD-equivalence of multiterms is connected with LD-equivalence of standard terms. This leads to a complete description of FLDM_X inside FLD_X, an example of the coadjoint construction of Section 2.

Lemma 3.8. *Let x be a fixed variable, and $\vec{t}$, $\vec{t}'$ be multiterms on X, Then the following are equivalent:*
(i) The multiterms $\vec{t}$ and $\vec{t}'$ are LD-equivalent;
(ii) The terms $\vec{t}\,x$ and $\vec{t}'\,x$ are LD-equivalent.

Proof. If $(\vec{t}, \vec{t}')$ is a pair of multiterms of type (i) or (ii), then the terms $\vec{t}\,x$ and $\vec{t}'\,x$ are LD-equivalent, and (i) implies (ii). Conversely, assume $t_1{^\wedge}\cdots{^\wedge}t_p{^\wedge}x =_{LD} t'_1{^\wedge}\cdots{^\wedge}t'_{p'}{^\wedge}x$. Necessarily, $p' = p$ holds. Since LD-equivalence is generated by basic LD-expansions, we may assume that $t'_1{^\wedge}\cdots{^\wedge}t'_p{^\wedge}x = (t_1{^\wedge}\cdots{^\wedge}t_p{^\wedge}x)\bullet\alpha$ holds for some address α. For $\alpha = 1^i0\beta$, t'_i is a basic LD-expansion of t_i, and, in this case, $(t_1{^\wedge}\cdots{^\wedge}t_p{^\wedge}, t'_1{^\wedge}\cdots{^\wedge}t'_p{^\wedge})$ is a pair of type (i). For $\alpha = 1^i$, we have $t'_i = t_i{^\wedge}t_{i+1}$ and $t'_{i+1} = t_i$, and $(t_1{^\wedge}\cdots{^\wedge}t_p{^\wedge}, t'_1{^\wedge}\cdots{^\wedge}t'_p{^\wedge})$ is a pair of type (ii). Hence (ii) implies (i). ∎

Proposition 3.9. (**realization**) *Assume that $(S, {^\wedge})$ is a free LD-system based on X. Let x_0 be a fixed element of X. Define on the left ideal $I_L(S, x_0)$ of S generated by x_0 two new operations by*

$$(a_1 \wedge \cdots \wedge a_p \wedge x_0) \cdot (b_1 \wedge \cdots \wedge b_q \wedge x_0) = a_1 \wedge \cdots \wedge a_p \wedge b_1 \wedge \cdots \wedge b_q \wedge x_0$$
$$(a_1 \wedge \cdots \wedge a_p \wedge x_0) * (b_1 \wedge \cdots \wedge b_q \wedge x_0) = c_1 \wedge \cdots \wedge c_q \wedge x_0,$$

with $c_\ell = a_1{^\wedge}\cdots{^\wedge}a_p{^\wedge}b_\ell$. Then $(I_L(S, x_0), \cdot, x_0, *)$ is a free LD-monoid based on $\{x{^\wedge}x_0\ ;\ x \in X\}$, and $\widetilde{\mathrm{ad}}_{x_0} : a \mapsto a{^\wedge}x_0$ embeds $(S, {^\wedge})$ into $(I_L(S, x_0), *)$.

Proof. The mapping $i_{x_0} : \vec{t} \mapsto \vec{t}\,x_0$ is an injection of $T^*_{(X,^\wedge)}$ into $T_{(X,^\wedge)}$, and, by Lemma 3.8, two multiterms in $T^*_{(X,^\wedge)}$ are LD-equivalent if and only if their images under i_{x_0} are LD-equivalent. Hence i_{x_0} induces a well-defined injective mapping of FLDM_X into S. By construction, the image of this mapping is the left ideal $I_L(S, x_0)$. Now, the product of two elements a and b represented by the multiterms $t_1{}^\wedge \cdots {}^\wedge t_p{}^\wedge$ and $s_1{}^\wedge \cdots {}^\wedge s_q{}^\wedge$ respectively is the element represented by $t_1{}^\wedge \cdots {}^\wedge t_p{}^\wedge s_1{}^\wedge \cdots {}^\wedge s_q{}^\wedge$. This means that $i_{x_0}(a \cdot b)$ is what we have defined to be $i_{x_0}(a) \cdot i_{x_0}(b)$. Similarly, the exponentiation $a^\wedge b$ is represented by $s'_1{}^\wedge \cdots {}^\wedge s'_q{}^\wedge$ with $s'_j = t_1{}^\wedge \cdots {}^\wedge t_p{}^\wedge s_j$. This means that $i_{x_0}(a^\wedge b)$ is what we have defined to be $i_{x_0}(a) * i_{x_0}(b)$. ■

The previous construction is the construction of the coadjoint LD-monoid of an LD-system as developed in Section 2. Instead of resorting to the free completion, we can check directly that free LD-systems satisfy the sufficient conditions of Proposition 2.17 (new operations).

Lemma 3.10. *Assume that $(S, {}^\wedge)$ is a free LD-system based on X. Then every element of X is generic and right cancellable in S, and S is nice.*

Proof. Assume $x \in X$. Proving that x is generic amounts to proving that, for all terms $t_1, \ldots, t'_{p'}$, $t_1{}^\wedge \cdots {}^\wedge t_p{}^\wedge x =_{LD} t'_1{}^\wedge \cdots {}^\wedge t'_{p'}{}^\wedge x$ implies $t_1{}^\wedge \cdots {}^\wedge t_p{}^\wedge t =_{LD} t'_1{}^\wedge \cdots {}^\wedge t'_{p'}{}^\wedge t$ for every t in $T_{(X,{}^\wedge)}$. Without loss of generality, we may assume that $t'_1{}^\wedge \cdots {}^\wedge t'_{p'}{}^\wedge x$ is a basic LD-expansion of $t_1{}^\wedge \cdots {}^\wedge t_p{}^\wedge x$, and the result is clear.

Proving that x is right cancellable in S amounts to proving that $t^\wedge x =_{LD} t'^\wedge x$ implies $t =_{LD} t'$, which is clear as every basic LD-expansion of a term $t^\wedge x$ has the form $t_1{}^\wedge x$ with t_1 a basic LD-expansion of t.

Finally, proving that S is nice amounts to proving that $t_1{}^\wedge \cdots {}^\wedge t_p{}^\wedge x =_{LD} t'_1{}^\wedge \cdots {}^\wedge t'_{p'}{}^\wedge x$ implies

$$(t^\wedge t_1)^\wedge \cdots {}^\wedge (t^\wedge t_p)^\wedge x =_{LD} (t^\wedge t'_1)^\wedge \cdots {}^\wedge (t^\wedge t'_{p'})^\wedge x$$

for every x. Again the result is clear in the case of a basic LD-expansion, and this is enough to conclude. ■

It follows that free LD-systems are eligible for the construction of the adjoint LD-monoid, and we obtain a second construction of free LD-monoids:

Proposition 3.11. (free LD-monoid as adjoint completion) *For every X, the adjoint LD-monoid of* FLD_X *is a free LD-monoid based on $\{\mathrm{ad}_x \,;\, x \in X\}$.*

Proof. For x in X, the evaluation mapping eval_x of $\mathrm{Ad}(S)$ into S is injective, and, therefore, the adjoint LD-monoid and the x-coadjoint LD-monoid are isomorphic. By Proposition 3.9 (realization), $\mathrm{Ad}(S)$ is a free LD-monoid. ■

LDM-equivalence

The description of free LD-monoids provided by Proposition 3.9 (realization) enables us to translate every question about free LD-monoids into a question about free LD-systems. However, we can address the question of describing free LD-monoids directly by considering convenient terms and the congruence relation associated with the identities (LDM). This approach is equivalent to the previous one, as will be seen now.

In order to represent LD-monoids, which involve two binary operations and one constant (the unit of the product), we consider terms involving two operators, denoted $^\wedge$ and $\cdot$, and one constant, denoted 1. We denote by $T_{(X,^\wedge,\cdot,1)}$ the set of all such terms constructed using variables in X. By construction, the set $T_{(X,^\wedge)}$ is a subset of $T_{(X,^\wedge,\cdot,1)}$. Now, we introduce on $T_{(X,^\wedge,\cdot,1)}$ the congruence relation $=_{LDM}$ generated by all $T_{(X,^\wedge,\cdot,1)}$-instances of the identities in (LDM). Then $T_{(X,^\wedge,\cdot,1)}/=_{LDM}$ is a free LD-monoid based on X, and we are left with the word problem of $=_{LDM}$, *i.e.*, with the question of deciding whether two given terms in $T_{(X,^\wedge,\cdot,1)}$ are LDM-equivalent.

As shown in Section V.1 in the easy case of free LD-quasigroups, a simple method for solving a word problem consists in introducing normal terms. Here, this approach will enable us to reduce the general problem of LDM-equivalence to the (already solved) case of LD-equivalence. Indeed, we shall introduce distinguished terms in $T_{(X,^\wedge,\cdot,1)}$ called prenormal, and show how to reduce LDM-equivalence of prenormal terms to LD-equivalence.

Definition 3.12. (prenormal, mapping i) Assume that t is a term in $T_{(X,^\wedge,\cdot,1)}$. We say that t is *prenormal* if we have either $t = 1$, or $t = t_1 \cdot \cdots \cdot t_p$ with $t_1, \ldots, t_p \in T_{(X,^\wedge)}$, *i.e.*, t_i involves neither the operator $\cdot$ nor the constant 1. In the former case, we define $i(t) = \varepsilon$; in the latter case, we define $i(t) = t_1{}^\wedge \cdots {}^\wedge t_p{}^\wedge$.

Definition 3.13. (reduct) For t in $T_{(X,^\wedge,\cdot,1)}$, the *reduct* of t is the term $\mathrm{red}(t)$ inductively defined by the rules:

- For $t = 1$ or t a variable, we put $\mathrm{red}(t) = t$;

- For $t = t_1 \cdot t_2$, with $\mathrm{red}(t_1) = t_{1,1} \cdot \cdots \cdot t_{1,p}$ and $\mathrm{red}(t_2) = t_{2,1} \cdot \cdots \cdot t_{2,q}$, we put $\mathrm{red}(t) = t_{1,1} \cdot \cdots \cdot t_{1,p} \cdot t_{2,1} \cdot \cdots \cdot t_{2,q}$; for $\mathrm{red}(t_1) = 1$, we put $\mathrm{red}(t) = \mathrm{red}(t_2)$, and, for $\mathrm{red}(t_2) = 1$, we put $\mathrm{red}(t) = \mathrm{red}(t_1)$;

- For $t = t_1{}^\wedge t_2$, with $\mathrm{red}(t_1) = t_{1,1} \cdot \cdots \cdot t_{1,p}$ and $\mathrm{red}(t_2) = t_{2,1} \cdot \cdots \cdot t_{2,q}$, we put $\mathrm{red}(t) = t'_1 \cdot \cdots \cdot t'_q$, with $t'_j = t_{1,1}{}^\wedge \cdots {}^\wedge t_{1,p}{}^\wedge t_{2,j}$; for $\mathrm{red}(t_1) = 1$, we put $\mathrm{red}(t) = \mathrm{red}(t_2)$, and, for $\mathrm{red}(t_2) = 1$, we put $\mathrm{red}(t) = 1$.

Example 3.14. (**reduction**) *Let* $t = (x_1 \cdot (x_2 \cdot x_3))^\wedge(x_1 \cdot x_2)$. *As a tree,* t *is represented by* 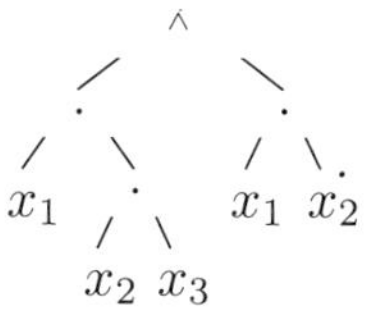 *Then* red(t) *is the term* $(x_1{}^\wedge x_2{}^\wedge x_3{}^\wedge x_1) \cdot (x_1{}^\wedge x_2{}^\wedge x_3{}^\wedge x_2)$, *i.e.,* 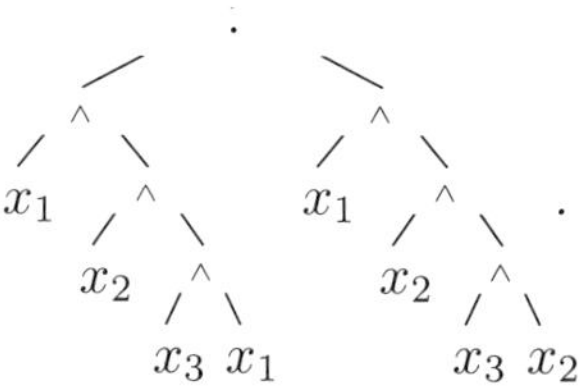.

Lemma 3.15. *For every* t *in* $T_{(X,^\wedge,\cdot,1)}$, *the term* red(t) *is prenormal, and it is LDM-equivalent to* t.

The easy verification is left to the reader. It suffices now to describe the restriction of LDM-equivalence to prenormal terms. Now, by definition, prenormal terms are very close to multiterms, and we have the following connection:

Lemma 3.16. *For all prenormal terms* t, t', *the following are equivalent:*
(i) The terms t *and* t' *are LDM-equivalent;*
(ii) The multiterms $i(t)$ *and* $i(t')$ *are LD-equivalent.*

Proof. By definition, the terms t and t' are LDM-equivalent if and only if they represent the same element of FLDM_X. By Proposition 3.7 (LD-equivalence), this happens if and only if the multiterms $i(t)$ and $i(t')$ are LD-equivalent. ■

Remark 3.17. Instead of resorting to Proposition 3.7, we could also prove the previous lemma by a direct verification. Proving that (i) implies (ii) is a little tedious, as one has to consider all possible identities of (LDM) and all possible terms: now prenormal terms are not closed under LDM-equivalence, so, the intermediate terms connecting t to t' in a sequence of elementary LDM-equivalence steps need not be prenormal, even if t and t' are.

Proposition 3.18. (**word problem**) *The word problem of* (LDM) *is decidable. More precisely, for every set* X *and every* x *in* X, *two terms* t, t' *in* $T_{(X,^\wedge,\cdot,1)}$ *are LDM-equivalent if and only if the terms* red$(t^\wedge x)$ *and* red$(t'^\wedge x)$ *are LD-equivalent.*

Proof. The terms t and t' are LDM-equivalent if and only if the prenormal terms $\mathrm{red}(t)$ and $\mathrm{red}(t')$ are LDM-equivalent, hence if and only if the multi-terms $i(\mathrm{red}(t))$ and $i(\mathrm{red}(t'))$ are LD-equivalent. By Lemma 3.8, the latter are LD-equivalent if and only if the terms $i(\mathrm{red}(t))\,x$ and $i(\mathrm{red}(t'))\,x$ are LD-equivalent. Now, by construction, $i(\mathrm{red}(t))\,x = \mathrm{red}(t^\wedge x)$ holds. ■

Example 3.19. **(word problem)** *Consider* $t = (x_1 \cdot x_2)^\wedge x_1$, $t' = x_1 \cdot x_2 \cdot x_1$. *We have* $\mathrm{red}(t) = x_1{}^\wedge x_2{}^\wedge x_1$, *and* $\mathrm{red}(t^\wedge x) = (x_1{}^\wedge x_2{}^\wedge x_1)^\wedge x$, *and, similarly,* $\mathrm{red}(t') = t'$, $\mathrm{red}(t'^\wedge x) = x_1{}^\wedge x_2{}^\wedge x_1{}^\wedge x =_{LD} (x_1{}^\wedge x_2{}^\wedge x_1)^\wedge x_1{}^\wedge x_2{}^\wedge x$. *The terms* $(x_1{}^\wedge x_2{}^\wedge x_1)^\wedge x$ *and* $(x_1{}^\wedge x_2{}^\wedge x_1)^\wedge x_1{}^\wedge x_2{}^\wedge x$ *have a variable clash, so we conclude that* t *and* t' *are not LDM-equivalent.*

Proposition 3.20. **(conservativity)** *Assume that* t, t' *are terms in* $T_{(X,^\wedge)}$. *Then* t *and* t' *are LDM-equivalent if and only if they are LD-equivalent.*

Proof. Let x be a fixed variable. For t in $T_{(X,^\wedge)}$, $\mathrm{red}(t^\wedge x) = t^\wedge x$ holds, and we have already noted that $t^\wedge x =_{LD} t'^\wedge x$ is equivalent to $t =_{LD} t'$. ■

Exercise 3.21. **(unique basis)** *Prove that a free LD-monoid admits only one basis.*

---

4. Orders on Free LD-Monoids

Free LD-systems are equipped with canonical linear orderings, and we have seen that free LD-monoids can be considered as subsets of free LD-systems. Hence they inherit order properties. However, the algebraic operations involved in the LD-monoid structure are not the operation of the LD-system, so compatibility questions arise. It turns out that everything works correctly. In particular, the embedding of FLD_X into FLDM_X is order preserving provided the fixed variable involved in the construction is chosen conveniently, and, then, the free LD-monoid FLDM_X can be described as an order completion of the free LD-system FLD_X.

Carrying the ordering

Assume that M is a free LD-monoid based on X. By the results of Section 3, there exists for each x in X an injection of M into the free LD-system based on X that maps 1 to x. It will be more convenient here to introduce a new variable x_0 not in X and to consider the injection of M into the free LD-system based on $X \cup \{x_0\}$ that maps 1 to x_0. This allows us to freely choose the position of x_0: the good choice here is to require x_0 to be minimal.

Definition 4.1. (injection) Assume that M is a free LD-monoid based on X. Fix a new variable x_0 not in X. Then i_0 denotes the injection of M into $\mathrm{FLD}_{X \cup \{x_0\}}$ defined in Proposition 3.9 (realization) that maps x to $x{}^\wedge x_0$ for x in X.

Thus, assuming $i_0(a) = a_1{}^\wedge \cdots {}^\wedge a_p{}^\wedge x_0$ and $i_0(b) = b_1{}^\wedge \cdots {}^\wedge b_q{}^\wedge x_0$, we have by definition $i_0(a \cdot b) = a_1{}^\wedge \cdots {}^\wedge a_p{}^\wedge b_1{}^\wedge \cdots {}^\wedge b_q{}^\wedge x_0$, and $i_0(a{}^\wedge b) = c_1{}^\wedge \cdots {}^\wedge c_q{}^\wedge x_0$, with $c_j = a_1{}^\wedge \cdots {}^\wedge a_p{}^\wedge b_j$.

Definition 4.2. (order) Assume that M is a free LD-monoid based on X, and $<$ is a linear order on X. For a, b in M, we say that $a < b$ holds if and only if $i_0(a) < i_0(b)$ holds in $\mathrm{FLD}_{X \cup \{x_0\}}$, when the order on X is extended by $x_0 < x$ for every x in X.

Example 4.3. (order) *Let us consider* FLDM_1. *We have* $i_0(x^2) = i_0(x \cdot x) = x{}^\wedge x{}^\wedge x_0$, *and* $i_0(x^{[2]}) = x^{[2]}{}^\wedge x_0$. *We have* $x{}^\wedge x{}^\wedge x_0 = x^{[2]}{}^\wedge x{}^\wedge x_0$ *and* $x_0 < x$ *holds by definition, so we find* $i_0(x^{[2]}) \ll i_0(x^2)$, *hence* $i_0(x^{[2]}) < i_0(x^2)$. *Therefore,* $x^{[2]} < x^2$ *holds in* FLDM_1.

On the other hand, we have $i_0(x^{[3]}) = x^{[3]}{}^\wedge x_0$. *Because* $x_0 < x$ *holds by definition, we have* $x{}^\wedge x{}^\wedge x_0 \ll x^{[3]}$. *Hence,* $i_0(x^2) < i_0(x^{[3]})$ *holds, and* $x^2 < x^{[3]}$ *holds in* FLDM_1.

As the mapping i_0 is injective, we have at once:

Lemma 4.4. *Assume that M is a free LD-monoid based on X, and $<$ is a linear order on X. Then the relation of Definition 4.2 is a linear order on M that extends $<$.*

Definition 4.5. (canonical) Under the above hypotheses, we say that the ordering of M given by Definition 4.2 is the *canonical extension* of $<$ to M. We say that an ordering $\prec$ on M is *canonical* if $\prec \restriction X$ is a linear ordering, and $\prec$ is the canonical extension of the latter.

Assume that M is a free LD-monoid based on X, and $<$ is a linear order on X. Then the sub-LD-system of M generated by X, which is free by Proposition 3.20 (conservativity), has been equipped with two linear orderings extending $<$ and involving the structures of LD-system and of LD-monoid respectively. These extensions coincide:

Proposition 4.6. (**coherence**) *Assume that M is a free LD-monoid based on X, and $<$ is a linear order on X. Let S be the sub-LD-system of M generated by X. Write $<^M$ for the canonical extension of $<$ to M as a free LD-monoid, and $<^S$ for the canonical extension of $<$ to S as a free LD-system based on X. Then $<^S$ is the restriction of $<^M$ to S.*

Notation 4.7. (**element $\bar{a}$**) In the above situation, S is isomorphic to the subsystem FLD_X of $\mathrm{FLD}_{X\cup\{x_0\}}$, and there is no danger in identifying every element of S with its image in $\mathrm{FLD}_{X\cup\{x_0\}}$ under the isomorphism that is the identity on X. However, we think reading is easier if the identification is not made; so we write $\bar{a}$ for the image of a under the previous isomorphism.

Proof. We claim that, for a, b in S, $a <^S b$ implies $\bar{a}\hat{\ }x_0 \ll \bar{b}\hat{\ }x_0$ in $\mathrm{FLD}_{X\cup\{x_0\}}$. Indeed, either $a \ll b$ holds in S, and, in this case, $\bar{a}\hat{\ }x_0 \ll \bar{b}\hat{\ }x_0$ holds as well. Or $a \sqsubset b$ holds. So there exist elements $a_1, \ldots, a_p$, with $p \geq 1$ and $b = (\cdots(a\hat{\ }a_1)\hat{\ }a_2\cdots)\hat{\ }a_p$. Then we have $\bar{b} = (\cdots(\bar{a}\hat{\ }\bar{a}_1)\hat{\ }\bar{a}_2\cdots)\hat{\ }\bar{a}_p$. No expression of a_1 involves x_0, hence $x_0 \ll \bar{a}_1$ holds, and we find $\bar{a}\hat{\ }x_0 \ll \bar{a}\hat{\ }a_1$, hence $\bar{a}\hat{\ }x_0 \ll \bar{b}$, and, *a fortiori*, $\bar{a}\hat{\ }x_0 \ll \bar{b}\hat{\ }x_0$.

We deduce that $a <^S b$ implies $\bar{a}\hat{\ }x_0 < \bar{b}\hat{\ }x_0$, *i.e.*, $i_0(a) < i_0(b)$, hence, by definition, $a <^M b$. As $<^S$ is a linear order, this is enough to conclude the argument. ∎

Proposition 4.8. (**characterization**) *Assume that $(M, \cdot, \hat{\ })$ is a free LD-monoid based on X and $<$ is a canonical order on M. Let S be the sub-LD-system of M generated by X. Then, for a, b in M, $a < b$ holds if and only if*

(i) either b admits a decomposition of the form

$$b = ((\cdots((a \hat{\ } a_1) \hat{\ } a_2) \hat{\ } \cdots) \hat{\ } a_p) \cdot a_{p+1} \tag{4.1}$$

with $p \geq 1$ and $a_1, \ldots, a_p \in S$,

(ii) or a and b admit decompositions of the form

$$\begin{cases} a = ((\cdots(c \hat{\ } x) \hat{\ } a_1) \hat{\ } \cdots) \hat{\ } a_p) \cdot a_{p+1}, \\ b = ((\cdots(c \hat{\ } y) \hat{\ } b_1) \hat{\ } \cdots) \hat{\ } b_q) \cdot b_{q+1} \end{cases} \tag{4.2}$$

with $a_1, \ldots, a_p, b_1, \ldots, b_q \in S$, and x, y in X satisfying $x < y$.

Proof. First, we prove an auxiliary claim. Assume $a, c \in M$, and $c = c_1 \cdots c_\ell$ with $c_1, \ldots, c_\ell \in S$. Then we claim that $\bar{c}_1 \hat{} \cdots \hat{} \bar{c}_\ell \sqsubset i_0(a)$ holds if and only if there exist $a_1, \ldots, a_p$ in S with $p \geq 0$ and a_{p+1} in M satisfying

$$a = ((\cdots((c \wedge x) \wedge a_1) \wedge \cdots) \wedge a_p) \cdot a_{p+1}. \tag{4.3}$$

Indeed, assume $\bar{c}_1 \hat{} \cdots \hat{} \bar{c}_\ell \hat{} x \sqsubset i_0(a)$. There exists a decomposition $i_0(a) = \bar{d}_1 \hat{} \cdots \hat{} \bar{d}_n \hat{} x_0$ satisfying $\bar{c}_1 \hat{} \cdots \hat{} \bar{c}_\ell \hat{} x \sqsubseteq \bar{d}_1$. So there exist $p \geq 0$ and $a_1, \ldots, a_p$ in S satisfying

$$\bar{d}_1 = (\cdots((\bar{c}_1 \wedge \cdots \wedge \bar{c}_\ell \wedge x) \wedge \bar{a}_1) \wedge \cdots) \wedge \bar{a}_p.$$

Back in M, we find $d_1 = (\cdots((c_1 \hat{} \cdots \hat{} c_\ell \hat{} x) \hat{} a_1) \hat{} \cdots) \hat{} a_p$, and we obtain (4.3) when we put $a_{p+1} = d_2 \cdots d_n$. Conversely, assuming (4.3), and decomposing a_{p+1} into $d_2 \cdots d_n$ with $d_2, \ldots, d_n$ in S, we obtain $i_0(a) = \bar{d}_1 \hat{} \cdots \hat{} \bar{d}_n \hat{} x_0$ with $d_1 = (\cdots((c_1 \hat{} \cdots \hat{} c_\ell \hat{} x) \hat{} a_1) \hat{} \cdots) \hat{} a_p$, and $\bar{c}_1 \hat{} \cdots \hat{} \bar{c}_\ell \sqsubset i_0(a)$ holds.

We come back to the proof of the proposition. Let a, b be arbitrary elements of M. Assume $a < b$. As was shown in the proof of Proposition 4.6 (coherence), $i_0(a) \ll i_0(b)$ holds in $\mathrm{FLD}_{X \cup \{x_0\}}$. By Proposition V.5.6 (order), this means that there exist elements $c_1, \ldots, e_n$ of S, and elements x, y in X satisfying $x < y$ such that the following equalities hold:

$$i_0(a) = (\cdots((\bar{c}_1 \wedge \ldots \bar{c}_\ell \wedge x) \wedge \bar{d}_1) \cdots) \wedge \bar{d}_m$$
$$i_0(b) = (\cdots((\bar{c}_1 \wedge \ldots \bar{c}_\ell \wedge y) \wedge \bar{e}_1) \cdots) \wedge \bar{e}_n.$$

Two cases may occur, according to the possible positions of the variable clash. If x is the final variable x_0 in $i_0(a)$, then we have $\bar{c}_1 \hat{} \cdots \hat{} \bar{c}_\ell \hat{} y \sqsubset i_0(b)$ by construction, and, by the previous claim, b admits a decomposition of the form (4.1)—actually, we may even assume $a_1 \in X$. If x is not the final x_0 in $i_0(a)$, applying the claim to the relations $\bar{c}_1 \hat{} \cdots \hat{} \bar{c}_\ell \hat{} x \sqsubset i_0(a)$ and $\bar{c}_1 \hat{} \cdots \hat{} \bar{c}_\ell \hat{} y \sqsubset i_0(b)$ yields two equalities of the type (4.2).

Finally, as the claim above is an equivalence, the computation can be reversed, and both (4.1) and (4.2) imply $a < b$. ■

Observe that requiring $a_1, \ldots, a_p \in S$ is necessary in (4.1): indeed we always have $1 = a \hat{} 1$, although $1 < a$ holds by definition. Actually, the element 1 is the only possible obstruction, and we can replace $a_i \in S$ with $a_i \neq 1$ in (4.1).

Definition 4.9. (relation $\sqsubset$ and $\ll$) Assume that M is a free LD-monoid based on X, and $<$ is a canonical order on M. For a, b in M, we say that $a \sqsubset b$ holds if a and b satisfy Condition (i) in Proposition 4.8, and that $a \ll b$ holds if they satisfy Conditions (ii).

These definitions are natural extensions of the case of free LD-systems.

Lemma 4.10. *Assume that M is a free LD-monoid based on X, and let S be the free sub-LD-system of M generated by X. Assume that $<$ is a canonical* *order on M. Then the relations $\sqsubset$ and $\ll$ of Definition 4.9 coincide on S with those defined in Section V.5.*

Proof. The involved decompositions are similar. If $a \sqsubset b$ holds in the sense of Section V.5, it holds as well in the current sense, and the same holds for $\ll$ and $\gg$. The converse implication is not obvious, as the elements occurring in the decompositions need not belong to S. However, one inclusion is enough to conclude, as we know that, on S, the union of $\sqsubset$ and $\ll$ as defined in Section V.5 is a linear ordering. ∎

As for free LD-systems, the relations $\sqsubset$ and $<$ coincide in the case of rank 1.

Proposition 4.11. (**uniqueness, rank** 1) *Assume that M is a free monogenic LD-monoid. Then the canonical ordering of M is the unique partial ordering on M that satisfies $a < a$^b and $a \le a \cdot b$ for every a and for every $b \ne 1$.*

Proof. Assume that $\prec$ is a partial ordering on M that satisfies the hypotheses of the proposition. Assume $a, b \in M$, with $a < b$. By Proposition 4.8 (characterization), there exists a decomposition $b = ((\cdots((a^\wedge a_1)^\wedge a_2)^\wedge\cdots)^\wedge a_p) \cdot a_{p+1}$. Then, by hypothesis, we have

$$\begin{aligned} &a \prec a \wedge a_1 \prec (a \wedge a_1) \wedge a_2 \prec \ldots \\ &\quad \prec (\cdots((a \wedge a_1) \wedge a_2) \wedge \cdots) \wedge a_p) \preceq ((\cdots((a \wedge a_1) \wedge a_2) \wedge \cdots) \wedge a_p) \cdot a_{p+1} = b, \end{aligned}$$

hence $a \prec b$ holds. As $\prec$ has no cycle, we conclude that it coincides with $\sqsubset$. ∎

Proposition 4.12. (**uniqueness**) *Assume that M is a free LD-monoid based on X, and $<$ is a linear ordering on X. Then the canonical extension of $<$ to M is the unique partial ordering on M such that*

(i) $a \prec a$^b and $a \preceq a \cdot b$ hold for every a, and every $b \ne 1$, and

(ii) if x and y belong to X and $x < y$ holds, then

$$((\cdots(c \wedge x) \wedge a_1)\cdots) \wedge a_p) \cdot a_{p+1} \prec ((\cdots((c \wedge y) \wedge b_1)\cdots) \wedge b_q) \cdot b_{q+1} \tag{4.4}$$

holds for every $a_1, \ldots, b_{q+1}, c$ in M.

Proof. By Proposition 4.8 (characterization), the canonical extension of $<$ satisfies the conditions. Conversely, assume that $\prec$ is a partial ordering that satisfies the conditions of the proposition. Let $a, b \in M$ satisfying $a < b$. Then $a \sqsubset b$ implies $a \prec b$ by the argument of the previous proof. Assume now $a \ll b$. By Proposition 4.8 again, a and b admit decompositions of the form

$$\begin{aligned} a &= ((\cdots((c_1 \wedge \cdots \wedge c_r \wedge x) \wedge a_1) \wedge \cdots) \wedge a_p) \cdot a_{p+1} \\ b &= ((\cdots((c_1 \wedge \cdots \wedge c_r \wedge y) \wedge b_1) \wedge \cdots) \wedge b_q) \cdot b_{q+1} \end{aligned}$$

with x, y in X satisfying $x < y$. Then $a \prec b$ holds by hypothesis. Hence $<$ is included in $\prec$, and, as the latter is an ordering, it must coincide with $<$. ∎

Criteria for freeness

As in the case of LD-systems, the existence of canonical orders characterizes free LD-monoids.

Proposition 4.13. (criterion I) *Assume that* $(M, \cdot, {}^\wedge)$ *is a monogenic LD-monoid with at least two elements. Then the following are equivalent:*

(i) The LD-monoid $(M, \cdot, {}^\wedge)$ *is free;*

(ii) No equality of the form

$$a = (\cdots((a \wedge a_1) \wedge a_2)\cdots) \wedge a_p \tag{4.5}$$

with $p \geq 1$ *and* $a_1, \ldots, a_p \neq 1$ *holds in* M*;*

(iii) The set $M \setminus \{1\}$ *is closed under* ${}^\wedge$*, and left division in the LD-system* $(M \setminus \{1\}, {}^\wedge)$ *admits no cycle;*

(iv) There exists a partial ordering $\prec$ *on* M *such that* $a \prec a^\wedge b$ *holds for every* a *and every* $b \neq 1$.

Proof. We know that (i) implies (iv). Then (iv) implies (iii). Indeed assume $a^\wedge b = 1$. Then we have $(a^\wedge b)^\wedge a = a$, so, if (iv) is true, we must have $a = 1$. Then, as for LD-systems, the existence of $\prec$ forbids any equality of the form $a = (\cdots((a^\wedge a_1)^\wedge a_2)^\wedge\cdots)^\wedge a_p$. Then (iii) implies (ii), and no equality of the form (4.5) is possible with $a \neq 1$ if $M \setminus \{1\}$ is closed under operation ${}^\wedge$. It remains to prove that (ii) implies (i). So assume that M satisfies (ii), and is generated by g. There exists a (unique) surjective homomorphism π of FLDM_1 onto M that maps x to g. We claim that π is injective. Indeed, assume that a, b are distinct elements of FLDM_1. Then $a < b$ or $b < a$ holds. Assume for instance $a < b$. Then, by Proposition 4.8 (characterization), b admits in FLDM_1 a decomposition of the form

$$b = ((\cdots((a \wedge a_1) \wedge a_2) \wedge \cdots) \wedge a_p) \cdot a_{p+1}$$

with $a_1, \ldots, a_p$ in FLD_1. Applying the homomorphism π, we find in M the equality

$$\pi(b) = ((\cdots((\pi(a) \wedge \pi(a_1)) \wedge \pi(a_2)) \wedge \cdots) \wedge \pi(a_p)) \cdot \pi(a_{p+1}),$$

which implies

$$\begin{aligned}\pi(b) \wedge \pi(a_1) &= (((\cdots((\pi(a) \wedge \pi(a_1)) \wedge \pi(a_2)) \wedge \cdots) \wedge \pi(a_p)) \cdot \pi(a_{p+1})) \wedge \pi(a_1)\\ &= (((\cdots((\pi(a) \wedge \pi(a_1)) \wedge \pi(a_2)) \wedge \cdots) \wedge \pi(a_p)) \wedge (\pi(a_{p+1}) \wedge \pi(a_1)).\end{aligned}$$

This gives an equality of type (4.5) in M, and it remains to prove that $\pi(a_2)$, $\ldots$, $\pi(a_p)$ are not 1. Actually, we claim that $\pi(a) \neq 1$ holds for every a in FLD_X. This is true for $a = x$, as M has at least two elements. Then, assuming $\pi(a), \pi(b) \neq 1$, we deduce $\pi(a^\wedge b) \neq 1$, for, otherwise, we would have, as above, $(\pi(a)^\wedge\pi(b))^\wedge\pi(a) = \pi(a)$, a forbidden equality of type (4.5). ∎

Proposition 4.14. (**criterion II**) *Assume that $(M, \cdot, ^\wedge)$ is an LD-monoid generated by X. Then the following are equivalent:*

(i) The LD-monoid $(M, \cdot, ^\wedge)$ is free based on X;

(ii) No equality of the form

$$a = (\cdots((a \,^\wedge\, a_1) \,^\wedge\, a_2)\cdots) \,^\wedge\, a_p \tag{4.6}$$

with $p \geq 1$ and $a_1, \ldots, a_p \neq 1$, or of the form

$$(\cdots((c \,^\wedge\, x) \,^\wedge\, a_1) \,^\wedge \cdots) \,^\wedge\, a_p = (\cdots((c \,^\wedge\, y) \,^\wedge\, b_1) \,^\wedge \cdots) \,^\wedge\, b_q \tag{4.7}$$

with x, y distinct elements of X and $a_1, \ldots, b_q \neq 1$ holds in M.

(iii) The set $M \setminus \{1\}$ is closed under $^\wedge$, left division in the LD-system $(M \setminus \{1\}, ^\wedge)$ has no cycle and X is quasifree in $(M \setminus \{1\}, ^\wedge)$;

(iv) There exists a partial ordering $\prec$ on M such that the restriction of $\prec$ to X is linear, $a \prec a^\wedge b$ and $a \prec a \cdot b$ hold for every a and every $b \neq 1$, and

$$(\cdots((c_1 \,^\wedge \cdots \,^\wedge\, c_r \,^\wedge\, x) \,^\wedge\, a_1) \,^\wedge \cdots) \,^\wedge\, a_p \prec (\cdots((c_1 \,^\wedge \cdots \,^\wedge\, c_r \,^\wedge\, y) \,^\wedge\, b_1) \,^\wedge \cdots) \,^\wedge\, b_q$$

holds whenever x, y are elements of X that satisfy $x \prec y$ and $a_1, \ldots, b_q \neq 1$.

Proof. As previously, we know that (i) implies (iv). That (iv) implies (ii) is clear, for the existence of the ordering $\prec$ shows that the first term of the equality is smaller than the second one both in (4.6) and (4.7). It is clear that (iii) implies (ii), and, conversely, in order to show that (ii) implies (iii), it suffices to observe as in the previous proof that $a^\wedge b = 1$ implies $(a^\wedge b)^\wedge a = a$, an equality of type (4.6). It remains to prove that (ii) implies (i). So, assume that M satisfies the conditions of (ii). Let π be the homomorphism of FLDM_X onto M that is the identity on X. Let a, b be distinct elements of FLDM_X. Assume for instance $a < b$. By Proposition 4.8 (characterization), two cases are possible.

Case 1: There exists a decomposition

$$b = ((\cdots((a \,^\wedge\, a_1) \,^\wedge\, a_2) \,^\wedge \cdots) \,^\wedge\, a_p) \cdot a_{p+1},$$

which projects on

$$\pi(b) = ((\cdots((\pi(a) \,^\wedge\, \pi(a_1)) \,^\wedge\, \pi(a_2)) \,^\wedge \cdots) \,^\wedge\, \pi(a_p)) \cdot \pi(a_{p+1}),$$

implying

$$\pi(b) \,^\wedge\, \pi(a_1) = ((\cdots((\pi(a) \,^\wedge\, \pi(a_1)) \,^\wedge\, \pi(a_2)) \,^\wedge \cdots) \,^\wedge\, \pi(a_p)) \,^\wedge\, (\pi(a_{p+1}) \,^\wedge\, \pi(a_1)),$$

and we deduce $\pi(a) \neq \pi(b)$ as in the previous proof.

Case 2: There exist decompositions of the form

$$\begin{cases} a = ((\cdots(c^\wedge x)^\wedge a_1)^\wedge\cdots)^\wedge a_p) \cdot a_{p+1}, \\ b = ((\cdots(c^\wedge y)^\wedge b_1)^\wedge\cdots)^\wedge b_q) \cdot b_{q+1}, \end{cases}$$

with x, y distinct elements of X, and $a_1, \ldots, a_p, b_1, \ldots, b_q$ in FLD_X. We project these equalities to

$$\begin{cases} \pi(a) = ((\cdots(\pi(c)^\wedge x)^\wedge\pi(a_1))^\wedge\cdots)^\wedge\pi(a_p))\cdot\pi(a_{p+1}), \\ \pi(b) = ((\cdots(\pi(c)^\wedge y)^\wedge\pi(b_1))^\wedge\cdots)^\wedge\pi(b_q))\cdot\pi(b_{q+1}). \end{cases} \tag{4.8}$$

As noted in the proof of Proposition 4.13 (criterion I), we have $\pi(a_1), \ldots, \pi(b_q) \neq 1$, and the only possible problem is with $\pi(a_{p+1})$ and $\pi(b_{q+1})$. Now let z be an arbitrary element of X. Multiplying (4.8) by z on the right, we obtain

$$\begin{cases} \pi(a)^\wedge z = ((\cdots(\pi(c)^\wedge x)^\wedge\pi(a_1))^\wedge\cdots)^\wedge\pi(a_p))^\wedge(\pi(a_{p+1}{}^\wedge z)), \\ \pi(b)^\wedge z = ((\cdots(\pi(c)^\wedge y)^\wedge\pi(b_1))^\wedge\cdots)^\wedge\pi(b_q))^\wedge(\pi(b_{q+1}{}^\wedge z)). \end{cases}$$

By construction, $a_{p+1}{}^\wedge z$ admits an expression where the product does not occur, *i.e.*, it belongs to FLD_X, and so does $b_{q+1}{}^\wedge z$. Hence $\pi(a_{p+1}{}^\wedge z) = 1$ and $\pi(b_{q+1}{}^\wedge z) = 1$ are impossible. Then the hypothesis of (ii) implies $\pi(a)^\wedge z \neq \pi(b)^\wedge z$, and, finally, $\pi(a) \neq \pi(b)$. ∎

Order and multiterms

The elements of the free LD-monoids are LD-equivalence classes of multiterms. Here, we characterize the order at the level of multiterms. To this end, we come back on the syntactic construction of terms in order to represent multiterms easily. The following lemma completes Lemma V.4.16.

Lemma 4.15. *Let X be a nonempty set. For every word w over the alphabet $X \cup \{\bullet\}$, the following properties are equivalent:*

(i) There exists a term t such that w is a prefix of t;

(ii) There exists a nonnegative integer p such that the word $w\bullet^p$ is a term;

(iii) There exist finitely many terms $t_1, \ldots, t_p$ in $T_{(X,^\wedge)}$ such that w is the concatenation $t_1 \cdots t_p$.

Moreover, in this situation, the decomposition of (iii) is unique.

Proof. For u a word on $X \cup \{\bullet\}$, let $\mu(u) = \#_X(u) - \#_\bullet(u)$. By Lemma V.1.7, the word w is the prefix of a term in $T_{(X,^\wedge)}$ if and only if $\mu(u) \geq 1$ holds for every prefix u of w, and, in this case, the word $w\bullet^p$ with $p = \mu(w) - 1$ is a term. So (i) and (ii) are equivalent. Clearly (iii) implies (ii), so it remains to show that (i) implies (iii) and the uniqueness. Assume $w = t_1 \ldots t_p$, where $t_1, \ldots, t_p$ are terms. By construction, the word $t_1 \ldots t_i$ is the longest prefix u of w satisfying $\mu(u) = i$. So the decomposition of w in a product of terms, if it exists, is unique. Assume that w satisfies Condition (i), and let $p = \mu(w)$. As the function μ cannot jump, there exists, for each $i \leq p$, a longest prefix u_i of w with $\mu(u_i) = i$. By construction, u_1 is a term, and, for $i \geq 2$, we have $u_i = u_{i-1}t_i$, where t_i is a term. So we have $w = t_1 \ldots t_p$. ∎

It follows that we may identify without ambiguity the multiterm $(t_1, \dots, t_p)$, also denoted $t_1{}^\wedge \cdots {}^\wedge t_p{}^\wedge$, with the word $t_1 \cdots t_p$, a prefix of the term $t_1 \cdots t_p \bullet^{p-1}$. Actually, $t_1 \cdots t_p$ can be seen as the right Polish expression of $t_1{}^\wedge \cdots {}^\wedge t_p{}^\wedge$, and we come back to the intuition of multiterms as incomplete terms already illustrated in the incomplete trees of Section 3.

As multiterms are now words over $X \cup \{\bullet\}$, the relations $\sqsubset$ and $\ll$ introduced for terms in Section V.4 extend to multiterms: $s \sqsubset t$ means that, as a word, s is a strict prefix of t, and $s \ll t$ means that there exist a word w and two variables x, y with $x < y$ such that wx is a prefix of s and wy is a prefix of t. The following result shows that this representation of multiterms and the relations so deduced are compatible with the previous constructions.

Proposition 4.16. (expressions) *Assume that $(M, \cdot, {}^\wedge)$ is a free LD-monoid based on X, and $<$ is a canonical ordering on M.*

(i) For all a, b in M, $a \sqsubset b$ holds if and only if $t \sqsubset t'$ holds for some multiterms t, t' representing a and b respectively.

(ii) For all a, b in M, $a \ll b$ holds if and only if $t \ll t'$ holds for some multiterms t, t' representing a and b respectively.

Proof. Assume $a \sqsubset b$. By Proposition 4.8 (characterization), there exists a decomposition of b under the form

$$b = ((\cdots((a \wedge a_1) \wedge a_2) \wedge \cdots) \wedge a_p) \cdot a_{p+1}$$

with $a_1, \dots, a_p$ in the sub-LD-system S of M generated by X. Let t be a multiterm that represents a, $t_1, \dots, t_p$ be terms that represent $a_1, \dots, a_p$ and t_{p+1} be a multiterm that represents a_{p+1}. By construction, the multiterm $tt_1 \bullet^{\mu(t)} t_2 \bullet \cdots t_p \bullet t_{p+1} \bullet^{\mu(t_{p+1})}$ represents b, and t is a prefix of this word. The case $a \ll b$ is similar, and the proof is complete as we know that the union of $\sqsubset$ and $\ll$ is a linear ordering on M. ∎

Properties of the canonical orders

We conclude with a few additional properties of the canonical orders on a free LD-monoid.

Proposition 4.17. (translations) *Assume that M is a free LD-monoid. Then the translations $x \mapsto c{}^\wedge x$ and $x \mapsto cx$ are increasing with respect to every canonical ordering of M.*

Proof. Assume $a < b$. Then $c{}^\wedge a < c{}^\wedge b$ holds. Indeed, the left translation ad_c is a homomorphism with respect to both product and exponentiation, and the order is defined by means of equalities involving product and exponentiation.

For instance, $a \sqsubset b$ holds if and only if b admits a decomposition of the form $b = ((\cdots((a^\wedge a_1)^\wedge\cdots)^\wedge a_p) \cdot a_{p+1}$, which implies

$$c \wedge b = ((\cdots(((c \wedge a) \wedge (c \wedge a_1)) \wedge \cdots) \wedge (c \wedge a_p)) \cdot (c \wedge a_{p+1}),$$

and therefore $c^\wedge a \sqsubset c^\wedge b$. As for the translations associated with the product, using multiterms as in the previous paragraph, we observe that an expression of ca is obtained by concatenating an expression of c and an expression of a. So, if $a \sqsubset b$ holds, some expression t of a is a proper prefix of some expression t' of b. If t_0 is any expression of c, t_0t is an expression of ca, t_0t is an expression of cb, and t_0t' is a proper prefix of t_0t. The argument is similar for $\ll$. ■

Proposition 4.18. (**cancellativity**) *In a free LD-monoid, both product and exponentiation admit left cancellatation.*

Proof. Fix some canonical ordering $<$. Assume $a \neq b$. Then one of $a < b$, $b < a$ holds. This implies $c^\wedge a < c^\wedge b$ or $c^\wedge b < c^\wedge a$, so $c^\wedge a \neq c^\wedge b$ in every case. The proof is similar for the product. ■

Let us finally consider the order type of a free LD-monoid equipped with a canonical ordering. Let X be a nonempty set, and $<$ be a linear order on X. The free LD-system FLD_X is a subset of the free LD-monoid FLDM_X, and the canonical extension of $<$ to FLD_X is the restriction of the canonical extension of $<$ to FLDM_X. We are going to describe the connection $(\mathrm{FLD}_X, <)$ and $(\mathrm{FLDM}_X, <)$ more precisely: we shall see that FLDM_X is obtained from FLD_X by adding a minimal element, namely 1, and upper bounds for some increasing sequences in FLD_X.

Proposition 4.19. (**density**) *Assume that X is a nonempty set. Then $1 \sqsubset a$ holds for every a in FLD_X, and, for $a_1, \ldots, a_p$ in FLD_X, $a_1 \cdots a_p$ is the $\sqsubset$-supremum of the $\sqsubset$-increasing sequence $\mathrm{iter}_n(a_1, \ldots, a_p)$ in FLDM_X.*

Proof. The relation $1 \sqsubset a$ holds by construction, as $i_0(1)$ is defined to be x_0, which is smaller than every variable in X. Assume $a_1, \ldots, a_p \in \mathrm{FLD}_X$. By Proposition V.5.16 (predecessors), the sequence $(\mathrm{iter}_n(a_1, \ldots, a_p)\ ;\ n \geq 0)$ is $\sqsubset$-increasing, and, for every x in X, $a_1{}^\wedge \cdots {}^\wedge a_p{}^\wedge x$ is a $\sqsubset$-supremum in FLD_X for this sequence.

Let $(a'_1, \ldots, a'_p) = \partial(a_1, \ldots, a_p)$. By construction, we have $a'_1 \cdots a'_p = a_1 \cdots a_p$. This implies $a'_1 \sqsubset a_1 \cdots a_p$, and, inductively, $\mathrm{iter}_n(a_1, \ldots, a_p) \sqsubset a_1 \cdots a_p$ for every n. On the other hand, assume $b < a_1 \cdots a_p$ in FLDM_X. Let $a = a_1{}^\wedge \cdots {}^\wedge a_p{}^\wedge x$, where x is an arbitrary element of X. Then we have $b < a$. We follow the proof of Proposition V.5.16. If $b \sqsubset a$ holds, there must exist some element y in X such that $b^\wedge y \sqsubset a$ holds, $b^\wedge y \in \mathrm{FLD}_X$, and $b^\wedge y$, hence b as well, is a prefix of $\mathrm{iter}_n(a_1, \ldots, a_p)$ for some n. If $b \ll a$ holds, $b^\wedge y \ll a^\wedge x$ holds for all x, y in X, and, again, we find n satisfying $b^\wedge y \ll \mathrm{iter}_n(a_1, \ldots, a_p)$. ■

(The previous result can also be established using multiterms.) When the basis has a minimal element, in particular when it is finite, we have the following simple result.

Proposition 4.20. (**predecessor**) *Assume $<$ is a linear order on X, and x is $<$-minimal. Then, for every $a_1, \ldots, a_p$ in FLD_X, $a_1 \cdots a_p$ is the immediate predecessor of $a_1{}^\wedge \cdots {}^\wedge a_p{}^\wedge x$ in FLDM_X.*

Proof. Let $a = a_1{}^\wedge \cdots {}^\wedge a_p{}^\wedge x$. We have $a = (a_1 \cdots a_p)^\wedge x$, hence $a_1 \cdots a_p \sqsubset a$. Conversely, assume $b < a$. Then either b can be represented by a multiterm that is a proper prefix of a term representing a, hence of a term of the form $\partial^n(t_1{}^\wedge \cdots {}^\wedge t_p{}^\wedge x)$ where t_i represents a_i, and therefore we have $b \sqsubset \mathrm{iter}_n(a_1, \ldots, a_p)$ for n large enough. Or $b \ll a$ holds, and, because x is minimal, the variable clash must occur before x, which means that, with the previous notations, $b \ll \mathrm{iter}_n(a_1, \ldots, a_p)$ holds for some n. In both cases, $b \sqsubset a_1 \cdots a_p$ follows. ∎

Thus, for instance, in FLDM_1, 1 is the immediate predecessor of x, x^2 is the immediate predecessor of $x^{[3]}$ and the supremum of the sequence $(\mathrm{iter}_n(x, x)\ ;\ n \geq 0)$, x^3 is the immediate predecessor of $x^{[4]}$ and the supremum of the sequence $(\mathrm{iter}_n(x, x, x)\ ;\ n \geq 0)$, *etc.*

Exercise 4.21. (**right cancellativity**) *Show that the equalities $x^\wedge x^{[2]} = x^{[2]\wedge} x^{[2]}$ and $x \cdot x = x^{[2]} \cdot x$ hold in every LD-monoid. Deduce that neither the product nor the exponentiation may admit right cancellatation in an LD-monoid unless exponentiation is idempotent.*

Exercise 4.22. (**sub-LD-monoid**) *Prove that every nontrivial monogenic sub-LD-monoid of a free LD-monoid is free.*

---

5. Extended Braids

Here we come back to the LD-monoid EB_∞ of extended braids. This LD-monoid was introduced in Section I.4 as a partial topological completion of the braid group B_∞. We shall show here that EB_∞ is also closely connected with the adjoint LD-monoid of the LD-system $(B_\infty, {}^\wedge)$. This will lead us to a realization of the free LD-monoid of rank 1, and to the construction of a new left self-distributive operation on EB_∞ from which both braid exponentiation and the product on EB_∞ can be recovered.

Free subsystems

A remarkable property of braid exponentiation is that every braid generates a free LD-system. The LD-monoid EB_∞ satisfies a similar property—provided we use operation $^\wedge$ rather than the direct extension $_\vee$ of conjugacy.

We recall that every element a of EB_∞ admits a decomposition $a = b\tau^q$ with b a braid and q a nonnegative integer. Such a decomposition is not unique, but the integer q is, and it is called the degree of a, and denoted $\deg(a)$.

Proposition 5.1. (**sub-LD-monoids free**) *The sub-LD-monoid of* $(EB_\infty, \cdot, ^\wedge)$ *generated by an extended braid of positive degree is free.*

Proof. By Proposition 4.13 (criterion I), it suffices to show that no equality of the form

$$a = (\cdots(a \,^\wedge\, b_1) \,^\wedge\, b_2)\cdots) \,^\wedge\, b_k \tag{5.1}$$

with $k \geq 1$ and $b_1, \ldots, b_k \neq 1$ is possible in the sub-LD-monoid M of EB_∞ generated by an extended braid of positive degree. Let us assume that (5.1) holds. By left self-distributivity, if the sequence $(a, b_1, \ldots, b_k)$ satisfies (5.1), so does the sequence $(a^\wedge b_1, b_2, \ldots, b_k, b_1)$. So, as $\deg(a^\wedge b_1) = \deg(b_1)$ holds, we may assume $\deg(a) \leq \deg(b_j)$ for every j. Let $p = \deg(a)$, and $q_j = \deg(b_j)$. By construction, $\deg(x) \geq 1$ holds for every element x in M except 1, and, as $a = 1$ is impossible in (5.1), we have $p \geq 1$. Now expanding (5.1) gives a relation of the form

$$\mathrm{sh}^p(c_0)\, \tau_{p,q_1}\, \mathrm{sh}^{q_1}(c_1)\, \tau_{q_1,q_2}\, \mathrm{sh}^{q_2}(c_2) \cdots \tau_{q_{k-1},q_k} \mathrm{sh}^{q_k}(c_k) \in B_p. \tag{5.2}$$

In other words, there exists a braid c in B_p such that the braid

$$c\, \mathrm{sh}^p(c_0)\, \tau_{p,q_1}\, \mathrm{sh}^{q_1}(c_1)\, \tau_{q_1,q_2}\, \mathrm{sh}^{q_2}(c_2) \cdots \tau_{q_{k-1},q_k} \mathrm{sh}^{q_k}(c_k) \tag{5.3}$$

is trivial. Now p is positive, and $q_j \geq p$ holds for every j. Hence, the braid occurring in (5.3) admits an expression where the letter σ_p occurs at least once, while the letter σ_p^{-1} does not occur. By Proposition III.2.18 (σ_i-positive), such a braid cannot be trivial. So (5.1) is impossible. ∎

The adjoint LD-monoid of $(B_\infty, ^\wedge)$.

We show now that there exists a close connection between EB_∞ and the adjoint LD-monoid of the LD-system $(B_\infty, ^\wedge)$.

Definition 5.2. (**monoid** EB_∞^+) We denote by EB_∞^+ the sub-LD-monoid of EB_∞ consisting of 1 and all elements of positive degree.

Thus EB_∞^+ is $EB_\infty \setminus B_\infty \cup \{1\}$. We shall prove the following result:

Proposition 5.3. (**adjoint**) *The adjoint LD-monoid of the LD-system* $(B_\infty, {}^\wedge)$ *exists, and*

$$\mathrm{ad}_{a_1} \circ \cdots \circ \mathrm{ad}_{a_p} \mapsto a_1\,\mathrm{sh}(a_2)\;\cdots\;\mathrm{sh}^{p-1}(a_p)\tau^p$$

defines an isomorphism of $\mathrm{Ad}(B_\infty)$ *onto* EB_∞^+.

Several verifications are needed. First, we check that the adjoint LD-monoid of $(B_\infty, {}^\wedge)$ exists, *i.e.*, that the LD-system $(B_\infty, {}^\wedge)$ is nice.

Lemma 5.4. *For* b, $b \in B_\infty$, *and* p, $p' \geq 0$, *the following are equivalent:*
(i) We have $b\,\tau_{p,1}\,\mathrm{sh}(b^{-1}) = b'\,\tau_{p',1}\,\mathrm{sh}(b'^{-1})$;
(ii) We have $p = p'$ *and* $b^{-1}b' \in B_p$, *i.e.*, $b\tau^p = b'\tau^{p'}$ *holds in* EB_∞.

Proof. Assume (i). We recall that, for b a braid, the exponent sum $\mathrm{es}(b)$ of b is the image of b under the group homomorphism of B_∞ in $\mathbf{Z}$ that maps every σ_i to 1. Then we have $\mathrm{es}(b\tau_{p,1}\mathrm{sh}(b^{-1})) = p$, and, therefore, (i) implies $p' = p$. Let $c = b^{-1}b'$. The hypothesis gives

$$\mathrm{sh}(c) = \tau_{p,1}^{-1}\,c\,\tau_{p,1}. \tag{5.4}$$

Let n be the least integer satisfying $c \in B_n$. Assume $n > p$. Then we have $\tau_{p,1} \in B_n$, and $\tau_{p,1}^{-1}\,c\,\tau_{p,1} \in B_n$. By hypothesis, $\mathrm{sh}(c) \notin B_n$ holds, so (5.4) is impossible. So we deduce $n \leq p$, hence $b^{-1}b' \in B_p$, and (i) implies (ii). Observe that the argument works for $p = 0$ and $p = 1$, and, in these cases, it simply gives $b = b'$. Conversely, $c \in B_p$ implies (5.4) by a direct computation, so $b' = bc$ with $c \in B_p$ implies (i). ∎

Lemma 5.5. *(i) Assume* $a_1, \ldots, a_p, a'_1, \ldots, a'_{p'} \in B_\infty$. *Then* $\mathrm{ad}_{a_1} \circ \cdots \circ \mathrm{ad}_{a_p} = \mathrm{ad}_{a'_1} \circ \cdots \circ \mathrm{ad}_{a'_{p'}}$ *holds if and only if we have* $p = p'$ *and* $a^{-1}a' \in B_p$ *for* $a = a_1\,\mathrm{sh}(a_2) \cdots \mathrm{sh}^{p-1}(a_p)$ *and* $a' = a'_1\,\mathrm{sh}(a'_2) \cdots \mathrm{sh}^{p'-1}(a'_{p'})$.
(ii) The LD-system $(B_\infty, {}^\wedge)$ *is nice, and* 1 *is generic in* $(B_\infty, {}^\wedge)$.

Proof. By definition, we have

$$\mathrm{ad}_{a_1} \circ \cdots \circ \mathrm{ad}_{a_p} : b \mapsto a\mathrm{sh}^p(b)\tau_{p,1}\mathrm{sh}(a^{-1}). \tag{5.5}$$

Assume $\mathrm{ad}_{a_1} \circ \cdots \circ \mathrm{ad}_{a_p}(1) = \mathrm{ad}_{a'_1} \circ \cdots \circ \mathrm{ad}_{a'_{p'}}(1)$. Using (5.5), we find $a\tau_{p,1}\mathrm{sh}(a^{-1}) = a'\tau_{p',1}\mathrm{sh}(a'^{-1})$, hence $p = p'$ and $a^{-1}a' \in B_p$ by Lemma 5.4. Conversely, assume the latter relations. Using (5.5) again and the equality $c\tau_{p,1} = \tau_{p,1}\mathrm{sh}(c)$ which holds for $c \in B_p$, we find

$$\begin{aligned}\mathrm{ad}_{a'_1} \circ \cdots \circ \mathrm{ad}_{a'_{p'}}(b) &= a'\mathrm{sh}^p(b)\tau_{p,1}\mathrm{sh}(a'^{-1})\\ &= a(a^{-1}a')\mathrm{sh}^p(b)\tau_{p,1}\mathrm{sh}(a'^{-1})\\ &= a\,\mathrm{sh}^p(b)(a^{-1}a')\tau_{p,1}\mathrm{sh}(a'^{-1})\\ &= a\,\mathrm{sh}^p(b)\tau_{p,1}\mathrm{sh}(a^{-1}a')\mathrm{sh}(a'^{-1})\\ &= a\,\mathrm{sh}^p(b)\tau_{p,1}\mathrm{sh}(a^{-1}) = \mathrm{ad}_{a_1} \circ \cdots \circ \mathrm{ad}_{a_p}(b).\end{aligned}$$

This proves Point (i), and that 1 is generic in $(B_\infty, {}^\wedge)$.

Assume $\mathrm{ad}_{a_1} \circ \cdots \circ \mathrm{ad}_{a_p} = \mathrm{ad}_{a'_1} \circ \cdots \circ \mathrm{ad}_{a'_{p'}}$, and let c be an arbitrary braid. With the notations of (i), we have $p = p'$ and $a^{-1}a' \in B_p$. Proving $\mathrm{ad}_{c^\wedge a_1} \circ \cdots \circ \mathrm{ad}_{c^\wedge a_p} = \mathrm{ad}_{c^\wedge a'_1} \circ \cdots \circ \mathrm{ad}_{c^\wedge a'_{p'}}$ amounts to proving the equality $p = p'$, which is in our assumptions, and the relation

$$((c \wedge a_1)\,\mathrm{sh}(c \wedge a_2) \cdots \mathrm{sh}^{p-1}(c \wedge a_p))^{-1}((c \wedge a'_1)\,\mathrm{sh}(c \wedge a'_2) \cdots \mathrm{sh}^{p'-1}(c \wedge a'_{p'})) \in B_p.$$

Expanding the latter expression gives

$$(c\,\mathrm{sh}(a)\,\tau_{1,p}\,\mathrm{sh}^p(c^{-1}))^{-1}(c\,\mathrm{sh}(a')\,\tau_{1,p}\,\mathrm{sh}^p(c^{-1})) \in B_p, \tag{5.6}$$

i.e., $\mathrm{sh}^p(c)\,\tau_{1,p}^{-1}\,\mathrm{sh}(a^{-1}a')\,\tau_{1,p}\,\mathrm{sh}^p(c^{-1}) \in B_p$. Now $a^{-1}a' \in B_p$ implies $\tau_{1,p}^{-1}\,\mathrm{sh}(a^{-1}a')\,\tau_{1,p} = a^{-1}a'$. As every braid in B_p commutes with every braid in the image of sh^p, we finally see that the braid in (5.6) is equal to $a^{-1}a'$, so (5.6) is true, and $(B_\infty, {}^\wedge)$ is nice. ∎

Hence the adjoint LD-monoid of $(B_\infty, {}^\wedge)$ exists. We can now complete the proof of Proposition 5.5 (adjoint).

Proof. Let f be the function of the free monoid B_∞^* of all finite sequences over B_∞ into EB_∞ defined by

$$f : (a_1, \dots, a_p) \mapsto a_1\,\mathrm{sh}(a_2)\ \cdots\ \mathrm{sh}^{p-1}(a_p)\tau^p. \tag{5.7}$$

By Lemma 5.5, $f(a_1, \dots, a_p) = f(a'_1, \dots, a'_{p'})$ is equivalent to $\mathrm{ad}_{a_1} \circ \cdots \circ \mathrm{ad}_{a_p} = \mathrm{ad}_{a'_1} \circ \cdots \circ \mathrm{ad}_{a'_{p'}}$. Hence f induces a well-defined injective mapping of $\mathrm{Ad}(B_\infty)$ into EB_∞. By definition, the image of this mapping consists of the unit braid, and of all extended braids of the form $a\tau^p$ with $a \in B_\infty$ and $p \geq 1$, *i.e.*, it is EB_∞^+. It remains to see that the operations are preserved. The result is obvious in the case of the product. As for exponentiation, owing to the LD-monoid identities, it suffices to consider the case of two braids, say a, b. By definition, ad_a is mapped to $a\tau$, ad_b is mapped to $b\tau$, and $\mathrm{ad}_a{}^\wedge\mathrm{ad}_b$, which is $\mathrm{ad}_{a^\wedge b}$ by definition, is mapped to $(a^\wedge b)\tau$. Now the exponentiation on EB_∞ has been constructed so that the mapping $a \mapsto a\tau$ is a homomorphism oif $(B_\infty, {}^\wedge)$ into $(EB_\infty, {}^\wedge)$. ∎

Using the construction of the coadjoint LD-monoid, we can apply the above results and define the structure of an LD-monoid on the left ideal $I_L(B_\infty, 1)$ of $(B_\infty, ^\wedge)$ generated by 1 by using the bijection

$$\mathrm{eval}_1 : \mathrm{ad}_{a_1} \circ \cdots \circ \mathrm{ad}_{a_p} \mapsto a_1 {}^\wedge \cdots {}^\wedge a_p {}^\wedge 1$$

of $\mathrm{Ad}(B_\infty)$ onto $I_L(B_\infty, 1)$. As $\mathrm{Ad}(B_\infty)$ is isomorphic to EB_∞^+, composing the two bijections gives a bijection of EB_∞^+ onto $I_L(B_\infty, 1)$, namely

$$I : b\tau^q \mapsto b\,\tau_{q,1}\,\mathrm{sh}(b^{-1}). \tag{5.8}$$

Using (5.8), we extend I to the whole of EB_∞, *i.e.*, we put $I(b) = b\,\mathrm{sh}(b^{-1})$ for $b \in B_\infty$. By Lemma 5.4, the extended mapping I is still an injection.

A new self-distributive operation on EB_∞

As EB_∞ is an LD-monoid, we can consider the companion self-distributive operations of Definition 1.9. Using the element 1 leads to the trivial operation

$$(a\tau^p) *_1 (b\tau^q) = b\tau^q.$$

Starting with τ gives the following more interesting operation.

Definition 5.6. (operation $*$) For a, $b \in B_\infty$, and p, $q \geq 0$, we put

$$(a\tau^p) * (b\tau^q) = a\,\tau_{p,1}\,\mathrm{sh}(a^{-1}b)\,\tau^{q+1}. \tag{5.9}$$

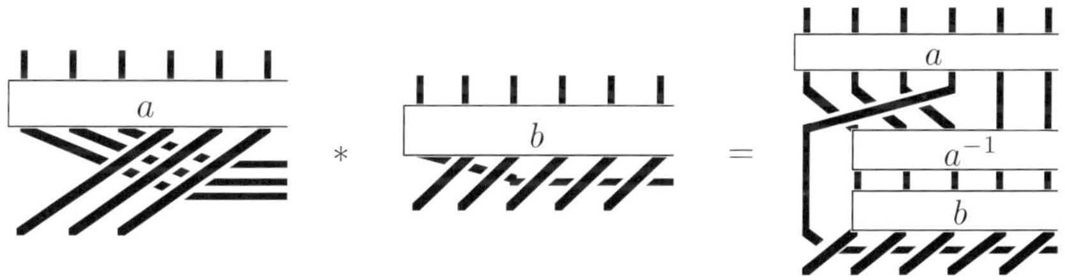

Figure 5.1. Operation $*$ applied to $a\tau^p$ and $b\tau^q$ (here $p = 3$, $q = 1$)

Proposition 5.7. (operation $*$) *(i) The system $(EB_\infty, *)$ is left self-distributive.*

(ii) Every monogenic sub-LD-system of $(EB_\infty, *)$ is free.

(iii) The mapping I of (5.8) is an embedding of $(EB_\infty, *)$ into $(B_\infty, ^\wedge)$.

(iv) The LD-system $(EB_\infty, *)$ is nice, and 1 is generic in $(EB_\infty, *)$.

(v) The LD-monoid operations on EB_∞ coincide with the 1-coadjoint operations of $(EB_\infty, *)$ on the left ideal of $(EB_\infty, *)$ generated by 1, i.e., we have

$$(a_1 * \cdots * a_p * 1) \cdot (b_1 * \cdots * b_q * 1) = a_1 * \cdots * a_p * b_1 * \cdots * b_q * 1,$$

$$(a_1 * \cdots * a_p * 1) \wedge (b_1 * \cdots * b_q * 1) = b'_1 * \cdots * b'_q * 1$$

with $b'_j = a_1 * \cdots * a_p * b_j$.

Proof. (i) By definition, $(EB_\infty, *)$ is the $\{\tau\}$-companion of the LD-monoid EB_∞.

(ii) By Laver's criterion (Proposition V.6.4), it suffices to show that no equality of the form

$$a\tau^p = (\cdots((a\tau^p * b_1\tau^{q_1}) * b_2\tau^{q_2})\cdots) * b_k\tau^{q_k}$$

with $k \geq 1$ is possible in EB_∞. Expanding this equality gives a relation of the form

$$\tau_{p,1}\,\mathrm{sh}(c_0)\,\tau_{q_1+1,1}\,\mathrm{sh}(c_1)\,\tau_{q_2+1,1}\ \cdots\ \tau_{q_k+1,1}\,\mathrm{sh}(c_k) \in B_p,$$

i.e.,

$$c^{-1}\,\tau_{p,1}\,\mathrm{sh}(c_0)\,\tau_{q_1+1,1}\,\mathrm{sh}(c_1)\,\tau_{q_2+1,1}\ \cdots\ \tau_{q_k+1,1}\,\mathrm{sh}(c_k) = 1$$

for some c in B_p. Now, for such a braid c, we have $c^{-1}\,\tau_{p,1} = \tau_{p,1}\mathrm{sh}^p(c)$. For $p = 0$, we must have $c = 1$, so, in every case, we have $\mathrm{sh}^p(c) \in \mathrm{sh}(B_\infty)$, and we obtain in B_∞ an equality of the form

$$\tau_{p,1}\,\mathrm{sh}(c'_0)\,\tau_{q_1+1,1}\,\mathrm{sh}(c_2)\,\tau_{q_2+1,1}\ \cdots\ \tau_{q_k+1,1}\mathrm{sh}(c_k) = 1.$$

The latter braid admits an expression where σ_1 occurs, but σ_1^{-1} does not, which contradicts Proposition I.3.19 (braids acyclic).

(iii) Let $a\tau^p$ and $b\tau^q$ be arbitrary extended braids. Let $c = a\tau_{p,1}\mathrm{sh}(a^{-1}b)$ and $r = q+1$. By definition, we have $c\tau^r = (a\tau^p) * (b\tau^q)$. On the other hand, we find

$$\begin{aligned} I(a\tau^p) \wedge I(b\tau^q) &= a\,\tau_{p,1}\,\mathrm{sh}(a^{-1})\,\mathrm{sh}(b)\,\mathrm{sh}(\tau_{q,1})\,\mathrm{sh}^2(b^{-1})\,\sigma_1\,\mathrm{sh}^2(a)\,\mathrm{sh}(\tau_{p,1}^{-1})\,\mathrm{sh}(a^{-1}) \\ &= a\,\tau_{p,1}\,\mathrm{sh}(a^{-1}b)\,\tau_{q+1,1}\,\mathrm{sh}^2(b^{-1}a)\,\mathrm{sh}(\tau_{p,1}^{-1})\,\mathrm{sh}(a^{-1}) \\ &= c\,\tau_{r,1}\,\mathrm{sh}(c^{-1}) = I(c\tau^r), \end{aligned}$$

which shows that I is a homomorphism.

(iv) Assume $a_1, \ldots, a'_{k'} \in EB_\infty$ and $a_1 * \cdots * a_k * 1 = a'_1 * \cdots * a'_{k'} * 1$ holds in EB_∞. As $I(1) = 1$ holds, we have in B_∞

$$I(a_1) \wedge \cdots \wedge I(a_k) \wedge 1 = I(a'_1) \wedge \cdots \wedge I(a'_{k'}) \wedge 1.$$

As 1 is generic in $(B_\infty, \wedge)$, this implies $I(a_1)\wedge \cdots \wedge I(a_k)\wedge b = I(a'_1)\wedge \cdots \wedge I(a'_{k'})\wedge b$ for every braid b, and, in particular, for every braid b in the image of I. As I is injective, this in turn implies $a_1 * \cdots * a_k * b = a'_1 * \cdots * a'_{k'} * b$ for every b in EB_∞, *i.e.*, 1 is generic in $(EB_\infty, *)$. Under the same hypothesis, we have

$$(c \wedge I(a_1)) \wedge \cdots \wedge (c \wedge I(a_k)) \wedge 1 = (c \wedge I(a'_1)) \wedge \cdots \wedge (c \wedge I(a'_{k'})) \wedge 1$$

for every braid c, and, in particular, for every c in the image of I. Hence we find

$$(c * a_1) * \cdots * (c * a_k) * 1 = (c * a'_1) * \cdots * (c * a'_{k'}) * 1$$

for every c, and $(EB_\infty, *)$ is nice.

(v) Apply Proposition 2.20 (companion *vs.* coadjoint). ∎

Exercise 5.8. **(colouring extended braids)** *(i) Generalize braid colourings to extended braids. Show that, for every extended braid b and every sequence of braids $\vec{a}$ in $B_\infty^{(\mathbf{N})}$, $\vec{a} \bullet b$ determines b.*

(ii) Define a special extended braid to be one that can be obtained from 1 using exponentiation. Show that an extended braid b is special if and only if the sequence $(1, 1, \ldots) \bullet b$ exists and it is equal to $(I(b), I(b), I(b), \ldots)$.

(iii) Deduce that $b\tau^q$ is special if and only if the braid b admits a decomposition $b = b_1\, \mathrm{sh}(b_2)\, \cdots\, \mathrm{sh}^{q-1}(b_q)$ with $b_1, \ldots, b_q$ special.

Exercise 5.9. **(image of I)** *(i) Show that the image of the mapping I of (5.8) is a left ideal in $(B_\infty, {}^\wedge)$.*

(ii) For $b \in B_\infty$ and $p \geq 1$, put $I_p(b) = b\tau_{p,1}\mathrm{sh}(b^{-1})$. Prove that $b \in \mathrm{Im}(I)$ holds if and only if $b \in \mathrm{Im}(I_{\mathrm{es}(b)})$ does, and if and only if the braids $b\sigma_{n,1}^{-1}$ and $\sigma_{n,\mathrm{es}(b)+1}^{-1}$ are B_n-conjugate, where $\mathrm{es}(b)$ is the exponent sum of b, and a and a' being B_n-conjugate means that $a' = cac^{-1}$ holds for some c in B_n.

(iii) For f a permutation, define $\nu(f)$ as the number of integers i satisfying $f(i+1) = i$. Prove that $b \in \mathrm{Im}(I)$ implies $\mathrm{es}(b) = \nu(\mathrm{perm}(b))$.

(iv) Show that, if b is a positive braid, then the necessary condition of (iii) is also sufficient. [Hint: Show that, if (ii) holds, then b must admit an expression of the form $\sigma_{i_1} \cdots \sigma_{i_p}$ with $i_1 > \cdots > i_p$, and compute $I(\sigma_{1,i_p-1}^{-1} \cdots \sigma_{1,i_1-1}^{-1})$.]

(v) Prove that $b \in B_n \cap \mathrm{Im}(I_p)$ implies $b^{[n-p]} = \tau_{n-1,1}$.

Exercise 5.10. **(right powers)** *(i) Prove that, for b a braid, one has $(b\tau^q)^{[m]} = b\tau_{q+m-1}$ in $(EB_\infty, *)$. Deduce that the solutions of $x^{[2]} = b\tau^q$ in $(EB_\infty, *)$ are the elements $bc\tau^{q-1}$ with $c \in B_q$.*

(ii) Prove that $b \in B_n$ implies $(b\tau^q)^{[m-q]} = 1^{[m]}$ for $m > n$. Deduce that the LD-system $(EB_\infty, *)$ includes no free LD-system of rank 2, and that the same property holds for the sub-LD-system $\mathrm{Im}(I)$ of $(B_\infty, {}^\wedge)$.

Exercise 5.11. **(distributive operation)** *Assume that* $\bullet$ *is a left self-distributive operation on* EB_∞ *and Identity* (LDM_1) *connects* $\bullet$ *and the product of* EB_∞. *Show that, for* a, b *braids, we must have*

$$(a\tau^p)\bullet(b\tau^q) = a\,\mathrm{sh}^p(b)\, f(p,q,a,b)\, sh^q(a^{-1}),$$

with $f(p,q,a,b) \in B_{p+q}$. *Prove that, if we assume* $f(p,q,a,b)$ *to depend on* p *and* q *only, the only possible choices are* $f(p,q) = \tau_{p,q}$ *and* $f(p,q) = \tau_{q,p}^{-1}$, *thus leading to the operations* $^\wedge$ *and* $_\vee$.

Exercise 5.12. **(left division)** *Assume that* a *and* c *are extended braids, say* $a = a_0\tau^p$, $c = c_0\tau^r$ *with* $a_0, c_0 \in B_\infty$. *Let* $d = \tau_{p,1}^{-1}a_0^{-1}c$. *Prove that there exists* b *in* EB_∞ *satisfying* $a * b = c$ *if and only if* $d \in \mathrm{sh}(B_\infty)\,B_r$ *holds. Prove that* $a*c = a^{[2]}*c$ *holds if and only if* $\mathrm{sh}(d^{-1})\sigma_1\mathrm{sh}(d) \in B_{r+1}$ *does. Compare the previous conditions.*

6. Notes

It seems that LD-monoids were considered first, under the name of "semi-abelian monoids", in (Dehornoy 86), as a way to axiomatize the algebraic properties of elementary embeddings in set theory (*cf.* Chapter XII). Actually the relevant axioms were part of the folklore of set theory, and it is certain that they were considered in the 80s by a number of set theorists, in particular R. Laver, in connection with elementary embeddings. The name "LD-algebra" has been used in (Dehornoy 92), (Drápal 94), (Drápal 95), but it seems unappropriate here as the structure has nothing to do with what is usually called an algebra. It can be observed that LD-monoids, considered as monoids equipped with an action on themselves, are reminiscent of crossed modules.

Exercise 1.19 (strong idempotent) was suggested by P. Ageron. Heyting algebras were introduced as an algebraic counterpart to the intuitionistic logic where negation is no longer involutive in the same way as Boolean algebras are an algebraic counterpart of classical logic—see for instance (Mac Lane & Moerdijk 92). In the case of a Boolean algebra (or of a Heyting algebra), the formula $a \wedge b = a$ defines an ordering. If we try to extend this construction to an arbitrary monoid, we must restrict ourselves to those elements that are idempotent, but such elements do not make a submonoid in general. In the case of a Heyting algebra, the ordering can be defined alternatively by the condition $a \Rightarrow b = 1$. Exercise 1.23 (shuffle) is from (Dehornoy 86).

The free completion of an LD-system into an LD-monoid appears in (Dehornoy 86). The adjoint LD-monoid is introduced with various examples, including free LD-systems, in (Dehornoy 92), which builds on (Zapletal *91*); the developments about coadjoint LD-monoids and companion multi-LD-systems improve partial results in the previous papers. The existence of an embedding of every free LD-monoid into a free LD-system is established in (Laver 92), where in particular the order properties and the criterion of freeness are proved.

The LD-monoid structure on the finite tables A_n was discovered by R. Laver. It naturally appears as the projection of composition when A_n is considered as a quotient of the iterations of an elementary embedding in set theory—see Chapter XII. Further results about the monoids $(A_n, \circ_n)$ and the LD-monoids $(A_n, \circ_n, *_n)$ can be found in (Drápal 95*a*) and (Drápal 97), where it is shown in particular that every monogenic LD-monoid can be constructed from the tables $(A_n, \circ_n, *_n)$—a preparatory result to the more general result that every monogenic LD-system can be constructed from the tables $(A_n, *_n)$ presented in Section X.2.

The LD-monoid EB_∞ is investigated in (Dehornoy 98) and (Dehornoy 99*a*). It seems that the LD-system $(EB_\infty, *)$ is an interesting structure. In particular, it embeds in the LD-system $(B_\infty, {}^\wedge)$ as a proper subsystem, and it seems easier to control things in $(EB_\infty, *)$ than in $(B_\infty, {}^\wedge)$, which could be too large in some sense. For instance, the results of Exercise 5.10 support such a feeling. Working with $(EB_\infty, *)$ is equivalent to working with its image under the homomorphism I, *i.e.*, working with the sub-LD-system $\mathrm{Im}(I)$ of $(B_\infty, {}^\wedge)$.

Question 6.1. *Describe the inverse mapping I^{-1} effectively, i.e., give an algorithm that, for a braid b in the image of I, returns a braid a and an integer p satisfying $b = a\,\tau_{p,1}\,\mathrm{sh}(a^{-1})$.*

By Exercise 5.9, we know that p must be the exponent sum of b, but we have no practical method for finding a (which is unique up to multiplying on the right by an element of B_q). In particular, we do not know whether the necessary condition of Exercise 5.9(v) is sufficient. R. Varela has noted that the necessary condition of Exercise 5.9(iii) is not sufficient.

Exercise 5.11 is from (Dehornoy 98). We conjecture that the operations ${}^\wedge$ and ${}_\vee$ are the only solutions even if we remove the restricting hypothesis about the function f. Exercise 5.12 is connected with the possible closedness of the LD-system $(EB_\infty, *)$, *i.e.*, with the question of whether $a * c = a^{[2]} * c$ implies that c is left divisible by a. Using the exercise, the problem reduces to

Question 6.2. *Is $d \in \mathrm{sh}(B_\infty)\,B_r$ equivalent to $\mathrm{sh}(d^{-1})\sigma_1\mathrm{sh}(d) \in B_{r+1}$ for every braid d?*

The condition is obviously sufficient. R. Fenn and R. Varela announced that it is also necessary (personal communication).

XII
Elementary Embeddings

Here, we describe some left self-distributive structures that appear in set theory when mappings called elementary embeddings are iterated. We try to give a self-contained exposition of the subject by extracting those minimal results necessary for the construction, and assuming no knowledge of set theory. However, the chapter can be used as a black box leading to exportable results like Proposition 4.11. Actually, it can even be skipped, for it only consists in one more example—but one that has played a crucial role in the subject, and still does, as some of the algebraic results it leads to have so far received no alternative proof, as we shall see in Chapter XIII.

The organization of the chapter is as follows. In Section 1, we introduce large cardinals in set theory. In Section 2, we define the notions of an elementary embedding and of a self-similar rank, and we explain why the existence of such objects cannot be proved in ordinary mathematics. We associate with every nontrivial elementary embedding a distinguished ordinal called its critical ordinal. In Section 3, we show how a left self-distributive operation arises on elementary embeddings associated with a self-similar rank. The main result here is the Laver–Steel theorem, a deep consequence of the fact that ordinals are well ordered: it asserts that the critical ordinals of a certain type of sequence of elementary embeddings have no upper bound. Using this result, one shows that left division in the LD-system constructed above has no cycle. Finally, we study in Section 4 the finite quotients of this LD-system, and the Laver tables of Chapter X appear naturally.

1. Large Cardinals

We begin with a quick review of some basic notions of set theory that will be used below. For more details we refer to standard textbooks, like (Levy 79) or (Jech 78)—or (Kanamori 94) at a more elaborate level.

Ordinals and ranks

Set theory is the study of those mathematical objects that are sets, and that of the membership relation between sets and their elements. Provided sufficiently strong formation rules are accepted for sets, it is well known that every mathematical object can be represented as a set—even those, like integers or functions, that are not obviously (or are obviously not) sets from an intuitive point of view. This means that the whole mathematical world can be considered as consisting exclusively of sets, and this clearly implies that the class of all sets is a very complicated structure.

Technically, the development of set theory can be organized as the study of a unique binary relation, namely the membership relation $\in$. In particular, we say that a mathematical formula $F(x)$ is a formula in the language of set theory, or, simply, a *set theoretical* formula, if the only symbols used in F are $\in$, variables, parentheses, and logical connectives $\forall$, $\exists$, $\Leftrightarrow$, $\Rightarrow$, $\&$, $\neg$. For instance $(\forall y)(y \notin x)$ is a set theoretical formula (with one free variable x). This however does not prohibit using derived notions, *i.e.*, other operations or relations that are themselves *definable* from $\in$. For instance, the inclusion relation $\subseteq$ is defined from $\in$ by the equivalence

$$x \subseteq y \Longleftrightarrow (\forall z)(z \in x \Rightarrow z \in y),$$

and, therefore, formulas containing the inclusion symbol can still be considered as set theoretical: in order to obtain an equivalent formula involving $\in$ only, it suffices to replace everywhere the symbol $\subseteq$ by its definition from $\in$. This procedure is very powerful, as we said that all usual mathematical objects can be represented by sets, thus becoming definable from $\in$, and therefore usable in set theoretical formulas: so are for instance the natural numbers and their set $\mathbf{N}$, the real numbers and their set $\mathbf{R}$, or the standard operations on these objects like addition or multiplication. So $p \in \mathbf{N}$ and $(\forall m \in \mathbf{N})(m|p \Rightarrow (m = 1 \text{ or } m = p))$ can be considered as set theoretical formulas too.

Certain sets play a crucial role. The *ordinals* are such sets. The important points here are that being an ordinal is a definable property—from now on definable always means "definable by a set theoretical formula"— and that there exists a definable relation $<$ on ordinals that is a well-ordering, *i.e.*, every set of ordinals has a least element with respect to $<$. Let $\mathrm{Ord}(\alpha)$ be a fixed definition for the property of being an ordinal, *i.e.*, a set theoretical formula with one free variable α such that $\mathrm{Ord}(\alpha)$ is true if and only if α is an ordinal. Easy arguments connected with Russell's paradox about the set of all sets show that there cannot exist a set of all ordinals. However it is convenient to introduce the name of *class* for a set theoretical formula with one free variable, and to use for such objects a vocabulary that imitates the terminology used for sets: for instance, one speaks of the class Ord of all ordinals, and one writes $\alpha \in \mathrm{Ord}$ for $\mathrm{Ord}(\alpha)$. Similarly, one speaks of a *functional* to refer to a set theoretical formula $F(x, y)$ with two free variables such that, for each set x, there exists exactly one set y satisfying $F(x, y)$, and, in this case, one uses a notation of the type $G(x)$ to denote the corresponding set y, speaking of the functional G to refer to F.

Ordinals should be thought of as a continuation of the sequence of natural numbers. Because the ordering of the ordinals is a well-ordering, there exists a least ordinal, which is denoted 0, and every ordinal α has a least successor, denoted $S(\alpha)$. The mapping $n \mapsto S^n(0)$ is injective, so there is no problem in identifying the ordinal $S^n(0)$ with the natural number n, and, under this identification, the natural numbers make an initial segment of the class Ord. This initial segment is strict, as there exists a least ordinal, denoted ω, that is strictly larger than every natural number, and the construction continues with the successors of ω. It is not hard to show that there exist definable operations $+$, $\cdot$ on Ord that extend the corresponding operations on natural numbers in a reasonable sense. In particular, the successor of the ordinal α is always the ordinal $\alpha + 1$, so that, for instance, the increasing enumeration of the ordinals begins with

$$0,\ 1,\ 2, \ldots, \omega,\ \omega{+}1,\ \omega{+}2, \ldots, \omega{+}\omega = \omega{\cdot}2,\ \omega{\cdot}2{+}1, \ldots, \omega{\cdot}3, \ldots, \omega^2,\ \omega^2{+}1, \ldots$$

An ordinal that is not a successor is called a *limit ordinal*: for instance ω, $\omega \cdot 2$ or ω^2 are limit ordinals, while 3 or $\omega^3 + \omega + 1$ are successor ordinals.

A major technical point about ordinals is the principle of induction, which directly mimics that of natural numbers:

Lemma 1.1. *Assume that $F(x)$ is a set theoretical formula such that, for every ordinal α, if $F(\beta)$ is true for $\beta < \alpha$, then $F(\alpha)$ is true as well. Then $F(\alpha)$ is true for every ordinal α.*

This enables one to construct objects using ordinal induction. In the case of natural numbers, the general scheme for an inductive construction consists in giving a function f, and defining a sequence $(x_n)_{n \in \mathbf{N}}$ by the clause

$$x_n = f((x_p)_{p<n}),$$

of which the more usual form $x_0 = a$, $x_n = g(x_{n-1})$ is a particular case. The extension to ordinals consists in starting with a functional f, and defining a sequence of sets indexed by the ordinals $(x_\alpha)_{\alpha\in\mathrm{Ord}}$ by

$$x_\alpha = f((x_\beta)_{\beta<\alpha}). \tag{1.1}$$

Again, a usual particular case corresponds to definitions of the form

$$x_0 = a, \quad x_\alpha = g(x_\beta) \text{ for } \alpha = \beta+1, \quad x_\alpha = h((x_\beta)_{\beta<\alpha}) \text{ for } \alpha \text{ a limit ordinal.} \tag{1.2}$$

Lemma 1.1 implies that the sequences defined in (1.1) and (1.2) do exist, and they are definable in the sense that there exists a set theoretical formula $F(x,y)$ such that, for every ordinal α and every set x, $F(\alpha,x)$ is equivalent to $x = x_\alpha$.

Definition 1.2. **(rank)** The sequence $(R_\alpha)_{\alpha\in\mathrm{Ord}}$ is constructed using the induction clauses

$$R_0 = \emptyset, \quad R_\alpha = \mathfrak{P}(R_\beta) \text{ for } \alpha = \beta+1, \quad R_\alpha = \bigcup_{\beta<\alpha} R_\beta \text{ for } \alpha \text{ a limit ordinal,}$$

where $\mathfrak{P}(X)$ is the powerset of X. The set R_α is called the *α-th rank*.

Example 1.3. **(rank)** *The first ranks are:*
$R_0 = \emptyset$;
$R_1 = \mathfrak{P}(R_0) = \{\emptyset\}$;
$R_2 = \mathfrak{P}(R_1) = \{\emptyset, \{\emptyset\}\}$;
$R_3 = \mathfrak{P}(R_2) = \{\emptyset, \{\emptyset\}, \{\{\emptyset\}\}, \{\emptyset, \{\emptyset\}\}\}$, *etc.*
Inductively, R_n has 2^{n-1} elements, and R_ω is a countable set.

Ranks are unintuitive objects, but they are relevant for our current purpose. In particular, they contain enough ordinals and are "broad" sets.

Proposition 1.4. **(rank)** *(i) For $\beta < \alpha$, we have $R_\beta \subsetneqq R_\alpha$.*
(ii) The ordinals contained in R_α are the ordinals below α.
(iii) Assume that λ is a limit ordinal, and $\alpha < \lambda$ holds. Then $x \subseteq R_\lambda$ implies $x \cap R_\alpha \in R_\lambda$. Similarly, for every function f of R_λ into itself, we have $f \cap R_\alpha^2 \in R_\lambda$.

Proof. (i) We first prove that $\alpha = \beta+1$ implies $R_\beta \subsetneqq R_\alpha$ using induction on β. We have $R_0 = \emptyset$ and $R_1 \neq \emptyset$, so the result is true for $\beta = 0$. Assume now $\beta = \gamma+1$. By induction hypothesis, we have $R_\gamma \subsetneqq R_\beta$, hence $R_\gamma \notin \mathfrak{P}(R_\gamma) = R_\beta$, but $R_\beta \in \mathfrak{P}(R_\beta)$, and we deduce $R_\beta = \mathfrak{P}(R_\gamma) \subsetneqq \mathfrak{P}(R_\beta) = R_\alpha$. Assume now that β is a limit ordinal and $x \in R_\beta$ holds. By definition, $x \in R_\gamma$ holds for some $\gamma < \beta$, hence, by induction hypothesis, we have $x \in R_{\gamma+1}$, *i.e.*, $x \subseteq R_\gamma$, hence $x \subseteq R_\beta$, *i.e.*, $x \in R_\alpha$. So $R_\beta \subseteq R_\alpha$ holds. Assume $R_\beta = R_\alpha$.

Then we have $R_\beta = R_\beta$, hence $R_\beta \in R_\gamma$ for some γ with $\gamma < \beta$. This implies $R_\beta \in R_{\gamma+1}$, hence $R_\beta \subseteq R_\gamma$, and, using the induction hypothesis, we obtain the contradiction

$$R_\beta \subseteq R_\gamma \subsetneqq R_{\gamma+1} \subseteq R_\beta.$$

Using an induction, we obtain $R_{\beta+1} \subseteq R_\alpha$ for $\alpha \geq \beta + 1$, and we deduce $R_\beta \subsetneqq R_\alpha$ for $\alpha > \beta$.

(ii) The property depends on the way the ordinals are constructed. In the standard way, ordinals are constructed so that 0 is the empty set, $S(\alpha)$ is $\alpha \cup \{\alpha\}$, and, for α a limit ordinal, α is the union of all smaller ordinals. This implies that every ordinal α is exactly the set of all ordinals smaller than α. Then, the property of the proposition is proved using induction on α. For $\alpha = 0$, R_α is empty, and the property is true. Assume $\alpha = \beta + 1$. By induction hypothesis, $\gamma < \beta$ implies $\gamma \in R_\beta$, hence $\gamma \in R_\alpha$. As $\gamma < \beta$ is equivalent to $\gamma \in \beta$, we also deduce $\beta \subseteq R_\beta$, and, therefore, $\beta \in R_\alpha$. On the other hand, the induction hypothesis $\beta \notin R_\beta$ implies $\{\beta\} \not\subseteq R_\beta$, hence $\alpha \notin R_\alpha$, and, therefore, $\gamma \notin R_\alpha$ for $\gamma \geq \alpha$. Assume finally that α is a limit ordinal. For $\beta < \alpha$, we have $\beta \in R_{\beta+1} \subseteq R_\alpha$, hence $\beta \in R_\alpha$. If we have $\alpha \in R_\alpha$, we deduce $\alpha \in R_\beta$ for some β with $\beta < \alpha$, hence $\beta \in R_\beta$ (as β is included in α), which contradicts the induction hypothesis. So $\alpha \notin R_\alpha$ holds, and, as above, this implies $\gamma \notin R_\alpha$ for $\gamma \geq \alpha$.

(iii) By definition, we have $x \cap R_\alpha \subseteq R_\alpha$, hence $x \cap R_\alpha \in R_{\alpha+1}$, and $x \cap R_\alpha \in R_\lambda$. Similarly, a function f being identified with the set of all pairs of the form $(x, f(x))$, and the pair (x, y) being itself identified with the set $\{\{x\}, \{x, y\}\}$, a pair (x, y) with x, y in R_α turns out to be an element of $R_{\alpha+2}$, and a function of R_α into itself is an element of $R_{\alpha+3}$. So, for f a function of R_λ into itself, we have $f \cap R_\alpha^2 \in R_{\alpha+3}$, and therefore $f \cap R_\alpha^2 \in R_\lambda$. ■

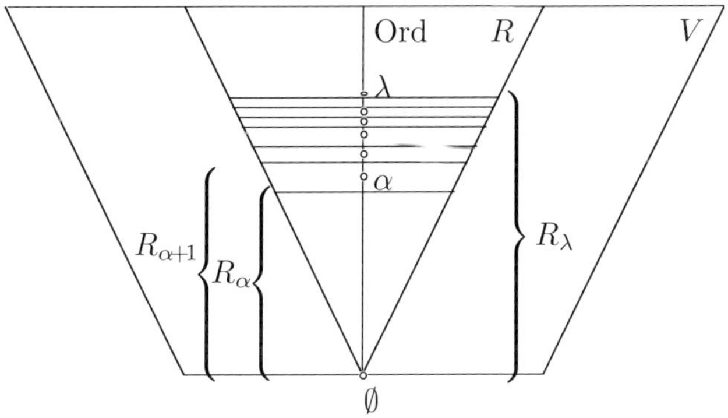

Figure 1.1: The "ice-cream" model of ranks

If λ is a limit ordinal, then, by construction, every set x in R_λ is the union of the sets $x \cap R_\alpha$ for $\alpha < \lambda$: so the previous result asserts that every subset of R_λ can be approximated by a sequence of sets each of which belongs to R_λ, and, similarly, that any function on R_λ can be approximated by a sequence of functions each of which belongs to R_λ. It is usual to represent ranks using the "ice-cream" model of Figure 1.1: here V represents the class of all sets, and Ord is the spine consisting of all ordinals. Ranks form an increasing family starting with the empty set, the rank R_α corresponds to the triangle just below the level of ordinal α. Limit ordinals are cluster points of the construction.

Large sets

From the point of view of set theory, infinite sets are involved in much deeper phenomena than finite sets. In particular, the assumption that at least one infinite set exists turns out to be a very powerful hypothesis, which enables one to establish a number of properties—even some properties that involve finite objects and procedures only. This point is so important that set theory from its beginnings can really be construed as a mathematical study of infinity.

Most recent developments of set theory involve *large sets*. The idea is to consider higher notions of infinity: the associated "very infinite" objects play with respect to smaller infinite objects the same role as infinite objects play with respect to finite objects. In order to introduce such higher notions of infinity, we can start from any property distinguishing the infinite sets from the finite sets, and repeat the process at a further level. Two classical examples are the following notions of an inaccessible cardinal, and of a measurable cardinal. A *cardinal* is defined to be an ordinal κ such that there is no bijection of κ onto a smaller ordinal. By Zermelo's theorem (which is equivalent to the axiom of choice), every set x is equipotent to a unique cardinal, its cardinality, denoted $\mathrm{card}(x)$, and the study of large sets turns out to be a study of *large cardinals*.

Let $\aleph_0$ denote the cardinal of countable sets: under the above definition, $\aleph_0$ coincides with the first infinite ordinal ω. On the other hand, every finite ordinal is also a cardinal, since, for p, q natural numbers with $p \neq q$, there exists no bijection between a set with p elements and one with q elements. Now $\aleph_0$ is bigger than all finite cardinals and it is inaccessible from the latter using the powerset operation, or the operation of taking the union of a sequence.

Definition 1.5. (inaccessible) Assume that κ is a cardinal. We say that κ is *inaccessible* if $\mathrm{card}(A) < \kappa$ implies $\mathrm{card}(\mathfrak{P}(A)) < \kappa$, and the conjunction of $\mathrm{card}(I) < \kappa$ and $\mathrm{card}(A_i) < \kappa$ for every i in I implies $\mathrm{card}(\bigcup_{i \in I} A_i) < \kappa$.

By definition, $\aleph_0$ is inaccessible, and inaccessible cardinals beyond $\aleph_0$ dominate their predecessors in the same way $\aleph_0$ dominates finite cardinals. This definition forces inaccessible cardinals to be very large.

Example 1.6. (inaccessible) *There exists an increasing enumeration $\aleph$ of the infinite cardinals by ordinals: $\aleph_0$ is the cardinal of countable sets, $\aleph_1$ is the next infinite cardinal, followed by $\aleph_2, \ldots, \aleph_\omega, \aleph_{\omega+1}$, etc. For instance, the continuum hypothesis is the assertion* $\mathrm{card}(\mathbf{R}) = \aleph_1$. *Now,* $\mathrm{card}(\mathfrak{P}(x)) > \mathrm{card}(x)$ *holds for every set x: hence we have* $\mathrm{card}(\mathfrak{P}(\aleph_\alpha)) \geq \aleph_{\alpha+1}$, *and therefore no cardinal of the form $\aleph_{\alpha+1}$ may be inaccessible. But $\aleph_\omega$ cannot be inaccessible either, since $\aleph_\omega$ is the union of the sequence, indexed by ω which is less than $\aleph_\omega$, of the cardinals $\aleph_n$, each of which is less than $\aleph_\omega$. The same argument works for every cardinal $\aleph_\alpha$ with $\alpha < \aleph_\alpha$. So, if κ is an inaccessible cardinal above $\aleph_0$, we must have $\kappa = \aleph_\kappa$.*

There is no end in the previous attempt to describe the first inaccessible cardinal above $\aleph_0$ because one *cannot* describe an inaccessible cardinal from below: this is the very meaning of the definition. Indeed, any such description would be a proof that inaccessible cardinals exist, contradicting the next result. The Zermelo-Fraenkel system, ZF for short, or ZFC if the axiom of choice is included, is a list of set theoretical statements, and there is a rather general agreement about this list correctly axiomatizing our intuition of what sets and membership are.

Proposition 1.7. (unprovability) *The existence of an inaccessible cardinal above $\aleph_0$ is not provable in the Zermelo-Fraenkel system.*

Proof. (Sketch) Saying that the system ZF axiomatizes the membership relation of sets means that the class V of all sets equipped with $\in$ satisfies all axioms in ZF, like a group satisfies the axioms of groups. Let R denote the union of all ranks R_α (this makes sense as R is a definable functional, *i.e.*, the relation $x = R_\alpha$ is a set theoretical formula). Whether every set lies in R, *i.e.*, whether the classes V and R coincide, depends on the initial choices in the construction of sets, but, in any case, the class R equipped with $\in$, or, more exactly, with the restriction of $\in$ to R, satisfies all axioms of ZF. Now the point is that, if κ is an inaccessible cardinal, then the rank R_κ, when equipped with the restriction of $\in$, enjoys enough closure properties to satisfy all axioms of ZF. More precisely, R_ω satisfies all axioms except the existence of an infinite set, while, for κ inacessible above ω, R_κ satisfies all axioms of ZF.

The rules for constructing sets are powerful enough to guarantee that sets include a copy of the integers with their operations. This enables one to code, using sets, every notion that can be coded in the integers, and, in particular, the notions of a mathematical formula and of a proof. It follows that there exists a formula of arithmetic, and therefore a formula of set theory, that codes the statement "there is no proof of $0 = 1$ in the system ZF". Now, let ZFI denote the system ZF enhanced with the statement "there exists an inaccessible cardinal above $\aleph_0$". Assume that κ is the least inaccessible cardinal above $\aleph_0$. Then the rank R_κ satisfies all axioms of ZF, and there is a proof of this statement in the system ZFI. By Gödel's completeness theorem for first order logic, this implies that the statement "there exist a set R and a binary relation E on R such that (R, E) satisfies all axioms of ZF" is provable from the axioms of ZFI. But this implies that ZFI also proves the statement "there is no proof of $0 = 1$ in the system ZF", since any proof of $0 = 1$ from a family of axioms would forbid any set satisfying these axioms to exist. Now Gödel's second incompleteness theorem tells us that no formal system including arithmetic can prove its own consistency. Assume that the existence of an inaccessible cardinal above $\aleph_0$ can be proved in ZF. It follows from the above argument that the consistency of ZF, which is the statement "there is no proof of $0 = 1$ in the system ZF", can be proved in ZF, contradicting Gödel's theorem, since arithmetic can be constructed in the system ZF. So the only possibility is that the statement "there exists an inaccessible cardinal above $\aleph_0$" has no proof in the system ZF. ∎

Remark 1.8. The previous result asserts that the existence of an inaccessible cardinal above $\aleph_0$ cannot be proved from the usual construction rules for sets. Actually, the same argument gives a little more, namely that even the consistency of the existence of an inaccessible cardinal cannot be proved. One knows that some statements, like the axiom of choice, the continuum hypothesis, or their negations, cannot be proved in the system ZF, but their relative consistency can be proved: these statements are not consequences of the axioms of ZF, but they do not contradict the latter, *i.e.*, no contradiction can appear when these statements are taken as additional axioms. This is not the case for the existence of an inaccessible cardinal above $\aleph_0$: not only is the latter an unprovable axiom, but it is one for which we cannot even hope to prove that it is consistent, and the only possibly provable result would be that inaccessibles do not exist. This situation is typical of the so-called *axioms of large cardinals*, of which the existence of an inaccessible cardinal above $\aleph_0$ is a standard example.

Let us mention a second example. Another property that distinguishes the infinite set ω from all finite sets is the existence of a non-principal measure.

Definition 1.9. (measurable) The cardinal κ is *measurable* if there exists a κ-complete non-principal two-valued measure on the set κ, *i.e.*, a mapping μ of $\mathfrak{P}(\kappa)$ into $\{0, 1\}$ such that $\mu(\kappa)$ is 1, singletons are μ-null, and, for any family $(A_\alpha)_{\alpha<\lambda}$ of μ-null sets indexed by an ordinal λ satisfying $\lambda < \kappa$, the union $\bigcup_{\alpha<\lambda} A_\alpha$ is μ-null.

The cardinal $\aleph_0$ is measurable, provided the axiom of choice is assumed: if U is any non-principal ultrafilter on $\aleph_0$ ($= \mathbf{N}$), *i.e.*, one that contains no singleton, we obtain an $\aleph_0$-complete measure on $\aleph_0$ by defining $\mu(A) = 1$ by $A \in U$.

Proposition 1.10. (**measurable** *vs.* **inaccessible**) *Every measurable cardinal is inaccessible.*

Proof. Assume that κ is measurable and μ is a κ-complete measure on κ. Assume that some set A satisfies $\text{card}(A) = \lambda < \kappa$ and $\text{card}(\mathfrak{P}(A)) \geq \kappa$. Fix an enumeration $(x_\beta)_{\beta<\lambda}$ of A. Let $(B_\alpha)_{\alpha<\kappa}$ be a sequence of pairwise distinct subsets of A. For $\beta < \lambda$, exactly one of the sets

$$\{\alpha < \kappa \; ; \; x_\beta \in B_\alpha\}, \qquad \{\alpha < \kappa \; ; \; x_\beta \notin B_\alpha\}$$

has μ-measure 1. Let X_β be that set. The intersection $\bigcap_{\beta<\lambda} X_\beta$ still has measure 1, but this set contains at most one element α, since the truth value of the statement $x_\beta \in B_\alpha$ is fixed for every β less than λ. This contradicts the definition of μ.

Similarly, assume that $(A_\alpha)_{\alpha<\lambda}$ is a sequence of sets with cardinality less than κ indexed by an ordinal less than κ, whose union has cardinality κ. By using a bijection between the latter set and κ itself, we obtain a new sequence $(B_\alpha)_{\alpha<\lambda}$ whose union is κ, and each set B_α has cardinality less than κ. Now $\text{card}(B) < \kappa$ implies $\mu(B) = 0$, as B is the union of less than κ singletons. Hence, $\mu(B_\alpha) = 0$ holds for each set in the previous sequence, and we deduce $\mu(\kappa) = 0$, a contradiction. ∎

It follows that the existence of a measurable cardinal above $\aleph_0$ is a large cardinal axiom, which cannot be proved more than the existence of an inaccessible cardinal. Actually, measurable cardinals (above $\aleph_0$) are much larger than inaccessible cardinals: by using the technique of ultrapowers, one can show that, if κ is a measurable cardinal above $\aleph_0$, and μ is a κ-complete two-valued measure on κ, then almost all ordinals below κ in the sense of μ are inaccessible.

Exercise 1.11. (**measurable implies cardinal**) *Assume that κ is a measurable ordinal. Show that κ must be a cardinal. [Hint: Assume that f is a bijection of some ordinal λ below κ onto κ, and consider the λ singletons $\{f(\alpha)\}$ for $\alpha < \lambda$.]*

---

2. Elementary Embeddings

Still another property distinguishing infinite objects from finite ones is that an infinite set is similar to some of its proper subsets. Here, similar means equipotent, *i.e.*, we refer to the property, often taken as a definition, that a set X is infinite if there exists an injection

$$j : X \to X \tag{2.1}$$

such that the image of j is a proper subset of X. For instance, this is the case for the injection $\mathrm{sh} : n \mapsto n+1$ on the set $\mathbf{N}$.

A natural idea for strengthening the above property and introducing a stronger notion of infinity is to require the similarity between the set and its proper subsets to involve an additional structure, *i.e.*, the injection of (2.1) to preserve some operation or relation on X.

Example 2.1. (**shift**) *The shift mapping* sh *of* $\mathbf{N}$ *preserves the ordered structure:* $n < n'$ *implies* $\mathrm{sh}(n) < \mathrm{sh}(n')$. *But it does not preserve the algebraic structure:* $\mathrm{sh}(n + n')$ *is not* $\mathrm{sh}(n) + \mathrm{sh}(n')$.

Elementary embeddings

In the context of set theory, where membership is the basic relation, it is natural to consider preservation of the structure provided by $\in$, *i.e.*, to consider $\in$-preserving homomorphisms. However, the technical developments of the theory constantly appeal to those objects that are defined from $\in$ using set theoretical formulas. So we are led to consider a stronger notion of preservation, involving not only the relation $\in$ itself, but also everything that is definable from $\in$.

Definition 2.2. (**elementary embedding**) Assume that S, S' are two structures made of a set equipped with relations or operations of the same type, and j is a mapping of S into S'. We say that j is an *elementary embedding* if, for every formula $F(\vec{x})$ in the language of S, and any sequence $\vec{a}$ in the domain of S, $F(\vec{a})$ is true in S if and only if $F(j(\vec{a}))$ is true in S'.

The existence of an elementary embedding of S in S' is a very strong similarity statement: in particular, it implies that every sentence (without free variables) that is true in S is true in S' as well, and conversely.

Lemma 2.3. *Assume that j is an elementary embedding of S into S'. Then j is an injective mapping, and it is a homomorphism with respect to every relation or operation that is definable in S and its counterpart in S'.*

Proof. By definition, the equality $a = b$ is true in S if and only if the equality $j(a) = j(b)$ is true in S', so j is injective. Now, assume that R is a definable binary relation of S: this means that there exists a formula $F(x, y)$ in the language of S such that $R(a, b)$ holds if and only if $F(a, b)$ is true in the domain of S. The counterpart R' of R in S' is defined similarly by the equivalence of $R'(a', b')$ and $F(a', b')$. Now the hypothesis that j is an elementary embedding implies that $F(a, b)$ is true in S if and only if $F(j(a), j(b))$ is true in S'. Similarly, assume that $*$ is a definable binary operation of S: this means that there exists a formula $F(x, y, z)$ in the language of S such that $c = a * b$ is true if and only if $F(a, b, c)$ holds. Again we deduce that $F(a, b, c)$ holds in S if and only if $F(j(a), j(b), j(c))$ holds in S'. ■

In the sequel, we consider the elementary embeddings of a structure into itself, with the intuition that the existence of such embeddings means that the considered structure is large. Observe that the identity mapping, or, more generally, any automorphism of a structure S, is an elementary embedding of S into itself. We shall say that an elementary embedding is *nontrivial* when it is not surjective. So, for instance, a set S is infinite if and only if there exists a nontrivial elementary embedding of S (the structure consisting of S alone) into itself.

Example 2.4. **(elementary embedding)** *Let us consider* $(\mathbf{N}, <)$ *again. The shift mapping* sh *is an injective homomorphism of* $(\mathbf{N}, <)$ *into itself. But this mapping is not an elementary embedding. Indeed, the formula*

$$F(x) = (\forall y)(x \leq y),$$

expresses that x *is a minimum for* $<$. *Now* $F(0)$ *is true in* $(\mathbf{N}, <)$, *while* $F(\mathrm{sh}(0))$ *is false. More generally, assume that* j *is any elementary embedding of* $(\mathbf{N}, <)$ *into itself. The previous argument shows that* 0 *has to be invariant under* j. *Now the integer* 1 *is also definable in* $(\mathbf{N}, <)$, *since it is the least element bigger than* 0. *Hence* 1 *must be invariant under* j, *and the same argument works for* 2, 3, *etc. So* j *has to be the identity mapping. In other words, there is no nontrivial elementary embedding of* $(\mathbf{N}, <)$ *into itself.*

We can formulate the previous argument as:

Lemma 2.5. *Assume that* j *is an elementary embedding of the structure* S *into itself. Then every element that is definable in* S *is invariant under* j.

Self-similar ranks

Coming back to the framework of set theory, and of ranks, it is natural to consider possible elementary embeddings between ranks.

Definition 2.6. (self-similar) We say that the rank R_α is *self-similar* if there exists a nontrivial elementary embedding of $(R_\alpha, \in)$ into itself; we define $\widehat{E}_\alpha$ and E_α respectively to be the set of all elementary embeddings, and of all nontrivial elementary embeddings, of $(R_\alpha, \in)$ into itself.

So, the rank R_α is self-similar if and only if the set E_α is not empty.

Ordinals are central objects in set theory. It is therefore natural that the images of ordinals under elementary embeddings play a significant role.

Definition 2.7. (critical) The *critical ordinal* $\mathrm{crit}(j)$ of the elementary embedding j is the least ordinal that is not invariant under j, if it exists.

Proposition 2.8. (critical) *Assume that j is a nontrivial elementary embedding of $(R_\alpha, \in)$ into itself.*

(i) The mapping j induces an increasing injection of the ordinals below α into themselves;

(ii) The critical ordinal of j exists. If $\mathrm{crit}(j)$ is κ, then $j(\kappa) > \kappa$ holds, and every set in R_κ is invariant under j. Moreover, κ is a measurable cardinal, and it is the κ-th measurable cardinal.

Proof. (i) This is a direct consequence of our definitions. Being an ordinal is expressed by some set theoretical formula $\mathrm{Ord}(x)$. If j is an elementary embedding of $(R_\alpha, \in)$ into itself, and γ is an ordinal below α, the formula $\mathrm{Ord}(\gamma)$ is true in $(R_\alpha, \in)$, and so is the formula $\mathrm{Ord}(j(\gamma))$. (The previous argument is inaccurate: it is not obvious that the formula $\mathrm{Ord}(\gamma)$ holding in the universe of all sets is equivalent to its holding in the substructure consisting of $(R_\alpha, \in)$; actually, there is no problem here, because the formula $\mathrm{Ord}(\gamma)$ can be constructed so that all quantifications it contains are bounded, and this implies that it is absolute, *i.e.*, its truth value is not altered by restricting the domain.) That the restriction of j to ordinals is increasing follows from the definability of the order of ordinals: by construction, $\gamma < \gamma'$ holds if and only if $\gamma \in \gamma'$ does.

(ii) Assume that j is the identity on the ordinals below θ. We show inductively on $\gamma < \theta$ that $j{\restriction}R_\gamma$ is the identity. The result is vacuously true for $\gamma = 0$. Assume $\gamma = \delta + 1$, and let $x \in R_\gamma$, *i.e.*, $x \subseteq R_\delta$. For $y \in x$, we have $y \in R_\delta$, hence $j(y) = y$ by induction hypothesis. Now $y \in x$ implies $j(y) \in j(x)$, and so we have proved $x \subseteq j(x)$. Conversely, $x \in R_{\delta+1}$ implies $j(x) \in R_{j(\delta)+1}$, hence $j(x) \in R_{\delta+1}$ by hypothesis. So, $y \in j(x)$ implies $y \in R_\gamma$, and, therefore, $y = j(y)$. Hence $y \in j(x)$ is also $j(y) \in j(x)$, which is equivalent to $y \in x$. So

$j(x) \subseteq x$ holds, and we conclude that $j(x) = x$ holds for every x in R_γ. Finally, assume that γ is a limit ordinal. The fact that j is the identity on each R_δ with $\delta < \gamma$ implies that j is the identity on R_γ, since R_γ is the union of the R_δ's.

By applying the previous result with $\theta = \alpha$, we see that the hypothesis of j being nontrivial on R_α implies that j moves at least one ordinal below α. Let κ be the least such ordinal: $j(\kappa) < \kappa$ is impossible, since j is injective and every ordinal below κ is its own image. Hence $j(\kappa) > \kappa$ is the only possibility. Finally, applying the above result with $\theta = \kappa$ shows that $j{\restriction}R_\kappa$ is the identity.

For $X \subseteq \kappa$, define $\mu(X) = 1$ for $\kappa \in j(X)$, and $\mu(X) = 0$ otherwise. We claim that μ is a measure and it shows that κ is measurable. Indeed let γ be any ordinal below κ. Then we have $j(\{\gamma\}) = \{j(\gamma)\}$: indeed, $\{\gamma\}$ is the unique set x that satisfies the set theoretical formula (with one parameter γ)

$$(\forall y)(y \in x \Leftrightarrow y = \gamma),$$

hence $j(\{\gamma\})$ is the unique set x' that satisfies the corresponding formula, where γ is replaced with $j(\gamma)$,

$$(\forall y)(y \in x' \Leftrightarrow y = j(\gamma)),$$

i.e., we have $j(\{\gamma\}) = \{j(\gamma)\}$. Now $\gamma < \kappa$ implies $j(\gamma) = \gamma$, and, therefore, $\kappa \notin j(\{\gamma\})$, *i.e.*, $\mu(\{\gamma\}) = 0$. On the other hand, $j(\kappa)$ is an ordinal strictly bigger than κ, so, by construction, it contains κ, which implies $\mu(\kappa) = 1$. Assume $\theta < \kappa$, and let $(X_\gamma)_{\gamma<\theta}$ be a family of sets all satisfying $\mu(X_\gamma) = 1$. So we have $\kappa \in j(X_\gamma)$ for every γ. Let $X = \bigcap_{\gamma<\theta} X_\gamma$. Write f for the function of θ to $\mathfrak{P}(\kappa)$ that maps γ to X_γ (*i.e.*, for the sequence $(X_\gamma)_{\gamma<\theta}$ itself). Then X is the only set x that satisfies the following formula, which involves the parameters θ and f,

$$(\forall y)(y \in x \Longleftrightarrow (\forall \gamma < \theta)(\exists z)((\gamma, z) \in f \text{ and } y \in z)).$$

It follows that $j(X)$ is the only set x' that satisfies the corresponding formula, where θ and f are replaced with their images:

$$(\forall y)(y \in x' \Longleftrightarrow (\forall \gamma < j(\theta))(\exists z)((\gamma, z) \in j(f) \text{ and } y \in z)).$$

Now we have $j(\theta) = \theta$, and, as j is elementary, $j(f)$ is a function of $j(\theta)$, *i.e.*, of θ, into $\mathfrak{P}(j(\kappa))$. This proves the equality

$$j(X) = \bigcap_{\gamma<\theta} j(X_\gamma).$$

By hypothesis, we have $\kappa \in j(X_\gamma)$ for every γ, and, therefore, it belongs to the intersection $j(X)$. This implies $\mu(X) = 1$, and completes the proof that κ is measurable. As was mentioned in Exercise 1.11, it is superfluous to check that κ is a cardinal, since this follows from the existence of the measure μ.

Finally, let X be the set of those ordinals below κ that are measurable cardinals. Then $j(X)$ is the set of all cardinals below $j(\kappa)$ that are measurable cardinals, since the property of being a measurable cardinal is defined by a set theoretical formula. Now κ belongs to the latter set, which implies $\mu(X) = 1$. Since every set with μ-measure 1 must have cardinality κ, we can conclude that κ is the κ-th measurable cardinal—and therefore also the κ-th inaccessible cardinal. ■

Corollary 2.9. *One cannot prove in the system ZF that a self-similar rank exists, nor can one prove it either in the systems "ZF + there exists an inaccessible cardinal" or "ZF + there exists a measurable cardinal".*

Proof. The argument is the same as for Proposition 1.7 (unprovability): assume that R_λ is a self-similar rank, and let κ be the critical ordinal of a nontrivial elementary embedding of $(R_\lambda, \in)$ into itself. Then there exists a set, namely the rank R_κ, that satisfies all the axioms of "ZF+ there exists a measurable cardinal" when equipped with the restriction of $\in$ (all the axioms of ZF being satisfied follows from κ being inaccessible, the axiom "there exists a measurable cardinal" being satisfied follows from κ not being the least measurable cardinal). So the system "ZF + there exists a self-similar rank" proves the consistency of the system "ZF + there exists a measurable cardinal", and, therefore, the consistency of the latter system cannot imply that of the former. ■

Another consequence of the proof of Proposition 2.8 (critical) is that the identity mapping is the only elementary embedding of a rank R_α into itself that is surjective on ordinals. Hence, in every case, the set $\widehat{E}_\alpha$ is the union of E_α and the singleton $\{\mathrm{id}\}$.

Kunen's bound

We conclude this section with a limitation result showing that the critical ordinal of a (nontrivial) elementary embedding of a self-similar rank into itself has to lie rather close to the top of that rank, or, more exactly, that the top of the rank has to be definable from the critical ordinal. This implies that such an elementary embedding has no fixed ordinal beyond the critical point.

Proposition 2.10. **(Kunen's bound)** *Assume that j is an elementary embedding of the rank R_α into itself, and that κ is the critical ordinal of j. Then we have either $\alpha = \sup_n j^n(\kappa)$, or $\alpha = \sup_n j^n(\kappa) + 1$.*

Proof. Write κ_n for $j^n(\kappa)$ (with $\kappa_0 = \kappa$). Let κ_ω be the supremum of the κ_n's. By definition, we have $\kappa_0 < \kappa_1$, hence $\kappa_n < \kappa_{n+1}$ for every n, as j is elementary. So the sequence of κ_n's is increasing. Moreover κ is a cardinal, and if follows inductively that κ_n is a cardinal for every n. This in turn implies that κ_ω is a cardinal. By construction, we have $\alpha \geq \kappa_\omega$. We assume $\alpha \geq \kappa_\omega + 2$, and derive a contradiction.

Every cardinal μ below κ_ω satisfies $\mu < \kappa_n$ for some n, and, therefore, the cardinal 2^μ of $\mathfrak{P}(\mu)$ is less than κ_n, since the latter is inaccessible. Fix an injection i_n of $\mathfrak{P}(\kappa_n)$ into κ_ω. Then the mapping $X \mapsto (i_n(X \cap \kappa_n))_{n \in \mathbf{N}}$ defines an injection of $\mathfrak{P}(\kappa_\omega)$ into the set $\kappa_\omega^{\mathbf{N}}$ of all $\mathbf{N}$-indexed sequences from κ_ω. For every set A, denote by $[A]^{\kappa_\omega}$ the set of those subsets of A that have cardinality κ_ω. We have $\mathrm{card}(\kappa_\omega \times [\kappa_\omega]^{\kappa_\omega}) \leq \mathrm{card}(\mathfrak{P}(\kappa_\omega))$, so, using the axiom of choice, we can fix an enumeration $(\gamma_\xi, X_\xi)_{\xi < \nu}$ of $\kappa_\omega \times [\kappa_\omega]^{\kappa_\omega}$, and then construct inductively on $\xi < \nu$ a sequence of elements s_ξ in $\kappa_\omega^{\mathbf{N}}$ such that s_ξ belongs to $[X_\xi]^{\kappa_\omega}$ and $\xi \neq \xi'$ implies $s_\xi \neq s_{\xi'}$: this is possible because the cardinality of $\kappa_\omega \times [\kappa_\omega]^{\kappa_\omega}$ is that of $\kappa_\omega^{\mathbf{N}}$. Now consider the function f of $\kappa_\omega^{\mathbf{N}}$ to κ_ω defined by $f(s) = \gamma_\xi$ for $s = s_\xi$, and $f(s) = 0$ for s not of the form s_ξ. Let $X \in [\kappa_\omega]^{\kappa_\omega}$. Then, for each $\gamma < \kappa_\omega$, there exists $\xi < \nu$ satisfying $(\gamma, X) = (\gamma_\xi, X_\xi)$. For this ordinal ξ, we have $s_\xi \in [X]^{\kappa_\omega}$ by hypothesis, and $f(s_\xi) = \gamma_\xi$. In other words, the function f has the property that, for every X in $[\kappa_\omega]^{\kappa_\omega}$, $f{\restriction}X^{\mathbf{N}}$ is still a surjection onto κ_ω.

By construction, $f \in R_{\kappa_\omega+2}$ holds, so it makes sense to consider the image $j(f)$. We have $\kappa_\omega = \sup_n \kappa_n$, hence $j(\kappa_\omega) = \sup_n \kappa_{n+1} = \kappa_\omega$. Hence $j(f)$ is a function of κ_ω into itself, and, as j is elementary, it must have the property that, for every X in $[\kappa_\omega]^{\kappa_\omega}$, $j(f){\restriction}X^{\mathbf{N}}$ is still a surjection onto κ_ω. Now, let $X = \{\theta < \kappa_\omega\ ;\ \theta \in \mathrm{Im}(j)\}$. For every s in $X^{\mathbf{N}}$, we have $s_n \in \mathrm{Im}(j)$ for every n, hence $s = j(s')$ holds for some s', and we find

$$j(f)(s) = j(f)(j(s')) = j(f(s')) \in X.$$

As, by construction, X is not equal to κ_ω, this contradicts the hypothesis of $j(f)$ being a surjection onto κ_ω. ∎

There remain at most two possibilities for the position of α with respect to the supremum of the iterated critical ordinals $j^n(\mathrm{crit}(j))$. The latter ordinal is always a limit cardinal (hence a limit ordinal), so we see that, if R_α is a self-similar rank, necessarily α is either a limit ordinal, or the successor of a limit ordinal. Actually, one possiblity subsumes the other.

Lemma 2.11. *Assume that j is a nontrivial elementary embedding of R_α into itself, and $\alpha = \lambda+1$ holds for some limit ordinal λ. Then $j{\restriction}R_\lambda$ is an elementary embedding of R_λ into itself.*

Proof. First, the image of the ordinal λ under j is necessarily λ (since it cannot be less), and, therefore, the image of an element of R_λ is an element of R_λ. This is not enough to conclude the proof, because the ranks R_λ and $R_{\lambda+1}$ do not satisfy the same set theoretical formulas: for instance, R_λ satisfies the formula "there is no maximum ordinal" (since λ is a limit ordinal), while $R_{\lambda+1}$ satisfies "there exists a maximum ordinal" (namely λ, which is an element of $R_{\lambda+1}$, but not of R_λ). Now, for $F(\vec{x})$ a set theoretical formula, define $F^{R_\lambda}(\vec{x})$ to be the formula inductively obtained from $F(\vec{x})$ by replacing all quantifications $(\forall y)$ by $(\forall y \in R_\lambda)$, and all quantifications $(\exists y)$ by $(\exists y \in R_\lambda)$. An inductive argument shows that, for $\vec{a}$ in R_λ, the formula $F(\vec{a})$ is true in R_λ if and only if the formula $F^{R_\lambda}(\vec{a})$ is true in $R_{\lambda+1}$. Then, for every formula $F(\vec{x})$, and every sequence $\vec{a}$ from R_λ, $F(\vec{a})$ is true in R_λ if and only if $F^{R_\lambda}(\vec{a})$ is true in $R_{\lambda+1}$, and if and only if $F^{R_{j(\lambda)}}(j(\vec{a}))$ is true in $R_{\lambda+1}$, hence if and only if $F(j(\vec{a}))$ is true in R_λ: this shows that $j{\restriction}R_\lambda$ is an elementary embedding of R_λ into itself. ∎

It follows that the existence of a self-similar rank of limit type—we shall simply say a limit self-similar rank in the sequel—is, from the point of view of logical strength, at most as strong as the existence of a self-similar rank of successor type. Actually it is weaker, as the existence of the latter proves that the existence of the former is consistent. Keeping this in mind, we shall always consider the weaker case in the sequel, *i.e.*, the case of a limit rank. In other words, we shall concentrate on the sets E_λ with λ a limit ordinal.

Proposition 2.12. (no fixed point) *Assume that j is an elementary embedding of a limit rank R_λ into itself. Then, for every ordinal γ below λ, we have either $\gamma < \mathrm{crit}(j)$, and then $j(\gamma) = \gamma$ holds, or $\gamma = \mathrm{crit}(j)$, or $\gamma > \mathrm{crit}(j)$, and then $\delta < \gamma \leq j(\delta) < j(\gamma)$ holds for some δ.*

Proof. Let κ denote the critical ordinal of j. Assume $\gamma > \kappa$, and let δ be the least ordinal satisfying $\gamma \leq j(\delta)$: since $\gamma \leq j(\gamma)$ always holds, δ exists, and we have $\delta \leq \gamma$. Assume $\delta = \gamma$. This means that $\xi < \gamma$ implies $j(\xi) < \gamma$. Now, by hypothesis, $\kappa < \gamma$ holds. So, we deduce $j(\kappa) < \gamma$, and, iteratively, $j^n(\kappa) < \gamma$ for every n. We deduce $\lambda \leq \gamma$, contradicting Kunen's bound. ∎

Exercise 2.13. (elementary embedding) *Assume that A and A' are sets without any additional structure. Show that $j : A \to A'$ is an elementary embedding if and only if j is an injection and either the sets have the same finite cardinality (and the injection is a bijection), or both sets are infinite.*

---

3. Operations on Elementary Embeddings

We turn to the main task of this chapter, namely constructing a left self-distributive structure on the sets E_λ. To this end, we use the idea of applying one elementary embedding to another one—an operation that is completely different from composing the embeddings. We establish two major results in this section, namely that left division for the left self-distributive operation on E_λ has no cycle, and a deep well-foundedness result, the Laver–Steel theorem, which turns out to have many consequences, in particular in terms of the finite Laver tables.

Applying an elementary embedding to another one

In the sequel, λ always refers to a limit ordinal. We recall that E_λ denotes the family of all nontrivial elementary embeddings of $(R_\lambda, \in)$ into itself, and that $\widehat{E}_\lambda$ is the union of E_λ and the singleton $\{\mathrm{id}_{R_\lambda}\}$. We are interested only in the case when the rank R_λ is self-similar, *i.e.*, the set E_λ is nonempty. The first remark is that $\widehat{E}_\lambda$ is closed under composition: if j_1 and j_2 are elementary embeddings of $(R_\lambda, \in)$ into itself, so is $j_1 \circ j_2$. Thus composition gives $\widehat{E}_\lambda$ the structure of a monoid. The special framework of set theory and ranks allows us to define another operation.

Definition 3.1. (**application**) Assume that j_1, j_2 are elementary embeddings of a limit rank R_λ into itself. Then the *application of* j_1 *to* j_2 is defined to be

$$j_1[j_2] = \bigcup_{\gamma<\lambda} j_1(j_2{\restriction}R_\gamma).$$

This definition makes sense, as $x \in R_\gamma$ implies $j_2(x) \in R_{j_2(\gamma)}$, and the mapping $j_2{\restriction}R_\gamma$, viewed as a set of pairs, is included in $R_\gamma \times R_{j_2(\gamma)}$, hence, according to the way pairs are constructed, in $R_{j_2(\gamma)+2}$, so $j{\restriction}R_\gamma$ belongs to $R_{j_2(\gamma)+3}$, hence to R_λ, and its image under j_1 is well defined.

Lemma 3.2. *Assume that* j_1, j_2 *are elementary embeddings of* $(R_\lambda, \in)$ *into itself. Then so is* $j_1[j_2]$.

Proof. When γ ranges over the ordinals below λ, the various mappings $j_2{\restriction}R_\gamma$ are compatible, since all of them are subsets of j_2 viewed as a set of pairs. Since j_1 is elementary, $j_1(j_2{\restriction}R_\gamma)$ is a partial mapping defined on $R_{j_1(\gamma)}$, and the partial mappings $j_1(j_2{\restriction}R_\gamma)$ and $j_1(j_2{\restriction}R_{\gamma'})$ associated with different ordinals γ, γ' must be coherent. Hence $j_1[j_2]$ is a mapping of R_λ into itself.

Assume that $F(\vec{x})$ is a set theoretical formula. Because j_2 is elementary, for every $\vec{a}$ in R_λ, $F(\vec{a})$ is true in R_λ if and only if $F(j_2(\vec{a}))$ is. If $\vec{a}$ belongs to the domain of the restriction $j_2{\restriction}R_\gamma$, $j_2(\vec{a})$ is also $(j_2{\restriction}R_\gamma)(\vec{a})$, and, therefore, the formula $G(j_2{\restriction}R_\gamma)$ is true for each γ, where $G(f)$ is the formula

$$(\forall \vec{x} \in \mathrm{Dom}(f))(F(\vec{x}) \Leftrightarrow F(f(\vec{x}))).$$

Since j_1 is elementary, the formula $G(j_1(j_2{\restriction}R_\gamma))$ must be true as well: this means that, for every $\vec{a}$ in the domain of $j_1(j_2{\restriction}R_\gamma)$, the equivalence

$$F(\vec{a}) \Leftrightarrow F(j_1(j_2{\restriction}R_\gamma)(\vec{a}))$$

holds. By coherence, this means that, for every $\vec{a}$ in R_λ, the formula $F(\vec{a})$ holds if and only if the formula $F(j_1[j_2](\vec{a}))$ does, *i.e.*, that $j_1[j_2]$ is an elementary embedding of $(R_\lambda, \in)$ into itself. ∎

For future use, let us notice that, if j_1 and j_2 are elementary embeddings of R_λ into itself, then, for $\gamma < \lambda$, the equality

$$j_1[j_2]{\restriction}R_{j_1(\gamma)} = j_1(j_2{\restriction}R_\gamma) \tag{3.1}$$

holds by construction.

Application is *not* composition. For instance, assume that j is an elementary embedding as above, and that κ is the critical ordinal of j. Then $j \circ j$ maps κ to $j(j(\kappa))$, and we have seen that $\kappa < j(\kappa) < j(j(\kappa))$ holds. On the other hand, by Proposition 2.8 (critical), $j{\restriction}R_\kappa$ is the identity mapping on R_κ, so $j(j{\restriction}R_\kappa)$ is the identity mapping on $R_{j(\kappa)}$, and, in particular, it leaves κ, which is an element of $R_{j(\kappa)}$, unchanged. So we have $j[j](\kappa) = \kappa \neq j(j(\kappa))$.

Lemma 3.3. *Assume that j_1, j_2 are elementary embeddings of R_λ into itself. Then, for every element x in the image of j_1, we have*

$$j_1[j_2](x) = j_1 j_2 j_1^{-1}(x). \tag{3.2}$$

Proof. Assume that f and x belong to R_λ, and, in addition, that f is a function and $x = f(y)$ holds. By construction, y belongs to R_λ as well. The point is that the equality $x = f(y)$ has a set-theoretical definition, for instance $(y, x) \in f$ if functions have been identified with their graph. Assume that j is an elementary embedding of $(R_\lambda, \in)$ into itself. Then, we have the equivalence

$$x = f(y) \quad \Leftrightarrow \quad j(x) = j(f)\,(\,j(y)\,). \tag{3.3}$$

Applying (3.3) to $j = j_1$ and $f = j_2{\restriction}R_\gamma$ for γ large enough gives $j_1[j_2](j_1(y)) = j_1 j_2(y)$, which is (3.2). ∎

The definition of the injection bracket on $\mathfrak{I}_\infty$ in Chapter I is reminiscent of Formula (3.2): this operation can be seen as a naive attempt to adapt the application operation of elementary embeddings to the framework of injections on $\mathbf{N}$. In the latter case, we have no way for defining the value of $f[g](x)$ for

$x \notin \mathrm{Im}(f)$, and it turns out that the default value $f[g](x) = x$ works. Notice that, in the case of an elementary embedding, (3.2) represents the trivial part of the application operation.

Proposition 3.4. (LD-system) *Assume that λ is a limit ordinal. Then $\widehat{E}_\lambda$ equipped with the application operation is an LD-system.*

Proof. Assume $j_1, j_2, j_3 \in \widehat{E}_\lambda$. Let γ be an arbitrary ordinal below λ. There exists β below λ such that $j_3 {\restriction} R_\gamma$ (as a set of pairs) belongs to R_β. From the definition, we have

$$j_2[j_3]{\restriction} R_{j_2(\gamma)} = (j_2 {\restriction} R_\beta)(j_3 {\restriction} R_\gamma). \tag{3.4}$$

Applying j_1 to this equality leads to

$$j_1(j_2[j_3]{\restriction} R_{j_2(\gamma)}) = j_1(j_2 {\restriction} R_\beta)[j_1(j_3 {\restriction} R_\gamma)]. \tag{3.5}$$

Now, the left-hand term in (3.5) is $j_1[j_2[j_3]]{\restriction} R_{j_1(j_2(\gamma))}$, while the right-hand side is $j_1[j_2][j_1[j_3]]{\restriction} R_{j_1(j_2(\gamma))}$, because $j_1(j_2(\gamma)) = j_1[j_2](j_1(\gamma))$ holds. As γ is arbitrary, we deduce $j_1[j_2[j_3]] = j_1[j_2][j_1[j_3]]$. ■

The set $\widehat{E}_\lambda$ is now equipped with two binary operations, namely composition and application. It is easy to verify that these two operations satisfy the mixed identities defining LD-monoids:

Proposition 3.5. (LD-monoid) *Assume that λ is a limit ordinal. Then the set $\widehat{E}_\lambda$ equipped with composition and application is an LD-monoid.*

Proof. We have already observed that $\widehat{E}_\lambda$ equipped with composition is a monoid. It is clear that $j[\mathrm{id}] = \mathrm{id}$ and $\mathrm{id}[j] = j$ holds for every j in $\widehat{E}_\lambda$. So it only remains to verify Identities (LDM_1) to (LDM_3). Assume $j_1, j_2, j_3 \in \widehat{E}_\lambda$. Let x be an arbitrary element of R_λ. For γ large enough, x belongs to the domain of $j_2 {\restriction} R_\gamma$. So, we have

$$\begin{aligned}(j_1 \circ j_2)(x) &= j_1(j_2(x)) = j_1((j_2 {\restriction} R_\gamma)(x))\\ &= j_1(j_2 {\restriction} R_\gamma)(j_1(x)) = j_1[j_2](j_1(x)) = (j_1[j_2] \circ j_1)(x).\end{aligned}$$

This shows (LDM_1). Let now γ be an arbitrary ordinal below λ. We have

$$(j_1 \circ j_2)[j_3] {\restriction} R_{j_1 j_2(\gamma)} = j_1(j_2(j_3 {\restriction} R_\gamma)) = j_1(j_2[j_3]{\restriction} R_{j_2(\gamma)}) = j_1[j_2[j_3]] {\restriction} R_{j_1 j_2(\gamma)},$$

so (LDM_2) holds. Finally, using the fact that the elementary embedding j_1 preserves the composition of functions, we have for (LDM_3)

$$\begin{aligned}j_1[j_2 \circ j_3]{\restriction} R_{j_1(\gamma)} &= j_1((j_2 \circ j_3){\restriction} R_\gamma) = j_1((j_2 {\restriction} R_{j_3(\gamma)}) \circ (j_3 {\restriction} R_\gamma))\\ &= (j_1[j_2]{\restriction} R_{j_1 j_3(\gamma)}) \circ (j_1[j_3]{\restriction} R_{j_1(\gamma)}) = (j_1[j_2] \circ j_1[j_3]){\restriction} R_\gamma.\end{aligned}$$ ■

Lemma 3.6. *Assume that* j_1, j_2 *are elements of* E_λ. *Then we have*

$$\operatorname{crit}(j_1[j_2]) = j_1(\operatorname{crit}(j_2)), \tag{3.6}$$
$$\operatorname{crit}(j_1 \circ j_2) = \inf(\operatorname{crit}(j_1), \operatorname{crit}(j_2)). \tag{3.7}$$

Proof. Let $\kappa_2 = \operatorname{crit}(j_2)$. Then, by definition, $j_2(\kappa_2) > \kappa_2$ holds, and so does $j_1[j_2](j_1(\kappa_2)) > j_1(\kappa_2)$, since j_1 is elementary and we have $j_1(j_2(\kappa_2)) = j_1[j_2](j_1(\kappa_2))$. On the other hand, the formula $(\forall \gamma < \kappa_2)(j_2(\gamma) = \gamma)$ holds, which implies $(\forall \gamma < j_1(\kappa_2))(j_1[j_2](\gamma) = \gamma)$. Hence the critical ordinal of $j_1[j_2]$ exists, and it is $j_1(\kappa_2)$.

Similarly, let $\kappa_1 = \operatorname{crit}(j_1)$. If γ is both below κ_1 and κ_2, we have $j_1(j_2(\gamma)) = j_1(\gamma) = \gamma$. Now we have $j_2(\kappa_2) > \kappa_2$, and therefore $j_1(j_2(\kappa_2)) > j_1(\kappa_2) \geq \kappa_2$. This shows $\kappa_1 \geq \operatorname{crit}(j_1 \circ j_2)$. Similarly, we have $j_2(\kappa_1) \geq \kappa_1$, and, therefore, $j_1(j_2(\kappa_1)) \geq j_1(\kappa_1) > \kappa_1$, which gives $\kappa_1 \geq \operatorname{crit}(j_1 \circ j_2)$, and, finally, $\inf(\kappa_1, \kappa_2) \geq \operatorname{crit}(j_1 \circ j_2)$. ∎

The following inequality is an easy consequence of the ordinals being well-ordered and of the definition of the application operation. We shall see in Chapter XIII that it leads to a nontrivial property of the Laver tables.

Proposition 3.7. **(inequality)** *Assume that* j *is an elementary embedding of* $(R_\lambda, \in)$ *into itself. Then* $j[j](\alpha) \leq j(\alpha)$ *holds for every ordinal* α *below* λ.

Proof. Let X be the set of all ordinals ξ such that $j(\xi) > \alpha$ holds. This set is well-ordered, so it possesses a least element, say β. So the following formula, whose parameters are α, β and j, holds

$$j(\beta) > \alpha \quad \text{and} \quad (\forall \xi < \beta)(j(\xi) \leq \alpha).$$

As j is elementary, we deduce that the formula

$$j[j](j(\beta)) > j(\alpha) \quad \text{and} \quad (\forall \xi < j(\beta))(j[j](\xi) \leq j(\alpha))$$

holds as well—we can make things rigorous by replacing the parameter j with some approximation of the form $j{\restriction}R_\gamma$ with γ large enough. As $\alpha < j(\beta)$ holds, we can take $\xi = \alpha$ in the second formula, which gives $j[j](\alpha) \leq j(\alpha)$. ∎

The Laver–Steel theorem

Assume that λ is a limit ordinal and j is an element of E_λ. By Lemma 3.6, $j^n(\operatorname{crit}(j))$ is the critical ordinal of $j^{[n+1]}$, which is also, by Lemma V.2.9, $j^{[n]}[j^{[n]}]$: so, in the sequence of the right powers j, $j^{[2]}$, $j^{[3]}$, ..., every term is a left divisor of the next one. Kunen's bound asserts that the supremum of the critical ordinals in the previous sequence is λ. Actually, this property has nothing to do with the particular choice of the elementary embeddings $j^{[n]}$, and it is a special case of a much stronger statement:

Proposition 3.8. (**Laver–Steel theorem**) *Assume that* $j_1, j_2, \ldots$ *is a sequence of nontrivial elementary embeddings of a limit rank* $(R_\lambda, \in)$ *into itself and that, for every* n, j_n *is a left divisor of* j_{n+1} *in the LD-system* $(E_\lambda, [\,])$. *Then the supremum of the ordinals* $\mathrm{crit}(j_n)$ *is* λ.

The proof relies on a preliminary result which we establish first.

Definition 3.9. (**representable**) Assume $j \in E_\lambda$, and $\gamma < \lambda$. We say that the ordinal α is *γ-representable by j* if it can be expressed as $j(f)(x)$ where f and x belong to R_γ and f is a mapping with ordinal values; $S_\gamma(j)$ denotes the set of all ordinals that are γ-representable by j.

The definition makes sense, for, since j is elementary, the set $j(f)$ is also a mapping with ordinal values.

Lemma 3.10. *Assume that* j_1 *is a left divisor of* j_2 *in the LD-system* $(E_\lambda, [\,])$ *and* γ *is an inaccessible cardinal satisfying* $\mathrm{crit}(j_1) < \gamma < \lambda$. *Then the order type of* $S_\gamma(j_1)$ *is larger than the order type of* $S_\gamma(j_2)$.

Proof. The point is to construct an increasing mapping of $S_\gamma(j_2)$ into some proper initial segment of $S_\gamma(j_1)$. By hypothesis, there exists an elementary embedding j in E_λ satisfying $j_2 = j_1[j]$, and the idea is that $S_\gamma(j_2)$ is (more or less) the image under j_1 of some set $S_\delta(j)$ with $\delta < \gamma$—a set that we can expect to be smaller than $S_\gamma(j_1)$ because $\delta < \gamma$ holds and γ is inaccessible.

By Proposition 2.12 (no fixed point), there exists an ordinal δ satisfying $\delta < \gamma \le j_1(\delta)$ since $\gamma > \mathrm{crit}(j_1)$ is assumed. Let G be the function that maps every pair (f, x) in R_δ^2 such that f is a function with ordinal values and x lies in the domain of $j(f)$ to $j(f)(x)$. By construction, the image of G is the set $S_\delta(j)$. The cardinality of this set is at most that of R_δ^2, hence it is strictly less than γ since γ is inaccessible. So the order type of the set $S_{j,\delta}$ is less than γ, and, by ordinal induction, we construct an order-preserving mapping H of $S_\delta(j)$ onto some ordinal β below γ—we recall that, by construction, β coincides with the set of those ordinals that are smaller than β. Let us apply now j_1: the mapping $j_1(H)$ is also order-preserving, and it maps $j_1(S_\delta(j))$, which is $S_{j_1(\delta)}(j_2)$, onto $j_1(\beta)$. By hypothesis, $j_1(\delta) \ge \gamma$ holds, so $S_{j_1(\delta)}(j_2)$ includes $S_\gamma(j_2)$. Let α be an ordinal in the latter set: by definition, there exist f, x in R_γ, f a mapping with ordinal values, x an element in the domain of $j_2(f)$, satisfying $\alpha = j_2(f)(x)$, and we have

$$j_1(H)(\alpha) = j_1(H)(j_2(f)(x)) = j_1(H)(j_1(G)((f, x))) = j_1(H \circ G)((f, x)). \quad (3.8)$$

Now H and G belong to R_γ, and therefore both $H \circ G$ and (f, x) are elements of R_γ. Thus (3.8) shows that the ordinal $j_1(H)(\alpha)$ is γ-representable by j_1, and the mapping $j_1(H)$ is an order-preserving mapping of $S_\gamma(j_2)$ into $S_\gamma(j_1)$. Moreover, the image of the mapping H is, by definition, the ordinal β, so the image of $j_1(H)$ is the ordinal $j_1(\beta)$, and, therefore, $j_1(H)$ is an order-preserving mapping of $S_\gamma(j_2)$ into $\{\xi \in S_\gamma(j_1); \xi < j_1(\beta)\}$. Now we have $j_1(\beta) = j_1(f)(0)$, where f is the mapping $\{(0, \beta)\}$. Since $\beta < \gamma$ holds, this shows that $j_1(\beta)$ is itself γ-representable by j_1, and the above set $\{\xi \in S_\gamma(j_1); \xi < j_1(\beta)\}$ is a proper subset of $S_\gamma(j_1)$. So the order type of $S_\gamma(j_2)$, which is that of $\{\xi \in S_\gamma(j_1); \xi < j_1(\beta)\}$, is strictly smaller than the order type of $S_\gamma(j_1)$. ∎

We can now prove the Laver–Steel theorem easily.

Proof. Assume for a contradiction that there exists an ordinal γ with $\gamma < \lambda$ satisfying $\gamma > \mathrm{crit}(j_n)$ for every n. We may assume that γ is an inaccessible cardinal: indeed, by Kunen's bound, there exists an integer m such that $j_1^m(\mathrm{crit}(j_1)) \geq \gamma$ holds, and we know that $j_1^m(\mathrm{crit}(j_1))$ is inaccessible. Now the previous lemma applies to each pair (j_n, j_{n+1}), and it gives an infinite descending sequence of ordinals

$$\text{order type}(S_\gamma(j_1)) > \text{order type}(S_\gamma(j_2)) > \cdots,$$

which is impossible since the ordering of the ordinals is a well-ordering. ∎

This result, which deeply relies on the well-ordering of the ordinals, is crucial, and we shall see several applications in the sequel. A first one is the following result.

Proposition 3.11. (embeddings acyclic) *Assume that R_λ is a self-similar limit rank. Then left division in the LD-system $(E_\lambda, [\ ])$ has no cycle.*

Proof. Assume that $j_1, \ldots, j_n$ is a cycle for left division in E_λ. Consider the infinite periodic sequence $j_1, \ldots, j_n, j_1, \ldots, j_n, j_1, \ldots$. The Laver–Steel theorem applies, and it asserts that the supremum of the critical ordinals in this sequence is λ. But, on the other hand, there are only n different embeddings in the sequence, and the supremum of finitely many ordinals below λ cannot be λ, a contradiction. ∎

Exercise 3.12. *(**no 2-cycle**) (i) Assume that j_1, j_2 are elements of E_λ. Show that $j_1[j_2] = j_1$ is impossible. [Use the critical ordinal.]*

(ii) Assume that j_1, j_2, j_3 are elements of E_λ satisfying $j_1[j_2][j_3] = j_1$. Prove that $\mathrm{crit}(j_1) > \mathrm{crit}(j_3)$ implies $j_1[j_2](\mathrm{crit}(j_3)) \in \mathrm{Im}(j_1)$, and get a contradiction. Prove that $\mathrm{crit}(j_1) < \mathrm{crit}(j_3)$ is also impossible. Applying

the previous argument to $j_1[j_2]$, j_3, *and* j_2, *prove* $\mathrm{crit}(j_1[j_2]) = \mathrm{crit}(j_2)$, *and deduce* $\mathrm{crit}(j_2) < \mathrm{crit}(j_1)$. *Deduce that left division in* $(E_\lambda, [\ ])$ *has no 2-cycle.*

---

4. Finite Quotients

In this section, we investigate the finite quotients of the monogenic subsystems of the LD-systems E_λ. The context and the tools of set theory prove especially relevant for this task: the possibility of approximating elementary embeddings by convenient restrictions leads to a natural family of finite quotients, and the latter turn out to be the Laver tables introduced in Section X.1.

Equivalence of two elementary embeddings

Definition 4.1. (**iterates**) Assume that j is a nontrivial elementary embedding of a limit rank R_λ into itself. We define $\mathrm{Iter}(j)$ to be the sub-LD-system of $(E_\lambda, [\ ])$ generated by j, and $\widehat{\mathrm{Iter}}(j)$ to be the sub-LD-monoid of $(\widehat{E}_\lambda, \circ, [\ ])$ generated by j. The elements of $\mathrm{Iter}(j)$ are called the *pure iterates* of j, while the elements of $\widehat{\mathrm{Iter}}(j)$ are called the *iterates* of j.

By construction, $\mathrm{Iter}(j)$ is a monogenic LD-system, and $\widehat{\mathrm{Iter}}(j)$ is a monogenic LD-monoid. Applying Proposition 3.11 (embeddings acyclic) and Laver's criterion, we immediately obtain:

Proposition 4.2. (**free**) *Assume that j is a nontrivial elementary embedding of a self-similar rank into itself. Then* $\mathrm{Iter}(j)$ *is a free LD-system of rank* 1, *and* $\widehat{\mathrm{Iter}}(j)$ *is a free LD-monoid of rank* 1.

In order to introduce possible quotients of the structures $\mathrm{Iter}(j)$ and $\widehat{\mathrm{Iter}}(j)$, a simple idea could be to concentrate on the critical ordinals, or, more generally, on the values that the elementary embeddings take on particular fixed ordinals (see Exercise 4.21). As subsequent results will explain, the previous naive approach is not relevant beyond the first levels, and we must instead consider equivalence relations that take into account more than a few scattered values. Another natural idea would be to consider the restrictions of the embeddings to a fixed rank, *i.e.*, to consider equivalence relations of the form $j{\restriction}R_\gamma = j'{\restriction}R_\gamma$. Such relations do not behave properly with respect to the application operation, and we are led to introduce the following slightly different relations.

Definition 4.3. (**equivalence**) Assume j, $j' \in E_\lambda$ and γ is a limit ordinal below λ. We say that j and j' are *γ-equivalent*, denoted $j \stackrel{\gamma}{=} j'$, if $j(x) \cap R_\gamma = j'(x) \cap R_\gamma$ holds for every x in R_γ (Figure 4.1).

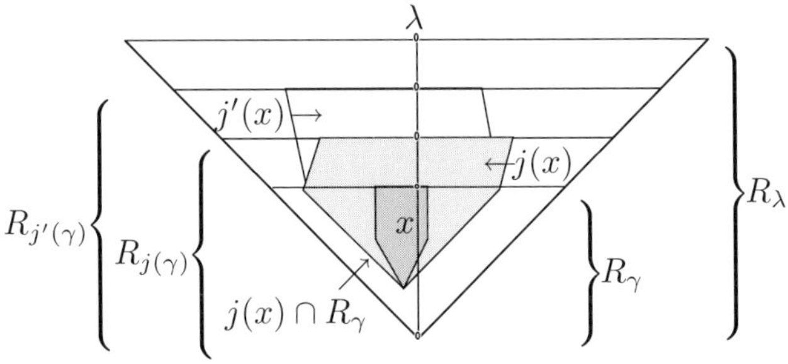

Figure 4.1. γ-equivalence of j and j'

By definition, γ-equivalence is an equivalence relation on $\widehat{E}_\lambda$. Note that, if j and j' are γ-equivalent, then $j(x) \cap R_\gamma = j'(x) \cap R_\gamma$ holds for every x in R_λ—and not only for x in R_γ—for, if y belongs to R_β with $\beta < \gamma$, $y \in j(x) \cap R_\gamma$ is equivalent to $y \in j(x \cap R_\beta) \cap R_\gamma$, and $x \cap R_\beta$ is an element of $R_{\beta+1}$, hence of R_γ since we assume γ to be a limit ordinal.

Lemma 4.4. *Assume $j \stackrel{\gamma}{=} j'$ and $\alpha < \gamma$. Then we have either $j(\alpha) < \gamma$, and then $j'(\alpha) = j(\alpha)$ holds, or $j(\alpha) \geq \gamma$, and then $j'(\alpha) \geq \gamma$ holds as well. So, in particular, we have either* $\mathrm{crit}(j) = \mathrm{crit}(j') < \gamma$, *or both* $\mathrm{crit}(j) \geq \gamma$ *and* $\mathrm{crit}(j') \geq \gamma$.

Proof. Since every ordinal is the set of all smaller ordinals, $j(\alpha) > \beta$ is equivalent to $\beta \in j(\alpha)$. So, if j' is γ-equivalent to j, $j(\alpha) > \beta$ is equivalent to $j'(\alpha) > \beta$ for α, $\beta < \gamma$. ∎

Lemma 4.5. *Assume j_1, $j_2 \in E_\lambda$. Then $j_1[j_2]$ and j_2 are* $\mathrm{crit}(j_1)$*-equivalent.*

Proof. Let $\gamma = \mathrm{crit}(j_1)$. Then γ is a limit ordinal (even a measurable cardinal), and, by Proposition 2.8 (critical), $z \in R_\gamma$ implies $j_1(z) = z$. So, for x, $y \in R_\gamma$, $y \in j_2(x)$ is equivalent to $j_1(y) \in j_1(j_2(x))$, hence to $y \in j_1[j_2](x)$, as we have $j_1(y) = y$, and $j_1(j_2(x)) = j_1[j_2](j_1(x)) = j_1[j_2](x)$. ∎

Proposition 4.6. (**compatibility**) *For every limit ordinal γ below λ, γ-equivalence is a congruence on the LD-monoid $\widehat{E}_\lambda$.*

Proof. Assume $j_1 \stackrel{\gamma}{=} j_1'$ and $j_2 \stackrel{\gamma}{=} j_2'$. Assume x, $y \in R_\gamma$ and $y \in j_1(j_2(x))$. Since γ is a limit ordinal, we have $x, y \in R_\beta$ for some β with $\beta < \gamma$. As y belongs to R_β, $y \in j_1(j_2(x))$ implies

$$y \in j_1(j_2(x) \cap R_\beta) \cap R_\gamma.$$

Now, by hypothesis, we have $j_2(x) \cap R_\beta = j_2'(x) \cap R_\beta$, and this set is an element of $R_{\beta+1}$, hence of R_γ. We deduce

$$j_1'(j_2'(x) \cap R_\beta) \cap R_\gamma = j_1(j_2'(x) \cap R_\beta) \cap R_\gamma = j_1(j_2(x) \cap R_\beta) \cap R_\gamma.$$

So $y \in j_1'(j_2'(x))$ holds, and this proves $j_1(j_2(x)) \subseteq j_1'(j_2'(x))$. The roles being symmetric, the compatibility of γ-equivalence with composition is proved. For the compatibility with application, we shall actually prove a stronger result stated below as a separate lemma. ■

Lemma 4.7. *Assume* $j_1 \stackrel{\gamma}{=} j_1'$ *and* $j_2 \stackrel{\delta}{=} j_2'$ *with* $j_1(\delta) \geq \gamma$. *Then we have* $j_1[j_2] \stackrel{\gamma}{=} j_1'[j_2']$.

Proof. Assume first $\mathrm{crit}(j_1) \geq \gamma$. By Lemma 4.4, we have also $\mathrm{crit}(j_1') \geq \gamma$. Moreover, $\delta \geq \gamma$ holds, for $\delta < \gamma$ would imply $j_1(\delta) = \delta < \gamma$, contradicting the hypothesis. Hence, $j_2 \stackrel{\delta}{=} j_2'$ implies $j_2 \stackrel{\gamma}{=} j_2'$. By Lemma 4.5, we have

$$j_1[j_2] \stackrel{\gamma}{=} j_2 \stackrel{\gamma}{=} j_2' \stackrel{\gamma}{=} j_1'[j_2'].$$

Assume now $\mathrm{crit}(j_1) < \gamma$, and, therefore, $\mathrm{crit}(j_1') = \mathrm{crit}(j_1)$. Since $j_2 \stackrel{\delta}{=} j_2'$ implies $j_2 \stackrel{\delta'}{=} j_2'$ for $\delta' \leq \delta$, we may assume without loss of generality that δ is minimal such that $j_1(\delta) \geq \gamma$ holds, which, by Proposition 2.12 (no fixed point), implies $\gamma > \delta$. Let $j \stackrel{*}{\cap} R_\alpha$ denote the set $\{(x, y) \in R_\alpha^2 \ ; \ y \in j(x)\}$. By definition, $j \stackrel{\alpha}{=} j'$ is equivalent to $j \stackrel{*}{\cap} R_\alpha = j' \stackrel{*}{\cap} R_\alpha$. We have

$$j_1[j_2] \stackrel{*}{\cap} R_\gamma = (j_1[j_2] \stackrel{*}{\cap} R_{j_1(\delta)}) \cap R_\gamma^2 = j_1(j_2 \stackrel{*}{\cap} R_\delta) \cap R_\gamma^2.$$

Now, $R_\gamma^2 = R_\gamma$ holds, since, by construction, an ordered pair of elements of R_α is an element of $R_{\alpha+2}$, and $\alpha < \gamma$ implies $\alpha + 2 < \gamma$ for γ a limit ordinal. Similarly, $j_2 \stackrel{*}{\cap} R_\delta$ is a set of ordered pairs of elements of R_δ, hence a set of elements of $R_{\delta+2}$, *i.e.*, an element of $R_{\delta+3}$, and, therefore, an element of R_γ. So, applying the hypotheses $j_2 \stackrel{*}{\cap} R_\delta = j_2' \stackrel{*}{\cap} R_\delta$ and $j_1(x) \cap R_\gamma = j_1'(x) \cap R_\gamma$ for $x \in R_\gamma$, we find

$$\begin{aligned} j_1[j_2] \stackrel{*}{\cap} R_\gamma &= j_1(j_2 \stackrel{*}{\cap} R_\delta) \cap R_\gamma \\ &= j_1'(j_2 \stackrel{*}{\cap} R_\delta) \cap R_\gamma = j_1'(j_2' \stackrel{*}{\cap} R_\delta) \cap R_\gamma = j_1'[j_2'] \stackrel{*}{\cap} R_\gamma. \end{aligned}$$ ■

Assume that j_0, j_1, j_2 are elements of E_λ. By left self-distributivity, the embeddings $j_0[j_1[j_2]]$ and $j_0[j_1][j_0[j_2]]$ coincide, but these embeddings need not be equal to $j_0[j_1][j_2]$: actually equality occurs if and only if $j_0[j_2] = j_2$ is true. Now, by Lemma 4.5, $j_0[j_2]$ and j_2 are $\mathrm{crit}(j_0)$-equivalent, and this implies that $j_0[j_1[j_2]]$ and $j_0[j_1][j_2]$ are $j_0[j_1](\mathrm{crit}(j_0))$-equivalent. Generalizing the argument, we obtain the following technical lemma.

Lemma 4.8. *Assume* $j_0, j_1, \ldots, j_p \in E_\lambda$, *and let* $\gamma = \mathrm{crit}(j_0)$.

(i) Assume $j_0[j_1[j_2] \cdots [j_\ell]](\gamma) \geq \gamma'$ *for* $1 \leq \ell \leq p-1$. *Then we have*

$$j_0[j_1][j_2] \cdots [j_p] \stackrel{\gamma'}{=} j_0[j_1[j_2] \cdots [j_p]]. \tag{4.1}$$

(ii) Assume $\mathrm{crit}(j_1[j_2] \cdots [j_\ell]) < \gamma$ *for* $1 \leq \ell \leq p-1$ *and* $\mathrm{crit}(j_1[j_2] \cdots [j_p]) \leq \gamma$. *Then we have*

$$\mathrm{crit}(j_0[j_1][j_2] \cdots [j_p]) = j_0(\mathrm{crit}(j_1[j_2] \cdots [j_p])). \tag{4.2}$$

Proof. (i) Use induction on p. For $p = 1$, the equivalence is an equality. Otherwise, we have, by induction hypothesis, $j_0[j_1][j_2] \cdots [j_{p-1}] \stackrel{\gamma'}{=} j_0[j_1[j_2] \cdots [j_{p-1}]]$, and, therefore,

$$j_0[j_1][j_2] \cdots [j_{p-1}][j_p] \stackrel{\gamma'}{=} j_0[j_1[j_2] \cdots [j_{p-1}]][j_p]. \tag{4.3}$$

Lemma 4.5 gives $j_p \stackrel{\gamma}{=} j_0[j_p]$, which implies

$$j_0[j_1[j_2] \cdots [j_{p-1}]][j_p] \stackrel{\gamma'}{=} j_0[j_1[j_2] \cdots [j_{p-1}]][j_0[j_p]], \tag{4.4}$$

since $j_0[j_1[j_2] \cdots [j_{p-1}]](\gamma) \geq \gamma'$ holds by hypothesis. Now the right-hand embedding in (4.4) is also $j_0[j_1[j_2] \cdots [j_p]]$ by left self-distributivity, so combining (4.3) and (4.4) gives (4.1).

(ii) For $p = 1$, the result is obvious. Assume $p \geq 2$, and let γ' be the smallest of the ordinals $j_0[j_1](\gamma)$, $j_0[j_1[j_2]](\gamma)$, ..., $j_0[j_1[j_2] \cdots [j_{p-1}]](\gamma)$. Then, by construction, we may apply (i), and we have

$$j_0[j_1][j_2] \cdots [j_p] \stackrel{\gamma'}{=} j_0[j_1[j_2] \cdots [j_p]]. \tag{4.5}$$

Let q be the least integer satisfying $\gamma' = j_0[j_1[j_2] \cdots [j_q]](\gamma)$. If we let $j = j_1[j_2] \cdots [j_q]$, we have $\gamma' = j_0[j](\gamma)$. Now, by hypothesis, $\mathrm{crit}(j) < \gamma$ holds, and, therefore, there exists an ordinal δ that satisfies $\delta < \gamma \leq j(\delta)$. From (4.5) we deduce

$$j_0(\gamma) \leq j_0(j(\delta)) = j_0[j](j_0(\delta)) = j_0[j](\delta) < j_0[j](\gamma) = \gamma'.$$

Hence $\mathrm{crit}(j_1[j_2] \cdots [j_p]) \leq \gamma$ implies $\mathrm{crit}(j_0[j_1[j_2] \cdots [j_p]]) \leq j_0(\gamma) < \gamma'$. Therefore the right-hand embedding in (4.5) has its critical ordinal below γ', and, by Lemma 4.4, so has the left-hand embedding, and the two critical ordinals are equal, which is the expected result. ∎

Approximation by left powers

For j a nontrivial elementary embedding of a limit rank into itself, both $\mathrm{Iter}(j)$ and $\widehat{\mathrm{Iter}}(j)$ consist of countably many elementary embeddings, each of which except the identity has a critical ordinal. So, we can associate with every elementary embedding j the countable family of all critical ordinals of iterates of j. Such a family is ordered by the canonical order of the ordinals, and it can be enumerated according to this ordering.

Definition 4.9. (ordinals crit_n) The ordinal $\mathrm{crit}_n(j)$ is defined to be the $(n+1)$-th element in the increasing enumeration of those ordinals that are the critical ordinal of some iteration of j.

Example 4.10. (ordinals crit_n) *The formulas* $\mathrm{crit}(j_1[j_2]) = j_1(\mathrm{crit}(j_2))$, $\mathrm{crit}(j_1 \circ j_2) = \inf(\mathrm{crit}(j_1), \mathrm{crit}(j_2))$ *and an obvious induction show that* $\mathrm{crit}(i) \geq \mathrm{crit}(j)$ *holds for every iterate* i *of* j. *Hence the ordinal* $\mathrm{crit}_0(j)$ *is the critical ordinal of* j. *The equalities* $\mathrm{crit}_1(j) = j(\mathrm{crit}(j))$ *and* $\mathrm{crit}_2(j) = j^2(\mathrm{crit}(j))$ *will be proved in Chapter XIII. Things become complicated subsequently. At the moment, we do not know (yet) that the sequence of the ordinals* $\mathrm{crit}_n(j)$ *exhaust all critical ordinals in* $\widehat{\mathrm{Iter}}(j)$: *it could happen that some nontrivial iterate* i *of* j *has its critical ordinal beyond all* $\mathrm{crit}_n(j)$*'s.*

The result we are to prove now completely describes the quotients of the structure $\mathrm{Iter}(j)$—or $\widehat{\mathrm{Iter}}(j)$—under γ-equivalence. It will be the basis of all developments of Chapter XIII.

Proposition 4.11. (quotients) *(i) Assume that* j *is a nontrivial elementary embedding of a limit rank into itself. Then* $\mathrm{crit}_n(j)$*-equivalence is a congruence on the LD-system* $\mathrm{Iter}(j)$, *and the associated quotient is isomorphic to the Laver table* $(A_n, *_n)$.

(ii) Similarly, $\mathrm{crit}_n(j)$*-equivalence is a congruence on the LD-monoid* $\widehat{\mathrm{Iter}}(j)$, *and the associated quotient is isomorphic to the Laver table* $(A_n, \circ_n, *_n)$.

The proof requires several preliminary results. We begin with a direct application of Lemma 4.8, whose statement is reminiscent from the Laver–Steel theorem, but which we shall see is rather different:

Lemma 4.12. *Assume that* $i_1, i_2, \ldots, i_{2^n}$ *are iterates of* j. *Then we have* $\mathrm{crit}(i_1[i_2]\cdots[i_p]) \geq \mathrm{crit}_n(j)$ *for some* p *with* $p \leq 2^n$.

Proof. We use induction on n. For $n = 0$, the result is the inequality $\mathrm{crit}(i_1) \geq \mathrm{crit}(j)$, which we have seen holds for every iterate i_1 of j. Otherwise, we apply the induction hypothesis twice. First, we find $q \leq 2^{n-1}$ satisfying

$$\mathrm{crit}(i_1[i_2] \cdots [i_q]) \geq \mathrm{crit}_{n-1}(j). \tag{4.6}$$

If the inequality is strict, we have $\mathrm{crit}(i_1[i_2] \cdots [i_q]) \geq \mathrm{crit}_n(j)$, and we are done. So, we can assume from now on that (4.6) is an equality. By applying the induction hypothesis again, we find $r \leq 2^{n-1}$ satisfying

$$\mathrm{crit}(i_{q+1}[i_{q+2}] \cdots [i_{q+r}]) \geq \mathrm{crit}_{n-1}(j).$$

If r is chosen to be minimal, we can apply Lemma 4.8(i) with $p = r$, $j_0 = i_1[i_2] \cdots [i_q]$, $j_1 = i_{q+1}$, ..., $j_p = i_{q+r}$, $\gamma = \mathrm{crit}_{n-1}(j)$, and $\gamma' = \mathrm{crit}_n(j)$. Indeed, with these notations, we have $\mathrm{crit}(j_1[j_2] \cdots [j_\ell]) < \gamma$ for $\ell < r$, hence

$$\mathrm{crit}(j_0[j_1[j_2] \cdots [j_\ell]]) = j_0(\mathrm{crit}(j_1[j_2] \cdots [j_\ell])) = \mathrm{crit}(j_1[j_2] \cdots [j_\ell]) < \gamma,$$

and, therefore, $j_0[j_1[j_2] \cdots [j_\ell]](\gamma) > \gamma$ holds, which gives $j_0[j_1[j_2] \cdots [j_\ell]](\gamma) \geq \gamma'$ by definition. So we have

$$j_0[j_1][j_2] \cdots [j_p] \stackrel{\gamma'}{=} j_0[j_1[j_2] \cdots [j_p]].$$

As we have $\mathrm{crit}(j_0[j_1[j_2] \cdots [j_p]]) \geq j_0(\gamma) \geq \gamma' = \mathrm{crit}_n(j)$, we deduce $\mathrm{crit}(j_0[j_1][j_2] \cdots [j_p]) \geq \mathrm{crit}_n(j)$ by Lemma 4.4, *i.e.*, $\mathrm{crit}(i_1[i_2] \cdots [i_q] \cdots [i_{q+r}]) \geq \mathrm{crit}_n(j)$, as was expected. ■

The main task is now to show that all iterates of j can be approximated by left powers of j up to $\mathrm{crit}_n(j)$-equivalence. We begin with approximating arbitrary iterates by pure iterates.

Lemma 4.13. *Assume that n is a fixed integer, and i is an iterate of j. Then there exists a pure iterate i' of j that is $\mathrm{crit}_n(j)$-equivalent to i.*

Proof. Let A be the set of those iterates of j that are $\mathrm{crit}_n(j)$-equivalent to some pure iterate of j. The set A contains j, and it is obviously closed under application. So, in order to show that A is all of $\widehat{\mathrm{Iter}}(j)$, it suffices to show that it is closed under composition, and, because $\mathrm{crit}_n(j)$-equivalence is compatible with composition, it suffices to show that, if i_1, i_2 are pure iterates of j, then some pure iterate of j is $\mathrm{crit}_n(j)$-equivalent to $i_2 \circ i_1$. To this end, we define inductively a sequence of pure iterates of j, say i_3, i_4, ... by the inductive clause $i_{p+2} = i_{p+1}[i_p]$. Then we have

$$i_3 \circ i_2 = i_2[i_1] \circ i_2 = i_2 \circ i_1,$$

and, inductively, $i_{p+1} \circ i_p = i_2 \circ i_1$ for every p. We claim that $\mathrm{crit}(i_p) \geq \mathrm{crit}_n(j)$ holds for at least one of the values $p = 2n$ or $p = 2n + 1$. If this is known, $i_2 \circ i_1$, which is $i_p \circ i_{p-1}$, is $\mathrm{crit}_n(j)$-equivalent to i_{p-1}, and we are done.

In order to prove the claim, we separate the cases $\mathrm{crit}(i_2) > \mathrm{crit}(i_1)$ and $\mathrm{crit}(i_2) \leq \mathrm{crit}(i_1)$. Assume the first inequality. Then we have

$$\mathrm{crit}(i_3) = i_2(\mathrm{crit}(i_1)) = \mathrm{crit}(i_1), \quad \text{and} \quad \mathrm{crit}(i_4) = i_3(\mathrm{crit}(i_2)) > \mathrm{crit}(i_2).$$

An immediate induction gives

$$\begin{cases} \mathrm{crit}(i_1) = \mathrm{crit}(i_3) = \mathrm{crit}(i_5) = \cdots, \\ \mathrm{crit}(i_2) < \mathrm{crit}(i_4) < \mathrm{crit}(i_6) < \cdots. \end{cases}$$

By definition, we have $\mathrm{crit}(i_1) \geq \mathrm{crit}_0(j)$, and, therefore, we have $\mathrm{crit}(i_2) \geq \mathrm{crit}_1(j)$, and, inductively, $\mathrm{crit}(i_{2n}) \geq \mathrm{crit}_n(j)$, as was claimed. Assume now $\mathrm{crit}(i_2) \leq \mathrm{crit}(i_1)$. Similar computations give

$$\begin{cases} \mathrm{crit}(i_1) < \mathrm{crit}(i_3) < \mathrm{crit}(i_5) < \cdots, \\ \mathrm{crit}(i_2) = \mathrm{crit}(i_4) = \mathrm{crit}(i_6) = \cdots, \end{cases}$$

and we find now $\mathrm{crit}(i_{2n+1}) \geq \mathrm{crit}_n(j)$. So the claim is established, and the proof is complete. ∎

Example 4.14. (approximation by pure iterates) *Consider $i = j \circ j$. We are in the case "$\mathrm{crit}(i_2) \leq \mathrm{crit}(i_1)$", and we know that the pure iterate denoted above i_{2n+1} is a $\mathrm{crit}_n(j)$-approximation of i. For instance, we find $i_3 = j_{[2]}$, $i_4 = j_{[3]}$, $i_5 = j_{[3]}[j_{[2]}] = j_{[4]}{}^{[2]}$, so $j \circ j$ and $(j_{[4]})^{[2]}$ are $\mathrm{crit}_2(j)$-equivalent. We shall see later that the critical ordinal of $(j_{[4]})^{[2]}$, i.e., $j_{[4]}(\mathrm{crit}(j_{[4]}))$, is larger than $\mathrm{crit}_2(j)$, namely it is $\mathrm{crit}_3(j)$, and the previous equivalence is actually a $\mathrm{crit}_3(j)$-equivalence.*

Proposition 4.15. (approximation) *Assume that j is a nontrivial elementary embedding of a limit rank into itself, n is a positive integer, and i is an iterate of j. Then $j_{[p]}$ is $\mathrm{crit}_n(j)$-equivalent to i for some p with $p \leq 2^n$.*

Proof. Owing to Lemma 4.13, we may assume that i is a pure iterate of j. The principle is to iteratively divide by j on the right, *i.e.*, we construct pure iterates of j, say i_0, i_1, ... such that i_0 is i, and i_p is $\mathrm{crit}_n(j)$-equivalent to $i_{p+1}[j]$ for every p. So, i is $\mathrm{crit}_n(j)$-equivalent to $i_p[j]\cdots[j]$, p times j, for every p. We stop the process when we have either $i_p = j$, in which case i is $\mathrm{crit}_n(j)$-equivalent to $j_{[p+1]}$, or $p = 2^n$: in this case, we have obtained a sequence of 2^n iterates of j and using Lemma 4.12 allows us to conclude the proof.

Let us go into details. In order to see that the construction is possible, let us assume that i_p has been obtained. If $i_p = j$ holds, we are done, as mentioned above. Otherwise, i_p has the form $i'_1[i'_2[\cdots[i'_r[j]]\cdots]]$, where i'_1, ..., i'_r are some uniquely defined pure iterates of j. Then, by (LDM_1), we have

$i_p = (i'_1 \circ \cdots \circ i'_r)[j]$, and we define i_{p+1} to be a pure iterate of j that is $\mathrm{crit}_n(j)$-equivalent to $i'_1 \circ \ldots \circ i'_r$.

Assume that the construction continues for at least 2^n steps, and let us consider the 2^n embeddings $i_{2^n}[j]$, $i_{2^n}[j][j]$, ..., $i_{2^n}[j][j] \cdots [j]$, 2^n times j. By Lemma 4.12, there must exist $p \leq 2^n$ such that the critical ordinal of $i_{2^n}[j][j] \cdots [j]$, p times j, is at least $\mathrm{crit}_n(j)$. Let i' be the latter elementary embedding. Then i is $\mathrm{crit}_n(j)$-equivalent to $i_{2^n}[j][j] \cdots [j]$, 2^n times j, which is also $i'[j][j] \cdots [j]$, $2^n - p$ times j, and, therefore, i is $\mathrm{crit}_n(j)$-equivalent to $\mathrm{id}[j][j] \cdots [j]$, $2^n - p$ times j, i.e., to $j_{[2^n-p]}$, and we are done as well. ∎

Example 4.16. (approximation by left powers) *The previous argument is effective. Starting with an arbitrary iteration i of j and a fixed level of approximation $\mathrm{crit}_n(j)$, we can find some left power of j that is $\mathrm{crit}_n[j]$-equivalent to i in a finite number of steps. However, the computation becomes quickly very intricate, and there is no uniform way to know how many steps are needed. For instance, let $i = j^{[3]}$, the simplest iterate of j that is not a left power. We write $i = (j \circ j)[j]$, and have to find an approximation of $j \circ j$. Now, $j \circ j$ is $\mathrm{crit}_3(j)$-equivalent to $j_{[3]}$, and, in this particular case, we obtain directly that $j^{[3]}$ is $\mathrm{crit}_3(j)$-equivalent to $j_{[3]}[j]$, i.e., to $j_{[4]}$. If we look for $\mathrm{crit}_4(j)$-equivalence, the computation is much more complicated. The results below will show that, if i is $\mathrm{crit}_3(j)$-equivalent to $j_{[4]}$, then it is $\mathrm{crit}_4(j)$-equivalent either to $j_{[4]}$ or to $j_{[12]}$. By determining the critical ordinal of $i[j][j][j][j]$, we could find that the final result is $j^{[3]}$ being $\mathrm{crit}_4(j)$-equivalent to $j_{[12]}$. We shall give in Chapter XIII an alternative way for proving such statements using the Laver tables.*

Proposition 4.17. (values) *Assume that j is a nontrivial elementary embedding of a limit rank into itself. Then, for every p, we have $\mathrm{crit}(j_{[p]}) = \mathrm{crit}_m(j)$, where m is the largest integer such that 2^m divides p.*

Proof. We establish inductively on $n \geq 0$ that $\mathrm{crit}(j_{[2^n]}) \geq \mathrm{crit}_n(j)$ holds, and that $\mathrm{crit}(j_{[p]}) = \mathrm{crit}(j_{[2^m]})$ holds for $p < 2^n$ with m the largest integer such that 2^m divides p. For $n = 0$, we already know $\mathrm{crit}(j) = \mathrm{crit}_0(j)$. Otherwise, let us consider the embeddings $j_{[2^n+p]}$ for $1 \leq p \leq 2^n$. By definition, we have

$$j_{[2^n+p]} = j_{[2^n]}[j] \cdots [j], \quad p \text{ times } j,$$

and, by induction hypothesis, we have $\mathrm{crit}(j_{[\ell]}) < \mathrm{crit}(j_{[2^n]})$ for $\ell < 2^n$, and $\mathrm{crit}(j_{[2^n]}) \geq \mathrm{crit}_n(j)$. By Lemma 4.8(ii) applied with $j_0 = j_{[2^n]}$ and $j_1 = \ldots = j_p = j$, we have

$$\mathrm{crit}(j_{[2^n+p]}) = j_{[2^n]}(\mathrm{crit}(j_{[p]})).$$

For $p < 2^n$, we deduce $\mathrm{crit}(j_{[2^n+p]}) = \mathrm{crit}(j_{[p]}) = \mathrm{crit}_m(j)$ where m is the largest integer such that 2^m divides p, which is also the largest integer such that 2^m divides $2^n + p$. For $p = 2^n$, we obtain

$$\mathrm{crit}(j_{[2^{n+1}]}) = j_{[2^n]}(\mathrm{crit}(j_{[2^n]}) > \mathrm{crit}(j_{[2^n]}) = \mathrm{crit}_n(j),$$

and we deduce $\mathrm{crit}(j_{[2^{n+1}]}) \geq \mathrm{crit}_{n+1}(j)$. So the induction is complete. Now, it follows from Proposition 4.15 (approximation) that the critical ordinal of any iterate of j is either equal to the critical ordinal of some left power of j, or is larger than all ordinals $\mathrm{crit}_m(j)$. Since the sequence of all ordinals $\mathrm{crit}(j_{[2^n]})$ is increasing, the only possibility is $\mathrm{crit}(j_{[2^n]}) = \mathrm{crit}_n(j)$. ∎

Lemma 4.18. *The left powers $j_{[p]}$ and $j_{[p']}$ are $\mathrm{crit}_n(j)$-equivalent if and only if $p = p' \bmod 2^n$ holds.*

Proof. We have $\mathrm{crit}(j_{[2^n]}) = \mathrm{crit}_n(j)$, so $j_{[2^n]}$ is $\mathrm{crit}_n(j)$-equivalent to the identity mapping, which, by Lemma 4.6, inductively implies that $j_{[p]}$ and $j_{[2^n+p]}$ are $\mathrm{crit}_n(j)$-equivalent for every p. Hence the condition of the lemma is sufficient. On the other hand, we prove inductively on $n \geq 0$ that $1 \leq p < p' \leq 2^n$ implies that $j_{[p]}$ and $j_{[p']}$ are not $\mathrm{crit}_n(j)$-equivalent. The result is vacuously true for $n = 0$. Otherwise, for $p' \neq 2^{n-1} + p$, the induction hypothesis implies that $j_{[p]}$ and $j_{[p']}$ are not $\mathrm{crit}_{n-1}(j)$-equivalent, and *a fortiori* they are not $\mathrm{crit}_n(j)$-equivalent. Now, assume $p' = 2^{n-1} + p$ and $j_{[p]}$ and $j_{[p']}$ are $\mathrm{crit}_n(j)$-equivalent. By applying Lemma 4.6 $2^{n-1} - p$ times, we deduce that $j_{[2^{n-1}]}$ and $j_{[2^n]}$ are $\mathrm{crit}_n(j)$-equivalent, which is impossible as we have $\mathrm{crit}(j_{[2^{n-1}]}) < \mathrm{crit}_n(j)$ and $\mathrm{crit}(j_{[2^n]}) \geq \mathrm{crit}_n(j)$. ∎

Lemma 4.19. *Assume that j is a nontrivial elementary embedding of a limit rank into itself. Then $\mathrm{crit}_n(j)$-equivalence is a congruence on the LD-monoid $\widehat{\mathrm{Iter}}(j)$, and the quotient LD-monoid has 2^n elements, namely the classes of j, $j_{[2]}, \ldots, j_{[2^n]}$, the latter being also the class of the identity mapping.*

Proof. The result is clear from Proposition 4.15 (approximation) and the previous lemma. That $j_{[2^n]}$ and the identity mapping are $\mathrm{crit}_n(j)$-equivalent follows from $\mathrm{crit}_n(j)$ being the critical ordinal of $j_{[2^n]}$. ∎

So, the quotient of $\widehat{\mathrm{Iter}}(j)$, which, by Lemma 4.13, is also the quotient of $\mathrm{Iter}(j)$, under $\mathrm{crit}_n(j)$-equivalence is a finite LD-system with 2^n elements. In addition, this LD-system consists of the first 2^n left powers of the class of j, and the class of the $2^n + 1$-th left power is the class of j. By Lemma X.1.5, such an LD-system is isomorphic to the Laver table A_n. We thus have completed the proof of Proposition 4.11(i) (quotients). The results with composition added follow immediately. Indeed, we have seen in Proposition XI.2.15 (Laver tables) that $\circ_n$ is the only associative operation on A_n that gives $(A_n, \circ_n, *_n)$ the structure of an LD-monoid.

Exercise 4.20. (equivalence *vs.* **restriction)** *Assume $j \in E_\lambda$, and $\theta < \lambda$. Prove that, if γ is the least ordinal satisfying $j(\theta) \geq \gamma$, then $j{\restriction}R_{\theta+1} = j'{\restriction}R_{\theta+1}$ implies that j and j' are γ-equivalent, and that j and j' being γ-equivalent implies $j{\restriction}R_\theta = j'{\restriction}R_\theta$.*

Exercise 4.21. (naive quotients) *(i) Assume $j \in E_\lambda$ and $\gamma_0 = \mathrm{crit}(j)$. Show that $i(\gamma_0) \geq \gamma_0$ holds for every pure iterate i of j. Prove that the partition of $\mathrm{Iter}(j)$ defined by $\mathrm{crit}(i) = \gamma_0$ vs. $\mathrm{crit}(i) > \gamma_0$ is compatible with the application operation, and that the corresponding quotient is the Laver table A_1.*

(ii) Let $\gamma_1 = j(\gamma_0)$. Show that there are four possibilities for the action of an iterate of j on $\{\gamma_0, \gamma_1\}$, namely

	j	$j_{[2]}$	$j_{[3]}$	$j_{[4]}$
γ_0	γ_1	γ_0	$>\gamma_1$	γ_0
γ_1	$>\gamma_1$	$>\gamma_1$	$>\gamma_1$	γ_1

Prove the associated equivalence relation is compatible with application. [Hint: The class of i can be described in terms of the critical ordinals of i and $i^{[2]}$]. Show that the corresponding quotient is A_2.

Exercise 4.22. (no cycle) *Assume $j_1, j_2, \ldots \in E_\lambda$. Using Lemma 4.8, prove that the supremum of the critical ordinals of j_1, $j_1[j_2]$, $j_1[j_2][j_3]$, ... is λ. Deduce that left division in E_λ has no cycle without using the Laver–Steel theorem.*

5. Notes

Most recent developments in set theory involve elementary embeddings and large cardinals of the type considered in this chapter. The fact that the existence of such objects cannot be proved does not diminish the interest of investigating them, but, on the contrary, it is the condition that makes their study significant. Indeed, Gödel's incompleteness theorems show that no axiomatic system can describe the universe of sets exhaustively. This means that, for each axiomatic system of set theory, there needs to remain open statements that are neither provable nor refutable in that system. It is well known for instance that the axiom of choice or Cantor's continuum hypothesis are unprovable in the usual axiomatic system ZF of Zermelo–Fraenkel. So the problem of comparing the possible extensions of ZF and of understanding the connection between various unprovable statements remains open. Large cardinal axioms, like the axiom asserting the existence of an inaccessible cardinal, or of a measurable cardinal, turn out to answer such questions. Very roughly speaking, there exists a hierarchy of increasingly powerful axioms of the type above, and these axioms provide a scale for evaluating the logical strength of many statements that are unprovable in the basic system ZF: in some sense, the point is to evaluate how much of infinity is needed to prove the considered property. A typical example in this direction is Lebesgue measurability of the PCA subsets of the real line, *i.e.*, of those subsets that are continuous images of complements of analytical sets (projections of Borel sets). The question is undecidable in ZF, and, more precisely, a dichotomy appears: always roughly speaking, either there exists at least one measurable cardinal, and then all PCA sets are Lebesgue measurable, or there exists no measurable cardinal, and then one can construct (in most cases) a non-measurable PCA set—there is absolutely no obvious connection between the Lebesgue measure for sets of reals and the discrete measure involved in the definition of a measurable cardinal. The point is that the above situation is in some sense generic, as shown by many recent works in set theory, *cf.* (Kanamori 94).

Thus large cardinals are central in set theory, and so are elementary embeddings, as most of the higher large cardinal axioms are stated in terms of elementary embeddings. The existence of a self-similar rank is just one example of such a type of axiom, actually a technically simple one since it involves only one rank and one elementary embedding. The study of iterates of elementary embeddings was initiated in the late 60s in (Gaifman 67) and (Kunen 70), and it became central from the beginning of the 80s when a major challenge was to find a large cardinal axiom implying the determinacy property for all sets in Lusin's projective hierarchy. Early results about the iterates of an elementary embedding of a rank into itself are mentioned in (Laver 86). In (Dehornoy 86) and (Dehornoy 88) the role of the algebraic properties of the iterates of an elementary embedding is emphasized—as opposed to their logical properties: the only fact that the iterates form an LD-monoid leads to a nontrivial consequence, namely the determinacy of all analytical sets.

Most of the results in Sections 3 and 4 are due to R. Laver. The fact that left division in the LD-system Iter(j) has no cycle, and therefore Iter(j) is a free LD-system, was proved in 1989, and it appears in (Laver 92). The original proof made no use of the Laver–Steel theorem, but it directly built on Lemma 4.8—this shows that the acyclicity result relies on the algebraic properties of the critical ordinals only, and not on the full power of the Laver–Steel theorem. The latter and the results on the quotients of Iter(j) appear in (Laver 95). The original argument involves the theory of extenders in set theory; the present exposition follows (Dougherty 96), which extracts a minimal self-contained version of these constructions. The crucial point is the well-foundedness of the ordering of ordinals.

As was mentioned in the preface, the LD-systems Iter(j) have been the first known examples of LD-systems with an acyclic left division, and they remained the only ones for several years. It had already been observed at that time both by Laver and in (Dehornoy 89*b*) that the existence of such LD-systems implies a number of consequences for free LD-systems. The strange point was that such corollaries were not really *proved* in this way: indeed, the existence of a self-similar rank is an unprovable statement, and resorting to Iter(j) in a proof amounts to introducing an unproved (and unprovable) logical hypothesis.

Even if alternative proofs have been discovered subsequently, the role of set theory remains crucial: in (Dehornoy 89*b*) the existence of an acyclic LD-system is introduced as a simple conjecture of algebra, while the construction of (Laver 92), even relying on an unprovable axiom, gives a much stronger evidence for this existence. The subsequent investigations for an alternative construction, which have led to braid exponentiation, would have never been undertaken if set theory had not given the first evidence that such a construction had to be possible.

As a final remark, let us mention without proof another result of R. Laver:

Proposition 5.1. (trace) *Assume that j is a nontrivial elementary embedding of a limit rank into itself, and i and i' are iterates of j satisfying $i(\mathrm{crit}_\ell(j)) = i'(\mathrm{crit}_\ell(j))$ for every ℓ. Then $i = i'$ holds.*

This result is a global result, and not a local one. If two iterates of j are $\mathrm{crit}_n(j)$-equivalent, then, by Lemma 4.4, they must behave similarly on $\mathrm{crit}_0(j)$, ..., $\mathrm{crit}_{n-1}(j)$ in the sense that either the values both are under $\mathrm{crit}_n(j)$ and are equal, or that both are above $\mathrm{crit}_n(j)$. However, the converse implication need not be true in general, as shown by the next example.

Example 5.2. (trace) *Let $\gamma_n = \mathrm{crit}_n(j)$. According to the above remark, if i, i' are γ_4-equivalent iterates of j, they must behave similarly on γ_0, ..., γ_3. But the converse is not true: by Lemma 4.18, j and $j_{[9]}$ are not γ_4-equivalent, yet their action on γ_0, ..., γ_3 are similar. Indeed, j maps γ_0 to γ_1, γ_1 to γ_2 and γ_2 to γ_4 (this will be proved in Chapter XIII). Now we have $\mathrm{crit}(j_{[8]}) = \gamma_3$, and therefore $j_{[9]}$ maps $j_{[8]}(\gamma_0)$, which is γ_0, to $j_{[8]}(\gamma_1)$, which is γ_1, and, similarly, it maps γ_1 to γ_2, and γ_2 to $j_{[8]}(\gamma_3)$, which is at least γ_4. This means that, in the framework of Exercise 4.21, both j and $j_{[9]}$ correspond to a column that we should denote as $(\gamma_1, \gamma_2, \infty, \infty)$.*

So the approach that we considered in Exercise 4.21 (naive quotients), which works in the case of the first two (or three) critical ordinals, cannot work in general.

XIII
More about the Laver Tables

In this chapter, we come back to the finite Laver tables introduced in Chapter X. We show how using the results of set theory established in Chapter XII gives new properties of A_n, in particular the result that the period in the first row of A_n goes to infinity with n. The unusual point here is that the result is established using the existence of a self-similar rank as an hypothesis, and we know that this is an unprovable logical statement. Thus the previous argument is not a proof in the usual sense, and another proof has to be found, typically a combinatorial argument involving only intrinsically finite objects. Unfortunately (or fortunately...), no such proof is known to date, and it is only known that such a proof, if it exists, has to be very complicated in some sense.

The chapter is organized as follows. In Section 1, we construct a dictionary between the iterations of an elementary embedding as described in Chapter XII and the finite tables A_n of Section X.1, and we use it to translate the Laver–Steel theorem into results about the systems A_n. In Section 2, we sketch partial combinatorial results toward those properties established in Section 1 using elementary embeddings. In Section 3, we come back to elementary embeddings, and we show that computing some critical ordinals requires using fast growing functions on the integers. Finally, in Section 4, we deduce from the previous results a huge lower bound for the least n such that the period in the first row of A_n goes beyond 16. The existence of such bounds suggests that no simple combinatorial proof of the results of Section 1 is likely to exist, which is confirmed by the result that no such proof can be formalized inside the logical system called Primitive Recursive Arithmetic.

1. A Dictionary

The results of Chapter XII about the iterates of an elementary embedding can be translated into results about the LD-systems A_n. We obtain in this way new properties of these systems, in particular by using the Laver-Steel theorem. The point is that the critical ordinals of the elementary embeddings are directly connected with the periods in the systems A_n. However, the existence of an elementary embedding of the required sort is an unprovable logical hypothesis. As a consequence, every property of the tables A_n proved using results on elementary embeddings remains an hypothetical statement in the framework of ordinary mathematics—but one for which a sort of technical evidence exists.

From elementary embeddings to Laver's tables

We recall from Chapter XII that, when j is an elementary embedding of a limit rank into itself, $\mathrm{Iter}(j)$ and $\widehat{\mathrm{Iter}}(j)$ respectively denote the LD-system and the LD-monoid of all pure iterates and of all iterates of j, *i.e.*, the family of those elementary embeddings obtained from j by using the application operation and, in the second case, composition. A critical ordinal $\mathrm{crit}(i)$ is associated with every iterate i of j, and the critical ordinals of the iterates of j make a countable sequence

$$\mathrm{crit}_0(j) < \mathrm{crit}_1(j) < \cdots.$$

We have introduced for every ordinal γ a congruence relation on $\widehat{\mathrm{Iter}}(j)$ called γ-equivalence. The connection with the finite tables A_n is provided by the following result, which is a restatement of Proposition XII.4.11 (quotients):

Proposition 1.1. (**isomorphism**) *Assume that j is a nontrivial elementary embedding of a limit rank into itself. Then the quotient of the LD-system $\mathrm{Iter}(j)$ under $\mathrm{crit}_n(j)$-equivalence is isomorphic to the LD-system $(A_n, *_n)$, and the quotient of the LD-monoid $\widehat{\mathrm{Iter}}(j)$ under $\mathrm{crit}_n(j)$-equivalence is isomorphic to the LD-monoid $(A_n, \circ_n, *_n)$.*

Under the previous isomorphism, the element a of A_n is the image of the class of the embedding $j_{[a]}$, and, in particular, 2^n is the image of the class of $j_{[2^n]}$, which is also the class of identity.

By construction, if S is an LD-system, and a is an element of S, there exists a well-defined evaluation for every term t in T_1 when the variable x is given the value a. We shall use here the generic notation $t(a)^S$ for the evaluation of $t(x)$ at $x = a$ in S. With these notations, every pure iterate of j can be written as $t(j)^{\mathrm{Iter}(j)}$ where $t(x)$ is some term; in the sequel, we shall simply write $t(j)$, since j is already explicit in the expression. Similarly, 1 is a generator of A_n, and the elements of A_n have the form $t(1)^{A_n}$—here we shall keep the superscript A_n, as several tables A_n may be considered simultaneously. With these notations, we have the following result:

Lemma 1.2. *For every term $t(x)$, the image of the $\mathrm{crit}_n(j)$-equivalence class of $t(j)$ in A_n under the isomorphism of Proposition 1.1 is $t(1)^{A_n}$.*

Example 1.3. (isomorphism) *The previous isomorphism can be used to obtain results about the iterations of an elementary embedding. For instance, we considered in Examples XII.4.14 and XII.4.16 the question of determining which left powers of j are $\mathrm{crit}_4(j)$-approximations of $j \circ j$ and of $j^{[3]}$. By looking at the table of the LD-monoid A_4, we obtain $1 \circ_4 1 = 11$, and $1^{[3]A_4} = 12$. We deduce that $j \circ j$ is $\mathrm{crit}_4(j)$-equivalent to $j_{[11]}$ and $j^{[3]}$ is $\mathrm{crit}_n(j)$-equivalent to $j_{[12]}$.*

The key to further results is the possibility of translating into the language of the finite tables A_n the values of the critical ordinals associated with the iterations of an elementary embedding.

Proposition 1.4. (dictionary I) *Assume that j is a nontrivial elementary embedding of a limit rank into itself. Then, for every term $t(x)$ and every integer n, the following relations are equivalent:*
(i) The critical ordinal of $t(j)$ is at least $\mathrm{crit}_n(j)$;
(ii) The value of $t(1)$ computed in A_n is 2^n.
Similarly, the following are equivalent:
(i′) The critical ordinal of $t(j)$ is $\mathrm{crit}_n(j)$;
(ii″) The value of $t(1)$ computed in A_{n+1} is 2^n.

Proof. By definition, (i) is equivalent to $t(j)$ being $\mathrm{crit}_n(j)$-equivalent to the identity mapping, hence to the class of $t(j)$ in A_n being the class of the identity mapping, which is 2^n.

For (i′) and (ii′), we observe that $\mathrm{crit}(t(j)) = \mathrm{crit}_n(j)$ holds if and only if $\mathrm{crit}(t(j)) \geq \mathrm{crit}_n(j)$ and $\mathrm{crit}(t(j)) \not\geq \mathrm{crit}_{n+1}(j)$ do. This is therefore equivalent to the conjunction of $t(1)^{A_n} = 2^n$ and $t(1)^{A_{n+1}} \neq 2^{n+1}$. Now, if $t(1)^{A_n} = 2^n$ holds, the only possible values for $t(1)^{A_{n+1}}$ are 2^n and 2^{n+1}, so $t(1)^{A_{n+1}}$ must be 2^n if it is not 2^{n+1}. Conversely, if $t(1)^{A_{n+1}} = 2^n$ holds, $t(1)^{A_n}$, which is its projection, must be 2^n. ■

Example 1.5. (dictionary) *We can check that $1^{[3]}$ computed in A_3 is 4, and $1^{[4]}$ computed in A_5 is 16. Using the dictionary, we deduce that the critical ordinal of $j^{[3]}$ is $\mathrm{crit}_2(j)$, while the critical ordinal of $j^{[4]}$ is $\mathrm{crit}_4(j)$.*

Proposition 1.6. **(dictionary** II) *Assume that j is a nontrivial elementary embedding of a limit rank into itself. Then, for every term $t(x)$, and for all nonnegative integers m, n with $m \leq n$, the following are equivalent:*

(i) The embedding $t(j)$ maps $\mathrm{crit}_m(j)$ to $\mathrm{crit}_n(j)$;

(ii) The equality $t(1)^{A_{n+1}} *_{n+1} 2^m = 2^n$ holds.

Proof. We know that $\mathrm{crit}_m(j)$ is the critical ordinal of $j_{[2^m]}$, and that the critical ordinal of $j_1[j_2]$ is $j_1(\mathrm{crit}(j_2))$. Hence we have

$$t(j)(\mathrm{crit}_m(j)) = \mathrm{crit}(t(j)[j_{[2^m]}]).$$

By Proposition 1.4 (dictionary I), the latter is $\mathrm{crit}_n(j)$ if and only if the value of $t(1) * 1_{[2^m]}$ computed in A_{n+1} is 2^n. Now, for $m \leq n+1$, $1_{[2^m]}$ evaluated in A_{n+1} is 2^m. ■

Example 1.7. **(dictionary)** *We have $4 *_4 4 = 8$ in A_4. This implies that, if j is a nontrivial elementary embedding, then $j_{[4]}$ maps $\mathrm{crit}_2(j)$ to $\mathrm{crit}_3(j)$—as can be established directly. Similarly, we have $1 *_5 4 = 16$ in A_5, which corresponds to $j(\mathrm{crit}_2(j)) = \mathrm{crit}_4(j)$.*

Proposition 1.8. **(dictionary** III) *Assume that j is a nontrivial elementary embedding of a rank into itself. Then, for a, $m \leq n$, the following are equivalent:*

(i) The embedding $j_{[a]}$ maps $\mathrm{crit}_m(j)$ to $\mathrm{crit}_n(j)$;

(ii) The period of a jumps from 2^m to 2^{m+1} between A_n and A_{n+1}.

Proof. The image of $j_{[a]}$ is a both in A_n and A_{n+1}, hence Proposition 1.6 (dictionary II) asserts that (i) is equivalent to $a *_{n+1} 2^m = 2^n$. If this holds, the period p of a in A_{n+1} is 2^{m+1}: indeed, $a *_{n+1} 2^m < 2^{n+1}$ implies $p > 2^m$, while $2 \times 2^n = 2^{n+1}$ implies $p \leq 2 \times 2^m$. Therefore, the period of 1 in A_n must be 2^m. Conversely, assume (ii). Then we have $a *_n 2^m = 2^n$ and $a *_{n+1} 2^m \neq 2^{m+1}$, so the only possibility is $a *_{n+1} 2^m = 2^m$. ■

Example 1.9. **(dictionary)** *We read from the tables that the period of 1 jumps from 4 to 8 between A_4 and A_5: we deduce that, if j is a nontrivial elementary embedding, then j maps $\mathrm{crit}_2(j)$ to $\mathrm{crit}_4(j)$. Similarly, the period of 3 jumps from 8 to 16 between A_5 and A_6: we deduce that $j_{[3]}$ maps $\mathrm{crit}_3(j)$ to $\mathrm{crit}_5(j)$.*

Periods in A_n

In the previous examples, we have used our dictionary between the iterates of an elementary embedding and the tables A_n to deduce information about elementary embeddings from explicit values in A_n. We can also use the correspondence in the other direction, and deduce results about the tables A_n from properties of the elementary embeddings.

Proposition 1.10. (inequality) *Assume that there exists a self-similar rank. Then, for every n, the period of 2 in A_n is at least equal to the period of 1.*

Proof. Assume that the period of 1 in A_n is 2^m. Let n' be the largest integer such that the period of 1 in $A_{n'}$ is 2^{m-1}. By construction, the period of 1 jumps from 2^{m-1} to 2^m between $A_{n'}$ and $A_{n'+1}$. Assume that j is a nontrivial elementary embedding of a rank into itself. By Proposition 1.8 (dictionary III), j maps $\mathrm{crit}_m(j)$ to $\mathrm{crit}_{n'}(j)$. By Proposition XII.3.7 (inequality), $j[j]$ maps $\mathrm{crit}_m(j)$ to some ordinal of the form $\mathrm{crit}_{n''}(j)$ with $n'' \leq n'$. This implies that the period of 2 jumps from 2^{m-1} to 2^m between $A_{n''}$ and $A_{n''+1}$. By construction, we have $n'' \leq n' < n$, hence the period of 2 in A_n is at least 2^m. ∎

The Laver–Steel theorem (Proposition XII.3.8) is probably the deepest well-foundedness result about elementary embeddings known to date. It implies the following statements.

Proposition 1.11. (cofinal) *Assume that j is a nontrivial elementary embedding of $(R_\lambda, \in)$ into itself.*

(i) The ordinals $\mathrm{crit}_n(j)$ are cofinal in λ, i.e., there exists no θ with $\theta < \lambda$ such that $\mathrm{crit}_n(j) < \theta$ holds for every n.

(ii) For every iterate i of j, we have $\mathrm{crit}(i) = \mathrm{crit}_m(j)$ for some integer m, and, therefore, i is not $\mathrm{crit}_m(j)$-equivalent to the identity mapping.

Proof. (i) By definition, every entry in the sequence j, $j_{[2]}$, $j_{[3]}$, ... is a left divisor of the next one, hence the Laver-Steel theorem implies that the critical ordinals of j, $j_{[2]}$, ... are cofinal in λ. By Proposition XII.4.17 (values), these critical ordinals are exactly the ordinals $\mathrm{crit}_n(j)$.

(ii) Proposition XII.4.15 (approximation) implies that either $\mathrm{crit}(i) > \mathrm{crit}_m(j)$ holds for every m, or there exists m satisfying $\mathrm{crit}(i) = \mathrm{crit}_m(j)$. By (i), the first case is impossible. ∎

Observe that the point in the previous argument is really the Laver–Steel theorem, because Proposition XII.4.15 (approximation) or Lemma XII.4.13 alone do not discard the possibility for the critical ordinal of some iterate i to lie above all $\mathrm{crit}_m(j)$'s.

If follows from the previous result that, for every m, the image under j of the critical ordinal $\mathrm{crit}_m(j)$ is again an ordinal of the form $\mathrm{crit}_n(j)$. Indeed, $\mathrm{crit}_m(j)$ is the critical ordinal of $j_{[2^m]}$, and, therefore, $j(\mathrm{crit}_m(j))$ is the critical ordinal of $j[j_{[2^m]}]$, hence the critical ordinal of an iterate of j and, therefore, an ordinal of the form $\mathrm{crit}_n(j)$ for some finite n. We deduce in this way an action of j, and, similarly, of each iterate of j, on $\mathbf{N}$. This enables us to introduce a sort of miniaturization of j which, instead of dealing with (huge) ordinals, involves natural numbers only.

Definition 1.12. **(micro-j)** Assume that j is a nontrivial elementary embedding of a rank into itself. We define μj ("micro-j") : $\mathbf{N} \to \mathbf{N}$ by

$$j : \mathrm{crit}_m(j) \mapsto \mathrm{crit}_{\mu j(m)}. \tag{1.1}$$

Then μj^* ("iterated micro-j") : $\mathbf{N} \to \mathbf{N}$ is defined by

$$\mu j^*(k) = \mu j^k(0) = \mu j(\mu j(\dots(0)\dots)), \quad k \text{ times } \mu j. \tag{1.2}$$

Example 1.13. **(micro-j)** *We have seen above in Example* 1.7 *(dictionary II) that j maps $\mathrm{crit}_2(j)$ to $\mathrm{crit}_4(j)$: by definition, this means that μj maps* 2 *to* 4.

Lemma 1.14. *The mappings μj and μj^* are increasing injections of* $\mathbf{N}$ *into itself. They do not depend on j.*

Proof. The function μj being increasing follows from its definition and from elementary embeddings being injective and preserving the order of ordinals. Its not depending on j follows from Proposition 1.6 (dictionary II): $\mu j(m) = n$ is equivalent to $1 *_{n+1} 2^m = 2^n$ holding in the finite table A_{n+1}, a condition that does not involve j. ■

Now, the existence of the function μj and, therefore, of its iterate μj^*, translates into the following asymptotic result about the periods in the tables A_n. We recall that $o_n(a)$ denotes the integer such that the period of a in A_n is $2^{o_n(a)}$.

Proposition 1.15. **(limit)** *Assume that there exists a self-similar rank. Then, for every a, the period of a in A_n tends to infinity with n. More precisely, if j is any nontrivial elementary embedding of a rank into itself, we have*

$$o_n(a) \leq \mu j^*(r) \qquad \text{if and only if} \qquad n \leq \mu j^*(r+1) \tag{1.3}$$

for $r \geq a$. In particular, (1.3) *always holds for $a = 1$.*

Proof. Let us first consider the case $a = 1$. By construction, j maps $\mathrm{crit}_{\mu j^*(r)}(j)$ to $\mathrm{crit}_{\mu j^*(r+1)}(j)$ for every r. By Proposition 1.8 (dictionary III), this implies that the period of 1 doubles from $2^{\mu j^*(r)}$ to $2^{\mu j^*(r)+1}$ between $A_{\mu j^*(r+1)}$ and $A_{\mu j^*(r+1)+1}$. So we have

$$o_{\mu j^*(r+1)}(1) = \mu j^*(r), \quad \text{and} \quad o_{\mu j^*(r+1)+1}(1) = \mu j^*(r) + 1,$$

which gives (1.3). Assume now $a \geq 2$. By Lemma V.2.9, we have $(j_{[a]})^{[r]} = j^{[r+1]}$ for $r \geq a$, so the critical ordinal of $(j_{[a]})^{[r]}$ is $\mathrm{crit}_{\mu j^*(r)}(j)$. Hence, for $r \geq a$, the embedding $j_{[a]}$ maps $\mathrm{crit}_{\mu j^*(r)}(j)$ to $\mathrm{crit}_{\mu j^*(r+1)}(j)$, and the argument is as in the case $a = 1$. ■

Exercise 1.16. (critical map) *Assume that S is an LD-system and ν is a mapping of S to $\mathbf{N}$ satisfying*

$\nu(a) = \nu(b) \Rightarrow \nu(ca) = \nu(cb)$ *for every* c,

$\nu(a) \leq \nu(b) \Rightarrow \nu(ab) > \nu(b)$,

$\nu(a) > \nu(b) \Rightarrow \nu(ab) = \nu(b)$.

Prove that left division in S has no cycle of length 2. [Hint: Follow the proof of Exercise XII.3.12.]

---

2. Computation of Rows

The logical status of Propositions 1.10 (inequality) and 1.15 (limit) is unusual: these statements involve only finite, effective objects, namely the LD-systems A_n, while the hypothesis under which they are proved is a logical statement involving huge objects, which we recall cannot be proved to exist in the ordinary mathematical framework. The LD-systems A_n have nothing to do with such huge objects of set theory, but, as long as the latter are used in the proof, the results do depend on their existence.

We are led to the question of eliminating the logical assumption. The aim of this section is to show how some periods can be computed in the tables A_n. The ultimate result would be a general formula for $o_n(a)$, or, at least, a direct proof that $o_n(1)$ tends to infinity with n. The results we shall mention here are still in progress, but they illustrate the existing techniques.

First, the inductive construction of the systems A_n's as described in Section X.1 gives some periods directly. We recall that $(b)_N$ denotes the unique integer equal to b *modulo* N in the interval $\{1, \ldots, N\}$.

Proposition 2.1. (**rows** I) *For $k \le n$, the row of $2^n - 2^k$ in A_n is given by*

$$(2^n - 2^k) *_n b = 2^n - 2^k + (b)_{2^k},$$

and, in particular, we have $o_n(2^n - 2^k) = k$.

Proof. For $2^n - 2^k + 1 \le a \le 2^n - 1$, the period of a in A_n is at most $2^k - 1$, so it divides 2^{k-1}, and therefore $2^n - 2^k$. So Lemma X.1.9 applies. ■

Example 2.2. (**table** A_6) *Let us consider the table of A_6 (64 elements). The previous result describes the following rows:*

	1	2	3	4	5	6	7	8	9	...	16	17	...	32	33	...
⋮																
32	33	34	36	37	38	39	40	41	42	...	48	49	...	64	33	...
48	49	50	51	52	53	54	55	56	57	...	64	49	...			
56	57	58	59	60	61	62	63	64	57	...						
60	61	62	63	64	61	...										
62	63	64	63	...												
63	64	...														

With a little more work, we can compute the rows of the elements $2^{n-1} - 2^k$ similarly. By Lemma X.1.9, we know that these rows are connected with those of the elements $2^n - 2^k$, *i.e.*, with those of the elements $2^{n-1} - 2^k$ in A_{n-1}. The problem is to compute the threshold, *i.e.*, to decide whether the period doubles. Actually, it always does in the current case, *i.e.*, the value of $(2^{n-1} - 2^k) *_n 2^k$, which can be either 2^{n-1} or 2^n, is 2^{n-1}.

Lemma 2.3. *For $0 \le k \le n - 1$, and $0 \le \ell \le k$, we have*

$$(2^{n-1} - 2^k) *_n 2^\ell = 2^{n-1} - 2^k + 2^\ell.$$

Proof. We use induction on k and ℓ. The formula is obvious for $k = 0$. Otherwise, we can write

$$2^\ell = 2^{\ell-1} (*_n 1)_{[2^{\ell-1}]},$$

where $a (*_n b)_{[m]}$ denotes the mixed left power $(\cdots((a *_n b) *_n b)\cdots) *_n b$, m times b. Applying left self-distributivity, we deduce

$$(2^{n-1} - 2^k) *_n 2^\ell = ((2^{n-1} - 2^k) *_n 2^{\ell-1}) (*_n (2^{n-1} - 2^k) *_n 1)_{[2^{\ell-1}]},$$

which, by induction hypothesis, is

$$2^{n-1} - 2^k + 2^{\ell-1} (*_n 2^{n-1} - 2^k + 1)_{[2^{\ell-1}]}.$$

So, the desired equality follows from

$$2^{n-1} - 2^k + 2^{\ell-1} (*_n (2^{n-1} - 2^k + 1))_{[m]} = 2^{n-1} - 2^k + 2^{\ell-1} + m, \qquad (2.1)$$

which we shall now show inductively for $0 \le m \le 2^{\ell-1}$. By definition, (2.1) is true for $m = 0$. Assume $m > 0$. Denoting the left member of (2.1) by c_m, we obtain by induction hypothesis

$$\begin{aligned} c_m &= c_{m-1} *_n (2^{n-1} - 2^k + 1) \\ &= (2^{n-1} - 2^k + 2^{\ell-1} + m - 1) *_n (2^{n-1} - 2^k + 1). \end{aligned}$$

Now, the point is that the period of $2^{n-1} - 2^k + 2^{\ell-1} + m - 1$ in A_n is at most 2^k, since the period of $2^n - 2^k + 2^{\ell-1} + m - 1$ is at most 2^{k-1}, as we have seen in Proposition 2.1 (rows I). Since 2^k divides $2^{n-1} - 2^k$, we deduce, as in the proof of Lemma X.1.9, the equality

$$\begin{aligned} &(2^{n-1} - 2^k + 2^{\ell-1} + m - 1) *_n (2^{n-1} - 2^k + 1) \\ &\qquad = (2^{n-1} - 2^k + 2^{\ell-1} + m - 1) *_n 1 = 2^{n-1} - 2^k + 2^{\ell-1} + m, \end{aligned}$$

which completes the induction. ∎

Proposition 2.4. (rows II) *For $k < n-1$, the row of $2^{n-1} - 2^k$ in A_n is given by*

$$(2^{n-1} - 2^k) *_n b = \begin{cases} 2^{n-1} - 2^k + (b)_{2^k} & \text{for } (b)_{2^{k+1}} \le 2^k, \\ 2^n - 2^k + (b)_{2^k} & \text{for } (b)_{2^{k+1}} > 2^k, \end{cases}$$

and, in particular, we have $o_n(2^{n-1} - 2^k) = k + 1$.

Proof. By Proposition 2.1 (rows I), the period of $2^{n-1} - 2^k$ in A_{n-1} is 2^k, so its period in A_n is either 2^k or 2^{k+1}. In order to discard 2^k, it suffices to prove $(2^{n-1} - 2^k) *_n 2^k \neq 2^n$: this follows from the previous lemma. ∎

Example 2.5. (**table** A_6) *Proposition 2.4 gives the following new rows in the table of A_6:*

	1	2	3	4	5	6	7	8	9	10	11	...	16	17	...	32	33	...
⋮																		
16	17	18	19	20	21	22	23	24	25	26	27	...	32	33	...	64	17	...
24	25	26	27	28	29	30	31	32	57	58	59	...	64	25	...			
28	29	30	31	32	61	62	63	64	29	...								
30	31	32	63	64	31	...												
31	32	64	32	...														

Going further requires more powerful methods. The principle of the subsequent computations is the following observation.

Lemma 2.6. *Assume that there exists an injective homomorphism f of A_m into A_n that maps 1 to $a+1$. Then the row of a in A_n is determined by*

$$a *_n b = f((b)_m).$$

In particular, we have $o_n(a) = m$.

Proof. By Proposition X.1.15 (basics), the homomorphism f maps b to $a *_n b$ for every b, and we know that it maps 2^m to 2^n. So we have $a *_n 2^m = 2^n$, which implies that the period of a in A_n is 2^m at most. Since f is injective, we have $a *_n b \neq 2^n$ for $b < 2^m$, hence

$$a *_n 1 < a *_n 2 < \cdots < a *_n 2^m = 2^n,$$

which shows that the period of a in A_n is 2^m exactly, and, moreover, gives a complete description of the row of a in the table of A_n. Observe that the hypotheses imply that f viewed as a mapping of $\{1, \ldots, 2^m\}$ into $\{1, \ldots, 2^n\}$ is increasing. ■

Example 2.7. (homomorphism) *Consider the application*

$$b \mapsto 2^{n-1} + b$$

of A_{n-1} into A_n. By Lemma X.1.8, this mapping is a homomorphism of A_{n-1} into A_n, and it is injective. We (re)obtain in this way the description of the row of 2^{n-1} in A_n, namely

	1	2	$\ldots$	2^{n-1}	$2^{n-1}+1$	$\ldots$
2^{n-1}	$2^{n-1}+1$	$2^{n-1}+2$	$\ldots$	2^n	$2^{n-1}+1$	$\ldots$

Finding similar homomorphisms that jump higher allows one to describe further rows in the table A_n. Here we state without proof some results in this direction. After the above homomorphism, which has the form $b \mapsto b+C$ and can therefore be called additive, it is natural to look for multiplicative homomorphisms of the form $b \mapsto Cb$. This amounts to looking for rows with a linear growth rate bigger than 1, like the rows of 3 in A_3 and A_4.

Proposition 2.8. (first level) *For every integer d, and for $0 \leq m \leq 2^d + 1$, the mapping*

$$b \mapsto 2^{2^d} b$$

defines an injective homomorphism of A_m into A_{m+2^d}.

Corollary 2.9. *For $2^d \le n \le 2^{d+1}+1$, the row of $2^{2^d}-1$ in A_n is given by*

$$(2^{2^d}-1) *_n b = 2^{2^d} b.$$

Proof. The injective mapping $b \mapsto 2^{2^d} b$ is a homomorphism of A_{n-2^d} into A_n. It maps 1 to 2^d, *i.e.*, to $(2^d-1) *_n 1$. We then apply Lemma 2.6. ∎

Example 2.10. (**first level**) *Take $d=2$: multiplication by 16 gives a homomorphism of A_{n-4} into A_n for $4 \le n \le 9$. This applies for instance to $n=6$, and gives the description of the row of $16-1=15$ in A_6:*

	1	2	3	4	5	...
15	16	32	48	64	16	...

The next step is to look for exponential homomorphisms of the form $b \mapsto C^b$ or similar.

Proposition 2.11. (**second level**) *For every integer d, and for $0 \le m \le 2^{2^{d+1}}$, the mapping f_d defined by*

$$f_d : 2^i \mapsto 2^{(i+1)2^d} - 2^{i2^d}$$

and $f_d(\sum b_i 2^i) = \sum b_i f_d(2^i)$ defines an injective homomorphism of A_m into A_{m2^d}.

Corollary 2.12. *For $0 \le n \le 2^{2^{d+1}+d}$ such that 2^d divides n, the row of $2^{2^d}-2$ in A_n is given by*

$$(2^{2^d}-2) *_n b = f_d(b).$$

Proof. The homomorphism f_d maps 1 to $2^{2^d}-1$, which is $(2^{2^d}-2) *_n 1$. ∎

Example 2.13. (**second level**) *For $d=0$, we obtain the identity mapping. For $d=1$, f_1 is a homomorphism of A_m into A_{2m} for $m \le 4$. Since f_1 maps 1 to 3, this enables us to compute the row of 2. For instance, we determine the row of 2 in A_6 by using f_1 from A_3, thus getting*

	1	2	3	4	5	6	7	8	9	...
2	3	12	15	48	51	60	63	64	3	...

Let us consider now the case $d=2$: f_2 is a homomorphism of A_m into A_{4m} for $m \le 16$. Since f_2 maps 1 to 15, which gives the description of the row *of 14. The result does not apply to A_6 directly, since 6 is not a multiple of 4, but we can use f_2 to compute the row of 14 in A_8, namely*

	1	2	3	4	5	...
14	15	240	255	256	15	...

and deduce the row of 14 in A_6 using projection modulo 64:

	1	2	3	4	5	...
14	15	48	63	64	15	...

Finally, let us mention a third family of superexponential homomorphisms.

Proposition 2.14. (third level) *Define inductively the integer e_n by $e_0 = 2$ and $e_{n+1} = e_n^{e_n}$. Then, for every integer d, and for $0 \le m \le e_d^2$, the mapping g_d defined by*

$$g_d : 2^i \mapsto 2^{d^{i+1}} - 2^{d^i}$$

and $g_d(\sum b_i 2^i) = \sum b_i g_d(2^i)$ is an injective homomorphism of A_m into $A_{e_d^m}$.

Since g_d maps 1 to $2^{2^d} - 2$, the previous proposition gives a description of the row of $2^{2^d} - 3$ in certain tables A_n, for instance in $A_{e_{d+1}}$.

Example 2.15. (third level) *Consider $d = 2$, the first nontrivial case. Then g_2 is a homomorphism of A_m to A_{2^m} for $m \le 4$, and it maps 1 to 2, thus determining the row of 1. In this way, we obtain that the row of 1 in A_8 is*

	1	2	3	4	5	6	7	8	9	...
1	2	12	14	240	242	252	254	256	2	...

and we deduce for instance that the row of 1 in A_6 is

	1	2	3	4	5	6	7	8	9	...
1	2	12	14	48	50	60	62	64	2	...

We shall not go further here.

Exercise 2.16. (translations) *(i) Assume $a < 2^{n-1}$. Prove that, for every b, $(a + 2^{n-1}) *_n b$ is the unique integer equal to $a *_n b$ modulo 2^{n-1} that lies between $2^{n-1} + 1$ and 2^n.*

(ii) Deduce that every row in the first half of A_n determines a row in the second half, and, more generally, that, for $2^n - 2^k < a < 2^n$ and $a' = a \pmod{2^k}$, the row of a' determines the row of a. Apply this the rows of 33, 34, 46, and 47 in A_6, then to those of 49 and 50, of 57 and 58, and finally, to that of 61.

---

3. Fast Growing Functions

The aim of the last two sections of this chapter is to show that a possible combinatorial proof of the results of Section 1 has to be very complicated. In this preparatory section, we prove that some fast growing function has to appear when the critical ordinals of the iterates of an elementary embedding are computed. The statement we prove here is as follows:

Proposition 3.1. (**not primitive recursive**) *Assume that j is a nontrivial elementary embedding of a limit rank into itself. Then the function μj^* grows faster than any primitive recursive function.*

The class of primitive recursive functions contains all usual arithmetic functions, including iterated exponentials of any kind, and, therefore, the above result means that the function μj^* has an extremely large growth rate.

First values of μj^*

We assume till the end of the section that j is a fixed nontrivial elementary embedding of a limit rank into itself. In order to get easier notations, the increasing sequence of critical ordinals associated with the iterates of j will be denoted $\gamma_0 < \gamma_1 < \cdots$, *i.e.*, γ_n stands for $\mathrm{crit}_n(j)$. With these notations, the function μj is determined by

$$\gamma_{\mu j(m)} = j(\gamma_m), \tag{3.1}$$

and its iterate μj^* is determined by

$$\gamma_{\mu j^*(r)} = j^r(\gamma_0). \tag{3.2}$$

(where j^r refers to composition). We are going to establish lower bounds for the values of the function μj^*—which, we recall, does not depend on j.

The first values of the function μj^* can be computed exactly. The principle is to determine sequences of iterated values for the left powers of j, and, in particular, to determine the *critical sequence* of these embeddings, defined as the sequence made of the critical ordinal and its successive images. We use the notation

$$i : \Vdash\!\!\to \theta_0 \mapsto \theta_1 \mapsto \cdots$$

to mean that θ_0 is the critical ordinal of i, that θ_1 is the image of θ_0 under i—hence θ_1 is the critical ordinal of $i^{[2]}$—*etc.* For instance, by definition of μj, we have

$$j : \gamma_0 \mapsto \gamma_{\mu j^*(1)} \mapsto \gamma_{\mu j^*(2)} \mapsto \gamma_{\mu j^*(3)} \mapsto \cdots . \tag{3.3}$$

Now, for each sequence of the form

$$i : \Vdash\!\!\to \theta_0 \mapsto \theta_1 \mapsto \theta_2 \mapsto \cdots ,$$

we deduce for each elementary embedding j_0 a new sequence

$$j_0[i] : \Vdash\!\!\to j_0(\theta_0) \mapsto j_0(\theta_1) \mapsto j_0(\theta_2) \mapsto \cdots .$$

Applying the previous principle to the sequence (3.3) with $j_0 = j$, and using $\mu j^*(1) = 1$, we obtain the sequence

$$j_{[2]} : \Vdash\!\!\to \gamma_1 \mapsto \gamma_{\mu j^*(2)} \mapsto \gamma_{\mu j^*(3)} \mapsto \cdots . \tag{3.4}$$

Applying the same principle to (3.3) with $j_0 = j_{[2]}$, we obtain

$$j_{[3]} : \Vdash\!\!\to \gamma_0 \mapsto \gamma_{\mu j^*(2)} \mapsto \gamma_{\mu j^*(3)} \mapsto \cdots . \tag{3.5}$$

We know that γ_2 is the critical ordinal of $j_{[4]}$, hence $\gamma_2 = j_{[3]}(\gamma_0)$ holds. Now (3.5) shows that the latter ordinal is $\gamma_{\mu j^*(2)}$, *i.e.*, we have proved the equality $\gamma_{\mu j^*(2)} = \gamma_2$, and, therefore we have $\mu j(1) = 2$ and $\mu j^*(2) = 2$.

Let us continue the previous computation, and try to find the value of $\mu j^*(3)$, *i.e.*, of $\mu j(2)$. From (3.5) and the fact that every elementary embedding induces an increasing function on ordinals, we deduce that $j_{[3]}(\gamma_1)$, which, by construction, is the critical ordinal of $j_{[3]}[j_{[2]}]$, lies strictly between γ_2 and $\gamma_{\mu j^*(3)}$. Hence $\mu j^*(3) \geq 4$ holds. By successively applying $j_{[4]}, \ldots, j_{[7]}$ to (3.3), we obtain the following sequences

$$\begin{array}{llllll}
j_{[4]} : & \Vdash\!\!\to \gamma_2 & \mapsto j_{[3]}(\gamma_1) & \mapsto \gamma_{\mu j^*(3)} & \mapsto \cdots , \\
j_{[5]} : & \Vdash\!\!\to \gamma_0 & \mapsto \gamma_1 & \mapsto j_{[3]}(\gamma_1) & \mapsto \cdots , \\
j_{[6]} : & \Vdash\!\!\to \gamma_1 & \mapsto j_{[3]}(\gamma_1) & \mapsto j_{[5]}(\gamma_2) & \mapsto \cdots , \\
j_{[7]} : & \Vdash\!\!\to \gamma_0 & \mapsto j_{[3]}(\gamma_1) & \mapsto j_{[7]}(\gamma_2) & \mapsto \cdots .
\end{array}$$

As above, we deduce $\gamma_3 = \mathrm{crit}(j_{[8]}) = j_{[7]}(\gamma_0) = j_{[3]}(\gamma_1)$.

At this point, the smallest possible value for $\mu j^*(3)$ is 4, and, in order to apply the previous scheme, we have to determine the beginnings of the critical sequences of $j_{[n]}$ for $n \leq 15$. However, applying the method is not sufficient for determining enough values. In the current case, the following specific result is needed.

Lemma 3.2. *Assume* $\mathrm{crit}(i) \geq \gamma_2$. *Then we have* $i(\gamma_3) = i[j][j][j](\gamma_1)$.

Proof. By hypothesis, i is γ_2-equivalent to identity, hence $i[j]$ is γ_2-equivalent to j and $i[j][j]$ is γ_2-equivalent to $j_{[2]}$. So $j(\gamma_1) \geq \gamma_2$ implies $i[j](\gamma_1) \geq \gamma_2$, and, similarly, $j_{[2]}(\gamma_1) \geq \gamma_2$ implies $i[j][j](\gamma_1) \geq \gamma_2$. Now we have

$$\begin{aligned} i[j](\gamma_2) = i[j](j(\gamma_1)) &= i[j][j](i[j](\gamma_1)) \\ &\geq i[j][j](\gamma_2) = i[j][j](j(\gamma_1)) = i[j][j][j](i[j][j](\gamma_1)) \\ &\geq i[j][j][j](\gamma_2) > i[j][j][j](\gamma_1) \,. \end{aligned}$$

The γ_2-equivalence of $i[j]$ and j gives

$$i[j][j][i[j]] \overset{i[j][j](\gamma_2)}{=} i[j][j][j] \qquad \text{and} \qquad i[j][i[j]] \overset{i[j](\gamma_2)}{=} i[j][j].$$

So we have

$$i[j_{[3]}] = (i[j])_{[3]} \overset{i[j](\gamma_2)}{=} i[j][j][i[j]] \overset{i[j][j](\gamma_2)}{=} i[j][j][j].$$

This implies that the embeddings $i[j_{[3]}]$ and $i[j][j][j]$ are $i[j][j][j](\gamma_2)$-equivalent, which completes the proof, since we have $\gamma_1 = i(\gamma_1)$, and, therefore, $i[j_{[3]}](\gamma_1) = i(j_{[3]}(\gamma_1)) = i(\gamma_3)$. ∎

The previous lemma shows that $j_{[7]}$ maps γ_1 to $\gamma_{\mu j^*(3)}$: indeed $j_{[4]}$ has critical ordinal γ_2, so $j_{[4]}[j][j][j]$, *i.e.*, $j_{[7]}$, maps γ_1 to $j_{[4]}(\gamma_3)$, which we have seen is $\gamma_{\mu j^*(3)}$. Similarly, we have $j_{[11]}(\gamma_1) = j_{[8]}(\gamma_3)$, and we construct inductively the sequences

$$\begin{aligned}
j_{[7]} &: \Vdash\!\!\to \gamma_0 \mapsto \gamma_3, \qquad \gamma_1 \mapsto \gamma_{\mu j^*(3)}, \\
j_{[8]} &: \Vdash\!\!\to \gamma_3 \mapsto \gamma_{\mu j^*(3)}, \\
j_{[9]} &: \Vdash\!\!\to \gamma_0 \mapsto \gamma_1 \mapsto \gamma_2, \\
j_{[10]} &: \Vdash\!\!\to \gamma_1 \mapsto \gamma_2, \\
j_{[11]} &: \Vdash\!\!\to \gamma_0 \mapsto \gamma_2, \qquad \gamma_1 \mapsto \gamma_{\mu j^*(3)}, \\
j_{[12]} &: \Vdash\!\!\to \gamma_2 \mapsto \gamma_{\mu j^*(3)}, \\
j_{[13]} &: \Vdash\!\!\to \gamma_0 \mapsto \gamma_1 \mapsto \gamma_{\mu j^*(3)}, \\
j_{[14]} &: \Vdash\!\!\to \gamma_1 \mapsto \gamma_{\mu j^*(3)}, \\
j_{[15]} &: \Vdash\!\!\to \gamma_0 \mapsto \gamma_{\mu j^*(3)}.
\end{aligned}$$

We deduce $\gamma_4 = \mathrm{crit}(j_{[16]}) = j_{[15]}(\gamma_0) = \gamma_{\mu j^*(3)}$. This proves $\mu j^*(3) = 4$, *i.e.*, that μj maps γ_2 to γ_4: γ_3 is the only critical ordinal between γ_2 and its image under j.

At this point, the computations can be summarized as follows:

Proposition 3.3. (**first values**) *The first values of the function* μj *are:* $\mu j(0) = 1$, $\mu j(1) = 2$, $\mu j(2) = 4$.

Equivalently, we have $\mu j^*(1) = 1$, $\mu j^*(2) = 2$, $\mu j^*(3) = 4$, which means that the critical ordinals of the right powers j, $j^{[2]}$, and $j^{[3]}$ are γ_1, γ_2, and γ_4 respectively. Using the dictionary of Section 1, we can also check the previous results using the explicit values in the tables A_n, *i.e.*, verify that the period of 1 jumps from 1 to 2 between A_1 and A_2, that it jumps from 2 to 4 between A_2 and A_3, and that it jumps from 4 to 8 between A_4 and A_5.

Proliferation of critical ordinals

We turn now to the proof of Proposition 3.1 (not primitive recursive). The above exact computations cannot be continued much further, as the function μj^* grows very rapidly after the first values. The origin of the phenomenon is the following simple observation.

Lemma 3.4. *Assume that some iterate i of j satisfies $i : \gamma_p \mapsto \gamma_q \mapsto \gamma_r$. Then we have $r - q \geq q - p$.*

Proof. As i is increasing on ordinals, $\gamma_p < \alpha < \alpha' < \gamma_q$ implies $\gamma_q < i(\alpha) < i(\alpha') < \gamma_r$. Moreover, if α is the critical ordinal of i_1, $i(\alpha)$ is that of $i[i_1]$, and, if i_1 is an iterate of j, so is $i[i_1]$. Hence the number of critical ordinals of iterates of j between γ_q and γ_r is at least the number of critical ordinals of iterates of j between γ_p and γ_q. The former is the number of γ_n's between γ_q and γ_r, *i.e.*, $r - q - 1$, while the latter is the number of γ_n's between γ_p and γ_q, *i.e.*, $q - p - 1$. ■

Definition 3.5. (realizable, base) We say that the sequence of ordinals $(\alpha_0, \ldots, \alpha_p)$ is *realizable* (with respect to j) if there exists an iterate i of j that satisfies $i : \Vdash\!\!\rightarrow \alpha_0 \mapsto \cdots \mapsto \alpha_p$. We say that the sequence $(\alpha_0, \ldots, \alpha_p)$ is a *base* for the sequence $\vec{\theta} = (\theta_0, \ldots, \theta_n)$ if, for each $m < n$, the sequence $(\alpha_0, \ldots, \alpha_p, \theta_m, \theta_{m+1})$ is realizable.

Observe that the existence of a base for $(\theta_0, \ldots, \theta_n)$ implies that this sequence is increasing, and that, if $(a_0, \ldots, a_p)$ is a base for a sequence $\vec{\theta}$, every final subsequence of the form $(a_m, \ldots, a_p)$ is still a base for $\vec{\theta}$: if i admits the critical sequence $\Vdash\!\!\rightarrow \alpha_0 \mapsto \cdots \mapsto \theta_m \mapsto \theta_{m+1}$, then $i^{[2]}$ admits the critical sequence $\Vdash\!\!\rightarrow \alpha_1 \mapsto \cdots \mapsto \theta_m \mapsto \theta_{m+1}$.

Lemma 3.6. *Assume that the sequence $(\theta_0, \theta_1, \ldots)$ admits a base. Then $\theta_n \geq \gamma_{2^n}$ holds for every n.*

Proof. Assume that (γ_p) is a base for $(\theta_0, \theta_1, \ldots)$. Define f by $\theta_n = \gamma_{f(n)}$. Lemma 3.4 gives $f(n+1) - f(n) \geq f(n) - p$ for every n. As $f(0) > p$ holds by definition, we deduce $f(n) \geq 2^n + p$ inductively. ■

Example 3.7. **(base)** *The embedding $j_{[2]}$ leaves γ_0 fixed, and it maps $\gamma_{\mu j^*(r)}$ to $\gamma_{\mu j^*(r+1)}$ for $r \geq 1$. So its $(r-1)$-th power with respect to composition satisfies*

$$(j_{[2]})^{r-1} : \Vdash\!\!\to \gamma_1 \mapsto \gamma_{\mu j^*(r)}, \quad \gamma_2 \mapsto \gamma_{\mu j^*(r+1)}.$$

Applying these values to the critical sequence of j, we obtain

$$(j_{[2]})^{r-1}[j] : \Vdash\!\!\to \gamma_0 \mapsto \gamma_{\mu j^*(r)} \mapsto \gamma_{\mu j^*(r+1)}.$$

This shows that (γ_0) is a base for the sequence $(\gamma_{\mu j^*(1)}, \gamma_{\mu j^*(2)}, \ldots)$. Lemma 3.6 gives $\mu j^*(n) \geq 2^{n-1}$. In particular, we find $\mu j^*(4) \geq 8$. This bound discards any hope to compute an exact value by applying the scheme used for the first values: indeed this would entail computing values until at least $j_{[255]}$. We shall see below that the value of $\mu j^*(4)$ is actually much larger than 8.

In order to improve the previous results, we use a trick that expands the sequences admitting a base by inserting many intermediate new critical ordinals.

Lemma 3.8. *Assume that the sequence $\vec{\theta}$ is based on (β) and it goes from γ to δ in n steps. Assume moreover that $(\alpha_0, \ldots, \alpha_p, \beta, \gamma)$ is realizable. Then there exists a sequence based on $(\alpha_0, \ldots, \alpha_p)$ that goes from β to δ in 2^n steps.*

Proof. We use induction on $n \geq 0$. For $n = 0$, the sequence (β, γ) answers the question, since $(\alpha_0, \ldots, \alpha_p, \beta, \gamma)$ being realizable means that (β, γ) is based on $(\alpha_0, \ldots, \alpha_p)$. For $n > 0$, let δ' be the next to last term of $\vec{\theta}$. The induction hypothesis gives a sequence $\vec{\tau}'$ based on $(\alpha_0, \ldots, \alpha_p)$ that goes from γ to δ' in 2^{n-1} steps. As (δ', δ) is based on (β), there exists an embedding i satisfying

$$i : \Vdash\!\!\to \beta \mapsto \delta' \mapsto \delta.$$

We define the sequence $\vec{\tau}$ by extending $\vec{\tau}'$ with 2^{n-1} additional terms

$$\tau_{2^{n-1}+m} = i(\tau'_m) \qquad \text{for } 1 \leq m \leq 2^{n-1}.$$

By hypothesis, we have $\tau'_{2^{n-1}} = \delta'$, hence $\tau_{2^n} = i(\delta') = \delta$. So $\vec{\tau}$ goes from β to δ in 2^n steps. Moreover, $(\alpha_0, \ldots, \alpha_p)$ is a base for $\vec{\tau}'$, so, for $0 \leq m < 2^{n-1}$, there exists i'_m satisfying

$$i'_m : \Vdash\!\!\to \alpha_0 \mapsto \cdots \mapsto \alpha_p \mapsto \tau'_m \mapsto \tau'_{m+1}.$$

As β is the critical ordinal of i and $\alpha_p < \beta$ holds, this implies

$$i[i'_m] : \Vdash\!\!\to \alpha_0 \mapsto \cdots \mapsto \alpha_p \mapsto i(\tau'_m) \mapsto i(\tau'_{m+1}),$$

which shows that $(\alpha_0, \ldots, \alpha_p)$ is a base for $\vec{\tau}$. Note that the case $m = 0$ works as $\tau'_0 = \beta$ implies $i(\tau'_0) = i(\beta) = \delta' = \tau_{2^n-1}$, as is needed. ■

By playing with the above construction one more time, we can obtain still longer sequences. In order to specify them, we introduce an *ad hoc* iteration of the exponential function. In the sequel, we use G_p for the function inductively defined by the clauses $G_0(n) = n$ and

$$G_{p+1}(n) = \begin{cases} 0 & \text{for } n = 0, \\ G_{p+1}(n-1) + G_p(2^{G_{p+1}(n-1)}) & \text{otherwise.} \end{cases}$$

Thus, G_1 is an iterated exponential. Observe that $G_p(1) = 1$ holds for every p.

Lemma 3.9. *Assume that the sequence $\vec{\theta}$ is based on (β_p, β_{p+1}) and it goes from γ to δ in n steps. Assume moreover that $(\beta_0, \ldots, \beta_{p+1}, \gamma)$ is realizable. Then there exists a sequence based on (β_{p+1}) that goes from γ to δ in $G_{p+1}(n)$ steps.*

Proof. We use induction on $p \geq 0$, and, for each p, on $n \geq 1$. For $n = 1$, the sequence (γ, δ) answers the question, since, if i satisfies $\Vdash\!\!\to \beta_p \mapsto \beta_{p+1} \mapsto \gamma \mapsto \delta$, then $i^{[2]}$ satisfies $\Vdash\!\!\to \beta_{p+1} \mapsto \gamma \mapsto \delta$. Assume $n \geq 2$. Let δ' be the next to last term of $\vec{\theta}$. By induction hypothesis, there exists a sequence $\vec{\tau}'$ based on (β_{p+1}) that goes from γ to δ' in $G_{p+1}(n-1)$ steps. As in Lemma 3.8, we complete the sequence by appending new terms, but, before translating it, we still fatten it one or two more times. First, we apply Lemma 3.8 to construct a new sequence $\vec{\tau}''$ based on (β_p, β_{p+1}) that goes from β_{p+1} to δ' in $2^{G_{p+1}(n-1)}$ steps and is based on (β_{p-1}, β_p) for $p \neq 0$ (*resp.* on (β_p) for $p = 0$). For $p \neq 0$, we are in position for applying the current lemma with $p - 1$ to the sequence of $\vec{\tau}''$. So we obtain a new sequence $\vec{\tau}'''$ based on (α_p), and going from β_{p+1} to δ' in $G_p(2^{G_{p+1}(n-1)})$ steps. For $p = 0$, we simply take $\vec{\tau}''' = \vec{\tau}''$: as $G_0(N) = N$ holds, this remains consistent with our notations. Now we make the translated copy: we choose i satisfying $\Vdash\!\!\to \beta_p \mapsto \beta_{p+1} \mapsto \delta' \mapsto \delta$, and complete $\vec{\tau}'$ with the new terms

$$\tau_{G_{p+1}(n-1)+m} = i(\tau'''_m) \qquad \text{for} \qquad 0 < m \leq G_p(2^{G_{p+1}(n-1)}).$$

The sequence $\vec{\tau}$ has length $G_{p+1}(n-1) + G_p(2^{G_{p+1}(n-1)}) = G_{p+1}(n)$, and it goes from γ to $i(\delta')$, which is δ. It remains to verify the base condition for the new terms. Now assume that i'''_m satisfy $\Vdash\!\!\to \beta_p \mapsto \tau'''_m \mapsto \tau'''_{m+1}$. As in the proof of Lemma 3.8, we see that $i[i'''_m]$ satisfies $\Vdash\!\!\to \beta_{p+1} \mapsto i(\tau'''_m) \mapsto i(\tau'''_{m+1})$, which completes the proof, since we notice, for continuity, that $i(\tau'''_0) = \delta'$ holds. ■

By combining Lemmas 3.8 and 3.9, we obtain:

Lemma 3.10. *Assume that the sequence $\vec{\theta}$ is based on (β_p, β_{p+1}) and it goes from γ to δ in n steps. Assume moreover that $(\beta_0, \ldots, \beta_{p+1}, \gamma)$ is realizable. Then there exists a sequence based on (β_0) that goes from β_1 to δ in $H_1(H_2(\ldots(H_{p+1}(n))\ldots))$ steps, where $H_q(m)$ is defined to be $2^{G_q(m)}$.*

Proof. We use induction on $p \geq 0$. In every case, Lemma 3.9 constructs from $\vec{\theta}$ a new sequence $\vec{\theta}'$ based on (β_{p+1}) that goes from β_{p+1} to δ in $G_{p+1}(n)+1$ steps. Then, Lemma 3.8 constructs from $\vec{\theta}'$ a new sequence $\vec{\theta}''$ that goes from (β_{p+1}) to δ in $2^{G_{p+1}(n)} + 1 = H_{p+1}(n) + 1$ steps, and that is based on (α_{p-1}, α_p) for $p \neq 0$, and on (α_p) for $p = 0$. For $p = 0$, $\vec{\theta}''$ answers the question. Otherwise, we are in position for applying the induction hypothesis to $\vec{\theta}''$. ■

Using the previous computational lemmas, we deduce the following lower bound for the function μj^*.

Proposition 3.11. (lower bound) *Assume that j is a nontrivial elementary embedding of a limit rank into itself. Then, for $r \geq 3$, we have*

$$\mu j^*(r) \geq 2^{H_1(H_2(\cdots(H_{r-2}(1))\cdots))}. \tag{3.6}$$

Proof. By definition, $(\gamma_{\mu j^*(r-1)}, \gamma_{\mu j^*(r)})$ is based on $(\gamma_{\mu j^*(r-3)}, \gamma_{\mu j^*(r-2)})$, and the auxiliary sequence $(\gamma_0, \ldots, \gamma_{\mu j^*(r-2)})$ is realizable. Indeed, j satisfies

$$j : \Vdash\!\!\to \gamma_{\mu j^*(0)} \mapsto \gamma_{\mu j^*(1)} \mapsto \gamma_{\mu j^*(2)} \mapsto \gamma_{\mu j^*(3)},$$

and, therefore, we have

$$j^{[r+1]} : \Vdash\!\!\to \gamma_{\mu j^*(r)} \mapsto \gamma_{\mu j^*(r+1)} \mapsto \gamma_{\mu j^*(r+2)} \mapsto \gamma_{\mu j^*(r+3)}$$

for every r. By applying Lemma 3.10, we find a new sequence based on (γ_0) that goes from γ_1 to $\gamma_{\mu j^*(r)}$ in $H_1(H_2(\cdots(H_{r-2}(1))\cdots))$ steps. We conclude using Lemma 3.6. ■

Example 3.12. (lower bound) *For $r = 4$, we obtain $\mu j^*(4) \geq 2^8 = 256$. We see that computing an exact value along the lines of the previous paragraph would require computing the beginning of the critical sequence of $j_{[n]}$ for $n < 2^{256}$ at least.*

Similarly, the value of $\mu j^*(5)$ is at least $2^{H_1(H_2(H_3(1)))}$, which is $2^{2^{G_1(16)}}$. It is easy to verify that $G_1(n+1)$ is more than the evaluation of a tower of exponentials with height n, so we see that $\mu j^*(5)$ is more than a tower of base 2 exponentials of height 17.

Observe that the constructions of Lemmas 3.8, 3.9 and 3.10 are effective: theoretically, we could write an explicit list of 256 iterates of j whose critical ordinals make an increasing sequence below $\gamma_{\mu j^*(4)}$.

It is not easy to figure out intuitively how large the lower bounds given by Proposition 3.11 (lower bound) are. A classical way to measure the growth rate of a function is to compare it with Ackermann's function f_ω^{Ack}.

Definition 3.13. (**Ackermann's function**) The function f_p^{Ack} is defined inductively by $f_0^{\mathrm{Ack}}(n) = n + 1$ and

$$f_{p+1}^{\mathrm{Ack}}(n) = \begin{cases} f_p^{\mathrm{Ack}}(1) & \text{for } n = 0, \\ f_p^{\mathrm{Ack}}(f_{p+1}^{\mathrm{Ack}}(n-1)) & \text{otherwise.} \end{cases}$$

Then we put $f_\omega^{\mathrm{Ack}}(n) = f_n^{\mathrm{Ack}}(n)$.

Each function f_p^{Ack} is constructed using recurrence from the previous one. The first three functions f_0^{Ack}, f_1^{Ack}, and f_2^{Ack} are linear, f_3^{Ack} is exponential, f_4^{Ack} is a tower of exponentials, *etc.* One of the interesting features of the construction is that every function on the integers definable from addition and multiplication using compositions and recurrences, *i.e.*, every so-called *primitive recursive function*, is dominated by some function f_p^{Ack}, *i.e.*, there exists p such that $f(n) < f_p^{\mathrm{Ack}}(n)$ holds for n large enough. This is not the case for the function f_ω^{Ack}, which eventually dominates all functions f_p^{Ack}, and, therefore, f_ω^{Ack} is an example of a non-primitive recursive function.

Using the similarity between the clauses that define Ackermann's functions and the functions G_p used above, it is easy to complete the proof of Proposition 3.1 (not primitive recursive).

Proof. As we know that the function f_ω^{Ack} grows faster than any primitive recursive function, it is enough to show the inequality $2^{H_1(H_2(\dots(H_{n-2}(1))\dots))} \geq f_\omega^{\mathrm{Ack}}(n-1)$ for $n \geq 5$. To this end, we first observe that $G_p(n+3) > f_p^{\mathrm{Ack}}(n)$ holds for all p, n. This is obvious for $p = 0$. Otherwise, for $n = 0$, we have, using $G_p(2) \geq 3$,

$$G_p(3) > G_{p-1}(2^{G_p(2)}) > f_{p-1}^{\mathrm{Ack}}(6) > f_{p-1}^{\mathrm{Ack}}(1) = f_p^{\mathrm{Ack}}(0).$$

For $n > 0$, we find

$$\begin{aligned} G_p(n+3) > G_{p-1}(2^{G_p(n+2)}) &> G_{p-1}(f_p^{\mathrm{Ack}}(n-1) + 3) \\ &> f_{p-1}^{\mathrm{Ack}}(f_p^{\mathrm{Ack}}(n-1)) = f_p^{\mathrm{Ack}}(n). \end{aligned}$$

Finally, we have $G_2(n) = n + 2$ for every n, and therefore

$$\begin{aligned} 2^{H_1(H_2(\cdots(H_{n+2}(1))\cdots))} &= 2^{H_1(H_2(\cdots(H_{n+1}(2))\cdots))} \\ &= 2^{H_1(H_2(\cdots(H_n(2^{n+3}))\cdots))} \\ &> G_n(2^{n+3})) \geq G_n(r+3) > f_n^{\mathrm{Ack}}(n). \end{aligned}$$

∎

Let us finally mention without proof the following strengthening of the lower bound for $\mu j^*(4)$:

Proposition 3.14. (**lower bound**) *Assume that j is a nontrivial elementary embedding of a limit rank into itself. Then we have $\mu j^*(4) \geq f_9^{\mathrm{Ack}}(f_8^{\mathrm{Ack}}(f_8^{\mathrm{Ack}}(254)))$.*

This shows that even the value of $\mu j^*(4)$ is inconceivably large, and cannot be described in terms of the usual operations on natural numbers. Observe that the above result, as well as Proposition 3.11, shows how powerful the Laver–Steel theorem is: indeed, we have seen that this theorem is the key point in the existence (*i.e.*, the finiteness) of $\mu j^*(r)$.

4. Lower Bounds

We shall see now how the results of the previous section imply lower complexity bounds for the computation of the periods in A_n. A remarkable point is that the negative results so obtained are true in ordinary mathematics, although we establish them using computations on elementary embeddings: the proofs remain valid even if no self-similar rank exists—contrary to the positive results of Section 1 of which no proof in ordinary mathematics is known.

Scales

We first restate the results of Section 1 in a slightly different form. In the sequel, we consider partial functions of $\mathbf{N}$ into itself. To make things precise, we define a *partial increasing injection* on $\mathbf{N}$ to be an increasing function of $\mathbf{N}$ into itself whose domain is either $\mathbf{N}$, or a finite initial segment of $\mathbf{N}$. As in the case of elementary embeddings, we shall say that a partial increasing injection f is nontrivial if it does not coincide with the identity mapping, in the sense that there exists at least one integer m such that $f(m)$ exists and is not m, in which case $f(m) > m$ necessarily holds.

Definition 4.1. (**critical**) Assume that f is a partial increasing injection on $\mathbf{N}$. We say that m is the *critical integer* of f, denoted $m = \mathrm{crit}(f)$, if we have $f(k) = k$ for $k < m$, and $f(m) \neq m$, *i.e.*, either $f(m) > m$ holds or $f(m)$ is not defined.

Notation 4.2. (**weak inequality**) For f a partial increasing injection on $\mathbf{N}$, and m, n in $\mathbf{N}$, we write $f(m) \geq^w n$ if either $f(m)$ is defined and $f(m) \geq n$ holds, or $f(m)$ is not defined; we write $\mathrm{crit}(f) \geq^w m$ for $(\forall k < m)(f(k) = k)$.

Lemma 4.3. *Assume that f is a partial increasing injection on $\mathbf{N}$. Then $f(m) = n$ is equivalent to the conjunction of $f(m) \geq^w n$ and $f(m) \not\geq^w n+1$, and $\mathrm{crit}(f) = m$ is equivalent to the conjunction of $\mathrm{crit}(f) \geq^w m$ and $\mathrm{crit}(f) \not\geq^w m+1$.*

Proof. By definition, $f(m) \not\geq^w n+1$ means that $f(m)$ is defined, and its value is n at most. Similatly, if $\mathrm{crit}(f) \geq^w m$ holds, then $\mathrm{crit}(f) \geq^w m+1$ holds if and only if $f(m)$ exists and equals m: so it fails if either $f(m) > m$ holds, or $f(m)$ does not exist. ■

Definition 4.4. **(scale)** Assume that S is a binary system. A *scale* on S is defined to be a family of partial increasing injections on $\mathbf{N}$ indexed by S, say $I = (I_a)_{a \in S}$, that satisfies

$$\mathrm{crit}(I_{ab}) = I_a(\mathrm{crit}(I_b)) \tag{4.1}$$

for all a, b in S. The scale is said to be *total* if it consists of total functions, *i.e.*, if the domain of I_a is $\mathbf{N}$ for each a.

Assume that j is a nontrivial elementary embedding of a rank into itself, and, as above, let $\gamma_0, \gamma_1, \ldots$ be the sequence of the iterated critical ordinals of j. By the Laver–Steel theorem, for each term t in T_1, the elementary embedding $t(j)$ maps the sequence $(\gamma_m)_m$ into itself, *i.e.*, there exists a well-defined injection I_t^j of $\mathbf{N}$ into itself satisfying

$$I_t^j(m) = n \quad \text{if and only if} \quad t(j) : \gamma_m \mapsto \gamma_n. \tag{4.2}$$

For instance, we have $I_x^j = \mu j$, where μj is the mapping considered in Section 3. For $t =_{LD} t'$, we have $t(j) = t'(j)$, hence $I_t^j = I_{t'}^j$, and the following definition is non-ambiguous:

Definition 4.5. **(scale I^j)** For a in FLD_1, we denote by I_a^j the common value of I_t^j for t representing a, and we define I^j to be the sequence of all (I_a^j)'s.

Lemma 4.6. *Assume that j is a nontrivial elementary embedding of a rank into itself. Then I^j is a total scale on FLD_1.*

Proof. By definition of an elementary embedding, the injection I_a^j is increasing, and it is defined everywhere by the Laver–Steel theorem. Every elementary embedding in $\mathrm{Iter}(j)$ admits a critical ordinal, and, therefore, every injection I_a^j admits a critical integer. Then Condition (4.1) follows from Lemma XII.3.6. ■

Thus, if there exists a self-similar rank, there exists a total scale on FLD_1. Let us now address the question of constructing a scale on FLD_1 without using elementary embeddings. A natural method could be to start with a left self-distributive operation on $\mathfrak{I}_\infty$ and to find an increasing injection i such that every injection in the subsystem generated by i is still increasing. For instance, we could consider the bracket operation of Chapter I. However, it can be checked that, for every injection i in $\mathfrak{I}_\infty$, the subsystem of $(\mathfrak{I}_\infty, [\ ])$ generated by i contains injections that are not increasing.

A more promising approach consists in using the tables A_n. Indeed, using the dictionary of Section 1, we know that the condition $t(j) : \gamma_m \mapsto \gamma_n$ of the definition of the injection I_t^j is equivalent to $t(1) * 2^m$ computed in A_{n+1} being 2^n. This makes the following definition natural.

Definition 4.7. (**scale** I^A) For a in FLD_1, we define I_a^A to be the partial mapping such that $I_a^A(m) = n$ holds if, for some term t representing a, $t(1) * 2^m$ computed in A_{n+1} is 2^n. We denote by I^A the family of all I_a^A's.

As A_{n+1} is an LD-system, the value of $t(1) * 2^m$ computed in A_{n+1} depends on the LD-class of t only, so the previous definition is non-ambiguous. By Proposition 1.6 (dictionary II), if j is a nontrivial elementary embedding of a rank into itself, then the injections I_a^A and I_a^j coincide for every a in FLD_1, and we deduce:

Proposition 4.8. (**total scale**) *Assume that there exists a self-similar rank. Then I^A is a total scale on FLD_1.*

At this point, our obvious task is to prove directly that the family I^A, which always exists, is a total scale on FLD_1.

Example 4.9. (**left powers**) *Assume that j is an elementary embedding of a limit rank into itself. Let p be a positive integer. We have seen in Section XII.4 that the critical ordinal of $j_{[p]}$ is $\mathrm{crit}_m(j)$, where m is the largest integer such that 2^m divides p. In terms of the injections I_a^j, this means that the critical integer of the injection $I_{x_{[p]}}^j$ exists and it is m. Proving the corresponding result for I^A amounts to proving $I_{x_{[p]}}^A(k) = k$ for $k < m$, and $I_{x_{[p]}}^A(m) > m$ or $I_{x_{[p]}}^A(m)$ does not exist. By definition, for $I_{x_{[p]}}^A$ to leave k fixed means that $1_{[p]} * 2^k = 2^k$ holds in A_{k+1}. This equality is true for $k < m$, as we have $1_{[p]} = 2^{k+1}$ in A_{k+1}, and $2^{k+1} *_{k+1} b = b$ for every b in A_{k+1}. On the other hand, we have $1_{[p]} = 2^m$ in A_{m+1}, and therefore $2^m *_{m+1} 2^m = 2^{m+1}$. We conclude that the critical integer of $I_{x_{[p]}}^A$ exists, and it is m. So, in this case, we have established, without using any set theoretical assumption, the property of I_a^A that is the counterpart of some property of I_a^j. Observe that the previous proof implies that every integer is the critical integer of some injection I_a^A, namely a left power of the generator.*

Lemma 4.10. *Assume that a is an element of* FLD_1, *and t represents a.*
(i) For $n \geq m$, $I_a^A(m) \succeq^w n$ is equivalent to $t(1)^{A_n} *_n 2^m = 2^n$;
(ii) The mapping I_a^A is a partial increasing injection;
(iii) The relation $\mathrm{crit}(I_a^A) \succeq^w n$ is equivalent to $t(1)^{A_n} = 2^n$.

Proof. (i) If $I_a^A(m) = p$ holds for some p with $p \geq n$, we have $t(1)^{A_{p+1}} *_{p+1} 2^m = 2^{p+1}$, which gives $t(1)^{A_n} *_n 2^m = 2^n$ by projecting from A_{p+1} to A_n. And $I_a^A(m)$ not being defined means that there exists no p satisfying $t(1)^{A_{p+1}} *_{p+1} 2^m < 2^{p+1}$: in other words $t(1)^{A_{p+1}} *_{p+1} 2^m = 2^{p+1}$ holds for $p+1 \geq m$, and, in particular, this equality holds for $p+1 = n$.

(ii) Assume $I_a^A(m+1) = n+1$. This implies $t(1)^{A_{n+2}} *_{n+2} 2^{m+1} = 2^{n+1}$, *i.e.*, $t(1)^{A_{n+2}}$ has period 2^{m+2} at least in A_{n+2}. By projecting from A_{n+2} to A_{n+1}, we deduce that $t(1)^{A_{n+1}}$ has period 2^{m+1} at least in A_{n+1}, hence $t(1)^{A_{n+1}} *_{n+1} 2^m \leq 2^n$ holds. If the latter relation is an equality, we deduce $I_a^A(m) = n$. Otherwise, by projecting, we find some integer $p < n$ satisfying $t(1)^{A_{p+1}} *_{p+1} 2^m = 2^p$, and we deduce $I_a^A(m) = p$. In both cases, $I_a^A(m)$ exists, and its value is at most n. This shows that the domain of I_a^A is an initial segment of $\mathbf{N}$, and that I_a^A is increasing.

(iii) Assume $\mathrm{crit}(I_a^A) \succeq^w n$, *i.e.*, $I_a^A(m) = m$ holds for $m < n$. We have $I_a^A(n-1) \not\succeq^w n$, hence $t(1)^{A_n} *_n 2^{n-1} \leq 2^{n-1}$, and, therefore, $t(1)^{A_n} = 2^n$, as $a *_n 2^{n-1} = 2^n$ holds for $a < 2^n$. Conversely, assume $t(1)^{A_n} = 2^n$, and $m < n$. By projecting from A_n to A_{n+1}, we obtain $t(1)^{A_{m+1}} = 2^{m+1}$, hence $t(1)^{A_{m+1}} *_{m+1} 2^m = 2^m < 2^{m+1}$, which gives $I_a^A(m) \not\succeq^w m+1$ by (i). As $I_a^A(m) \succeq^w m$ holds by (i), we conclude that $I_a^A(m)$ is defined and $I_a^A(m) = m$ holds. ∎

At this point, we can complete the first part of our program:

Proposition 4.11. (**scale**) *The family I^A is a scale on* FLD_1.

Proof. We have seen above that I_a^A is a partial increasing injection for every a, and we observed in Example 4.9 (left powers) that I_x^A, and, more generally, $I_{x_{[p]}}^A$, is never trivial. It remains to verify (4.1). So, let a, b be two elements of FLD_1 represented by t_1 and t_2 respectively.

Assume first $I_a^A(p) \succeq^w n$ and $\mathrm{crit}(I_b^A) \succeq^w p$. By Lemma 4.10, the hypothesis is $t_1(1)^{A_n} *_n 2^p = 2^n$, and $t_2(1)^{A_p} = 2^p$. By projecting from A_n to A_p, we deduce that $t_2(1)^{A_n}$ is a multiple of 2^p. Hence, the hypothesis $t_1(1)^{A_n} *_n 2^p = 2^n$ implies $t_1(1)^{A_n} *_n t_2(1)^{A_n} = 2^n$, *i.e.*, $(t_1{\cdot}t_2)(1)^{A_n} = 2^n$, which, by Lemma 4.10 again, gives $\mathrm{crit}(I_{ab}^A) \succeq^w n$.

Assume now $I_a^A(p) \not\succeq^w n+1$ and $\mathrm{crit}(I_b^A) \not\succeq^w p+1$. The hypothesis is now $t_1(1)^{A_{n+1}} *_{n+1} 2^p \neq 2^{n+1}$ in A_{n+1}, *i.e.*, the period of $t_1(1)^{A_{n+1}}$ in A_{n+1} is 2^{p+1} at least, and $t_2(1)^{A_{p+1}} \neq 2^{p+1}$, *i.e.*, $t_2(1)^{A_{p+1}} \leq 2^p$. If we had $t_2(1)^{A_{n+1}} \geq 2^{p+1}$, then, by projecting from A_{n+1} to A_{p+1}, we would obtain $t_2(1)^{A_{p+1}} = 2^{p+1}$, which contradicts our hypothesis. Hence we have $t_2(1)^{A_{n+1}} \leq 2^p$, and the hypothesis that the period of $t_1(1)^{A_{n+1}}$ in A_{n+1} is 2^{p+1} at least implies $t_1(1)^{A_{n+1}} *_{n+1} t_2(1)^{A_{n+1}} \leq 2^n$, hence $\mathrm{crit}(I^A_{ab}) \not\geq^w n+1$.

By comparing the previous implications, we deduce that the conjunction of $I^A_a(p) = n$ and $\mathrm{crit}(I^A_b) = p$ implies $\mathrm{crit}(I^A_{ab}) = n$. Hence, the family I^A satisfies the defining relation (4.1) of a scale. ■

Thus the only point we have not proved so far is the scale I^A being total. Before going further, let us observe that the latter property is connected with the asymptotic behaviour of the periods in the tables A_n, as well as with several equivalent statements:

Proposition 4.12. (**equivalence**) *The following statements are equivalent:*

(i) The scale I^A is total;

(ii) For every term t, the period of $t(1)^{A_n}$ in A_n goes to infinity with n—so, in particular the period of every fixed a in A_n goes to infinity with n;

(iii) The period of 1 in A_n goes to infinity with n;

(iv) For every r, there exists n satisfying $1^{[r]A_n} < 2^n$;

(v) The sub-LD-system of $\underleftarrow{\lim} A_n$ generated by $(1, 1, \ldots)$ is free.

Proof. Let t be an arbitrary term in T_1, and a be its class in FLD_1. Saying that the period of $t(1)^{A_n}$ in A_n goes to infinity with n means that, for every m, there exists n satisfying $t(1)^{A_n} *_n 2^m < 2^n$, *i.e.*, $I^A_a(m) \not\geq^w n$. If I^A is a total scale, the function I^A_a is defined everywhere, and the previous relation holds. Conversely, the relations imply that the value of $I^A_a(m)$ exists for every m. Hence, (i) and (ii) are equivalent, and they imply (iii), which is the special case $t = x$ of (ii).

Assume now (iii). By the previous argument, the injection I^A_x is defined everywhere. If I^A_a and I^A_b are defined everywhere, $\mathrm{crit}(I^A_{b^{[k]}})$ exists for every k, and so does $I^A_a(\mathrm{crit}(I^A_{b^{[k]}}))$, which is $\mathrm{crit}(I^A_{(ab)^{[k]}})$. This proves that $I^A_{ab}(m)$ exists for arbitrary large values of m, and this is enough to conclude that I^A_{ab} is defined everywhere. So, inductively, we deduce from I^A_x being defined everywhere that I^A_a is defined everywhere for every a in FLD_1, which gives (ii).

Then, we prove that (ii) implies (iv) using induction on $r \geq 1$. The result is obvious for $r = 1$. Otherwise, let p be the largest integer satisfying $1^{[r-1]A_p} = 2^p$, which exists by induction hypothesis. By (ii), there exists $n > p$ satisfying $1 *_n 2^p < 2^n$, so the period of 1 in A_n is a multiple of 2^{p+1}. By hypothesis, we have $1^{[r-1]A_{p+1}} = 2^p$, hence $1^{[r-1]A_n} = 2^p \bmod 2^{p+1}$, so 2^p is the largest power of 2 that divides $1^{[r-1]A_n}$. As the period of 1 in A_n is a multiple of 2^{p+1}, $1 *_n 1^{[r-1]A_n} = 2^n$ is impossible, so we have $1^{[r]A_n} = 1 *_n 1^{[r-1]A_n} < 2^n$.

Assume now (iv), and let t be an arbitrary term. By Proposition V.2.12 (attraction), there exist q, r satisfying $t^{[r]} =_{LD} x^{[q]}$. By (iv), there exists n satisfying $1^{[q]A_n} = t(1)^{[r]A_n} < 2^n$. This implies $t(1)^{A_n} < 2^n$, since every right power of 2^n in A_n is 2^n. Hence (iv) implies (iii).

Assume (i), and let t, t_1, ..., t_k be arbitrary terms. By (ii), we can find n such that none of the terms t, $t{\cdot}t_1$, $(t{\cdot}t_1){\cdot}t_2$, ..., $(\ldots(t{\cdot}t_1)\ldots){\cdot}t_k$ evaluated at 1 in A_n is 2^n: this is possible since $t(1)^{A_n} \neq 2^n$ implies $t(1)^{A_p} \neq 2^p$ for $p \geq n$. So we have

$$t(1)^{A_n} < (t \cdot t_1)(1)^{A_n} < ((t \cdot t_1) \cdot t_2)(1)^{A_n} < \cdots$$

and, in particular, $t(1)^{A_n} \neq (\cdots(t{\cdot}t_1){\cdot}) \cdots {\cdot}t_k)(1)^{A_n}$. This implies that left division in the sub-LD-system of $\varprojlim A_n$ generated by $(1, 1, \ldots)$ has no cycle, and, therefore, this LD-system is free. Conversely, assume that (i) fails, *i.e.*, there exists $p \geq 1$ satisfying $1 *_n 2^p = 2^n$ for every n. Let α denote the sequence $(1, 1, \ldots)$ in $\varprojlim A_n$. Then we have $\alpha_{[2^p]} = (1, 2, \ldots, 2^p, 2^p, \ldots)$ and

$$\alpha * \alpha_{[2^p+1]} = (\alpha * \alpha_{[2^p]}) * (\alpha * \alpha) = \alpha * \alpha.$$

The sub-LD-system generated by α is not free, since $g * g = g * g_{[2^p+1]}$ does not hold in a free LD-system generated by g. So (v) is equivalent to (i)–(iv). ■

Unprovability results

The status of the equivalent statements of Proposition 4.12 (equivalence) remains currently open. However, the results of Section 3 enables us to say more about them. We know that some fast growing function is connected with the critical ordinals of the iterations of an elementary embedding. More precisely, we have seen that the function μj^* grows faster than any primitive recursive function. In terms of the scale I^j, the function μj^* is the function on $\mathbf{N}$ that maps k to $(I^j_x)^k(0)$. Now, the scales I^A and I^j coincide when the latter exists. So, a natural idea is to look at the (possibly partial) function $r \mapsto (I^A_x)^r(0)$. The remarkable fact is that we can obtain for this function the same lower bound as for its I^j-counterpart μj^* without using any set theoretical hypothesis:

Proposition 4.13. **(fast growing)** *If the scale I^A is total, the function $r \mapsto (I^A_x)^r(0)$ grows faster than any primitive recursive function.*

Proof. Assume that every injection I^A_a is total. Let us consider the proof of Proposition 3.1 (not primitive recursive) in Section 3, and try to copy it using I^A_a and critical integers instead of $t(j)$ and critical ordinals. This is possible, because the only properties used in Section 3 are the left self-distributivity identity and Relation (4.1) about critical ordinals. First, the counterpart of Lemma 3.3 is true since every value of I^A_a is an increasing injection and its domain is an initial interval. Then the definitions of a base and of a realizable sequence can be translated without any change. Let us consider Lemma 3.8. The point is to be able to deduce from the hypothesis

$$I^A_b : \Vdash\!\!\!\rightarrow m_0 \mapsto m_1 \mapsto \cdots \mapsto m_p \tag{4.3}$$

the conclusion

$$I^A_{ab} : \Vdash\!\!\!\to I^A_a(m_0) \mapsto I^A_a(m_1) \mapsto \cdots \mapsto I^A_a(m_p). \tag{4.4}$$

We establish the equality

$$(I^A_a)^r(\mathrm{crit}(I^A_a)) = \mathrm{crit}(I^A_{a^{[r+1]}})$$

using induction on $r \geq 0$. The result is true by definition for $r = 0$. Otherwise, we have

$$\begin{aligned}(I^A_a)^r(\mathrm{crit}(I^A_a)) &= I^A_a((I^A_a)^{r-1}(\mathrm{crit}(I^A_a)))\\ &= I^A_a(\mathrm{crit}(I^A_{a^{[r]}})) = \mathrm{crit}(I^A_{aa^{[r]}}) = \mathrm{crit}(I^A_{a^{[r+1]}}).\end{aligned}$$

So (4.3) can be restated as

$$\mathrm{crit}(I^A_b) = m_0, \quad \mathrm{crit}(I^A_{b^{[2]}}) = m_1, \ \ldots, \quad \mathrm{crit}(I^A_{b^{[r+1]}}) = m_r.$$

By applying I^A_a and using (4.1), we obtain

$$\mathrm{crit}(I^A_{ab}) = I^A_a(m_0), \ \ldots, \quad \mathrm{crit}(I^A_{ab^{[r+1]}}) = I^A_a(m_r).$$

By (LD), we have $I^A_{ab^{[r]}} = I^A_{(ab)^{[r]}}$, and therefore (4.3) implies (4.4).

So the proof of Lemma 3.8 goes through in the framework of an arbitrary total scale, and so do those of the other results of Section 3. We conclude that, for $r > 3$, there are at least $2^{H_1(H_2(\cdots(H_{r-2}(1))\cdots))}$ critical integers below the number $(I^A_x)^r(0)$, where H_p are the fast growing functions of Section 3. As in the latter section, we conclude that the function $r \mapsto (I^A_x)^r(0)$ grows at least as fast as Ackermann's function f^{Ack}_ω, and therefore it grows faster than any primitive recursive function. ■

Remark 4.14. In the previous argument, the hypothesis that the injections are defined everywhere is not really used. Indeed, we establish lower bounds for the values, and the precise result is an alternative: for each r, either the value of $(I^A_x)^r(0)$ is not defined, or this value is at least equal to some explicit value. In particular, the result is local, and the lower bounds remain valid for the small values of k even if $(I^A_x)^r(0)$ is not defined for some large value of k.

A consequence of the above result is that a total scale on FLD_1 is a very complicated object, and there is little hope to describe one explicitly. On the other hand, as the definition of the functions I^A_a is simple, proving that the scale I^A is total is likely to be a very difficult task. In order to make such an assertion more precise, we can consider a logical system similar to the Peano axiomatization of the integers but with an induction scheme restricted to primitive recursive functions. Such a system is known as Primitive Recursive Arithmetic, PRA for short. On the one hand, PRA is a rather weak system, but, on the other hand, most of the usual statements of arithmetic can be proved inside PRA. The fact that a non-primitive recursive function necessarily occurs when a total scale on FLD_1 is given implies the following result:

Proposition 4.15. (**unprovability**) *It cannot be proved in PRA that the scale I^A is total. More generally, the existence of a total scale on* FLD_1 *cannot be proved in Primitive Recursive Arithmetic, nor can any of the statements mentioned in Proposition 4.12: in particular, it is impossible to prove in PRA that the period of 1 in the table A_n goes to infinity with n.*

Proof. (Sketch) In PRA, one can prove the existence of primitive recursive functions only. Now, being able to prove that a given scale I is total would in particular enable one to prove that a non-primitive recursive function exists, since we could then prove in PRA that the function $r \mapsto (I_x)^r(0)$ grows faster than any primitive recursive function, as the counterparts of the proofs of Section 3 all can be formulated inside PRA. ∎

Let us come back finally to the first values of the function μj^*. We have proved in Section 3 that, if j is a nontrivial elementary embedding of a rank into itself, then $\mu j^*(4) \geq 256$ holds, which, when translated into the language of A_n, means that the period of 1 in A_n is 16 for every n between 9 and 256 at least. With the arguments of the current section, we deduce that this lower bound remains valid even if no elementary embedding exists. Now, we have also mentioned without proof the stronger result $\mu j^*(4) \geq f_9^{\mathrm{Ack}}(f_8^{\mathrm{Ack}}(f_8^{\mathrm{Ack}}(254)))$. Again, this bound remains valid in every case, and we can state the last result of this text:

Proposition 4.16. (**lower bound**) *The least integer n such that the period of 1 in A_n reaches 32 is larger than the number $f_9^{\mathrm{Ack}}(f_8^{\mathrm{Ack}}(f_8^{\mathrm{Ack}}(254)))$.*

5. Notes

The dictionary of Section 1 is due to R. Laver, who already knew around 1990 a number of results about the finite tables A_n.

There remains a widely open question to describe the tables A_n, so as to be able to compute the periods or to give an explicit formula for the rows. As the results of Section 4 show, the question is probably extremely difficult. The results of Section 2 appear in (Drápal 95), (Drápal 95*a*), (Drápal 97). These results are conceived as the first steps in an infinite sequence where the p-th level should describe homomorphisms from A_n to $A_{F_p(n)}$ where F_p lies approximately on the p-th level of Ackermann's hierarchy of fast growing functions. One could expect to control in this way the periods of the rows with indices of the form $2^N - p$. Let us mention that A. Drápal has obtained further results ("the fourth and the fifth level homomorphisms", unpublished). R. Dougherty has obtained similar results (corresponding to the first two levels) in the context of critical ordinals and elementary embeddings (Dougherty 96).

The computations of critical ordinals in Section 3 are from (Dougherty 93). Section 4 was inspired by (Dougherty & Jech 97) and (Dougherty & Jech 97*a*). In these papers, the notion of an embedding algebra is introduced as a miniaturized version of the LD-monoid $\widehat{\mathrm{Iter}}(j)$, of which it captures a number of properties. The current notion of a scale is a weakening of the notion of an embedding algebra, as we do not require the value of I_{ab} to depend on I_a and I_b only, and we retain only Formula (4.1) from the properties of composition. The existence of an embedding algebra is equivalent to the statements of Proposition 4.12 (equivalence): as being an embedding algebra is a stronger condition than being a total scale, the proof of the equivalence is combinatorially much more difficult than the argument we have given here. In particular, it requires some counterpart of the ordinals to be constructed.

The last statement, namely the huge lower bound for the first n for which the period of 1 in A_n reaches 32 is stated in (Dougherty 96). It is not clear that all arguments of this paper can be automatically translated into computations in A_n. However, the bound for $\mu j^*(4)$ can be established using translatable results only, and, therefore, it is valid without any assumption (Dougherty, private communication). (Jech 97) contains further results about the critical ordinals and their counterparts in embedding algebras.

Let us mention here a simple question. If j is an elementary embedding, then every iterate of j admits a critical ordinal which is one of the ordinals called $\mathrm{crit}_n(j)$ here. As $\mathrm{Iter}(j)$ is isomorphic to FLD_1, we obtain in this way a well-defined mapping, say ν, of FLD_1 into $\mathbf{N}$, namely the mapping defined by $\mathrm{crit}(t(j)) = \mathrm{crit}_{\nu(t(x))}(j)$ for t in T_1. By definition of the critical ordinals, we have then

$$\begin{cases} \nu(a) = \nu(b) & \text{implies} \quad \nu(ca) = \nu(cb), \\ \nu(a) \le \nu(b) & \text{implies} \quad \nu(ab) > \nu(b), \\ \nu(a) > \nu(b) & \text{implies} \quad \nu(ab) = \nu(b). \end{cases} \tag{5.1}$$

Question 5.1. *Can the existence of a mapping ν on* FLD_1 *satisfying* (5.1) *be proved without assuming the existence of a self-similar rank?*

If the scale I^A is total, then the mapping ν defined by $\nu(a) = \mathrm{crit}(I^A_a)$ satisfies (5.1). Now, as we consider here some properties of the critical ordinals only, a positive solution to Question 5.1 could be easier to obtain than the existence of a total scale on FLD_1.

Applications of set theory

The current status of the statement

$$\text{“}\textit{The period of } 1 \textit{ in } A_n \textit{ goes to infinity with } n\text{”} \tag{S}$$

is strange: this combinatorial statement involves only finite, effective objects, and we know that it is true if we assume the strong axiom of infinity "There exists a self-similar rank". On the other hand, no proof of (S) in ordinary mathematics is known, and, on the contrary, we know that (S) is not provable in the weak system PRA.

It might happen that (S) is not provable in ordinary mathematics, say in Peano arithmetics, or, even, in Zermelo–Fraenkel set theory: this would be the case if one could for instance strengthen the methods of Section 4 so as to construct, starting from (S), hence from a total scale on FLD_1 or from an embedding algebra, an elementary embedding of a rank into itself. By Gödel's incompleteness theorem, we know that the existence of such an object cannot be proved in the Zermelo–Fraenkel system, and, therefore, we would deduce that (S) itself cannot be proved.

However, it seems more likely that a purely combinatorial proof of (S) will be discovered eventually. In this case, set theory will be no longer needed. This is what happened for the existence of an LD-system in which left division has no cycle: as we mentioned in Chapter XII, the first proof of this statement was the proof by R. Laver that left division in the LD-system Iter(j) has no cycle, and, for several years, the statement

"There exists an LD-system in which left division has no cycle" (S')

—as well as its corollaries, such as the existence of a linear ordering satisfying $a < ab$ on free LD-systems or the decidability of the word problem for LD-equivalence—has had the same logical status as the statement (S) currently has.

It seems to us that the role of set theory in such cases is quite similar to the role of physics when the latter gives heuristic evidence for some statements that mathematicians are to prove subsequently. In both cases, the statements are first established rapidly but at the expense of admitting some additional hypotheses or approximative proof methods—observe that adding a set theoretical axiom is nothing but adding a new proof method—and the subsequent task is to give a proof that does not use the additional hypotheses any longer. However, the new proofs are likely to be more complicated than the first ones, and the seminal role played by set theory or physics should not be forgotten even if they vanish in the final picture. So, definitely, and despite the fact that nothing of set theory remains in the constructions of Chapter III or VIII, we think that the braid order is an application of set theory, and, more precisely, of this most speculative part of set theory that investigates large cardinals.

The ultimate questions

The question of whether the periods in the Laver tables A_n go to infinity may superficially appear as a game rather than as a deep mathematical question, but everyone should agree that these tables are fascinating objects that capture a significant part of the intrinsic complexity of the algebraic structure of elementary embeddings and of left self-distributivity. In particular, they can be seen as increasingly complicated approximations of the free LD-system of rank 1, which we have seen is essentially their inverse limit.

On the other hand, we have seen that letting the braid group B_n act on free LD-systems leads to new results about braids. An obvious problem now is

Question 5.2. *How to use the action of braids on the Laver tables?*

Owing to the intrinsic complexity of these tables, it is likely that deep results about braids or knots could be obtained in this way. In particular, the fact that the tables are finite, even if intricate, should make it possible to obtain effective results. On the other hand, the Laver tables are not left cancellative, and it is not even clear how to obtain a well defined action.

This goal may remain out of reach for a long time, but it is surely worth keeping in mind. To conclude with more realistic problems, let us come back to the construction of A_n as a monogenic LD-system. As the braid group B_∞ includes a copy of the free LD-system of rank 1, namely the set B_{sp} of all special braids, the LD-system A_n is a quotient of the LD-system $(B_{sp}, {}^\wedge)$. This means that there must exist a congruence $\approx_n$ on B_{sp} such that A_n is the quotient $B_{sp}/\approx_n$.

Question 5.3. *Can the congruence $\approx_n$ on B_{sp} be defined naturally (geometrically)? Does it extend to all of B_∞?*

Nothing is known about the question, except the trivial result that the congruence $\approx_1$ partitions braids according to whether the associated permutation leaves 1 fixed or not.

Finally, it is extremely natural to think of possible linear representations of LD-systems. The Burau representation gives a first approach.

Question 5.4. *Does there exist a linear representation of the Laver tables, i.e., does there exist a left self-distributive operation on matrices together with a (injective?) homomorphism of A_n into the matrices equipped with this possible operation?*

Bibliography

ADHÉMAR, C. D'

1970 Quelques classes de groupoïdes non-associatifs, *Math. Sci. Humaines*, **31**, 17–31.

ADJAN, S.I.

1984 Fragments of the word Delta in a braid group, *Mat. Zam. Acad. Sci. SSSR*, **36-1**, 25–34.

ARTIN, E.

1947 Theory of braids, *Ann. of Math.*, **48**, 101–126.

BELOUSOV, V.D.

1958 Tranzitivnyje distributivnyje kvazygruppy, *Ukr. Mat. Ž.*, **10**, 13–22.
1960 О стрыктуре дистрибутивных квазигрупи, *Math. Sbornik*, **50**, 267–298.
1963 Об одном класс лево-дистрибутивных квасигруп, *Isvestia Vys. Uč. Zav.*, **32**, 16–20.
1967 *Osnovy teorii kvazigrupp i lup*, Nauka, Moscow.

BÉNÉTEAU, L.

1975 Boucles de Moufang commutatives d'exposant 3 et quasigroupes de Steiner distributifs, *C. R. Acad. Sci. Paris*, **281**, 75–76.

BESSIS, D., F. DIGNE, & J. MICHEL

2000 Springer theory in braid groups and the Birman-Ko-Lee monoid, Preprint.

BIRKENMEIER, G.

1983 Exponential without associativity, *Acta Univ. Carolinae Math. Phys.*, **24.2**, 3–7.
1985 Exponentiation and the identity $x^y = (xy)^2$, *Comm. Algebra*, **13**, 681–695.
1986 Right ideals in a right distributive groupoid, *Algebra Universalis*, **22**, 103–108.

BIRKENMEIER, G., & H. HEATHERLY

1989 Polynomial identity properties for near-rings on certain groups, *Near-ring newsletter*, **12**, 5–15.
1990 Left self distributive near-rings, *J. Austral. Math. Soc.*, **49**, 273–296.

BIRKENMEIER, G., H. HEATHERLY & T. KEPKA

1992 Rings with helft self distributive multiplication, *Acta Math. Hung.*, **60**, 107–114.

BIRMAN, J.

1975 *Braids, links, and mapping class groups*, Annals of Math. Studies **82** Princeton Univ. Press, Princeton.

BIRMAN, J., K.H. KO & S.J. LEE

1998 A new approach to the word problem in the braid groups, *Advances in Math.*, **139-2**, 322-353.

BOL, G.

1937 Gewebe und Gruppen, *Math. Annalen*, **114**, 414–431.

BRIESKORN, E.

1988 Automorphic sets and braids and singularities, *Braids, Contemporary Maths AMS*, **78**, 45–117.

BRIESKORN, E., & K. SAITO

1972 Artin-Gruppen und Coxeter-Gruppen, *Invent. Math.*, **17**, 245–271.

BRUCK, R. H.

1966 *A survey of binary systems*, Springer, Berlin.

BURCKEL, S.

1994 L'ordre total sur les tresses positives, Doctorat Université de Caen.
1997 The wellordering on positive braids, *J. Pure Appl. Alg.*, **120-1**, 1–17.
1996 The rank of a positive braid, Preprint.

BURDE, G., & H. ZIESCHANG

1985 *Knots*, de Gruyter, Berlin.

BURSTIN, C., & W. MAYER

1929 Distributive Gruppen von endlicher Ordnung, *J. reine u. angewandte Math.*, **160**, 111–130.

CANNON, J.W., W.J. FLOYD, W.R. PARRY

1996 Introductory notes on Richard Thompson's groups, *Enseign. Math.*, **42-2**, 215–256.

Chein, O., H.O. Pflugfelder & J.D.H. Smith

1990 *Quasigroups and loops, theory and applications*, Sigma Series in Pure Mathematics vol. 8, Heldermann, Berlin.

Crowell, R., & R. Fox

1963 *Introduction to knot theory*, Blaisdell, New York..

Dehornoy, P.

1986 Infinite products in monoids, *Semigroup Forum*, **34**, 21–68.
1988 Π^1_1-complete families of elementary sequences, *Ann. P. Appl. Logic*, **38**, 257–287.
1989 Algebraic properties of the shift mapping, *Proc. Amer. Math. Soc.*, **106-3**, 617–623.
1989*a* Free distributive groupoids, *J. P. Appl. Algebra*, **61**, 123–146.
1989*b* Sur la structure des gerbes libres, *C. R. Acad. Sci. Paris*, **309**, 143–148.
1992 The adjoint representation of a left distributive magma, *Comm. in Algebra*, **20-4**, 1201–1215.
1992*a* Problème de mots dans les gerbes libres, *Theor. Comp. Sci.*, **94**, 199–213.
1992*b* Preuve de la conjecture d'irréflexivité pour les structures distributives libres, *C. R. Acad. Sci. Paris*, **314**, 333–336.
1992*c* Deux propriétés des groupes de tresses, *C. R. Acad. Sci. Paris*, **315**, 633–638.
1993 Structural monoids associated to equational varieties, *Proc. Amer. Math. Soc.*, **117-2**, 293–304.
1993*a* About the irreflexivity hypothesis for LD-magmas, *Logic Colloquium'90, J. Oikkonen & al. eds, Lect. Notes in Logic*, **2**, 46–61.
1994 A canonical ordering for free LD systems, *Proc. Amer. Math. Soc.*, **122-1**, 31–37.
1994*a* Braid Groups and Left Distributive Operations, *Trans. Amer. Math. Soc.*, **345-1**, 115–151.
1994*b* A Normal Form for the Free Left Distributive Law, *Int. J. for Algebra & Computation*, **4-4**, 499–528.
1994c Construction of left distributive operations and charged braids, Int. J. Algebra & Computation, to appear.
1995 From large cardinals to braids via distributive algebra, *J. Knot Theory & Ramifications*, **4-1**, 33–79.
1996 The structure group for the associativity identity, *J. Pure Appl. Algebra*, **111**, 59–82.
1996*a* Weak faithfulness properties for the Burau representation, *Topology and its Applic*, **69**, 121–143.
1996*b* Another use of set theory, *Bull. Symb. Logic*, **2-4**, 379–391.
1999 Three-dimensional realizations of braids, *J. London Math. Soc.*, **60-2**, 108–132.
1997 Groups with a complemented presentation, *J. Pure Appl. Algebra*, **116**, 115–137.
1997*a* On the syntactic algorithm for the word problem of left distributivity, *Algebra Univers.*, **37**, 191–222.
1997*b* A fast method for comparing braids, *Advances in Math.*, **125**, 200–235.
1997*c* Multiple left distributive systems, *Comm. Math. Univ. Carol.*, **38,4**, 615–625.

1998 Transfinite braids and left distributive operations, *Math. Zeitschr.*, **228**, 405–433.

1998*a* Free zeropotent left distributive systems, *Comm. in Algebra*, **26,6**, 1967–1978.

1998*b* Gaussian groups are torsion free, *J. of Algebra*, **210**, 291–297.

1998c On completeness of word reversing, Discrete Math., to appear.

1999 A conjecture about conjugacy in free group, *Discuss. Math.*, **19**, 75–112.

1999*a* Strange questions about braids, *J. Knot Th. and its Ramifications*, **8-5**, 589–620.

2000 Petits groupes gaussiens, Preprint.

2000a Study of an identity, Preprint.

DEHORNOY, P. & L. PARIS

1999 Gaussian groups and Garside groups, two generalizations of Artin groups, *Proc. London Math. Soc.*, **79-3**, 569–604.

DELIGNE, P.

1972 Les immeubles des groupes de tresses généralisés, *Invent. Math.*, **17**, 273–302.

DERSHOWITZ, N. & J.P. JOUANNAUD

1994 *Rewrite systems, in* Handbook of Theor. Comp. Sc. vol. B, J. van Leeuwen *ed.*, Elsevier, Amsterdam.

DEVINE, D. J.

1992 Alexander invariants of links and the fundamental rack, PhD university of Sussex (G. B.).

DOUGHERTY, R.

1993 Critical points in an algebra of elementary embeddings, *Ann. P. Appl. Logic*, **65**, 211–241.

1996 Critical points in an algebra of elementary embeddings, II, Logic: From Foundations to Applications (W. Hodges, ed.), Oxford, 103–136.

DOUGHERTY, R., & T. JECH.

1997 Finite left-distributive algebras and embedding algebras, *Advances in Math.*, **130**, 201–241.

1997*a* Left-distributive embedding algebras, *Electr Research Announcements. of the Amer. Math. Soc.*, **3**, 28–37.

DRÁPAL, A.

1994 Homomorphisms of primitive left distributive groupoids, *Comm. in Algebra*, **22-7**, 2579–2592.

1995 Persistency of cyclic left distributive algebras, *J. Pure Appl. Algebra*, **105**, 137–165.

1995a On the semigroup structure of cyclic left distributive algebras, *Semigroup Forum*, **51**, 23–30.

1996 The third level of homomorphisms in finite cyclic left distributive algebras, Preprint.
1997 Finite left distributive algebras with one generator, *J. Pure Appl. Algebra*, **121**, 233–251.
1997*a* Finite left distributive groupoids with one generator, *Int. J. Algebra & Computation*, **7-6**, 723–748.
1997*b* Low level computations in finite left distributive algebras, *Comm. in Algebra*, **25-7**, 2071–2084.

DRÁPAL, A., & R. EL BASHIR

1994 Quasitrivial left distributive groupoids, *Comment. Math. Univ. Carolinae*, **35-4**, 597–606.

DRÁPAL, A., T. KEPKA & M. MUSÍLEK

1994 Group Conjugacy Has Non-trivial LD-Identities, *Comment. Math. Univ. Carolinae*, **35-2**, 219–222.

DUDEK, J.

1981 Some remarks on distributive groupoids, *Czech. Math. J.*, **31-106**, 451–456.

ENDRES, N.

1991 Semidirect products of symmetric groupoids, *Proc. of 'Combinatorics 88', Research and Lecture Notes in Math., Mediterranean Press*, **2**, 39–68.
1991*a* Group related symmetric groupoids, *Demonstratio Math.*, **24**, 63–74.
1993 Die algebraische Struktur des Homotopiemenge von Produktabbildungen auf Sphären, *Geom. Dedicata*, **48**, 267–294.
1994 Various applications of symmetric groupoids, *Demonstratio Math.*, **27**, 673–685.
1995 Homotopiemenge von Multiplikationen auf Sphären und ilre algebraische Struktur, *Geom. Dedicata*, **54**, 57–85.
1996 Idempotent and distributive group related groupoids (I and II), *Demonstratio Math.*, **29**, 270–308.

ELRIFAI, E. A. & H. R. MORTON

1994 Algorithms for positive braids, *Quart. J. Math. Oxford*, **45-2**, 479–497.

ENDRES, N., & M. LEISCHNER

1994 Powers and iteration processes on modules, *Demonstratio Math.*, **27**, 427–447.

EPSTEIN, D., J.W. CANNON, D.F. HOLT, S.V.S. LEVY, M.S. PATERSON, & W.P. THURSTON

1992 *Word Processing in Groups*, Jones & Barlett, Boston.

FAJTLOWICZ, S, & J. MYCIELSKI

1974 On convex linear forms, *Alg. Univ.*, **4**, 244–249.

FENN, R. A., R. RIMÁNYI, & C. ROURKE

1997 The braid-permutation group, *Topology*, **36-1**, 123–135.

FENN, R. A., & C. ROURKE

1992 Racks and links in codimension 2, *J. of Knot Theory and its Ramifications*, **1-4**, 343–406.

FENN, R. A., M.T. GREENE, D. ROLFSEN, C. ROURKE & B. WIEST

1999 Ordering the braid groups, *Pacific J. of Math.*, **191**, 49–74.

FENN, R. A., D. ROLFSEN & J. ZHU

1996 Centralisers in braid groups and singular braid monoids, *Enseignement Math.*, **42**, 75–96.

FISCHER, B.

1964 Distributive Quasigruppen endlicher Ordnung, *Math. Z.*, **83**, 267–303.

GAIFMAN, H.

1967 Pushing up the measurable cardinal, Proc. 1967 UCLA Summer Institute.

GARSIDE, F.A.

1969 The braid group and other groups, *Quart. J. Math. Oxford*, **20** No.78, 235–254.

GHYS, E., & V. SERGIESCU.

1987 Sur un groupe remarquable de diffomorphismes du cercle, *Comment. Math. Helverici*, **62**, 185–239.

HANSEN, V.L.

1989 *Braids and coverings, selected topics*, London Math. Soc. Student texts 18, Cambridge Univ. Press, Cambridge.

HOSSZÚ, M.

1961 Homogeneous groupoids, *Ann. Univ. Sci. Budapest*, **3–4**, 95–98.

HURWITZ, A.

1891 Über Riemann'schen Flächen mit gegebenen Verzweigungspunkten, *Math. Ann.*, **39**, 1–61.

HUET, G.

1980 Confluent reductions: Abstract properties and applications to term rewriting systems, *Journal of the ACM*, **27**, 797–821.

JECH, T.

1978 *Set Theory*, Academic Press, New York.
1997 Large ordinals, *Advances in Math.*, **125**, 155–170.

JEŽEK, J.

1997 Enumerating left distributive groupoids, Czech. Math. J., to appear.

JEŽEK, J., & T. KEPKA

1974 Atoms in the lattice of varieties of distributive groupoids, *Coll. Math. Soc. János Bolyai, Szeged*, **14**, 185–194.
1975 The lattice of varieties of commutative abelian distributive groupoids, *Algebra Universalis*, **5**, 225–237.
1976 Free commutative idempotent abelian groupoids and quasigroups, *Acta Univ. Carol. Math. Phys.*, **17/2**, 13–19.
1978 Quasitrivial and nearly quasitrivial distributive groupoids and semigroups, *Acta Univ. Carol. Math. Phys.*, **19–2**, 25–44.
1983 Notes on distributive groupoids, *Comm. Math. Univ. Carol.*, **24**, 237–249.
1984 Distributive groupoids and symmetry-by-mediality, *Alg. Universalis*, **19**, 208–216.
1999 Selfdistributive groupoids, Part A1: Non-idempotent left distributive groupoids, Rapport de recherche SDAD 1999–9, Caen.
1999*a* Selfdistributive groupoids, Part D1: Left distributive semigroups, Rapport de recherche SDAD 1999–26, Caen.

JEŽEK, J., T. KEPKA & P. NĚMEC

1981 Distributive groupoids, *Rozpravy Ceskoslov. Akad. Ved Rada Mat. Přirod. Ved*, **91**, 1–95.

JOYCE, D.

1982 A classifying invariant of knots: the knot quandle, *J. of Pure and Applied Algebra*, **23**, 37–65.
1982*a* Simple quandles, *J. of Algebra*, **79**, 307–318.

KANAMORI, A.

1994 The higher infinite, Springer, Berlin.

KASSEL, C.

1999 L'ordre de Dehornoy sur les tresses, Séminaire Bourbaki, exposé 865.

KAUFFMAN, L. H.

1991 Knots and physics, World Scientific, Singapore.

KAUFFMAN, L. H. & H. SALEUR

1991 Free fermions and the Alexander-Conway polynomial, *Comm. in Math. Physics*, **141**, 293–327.

KEPKA, T.

1978 Commutative distributive groupoids, *Acta Univ. Carol.*, **19–2**, 45–58.
1979 Distributive division groupoids, *Math. Nachr.*, **87**, 103–107.
1981 Notes on left distributive groupoids, *Acta Univ. Carol.*, **22–2**, 23–37.
1984 Varieties of left distributive semigroups, *Acta Univ. Carol.*, **25-1**, 3-18.
1994 Non-idempotent left symmetric left distributive groupoids, *Comm. Math. Univ. Carol.*, **35**, 181–186.
1994*a* Ideals in selfdistributive groupoids, *Comm. Math. Univ. Carol.*, **35**, 187-191.

KEPKA, T., & P. NĚMEC

1977 A note on left distributive groupoids, *Coll. Math. Soc. János Bolyai*, **29**, 467–471.
1979 Notes on distributive groupoids, *Math. Nachr.*, **87**, 93–101.
1981 Distributive groupoids and the finite basis property, *J. of Algebra*, **70-1**, 229–237.

KIM, D.M., & D. ROLFSEN

1998 An ordering of pure braids and hyperplane arrangements, Preprint.

KRÜGER, B.

1989 Automorphe Mengen und die Artinschen Zopfgruppen, Dissertation Univ. Bonn.

KUNEN, K.

1970 Some applications of iterated ultrapowers in set theory, *Ann. Math. Logic*, **1**, 179–227.

LARUE, D.

1994 On braid words and irreflexivity, *Algebra Univ.*, **31**, 104–112.
1994a Left-distributive and left-distributive idempotent algebras, Ph D Thesis, University of Colorado, Boulder.
1999 Left-distributive idempotent algebras, *Comm. in Algebra*, **27-5**, 2003-2029.

LAVER, R.

1986 Elementary embeddings of a rank into itself, *Abstracts Amer. Math. Soc.*, **7**, 6.
1992 The left distributive law and the freeness of an algebra of elementary embeddings, *Advances in Math.*, **91-2**, 209–231.
1993 A division algorithm for the free left distributive algebra, *Oikkonen & al. eds, Logic Colloquium '90, Lect. Notes Logic*, **2**, 155–162.
1995 On the algebra of elementary embeddings of a rank into itself, *Advances in Math.*, **110**, 334–346.
1996 Braid group actions on left distributive structures and well-orderings in the braid group, *J. Pure Appl. Algebra*, **108-1**, 81–98.

LEVY, A.

1979 Basic set theory, Springer, Berlin.

LONG, D. & M. PATON

1993 The Burau representation is not faithful for $n \geq 6$, *Topology*, **32-2**, 439–447.

LYNDON, R. C. & P. E. SCHUPP

1977 Combinatorial group theory, Springer, Berlin.

MACLANE, S.

1963 Natural associativity and commutativity, *Rice Univ. Studies*, **49**, 28–46.

MACLANE, S.; & I. MOERDIJK

1992 Sheaves in geometry and logic: a first introduction to topos theory, Springer, Berlin.

MANIN, YU.

1972 Kubičeskije formy (algebra, geometrija, arifmetika), Nauka, Moscow (English translation, Cubic forms, 1974).

MATVEEV, S. V.

1982 Distributive groupoids in knot theory, *Math. Sbornik*, **119, 1-2**, 73–83.

MCKENZIE, R.N., G.F. MCNULTY & W.F. TAYLOR

1986 Algebras, Lattices, Varieties, vol.1, Wadsworth & Brooks, Monterey.

MCKENZIE, R.N., R.J. THOMPSON

1973 An elementary construction of unsolvable word problems in group theory, *in* Word Problems, Boone & *al.* eds., North Holland, Studies in Logic vol. 71.

MOODY, J.

1991 The Burau representation of the group B_n is unfaithful for large n, *Bull. Americ. Math. Soc.*, **25-2**, 379–384.

MURDOCH, D.C.

1941 Structure of abelian quasigroups, *Trans. Americ. Math. Soc.*, **49**, 392–409.

NIELSEN, J.

1927 Untersuchungen zur Topologie des geschlossenen zweiseitigen Flächen, *Acta Math.*, **50**, 289–358.

OREVKOV, S.

1998 Quasipositivity and right-invariant order on braid groups, Preprint.

PASSMAN, D.S.

1977 *The algebraic structure of group rings*, Pure and Applied Mathematics, Wiley, New York.

PATERSON, A.L.T.

1999 Groupoids, inverse semigroups, and their operator algebras, Progress in Mathematics vol. 170, Birkhäuser, Basel.

PATERSON, M.S., & A.A. RAZBOROV

1991 The set of minimal braids is co-NP-complete, *J. of Algorithms*, **12**, 393–408.

PEIRCE, C. S.

1880 On the algebra of logic, *Amer. J. of Math.*, **III**, 15–57.

PICANTIN, M.

1999 The conjugacy problem in small Gaussian groups, Comm. Algebra, to appear.
2000 The center of small Gaussian groups, Preprint.

PIERCE, R. S.

1978 Symmetric groupoids, *Osaka J. Math.*, **15**, 51–76.
1979 Symmetric groupoids II, *Osaka J. Math.*, **16**, 317–348.

PRASOLOV, V.V., & A.B. SOSSINSKY

1997 Knots, links, braids, and 3-manifolds, Translation of mathematical monographs, **154**, Amer. Math. Soc., Providence.

RHEMTULLA, A. & D. ROLFSEN

1999 Incompatible group orderings, local indicability and braids, Preprint.

ROLFSEN, D.

1997 Braid subgroup normalisers, commensurators and induced representations, *Invent. Math.*, **130**, 575–587.

ROLFSEN, D., & J. ZHU

1998 Braids, orderings and zero divisors, *J. Knot Th. and its Ramifications*, **7**, 837–841.

ROURKE, C., & B. WIEST

1998 Order automatic mapping class groups, Pacific J. of Math., to appear.

RUEDIN, J.

1966 Structure des demi-groupes et anneaux distributifs, *C. R. Acad. Sci. Paris*, **262**, 985–988.

SCHRÖDER, E.

1887 Über Algorithmen und Calculn, *Archive der Math. und Phys.*, **5**, 225–278.

SESBOÜÉ, A.

1996 Finite monogenic distributive systems, *Czech. Math. J.*, **121**, 697–719.

SHORT, H. & B. WIEST

1999 Orderings of mapping class groups, Preprint.

SMITH, J.D.H.

1976 Finite distributive quasigroups, *Math. Proc. Cam. Phil. Soc.*, **80**, 37–41.
1992 Quasigroups and quandles, *Discrete Math.*, **109**, 277–282.

SOUBLIN, J. P.

1971 Etude algébrique de la notion de moyenne, *J. Math. Pures et Appl.*, **50**, 53–264.

SQUIER, C.

1984 The Burau representation is unitary, *Proc. Amer. Math. Soc.*, **90-2**, 199–202.

STASHEFF, J.

1963 Homotopy associativity of H-spaces, *Trans. Amer. Math. Soc.*, **108**, 275–292.

STEIN, S. K.

1957 On the foundation of quasigroups, *Trans. Amer. Math. Soc.*, **85**, 228–256.
1959 Left-distributive quasigroups, *Proc. Amer. Math. Soc*, **10**, 577–578.
1959*a* On a construction of Horszú, *Publ. Math.*, **6**, 10–14.

SUŠKEVIČ, A. K.

1937 Теория общенных груп, Гос. Хауч.-техн. изд. Укпаины, Хар - ков, Киев.

TAKASAKI, M.

1943 Abstractions of symmetric functions, *Tôhoku Math. J.*, **49**, 143–207.

TATSUOKA, K.

1993 An isoperimetric inequality for Artin groups of finite type, *Trans. Amer. Math. Soc.*, **339–2**, 537–551.

WADA, M.

1992 Group invariants of links, *Topology*, **31-2**, 399–406.

WEHRUNG, F.

1991 Gerbes primitives, *C. R. Acad. Sci. Paris*, **313-I**, 357–362.

WIEST, B.

1998 Dehornoy's ordering of the braid groups extends the subword ordering, Pacific J. of Math., to appear.

WINKER, S.

1984 Quandles, knot invariants and the n-fold branched cover, PhD Thesis University of Illinois.

ZAPLETAL, J.

1991 Completion of a free distributive groupoid, Circulated notes.

List of Symbols

Index